C0-AVC-650

Methods in Protein Structure Analysis

Methods in Protein Structure Analysis

Edited by

M. Zouhair Atassi

Baylor College of Medicine
Houston, Texas

and

Ettore Appella

National Cancer Institute
National Institutes of Health
Bethesda, Maryland

Plenum Press • New York and London

Library of Congress Cataloging-in-Publication Data

Methods in protein structure analysis / edited by M. Zouhair Atassi and Ettore Apella.
p. cm.
Includes bibliographical references and index.
ISBN 0-306-45124-7
1. Proteins--Analysis. 2. Amino acid sequence. I. Atassi, M. Z. II. Appella, Ettore.
QP551.M398 1995
574.19'245--dc20 95-36854
CIP

Proceedings of the Tenth International Conference on Methods in Protein Structure Analysis, held September 8–13, 1994, in Snowbird, Utah

ISBN 0-306-45124-7

A Division of Plenum Publishing Corporation
233 Spring Street, New York, N. Y. 10013

10 9 8 7 6 5 4 3 2 1

Printed in the United States of America

CONFERENCE ORGANIZERS

Chairman: M. Zouhair Atassi (Houston, TX)

Secretary General: Ettore Appella (Bethesda, MD)

Organizing Committee: Carl W. Anderson (Upton, NY); Ettore Appella (Bethesda, MD); Eugene M. Barnes (Houston, TX); Agnes Henschen (Irvine, CA); Michael W. Hunkapiller (Foster City, CA); Richard A. Laursen (Boston, MA); C. Dale Poulter (Salt Lake City, UT); Kenneth A. Walsh (Seattle, WA); Chao-Yuh Yang (Houston, TX)

Scientific Advisory Committee: Ettore Appella (Bethesda, MD); M. Zouhair Atassi (Houston, TX); Klaus Biemann (Cambridge, MA); Jui-Yoa Chang (Basel, Switzerland); Maria A. Coletti-Previero (Montpellier, France); Marshall Elzinga (Staten Island, NY); John B.C. Findlay (Leeds, UK); Gerhard Frank (Zurich, Switzerland); Agnes Henschen (Irvine, CA); Michael W. Hunkapiller (Foster City, CA); Hans Jörnvall (Stockholm, Sweden); Richard A. Laursen (Boston, MA); Hans Neurath (Seattle, WA); Richard N. Perham (Cambridge, MA); Fumio Sakiyama (Osaka, Japan); Richard J. Simpson (Melbourne, Australia); Harald Tschesche (Bielefeld, FRG); A. Tsugita (Tokyo, Japan); Joel Vanderkerckhove (Ghent, Belgium); Kenneth A. Walsh (Seattle, WA); Brigitte Wittmann-Liebold (Berlin, FRG)

CONFERENCE SPONSORSHIP AND SUPPORT

Sponsorship and Organization: This conference was sponsored by the International Symposium on the Immunobiology of Proteins and Peptides, Inc., and organized by the organizing committee of the 10th MPSA.

Support for this conference was provided by the following organizations.

Leading Contributor

- Perkin Elmer, Applied Biosystems Division

Major Contributors

- Beckman Instruments
- Finnigan Mat
- United States Army Medical Research & Development Command, Department of the Army
- Millipore Corporation

Contributors

- Amgen
- Hewlett-Packard
- PerSeptive Biosystems
- Pierce Chemical Company
- Shimadzu Corporation

MPSA CONFERENCES

The International Conference on Methods in Protein Structure Analysis was established in 1975 and has usually been held biannually. The following are the dates, locations, and Chairs of the past conferences as well as of the forthcoming 11th MPSA:

Conference	Year	Venue	Chair or Co-Chairs
I	1975	Boston, MA	Richard A. Laursen
II	1977	Montpellier, France	M.A. Coletti-Previero
III	1979	Heidelberg, France	Christian Birr
IV	1981	Brookhaven, NY	Marshall Elzinga
V	1984	Cambridge, UK	John Walker
VI	1986	Seattle, WA	Kenneth A. Walsh
VII	1988	Berlin, FRG	Brigitte Wittmann-Liebold
VIII	1990	Kiruna, Sweden	Hans Jörnvall
IX	1992	Otsu, Japan	Kazutomo Imahori and Fumio Sakiyama (Co-Chairs)
X	1994	Snowbird, UT	M. Zouhair Atassi
XI	1996	Annecy, France	Michel van der Rest and Joël Vadekerckhove (Co-Chairs)

PREFACE

The MPSA international conference is held in a different country every two years. It is devoted to methods of determining protein structure with emphasis on chemistry and sequence analysis. Until the ninth conference, MPSA was an acronym for Methods in Protein Sequence Analysis. To give the conference more flexibility and breadth, the Scientific Advisory Committee of the 10th MPSA decided to change the name to *Methods in Protein Structure Analysis;* however, the emphasis remains on "**methods**" and on "**chemistry**." In fact, this is the only major conference that is devoted to methods.

The MPSA conference is truly international, a fact clearly reflected by the composition of its Scientific Advisory Committee. The Scientific Advisory Committee oversees the scientific direction of the MPSA and elects the chairman of the conference. Members of the committee are elected by active members, based on scientific standing and activity. The chairman, subject to approval of the Scientific Advisory Committee, appoints the Organizing Committee. It is this latter committee that puts the conference together. The lectures of the MPSA have traditionally been published in a special proceedings issue. This is different from, and more detailed than, the special MPSA issue of the *Journal of Protein Chemistry* in which only a brief description of the talks is given in short papers and abstracts. In the 10th MPSA, about half the talks are by invited speakers and the remainder were selected from submitted short papers and abstracts. Inclusion of submitted contributions in the oral program is an important mechanism for bringing new discoveries and innovations to the forefront. These proceedings are divided into eight topics: (1) preparation of proteins and peptides for microsequence analysis; (2) N-terminal sequence analysis; (3) C-terminal sequence analysis; (4) mass spectrometry; (5) new strategies for protein and peptide characterization; (6) immunological recognition, phage and synthetic libraries; (7) analysis of protein structures of special interest; (8) database analysis, protein folding, and three-dimensional structures of proteins. We believe that the different chapters in this book will provide a timely resource for the analysis of protein structures, which constitutes an indispensable part of contemporary biochemistry and molecular biology. Protein structure analysis continues to progress in line with other developments in modern biochemistry, molecular biology, and biophysics, and is essential for the design of therapeutic agents useful for the control of human diseases.

Conferences of this size (493 participants) require considerable funding. We would like to express our gratitude to our sponsors. Without their support and generosity this meeting would not have been possible. We especially would like to thank Millipore, Inc. for supporting the Edman Award. Because of this wonderful support and the excellent registration we were able to offer assistance for attending the conference to six junior scientists and thirteen students. Together with the International Science Foundation, we co-sponsored six scientists from the former Soviet Republics of Russia and Uzbekistan.

The editors wish to thank Marie Pellum, Kella Kunz, Priscilla Igori, and Shana Atassi for their invaluable assistance in the organization of the conference and especially Priscilla Igori and Shana Atassi for their help in the organization and preparation of this book.

M. Zouhair Atassi
Ettore Appella

THE EDMAN AWARDS 1994

On behalf of the Pehr Edman Award Selection Committee of the 10th MPSA and Millipore Corporation, it is with great pleasure that I have been asked to write a few words on the 1994 awardees and on the prize itself. To understand the contributions that Dr. Reudi Aebersold and Dr. Joel Vanderkerckhove have made to this field, the pioneering work of Pehr Edman should be acknowledged.

The field which has immortalized his name got its early roots with Pehr Edman's studies in 1946 on bovine angiotensin at the Karolinska Institute in Stockholm. He observed that neither the molecular mass nor the amino acid composition were giving information on the biological activity of the small, physiologically important molecule. During his stay at the Rockefeller Institute at Princeton from 1946-47, he looked at different reagents that could react with the amino group of peptides. This led him to introduce the use of phenylisothiocyanate for carbamylation of reactive amines on model peptides. With this tool the door was opened for the systematic elucidation of the primary structure of polypeptides. It is evident from the proceedings of the 10th MPSA that this chemistry is still alive as we move the analytical capabilities into new directions all aimed at accelerating sequencing. However, the basic chemical approach has not changed since Pehr Edman recognized the reactivity of the amino group in peptides and proteins. He also recognized that the chemical degradation of the amino terminal is a three-stage process in which it is critical that the thiazolinone derivative be removed from the parent peptide and converted to the stable PTH derivative in a separate vessel. This was the only way to achieve high repetitive yields, a prerequisite for unraveling long peptide sequences.

Since the early work of Pehr Edman there have been many contributions to its emergence as the primary protein structural characterization tool that it is today. The selection of recipients of the 1994 Edman prize was a difficult one. Two individuals are recognized this year; Dr. Joel Vanderkerckhove and Dr. Reudi Aebersold. Their contributions have helped to make it possible to sequence samples from electrophoretic separations after electroblotting to solid phases. In their earlier work they both recognized the importance of the solid phase in presenting the sample to the instrumentation and chemistry of the modern sequencer. At Millipore we have seen many applications develop in this field with the introduction of the polyvinylidene fluoride (PVDF) membrane in 1986 and its recognition as a sequencing support in 1987. At the 10th MPSA, clearly, we were still learning more about the importance of the solid phase in this approach. Drs. Vanderkerckhove and Aebersold are both actively moving this field forward and were selected by the 10th MPSA Award Selection Committee to receive the Edman Award, sponsored by Millipore.

To learn more about the awardees who are being honored in memory of Pehr Edman, the accomplishments and short biographies of Dr. Joel Vanderkerckhove and Dr. Reudi Aebersold will be outlined here briefly.

Dr. Pehr Edman

JOEL VANDERKERCKHOVE

Dr. Joel Vankderkerckhove was born in Belgium and obtained a degree in Chemistry at the University of Ghent in Belgium. In 1967, he started his Ph.D. work on the sequence determination of the coat protein of bacteriophage MS2, work which brought him directly to the heart of protein chemistry. This work brought him in contact with Dr. Klaus Weber at the Max Plank Institute of Biophysical Chemistry in Gottingen. There, he started a second period in his research career, studying the molecular basis of the different isoforms of actin and analyzing their expression in different cells and tissues.

In 1981, Vanderkerckhove returned to his old "mews" where he continued the work on the isoactin expression. That work became a classic in cell biology. In order to solve some of the problems for this work, Vanderkerckhove introduced the protein electroblotting techniques; the original method employing polybased-coated glass-fiber membranes as immobilizing support. Similar techniques were at that time developed in the laboratory of Professor Lee Hood at California Institute of Technology (Cal Tech), together with Drs. Steven Kent and Reudi Aebersold. The success of this method is now well known.

Close contact with the successful plant-engineering group of Professor M. Van Montagu at the University of Ghent, enticed Joel Vanderkerckhove to use transgenic plants for the production of bioactive peptides as part of hybrid seed storage proteins. This was the first step towards molecular farming offering interesting prospectives for increasing the nutritional value of seeds.

The protein-chemical micropreparation techniques which were meanwhile developed were applied in the development of a 2D gel database in collaboration with the group of Professor J. Celis. This is one of the largest databases of its kind.

In 1990, Vanderkerckhove became head of the Department of Biochemistry at the Medical Faculty of the University of Ghent (an unusual appointment for a nonphysician).

Dr. Joel Vanderkerckhove

Since then, his research has concentrated on the molecular mechanisms underlying the organization of the microfilament system in the cell. In particular, he has addressed the problem of the multiple interactions between actin and actin-binding proteins, a complex protein-protein docking problem.

RUEDI AEBERSOLD

Dr. Aebersold is a native of Switzerland, where he grew up and received his education. He graduated in 1983 with a doctoral degree in cell biology from the Biocenter, University of Basel. His thesis work, which was carried out in the laboratory of Dr. J.Y. Chang at Ciba-Geigy in Basel, involved the sequence analysis of monoclonal antibodies directed against streptococcal group carbohydrates.

After this, he came to the U.S. to do a postdoctoral fellowship with Dr. Lee Hood at Cal Tech. He stayed at Cal Tech from 1984-88 where he and his co-workers worked out several protein analytical techniques. The most notable ones might be the isolation of proteins for N-terminal sequencing by electroblotting from polyacrylamide gels, a procedure for internal sequence analysis of small amounts of gel separated and electroblotted proteins by *in situ* digestion on nitrocellulose membranes and the application of these techniques to proteins separated by 2D gel electrophoresis. With this work, they attempted to make the technique of protein sequencing compatible with techniques most commonly used in the laboratory of the biochemist.

In 1988, Aebersold moved to the Biomedical Research Center at the University of British Columbia in Vancouver to take an Assistant Professorship at the Department of Biochemistry. In Vancouver, he started working on the delineation of signal transduction pathways inside cells using a protein analytical approach. Initially, he characterized several signaling proteins by sequencing. He developed protocols for the determination of sites of Ser, Thr as well as Tyr phosphorylation by solid-phase sequencing. These techniques were used to determine the sites of protein phosphorylation.

More recently, he has worked on developing combined chemical and mass spectrometric protocols for the determination of the sites of protein phosphorylation at high sensitivities. He has demonstrated that the interaction between the kinase Zap 70 and

Dr. Ruedi Aebersold

phosphorylated TCR zeta chain is mediated by tyrosine phosphorylation, is essential for the T-cell receptor signaling, and can be interrupted with a synthetic analog modeled after the phosphopeptide sequence.

In 1993, he moved to the Department of Molecular Biotechnology at the University of Washington in Seattle to take a position as Associate Professor and Associate Director of the NSF Science and Technology Center on Molecular Biotechnology. His work continued to focus on the development and application of protein analytical technology.

The general aim of his work is, therefore, the development of analytical technology thats can be directly interfaced with experiments in the typical biochemistry laboratory to answer biological problems.

Dr. Malcolm G. Pluskal
Senior Consulting Scientist
Millipore Corporation

CONTENTS

Protein and Peptide Preparation for Microsequence Analysis

N-Terminal Sequence Analysis

C-Terminal Sequence Analysis

Mass Spectrometry

New Strategies for Protein and Peptide Characterization

Immunological Recognition, Phage and Synthetic Libraries

Analysis of Protein Structures of Special Interest

Database Analysis, Protein Folding and Three-Dimensional Structures of Proteins

PROTEIN AND PEPTIDE PREPARATION FOR MICROSEQUENCE ANALYSIS

1

CHARACTERIZATION OF PROTEINS SEPARATED BY GEL ELECTROPHORESIS AT THE PRIMARY STRUCTURE LEVEL

Ruedi Aebersold,[1*] Lawrence N. Amankwa,[2] Heinz Nika,[1†] David T. Chow,[2] Edward J. Bures,[1] Hamish D. Morrison,[1] Daniel Hess,[1‡] Michael Affolter,[1**] and Julian D. Watts[1]

[1] Department of Molecular Biotechnology
University of Washington
Seattle, Washington.
[2] The Biomedical Research Centre
University of British Columbia
Vancouver, B.C., Canada

INTRODUCTION

The investigation of cell differentiation, development, and signal transduction pathways are examples of current research projects which have in common the focus on complex, highly regulated systems consisting of numerous interacting elements. A complete understanding of such processes can only be achieved if the problem is approached globally, considering the temporal and spatial interactions of all the elements involved. This task is supported by large amounts of data stored and annotated in databases such as nucleic acid and amino acid sequence databases and two dimensional (2D)[††] protein databases.

---

[*] Correspondence address: Dr. Ruedi Aebersold, Department of Molecular Biotechnology, University of Washington FJ-20, Seattle, WA 98195 USA. Phone 206 685 4235 Fax 206 685 6932.

[†] Current address: Hewlett-Packard Company, Palo Alto, California.

[‡] Current address: Friedrich-Miescher Institute, Basel, Switzerland.

[**] Current address: Nestec Ltd. Research Center, Lausanne, Switzerland.

[††] ABBREVIATIONS: 2D: two dimensional; 2DE: 2D gel electrophoresis; IEF: isoelectric focusing; SDS-PAGE: sodium dodecyl sulfate - polyacrylamide gel electrophoresis; RP-HPLC: reverse-phase high performance liquid chromatography; ESI-MS: electrospray ionization mass spectrometer/try; PITC 311: 4-(3-pyridinylmethylaminocarboxypropyl)phenyl isothiocyanate; 311 PTH: 4-(3-pyridinylmethylamino-carboxypropyl)phenyl thiohydantoin; CE: capillary electrophoresis; PTM: post translational modification; ER: enzyme reactor.

Methods in Protein Structure Analysis, Edited by M. Z. Atassi and E. Appella
Plenum Press, New York, 1995

The number of sequences entered in sequence databases is growing exponentially at least in part due to coordinated large scale sequencing programs (e.g. Dujon et al., 1994). Such genome sequencing efforts will result in the determination of the complete genome sequence for a few species within the current decade (Collins and Galas, 1993). In addition, systematic cDNA sequencing projects generate increasingly complete databases of the genes expressed in specific tissues. Although access to the most advanced cDNA databases is currently limited, it is expected and hoped that this resource will eventually become generally accessible.

While sequence databases are useful for answering specific questions, the linear structure of the stored information lacks important dimensions which are essential for biologists. Spatial and temporal expression profiles and expression levels, regulatory features including post-translational modifications and polypeptide processing, protein trafficking and turnover, information on interactions with other elements and integration of numerous components into complex pathways are examples of the types of information which are not directly coded for in the DNA sequence and are therefore not extractable from sequence databases alone.

2D protein databases, the display and annotation in a 2D pattern of the expressed and fully processed protein components of a cell or tissue, represents an alternative format to globally store and display information. While the term 2D in this context usually refers to the two dimensions [isoelectric focusing (IEF) and SDS polyacrylamide gel electrophoresis (SDS-PAGE)] used in a gel electrophoresis experiment to separate the hundreds or thousands of proteins in an extract into a 2D pattern, the information content in a 2D protein database is in fact multidimensional. Data dimensions which can be obtained by simple experiments and/or subtractive pattern analysis and can be easily integrated into 2D protein databases include temporal and spatial expression levels, information on regulatory features mediated by covalent protein modifications, protein trafficking and turnover and information on the interaction of polypeptides with other components to form functional protein complexes. 2D gel electrophoresis by itself does however not provide any structural information on the separated species.

The information contained in sequence databases and in 2D protein databases are therefore complementary but not easily linked. Here we describe our approach towards a rapid, sensitive and conclusive analysis of the complete covalent structure of gel separated proteins and illustrate with selected results the current status of these projects. We are pursuing two main objectives: i) characterization of gel separated proteins by their amino acid sequence with the aim of correlating a protein spot in a 2D protein database with the corresponding entry in DNA sequence database and ii) characterization of protein modifications with the aim of understanding regulatory features and protein processing pathways.

IDENTIFICATION OF PROTEINS SEPARATED BY GEL ELECTROPHORESIS

It is frequently required in biological research projects that protein species represented as protein bands or spots separated by gel electrophoresis be further characterized. Such proteins may be detected by comparative protein pattern analysis, western blotting, or as dominant species in samples prepared by protein purification. Characterization of such proteins by their amino acid sequence not only represents the most conclusive criterion for protein identification but also provides a unique basis for further experimentation such as cloning of the corresponding gene, altering specific activities by site-directed mutagenesis and modulating temporal and spatial protein expression patterns. In addition, limited

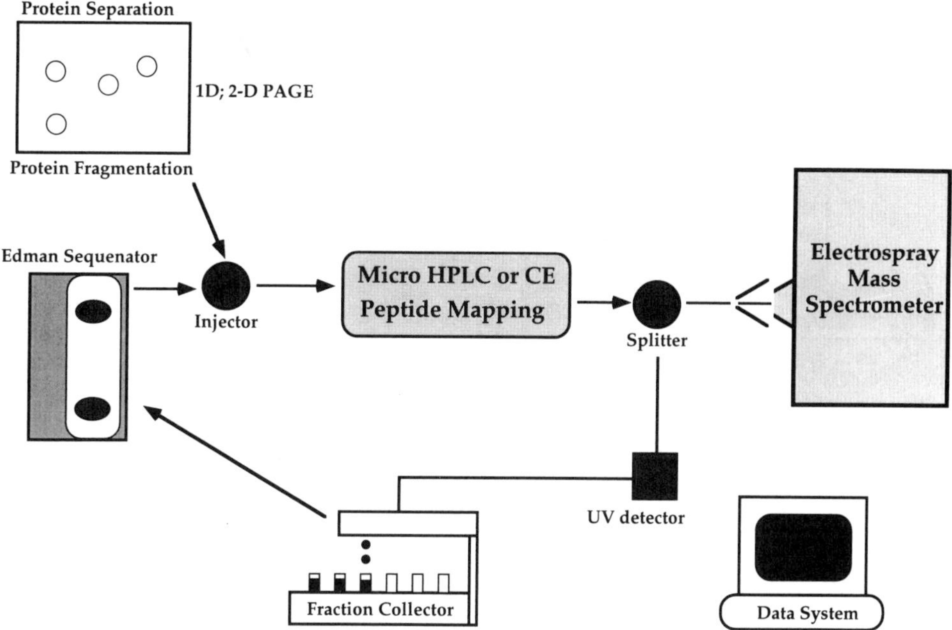

Figure 1. The protein chemistry workstation.

sequence information is suitable to unambiguously establish structural relationships between proteins of comparable electrophoretic mobilities and to characterize protein isoforms with different electrophoretic mobilities generated by differential protein processing or modification.

To rapidly, sensitively and reliably characterize proteins separated by gel electrophoresis at the level of the amino acid sequence we have assembled the protein chemistry workstation shown in Fig. 1. The system consists of a gel electrophoresis unit, a reverse-phase high performance liquid chromatography (RP-HPLC) system, an electrospray ionization mass spectrometer (ESI-MS), a fraction collector, a protein sequencer and a datasystem.

The system was operated in the following way. Proteins separated by gel electrophoresis, SDS-PAGE, IEF or 2D gel electrophoresis (2DE) were electrophoretically transferred from the gel either onto nitrocellulose or onto a membrane with a cationic surface (Immobilon CD) (Millipore Corp.) detected by staining and enzymatically cleaved as described (Aebersold et al.. 1987); Patterson et al.. 1992). Recovered peptides were separated by RP-HPLC and analyzed by on-line ESI-MS. Between the outlet of the chromatography column and the MS ion source we inserted a flow splitting device which split approximately 10% of the column effluent into the mass spectrometer and the remaining 90% of the sample was collected for further analysis such as peptide sequencing. (Hess et al., 1993). Since the ESI-MS is essentially a concentration dependent detector, splitting of the column effluent did not significantly reduce the sensitivity of peptide detection. The LC-MS peptide mapping experiment therefore yields the masses of peptides derived from the protein under investigation with minimal sample loss. In addition to indicating the degree of purity and homogeneity of the collected peptide fractions, as illustrated in Fig. 2, peptide masses can be used to identify a protein in a sequence database using any one of a number of peptide mass search algorithms which were developed independently in the last two years (Henzel et al. 1993; James et al. 1993; Mann et al. 1993; Pappin et al. 1994; Yates et al. 1993).

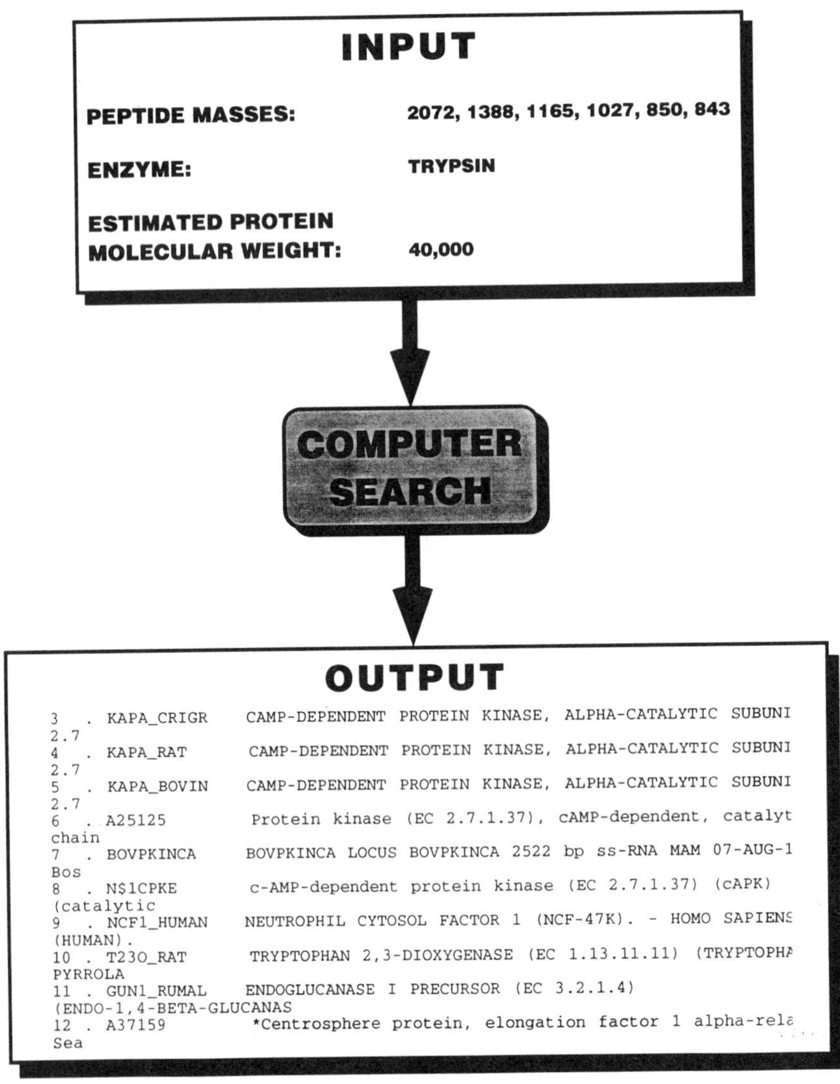

Figure 2. Protein identification by peptide mass search.

Illustration of the peptide mass database search procedure. Required entries for a search are the experimentally determined peptide masses, the estimated mass of the intact polypeptide (as determined by gel electrophoretic mobility) and the type of enzyme used for protein cleavage. Search results specify protein name, originating species, sequence database access code and more. The displayed example used the MOWSE search algorithm (Pappin et al. 1993) to search the OWL sequence database (Akrigg et al. 1988; Bleasby et al. 1990).

In cases in which the peptide mass database search could not or could not conclusively identify the protein, collected, homogeneous peptide fractions were subjected to automated peptide sequencing. To enhance the sensitivity of peptide sequencing we have developed a new degradation chemistry which uses ESI-MS for the detection of the degradation products. The chemistry is based on the reagent 4-(3-pyridinylmethylaminocarboxypropyl)phenyl isothiocyanate (PITC 311) (Bures et al. 1994, Hess et al. 1994) and is further detailed in an article by Bures et al. in this issue.

The chemistry is compatible with commonly used absorptive protein and peptide sequencing protocols. ESI-MS detection of the generated 4-(3 pyridinylmethylaminocarboxypropyl)phenyl thiohydantoins (311 PTH's) affords detection limits at the low femtomole level and provides significant data enhancement by selected ion monitoring. Fig. 3 shows results from a high sensitivity sequencing experiment using PITC 311 and illustrates the value of integrating different types of data obtained by the protein chemistry workstation. 3.2 pmole of bovine carbonic anhydrase (calibrated by quantitative amino acid composition analysis) was cleaved by trypsin, the resulting peptide fragments were separated by RP-HPLC and manually collected. 10% of the eluting peptide sample was split into the on-line ESI-MS system. A peptide of 973.5 Da was subjected to automated sequencing using the PITC 311 chemistry, the 311 PTH's were analyzed by ESI-MS and the data are represented in histogram format in Fig. 3. The specific residue determined in each sequencing cycle (marked with * in Fig. 3) was easily determined. Furthermore, the sum of the molecular weights of the determined amino acids and the experimentally determined molecular weight of the intact peptide helped confirm the determined amino acid sequence.

SUMMARY: PROTEIN IDENTIFICATION

We have developed a protein chemistry workstation for the rapid, sensitive and conclusive identification of proteins separated by gel electrophoresis. The system operates on a two-pass basis. In the first pass proteins are enzymatically fragmented and peptide molecular weights are determined by LC-ESI-MS or CE-ESI-MS. These peptide masses are used to search sequence databases for corresponding protein sequences. The second pass, required for protein identification in cases in which the peptide mass database search is inconclusive, consists of automated sequencing of the collected peptides using PITC 311 and detection of 311 PTH's by ESI-MS. Protein identification by the first pass is fast, simple, does not require any peptide sequencing and is growing in importance with increasing size of sequence databases. Protein identification by the second pass, PITC 311 sequencing is currently slightly more sensitive than peptide sequencing using PITC and yields less ambiguous results due to accurate mass analysis of the 311 PTH's. We anticipate that future developments in the PITC 311 degradation chemistry will make chemical peptide sequencing more sensitive by at least one order of magnitude, faster and more robust. Finally, the data obtained in the first pass and the second pass are synergistic. Integration of these data is useful for the selection of peptides for sequencing, for confirmation of the obtained peptide sequences and for minimizing the chance of sequencing in homogeneous peptide fractions or peptides derived from autocatalysis of proteolytic enzymes.

DETERMINATION OF THE MOLECULAR BASIS OF DIFFERENTIAL ELECTROPHORETIC MOBILITIES OF PROTEINS

There are numerous indications that the products of a single gene are translated and processed into different molecular species which frequently can be resolved by high-resolution gel electrophoresis. Such sets of closely related polypeptides are typically suspected if proteins with comparable mobilities by SDS-PAGE can be resolved by IEF (charge trains in 2DE). While the close structural relatedness can quite easily be verified by 2D western blotting experiments, the molecular basis for the differential mobility is more difficult to

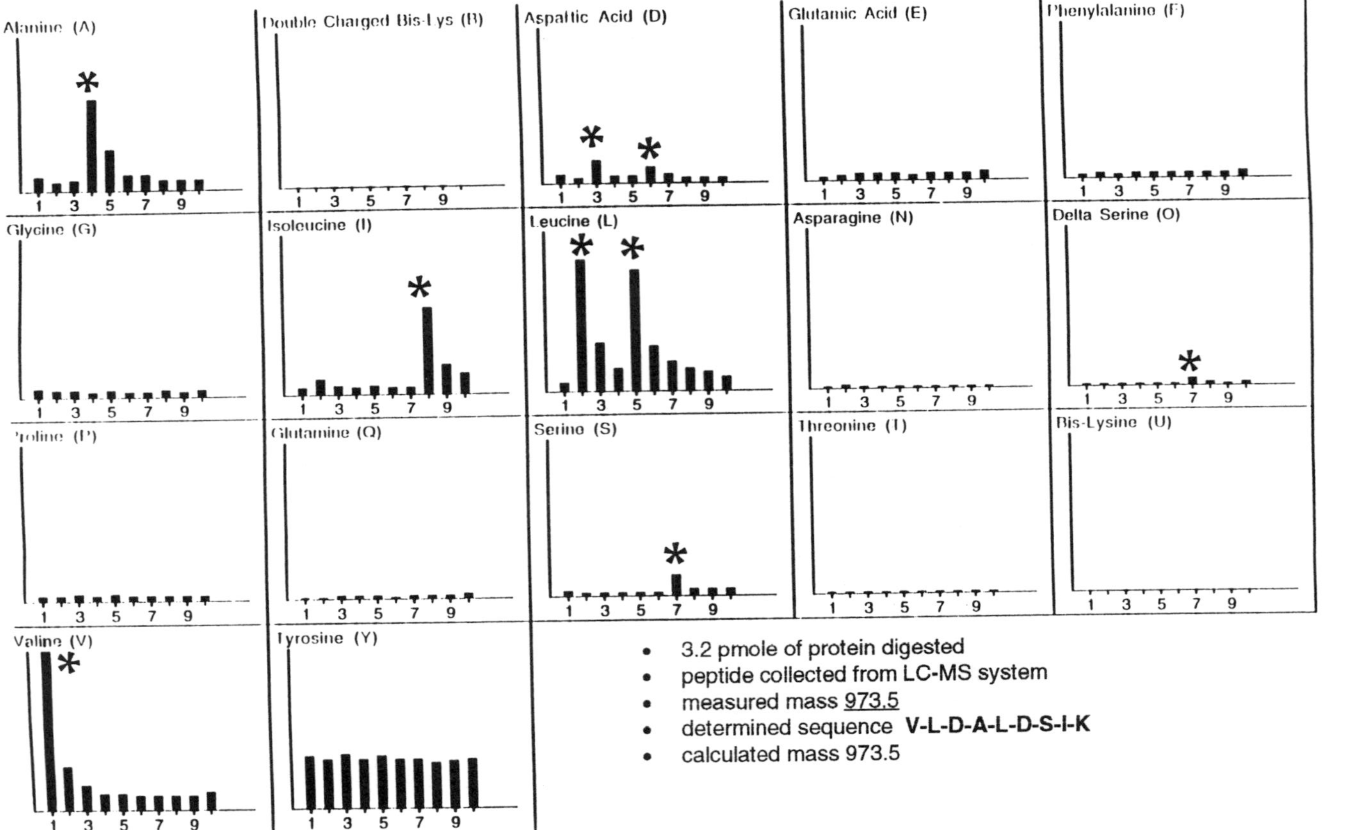

Figure 3. High sensitivity PITC 311 sequencing. A 3.2 pmole sample of bovine carbonic anhydrase was cleaved with trypsin and the resultant peptide fragments were separated by RP-HPLC and collected manually. One of these collected peptides (mass 973.5 Da) was subjected to automated sequence analysis using PITC 311 and the resultant 311 PTH's were monitored by ESI-MS operated in the multiple ion monitoring mode. Each panel depicts a cycle-by-cycle histogram of the abundance of ESI-MS signal corresponding to the amino acid indicated in each graph. The amino acid residue specific for the respective sequencing cycle are marked with *. Delta-S: 311 PTH dehydro serine.

elucidate. Since most sequences are determined by DNA sequencing which does not account for post translational processing and modifications, there is a definitive need for advanced technology to investigate the complete covalent structure of fully processed proteins. Localization of a known modification within a polypeptide sequence and "de-novo" determination and localization of modified residues, respectively, represent two predominant technological challenges.

LOCALIZATION OF MODIFIED RESIDUES WITHIN A POLYPEPTIDE SEQUENCE

Among the many types of protein modifications described to date (Wold 1981), (reversible) protein phosphorylation, mainly on serine, threonine and tyrosine residues is intensely studied for its essential regulatory role in many physiological processes. To localize the sites of protein phosphorylation we developed a post translational modification (PTM) analyzer as shown in Fig. 4.

The system consists of a micro enzyme reactor (ER), a separation instrument [HPLC or capillary electrophoresis] and an ESI-MS system. All three components are connected on-line. A data system controls the operation of the analyzer and integrates the generated data.

The use of a phosphatase ER for the determination of sites of protein tyrosine phosphorylation illustrates the operation of the PTM analyzer. Differentially phosphorylated polypeptides were prepared and enzymatically cleaved as described above. An aliquot of the recovered peptide mixture was subjected to enzymatic de-phosphorylation in the phosphatase ER, and the reaction products were separated and analyzed by CE-ESI-MS or by

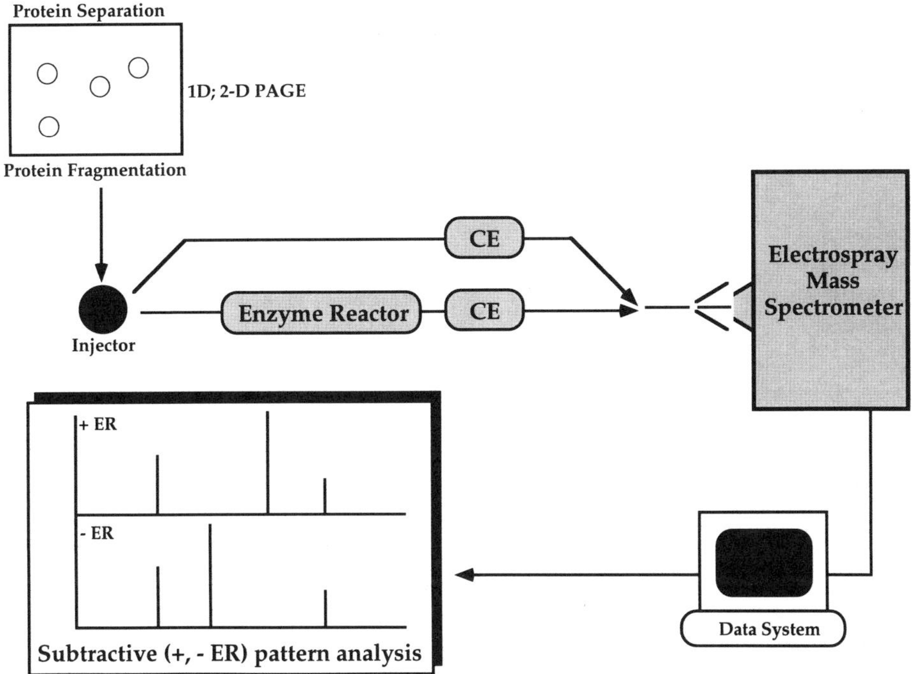

Figure 4. The post-translational modification analyzer

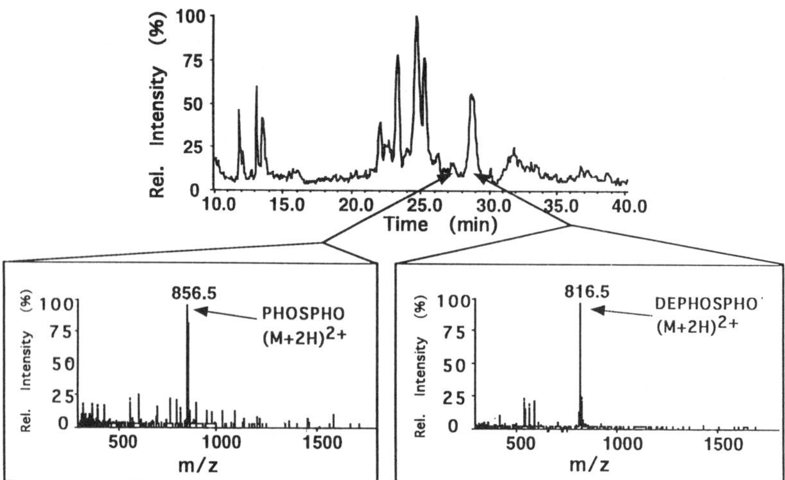

Figure 5. Determination of the site of protein tyrosine phosphorylation using the PTM analyzer. TCR/CD3 ζ chain, phosphorylated in vitro with the tyrosine kinase $p56^{lck}$ was cleaved with trypsin. Resulting peptides were analyzed in a PTM analyzer as described in Fig. 4, consisting of the following components: ER: human tyrosine phosphatase β immobilized on biotin-avidin-methacrylate; RP-HPLC: C18 capillary column; ESI-MS: Model API III mass spectrometer (PE/SCIEX). Total ion current (300-2000 Da) of peptides after ER exposure are shown. Inserts show the mass spectra of the peaks indicated by the arrows. The characteristic differences in elution time and measured mass identifies the two analyzed species as the phospho- and dephospho form, respectively, of the same peptide.

LC-ESI-MS, depending on the configuration of the system. A second aliquot of the same sample was analyzed in a similar way, except that it was not subjected to enzymatic dephosphorylation. The MS data from both samples were compared in the datasystem and phosphopeptides were identified by the characteristic change in mobility in the separation system as well as by the characteristic change in peptide mass induced by enzymatic dephosphorylation (reduction in peptide mass by 80 Da) per phosphate group removed. The data shown in Fig. 5 further illustrate the procedure. The cytoplasmic part of the T cell receptor ζ chain was phosphorylated in vitro with the tyrosine kinase $p56^{lck}$ as described (Watts et al. 1992), the phosphoprotein was cleaved with trypsin and the resulting peptide mixture was analyzed as described above.

The total ion current shown in Fig. 5 represents all the ions detected within the mass range 300-2000Da from the sample that was dephosphorylated in a ER consisting of immobilized human tyrosine phosphatase β and separated by capillary RP-HPLC (Amankwa et al., 1995). The ER was constructed by immobilizing a metabolically biotinylated human tyrosine phosphatase β fusion protein on avidin-modified sepharose packed in a capillary column. Mass spectral analysis of two peaks eluting around 28 min. indicated that the later eluting peptide [$(M+2H)^{2+}$=816.5] was the phosphatase product (the dephospho form) of the slightly earlier eluting phosphopeptide [$(M+2H)^{2+}$=856.5]. Comparative, systematic analysis of all the peptides present in the samples with or without phosphatase treatment allowed us to localize 5 additional phosphorylation sites in the ζ chain sample. ER's for CE-MS were constructed in a similar way by immobilizing the same phosphatase fusion protein onto the inner surface of fused silica capillaries coated with avidin as described (Amankwa and Kuhr, 1992). We used this type of ER to identify a site of tyrosine phosphorylation on the human platelet derived growth factor receptor (Amankwa et al., 1995).

The system has the following advantages and limitations. i) Two independent criteria, a shift in retention time and the characteristic mass reduction by 80 Da conclusively identify phosphopeptides in a peptide mixture. ii) If the sequence of the polypeptide and the specificity of the proteolytic enzyme used are known, the mass of the un-phosphorylated peptide is usually sufficient to unambiguously identify the phosphopeptide within the polypeptide sequence. iii) The system operates in an automated manner and at a low picomole sensitivity level and does not require radiolabeling of the sample. iv) The substrate specificity of the immobilized phosphatase distinguishes between serine, threonine and tyrosine phosphorylated peptides. v) In peptides which contain more than one site which could be phosphorylated, the exact sites of phosphorylation need to be determined by phosphopeptide sequencing (Aebersold et al., 1991; Meyer et al., 1991; Wettenhall et al., 1991) vi) The approach is general and can be extended to the analysis of essentially any protein modification for which a distinguishing enzyme reaction is known, and vii) The approach is easily adaptable to other types of analytical instrumentation such as matrix assisted laser-desorption time-of-flight MS.

DE-NOVO CHARACTERIZATION AND LOCALIZATION OF MODIFIED RESIDUES

De-novo characterization of modified amino acid residues is an important task in analytical protein biochemistry. Basic research into the structure and function of proteins continues to uncover novel types of protein modifications, the structures of which need to be analyzed. In the biotechnology industry the covalent structure of recombinant, overexpressed proteins needs to be documented. Frequently, such proteins carry modified residues. Metabolic products of pharmaceuticals potentially modify selected proteins and interfere with their function. Characterization of modified residues is therefore an important aspect of pharmacology and toxicology. Current chemical protein sequencing technology is of very limited value for the structural characterization of modified residues. We have therefore evaluated the potential of automated peptide sequencing using PITC 311 and ESI-MS detection of the resulting 311 PTH's.

The procedure is illustrated by the results shown in Fig. 6. A synthetic polypeptide containing the regulatory tyrosine residue in the protein tyrosine kinase p60^{src} was cleaved with trypsin and the phosphopeptide was isolated by RP-HPLC as described above. To allow for maximum flexibility in the extraction conditions during automated sequencing, the phosphopeptide was covalently attached to an arylamine-modified polyvinylidene fluoride membrane (Millipore Corp.) (Coull et al. 1991) applied to the protein sequencer cartridge and sequenced using PITC 311 following degradation protocols optimized for this chemistry (Nika et al. , in prep.). The samples extracted from the sequencer cartridge were scanned by on-line ESI-MS for the presence of an ion corresponding to 311 PTH phosphotyrosine $[(M+H)^{+}=555]$. Fig. 6 illustrates the presence of the expected mass in cycles 7 and 8, whereas the signal is absent in the cycles preceding cycle 7 and in later cycles. A phosphotyrosine residue was therefore positively identified in cycle 7 of the peptide.

We have used to same approach to structurally characterize additional types of modified residues including catalytic nucleophiles in the active site of glycosidases which were covalently modified with mechanism-based enzyme inhibitors (S. Lawson, D. Tull, S. Withers, University of British Columbia, unpublished). In these cases the samples extracted from the sequencer were analyzed by ESI-MS scanning an appropriate extended mass range for the presence of derivatives of modified amino acid residues (data not shown). In cases in which the 311 PTH mass did not unambiguously identify the modified residue, the structure of the derivative was further analyzed by collision-induced fragmentation in a tandem MS experiment (data not shown).

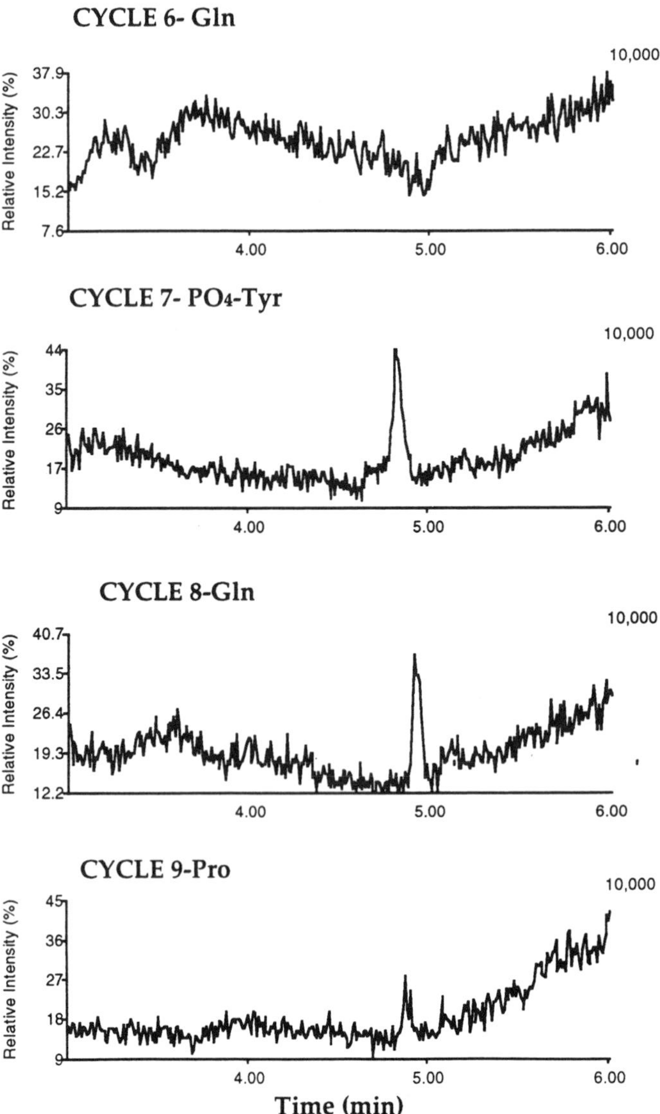

Figure 6. Determination of site of protein tyrosine phosphorylation by automated peptide sequencing/ESI-MS detection. A tyrosine phosphorylated peptide derived from p60^{src} was applied to automated protein sequencer and sequenced as described in the text. The sample extracted from the sequencer after each sequencing cycle was injected onto a C18 RP-HPLC column (Reliasil C18; 1x50 mm) (Michrom Bioresources) and the column effluent was analyzed by ESI-MS. The MS was operated in the multiple ion monitoring mode. Data corresponding to the extracted mass of 311 PTH phosphotyrosine [$(M+H)^+$=555] are displayed for the sequencing cycles around the detected phosphorylation site.

CONCLUSIONS

In this manuscript we describe a suite of complementary techniques for the high sensitivity determination of the covalent structure of proteins separated by gel electrophoresis. The described techniques focus on the two tasks which in our view will dominate the work of analytical protein chemists. The first is the identification of proteins relevant to a biological system or process at the level of the primary structure and the second is the determination and localization of modified or unnatural residues within an amino acid sequence. For both tasks we have developed two-pass processes which consist of a rapid, sensitive and simple initial screen which is followed, if necessary, by a more detailed, slower, less sensitive but general and conclusive secondary analysis.

Our strategy to identify proteins at the sequence level takes advantage of the power of rapidly growing sequence databases, the rapidly evolving capabilities in protein and peptide mass spectrometry and includes a new sequencing chemistry for amino acid sequencing at enhanced sensitivity.

The technique for the localization and identification of modified residues within polypeptides combines the specificity of enzyme reactions with the sensitivity and reliability of LC-MS and CE-MS analysis. For de-novo identification of modified residues we rely on chemical stepwise peptide degradation using a novel protein sequencing chemistry.

The described systems are modular. They consist of several components which are easily interfaced. Individual components can be easily exchanged without interfering with the performance of the system. For example, the described systems are compatible with a variety of gel electrophoresis techniques, with essentially any separation technique and with different MS techniques and instruments.

The central part of all the described instruments is a data system which stores and integrates the data generated by the subsystems and allows for data analysis and interpretation in a synergistic manner. Clearly it is these data interpretation and analysis aspects, together with systems integration aspects will require significant research and development efforts to make protein analytical technology even more powerful.

ACKNOWLEDGMENTS

This work was funded in part by the Department of Industry, Science and Technology, Canada (ISTC), the Canadian Human Genetics Network of Centres of Excellence, and the National Science Foundation Science and Technology Center for Molecular Biotechnology. R.A. was the recipient of a Medical Research Council (MRC) of Canada scholarship. We thank Drs. Tom Covey and Bori Shushan for stimulating discussions which lead up to some of the concepts described here. The excellent editorial assistance of Inge van Oostveen is gratefully acknowledged.

REFERENCES

Aebersold, R., Leavitt, J., Saavedra, R., Hood, L.E., Kent, S.B.H., 1987, Internal amino acid sequence analysis by *in situ* protease digestion on nitrocellulose of proteins separated by one- or two-dimensional gel electrophoresis, *Proc. Natl. Acad. Sci. USA* , 84, 6970-6974.

Aebersold, R., Watts, J.D., Morrison, H.D., Bures, E.J., 1991, Determination of the site of tyrosine phosphorylation at the low picomole level by automated solid-phase sequence analysis. *Anal. Biochem.*, 199, 51-60.

Akrigg, D., Bleasby, A.J., Dix, N.I.M., Findlay, J.B.C., North, A.C.T., Parry-Smith, D., Wootton, J.C., Blundell, T.I., Gardner, S.P., Hayes, F., et al., 1988, A protein sequence/structure database, *Nature*, 335, 745-746.

Amankwa, L.N., Kuhr W.G. 1992, Trypsin-modified fused-silica capillary microreactor for peptide mapping by capillary zone electrophoresis. *Anal. Chem.* 64, 1610-1613.

Amankwa, L.N., Harder, K., Jirik, F., Aebersold, R., High sensitivity determination of sites of protein tyrosine phosphorylation by on-line capillary electrophoresis and electrospray ionization mass spectrometry. *Protein Science* (1995), 4:113-125.

Bleasby, A.J., Wootton, J.C.,1990, Construction of validated, non-redundant composite protein databases, *Protein Eng.*, 3, 153-159.

Bures, E.J., Nika, H.,Chow, D.T., Hess, D., Morrison, H.D., Aebersold, R.,1994, Synthesis of the protein sequencing reagent 4-(3-pyridinylmethyl-aminocarboxypropyl) phenyl isothiocyanate and characterization of 4-(3-pyridinylmethylaminocarboxypropyl) phenylthiohydantoins, *Anal.Biochem*, (1995), 244:364-372.

Collins, F., Galas, D.,1993, A new five-year plan for the U.S. Human Genome Project, *Science*, 262, 43-46

Coull, J.M., Pappin, D.J.C., Mark, J., Aebersold, R., Koester, H. 1991. Functionalized membrane supports for covalent protein microsequence analysis, *Anal. Biochem.* 194, 110-120.

Dujon, B., Alexandraki, D., Andr'e, B., Ansorge, W., Baladron, V., Ballesta, J.P., Banrevi, A., Bolle, P.A., Bolotin-Fukuhara, M., Bossier, P., et al., 1994, Complete DNA sequence of yeast chromosome XI, *Nature*, 369, 371-378.

Henzel, W.J., Billeci, T.M., Stults, J.T., Wong, S.C., Grimley,C., Watanabe, C., 1993, Identifying proteins from two-dimensional gels by molecular mass searching of peptide fragments in protein sequence databases, *Proc.Natl.Acad.Sci.U.S.A.*, 90, 5011-5.

Hess, D., Covey, T.C., Winz, R., Brownsey, R., Aebersold, R., 1993, Analytical and micropreparative peptide mapping by high-performance liquid chromatography/ion spray mass spectrometry of proteins purified by gel electrophoresis, *Prot.Sci.* , 2, 1342-1351.

Hess, D., Nika, H., Chow, D.T., Bures, E.J., Morrison, H.D., Aebersold, R., 1994, Liquid chromatography-electrospray ionization mass spectrometry of 4-(3-pyridinylmethylaminocarboxypropyl) phenylthiohydantoins,*Anal.Biochem*, (1995), 224:373-381.

James, P., Quadroni, M. , Carafoli, E., Gonnet, G., 1993, Protein identification by mass profile fingerprinting, *Biochem.Biophys.Res.Commun..*, 195(1), 58-64.

Mann, M., Hojrup, P., Roepstorff, P., 1993, Use of mass spectrometric molecular weight information to identify proteins in sequence databases, *Biol.Mass.Spectrom.* 22, 338-45.

Meyer, H.E., Hoffmann-Posorske, E., Donella-Deana, A., Korte, H., 1991 Sequence analysis of phosphotyrosine-containing peptides *in* Methods in Enzymology (Hunter, T., Sefton, B.M., Eds.), Vol. 201, Academic Press, Orlando, FL.

Pappin, D.J.C., Hojrup, P., Bleasby, A.J., 1993, Rapid identification of proteins by peptide-mass fingerprinting, *Current Biology*, 3, 327-332.

Patterson, S.D., Hess, D., Yungwirth, T., Aebersold, R., 1992, High yield recovery of electroblotted proteins and cleavage fragments from a cationic polyvinylidine fluoride based membrane, *Anal. Biochem.*, 202, 193-203.

Watts, J.D., Wilson, G., Ettehadieh, E., Kubanek, C.A., Astell, C.R., Marth, J.D., Aebersold, R., 1992, Purification and initial characterization of the lymphocyte-specific protein tyrosine kinase $p56^{lck}$ from a baculovirus expression system, *J. Biol. Chem.*, 267, 901-907.

Wettenhall, R.E.H., Aebersold, R., Hood, L.E., Kent, S.B.H., 1991, Solid-phase sequencing of ^{32}P labeled phosphopeptides at picomole and subpicomole levels *in* Methods in Enzymology (Hunter, T., Sefton, B.M., Eds.), Vol. 201, Academic Press, Orlando, FL.

Wold, F., In vivo chemical modification of proteins, 1981, *Ann.Rev. Biochem.*, 50, 783-814.

Yates, J.R. III, Speicher, S., Griffin, P.R., Hunkapiller, T., 1993, Peptide mass maps: a highly informative approach to protein identification, *Anal.Biochem.* , 214, 397-408.

2

NEW STRATEGIES IN HIGH SENSITIVITY CHARACTERIZATION OF PROTEINS SEPARATED FROM 1-D OR 2-D GELS

Kris Gevaert,[1] Marc Rider,[2] Magda Puype,[1] Jozef Van Damme,[1] Stefaan De Boeck[1] and Joël Vandekerckhove[1]

[1] Laboratory of Physiological Chemistry
University of Gent
Ledeganckstraat 35, B-9000 Gent, Belgium
[2] Hormone and Metabolic Research Unit
International Institute of Cellular and Molecular Pathology
Avenue Hippocrate 75, B-1200 Brussels, Belgium

INTRODUCTION

One-dimensional (1-D) or two-dimensional (2-D) polyacrylamide gel electrophoresis is a convenient technique for purifying small amounts of proteins from very complex mixtures (O'Farrell, 1975; Celis & Bravo, 1984). For structural analysis, proteins are electrotransferred from the gels onto immobilizing membranes for subsequent NH_2-terminal sequence analysis (Vandekerckhove *et al.*, 1985; Aebersold *et al.*, 1986; Matsudaira, 1987). Alternatively, proteins can be cleaved either as membrane-bound molecules (Aebersold *et al.*, 1987; Bauw *et al.*, 1988) or when still present in the gel matrix (Rosenfeld *et al.*, 1992). The resulting peptides are then separated for further characterization. Recently, computer searching algorithms have been developed that use peptide mass fingerprinting to identify proteins whose sequences are stored in databases (Mann *et al.*, 1993; Pappin *et al.*, 1993; Yates *et al.*, 1993). Such peptide mass information can be obtained from previously unseparated mixtures using matrix assisted laser desorption ionization time of flight mass spectrometry (MALDI-TOF-MS) (Mann *et al.*, 1993, Zhang *et al.*, 1994) or from a reversed phase column eluate on-line connected with an electrospray ionization mass spectrometer (ESI-MS). Conventional automated Edman degradation techniques or mass spectrometric-based methods allow sample analysis in the low picomole or even femtomole range. Unfortunately, when only such small amounts are present in the starting mixture it is difficult to purify and digest the protein with high peptide recoveries. This is probably due to adsorption of a fraction of the protein within the pores of the immobilizing membrane by which they are trapped, out of reach of the proteases. In case of *in-gel* cleavage a fraction of the protein or its fragments stay inside the gel matrix. The problems mentioned above can be reduced by working in small but highly concentrated protein spots, thus reducing the membrane surface or gel matrix volume. When only small amounts of

Methods in Protein Structure Analysis, Edited by M. Z. Atassi and E. Appella
Plenum Press, New York, 1995

proteins are present in the gels, these conditions are seldomly met and therefore it is necessary to combine several protein spots into a small volume. In this paper we describe a method that allows to reproducibly elute and concentrate proteins from combined gel pieces. The concentration factor can be larger then 50 and the protein is concentrated into an agarose gel. The protein is then melted out of the agarose gel prior to proteolytic cleavage so that the digestion proceeds in a soluble phase. The overall peptide yields are at least 70% of those obtained from direct cleavage in free solution. This approach does not suffer from the problems of adsorption or *in-gel* trapping and should therefore be a better procedure. In addition the technique is amenable for miniaturization. Here we report our initial experiences with this technique and show that it can serve as an efficient link between polyacrylamide gel purification and protein identification by microsequencing or mass spectrometry in the very low picomole range.

MATERIALS AND METHODS

Tosyl-L-phenylalanine chloromethylketone-treated trypsin was obtained from Sigma Chemical Company, St. Louis, USA. Agarose (ultrapure, electrophoresis grade) was from BRL Life Technologies Inc., Gaithersbury, USA. Recombinant rat liver PFK-2/FBPase-2 was given to us by Dr. Crépin (International Institute of Cellular and Molecular Pathology; Brussels) and rabbit skeletal muscle actin was prepared as described by Spudich and Watt (1971). All other chemicals were from Janssen Chimica, Beerse, Belgium; Merck, Darmstadt, Germany or Serva, Heidelberg, Germany.

SDS-Polyacrylamide-Gel Electrophoresis

Proteins were separated by SDS-PAGE using the mini-gelelectrophoresis design originally published by Matsudaira and Burgess (1978). The gels were 0.5 mm thick. Proteins were detected by staining for 20 min. in a solution of 0.25% (w/v) Coomassie Brilliant Blue R250 (Serva, Heidelberg, Germany) in 45% (v/v) methanol/9% (v/v) acetic acid. Destaining was carried out in 5% (v/v) methanol/7.5% (v/v) acetic acid for 2 hours. Protein bands were excised, collected in Eppendorf tubes and washed in water for 1 hour by agitation. The washed protein bands were cut in ± 1 mm x 1 mm pieces and equilibrated for at least 1 h in 100-200 μl of sample buffer (1% (w/v) SDS, 10% (v/v) glycerol, 50 mM dithiothreitol, 12 mM Tris/HCl pH 7.1). Note that the sample buffer does not contain Bromophenol Blue. This sample is ready for elution and concentration (see below).

Narrowbore Reverse Phase HPLC and Microsequencing

Peptides were separated by reverse phase HPLC on a C18 Vydac 2.1 mm x 250 mm column (Separations Group, Hesperia, CA, USA), equilibrated in 0.1% (v/v) trifluoracetic acid/5% (v/v) acetonitrile (solvent A). A linear gradient program of 5-100% B in 100 min, where B = 0.1% (v/v) trifluoroacetic acid/70% (v/v) acetonitrile, was used to elute the peptides at a flow rate of 80 μl/min, and the absorbance was monitored at 214 nm using an Applied Biosystems 759 A absorbance detector. Peptides were collected by hand in Eppendorf tubes and directly applied on the gas-phase sequenator (Applied Biosystems model 470A or 477A) equipped with a 120A phenylthiohydantoin-amino acid analyzer.

On-Line Electrospray Ionization Mass-Spectrometry

The outlet of the narrowbore HPLC (run as described above) was connected to a solvent splitter which directed 80% of the eluate into the absorbance detector and 20% into the mass spectrometer (Fisons/VG Platform, Manchester, UK) via an interface employing a 0.0025 inch i.d. microbore polyetheretherketone tubing inlet. The instrument is equipped with an electrospray ion source and the m/z ratios were measured by a quadrupole analyzer. The flow rate of the peptide carrier solvent at the inlet of the interface was 16 µl/min. Droplet evaporation was achieved by a flow of warm (60°C) dried nitrogen gas. The mass spectrometer was scanned from m/z 300 to 1300 at a rate of 6s per scan during the first part of the chromatogram (between 15 and 45 min after injection). In the second part of the chromatogram (from 44 min to the end of the run) scans were made every 6s from m/z 500 to m/z 1600.

Matrix Assisted Laser Desorption Ionization Time of Flight Mass Spectrometry

Actin, concentrated in a volume of ± 6µl agarose gel (1.45% agarose, 0.1% SDS, 0.36 M Tris-HCl pH 8.7) was heated at 80°C for 5 min. The molten gel was mixed with an equal volume of digestion buffer, containing 0.1 M Tris-HCl pH 8.5 and 0.1 µg trypsin and kept at 0°C. Rapid mixing produces a fast drop in the sample temperature, limiting the rate of trypsin autodigestion. The sample was kept at 37°C overnight as a solid gel and the digestion was terminated by melting the gel at 80°C. An aliquot of 5 µl was removed from the molten gel, acidified with 1 µl of 50% trifluoroacetic acid and mixed with an equal volume of a saturated solution of α-cyano-4-hydroxycinnamic acid in acetonitrile and trifluoroacetic acid (1:2 v/v) (Beavis *et al.*, 1992). 2 µl of this mixture was applied to the sample support and dried. The MALDI mass spectra were recorded in the linear mode with a Bruker Bioflex (Bruker Instruments Inc., Bremen, Germany). The spectra shown represent the accumulation of 150 sample shot spectra taken with a conventional UV laser (nitrogen, 337 nm) set at an attenuation of 50-35. The acceleration voltage was at 28.5 kV and low molecular weight ions were deflected with a 1.7 kV puls of 500 nanoseconds.

RESULTS AND DISCUSSION

Protein *in-Gel* Concentration Procedure

The construction of the mini-agarose concentration gel is shown in Fig. 1. The gel is cast between two glass plates 10 cm x 9 cm, separated by spacers 1 cm wide and 0.75 mm thick and clamped together. They are sealed at the vertical edges with molten 2% agarose. A strip of Whatman 3 MM paper is inserted at the bottom and serves as support for the lower agarose gel stopping it from slipping during electrophoresis. The sample well is formed by a 1 cm wide x 0.75 mm thick spacer set between two parallel spacers each 0.5 cm wide x 0.75 mm thick inserted at the center of the glass plates and attached with adhesive tape at the top edge of the back plate. The volume of the sample well can be varied by changing the depth of the slot forming spacers and the height of the agarose concentration gel. The most convenient construction is shown in figure 1.

The lower gel is an agarose gel, 2 cm deep, 1.45% w/v agarose in 0.36 M Tris/HCl, pH 8.7, containing 0.1% SDS (w/v), poured as a freshly prepared hot molten solution. Once the agarose has set, it is overlayed with the polyacrylamide stacking gel, composed of 5.45% acrylamide, 0.13% bisacrylamide, 0.12 M Tris/HCl pH 6.8 and 0.1% SDS. The level of the stacking gel reaches up to 1 cm from the top edge of the front plate. When the stacking gel has set, the central spacer is removed, leaving a well ± 2 cm high, 1 cm wide and 0.75 mm thick. The mini concentration gel is then mounted on a small electrophoresis tank (LKB-

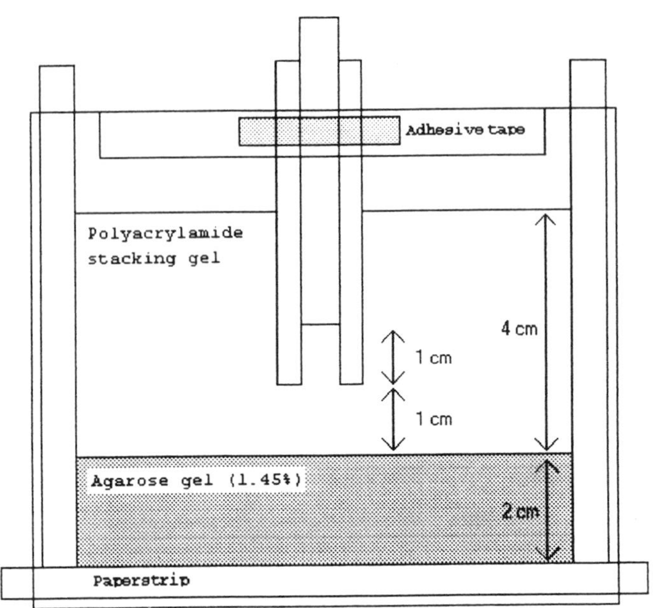

Figure 1. Construction of the mini agarose concentration gel.

Produkter AB, Bromma, Sweden) and the slot is filled with gel pieces collected from Coomassie Blue stained protein bands cut from primary mini-polyacrylamide gels and equilibrated with sample buffer. The sample well can accommodate approximately fifty 1 x 1 mm gel pieces. The remaining volume is further filled with blank gel pieces also equilibrated in sample buffer and if necessary with additional sample buffer, such that the level is 5 mm above that of the edge of the stacking gel. The electrophoresis run is started at 100 V, allowing the proteins to elute out of the combined gel pieces. The concentration effect is obtained by a combination of protein stacking and a horizontal compression of the stacked proteins. This is illustrated in figure 2 showing a series of time-lapse photographs of the migration of a coloured protein through the mini-concentration gel. In this particular experiment, the sample well was filled up with blank gel pieces equilibrated in sample buffer to which a cytochrome c solution was added. After running at 100 V, the protein enters the stacking gel as a sharp band between the two vertical sample spacers (Fig. 2A). At this point the central spacer is re-inserted into the sample well and the voltage is increased to 150-200 V. The migration of the protein between the spacers is thus retarded with respect to the migration of the solvent front down the sides of the spacers (Fig. 2B). The slot forming spacer is again removed when the solvent front on the outside has passed the end of the two vertical spacers and moves inwards (Fig. 2C). The protein band which has been retarded is now compressed into a small spot in the stacking gel (Fig. 2D). This spot moves further into the agarose gel (Fig 2E). The electrophoresis procedure is stopped when the protein has migrated in the agarose gel over a distance of approximately 5 mm (Fig 2F). Note that the sample buffer used in this experiment did not contain the tracking dye Bromophenol Blue. When Coomassie Blue is present in the protein samples it is well separated from the protein probably by an isotachophoretic effect, even though the agarose has no separating capacity. Several points are important in order to obtaining good protein concentration.

1. Protein concentration is controlled by re-inserting the slot forming spacer during electrophoresis, so that the migration of the protein between the central spacers

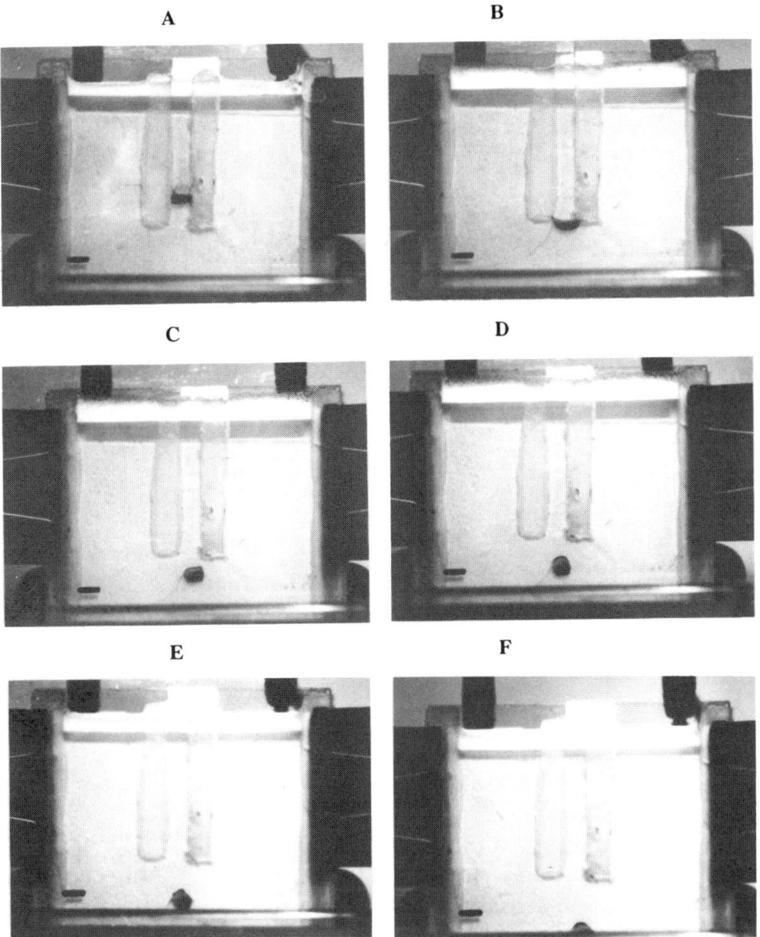

Figure 2. Time-lapse photographs of the migration of a protein through the agarose concentration gel.

acquires a ± 1 cm handicap versus the front in the stacking gel before it leaves the exit of the sample well (see below).

2. Control of the pH difference between the sample and the stacking gel is important for obtaining good sample concentration. The sample pH is 7.1, that of the stacking gel is 6.8. Therefore, the polyacrylamide gel pieces must be washed extensively with distilled water to remove remaining acetic acid of the destain before equilibrating in sample buffer.
3. The gel pieces should be kept as small blocks and not crushed in order to avoid clogging the sample slot and trapping air bubbles formed during electrophoresis.
4. It is important that the sample well is filled completely with gel pieces. If the protein-containing gel pieces do not completely fill the well; blank gel pieces can be added to build up the level.

Following fixation and staining with Coomassie Blue, the protein is seen as a ± 2 mm^2 spot, representing a concentration factor of about 50 fold. The gel is carefully washed with distilled water to remove excess of acid. The spot is then excised in a minimal volume of agarose gel (± 5 μl) and transferred into an Eppendorf tube. The agarose gel is melted in

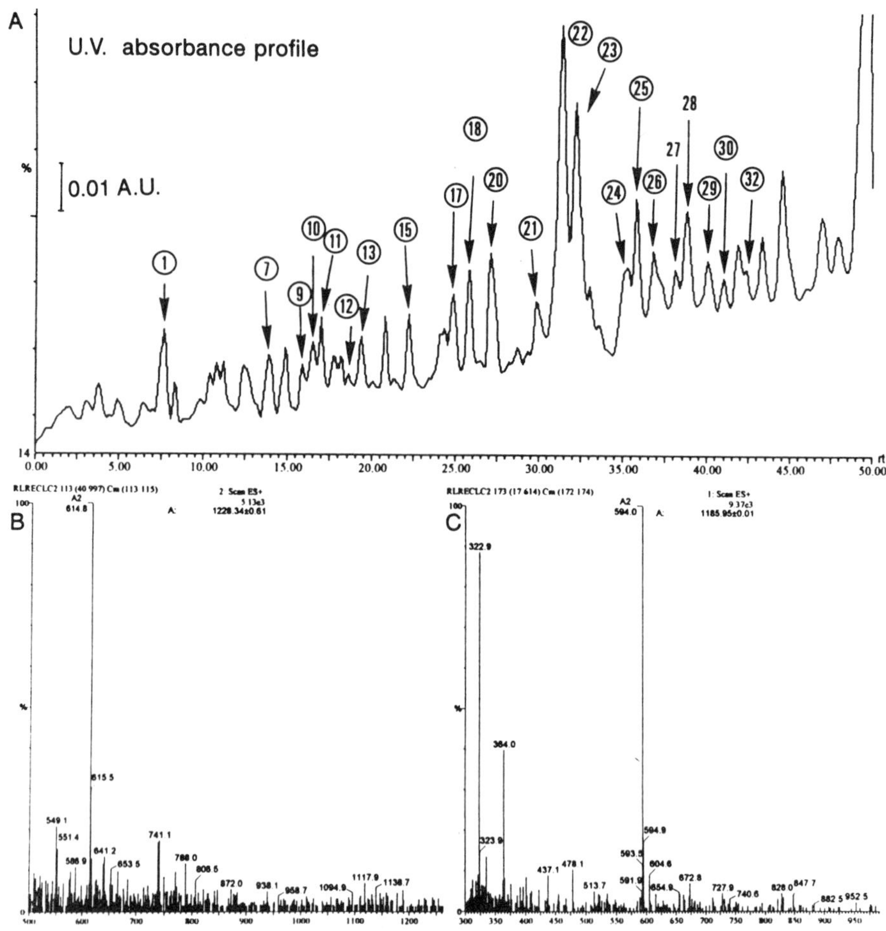

Figure 3. Tryptic peptide profile of recombinant rat PFK-2/FBP-ase-2 analysed by on-line narrowbore HPLC/ESI-MS.

50 μl of digestion buffer (0.1 M Tris/HCl, pH 8.7, 0.2% n-octyl-ß-glucoside) by heating at ± 80°C for ± 2 min and cooled to 37°C. Trypsin (0.1 μg in 1 μl) is added and the digestion is carried out overnight at 37°C. Following digestion, the peptide mixture is frozen at -80°C for at least 1 h, thawed and centrifuged at full speed during 5 min in an Eppendorf centrifuge to remove the precipitated agarose. The supernatant is used for mass spectrometry and narrowbore HPLC analysis.

Characterization of *in-Gel* Concentrated Proteins

The procedure described above is illustrated using two proteins: recombinant rat 6-phosphofructose-2-kinase/fructose 2,6-bisphosphophatase (PFK-2/FBPase-2) and skeletal muscle actin. In the first experiment recombinant PFK-2/FBPase-2 was passed over mini SDS-PAGE. A total amount of 3.5 μg or 63 pmol was divided over five wells of the primary gel. After Coomassie Blue staining, protein bands were excised, combined, concentrated in an agarose gel and digested with trypsin in molten gel. The peptide solution was frozen at -80°C and thawed. The agarose in the pellet fraction was separated from the peptide mixture

Tryptic peptides of PFK-2/FBP-ase-2

Number	Position	Theoretical mass (Da)	Found mass (Da)
1	231-238	968.0	967.4
7	55-60	711.8	712.9
9	357-360	656.8	658.6
10	269-278	932.2	932.0
11	448-457	1186.3	1186.0
12	292-299	818.9	818.7
13	3-11	1064.2	1063.0
15	258-266	1054.1	1054.5
17	122-136	1646.7	1646.1
17	364-373	1195.2	1195.8
18	74-82	1152.3	1153.3
20	458-470	1444.5	1444.0
21	407-415	1051.1	1050.4
22	15-28	1596.8	1596.0
23	153-171	2194.5	2193.0
24	64-73	1192.4	1191.7
25	308-323	1834.1	1834.2
26	89-104	2013.1	2012.4
29	374-383	1228.5	1228.3
30	31-52	2303.7	2305.6
32	281-291	1329.5	1328.7

Figure 3. *Continued.*

in the supernatant by centrifugation in an Eppendorf centrifuge. Peptides were separated by narrowbore reversed phase HPLC with a solvent splitter directing 80% of the eluting peptides into the UV detector and 20% into the mass spectrometer. The UV absorbance profile is shown in figure 3A. The m/z spectra of some selected peptides are shown in figures 3B and 3C. Peptides whose molecular weights could be deduced from the scans are listed in Fig. 3. Of the 23 peptides measured, 21 were assigned as fragments of the recombinant protein (Crepin *et al.*, 1989). This was further confirmed by NH_2-terminal sequence analysis (results not shown). The two remaining peptides are probably trypsin autodegradation products.

This experiment provides important points of information necessary for future development and application of this technique. First, proteins available in the 50 picomole range (here we refer to total amounts loaded on the primary gel) can be recovered from combined gel pieces and digested in the agarose solution with a high overall recovery. In the example shown, peptides are recovered with ± 75% yield. Second, the resolution of the HPLC peptide separation is not significantly affected by small amounts of agarose polymers left over in the peptide supernatant after agarose precipitation. Third, in a similar way, on-line electrospray mass spectrometry is not strongly affected by possible remaining agarose components. It should be mentioned however that some low molecular components are regularly observed in the mass spectra of some peptides (see for instance Fig. 3C). This may impose a lower detection limit when ESI-MS analysis is used. The use of capillary chromatography will undoubtly improve the sensitivity. However, when sufficient peptide has to be saved for subsequent sequence analysis, the practical lower limit of the procedure is expected to be around 30-50 pmol. Fourth, the presence of the agarose in the digestion mixture does not result in NH_2-terminal blocking of the peptides. This follows from a comparison of the initial yields of the phenylthiohydantoin amino acids of the peptides which did not differ in

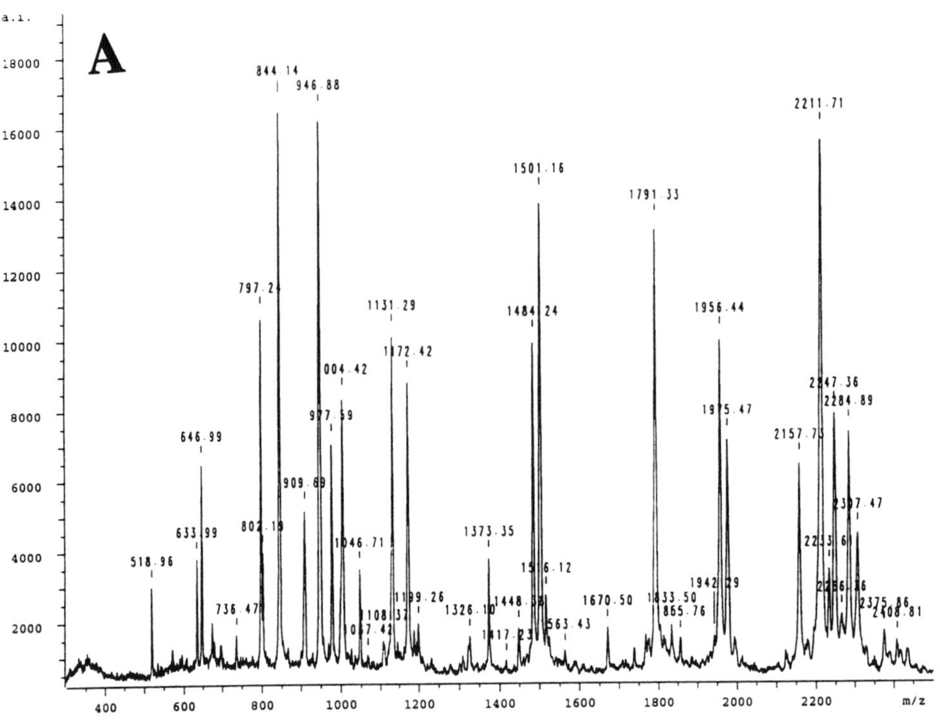

Position	Theoretical mass (Da)	Found mass (Da)	Mass difference (Da)	Position	Theoretical mass (Da)	Found mass (Da)	Mass difference (Da)
Tryptic peptides of skeletal muscle actin							
209-212	515.6	518.0	+ 2.4	31-41	1198.4	1198.3	- 0.1
194-198	630.7	633.0	+ 2.3	362-374	1500.6	1500.2	- 0.4
180-185	643.7	646.0	+ 2.3	87-97	1515.7	1515.1	- 0.6
287-292	733.8	735.5	+ 1.7	241-256	1790.9	1790.3	- 0.6
331-337	794.9	796.2	+ 1.3	98-115	1956.2	1955.4	- 0.8
64-70 *	800.0	801.2	+ 1.2	21-41 *	2156.4	2156.7	+ 0.3
21-30	976.0	976.6	+ 0.6	294-314	2246.5	2246.4	- 0.1
199-208	1130.2	1130.3	+ 0.1	293-314 *	2374.7	2374.9	+ 0.2
42-52	1171.4	1171.4	+ 0.0				
Autodigestion products of trypsin							
108-115	842.0	843.1	+ 1.1				
98-107	1045.1	1045.7	+ 0.6				
158-178	2158.5	2156.7	- 1.8				
58-77	2211.4	2210.7	- 0.7				
78-97	2283.6	2283.9	+ 0.3				

Figure 4. MALDI mass spectrometric peptide map of a tryptic digest of actin.

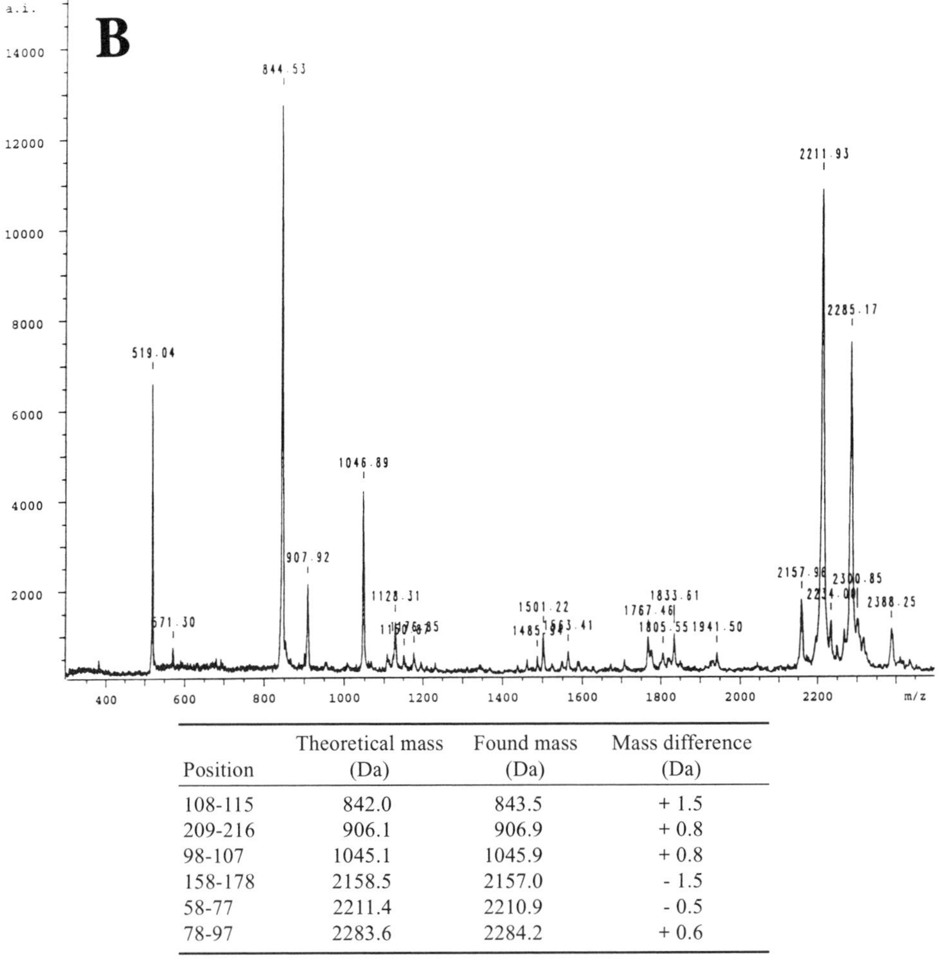

Position	Theoretical mass (Da)	Found mass (Da)	Mass difference (Da)
108-115	842.0	843.5	+ 1.5
209-216	906.1	906.9	+ 0.8
98-107	1045.1	1045.9	+ 0.8
158-178	2158.5	2157.0	- 1.5
58-77	2211.4	2210.9	- 0.5
78-97	2283.6	2284.2	+ 0.6

Figure 4. *Continued*

the molten agarose from those obtained in free solution (results not shown). In the second experiment we have mainly used the agarose concentration gel approach with respect to the problem of protein identification by tryptic peptide mass fingerprinting using as few protein as possible. Now the protein was digested in a smaller volume and the peptide mixture was not separated from the agarose for subsequent MALDI-TOF-MS analysis. Ten pmol (0.42 μg) of actin were recovered in 6 μl agarose gel (1.45% agarose). This was melted at 80°C and mixed with an equal volume of digestion buffer containing 0.1 μg of trypsin kept at 0°C. By rapid mixing the temperature is immediately shifted to about 40°C, reducing heat denaturation of the trypsin. The agarose (final concentration 0.7%) stays solid at this temperature. The digestion is continued overnight at 37°C and terminated by heating at 80°C and acidification with 1 μl of 50% TFA in water. An aliquot (5 μl) of the molten phase is removed, mixed with 5 μl of the matrix solution and 2 μl is taken for analysis. The amount analysed corresponds with 770 fmol of peptide, assuming complete cleavage of the protein. The corresponding MALDI-TOF-MS spectrum is shown in figure 4A. When compared with the blank analysis of trypsin (Fig 4B) we notice 17 peptides which could be assigned to the actin sequence (3 of these peptides resulted from partial digestion). This information is sufficient to identify the protein by molecular weight search in the tryptic peptide mass database. The high amount of trypsin autodigestion fragments is not surprising since the substrate to trypsin ratio used in this experiment was 4/1 and trypsin denaturation is likely in view of the use of 0.05% SDS during the digestion and the possible heat denaturation resulting from sample mixing.

In the example given here, 10 pmol of actin could be readily identified, however the quality of the MALDI-TOF spectra suggest that the starting amount of protein could be further decreased. The way this can be done may be directed by the finding that protein digestion with trypsin can proceed with the same efficiency in the gel phase as in solution - the agarose can be melted at high temperature either to obtain adequate mixing between the substrate and trypsin, or to take aliquots of the digestion mixture for analysis. Secondly and probably more importantly, MALDI-TOF-MS analysis can be successfully carried out on a peptide mixture still embedded in the agarose gel. This gives us the possibility to avoid the agarose/peptide separation step which takes place in a large volume and to work in small volumes.

CONCLUSIONS AND PERSPECTIVES

A feasibility study was made to use an agarose gel as protein holding matrix in which proteolytic cleavage can be carried out or from which mass spectrometric peptide analysis could be started. This approach could serve as an alternative for polyacrylamide *in-gel* digestion or *on-membrane* digestion procedures (Rosenfeld *et al*, 1992; Aebersold *et al*, 1986; Bauw *et al.*, 1988). The reason for such a study is based on potential advantages of using agarose versus other procedures. First, the agarose can be melted, converted into a liquid phase and therefore render the protein much more accessible for digestion while the recovery of the peptides should be more efficient. A second advantage is the finding that MALDI-TOF-MS analysis can be performed on peptide mixtures in the presence of high concentrations of agarose, opening the possibility for high sensitivity mass spectrometric protein identification. We have described in detail a procedure to transfer proteins from combined pieces of stained polyacrylamide gels into an agarose gel. By deformation of the electrical field, the eluting protein can be directed into a highly concentrated spot. In one strategy, the protein containing agarose gel is melted, followed by dilution with digestion buffer. Protein cleavage now proceeds in the liquid phase at 37°C. The example shown starts from 63 pmol of protein and peptides are subsequently separated by narrowbore HPLC on-line connected with an electrospray ion source. Peptide yields are sufficient for both peptide mass fingerprinting and individual amino acid sequence analysis. The sensitivity of the procedure described here could further be increased, e.g. by trying to concentrate the protein in smaller volumes and by using capillary chromatographic methods. Although these modifications seem possible, the lower limits will probably be set by contaminants derived from the incomplete agarose-peptide separation, interfering with ESI-MS. In the second approach we have digested the protein in solid agarose. This was achieved by melting the agarose gel piece at 80°C and consecutively mixing it with an equal volume of cold buffer containing trypsin. The digestion now proceeds in solid 0.75% agarose. At the end of the digestion the matrix is melted again for sample preparation. MALDI-TOF-MS peptide mass analysis is done in the presence of agarose, avoiding sample dilution or the agarose-peptide separation step. This second approach is particularly attractive because it should give us the opportunity to characterize very small amounts of gel-separated proteins. The example shown starts from 10 pmol of a 42 kDa protein. Further miniaturization of the system (e.g. concentrating in 1 μl volumes) will allow us to reduce the starting amount of protein by a factor of five or more. At this stage the lower limit will probably not be set by the limits of the miniaturization but probably by yet unknown parameters related with interactions with or modifications by either the primary polyacrylamide gel or the secondary agarose gel which may occur at these extreme low protein amounts. Other problems may be related to staining and destaining procedures and have to be considered in future experiments. The use of agarose as a holding matrix has clearly some interesting perspectives. While it allows to obtain peptides in high yields for further HPLC separation, the combination with *in-gel* MALDI-TOF mass spectrometric analysis of the peptide mixture opens the possibility of protein characterization on quantities which could never be reached previously.

ACKNOWLEDGMENTS

This work was supported by grants from the Belgian National Fund for Scientific Research (BNFSR) to J. V., the Commision of the European Union, the programs Concerted Research Actions and Centre of Emerging Technology of the Flemish Community. K. G. is supported by the Flemish Institute for Science and Technology. M. R. is Chercheur Qualifié of the BNFSR.

REFERENCES

Aebersold, R.H., Teplow, D.B., Hood, L.E. and Kent, S.B.H., 1986, Electroblotting onto activated glass. High efficiency preparation of proteins from analytical sodium dodecyl sulphate-polyacrylamide gels for direct sequence analysis. *J. Biol. Chem.* 261:4229-4238.

Aebersold, R.H., Leavitt, J., Saavedra, R.A., Hood, L.E. and Kent, S.B.H., 1987, Internal amino acid sequence analysis of proteins separated by one- or two-dimensional gel electrophoresis after *in situ* protease digestion on nitrocellulose. *Proc. Natl. Acad. Sci. USA* 84:6970-6974.

Bauw, G., Van Den Bulcke M., Van Damme, J., Puype, M. Van Montagu, M. and Vandekerckhove, J., 1988, Protein electroblotting on polybase-coated glassfiber and polyvinylidene difluoride membranes: an evaluation *J. Prot. Chem.* 7:194-196.

Beavis, R.C., Chaudhary, T. and Chait, B.T., 1992, α-Cyano-4-hydroxycinnamic acid as a matrix for matrix assisted laser desorption mass spectrometry. *Organ. Mass Spectrom.* 27:156-158.

Celis, J. and Bravo, R., 1984, in: Two-dimensional electrophoresis of proteins: methods and applications, Academic Press, New York (eds. Celis, J. and Bravo, R.).

Crepin, K.M., Darville, M.I., Michel, A., Hue, L. and Rousseau, G.G., 1989, Cloning and expression in *Escherichia Coli* of a rat hepatoma cell cDNA coding for 6-phosphofructo-2-kinase/fructose-2,6-bisphosphatase. *Biochem. J.* 264:151-160.

Mann, M., Højrup, P. and Roepstorff, P., 1993, Use of mass spectrometric molecular weight information to identify proteins in sequence databases. *Biol. Mass Spectrom.* 22:338-345.

Matsudaira, P.J., 1987, Sequence from picomole quantities of proteins electroblotted onto polyvinylidene difluoride membranes. *J. Biol. Chem.* 262:10035-10038.

Matsudaira, P.J. and Burgess, D.R., 1978, SDS microslab linear gradient polyacrylamide gel electrophoresis. *Anal. Biochem.* 87:386-396.

O'Farrell, P.H., 1975, High resolution two-dimensional electrophoresis of proteins. *J. Biol. Chem.* 250:4007-4021.

Pappin, D.J.C., Højrup, P. and Beasby, A.J., 1993, Rapid identification of proteins by peptide-mass fingerprinting. *Current Biology* 3:327-332.

Rosenfeld, J., Capdevielle, J., Guillemot, J.C. and Ferrara, P., 1992, In-gel digestion of proteins for internal sequence analysis after one- or two-dimensional gel electrophoresis. *Anal. Biochem.* 203:173-179.

Spudich, J.A. and Watt, S., 1971, The regulation of rabbit skeletal muscle contraction. *J. Biol. Chem.* 246:4866-4871.

Vandekerckhove, J., Bauw, G., Puype, M., Van Damme, J., and Van Montagu, M., 1985, Protein-blotting on polybrene-coated glass-fiber sheets. A basis for acid hydrolysis and gas-phase sequencing of picomole quantities of protein previously separated on sodium dodecyl sulfate/polyacrylamide gel. *Eur. J. Biochem.* 152:9-19.

Yates, J.R., Speicher, S., Griffin, P.R. and Hunkerpillar, T., 1993, Peptide mass maps: a highly informative approach to protein identification. *Anal. Biochem.* 214:397-408.

Zhang, W., Czernik, A.J., Yungwirth, T., Aebersold, R. and Chait, B.T., 1994, Matrix-assisted laser desorption mass spectrometric peptide mapping of proteins separated by two-dimensional gel electrophoresis: Determination of phosphorylation in synapsin I. *Prot. Sci.* 3:677-686.

3

HIGH-SPEED CHROMATOGRAPHIC SEPARATION OF PROTEINS AND PEPTIDES FOR HIGH SENSITIVITY MICROSEQUENCE ANALYSIS

Robert L. Moritz, James Eddes, and Richard J. Simpson

Joint Protein Structure Laboratory
Ludwig Institute for Cancer Research (Melbourne Branch) and
The Walter and Eliza Hall Institute for Medical Research
Parkville, Victoria 3050 Australia

INTRODUCTION

Reversed-phase high-performance liquid chromatography (RP-HPLC) and polyacrylamide gel electrophoresis (PAGE) are two of the most widely-used high-resolution techniques for isolating proteins and peptides for structural analysis (Simpson et al., 1988, 1989; Matsudaira, 1993; Patterson, 1994). In recent years, the importance of these two technologies has been further enhanced by using them in tandem. For example, proteins from complex mixtures such as total cell lysates can be resolved by two-dimensional gel electrophoresis (2-DE) and, following proteolytic digestion, the generated peptides separated by microbore column RP-HPLC. Proteolytic digestion of 2-DE gel spots can be accomplished either in the gel matrix (Ward et al., 1990, Eckerskorn et al., 1990, Rosenfeld et al, 1992, Ji et al., 1994, and Hellman, U. personal communication) or on immobilizing membranes such as polyvinylidine difluoride (Fernandez et al., 1992) and nitrocellulose (Aebersold et al., 1987), following electrotransfer from the gel. Protein identification can be achieved by microsequence analysis of the isolated peptides using either automated Edman degradation or tandem mass spectrometry (Hunt et al., 1986; Burlingame et al., 1994). More recently, alternative means of protein identification such as peptide mass fingerprinting (Pappin et al., 1993; Mann et al., 1993; Henzel et al., 1993, James et al., 1994) and amino acid compositional analysis (Sibbald et al., 1991, Shaw, 1993) have emerged. Since these latter techniques have the potential for generating large quantities of data rapidly, there is now an increasing need for rapid protein and peptide isolation procedures.

Although the concept of high-speed RP-HPLC using linear velocities in the range 1000-5000 cm/hr (i.e., 0.5-3.0 ml/min for a 2.1-mm ID column) has been previously documented (Kalghatgi and Horváth, 1987, 1988; Nugent, 1990, Fulton et al., 1991, Regnier, 1991), this technology has been slow to gain general acceptance in the bios-

Methods in Protein Structure Analysis, Edited by M. Z. Atassi and E. Appella
Plenum Press, New York, 1995

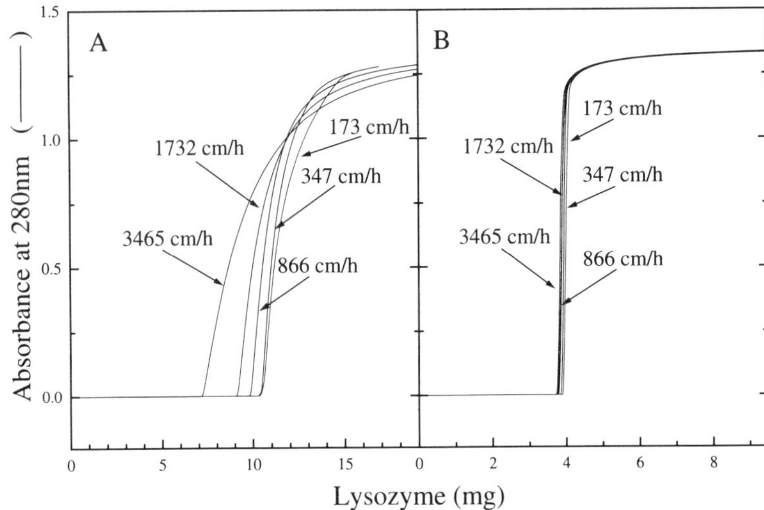

Figure 1. Frontal loading adsorption isotherms of conventional "wide pore" derivatised silica (Brownlee RP-300) and macroporous divinylbenzene crosslinked **polystyrene (Poros RII/H) supports.** Protein: 1 mg/ml solution in aqueous 0.1% TFA. Superficial linear flow velocities: 173, 347, 866, 1732 and 3465 cm/h. Temperature: 25°C. (**A**) RP-300 2.1 mm ID cartridge. (**B**) Poros RII/H 2.1 mm ID column. Adapted and reproduced with permission from Moritz *et al.*, 1994.

ciences. This has been due to a number of shortcomings in the methodology, foremost of these being stationary phase design. Initial attempts at designing stationary phase materials that would meet the fundamental criteria of fast protein chromatography, such as good solvent permeability and constant retention behaviour, led to the development of non-porous stationary phase packings (Unger et al., 1986; Kalghatgi and Horváth, 1987, Yamasaki et al., 1989). With these packings, "diffusion" involving the intraparticle pores is eliminated and the solvent passage is restricted to interparticle "convective flow". Since "diffusion", which results in slow mass transfer of analyte within particle pores (into and out of stagnant pools), is the major contributing factor to peak broadening (Fig. 1), non-porous packings exhibit minimal reduction in peak resolution over a broad range of flow rates. However, the reduction in particle surface area of these packings, due to the elimination of pores, results in their inferior binding capacity (~ 1.0 mg/g) compared to the conventional wide-pore (300Å) silica-based packings (~36 mg/g). Moreover, packed beds of non-porous packings are stable to very high pressure environments (typically, 6000 psi). However, these packings exhibit very high pressure drops across the column thereby restricting their use at very high linear flow velocities (Yamasaki et al., 1989; Nice and Simpson, 1989; Rozing and Goetz, 1989).

Many of the problems encountered with non-porous packings have been circumvented with the design of macroporous packings, such as "perfusive" stationary phases (Afeyan et al., 1990) and "hyperdiffusive" packings (Boschetti, 1994). The salient features of these packings are (i) the very large pore diameters (≥ 8000Å), (ii) the high binding capacity (compared to the non-porous packings), and (iii) maintenance of resolution (i.e., chromatographic efficiency) over a broad range of linear flow velocities (1000-5000 cm/h; i.e., 0.5-3.0 ml/min for a 2.1-mm ID column). While the "perfusive" packings are derived from divinylbenzene cross-linked polystyrene (PS-DVB), the "hyperdiffusive" packings are a soft agarose gel encased in a rigid PS-DVB spherical lattice. Both of these packings are robust in the practical operating range of flow rates, but at very high flow rates their utility, compared to some of the widely-used mesoporous silica packings (e.g., 300Å), is limited due to their fragile nature.

The principal advantages that macroporous packings afford over conventional silica-based reversed-phase packings are purported to be their ability to operate at very high linear flow velocities (1000 cm/h) while maintaining both high sample loading capacity and increased chromatographic resolution (Kassel et al., 1994). Here, we report a protocol for fast chromatographic analysis (<12 min) of proteins and peptides using a conventional 300Å, 7-μm silica-based support and standard liquid chromatographs. Using an inexpensive conventional 2.1-mm "wide-pore" reversed-phase cartridge and rapid linear flow velocities of 500-1000 cm/h (0.3-0.6 ml/min), highly reproducible separations can be achieved in 10-12 min, almost a magnitude faster than standard chromatographic conditions, without any serious compromise in chromatographic efficiency.

MATERIALS AND METHODS

Bovine serum albumin, bovine pancreatic ribonuclease-B, hen egg lysozyme, horse heart myoglobin, hen egg albumin (ovalbumin), rabbit muscle phosphorylase-b and bovine erythrocyte carbonic anhydrase were obtained from Sigma (St Louis, MI). Sequence grade trypsin was purchased from Promega. Coomassie Brilliant Blue R250 (CBR-250) was from LKB-Pharmacia (Uppsala, Sweden), 10% Tris-glycine acrylamide gels from Novex, trifluoroacetic acid (TFA) from Pierce and HPLC grade solvents were obtained from Mallinckrodt (Melbourne). High purity, deionized water was obtained from a tandem Milli-RO15 and Milli-Q system (Millipore, Bedford, MA). All other reagents used were of analytical grade quality.

High-Performance Liquid Chromatography

Instrumentation. Protein and peptide mixtures were fractionated by RP-HPLC on a Hewlett Packard model 1090A liquid chromatograph fitted with a model 1040A diode-array detector. Samples were injected either by an integrated autoinjector or by using a Rheodyne model 7125 injector equipped with a 2-ml injection loop installed in the heated column compartment. Fractions were collected manually in 1.5-ml polypropylene tubes (Eppendorf) and stored at -20°C.

Column Supports. The following columns were used in this study: (a) Brownlee RP-300 (300Å pore size, 7-μm particle diameter, octylsilica packed into a 100 x 2.1 mm ID cartridge, Applied Biosystems, Foster City, CA); (b) POROS RII/H (10-μm divinylbenzene crosslinked polystyrene packed into a 100 x 2.1 mm ID stainless steel column, Perseptive Biosystems, Cambridge, MA).

SDS-Polyacrylamide Gel Electrophoresis

Phosphorylase-b (97,000 M_r) was separated in 1.0 mm thick 10% SDS-gels (Novex). Two-dimensional gel electrophoresis (2-DE) of total cell lysates from cultured human colonic LIM1863 cells (Ji et al., 1993, 1994), with isoelectric focussing (IEF) using precast immobilized pH gradients (Pharmacia) in the first dimension, and SDS-PAGE in the second dimension were performed as described (Ji et al., 1994). Proteolytic digestion of gel-resolved proteins was performed *in-situ*, essentially as described (Ward et al., 1990a, b) with modifications based on the methods of Rosenfeld et al., (1992) and Hellman, U et al. (1994).

Step 1. *Visualization of proteins with CBR-250.* Gel staining conditions: 50% methanol / 10% acetic acid / 0.1% CB-R250 (~ 5-10 min). Destaining conditions: 12% methanol / 7% acetic acid for 1-1.5 h (with ~ 3 changes).

Step 2. *In-situ proteolysis:* (i) excise stained protein gel band; (ii) wash twice (~ 200 μl) for 30 min at 30°C with 1% ammonium bicarbonate / 50% acetonitrile; (iii) dry gel band completely by centrifugal lyophilisation (Savant, ~ 30 min); (iv) rehydrate gel band twice with trypsin-containing solution (~ 0.5-1.0 μg trypsin in 10 μl 1% ammonium bicarbonate / 0.5 mM $CaCl_2$) for 15-30 min; (v) add 150 μl 1% ammonium bicarbonate containing 0.5 mM $CaCl_2$ and incubate at 37°C for ~ 16 h.

Step 3. *Peptide extraction*: (i) collect enzymatic digestion buffer; (ii) add 200 μl of 1% TFA, sonicate the gel mixture for ~ 30 min (35-40°C) and collect the extract; (iii) add 200 μl of 0.1% TFA / 60% acetonitrile and sonicate the mixture for ~ 30 min at 35-40°C and collect the extract; (iv) concentrate the pooled extracts by centrifugal lyophilization to a final volume of 10-20μl for RP-HPLC.

RESULTS AND DISCUSSION

Protein Binding Capacity and Mass Transfer Kinetics of Conventional and Macroporous Packings

A conventional wide-pore silica-based reversed-phase column (e.g., Brownlee RP-300, C8, 7-μm, 300 Å particles) was evaluated for its binding capacity and mass transfer kinetics at varying superficial linear flow velocities and compared directly with a macroporous column (e.g., POROS RII/H, divinylbenzene cross-linked polystyrene, 10-μm, ≥ 8000Å particles). Using frontal analysis chromatography, it can be seen that the total protein binding capacity (saturation level) for lysozyme was significantly greater (~ 3-fold) for the conventional packing (~ 11.5 mg) than the macroporous support (~ 4 mg) (Fig. 1). The protein saturation levels for both conventional and macroporous packings are independent of linear velocity over the range 173-3465 cm/h. It should be noted for the conventional packing that the initial binding (or breakthrough), however, does depend on linear flow velocity. In contrast to the macroporous packing, marked variation in the frontal curve shape is observed for the conventional silica-based packing. For example, protein breakthrough at 173 cm/h occurs with loads > 11 mg, while at a 20-fold higher flow rate protein breakthrough occurs at ~ 7 mg. This variation in frontal curve shape for the conventional support is indicative of "stagnant mobile phase mass transfer" (attributable to the large number of inaccessible pockets in these particles where slow or unmoving mobile phase accumulates and mass transfer occurs by extended-path-length diffusion) (Fig. 2). This observation has been previously reported for a wide range of other silica-based supports (Snyder and Kirkland, 1973). The lack of variation in frontal curve shape for the macroporous packing are due to minimal stagnant mobile phase pool formation (Fig. 1B), a feature of this packing design.

The amount of protein that one can load on a reversed-phase packing is dependant largely on the total surface area of the packing. Narrow-pore silica-based porous packings (60Å) exhibit very high surface areas (300 m^2/g) compared to non-porous packings (< 5 m^2/g) (Yamasaki et al., 1989, Esser and Unger, 1991). Conventional wide-pore packings

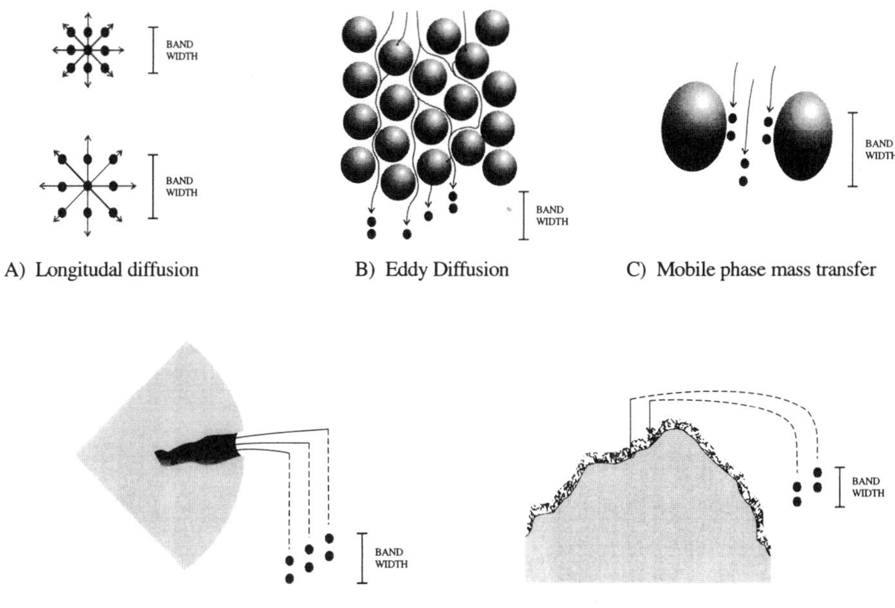

Figure 2. Physical factors which contribute to peak broadening in porous HPLC packings. (A) *Longitudinal diffusion* - results from normal diffusion of molecules in liquid medium and is more pronounced in slow moving fluid flow. (B) *Eddy diffusion* - results from multiple differentially distanced flowpaths between particles in a column. (C) *Mobile-phase mass transfer* - results from differing flow rates for different parts of a single flow stream between particles in a column. (D) *Stagnant mobile-phase mass transfer* - results from stagnant or unmoving mobile phase within the pores of a particle. Diffusion resulting from this is thought to be the major contributor to band broadening. (E) *Stationary-phase mass transfer* - results from molecules penetrating the stationary phase covering the surface of the particle by diffusion to varying extents. Adapted from Snyder and Kirkland, 1979.

(300 Å), originally designed for protein and peptide separations exhibit surface areas of 50-100 m^2/g and protein loads of ~ 35 mg/g. Macroporous packings (≥ 8000 Å) such as the POROS R series exhibit low protein binding capacity (i.e. ~ 5 mg/g) due to a smaller surface area. Later attempts to increase the binding capacity by the re-introduction of short-path-length pores led to a modest (2-fold) increase in the total protein binding capacity (~ 10 mg/g, data obtained from the manufacturer). With the exception of chromatographic techniques such as displacement chromatography, to achieve efficient chromatographic separation of proteins and peptides on reversed-phase packings, sample loads of ≤ 5% of the total capacity are typically used. Under these conditions, deleterious slow mass transfer kinetics, due to extended-pathlength diffusion into stagnant pools of mobile phase, become less pronounced; with higher sample loads deleterious peak shape can result from column overloading (Snyder and Kirkland, 1973)

Another aspect of rapid chromatography that warrants careful consideration is the instruments liquid pumping and data collection capabilities. For rapid microbore RP-HPLC, a binary pumping system capable of producing precise gradient formation at low flow rates with minimal system dead volume must be used. To achieve similar chromatographic efficiency within a reduced time frame, linear flow velocities are increased whilst maintaining the same gradient volume as formed at lower linear flow velocities. Pumping systems that are unable to produce precise gradient formation at low linear flow velocities and which incorporate large system dead volumes will not perform less efficiently at the higher linear flow velocities.

With respect to UV detection, to obtain a true representation of the chromatographic separation, the collected data must not be compromised by an erroneous data set. As linear flow velocities are increased whilst maintaining gradient volumes, proteins and peptides will elute in the same solvent fraction volume of the organic modifier as in slow linear velocities. This results in the analytes passing through the detector far more rapidly. If the operating parameters of the UV detector are similar to those used at conventional low linear flow velocities then there is an increased risk of false chromatographic separation representations. To compensate for this, data collection rates must be increased, (e.g., ~ 100 msec for linear flow velocities ≥ 3500 cm/h), accordingly.

Effect of Linear Flow Velocity on Resolution and Recovery of Proteins and Peptides

Chromatographic separation of a mixture of six proteins at varying flow rates on conventional and macroporous packings is shown in Fig. 3. It would appear that the resolution of these standard proteins varies little over the range of flow velocities examined. However, upon close inspection of the conventional reversed-phase packing (compare Fig. 3A and E), there is a discernible loss of resolution upon increasing the flow rate from 0.1 to 2.0 ml/min (i.e., 173 to 3465 cm/h). However, a loss of resolution is also evident for the macroporous packing, but to a lesser extent (compare Fig. 3F and J). It should be noted that the seemingly lower recoveries at the higher flow velocities are due to peak broadening. Sample recoveries for a glycoprotein (ribonuclease-b) by both stationary phases examined in this study are shown in Fig. 4. Good recoveries from the alkyl silica of ~ 98% per iterative step with an overall recovery of ~ 96% after two reinjections is shown in Fig 4A. For perfusive stationary phases however, lower recoveries of 94% per iterative reinjection and overall recoveries of 86% after two re-injections are found (Fig. 4B). For multidimensional

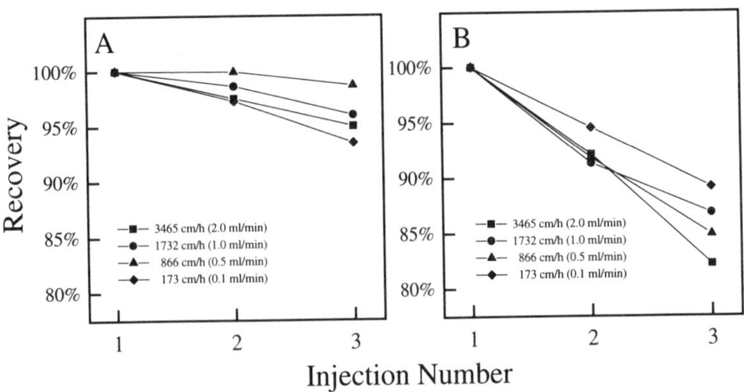

Figure 3. Rapid reversed-phase chromatography of standard proteins. Supports: RP-300 (panels **A** - **E**); Poros RII/H (panels **F** - **J**). Chromatographic conditions: linear 6-ml gradient of 0-100%B. Solvent A: aqueous 0.1% TFA, Solvent B: aqueous 0.1% TFA containing 60 % acetonitrile. Temperature: 45°C. Chromatographic runs performed at superficial linear flow velocities of 173 cm/h (0.1 ml/min) (**A**, **F**), 347 cm/h (0.2 ml/min) (**B**, **G**), 866 cm/h (0.5 ml/min) (**C**, **H**), 1732 cm/h (1.0 ml/min) (**D**, **I**) and 3465 cm/h (2.0 ml/min) (**E**, **J**). Proteins (5 μg): 1, ribonuclease-B; **2**, chick lysozyme; **3**, bovine serum albumin; **4**, myoglobin; 5, carbonic anhydrase; **6**, ovalbumin.

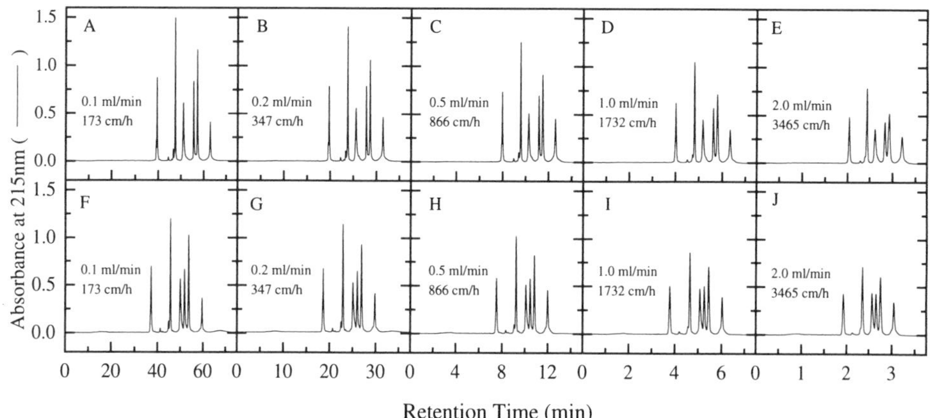

Figure 4. Protein recovery by rapid reversed-phase liquid chromatography. Supports: RP-300 (panel **A**); Poros RII/H (panel **B**). Chromatographic conditions are as described in Fig. 3. Sample: 5 μg Ribonuclease-B initially injected onto the respective columns. Once eluted the sample was collected into a 1.5ml polypropylene tube. The column was then re-equilibrated with at least 20 column volumes of Buffer A, and the collected peak diluted 1:1 with Buffer A and reapplied. Recovery measurements were calculated from peak heights. All experiments were performed in duplicate. Flow rate symbols: (◆) ; 173 cm/h, 0.1 ml/min, (▲) 866 cm/h, 0.5 ml/min (●) 1732 cm/h, 1.0 ml/min (■) 3465 cm/h, 2.0 ml/min.

purification strategies, low recoveries of protein obtained from perfusive packings would be of some concern.

The effect of linear flow velocity on the chromatographic separation of a tryptic digest of cytochrome-*c* on conventional and macroporous reversed-phase packing are compared in Fig. 5, panels A-J. It is apparent from the profiles at 0.1 ml/min that the chromatographic efficiency of the conventional silica-based packing exceeds that of the macroporous packing. It is also evident from the greater (~ 30%) "peak capacity" (i.e., the number of peaks that it is possible to resolve in a given chromatographic separation) that the silica-based packing exhibits a greater chromatographic efficiency than the macroporous

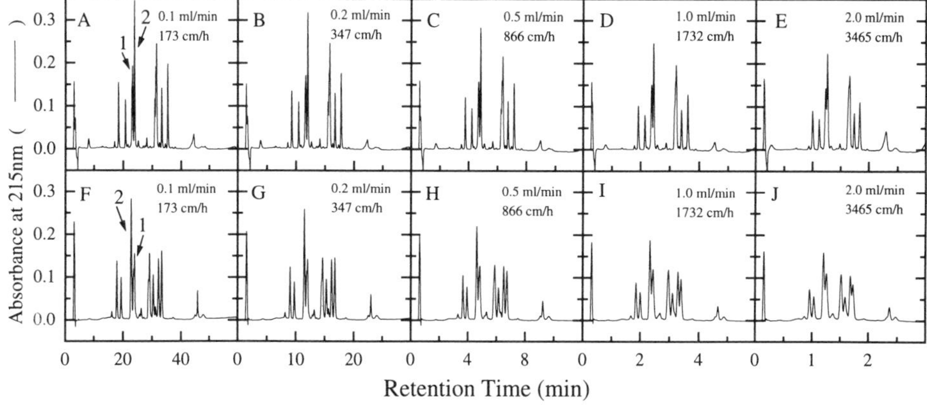

Figure 5. Rapid reversed-phase HPLC peptide mapping. Sample: 20 μg tryptic digest of cytochrome-c. Columns: RP-300 2.1mm ID cartridge (panels **A - E**); Poros RII/H 2.1mm ID (panels **F - J**). Chromatographic conditions are as described in Fig. 3.

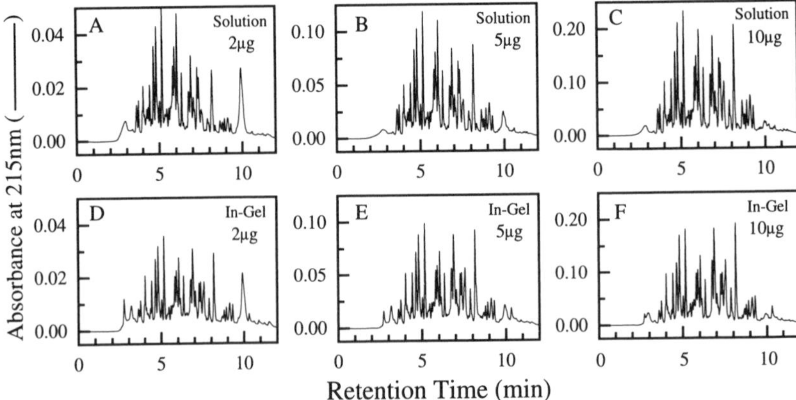

Figure 6. In-gel versus in-solution tryptic digestion of phosphorylase-b. Peptide maps were obtained by fast chromatography RP-HPLC using a Brownlee RP-300 100mm x 2.1mm I.D. column. Chromatographic conditions: a linear 6 ml gradient from 0-100% B; solvent A, aqueous 0.1% TFA; solvent B, aqueous 0.1% TFA / 60% CH_3CN. Flow: 0.5 ml/min (866 cm/h). Panels **A**, **B** & **C**: control tryptic digests in-solution using 2, 5 and 10 μg phosphorylase-b, respectively. Panels **D**, **E** & **F:** in-situ gel tryptic digests of 2, 5 and 10 μg of phosphorylase-b, respectively. Reproduced with permission from Moritz *et al.*, 1994.

packing. In contrast to earlier reports (Kassel et al., 1994), the chromatographic efficiency of the silica-based support exceeds that of the macroporous support at high flow rates (e.g. 3500 cm/h, compare Fig. 5E and J). The selectivity differences between the two packings used in this study (compare peaks 1 and 2 in Fig. 5A and F) indicate that they could be used in series in a multidimensional peptide purification strategy.

Rapid Peptide-Mapping of Acrylamide Gel-Resolved Proteins

Several internal amino acid sequencing strategies for electrophoretically separated proteins have been developed over the past few years (see Ward et al, 1990a, Rosenfeld et al, 1992; Patterson, 1994 and references therein). An excellent practical assessment of these methods by the Association of Biomedical Resource Facilities (ABRF) was published in 1992 (Stone, 1992) and 1993 (Williams et al., 1993). In an earlier report (Ward et al., 1990a) we described our *in-gel* digestion strategy which relies on first removing SDS from the CBR-250 stained gel prior to *in-gel* enzymatic digestion and an extensive acid extraction of generated peptides. In an effort to further reduce the overall time of the procedure, we omitted some of the TFA extraction steps (see Materials and Methods) without compromising the overall yield of recovered peptides. Additionally, we have replaced the initial SDS removal step with a dilute 1% ammonuim bicarbonate / acetonitrile extraction step. Using varying amounts of a tryptic digest of phosphorylase-*b* (M_r ~ 97000) and pre-cast 10% gels (Novex), we compared peptide recoveries from *in-gel* derived peptide maps with those obtained in-solution (Fig. 6A-F). It can be seen that the peptide map profiles of the *in-gel* and solution digests compare favourably, even with 20 pmol (2 μg) amounts of protein. The recovery of peptides from the *in-gel* proteolysis, based upon absorbance at 214 nm, is ~ 80% compared to the control *in-solution* digests. Comparable data was obtained using standard proteins of lower M_r such as lysozyme and β-lactoglobulin (data not shown). In an effort to minimize possible interference by detergent with electrospray ionization, we evaluated peptide recoveries in the absence of any added Tween 20. The data shown in Fig. 7A-F reveals that

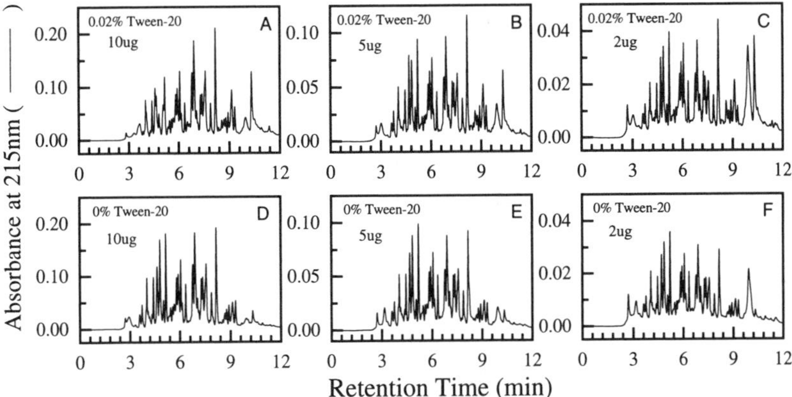

Figure 7. In-gel tryptic digestion of phosphorylase-b, effect of Tween-20. Peptide maps were obtained by fast chromatography RP-HPLC using a Brownlee RP-300 100mm x 2.1mm I.D. column. Chromatographic conditions: a linear 6 ml gradient from 0-100% B; solvent A, aqueous 0.1% TFA; solvent B, aqueous 0.1% TFA / 60% CH_3CN. Flow: 0.5 ml/min (866 cm/h). Panels **A**, **B** & **C**: in-situ gel tryptic digests containing 0.02% Tween-20 during digestion and extraction of 10, 5 and 2 μg of phosphorylase-b, respectively. Panels **D**, **E** & **F**: control in-situ gel tryptic digests using 10, 5 and 2 μg phosphorylase-b, respectively.

omission of Tween does not seriously affect peptide recoveries even at low quantities of protein (Fig. 7F).

Examples of the Application of Rapid Peptide-Mapping of 2-DE-Resolved Proteins

Over the past few years, we have been identifying 2-DE separated proteins from various human colorectal cancer cell lines by sequence and mass analysis as part of an ongoing program directed towards identifying specific colon tumour markers (Ward et al., 1990c; Ji et al., 1993, 1994). Examples of the power of this rapid peptide mapping approach are given in Figs. 8 and 9) for proteins #1 and #4 isolated from the colorectal cell line LIM 1863. Four CBR-250-stained protein spots from identical gels were digested with trypsin and the digest mixtures were chromatographed at 0.5 ml/min on a conventional 2.1 mm ID reversed-phase cartridge using a 6.0-ml linear gradient of acetonitrile in 0.1% TFA. In the case of protein spot #1, the partial sequence data obtained (Fig. 8), at 48-55 pmol levels, was used to search the available protein sequence databases and permit the unambiguous identification of this protein as thioredoxin. For protein spot #4 (Fig. 9), peptide T4 was sequenced directly while the peptide fraction containing peptides T1-3 was further resolved by rapid second dimensional chromatography on the same column, but utilising a modified mobile phase of 1% NaCl / 50% acetonitrile (Fig. 9B), prior to subjecting the component peptides to sequence analysis. The partial sequence data obtained (data not shown), at 5-17 pmol levels, was used to identify this protein as heat shock protein 60 (HSP-60).

SUMMARY

This report describes a rapid (~ 10 min) chromatographic approach for separating proteins and peptides on conventional silica-based reversed-phase packings employing a

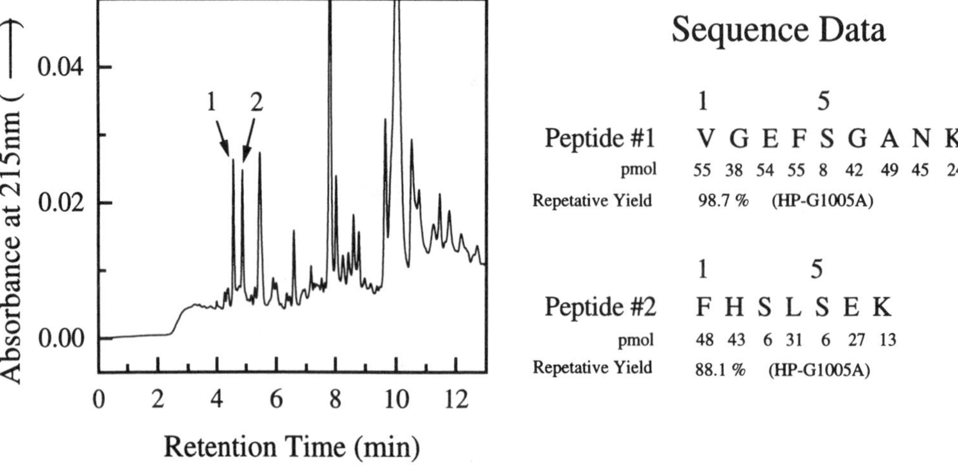

Figure 8. Rapid peptide mapping of colorectal cancer cell line LIM 1863 protein #1. Coomassie blue stained protein #1 from 4 identical 2-D gels was digested *in-gel* with trypsin, as described in Materials and Methods, and chromatographed on a conventional silica-based support (Brownlee RP-300) as described in Fig.3. First chromatographic dimension (Panel **A**): linear 6 ml gradient 0-100% B; solvent A, aqueous 0.1% TFA; solvent B, aqueous 0.1% TFA / 60% CH_3CN, Flow, 0.5 ml/min (866 cm/h). Sequence information obtained from T1 and T2 are shown. Protein identified as Thioredoxin.

standard liquid chromatograph. An improved in-gel enzymatic digestion strategy is described. Examples are given for peptide maps of phosphorylase-b from a 1-D gel and 2-DE protein spots from colorectal cancer cell line LIM 1863. In conjunction with microsequencing and mass spectrometric peptide-mass fingerprinting technologies, this approach may facilitate a rapid expansion of 2-DE gel protein databases.

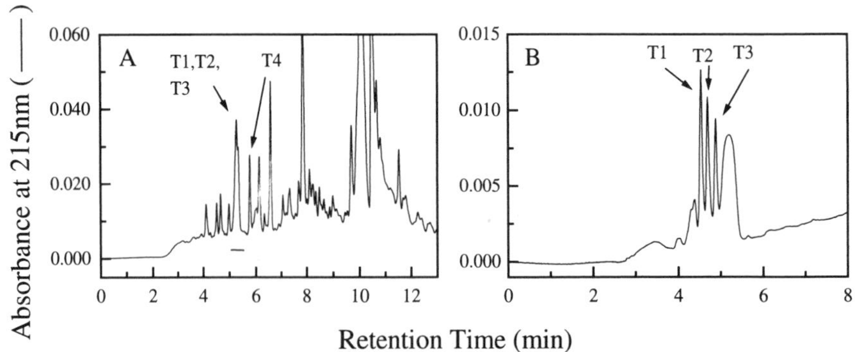

Figure 9. Rapid peptide mapping of colorectal cancer cell line LIM 1863 protein #4. Coomassie blue stained protein #4 from 4 identical 2-D gels was digested *in-gel* with trypsin, as described in Materials and Methods, and chromatographed on a conventional silica-based support (Brownlee RP-300) as described in Fig.3. First chromatographic dimension (Panel **A**): linear 6 ml gradient 0-100% B; solvent A, aqueous 0.1% TFA; solvent B, aqueous 0.1% TFA / 60% CH_3CN, Flow, 0.5 ml/min (866 cm/h). (B) Second chromatographic dimension (Panel **B**); peptide fraction containing peptides T1, T2 & T3 (see collection bar) from Fig.7A were rechromatographed on the same column but using a linear 5 ml gradient from 0-50% B; solvent A was aqueous 1% NaCl, pH 6.5 and solvent B was acetonitrile. Flow rate, 0.5 ml/min (866 cm/h). protein identified as heat-shock protein (HSP-60, data not shown). Reproduced with permission from Moritz *et al.*, 1994.

REFERENCES

Aebersold, R.H., Leavitt, J., Saavedra, R.A., Hood, L.E. and Kent, S.B. (1987) Internal amino acid sequence analysis of proteins separated by one- or two-dimensional gel electrophoresis after *in situ* protease digestion on nitrocellulose *Proc. Natl. Acad.* Sci. USA **84**: 6970-6974.

Afeyan, N.B., Gordon, N.F., Mazsaroff, I., Varady, L., Fulton, S.P., Yang, Y.B. and Regnier, F.E., (1990) Flow-through particles for the high-performance liquid chromatographic separation of biomolecules: perfusion chromatography *J. Chromatogr.* **519**, 1-29.

Boschetti, E., (1994) Advanced sorbents for preparative protein separation purposes *J. Chromatogr.* **658**, 207-236.

Burlingame, A.L., Boyd, R.K. and Gaskell, S.J. (1994) Mass spectrometry *Anal. Chem.* **66**: 634-683.

Eckerskorn, C., Strahler, S. and Lottspeich, F. (1990) in Two-dimensional electrophoresis (Endler, I. and Honash, S., Eds.), VCH, Weinheim.

Esser, U and Unger, K.K. (1991) Reversed-phase packings for the separation of peptides and proteins by means of gradient elution high-performance liquid chromatography. in High-performance liquid chromatography of peptides and proteins: separation, analysis and confirmation (Mant, C.T. and Hodges, R.S.), CRC Press Inc, Florida.

Fernandez, J., DeMott, M., Atherton, D. and Mische, S.M. (1992) Internal protein sequence analysis: Enzymic digestion for less than 10µg of protein bound to polyvinylidene difluoride or nitrocellulose membranes *Anal. Biochem.* **201**: 255-264.

Fulton, S.P., Afeyan, N.B., Gordon, N.F. and Regnier, F.E. (1991) Very high speed separation of proteins with a 20-µm reversed-phase sorbent. *J. Chromatogr.* **547,** 452-456.

Hellman, U., Wernstedt, C., Gonez, J. and Heldin, C-H. (1995) Improvements of an in-gel digestion procedure for the micropreparation of internal protein fragments for amino acid sequencing. *Anal. Biochem.* 224:451-455.

Henzel, W.J., Billeci, T.M., Stults, J.T., Wong, S.C., Grimley, C. and Watanabe, C. (1993) Identifying proteins from two-dimensional gels by molecular mass searching of peptide fragments in protein sequence databases *Proc. Natl. Acad. Sci.* USA **90**: 5011-5015.

Hunt, D.F., Yates, J.R., Shabanowitz, J., Winston, S. and Hauer, C. R. (1986) Protein sequencing by tandem mass spectrometry *Proc. Natl. Acad. Sci.* USA **83**: 6233-6237.

James, P., Quadroni, M., Carafoli, E., Gonnet, G. (1994) Protein identification in DNA databases by peptide mass fingerprinting *Protein Sci.* **3**, 1347-1350

Ji, H., Baldwin, G.S., Burgess, A.W., Moritz, R.L., Ward, L.D. and Simpson, R.J., (1993) Epidermal growth factor induces serine phosphorylation of stathmin in a human colon carcinoma cell line (LIM 1215), *J. Biol. Chem.* 286: 13396-13405.

Ji, H., Whitehead, R.H., Reid, G.E., Moritz, R.L., Ward, L.D. and Simpson, R.J. (1994) Two-dimensional electrophoretic analysis of proteins expressed by normal and cancerous human crypts: Application of mass spectrometry to peptide-mass fingerprinting, *Electrophoresis* **15**: 391-405.

Kalghatgi, K. and Horváth, C. (1987) Rapid analysis of proteins and peptides by reversed-phase chromatography *J. Chromatogr.* **398**, 335-339.

Kalghatgi, K. and Horváth, C., (1988) Rapid peptide mapping by high-performance liquid chromatography *J. Chromatogr.* **443**, 343-354.

Kassel, D.B., Shushan, B., Sakuma,T. and Salzman, J-P. (1994) Evaluation of packed capillary column HPLC/MS/MS for the rapid mapping and sequencing of enzymatic digests. *Anal. Chem.* **66**: 236-243.

Mann, M., Højrup, P., and Roepstorff, P. (1993) Use of mass spectrometric molecular weight information to identify proteins in sequence databases *Biol. Mass Spec.* **22**, 338-345.

Matsudaira, P. (1993) A practical guide to protein and peptide purification for microsequencing, *Academic Press*, San Diego.

Moritz, R.L. and Simpson, R.J. (1992a) Application of capillary reversed-phase high performance liquid chromatography to high-sensitivity protein sequence analysis *J. Chromatogr.* **599**, 119-130.

Moritz, R.L. and Simpson, R.J. (1992b) Purification of proteins and peptides for sequnce analysis using microcolumn liquid chromatography *J. Microcol. Sep.* **4**, 485-489

Moritz, R.L., Eddes, J., Ji, H., Reid, G.E. and Simpson, R.J. (1995) Rapid Separation of Proteins and Peptides using Conventional Silica-Based Supports: Identification of 2-D Gel Proteins Following In-Gel Proteolysis in Techniques in Protein Chemistry VI, (Crabb, J.W., ED.), *Academic press*, San Diego, pp. 311-322.

Nice, E.C. and Simpson, R.J. (1989) Micropreparative high-performance liquid chromatography of proteins and peptides *J. Phar. Biomed. Anal.* 7; 9, 1039-1053.

Nugent, K.D. (1990) Ultrafast protein analysis: A powerful technique for recombinant protein process monitoring, *In* "Current Research in Protein Chemistry: Techniques, Structure and Function" (Villafranca, J.J., ed.) *Academic Press*, pp. 233-244.

Pappin, D.J., Højrup, P. and Bleasby, A.J. (1993) Rapid identification of proteins by peptide-mass finger printing *Current Biology* **3**, 327-332.

Patterson, S. D. (1994) From electrophoretically separated protein to identification: strategies for sequence and mass analysis. *Anal. Biochem.* **221**: 1-15.

Rosenfeld, J., Capdevielle, J., Guillemot, J.C. and Ferrara, P. (1992) In-gel digestion of proteins for internal sequence analysis after one- or two-dimensional gel electrophoresis. *Anal. Biochem.* **203**: 173-179.

Rozing, G.P. and Goetz, H. (1989) Fast separation of biological macromolecules on non-porous, microparticulate columns *J. Chromatogr.* **476**, 3-19.

Shaw, G. (1993) Rapid identification of proteins *Proc. Natl. Acad. Sci. U.S.A.* 90, 5138-5142

Sibbald, P.R., Sommerfeldt, H. and Argos, P. (1991) Identification of proteins in sequence databases from amino acid composition data *Anal. Biochem.* **198**, 330-333.

Simpson, R.J., Moritz, R.L., Rubira, M.R. and Van Snick, J., (1988) Murine hybridoma/plasmacytoma growth factor: Complete amino acid sequence and relation to human interleukin-6. *Eur. J. Biochem.* **176**, 187-197.

Simpson, R.J., Moritz, R.L., Begg, G.S., Rubira, M.R., and Nice, E.C. (1989) Micropreparative procedures for high sensitivity sequencing of peptides and proteins. *Anal. Biochem.* **177**, 221-226.

Snyder, L.R. and Kirkland, J.J. (1979) Introduction to Modern Liquid Chromatography. Wiley Interscience, New York.

Stone, K. (1992) ABRF Workshop: Digestion of proteins from blots and gels *ABRF News* **3-3**, 8-9.

Unger, K.K., Jilge, G., Kinkel, J.N. and Hearn, M.T.W. (1986) Evaluation of advanced silica packings for the separation of biopolymers by high-performance liquid chromatography. II. Performance of non-porous monodisperse 1.5 μm silica beads in the separation of proteins by reversed-phase gradient elution high-performance liquid chromatogrraphy, *J. Chromatogr.* **359**, 61-72.

Ward, L.D., Reid, G.E., Moritz, R.L. and Simpson, R.J. (1990a) Strategies for internal amino acid sequence analysis of proteins separated by polyacrlyamide gel electrophoresis. *J. Chromatogr.* **519**: 199-216.

Ward, L.D., Reid, G.E., Moritz, R.L. and Simpson, R.J. (1990b) Peptide mapping and internal sequencing of proteins from acrylamide gels. . *In* "Current Research in Protein Chemistry: Techniques, Structure and Function" (Villafranca, J.J., ed.) Academic Press, pp. 179-190.

Ward, L.D., Ji, H., Whitehead, R.H. and Simpson, R.J., (1990c) Development of a database of amino acid sequences for human colon carcinoma proteins separated by two-dimensional polyacrylamide gel electrophoresis. *Electrophoresis* **11**, 883-891.

Williams, K., Kobayashi, R., Lane, W. and Tempst, P. (1993) Internal amino acid sequencing: Observations from four different laboratories *ABRF News* **4-4**, 7-12.

Yamasaki, Y., Kitamura, T., Nakatani, S. and Kato, Y. (1989) Recovery of proteins and peptides with nanogram loads on non-porous packings. *J. Chromatogr.* **481,** 391-396.

4

PRE-ELECTROPHORETIC LABELLING OF PROTEINS WITH A COLOURED WATER SOLUBLE EDMAN REAGENT

Keith Ashman*

Centre for Animal Biotechnology
School of Veterinary Science
The University of Melbourne
Parkville 3052, Melbourne, Victoria, Australia

INTRODUCTION

Sodium dodecyl sulphate polyacrylamide gel electrophoresis (Laemmli 1970) (SDS PAGE) is still the most powerful method of resolving a complex mixture of proteins. However, it is generally used as an analytical rather than a preparative tool. With the advent of PVDF membranes which are stable under the conditions employed by the Edman degradation, it has become common to try and obtain N-terminal sequence data from proteins separated by SDS PAGE followed by electrophoretic transfer to a PVDF membrane (Matsudaira 1987). It is also possible to electroelute proteins out of gel slices for sequencing or enzymic digestion. Alternatively the proteins can be enzymically digested within the gel matrix or directly on the membrane after transfer and the peptides eluted for subsequent HPLC purification (Bauw 1989). Both these procedures require the proteins to be stained after electrophoresis in order to locate their position in the gel or on the membrane. The staining process generally fixes the proteins and leads to significant loss of material.

Some of these problems may be overcome by pre-electrophoretic labelling (Kraft 1988). A simple method of pre-labelling proteins with a water soluble Edman reagent S-DABITC (see figure 1) (Chang 1989) which couples to the N-terminal amino acid and the epsilon amino group of lysine has been developed. The reaction takes place under very mild conditions and the reagent has been described for its use in the identification of reactive lysines on the surface protein molecules (Chang 1992). By denaturing proteins in the presence of SDS it is possible to label most of the available sites on a molecule. The method provides a simple method of generating coloured marker proteins for electrophoresis which can be used in preparative electrophoresis apparatus or on SDS PAGE gels. More importantly

* Current Address The European Molecular Biology Laboratory (EMBL), Meyerhof Strasse 1, Heidelberg D6900 Germany.

Methods in Protein Structure Analysis, Edited by M. Z. Atassi and E. Appella
Plenum Press, New York, 1995

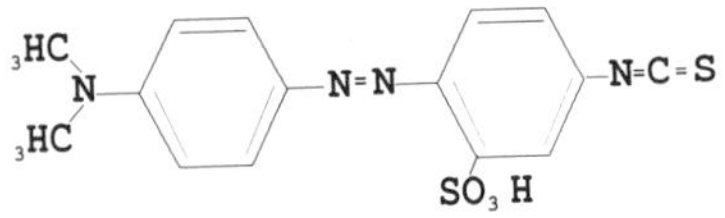

S-DABITC

$$(H_3C)_2N-C_6H_4-N{=}N-C_6H_3(SO_3H)-N{=}C{=}S$$

Figure 1. The structure of 4-N, N-dimethylaminobenzene-4′-isothiocyanate-2′-sulphonic acid, S-DABITC.

the labelled proteins can still be sequenced after the labelling procedure and electrophoretic separation. The N-terminal label is removed during the first cycle of Edman degradation. The labelled molecules can either be transferred to a suitable membrane for direct sequencing or passively eluted from the gel, since no fixing or further staining is required and collected on a Prospin cartridge or similar device. Passive elution is especially useful for high molecular proteins where it is often necessary to collect material from several gels to obtain enough for sequencing. The fact that the proteins carry a coloured label makes it easier to keep track of them. Further, the label does not interfere with enzymic or chemical digestion and lysine containing peptides are readily identified during HPLC separation because they have a characteristic absorption at 450nm. The procedure has been tested on several proteins and found to be a practical Method of labelling and recovering proteins and peptides for sequencing.

MATERIALS AND METHODS

ß-lactoglobulin, Problott and Prospin devices were from Applied Biosystems. bovine serum albumin (BSA), ribonuclease and carbonic anhydrase were from Sigma. S-DABITC was from Protein Institute, P.O. Box 550, Broomall PA, 19008-0550 U.S.A. The reaction buffer for labelling reactions was 20mM sodium bicarbonate pH 8.3, with or without 0.1%SDS. Proteins were labelled for 30 minutes at 60ºC. Electrophoretic transfer to Problot was performed in a Biorad mini-gel blotting apparatus using either 25mM tris glycine pH 8.5 or 10mM CAPS buffer pH 11, both buffers containing 10% methanol. Proteins were transferred at 100 volts constant voltage. Cyanogen bromide (CnBr) cleavage was performed by incubating the protein with 100µl of a saturated solution of CnBr in 70% formic acid, overnight, in the dark and at room temperature. Protein sequencing was performed on an Applied Biosystems 476a protein sequencer using the FSTBLT cycle. The instrument was equipped with a model 610a Data collection system.

RESULTS AND DISCUSSION

Presented here is a simple method of labelling proteins with a coloured water soluble Edman type reagent. The labelled proteins can be used as coloured markers during SDS PAGE and show a small increase in apparent molecular weight, probably due to the increased mass contributed by the label. Figure 2 shows the behaviour of several S-DABITC labeled proteins on SDS PAGE compared to coomassie blue stained marker proteins. The labelled proteins are yellow in colour at pH values greater than 7, but will appear red if subjected to acid conditions.

A major advantage of labelling proteins with S-DABITC is that they can still be sequenced after electroblotting and there is no requirement to stain the blot to locate the

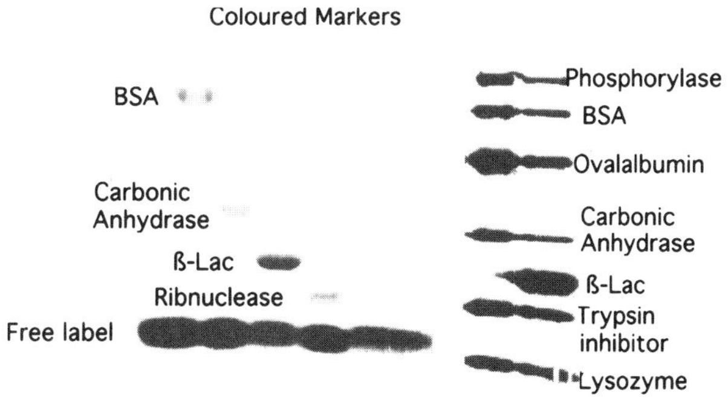

Coloured Markers

BSA

Carbonic Anhydrase

ß-Lac

Ribnuclease

Free label

Phosphorylase

BSA

Ovalalbumin

Carbonic Anhydrase

ß-Lac

Trypsin inhibitor

Lysozyme

Figure 2. SDS PAGE showing individual labelled proteins and coomassie blue stained marker proteins.

proteins. This is illustrated in figure 3 where the degradation of 50pmol of elctroblotted S-DABITC ß-lactoglobulin is shown. The N-terminal leucine is not visible, since the HPLC conditions are not designed for the detection of the S-DABPth derivative. However, there is no Pth-Leucine visible, showing that the N-terminus of the protein was completely labelled. There is a small amount of Pth-Isoleucine (the second amino acid in the ß-lactoglobulin sequence) present in cycle 1 which is probably due to the partial removal of the N-terminal

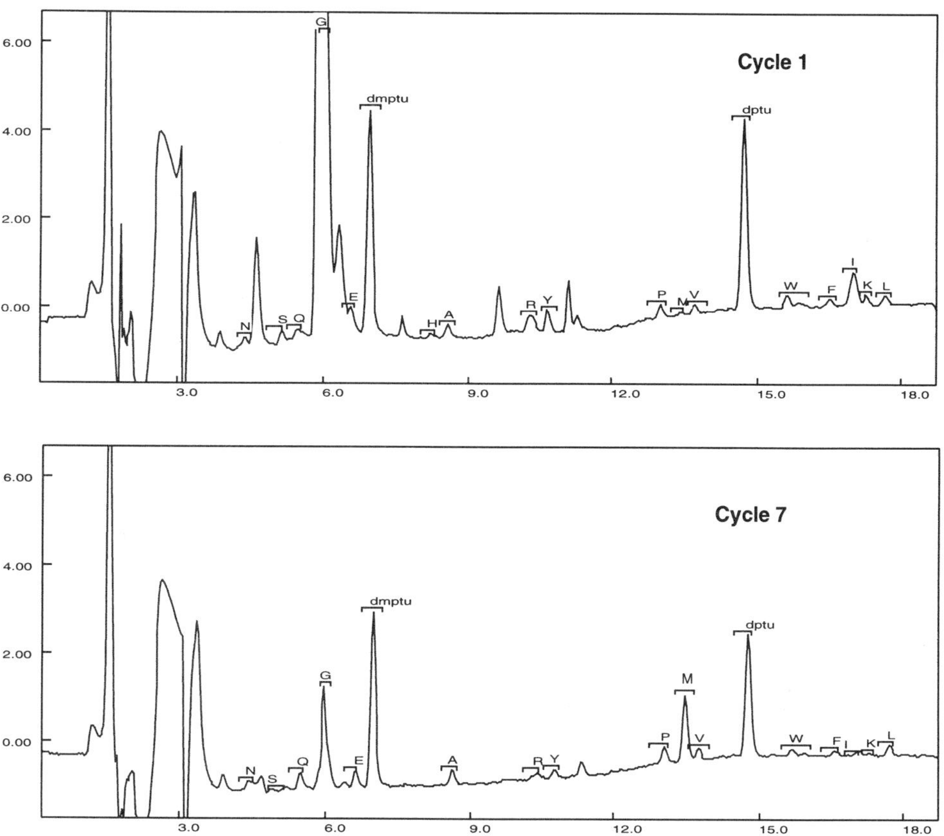

Figure 3. Edman degradation of 50pmol of ß-lactoglobulin labelled with S-DABITC before SDS PAGE and transfer to Problot.

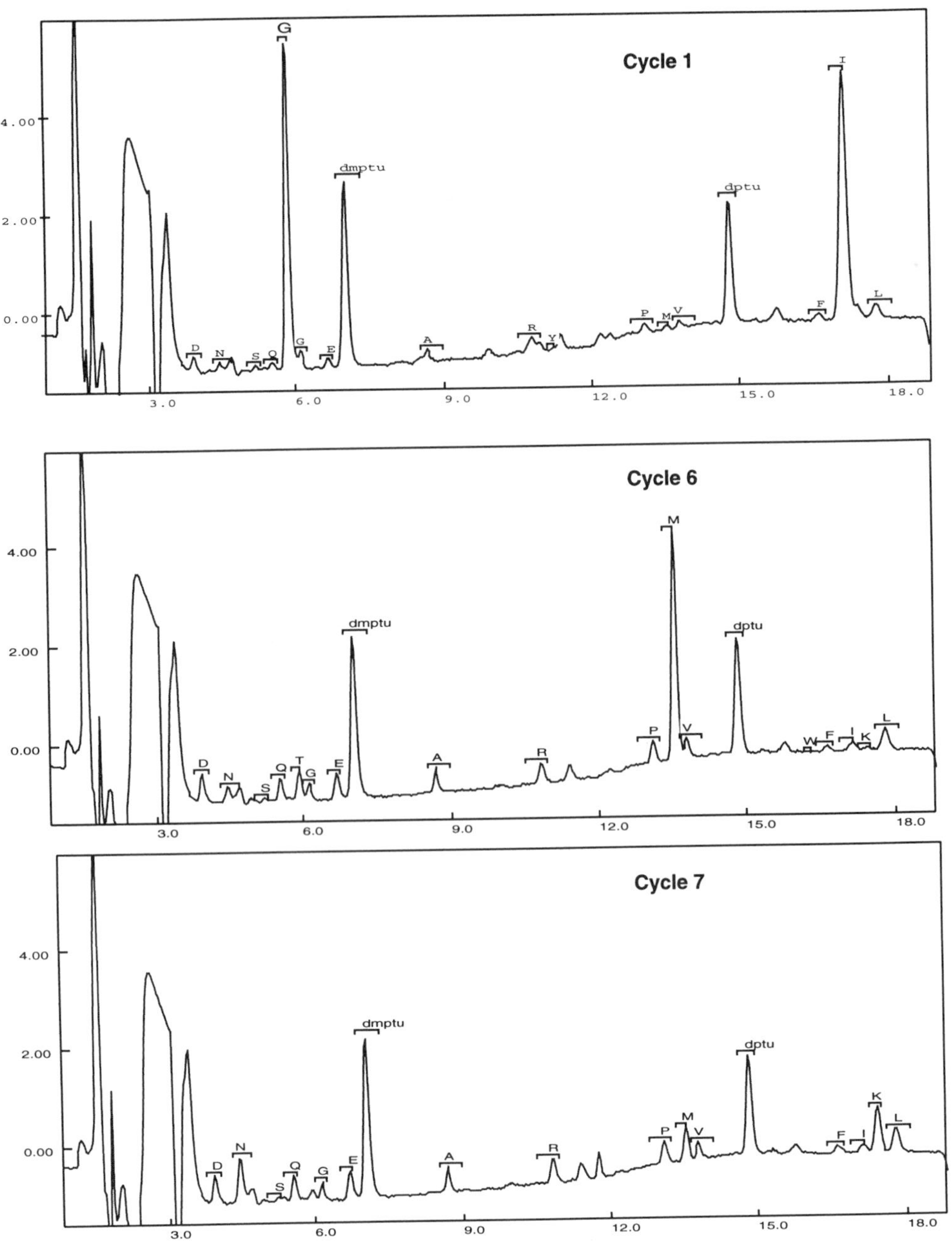

Figure 4. Edman degradation of 50pmol of ß-lactoglobulin labelled with S-DABITC before SDS PAGE and transfer to Problot. Prior to loading into the sequencer the blot was treated with 20 µl of TFA, dried in vacuo and washed with ethyl acetate in order to remove the derivatised N-terminal amino acid.

leucine during the Fstbgn cycle of the sequencer run. Pretreatment of the blotted sample with TFA prior to loading in the sequencer overcomes this preview problem and as shown in Figure 4 only the second residue of ß-lactoglobulin, isoleucine is visible on the chromatogram again indicating the labelling of the N-terminus was complete. Alternatively the fstbgn

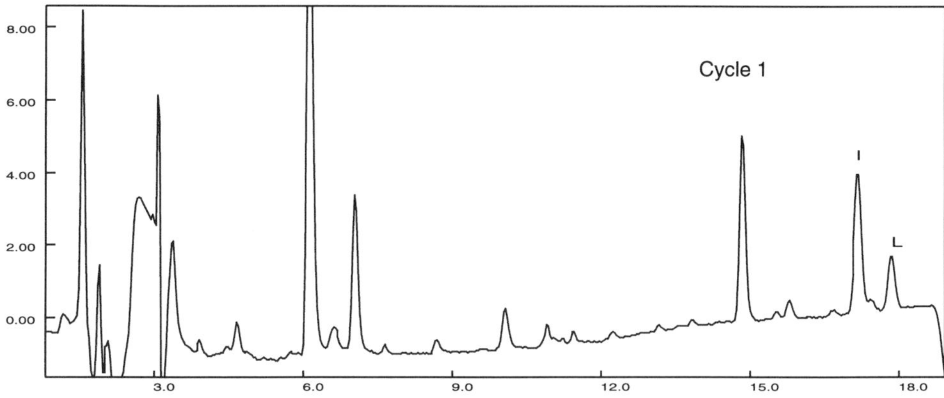

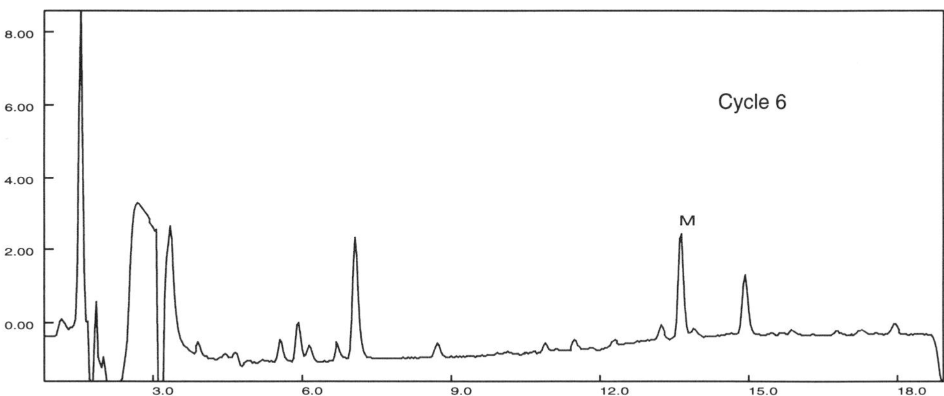

Figure 5. Edman degradation of 50pmol of ß-lactoglobulin labelled with S-DABITC before SDS PAGE. The protein was passively eluted into water and then captured on an Applied Biosystems Prospin device.

cycle could be modified. Cycle 7 in figure 2 shows the presence of a small amount of Pth lysine, which indicates the epsilon amino group of this residue was not completely derivatised with S-DABITC in this sample.

A further advantage of pre-electrophoretic labelling is that there is no requirement to stain the gels to visualise the proteins, therefore the proteins are not contaminated by the staining procedures nor are they fixed in any way and can be passively eluted from the excised bands by placing them in water. Figure 5 shows the results of sequencing passively eluted S-DABITC labelled ß lactoglobulin. After elution the protein was captured on a prospin cartridge for sequencing.Prior to sequencing the membrane was treated with TFA to cleave off the N-terminal leucine. There is a small amount of Pth leucine present in cycle one, indicating that the derivatisation of the N-terminus was not 100% in this case.

In order to demonstrate that the label can be used to isolate labelled peptides 200 pmol of BSA were labelled, digested with cyanogen bromide, the fragments separated by SDS PAGE and finally electro-blotted onto problot. The result of the blot is shown in figure 6. A parallel reaction was set up with unlabelled BSA and also separated and transferred to problott from the same gel. The S-DABITC labelled blot and the ponceau S stained unlabelled blot are shown in figure 6, and demonstrate similar senstivity. The results of

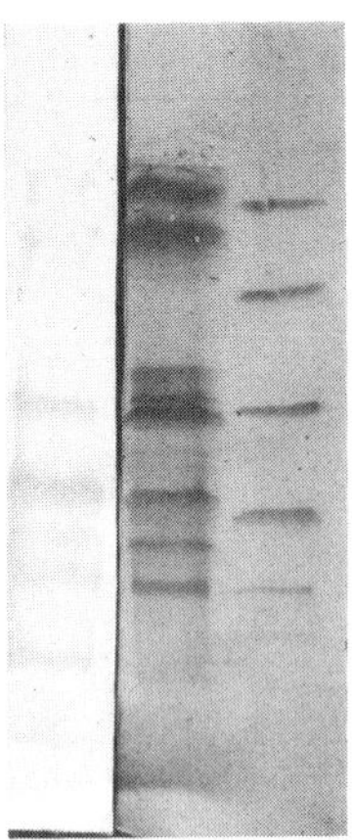

Figure 6. The SDS PAGE separation of the cyanogen fragments generated from BSA labelled with S-DABITC prior to digestion. Shown are the fragments after electrophoretic transfer to problot.

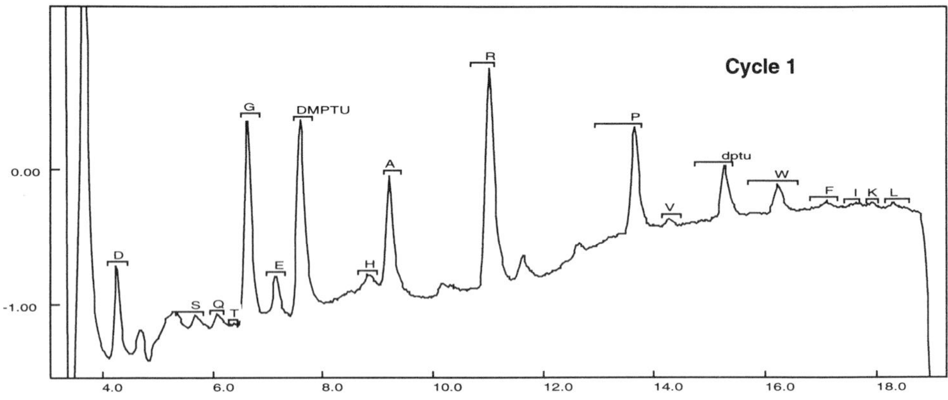

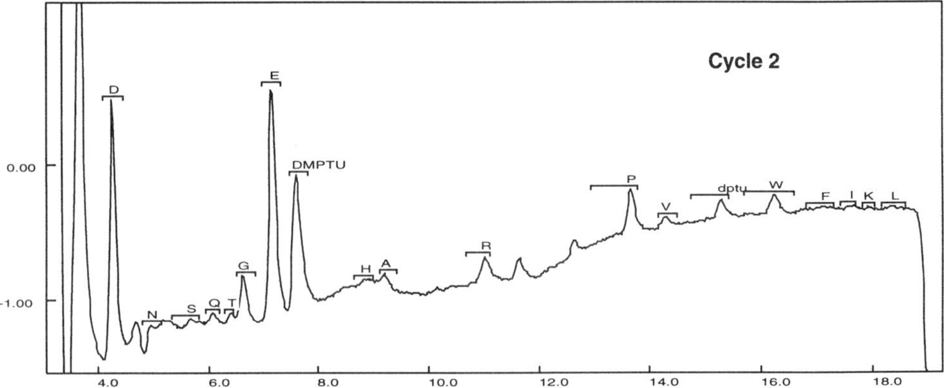

Figure 7. Edman degradation of a cyangen bromide fragment of BSA. The BSA was labelled with S-DABITC prior to digestion. The fragments were separated by SDS PAGE and electrophoretically transferred to Problot for sequencing.

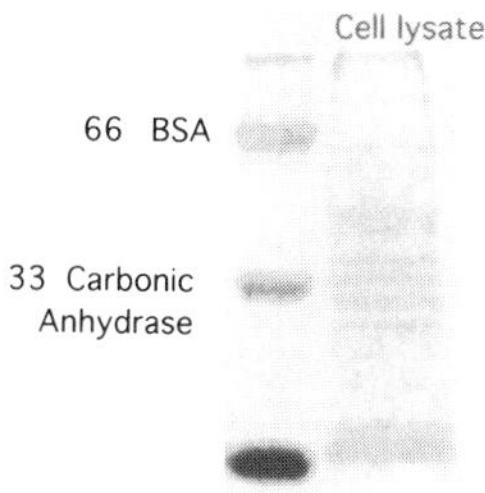

Figure 8. SDS PAGE showing the S-DABITC labelling of a cell lysate of Osteragia circumcincta.

Table 1. Some uses of pre-electrophoretic labelling

- To make PAGE Markers visible during electrophoresis
- Coloured proteins are easy to track
- Edman degradation not hindered
- No additional staining or fixing allows easy passive elution
- Tracking proteins for preparative electrophoresis eg HPEC, Prep Cell
- Isolation of labelled peptides by HPLC

sequencing of one of the bands on the S-DABITC labelled blot are shown in figure 7. Two sequences AD and RE, corresponding to the N-terminus of 2 BSA CNBR fragments are present.

The ability of S-DABITC to label a complex was tested by reacting a cell lysate of *Osteragia circumcincta* with the reagent. The labelled protein mixture was separated by SDS PAGE and the result is shown in figure 8. There are many labelled bands visible clearly showing that it is possible to label complex mixtures.

The results presented here demonstrate that it is possible to efficiently label proteins with a water soluble Edman reagent S-DABITC. The labelled molecules are visible during SDS PAGE separation and can be sequenced either after electro-blotting or passive elution from the gel. The latter method may be useful for accumulating and concentrating quantities of protein for sequencing or digestion, especially in the case of high molecular weight proteins where electro-transfer is often difficult. The label is stable to cyanogen bromide digestion and the labelled peptides can be isolated. The potential of the reagent as a tool in protein sequence analysis is clearly great, it may be a useful alternative to PITC for sequencing and work is in progress to assess this possibility.

REFERENCES

1. Bauw, G. et al (1989) Prot. Nat. Acad. Sci. USA 86, 7701-7705
2. Chang, J.Y. (1989) J.Biol. Chem. **264,** 3111-3115
3. Chang, J.Y. et al (1992) Biochemistry **31,** 2874-2878
4. Kraft, R. et al (1988) Biol. Chem. Hoppe Seyler **369**, 87-91
5. Laemmli, U. k. (1970) Nature **227**, 680-685
6. Matsudaira, P. (1987) J. Biol. Chem. **262**, 10035-10038

N-TERMINAL SEQUENCE ANALYSIS

5

HIGH SENSITIVITY ANALYSIS OF PTH AMINO ACIDS

Karen C. Waldron, Michael Carpenter, Ian Ireland, Darren Lewis, Xing Fang Li, and Norman J. Dovichi

Department of Chemistry
University of Alberta
Edmonton, Alberta T6G 2G2, Canada

ABSTRACT

Capillary electrophoresis and laser-based photothermal detection are used to analyze minute amounts of PTH amino acids. This technology is demonstrated for analysis of manual Edman degradation reactions. This technology is also used to analyze the products generated by a highly miniaturized automated protein sequencer.

INTRODUCTION

The determination of the primary amino acid sequence of minute amounts of proteins remains important in biology. Current technology relies on the repetitive application of the Edman degradation reaction (1). In this reaction, the N-terminal amino group of the peptide reacts with phenylisothiocyanate (PITC) under basic conditions to form the phenylthiocarbamyl (PTC) derivative. After excess reagent is extracted, the PTC- peptide is treated with anhydrous acid to cleave the cyclic phenylthiazolinone amino acid. In the process, the peptide is truncated by one amino acid residue. Last, the thiazolinone is extracted from the truncated peptide and treated with aqueous acid to produce the stable phenylthiohydantoin amino acid (PTH). There are also two common side products produced in the sequencing reaction, diphenylthiourea (DPTU) and dimethylphenylthiourea (DMPTU). Cysteine does not survive the Edman degradation reaction. As a result, there are 19 possible PTH amino acid products for unmodified amino acids, plus the two main interfering products.

Manual protein sequencing involves a laborious series of reactions and extractions to isolate the PTH amino acid products from each cycle of the Edman degradation reaction. The development of automated protein sequencers lead not only to a significant increase in efficiency and reproducibility but also allowed the use of smaller amounts of reagents, which allows the study of smaller amounts of peptide (2-4). Miniaturization of the automated sequencer has lead to improved sequencing sensitivity. This approach has been quite

Methods in Protein Structure Analysis, Edited by M. Z. Atassi and E. Appella
Plenum Press, New York, 1995

successful for one fundamental reason. The major impediment to sequence determination is reagent impurity and contamination. By reducing consumption of reagents, contamination is reduced; smaller amounts of protein may be sequenced.

In all examples of protein sequencing, either thin layer chromatography or high performance liquid chromatography have been used to identify the PTH amino acid products. In general, detection of less than one picomole of PTH amino acid is difficult. While the use of microbore columns may offer some sensitivity advantages, the limits of liquid chromatography appear to be in sight.

We have reported an alternative technology for detection of minute amounts of PTH amino acids. This technology is based on micellar capillary electrophoresis for separation and a laser-based thermo-optical absorbance technique for detection (5). Micellar capillary electrophoresis relies on addition of surfactant to the separation buffer in zone electrophoresis. The technique has been used to separate 22 PTH amino acids in 28 minutes (6).

Recently, we have studied the effect of SDS concentration, buffer concentration and pH on the separation of a mixture of nineteen PTH amino acids (PTH-cysteine was excluded) and two common by-products formed during Edman degradation: diphenylthiourea (DPTU) and dimethylphenylthiourea (DMPTU) (7). Many of the components in this mixture are sensitive to their immediate environment, which is a similar problem encountered in HPLC. PTH-histidine (pKa~6) is especially sensitive to pH during the separation . We have achieved baseline separation of the 19 PTHs and DPTU and DMPTU within 10 minutes with a pH 6.7 buffer consisting of 10.7 mM sodium phosphate, 1.8 mM sodium tetraborate and 25 mM SDS, at ambient temperature. Thermo-optical absorbance provides detection limits (3σ) for the PTH amino acids that range from 0.2 to 5 fmol injected onto the column. This limit is almost 1,000 times better than currently used methods for HPLC.

While these separation and detection capabilities are outstanding, it is important to understand one property of the technology: samples must be injected in small volumes, on the order of a few nanoliters; injection of larger volumes leads to unacceptable band-broadening, which destroys the separation. As a result, there is a fundamental mismatch between the volumes produced by commercial protein sequencers and the volumes required for capillary electrophoresis.

In this paper, we demonstrate the performance of capillary electrophoresis for analysis of the products generated by manual Edman degradation reaction. The electrophoretic analysis is much faster and much easier to perform than gradient elution reversed phase liquid chromatography. We also demonstrate the use of capillary electrophoresis for analysis of the products generated by a highly miniaturized protein sequencer.

EXPERIMENTAL

Manual Edman Degradation

The method for manual protein sequencing is described in detail elsewhere (7).

Micellar Capillary Electrophoresis

Determination of PTH amino acids was performed using the CE/thermo-optical absorbance instrument described in detail elsewhere (4, 7), with a few changes: the pump laser was operated at 625 Hz, a neutral density filter (O.D. = 0.3) was placed in the beam path to reduce the beam intensity, the probe beam intensity was detected by a variable gain/variable bandwidth Model 2001 Front-end Optical Receiver (New Focus, Inc., Mountain view, California, USA), and data were collected at 3 Hz directly from the lock-in amplifier to the PC via an RS232 interface. A program was written in BASIC for data collection and display.

Automated Protein Sequencing

Sequencing grade 12.5% trimethylamine in water and 5% phenylisothiocyanate in heptane were purchased from Applied Biosystems. Anhydrous trifluoroacetic acid, Polybrene, toluene (redistilled) and oxidized insulin chain B were purchased from Sigma. Acetonitrile (HPLC grade) was purchased from BDH Chemicals Canada. Argon gas was bubbled through the trimethylamine and trifluoroacetic acid solutions to deliver the reagent in the vapor form to the reactor. Acetonitrile and toluene were mixed in a ratio of 15:85. The acetonitrile:toluene mixture and phenylisothiocyanate were pumped as liquids through the reaction chamber with argon gas pressure. Insulin chain B was dissolved in 8.3% trimethylamine in propanol:water, 3:2 (V/V) adjusted to pH 9.5 with trifluoroacetic acid and loaded into the reaction chamber with a 1-µL syringe.

The reaction chamber was constructed from fused silica capillaries; the outside of the capillaries were supplied with a polyimide coating. A 350-µm inner diameter and 500-µm outer diameter capillary was inserted about 1-cm into a 530-µm inner diameter, 700-µm outer diameter and 5-cm long capillary; the two pieces were epoxied together. A 4-mm long bed of Porasil-T coated with 20% (w/w) Polybrene was placed in the larger capillary. A Zitex plug was inserted at either end of the bed to hold the Posasil packing in place. The reaction chamber was precycled once using the reaction protocol described below. The reaction chamber was flushed with argon gas and 890 picomoles of insulin chain B was loaded.

The reaction chamber was heated by Peltier devices purchased from Melcor. A Digi-sense thermocouple was used to monitor the temperature of the reaction chamber.

Prepurified argon was passed through an oxygen trap and distributed by a gas manifold. The argon was used to pressurize all reagents to 3.7 psig. The reagent bottles were connected to a valve block originally designed for use with the Beckman spinning cup sequencer. The valve block has three single position valves and one seven position valve; only the latter valve was used in this experiment. Reagents were delivered by opening the appropriate valve inlet and allowing the pressurized argon to push the reagent to the valve block. The liquid reagents were then pushed through the reaction chamber by a stream of argon.

The conditions used for the sequencing reaction are listed in Table 1.

Table 1.

Step	Reagent	Time (s)	Step	Reagent	Time (s)
coupling, 57 °C	12.5% TMA	60	Cleavage	TFA	180
	5% PITC	5		Argon	180
	12.5% TMA	300			
	5% PITC	5	Extraction	Acetonitrile/toluene	5
	12.5% TMA	300		argon	60
	5% PITC	5		acetonitrile/toluene	5
	12.5% TMA	300		argon	60
	5% PITC	5			
	Argon	120			
Wash	Acetonitrile/toluene	5			
	argon	60			
	acetonitrile/toluene	5			
	argon	60			

The extracted anilinothiazolinone was collected into a 600-μl microcentrifuge tube containing 10-μl of 25% trifluoroacetic acid in water. The extract was then dried in a Speed-Vac. Conversion to the PTH amino acid was performed by dissolving the extract in 50-μL of 25% trifluoroacetic acid and heating at 62 °C for 30-35 minutes. The product was dried in the Speed-Vac and stored at -20 °C.

RESULTS AND DISCUSSION

Separation of Manual Edman Degradation Products

SP-5 is pentapeptide with the sequence: NH_2-arginine-lysine-glutamic acid-valine-tyrosine-COOH (NH_2-R-K-E-V-Y-COOH). Figure 1 shows the electropherograms for the sequence analysis of 865 nmol of SP-5. The standard contains approximately 20 fmol for each PTH amino acid, DPTU and DMPTU. No PTH-cysteine is present in the standard. The electropherograms show good signal-to-noise for the analyte PTH residue because of the very large amount of starting material used. In the first cycle, a few unidentified peaks are seen besides the DPTU by-product peak. Extensive wash steps apparently completely remove DMPTU. Each cycle shows a slight amount of lag, PTH product from the previous cycle present in the current cycle. Only cycle 2 shows evidence of preview, PTH product from the following cycle present in the current cycle.

Separation of Edman Degradation Products from the Highly Miniaturized Sequencer

Figure 2 presents a set of electropherograms generated from 890 pmol of insulin chain B. Tyrosine (Y) was added to each electropherogram as an internal marker of retention time. The retention times were normalized to the DMPTU and tyrosine peaks. The first cycle shows a strong peak for phenyalinine, which demonstrates a small amount of lag in subsequent cycles. The second cycle shows a strong peak for valine (V), which again shows lag in subsequent cycles. The third cycle shows a medium-size peak for asparagine (N); although this peak is easily identified as the terminal residue, lag from previous cycles begin to confound the interpretation of the data. There are also two anomalous peaks in this electropherogram. The first appears at about 5.3 minutes and appears to be due to the passage of a bubble through the detection chamber. The broad peak at 5.7 minutes is of unknown origin. By the fourth cycle, the peak from glutamine (Q) is present, although peaks from N and V, due to lag from previous cycles, dominate the electropherogram.

It is clear that the automated sequencer produces very clean electropherograms, with relatively little spurious signals from reagent impurities. However, lag from previous cycles is a serious problem with the current instrument. We have investigated a number of experimental parameters, and the lag does not appear to be associated with low coupling efficiency. Instead, the instrument appears to suffer from modest cleavage efficiency. The instrument is being modified to improve the efficiency of the cleavage step.

CONCLUSIONS

We have reported the use of capillary electrophoresis for identification of PTH amino acid residues produced by both manual and automated protein sequencing. The electrophoresis system requires about 11 minutes to separate and identify the PTH amino acids.

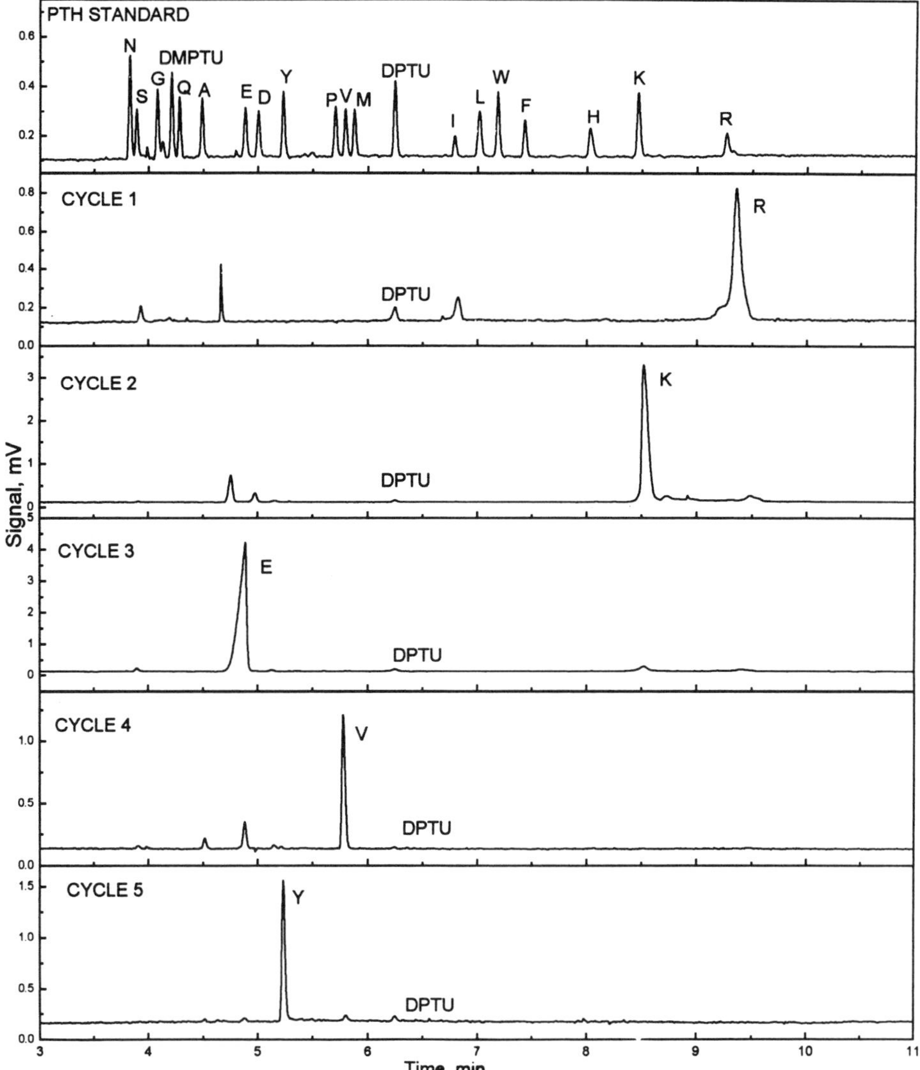

Figure 1. Electropherograms for the manual sequence analysis of 865 nmol of a pentapeptide.

Furthermore, because the system does not require re-equilibration, a new sample may be analyzed immediately after completion of an electropherogram. The capillary electrophoresis system is much faster and simpler than gradient elution high performance liquid chromatography.

In addition to highly efficient separations, our use of photothermal absorbance detection produces high sensitivity analysis. The laser-based detector generates sub-femtomole detection limits for the PTH amino acids. However, the system suffers from one important limitation. Only a few nanoliters of analyte may be injected onto the capillary without introducing an unacceptable amount of band broadening.

We report the development of a highly miniaturized protein sequencer, which is matched in volume to the volume required by capillary electrophoresis. The highly mini-

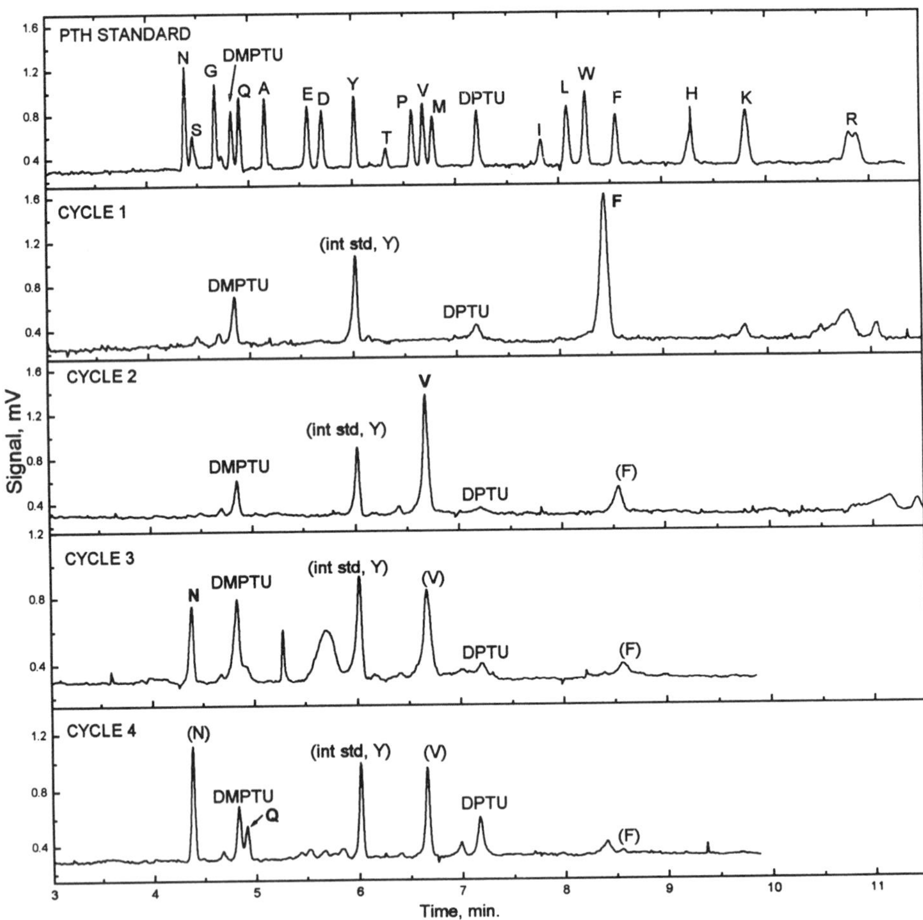

Figure 2. Electropherograms for the automated sequence analysis of 890 pmol of insulin chain B. Tyrosine is added to each sample as an internal standard.

aturized instrument is much smaller than conventional technology. It is based on a 400-μm diameter reaction mat, which has about 1/1000 the cross-sectional area of a conventional sequencer. This minute size allows a three order of magnitude reduction in reagent consumption, with a concomitant reduction in contamination. Two important steps remain in our instrumentation development program. First, we must improve the cleavage step in the miniaturized sequencer to reduce the amount of lag and to improve the overall conversion efficiency. Second, we need to couple directly the reaction chamber with the electrophoresis system. By achieving these two goals, we should be able to sequence routinely femtomole amounts of proteins.

ACKNOWLEDGMENTS

This work was funded by the Natural Sciences and Engineering Research Council of Canada and Sciex, Inc. Many helpful suggestions were provided by Ruedi Aebersold of the Department of Molecular Biotechnology, University of Washington, USA.

REFERENCES

1. P. Edman, *Acta Chem. Scand.* **4**, 283-293 (1950).
2. R. M. Hewick, M. W. Hunkapiller, L. E. Hood, and W. J. Dreyer (1981). *J. Biol. Chem.*, 256(15), 7990-7997.
3. M. Haniu, and J. E. Shively (1988). *Anal. Biochem.*, 173, 296-306.
4. N. F. Totty, M. D. Waterfield, and J. J. Hsuan, (1992). *Protein Science*, 1, 1215-1224.
5. K. C. Waldron , and N. J. Dovichi, (1992). *Anal. Chem.*, 64(13), 1396-1399.
6. K. Otsuka, S. Terabe, and T. Ando, (1985). *J. Chromatogr.*, 332, 219-226.
7. M. Chen, K.C. Waldron, Y. Zhao and N.J. Dovichi *Electrophoresis* in press.

6

SYNTHESIS, EVALUATION AND APPLICATION OF A PANEL OF NOVEL REAGENTS FOR STEPWISE DEGRADATION OF POLYPEPTIDES

Edward J. Bures,[1] Heinz Nika,[1] David T. Chow,[1] Daniel Hess,[1] Hamish D. Morrison,[1] Michael Bartlet-Jones,[2] Darryl J. C. Pappin,[2] and Ruedi Aebersold[1*]

[1] Biomedical Research Centre
University of British Columbia
Vancouver, Canada
[2] Imperial Cancer Research Fund
London WC2A 3PX, United Kingdom

INTRODUCTION

The Edman degradation (Edman, 1949) has been the most successful, general and widely used technique for the determination of the amino acid sequence of proteins and peptides. As a benefit of this distinction, over the last four decades the method has been refined to a high degree of perfection. Nevertheless, sequencing with phenyl isothiocyanate (PITC)[†] suffers from a few practical limitations. First, the extinction coefficient of the phenylthiohydantoins (PTH's) limits sequencing sensitivity. Currently, routine sequencing in most laboratories requires low picomole amounts of sample applied to the sequencer. Second, UV-absorbing products which may co-elute with PTH's during high performance liquid chromatography (HPLC) separation have a tendency to obscure the specific PTH signals during high sensitivity sequencing. Third, with the exception of select cases (Wettenhall et al, 1991; Meyer et al, 1990, 1991; Aebersold et al, 1991; Gooley et al, 1991; Pisano et al, 1993), modified and unnatural amino acids of known structure are difficult to identify

[*] Correspondence address: Department of Molecular Biotechnology, University of Washington, FJ-20, Seattle, WA 98195.

[†] ABBREVIATIONS: PITC: phenyl isothiocyanate; PTH: phenylthiohydantoin; HPLC: high-performance liquid chromatography; ESI-MS: electrospray ionization mass spectrometer/metry; MS/MS: tandem mass spectrometer/metry; PETMA-PITC: 3-[4′(ethylene-N,N,N-trimethylamino)-phenyl]-2-isothiocyanate; PITC-311: 4-(3 pyridylmethylaminocarboxypropyl)-phenyl isothiocyanate; RP-HPLC: reverse-phase high-performance liquid chromatography; TFA: trifluoroacetic acid; MeCN: acetonitrile.

Methods in Protein Structure Analysis, Edited by M. Z. Atassi and E. Appella
Plenum Press, New York, 1995

and de-novo characterization of such residues by UV absorbance detection alone is extremely difficult.

To overcome these limitations we have attempted to develop a new protein degradation chemistry. In particular, the aims of this new chemistry were to achieve higher sensitivity, to provide enhanced selectivity for detecting the specific signal in the products of a chemical sequencing cycle and to provide the possibility of structural characterization of modified residues. To this end we endeavored to design a sequencing reagent which generated derivatives that are detectable by electrospray ionization mass spectrometry (ESI-MS). Femtomole level detection sensitivity of ESI-MS is well documented and mass analysis of the cleaved and extracted residues is expected to enhance the ability to identify modified residues and to extract the specific signal out of the complex chemical mixture generated by the protein sequencer. An additional intrinsic capability of an ESI-MS-based sequencing chemistry is the potential for de-novo structure determination of modified residues by analysis of tandem MS (MS/MS) fragmentation patterns of amino acid derivatives.

As part of ongoing efforts in our group to develop and improve methods of protein structure analysis, we report the synthesis, evaluation and application of a panel of reagents for stepwise degradation of polypeptides and analysis of the resultant derivatives by ESI-MS. We describe the process by which the reagents were designed, the difficulties that arose with specific compounds, and the evolution toward a structure that met the intricate requirements.

A NEW PROTEIN SEQUENCING REAGENT: EVOLUTION OF THE DESIGN

The first reagent we synthesized and reported on was 3-[4′(ethylene-N,N,N-trimethylamino)phenyl]-2-isothiocyanate (PETMA-PITC) (Aebersold et al, 1992) as shown in Fig. 1, structure 1. The molecule was designed to include three specific components, each of which was to serve a distinct purpose. The PITC moiety was included to ensure the optimal coupling and cleavage kinetics which distinguish the Edman chemistry. The strongly basic functional group, a quaternary amine in the case of PETMA-PITC, was added to mediate efficient ionization for high sensitivity detection by ESI-MS. The bridging section consisting of an ethyl group in the case of PETMA-PITC was added to ensure steric and electronic separation of the PITC and the basic groups. Such separation was desirable to minimize interference between the two functional groups.

Experiments using PETMA-PITC demonstrated that the resulting amino acid derivatives could be detected at low femtomole sensitivities by ESI-MS and that the kinetic properties of the reagent were comparable with those of PITC. However, when PETMA-PITC was tested in an automated protein sequencer we observed two limitations. First, the reagent was too polar to be compatible with common absorptive sequencing protocols. This prevented the use of this chemistry in the majority of sequencers currently in use without significant modifications to hardware and protocols. Second, we discovered that it would be preferable for the sequencing reagent to have a higher molecular weight to ensure that the derivatives would appear in an area of the mass spectrum that was less abundant in interfering background contaminants.

To reduce polarity and to enhance the molecular weight the reagent C10-PETMA-PITC (Fig.1, structure 2) was synthesized. While the PITC and quaternary amine functional groups were maintained, this compound differed from PETMA-PITC by a ten-carbon chain extension which was attached with an amide bond linkage. With this reagent, the formation of amino acid derivatives detectable by ESI-MS was effected without difficulty, however,

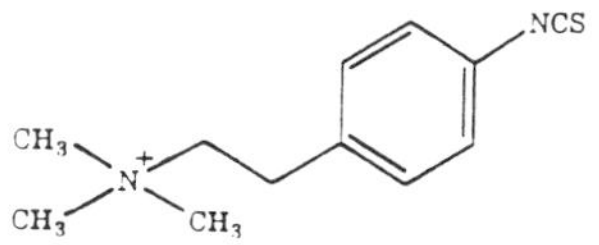

Figure 1. Molecular structure of reagents synthesized and evaluated. **1**: PETMA-PITC **2**: C10-PETMA-PITC (n = 10) **3**: C5-PETMA-PITC (n = 5) **4**: pyridylmethylisothiocyanate **5**: nicotinic phenylisothiocyanate **6**: pyridyl-methylphenylisothiocyanate **7**: 4-(3-pyridinylmethylaminocarboxypropyl) phenyl isothiocyanate (PITC 311).

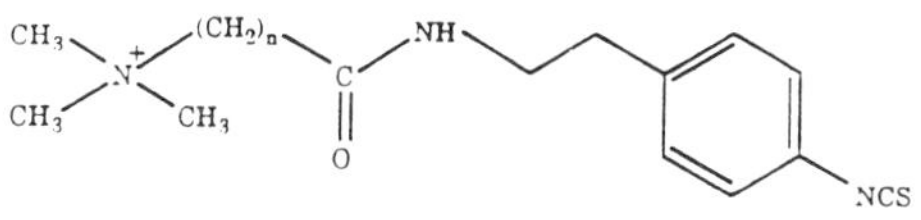

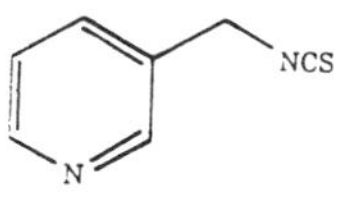

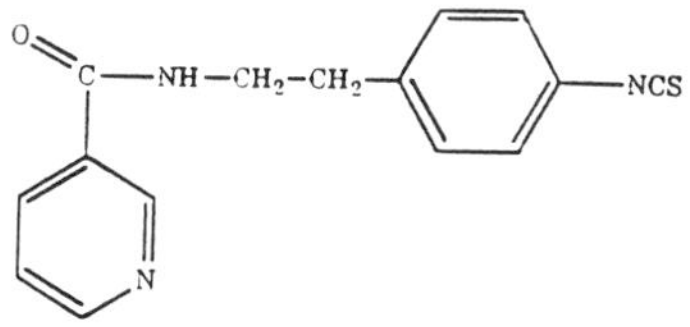

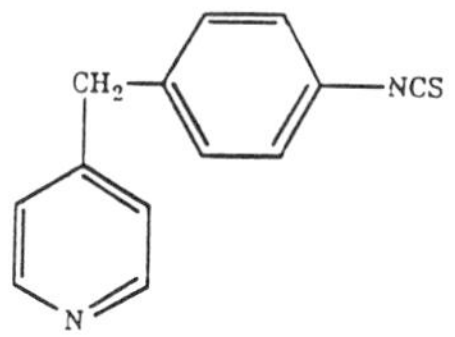

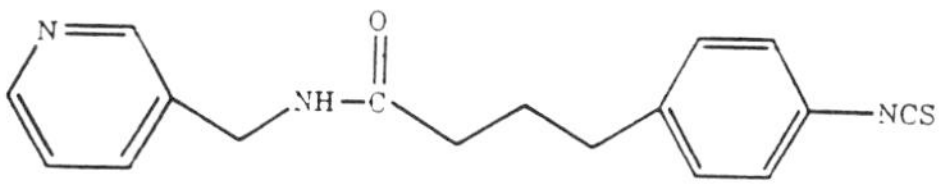

the products were found to chromatograph very poorly under typical reverse-phase high performance liquid chromatography (RP-HPLC) conditions. We attributed this occurrence to the formation of micelles, a structure common to amphipatic molecules. To reduce the potential for micelle formation we next shortened the length of the chain extension to five carbons to form the reagent C5-PETMA-PITC (Fig. 1., structure 3). This was achieved by using the same synthetic steps as with C10-PETMA-PITC, except that a starting material of different size carbon chain was used. This "cassette-style" process to create the reagents expedited synthesis considerably. Unfortunately, as was the case with C10-PETMA-PITC, the C5-PETMA-PTH's generated by sequencing polypeptides with C5-PETMA-PITC were difficult to resolve and recover by RP-HPLC. To overcome the difficulties associated with the strongly polar quaternary amine one of us (DJCP) suggested the use of a pyridyl group as a mediator of ionization in ESI-MS. In contrast to quaternary amines the pyridyl group is not formally charged under typical RP-HPLC conditions, suggesting that pyridyl-based reagents could be more suitable to chromatographic separation and absorptive sequencing conditions. Using the reagent 3-pyridylmethyl isothiocyanate (Fig. 1, structure 4) we showed that pyridyl-containing amino acid derivatives were detectable at sensitivities comparable to those derived from quaternary amine-based reagents. Unfortunately this reagent had the disadvantage of a small molecular weight, thus generating amino acid derivatives that would appear in a region of the mass spectrum which was obscured to a significant degree by low

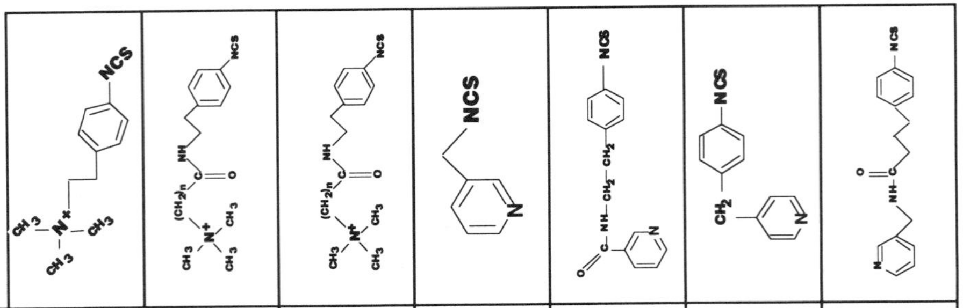

STRUCTURE #	1	2	3	4	5	6	7
POLARITY				●	●	●	●
STABILITY	●	●	●	●		●	●
REACTIVITY	●	●	●	●	●		●
MASS		●	●		●	●	●
HPLC CHROMATOGRAPHY	●			●	●	●	●
DETECTION SENSITIVITY	●			●		●	●

Figure 2. Evaluation of the reagents synthesized. The characteristics indicated with a black dot denote a favorable result was observed for the corresponding reagent.

molecular weight contaminants of unknown origin and nature. Thus our proposed course at this stage was to create a molecule that would concentrate on two main components: i) a pyridine ring for ESI-MS detection and reduced polarity compared to a positively charged quaternary amine, and ii) a higher molecular weight to increase the organic character of the compound and to produce derivatives to appear in a cleaner area of the mass spectrum.

The first reagent incorporating these insights was nicotinic phenylisothiocyanate (Fig. 1, structure 5). This compound was developed from the amide linkage of nicotinic acid and p-nitrophenethylamine. While preliminary results demonstrated desirable coupling and cyclization/cleavage kinetics as well as good detection sensitivity by ESI-MS, application of this reagent for automated sequencing revealed chemical instability during the sequencing process. We learned that the amide bond tended to cleave (at roughly 50% yield) upon exposure to trifluoroacetic acid (TFA). Adjustment of the sequencing conditions to limit TFA exposure and reduce temperature did not yield significant improvement.

To arrive at a more stable structure while maintaining the desirable characteristics of nicotinic phenylisothiocyanate we synthesized and evaluated 4-pyridylmethyl phenylisothiocyanate (Fig. 1, structure 6). Amino acid derivatives prepared with this reagent showed favorable chromatography and ESI-MS detection characteristics. However, the reagent showed poor coupling kinetics in manual "bench-top" coupling reactions as well as in automated peptide sequencing. We attributed this observation to the characteristic of possessing only one carbon atom in the spacer group.

At this point we decided to return to using an amide bond linking group similar to nicotinic phenylisothiocyanate, but with two modifications aimed at arresting the cleavage problem. The first change was to insert a spacer between the amide bond and the pyridine ring to attenuate the electron withdrawing effects of the ring on the amide bond, the effect believed to be responsible for weakening the amide in nicotinic phenylisothiocyanate. The second change was to reverse the sense of the amide bond so that the carbonyl would be even further removed from the pyridine ring. With this rationale, the reagent 4-(3-pyridylmethylaminocarboxypropyl)phenyl isothiocyanate (Fig. 1, structure 7) was synthesized. While the name 4-(3 pyridylmethylaminocarboxypropyl)phenyl isothiocyanate is a chemically accurate description for the reagent we use the simpler name PITC 311 to reflect the molecular weight of the compound in the name. It was with this reagent that we observed superior results with respect to chemical stability, reactivity and chromatography and we therefore proceeded to a detailed characterization of this reagent. The preliminary evaluation of the panel of compounds described above with respect to molecular weight, polarity, reaction kinetics, HPLC chromatography, chemical stability, and mass spectral detectability are summarized in Fig. 2.

ANALYSIS OF 311 PTH AMINO ACID DERIVATIVES BY ESI-MS

Initially we synthesized thiohydantoins of the 20 naturally occurring amino acids and analyzed the products by ESI-MS (Bures, 1994). The mass spectrum of 311 PTH Val shown in Fig. 3 is representative of a typical result obtained with such compounds. The measured mass of $[M+H]^+ = 411.0$ corresponded to the calculated mass for the molecule and it is apparent that the molecule displayed only very limited fragmentation under the ionization conditions used.

We next evaluated the detection sensitivity, linearity of detector response and the dynamic range of the detector. Different amounts of 311 PTH's ranging from 50 fmole to 10 pmole were applied to a 1mm i.d. column and subjected to LC-ESI-MS analysis. The results for residues with acidic, unpolar and neutral-polar side chains shown in Fig. 4. demonstrate that 311 PTH's are detectable by ESI-MS at a sensitivity below 50 femtomoles and that the

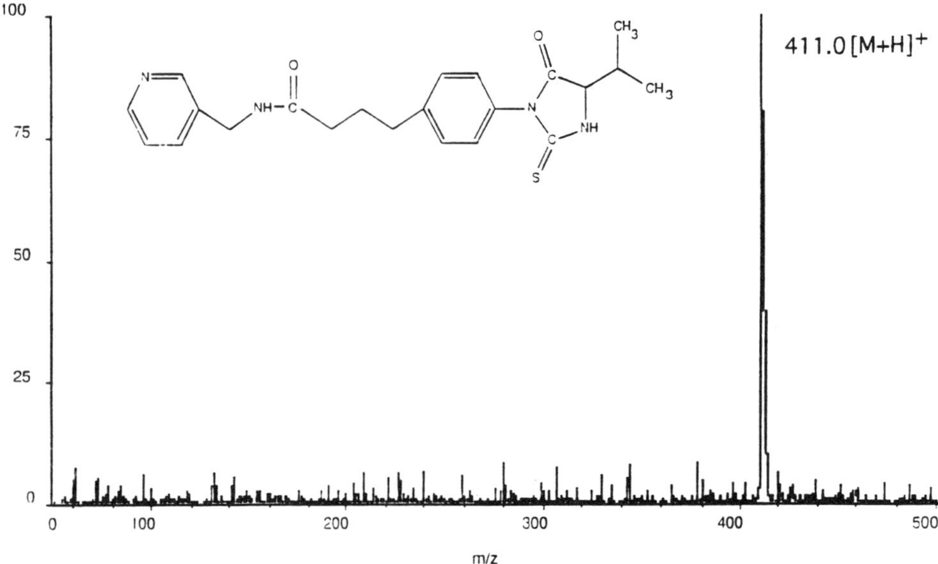

Figure 3. Structure and mass spectrum of 311 PTH Val. One pmole of purified 311 PTH Val was analyzed by LC-MS. The sample was chromatographed over a 1x50 mm Reliasil BDS C-18 column at a flow rate of 50 µl/min using a TFA/acetonitrile (MeCN) solvent system. The inset shows the chemical structure of the compound.

detection sensitivity is comparable for amino acid derivatives with acidic, unpolar and neutral-polar side chains. This sensitivity level was comparable to values achieved previously with PETMA-PITC and supported the potential for protein sequencing at enhanced sensitivity using PITC 311 and ESI-MS detection of 311 PTH's. Furthermore, the detector response was linear in the range of 50 fmole to several pmole and the dynamic range of detection covered three orders of magnitude. Finally, it is important to note, that the instrumental conditions employed for these experiments were such that they would emulate those to be used in a sequencing run in an automated sequencer. In particular, the system was compatible with the injection of sample volumes of up to 100 µl without loss of resolution and sensitivity (Hess et al, 1994).

AUTOMATED SEQUENCING WITH PITC 311

Given that the preliminary testing of PITC 311 showed promise, the next goal was to append an ESI-MS to an automated polypeptide sequencer to attempt "real" microsequencing conditions. The system that was employed is schematically presented in Fig. 5. A commercially available sequencer (Applied Biosystems model 477A) was interfaced on-line with the LC-ESI-MS configuration used for the PITC 311 evaluations described above.

To assess the potential of the ESI-MS system for detection of 311 PTH's generated by automated sequencing we applied an aliquot of a synthetic peptide to the cartridge of the protein sequencer, subjected the sample to automated sequencing using PITC 311 and monitored the degradation products sequentially by UV absorbance detection and by ESI-MS. The data shown in Fig. 6 compare the UV absorbance signals in the first 2

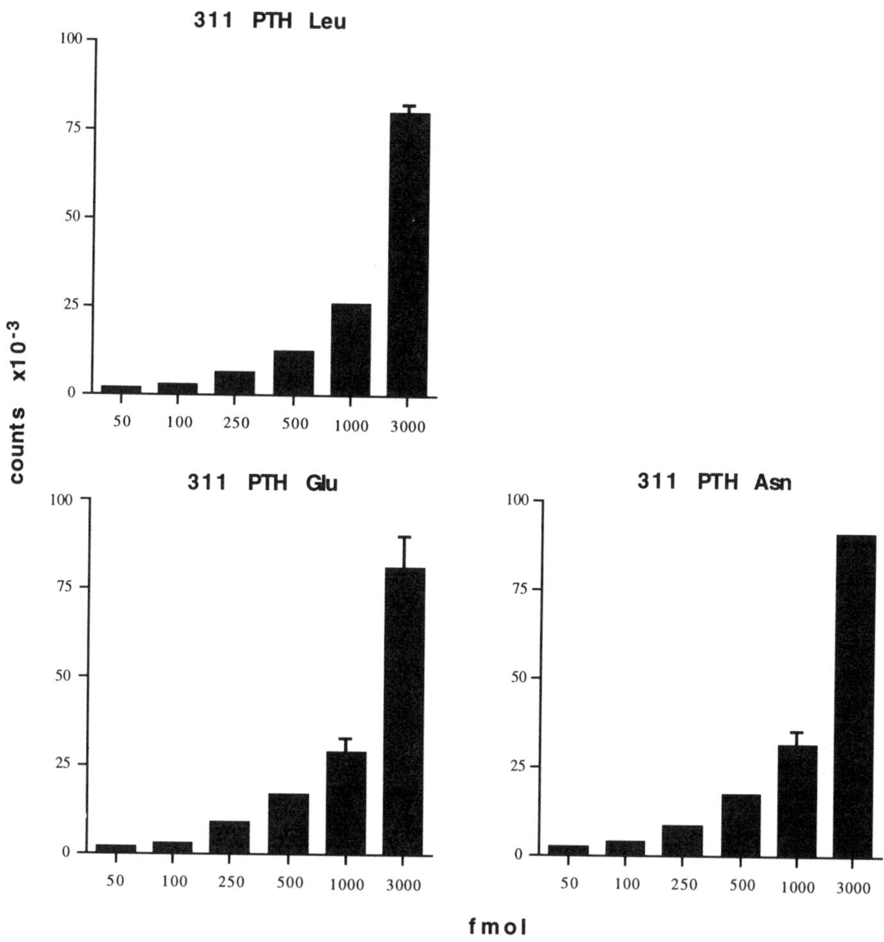

Figure 4. LC-ESI-MS of 311 PTH's. Samples of the amounts indicated and the derivatives indicated were chromatographed over a 1x50 mm Reliasil BDS C-18 column at a flow rate of 50 μl/min and analyzed by multiple ion monitoring LC-ESI-MS. Samples were injected with a 50 μl loop and subject to a 14 minute gradient using a TFA/MeCN solvent system. Integrated peak values are shown. The system consisted of a Michrom UMA HPLC system (Michrom Bioresources) and an API III triple quadrupole MS (PE/SCIEX).

sequencing cycles (panels A1, A2), the total ion current representing all the ions detected by ESI-MS in the mass range from 365-755 Da of the same cycles (panels B1,B2), and the enhanced MS signal generated by selected ion extraction (panels C1, C2). Comparison of panels A and B in Fig. 6 illustrates that most of the contaminants which were detected at a relatively constant level by UV absorbance detection during the sequencing experiment were also detected by the ESI-MS. In general, the ESI-MS results largely resemble the UV data in this form. Selective monitoring of the ions corresponding to the expected 311 PTH's and their adducts (panels C1, C2) dramatically enhanced signal levels, suggesting that the use of MS detection for the products of chemical peptide degradation will be advantageous for high sensitivity sequencing experiments.

In an experiment designed to evaluate the level of sequencing sensitivity achievable using ESI-MS detection of 311 PTH's, we applied decreasing amounts (calibrated by

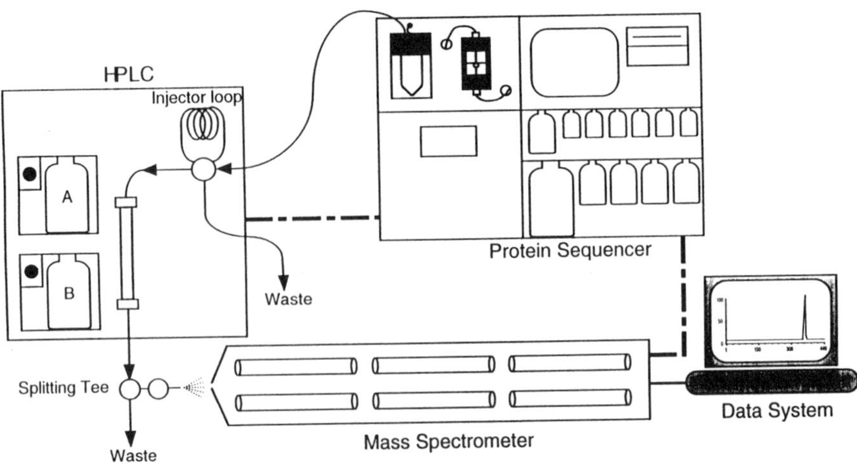

Figure 5. Schematic of system for PITC 311 sequencing. The operation of the sequencer, HPLC and MS was controlled by their respective controllers. Synchronization was achieved by using event A and B of the model 477A sequencer to start the HPLC and the MS respectively. The 311 PTH's were transferred by argon pressure from the sequencer through a standard teflon injector line into the injection loop of the HPLC using the manual injection port. The transfer delay was approximately 8 sec. at which time the injection of the HPLC was triggered and subsequently the data acquisition of the MS was started by contact closure signals. A fused silica capillary with an inner diameter of 75 μm was used to connect the UV-cell of the HPLC with the ESI-MS ion source. Liquid connections between instruments are in solid lines. Electrical connections are represented by broken lines. The flow splitter between the HPLC unit and the electrospray ionization interface was optional and did not affect the performance of the system.

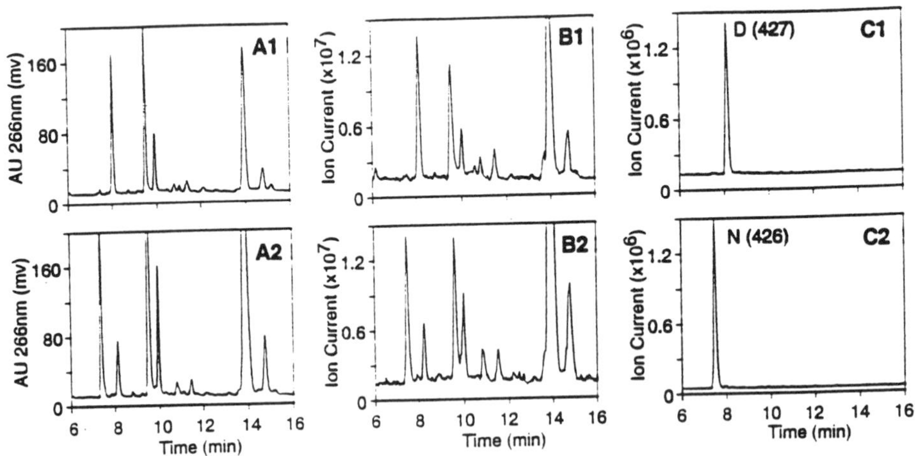

Figure 6. Signal enhancement by ESI-MS detection of 311 PTH's. A synthetic peptide was subjected to automated sequence analysis using PITC 311 and the resulting 311 PTH's were monitored sequentially by UV absorbance and ESI-MS. Results from the first two sequencing cycles (1,2) are displayed. *Row A:* UV absorbance detection of 311 PTH's. *Row B:* Total ion current of 311 PTH detection. Mass range displayed is 365-755 Da. *Row C:* Extraction of acquired MS data for the masses corresponding to 311 PTH's of naturally occurring amino acids. Peaks are designated with the one letter code of the corresponding amino acid and the mass of the compound.

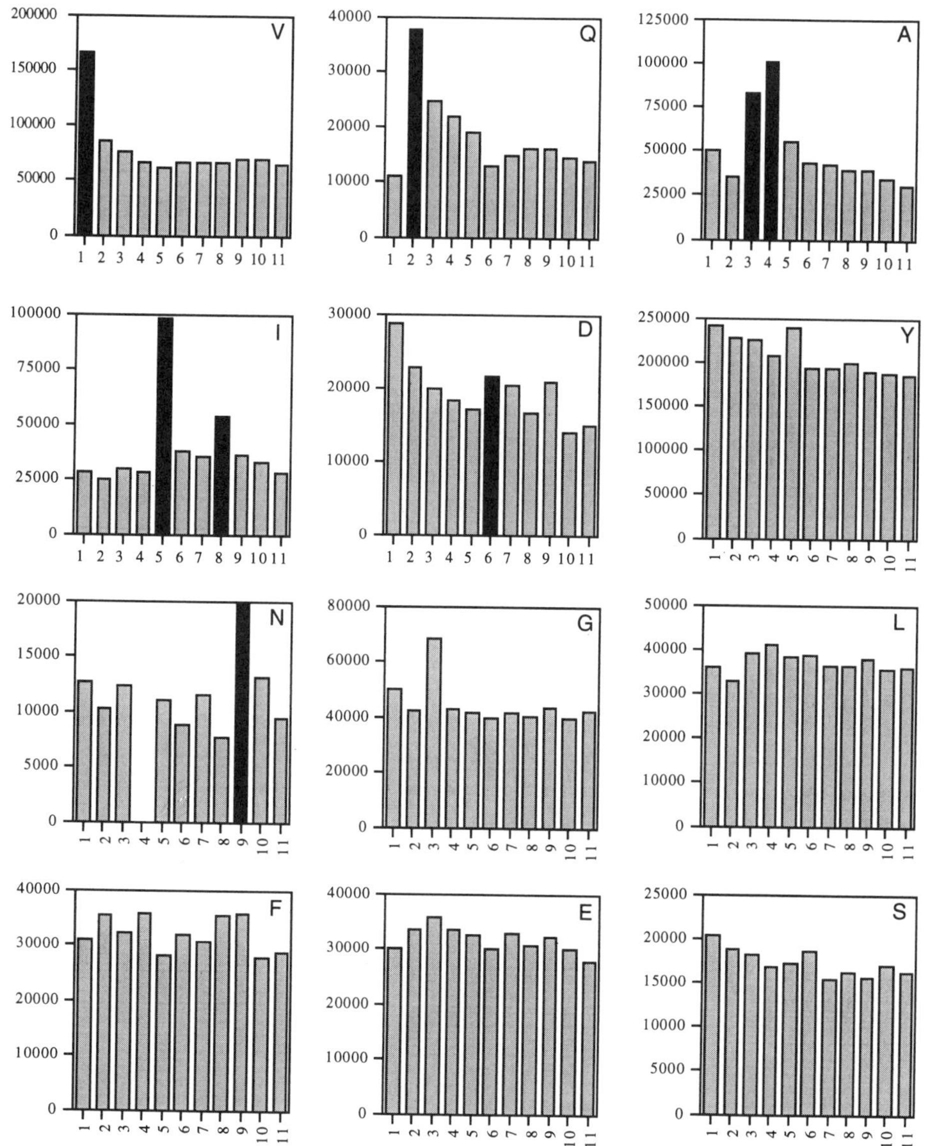

Figure 7. Subpicomole peptide sequencing with PITC 311. A 500 fmole amount of a synthetic decapeptide with the sequence NH_2-Val-Gln-Ala-Ala-Ile-Asp-Tyr-Ile-Asn-Gly-CO_2H was subjected to automated sequence analysis using PITC 311 and the resultant 311 PTH's were monitored by ESI-MS operated in the multiple ion monitoring mode. Each panel depicts a cycle-by-cycle histogram of the abundance of ESI-MS signal corresponding to the amino acid indicated by the single letter code in the upper right hand corner of the graph. The darkly shaded bars indicate the amino acid residue specific for the respective sequencing cycle.

quantitative amino acid composition analysis) of a synthetic decapeptide with the sequence NH_2-Val-Gln-Ala-Ala-Ile-Asp-Tyr-Ile-Asn-Gly-CO_2H to the cartridge of the protein sequencer and sequenced the peptides with PITC 311 under the conditions described above. The results of an experiment in which 500 fmole of peptide was covalently attached to an Arylamine Immobilon disc (Millipore) (Coull, 1991) and applied to the sequencer is shown in Fig. 7.

The 311 PTH mass provided a third data dimension in addition to the RP-HPLC retention time and the signal intensity. To reduce the complexity we displayed the sequencing data as a series of two-dimensional histograms, each histogram representing the abundance of ESI-MS signal corresponding to a particular amino acid derivative as a function of the sequencing cycle. The data in Fig. 7 illustrate that femtomole amounts of peptide can be sequenced by this system. Whereas the background signal for most 311 PTH's is relatively constant and low we observed a contaminant isobaric to 311 PTH Tyr which almost co-eluted with 311 PTH Tyr, obscuring the specific signal in cycle 7. We determined that this contaminant was associated with certain batches of PITC 311 and could be removed by additional purification of the reagent. The aspartic acid (D) signal in cycle 6 was suppressed as a result being covalently attached to the support membrane during sequencing. While covalent sample attachment is not a requirement for the PITC 311 chemistry it was advantageous to use covalent sample attachment in the early sequencing experiments prior to optimization of absorptive sequencing protocols.

While these results are an encouraging demonstration of the potential of PITC 311 chemistry for high sensitivity sequencing the use of a synthetic peptide substrate in tightly controlled test experiments precludes an assessment of the generality of the procedure. The key information pertinent to scientists that are working with proteins is how much material is needed to obtain N-terminal or internal sequence data. To address this practical issue we used trypsin to cleave decreasing amounts (calibrated by quantitative amino acid composition analysis) of bovine carbonic anhydrase , separated the resultant fragments by microbore RP-HPLC and sequenced selected collected peptides using PITC 311 as described above. The resultant degradation products were analyzed by ESI-MS and the data are displayed as described above for Fig. 7. The data shown in Fig. 8 represent the sequence of a peptide obtained by cleavage of 1.2 pmole of protein. Considering the losses associated with HPLC purification, collection and transfer of the peptide into the sequencer clearly only femtomole amounts of peptide were sequenced. The darkly shaded bars in Fig. 8 indicate that the sequence could be easily and unambiguously called even at that sensitivity level.

SUMMARY

We have synthesized and evaluated a panel of novel protein sequencing reagents designed to yield amino acid derivatives detectable at the low femtomole level by ESI-MS. Polypeptide degradation with these reagents is based on the phenylisothiocyanate functionality introduced by Edman (Edman). The chemistries were therefore easily adapted to automated stepwise degradation. Through a systematic process, we have arrived at a new reagent, PITC 311, that permits a sequencing approach that incorporates ESI-MS detection. Employing this approach, we have shown that PITC 311 is compatible with femtomole level peptide sequencing. Additionally, we have demonstrated that mass information provided by ESI-MS detection enhances confidence level in data interpretation. Similarly, mass information available by ESI-MS analysis of 311 PTH's assists in characterization of modified and unnatural amino acid residues. In future work with this reagent, our aim is to optimize automated sequencing cycles for high sensitivity protein sequencing. We also endeavor to develop a methodology to apply PITC 311 for high sensitivity absorptive sequencing, and to create rapid sequencing protocols.

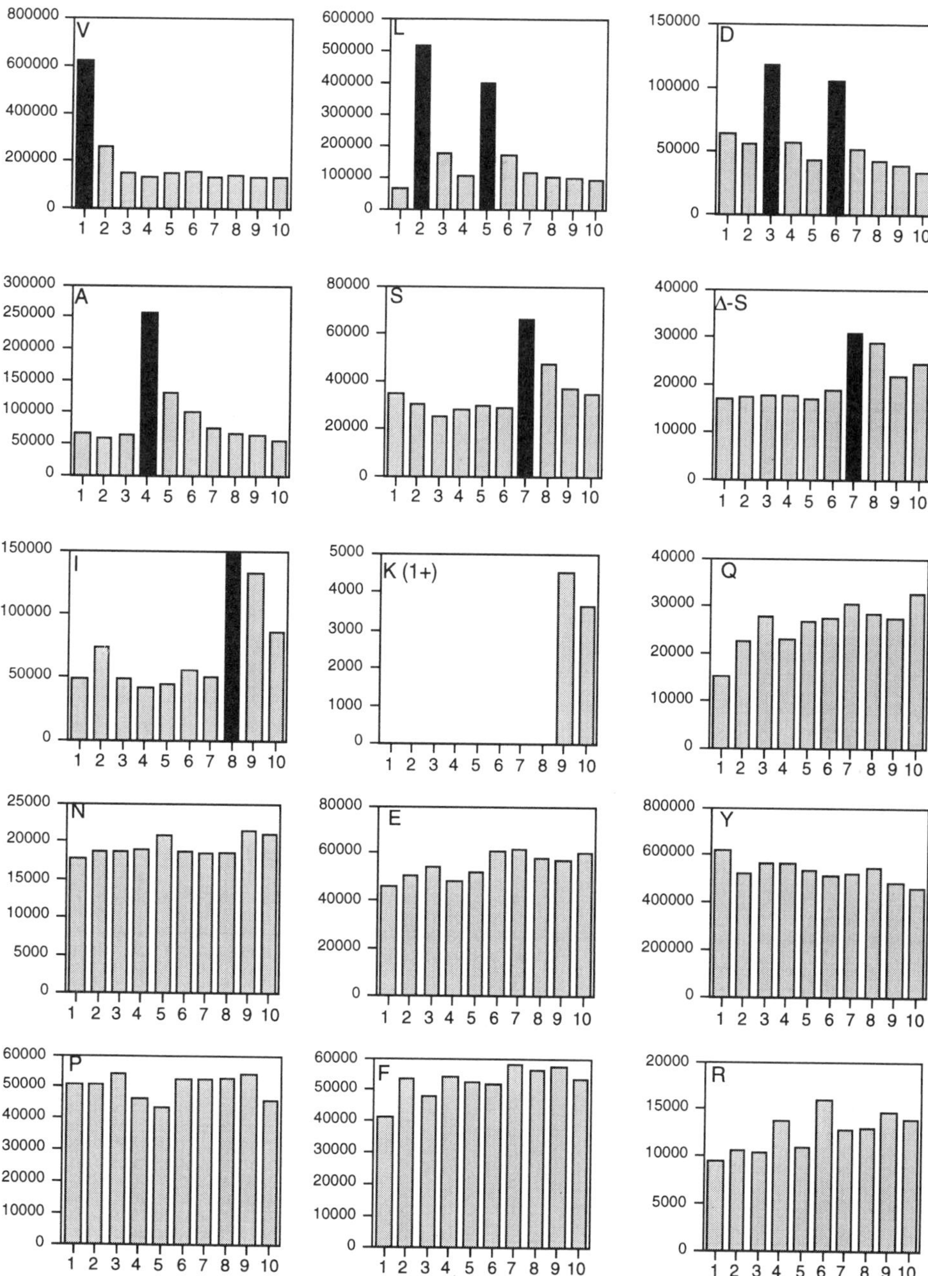

Figure 8. High sensitivity PITC 311 sequencing. A 1.2 pmole sample of bovine carbonic anhydrase was cleaved with trypsin and the resultant peptide fragments separated by RP-HPLC and collected manually. One of these collected peptides was subject to automated sequence analysis using PITC 311 and the resultant 311 PTH's were monitored by ESI-MS operated in the multiple ion monitoring mode. Each panel depicts a cycle-by-cycle histogram of the abundance of ESI-MS signal corresponding to the amino acid indicated by the single letter code in the upper left hand corner of the graph. The darkly shaded bars indicate the amino acid residue specific for the respective sequencing cycle. Δ -S: 311 PTH dehydro serine.

ACKNOWLEDGMENTS

This work was funded in part by the Department of Industry, Science and Technology, Canada(ISTC) and by the Canadian Human Genetic Disease Network. R.A. was the recipient of a Medical Research Council(MRC) of Canada scholarship.

REFERENCES

Aebersold, R., Watts, J.D., Morrison, H.D., Bures, E.J. 1991 Determination of the site of tyrosine phosphorylation at the low picomole level by automated solid-phase sequence analysis. Anal. Biochem. 199:51-60.

Aebersold, R., Bures, E.J., Namchuk, M., Goghari, M.H., Shushan, B., Covey, T.C., 1992 Design, synthesis, and characterization of a protein sequencing reagent yielding amino acid derivatives with enhanced detectability by mass spectrometry. Protein Sci. 1:494-503.

Bures, E.J., Nika, H., Chow, D.T., Morrison, H.D., Aebersold, R. 1994 Synthesis of the protein-sequencing reagent 4-(3-pyridinylmethylamino-carboxypropyl) phenyl isothiocyanate and characterization of 4-(3-pyridinyl-methylaminocarboxypropyl) phenylthiohydantoins. Anal. Biochem. in press.

Coull, J.M., Pappin, D.J.C., Mark, J., Aebersold, R., Koester, H. 1991 Functionalized membrane supports for covalent protein microsequence analysis. Anal. Biochem. 194:110-120.

Edman, P. 1949 A method for the determination of the amino acid sequence in peptides. Arch. Biochem. 22:475-476.

Gooley, A.A., Classon, B.J., Marschalek, R.,, Williams, K.L. 1991 Glycosylation sites identified by detection of glycosylated amino acids released from Edman degradation: the identification of Xaa-Pro-Xaa-Xaa as a motif for Thr-O-glycosylation. Biochem. and Biophys. Res. Commun. 178:1194-1201.

Hess, D., Nika, H., Chow, D.T., Bures, E.J., Morrison, H.D., Aebersold, R. 1994 Liquid chromatography-electrospray ionization mass spectrometry of 4-(3-pyridinylmethylaminocarboxypropyl) phenylthiohydantoins. Anal. Biochem. accepted for publication.

Meyer, H.E., Hoffmann-Posorske, E., Korte, H., Donella-Deana, A., Brunati, A.M., Pinna, L.A., Coull, J., Perich, J., Valerio, R.M., Johns, R.B. 1990 Determination and location of phosphoserine in proteins and peptides by conversion to S-ethylcysteine. Chromatographia 30:691-695.

Meyer, H.E., Hoffmann-Posorske, E., Donella-Deana, A., Korte, H., 1991 Sequence analysis of phosphotyrosine-containing peptides *in* Methods in Enzymology (Hunter, T., and Sefton, B.M., Eds.), Vol. 201, Academic Press, Orlando, FL.

Pisano, A., Redmond, J.W., Williams, K.L., Gooley, A.A. 1993 Glycosylation sites identified by solid-phase Edman degradation: O-linked glycosylation motifs on human glycophorin A. Glycobiology 3:429-435.

Wettenhall, R.E.H., Aebersold, R., Hood, L.E., Kent, S.B.H., 1991 Solid-phase sequencing of ^{32}P labeled phosphopeptides at picomole and subpicomole levels *in* Methods in Enzymology (Hunter, T., and Sefton, B.M., Eds.), Vol. 201, Academic Press, Orlando, FL.

7

IDENTIFICATION AND CHARACTERIZATION OF GLYCOSYLATED PHENYLTHIOHYDANTOIN AMINO ACIDS

Anthony Pisano, Nicolle H. Packer, John W. Redmond, Keith L. Williams, and Andrew A. Gooley

Macquarie University Centre for Analytical Biotechnology (MUCAB)
Macquarie University
Sydney, N.S.W., 2109 Australia

SUMMARY

The three major groups of glycosylated phenylthiohydantoin (PTH) derivatives Asn(Sac), Ser(Sac) and Thr(Sac), can be clearly resolved and separated from the other 20 commonly occurring PTH-amino acids using a new 5 mM triethylammonium formate (TEAF) buffer, pH 4·0 with an acetonitrile gradient. The glycosylated amino acids elute early in a 1·5 min "glycosylation window" between 6·5–8 min, while all the other PTH-amino acids elute between 8–15 min. This buffer system was developed principally for its ability to separate all PTH-amino acids and glycoamino acids at low ionic strength. The low buffer concentration is necessary to minimize glucose contamination for monosaccharide analysis of the PTH-glycoamino acids.

We demonstrate that: (a) a TEAF buffer system is compatible with monosaccharide analysis of the PTH-glycoamino acid and, in principle, the volatile nature of the buffer makes it suitable for ionspray mass spectrometric analysis of recovered PTH-glycoamino acids. (b) the "glycosylation window" is important for the detection of site-specific partial glycosylation and for identifying different forms of PTH-glycoamino acids.

INTRODUCTION

Bioactive proteins are commonly glycosylated and in many cases the glycosylation is important for stability, secretion, biological activity, recognition and cell-cell interactions (Williams and Barclay, 1988; Mallett and Barclay, 1991). However, in many other instances the glycosylation has not been assigned to specific residues and usually only the pooled oligosaccharides from the protein are characterized (Dwek *et al.*, 1993). The control of glycosylation is also becoming increasingly important in the biotechnology industry with

Methods in Protein Structure Analysis, Edited by M. Z. Atassi and E. Appella
Plenum Press, New York, 1995

the use of eukaryotic expression systems where the rules for *in vivo* glycosylation, especially *O*-linked oligosaccharides are only now being understood (Gooley and Williams, 1994).

The traditional approach to the study of carbohydrates attached to proteins and peptides has been to release them from the polypeptide chain by chemical or enzymatic treatment and then characterize the oligosaccharides/monosaccharides separately by: high performance anion exchange chromatography with pulsed amperometric detection (HPAEC-PAD) (Townsend *et al.*, 1989; Townsend and Hardy, 1991), fluorophore assisted carbohydrate electrophoresis (FACE™) (Jackson, 1991) or mass spectrometry (Carr *et al.*, 1993). However glycoproteins often contain a heterogeneous collection of both *N*-linked and *O*-linked oligosaccharides and the release of glycans of a particular class provides no specific information concerning site-specific glycosylation.

Solid-phase Edman degradation is a powerful tool in the identification and quantitation of sites of glycosylation as individual glycoamino acids are recovered for monosaccharide analysis and mass analysis (Gooley *et al.*, 1994a; Gooley *et al.*, 1994b). It is also the only method available for the characterization of clustered sites of glycosylation found in glycoproteins including, mucins, the extracellular domain of human glycophorin A and the macroglycopeptide of bovine κ-casein (Pisano *et al.*, 1993; Pisano *et al.*, 1994).

Phenylthiohydantoin (PTH)-amino acids are easily separated by reversed-phase HPLC and the most popular buffer systems use sodium acetate in tetrahydrofuran. This provides a compact chromatogram without a suitable "window" for the early elution of polar modified amino acids such as glycosylated Asn, Ser and Thr. Two alternative solvent systems have been proposed recently. The first is a 35 mM ammonium acetate, pH 4·9 /acetonitrile system recommended by Millipore/BioSearch for the MilliGen ProSequenator™ and used by Gooley *et al.*, (1991) and Pisano *et al.*, (1993) for the identification of glycosylated PTH-amino acids. This system was found to be unsuitable for monosaccharide analysis because of a high glucose contamination on hydrolysis (Gooley *et al.*, 1994a). It also does not allow for the unambiguous assignment of glycosylated residues because of insufficient resolution or the detection of low levels of glycosylation. The second solvent system (Strydom, 1994) which involves a mixture of triethylamine-phosphate buffered/methanol/acetonitrile as solvent A and a mixture of methanol/isopropanol/water as solvent B has not been systematically evaluated for extraneous sugar content, nor have the elution positions of glycosylated Ser/Thr residues been established.

For the detection, correct assignment and characterization of glycosylated PTH-amino acids by on-line reversed-phase C_{18} HPLC during routine N-terminal sequence analysis, it is advantageous to have a chromatographic system that:

a. identifies each type of glycosylated amino acid
b. separates the PTH-glycoamino acids from the PTH-amino acids to allow the detection of partially glycosylated sites
c. separates the 20 non-glycosylated PTH-amino acids
d. uses a solvent system which is low in glucose contamination to enable monosaccharide analysis and
e. is compatible with ionspray mass spectrometry.

Here we propose the use of a simple, low-molarity triethylammonium formate (TEAF)/acetonitrile system for the routine detection/characterization of glycosylated PTH-amino acids, which meets these criteria. We also show that the PTH chromophore conjugated to a glycoamino acid can be used as an effective "tag" for obtaining structural information on specific sites of glycosylation following exoglycosidase(s) treatment of glycopeptides.

MATERIALS AND METHODS

Materials

PTH-amino acid standards were from Applied Biosystems (Div. of Perkin-Elmer, CA); analytical grade formic acid, 90%, was from Ajax Chemicals (Australia); triethylamine, sequencing grade, Cat no. 25108, was from Pierce; human glycophorin A (Cat. no. G-9511), Ovomucoid (Cat No. T-2011) and bovine κ-casein macroglycopeptide (Cat no. C-7278) were purchased from Sigma. Recombinant PsA (rPsA), a secreated form of cell surface glycoprotein of *Dictyostelium discoideum* was prepared by the method of Zhou-Chou *et al* (1994). β-galactosidase was from *Diplococcus pneumoniae* (Cat no. 188718, Boehringer Mannheim). Trifluoroacetic acid (TFA) was obtained from Sigma-Aldrich.

Preparation of Tryptic Glycopeptides

The tryptic glycopeptides from the human serum albumin mutant Casebrook (Arg485–Lys500) and rPsA (Ile88–Lys122) were prepared according to the method of Gooley *et al.*, (1994a) except that cysteines were alkylated with 4-vinylpyridine according to Tarr (1986).

Preparation of Glycopeptide from Bovine κ-Casein Macroglycopeptide

The bovine κ-casein Met106–Thr124 glycopeptide was prepared by C_{18} (Sephasil™, Pharmacia-Biotech) reversed-phase chromatography (SMART™ system, Pharmacia-Biotech) from bovine κ-casein macroglycopeptide according to Pisano *et al.*, (1994).

Desialylation of Glycopeptides/Glycoproteins

Between 0·5–3 nmol (20-100 μl) of glycopeptide or glycoprotein in 20% (v/v) acetonitrile, was mixed with an equal volume of 0·2 M TFA and incubated at 80°C for 1h to remove sialic acids. The sample was then diluted 1 in 10 (v/v) with 20% (v/v) acetonitrile and concentrated to ≈ 10 μl in a vacuum centrifuge. The process of dilution and concentration was repeated once.

β-Galactosidase Digestion of Desialylated Albumin Casebrook Tryptic Glycopeptide

Approximately 3 nmol of desialylated albumin Casebrook (Arg485–Lys500) were dissolved in 150 μl of 40 mM acetate buffer, pH 6 with 10% (v/v) acetonitrile. The glycopeptide solution was then divided into three 50 μl aliquots and 2 mU of β-galactosidase was added to one aliquot and incubated for 2 hr at 37°C to obtain a partial digest. To the second glycopeptide aliquot, 2 mU of β-galactosidase was added and the third aliquot was incubated without enzyme as a control. These two samples were incubated at 37°C for 24 hr. The glycopeptide was separated from the β-galactosidase enzyme by C_{18} (Sephasil™, Pharmacia-Biotech) reversed-phase chromatography (SMART™ system, Pharmacia-Biotech) using a 30 min linear gradient: 0·05% (v/v) trifluoroacetic acid (TFA) as solvent A, 85% (v/v) acetonitrile + 0·045% (v/v) TFA as solvent B.

Covalent Attachment and Solid-Phase Edman Degradation

Human Glycophorin A (GpA). Between 0·5–2 nmol of desialylated human glycophorin A was dissolved in 20% (v/v) acetonitrile and covalently attached to Sequelon AA™ membranes via the side-chain carboxyl groups using water soluble *N*-ethyl-*N*′-dimethylaminopropylcarbodiimide (EDC). The coupling reaction was carried out by the addition of 5 µl of coupling buffer (0·2 mg EDC/µl), at 4°C for 15 min as described by Liang and Laursen, (1990). The coupling buffer used was that supplied by MilliGen/BioSearch in the Sequelon AA™ attachment kit. The coupling reaction was terminated by vortexing the Sequelon AA™ membranes in 1 ml of 50% (v/v) methanol, followed by 1 ml of methanol then drying the membranes at 55°C.

Bovine κ-Casein Glycopeptides and Tryptic Peptides of Serum Albumin Casebrook and rPsA. Between 0·2–1 nmol of desialylated glycopeptides were covalently attached to Sequelon AA™ membranes by using the manufacturer's recommended procedure (See Sequelon AA™ attachment kit User's guide) and the incubation was carried out at 4°C to increase coupling yield as recommended by Laursen *et al.*, (1991).

Sequelon AA™ coupled protein/peptide membranes were subjected to automated solid-phase Edman degradation using a MilliGen ProSequencer™ 6600 with the standard 6600B method supplied by the manufacturer. The PTH-glycoamino/amino acid derivatives were transferred directly from the conversion flask to the on-line HPLC system.

On line HPLC

The on-line HPLC system consisted of a Waters 600 multisolvent pump delivery system supported by a Waters 600-MS system controller and a Waters 490E programmable multiwavelength detector set at 269 nm and 313 nm. The PTH-amino acids were separated by on-line reversed-phase chromatography using a 3·9 mm x 300 mm C_{18} Nova-Pak™ (Waters) column.

Solvent A: 5 mM TEAF buffer was prepared by the addition of 300 µl of formic acid to 1·2 l of degassed MilliQ water and the pH was adjusted to pH 4·0 with the addition of triethylamine (620 µl). Solvent B: 100% acetonitrile (Ajax chemicals, Australia); both solvent A and solvent B reservoirs were kept under constant helium head pressure of approximately 20 kPa during HPLC operation. Optimal separation of PTH-glycoamino/amino acids was achieved by modifying the manufacturer's gradient (see Table 1).

Table 1. Gradient conditions for PTH–amino/glycoamino acid separation*

Time (min)	Solvent A	Solvent B
Initial	95	5
0.7	80	20
1.4	73	27
2.8	73	27
5.7	55	45
7.4	55	45
8.1	53	47
12	20	80
20	95	5

*Flow rate of 0.7 ml/min.

Calculation of Corrected Yields

The combined peak areas of a completely glycosylated PTH-Thr(Sac)142 site in bovine κ-casein macroglycopeptide was found to be equivalent to 0·83 of the equivalent yield of PTH-Val (Pisano *et al.*, 1994). Therefore a correction factor of 1·2 was used to convert the area of the PTH-Thr(Sac) to pmol and this was applied to the Met106–Thr124 glycopeptide to determine the amount of glycosylation on Thr121.

Monosaccharide Compositional Analysis of PTH-Ser(Sac)

The PTH-glycoamino acids (≈ 400 pmol) were collected from the MilliGen ProSequencer™ on-line HPLC. An equal volume of 4 M TFA was added and the sample was hydrolysed at 100°C for 4 h. After evaporation of the acid, the liberated monosaccharides were analysed by HPLC using a Dionex CarboPac PA1™ column (4 mm x 250 mm) with a waters 600 LC system and 464 pulsed amperometric electrochemical detector. The sugars were eluted isocratically with 15 mM NaOH and were identified by comparison with standards. An internal standard of 2-deoxyglucose was used for quantitation.

RESULTS

Solid-phase sequence analysis of glycophorin A, bovine κ-casein, ovomucoid and human mutant albumin Casebrook was used to characterize their glycosylated amino acids by their retention time, peak distribution pattern, chromatographic mobility shift following exoglycosidase treatment and monosaccharide composition. These results were made possible using low molarity acidic solvents as solvent A and acetonitrile as solvent B with on-line C_{18} reversed-phase HPLC analysis of the PTH-glycoamino/amino acids in either preparative or analytical modes.

Separation of PTH-Amino Acids in TEAF Buffer

The gradient of 5 mM TEAF pH 4·0 buffer in acetonitrile (Table 1) effectively resolves the 20 amino acids in an 8 min window (Fig. 1). The pattern of elution for the first 11 amino acids (Asp, Asn, Ser, Gln, Thr, Glu, Gly, His, Ala, Tyr and Arg) is similar to the sodium acetate/tetrahydrofuran buffer system except that Glu elutes before Gly. The final 9 amino acids (Pyridylethyl(PE)-Cys, Met, Val, Pro, Trp, Lys, Phe, Ile and Leu) have an elution profile typical for chromatogram using the MilliGen ProSequencer™ system equipped with a Waters PicoTag™ C_{18} column and ammonium acetate/acetonitrile gradient (Gooley *et al.*, 1991). Dehydroalanine (from Ser) and dehydro-α-aminobutyric acid (from Thr) elute at 11·8 and 12·1 min, respectively and are monitored by simultaneous detection at 269 nm and 313 nm (data not shown). The Edman degradation by-products dimethylphenylthiourea (DMPTU, 10·4 min) and diphenylthiourea (DPTU, 14 min, which co-elutes with Trp) are not problematic in solid-phase sequencing and do not interfere with amino acid assignments. The only disadvantage of the low molarity TEAF buffer system is that His and Arg are very sensitive to variations in pH and ionic strength. However, careful titration of the buffer provides a satisfactory elution position for both these amino acids (Fig. 1).

The elution positions of all 20 amino acids were confirmed by the known N-terminal amino acid sequence analysis of human glycophorin A and ovomucoid.

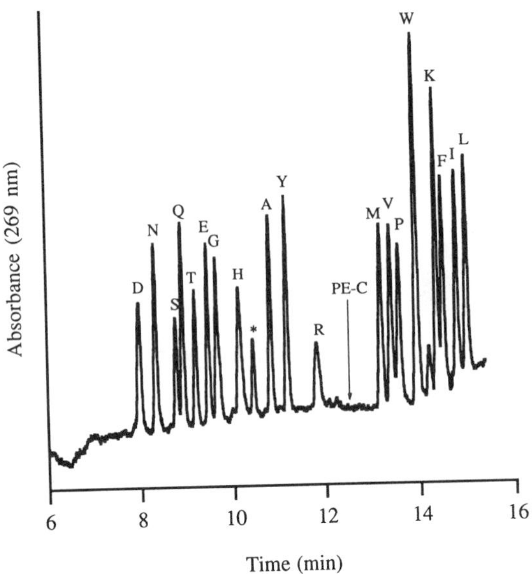

Figure 1. A C_{18} HPLC chromatogram with an elution profile of 19 PTH-amino acid standards routinely encountered in N-terminal sequence analysis. The PTH-amino acids are (in order of elution): Asp (D), Asn (N), Ser (S), Gln (Q), Thr (T), Glu (E), Gly (G), His (H), DMPTU at 10·4 min, Ala (A), Tyr (Y), Arg (R), Met (M), Val (V), Pro (P), Trp (W) which co-elutes with DPTU, Lys (K), Phe (F), Ile (I) and Leu (L). PE-Cys(PE-C) is not routinely included in our PTH-amino acid standards mixture and subsequently its elution time was identified separately. The elution position of PE-Cys is indicated on the elution profile by a arrow. The 20 PTH-amino acids were separated using 5 mM TEAF, pH 4·0 as solvent A and acetonitrile as solvent B.

Separation of PTH-Glycoamino Acids in TEAF Buffer

Reversed-phase elution of PTH-amino acids with 5 mM TEAF, pH 4·0, provides a clear chromatographic window for the elution of PTH-Asn(Sac), PTH-Ser(Sac) and PTH-Thr(Sac) which all elute prior to PTH-Asp (Fig. 2, Table 2). PTH-Asn(Sac) from Casebrook tryptic glycopeptide Arg485–Lys500 elutes first off the column as two peaks, one of which is heterogeneous [Asn$(Sac)_1$, at 6·9 min], and a small peak, Asn$(Sac)_2$, at 7·65 min (Fig.2d). PTH-Ser(Sac) from human glycophorin A consistently elutes as two peaks: Ser$(Sac)_1$ at 7·25 min and Ser$(Sac)_2$ at 7·85 min. The first peak elutes as a poorly resolved doublet and Ser$(Sac)_2$ co-elutes with Thr$(Sac)_2$ (Fig. 2b and c , Table 2). Thr(Sac) elutes as two major peaks Thr$(Sac)_1$ at 7·5 min and Thr$(Sac)_2$ at 7·85 min and two minor peaks Thr$(Sac)_3$ at 8·25 min and Thr$(Sac)_4$ at 8·65 min (Fig. 2b, Table 2).

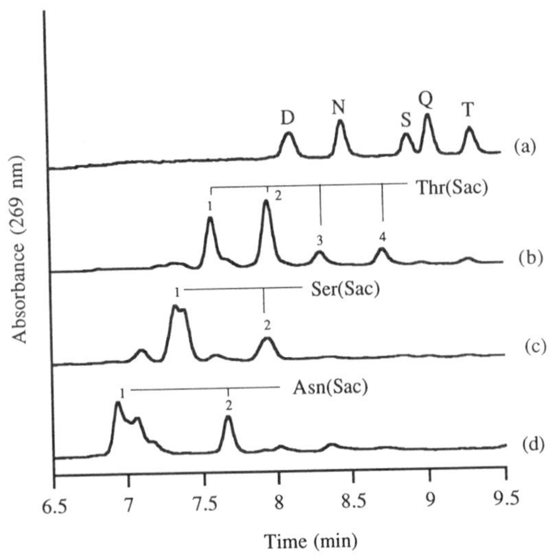

Figure 2. A comparison of the C_{18} HPLC elution profiles for the three main groups of PTH-glycoamino acids: Asn(Sac), Ser(Sac) and Thr(Sac) separated with solvent A as 5 mM TEAF buffer, pH 4·0. (a) First five PTH-amino acids to elute from the column: Asp (at 8·0 min); Asn (at 8·3 min); Ser (8·75 min), Gln (8·89 min) and Thr (at 9·2 min). (b) PTH-Thr(Sac) and (c) PTH-Ser(Sac) from the N-terminal sequence of human glycophorin A after 4 and 2 cycles of Edman degradation, respectively. (d) PTH-Asn(Sac) from the albumin Casebrook tryptic glycopeptide following 10 cycles of Edman degradation.

Table 2. Retention times of PTH-glycoforms and parent PTH-amino acids

PTH-(glyco)amino acids	PTH-oligosaccharide	Retention time (R_t, min)
Asn(Sac)$_1$	GlcNAc$_2$Man$_3$GlcNAc$_2$Gal$_2$	6.85
Asn(Sac)$_2$	GlcNAc$_2$Man$_2$*	7.65
Ser(Sac)$_1$	GalNAc$_1$Gal$_1$	7.25
Ser(Sac)$_2$		7.85
Thr(Sac)$_1$	GalNAc$_1$Gal$_1$	7.50
Thr(Sac)$_2$	GalNAc$_1$Gal$_1$?	7.85
Thr(Sac)$_3$	GlcNAc or GalNAc	8.25
Thr(Sac)$_4$	GlcNAc or GalNAc?	8.65
Asn		8.30
Ser		8.75
Thr		9.16

*Previously reported as GlcNAc$_2$Man$_1$ (Gooley *et al.*,1994a).

One major advantage of a unique chromatographic window for the glycosylated amino acids is the detection of partially glycosylated amino acids free from the background of the other amino acids. This is best demonstrated by the sequence analysis of the bovine κ-casein glycopeptide Met106–Thr124. Figure 3a shows the corrected yield for this peptide from Asn114–Asn123. Normal PTH-Thr/Ser are recovered in low yield due to the production of the dehydro forms of Ser/Thr during the coupling and cleavage reactions. Hence, these two amino acids are rarely quantitated and it generally suffices to assign these on the basis of detection of their dehydro forms which are detected at 313 nm.

However, the glycosylated forms of the Ser and Thr are recovered in high yield and it is possible to determine how much of the amino acid is modified (see Materials and Methods). It was estimated that 5% of the Thr121 is glycosylated. An enlarged section of the Thr121 chromatogram is shown in Fig. 3b with the previous cycle Pro120 overlaid. The solid line is the 5% glycosylated Thr121 and is clearly visible after 16 cycles of Edman degradation. The two major glycoforms of Thr(Sac): (Thr(Sac)$_1$ and Thr(Sac)$_2$ elute at 7·9 and 8·2 min respectively, while Thr elutes at 9·1 min) are easily distinguishable above the preceding cycle overlay (dotted line, Fig. 3b).

Monosaccharide Analysis of Glycosylated PTH Amino Acids

PTH-glycoamino acids released by solid-phase Edman degradation of the N-terminal extracellular domain of human GpA (PTH Ser(Sac)$_{1and2}$ at position 2 and PTH Thr(Sac)$_{1and2}$ at position 4), of the spacer domain of PsA of *Dictyostelium discoideum* (PTH Thr(Sac)$_{3and4}$ at position 91) and of the tryptic peptide of albumin Casebrook (PTH Asn(Sac)$_{1and2}$ at position 494) were collected and characterised by monosaccharide compositional analysis (Table 2). The low molarity elution buffers allowed the glucose contamination to be kept to a level able to be completely resolved from the component sugars by HPAEC.

Chromatographic Shift of PTH-Asn(Sac) following β-Galactosidase Treatment

Human mutant albumin Casebrook has a single biantennary *N*-linked oligosaccharide at Asn494. The PTH glycoamino acid released by Edman degradation showed chromatographic heterogeneity after 10 cycles (Fig. 2d) The intact tryptic glycopeptide (Arg485–

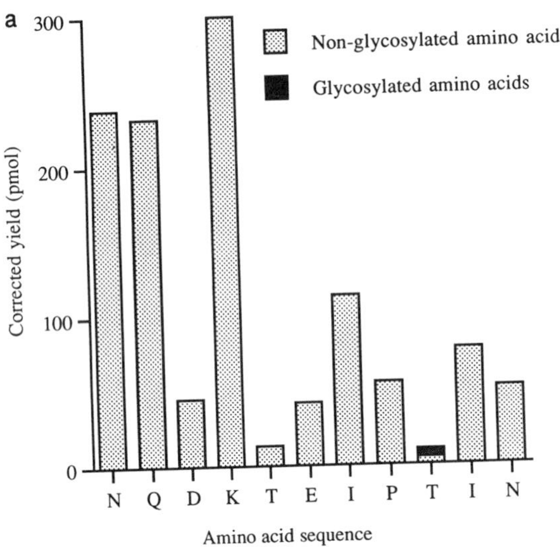

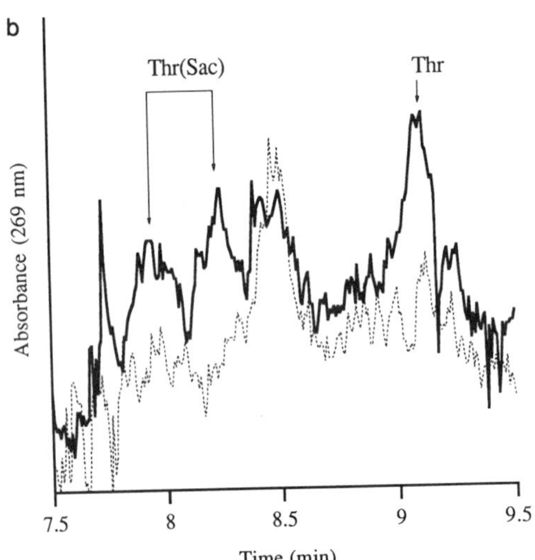

Figure 3. (a) Corrected yields for the solid-phase Edman degradation from Asn114–Asn123 from the glycopeptide Met106–Thr124 from bovine κ-casein macroglycopeptide. The recovery of both Glu (E) and Asp (D) was low since they remain covalently attached to the Sequelon AA membrane following carbodiimide activation of the side-chain carboxyl groups. Non-glycoamino acids are shown as shaded bars and glycoamino acids are solid black bars. (b) A chromatogram overlay of cycle 15 (dotted line) and cycle 16 (solid line) focused on the elution positions of both PTH-Thr(Sac)121 and PTH-Thr121 from the partially glycosylated peptide Met106–Thr124 following solid-phase Edman degradation. The peak at 8·5 min is a lag from Asn114. The chromatography is as described in the Materials and Methods section with the exception that Solvent A was 2 mM acetic acid.

Table 3. Resolution of glycoforms on PTH-Asn494 from albumin Casebrook

PTH-Oligosaccharide	Retention time (Rt, min)
$GlcNAc_2Man_3GlcNAc_2Gal_2$ $(Asn(Sac)_1)$	6.85
$GlcNAc_2Man_3GlcNAc_2Gal_1$	7.05
$GlcNAc_2Man_3GlcNAc_2$	7.15
$GlcNAc_2Man_2$ $(Asn(Sac)_2)$	7.40

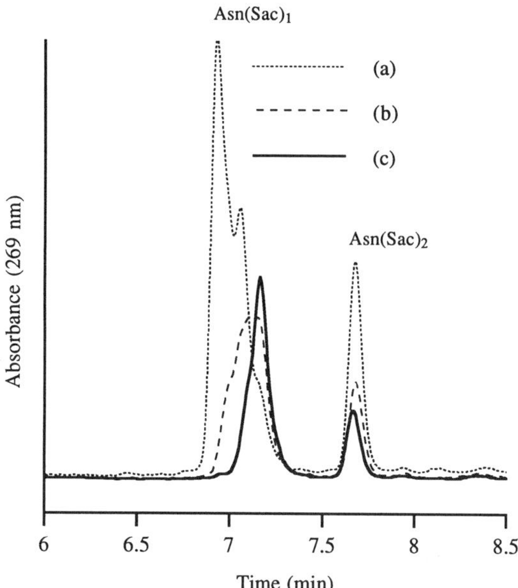

Figure 4. Solid-phase Edman degradation of human mutant albumin Casebrook tryptic glycopeptide following digestion by ß-galactosidase. Chromatograms show PTH-Asn(Sac) glycoamino acids separated after 10 cycles of Edman degradation of the glycopeptide(a) which had been incubated with enzyme for(b) 2 h and (c) 24 h.

Lys500) was subjected to a time course of digestion with β-galactosidase to see if the heterogeneity observed was due to the sequential loss of galactose. There was a progressive shift of the major peak of PTH Asn(Sac)$_1$ (6·85 min) to a longer retention time with a peak at 7·05 min appearing after 2 h and the final product after 24 h of incubation eluting at 7·15 min (Fig. 4). Monosaccharide analysis of this peak showed that there was no galactose present. The heterogeneity of the PTH-glycoamino acid could thus be attributed to loss of terminal galactose residues during the Edman chemistry. These glycoforms were separated by the chromatographic conditions used (Fig. 3, Table 3).

DISCUSSION

The accepted practice of ignoring modified amino acids during routine sequence analysis is no longer necessary as glycoamino acids can be identified as part of solid-phase Edman degradation. This is particularly important in the for quality control of products from eukaryotic expression. These modified amino acids hold key insights into the primary structure motifs that confer modifications. We have developed a method for detecting and characterizing glycosylated Asn, Ser and Thr (Gooley *et al.*, 1994b; Pisano *et al.*, 1994). However, in routine microsequencing our methodology was incompatible with core facilities where an instrument cannot be dedicated solely to sequencing known glycosylated proteins.

Here we describe the development of a triethylammonium formate (TEAF) buffered system for use in the routine sequence analysis of PTH-glycoamino acids extracted by N-terminal solid-phase Edman degradation with on-line C_{18} reversed-phase HPLC chromatography. This system can separate both the glycosylated and non-glycosylated PTH-amino acids in one chromatographic run. We also show that it is possible to obtain saccharide sequence/structure information by monitoring the retention times of PTH-glycoamino acids, following exoglycosidase(s) treatment.

PTH-Amino Acid/Glycoamino Acid Chromatography

The new TEAF system resolves all 20 amino acids (Fig. 1) and separates the three major groups of glycoamino acids (Fig. 2 b and d), into their own chromatographic space. The TEAF buffer system provides a 1·5 min "glycosylation window" for the unambiguous detection of glycosylated Asn, Ser, and Thr residues, including partially glycosylated forms (Fig. 3). This has been demonstrated by the N-terminal sequence analysis of Asn(Sac) at position 494 from albumin Casebrook tryptic glycopeptide (Arg485–Lys500), Ser(Sac) at position 2 and Thr(Sac) at position 4 from the N-terminus of human GpA and Thr(Sac) in the spacer domain of PsA of *Dictyostelium discoideum.* The elution profiles of Asn(Sac), Ser(Sac) and Thr(Sac) in the TEAF system were found to be consistent with data obtained from C_{18} reversed-phase chromatography with 2 mM acetic acid or 2 mM formic acid as solvent A (Pisano *et al.*, 1994; Gooley, *et al.*, 1994b).

The separation between $Asn(Sac)_1$ and $Asn(Sac)_2$ has been improved with TEAF so that $Asn(Sac)_2$ no longer co-elutes with the major Ser(Sac) glycoform $Ser(Sac)_1$. Also the separation between $Ser(Sac)_1$ and $Thr(Sac)_1$ has increased by 0·35 min (Table 2; Pisano *et al.*, 1994). The pattern of glycosylated Thr(Sac) peaks (Fig. 2b) is identical to that observed in the rat CD8α hinge peptide (Gooley *et al.*, 1991) and human GpA (Pisano *et al.*, 1993), with two major peaks, $Thr(Sac)_1$ and $Thr(Sac)_2$, eluting well before Thr (Fig. 2a and b). This characteristic pair of peaks probably represents diastereomeric forms of PTH-Thr(Sac), a similar pattern to that obtained for PTH-β-methyl-S-ethyl-cysteine, the ethanethiol adduct of β-eliminated phosphothreonine (Meyer *et al.*, 1993)

The separation of $Thr(Sac)_3$ and $Thr(Sac)_4$ from Asp, Asn and Ser also has significance because these peaks have been shown to co-elute with the diasteromeric forms of PTH-Thr(Sac) from the spacer domain of recombinant PsA, a cell surface glycoprotein, from *Dictyostelium discoideum* (Gooley *et al.*, 1991). The PTH-Thr(Sac) from recombinant PsA is believed to be a single GlcNAc residue conjugated to Thr (Gooley *et al.*, in preparation). Single GlcNAc residues have recently been reported with increased frequency on many intracellular proteins such as nuclear pore proteins (Hart *et al.*, 1989), neurofilaments (Dong *et al.*, 1993) and keratins (Ku and Omary, 1994).

The ability to detect partially glycosylated residues is dependent on the PTH-glycosylated residues eluting in a region on the chromatogram with low background noise. PTH-glycoamino acids separated by 2 mM acetic acid, formic acid and TEAF buffer all have very low PTH-amino acid background in the "glycosylation windows". Partially glycosylated glycoforms are typical of κ-casein (Pisano *et al.*, 1994) and their detection was made possible by their early elution.

Monosaccharide Analysis

Monosaccharide analysis is an important first step in the characterization of oligosaccharides since mass spectrometry alone cannot distinguish between isomeric sugars.

We have shown that it is possible to collect a PTH-glycoamino acid directly off the sequenator in TEAF buffer/acetonitrile and subject it to monosaccharide analysis (Table 2). TEAF, like 2 mM acetic acid and 2 mM formic acid, is volatile and compatible with ion-spray mass spectrometry.

β-Galactosidase Digest of Albumin Casebrook Tryptic Glycopeptide

Not all protein analysis core facilities may be equipped with HPAEC and ion-spray capabilities. The technique of on-line Edman degradation described in this paper permits the determination of the sequence of the oligosaccharide(s) by using exoglycosidase treatments and subsequent monitoring of the products .

The *N*-linked glycosylation site at Asn494 on serum albumin Casebrook has been well characterized by both monosaccharide and NMR analysis and was found to consist of a single $NeuAc_2Gal_2GlcNAc_2Man_3GlcNAc_2$ linkage (Haynes *et al.*, 1992). Recently, the desialylated PTH-Asn$(Sac)_1$ was collected after 10 cycles of Edman degradation from the Casebrook tryptic glycopeptide (Arg485–Lys500) and subjected to compositional analysis and ion-spray mass spectrometry. The major mass was found to be consistent with a PTH-Asn-$GlcNAc_2Man_3GlcNAc_2Gal_2$ (Gooley, *et al.*, 1994). There were also secondary masses which correspond to the loss of one and two hexose residues in the mass spectra for PTH-Asn(Sac)494 from the tryptic glycopeptide. The chromatography after exoglycosidase treatment shows that these were Edman degradation by-products and not fragmented parent ions that arose from the mass analysis process (Gooley *et al.*, 1994a). Treatment with β-galactosidase (Fig. 4) show that the degradation was due to the loss of terminal galactose residues. We have defined the retention times of a PTH-Asn *N*-linked biantennary oligosaccharide with the loss of one galactose as 7·05 min and loss of two galactoses as 7·15 min (Table 3).

Reversed-phase chromatography of PTH-glycoamino acids obtained by solid-phase Edman degradation was first demonstrated by Gooley *et al.*, (1991). We have now shown that it is possible to monitor single saccharide changes in PTH-glycoamino acids by using a TEAF/acetonitrile buffer system.

By using this system it is possible:

1. to create a retention time data base for the on-line detection of different oligosaccharide-amino acid linkages as a routine first step in the characterization of site-specific glycosylation sites released by solid-phase Edman degradation.

2. to subject either the glycopeptide or PTH-glycoamino acid to exoglycosidase(s) and monitor the change in chromatographic profile of the PTH-glycoamino acid.

CONCLUSION

Solid-phase Edman degradation with on-line C_{18} reversed-phase chromatography using TEAF/acetonitrile has been developed for the routine detection of PTH-glycoamino acids. This system is useful for analytical and preparative scale PTH-glycoamino acid analysis. The PTH-chromophore is a convenient “tag” for monitoring site-specific changes in glycosylation.

ACKNOWLEDGMENTS

KLW and AAC acknowledge support from Australia Research Grants A09330428 and a program grant as well as grants from Macquarie University. We thank Dr. S. Brennan who kindly donated the Human serum albumin Casebrook.

REFERENCES

Carr, S. A., Huddleston, M. J. and Bean, M. F. (1993) Selective identification and differentiation of N-and O-linked oligosaccharides in glycoproteins by liquid chromatography-mass spectrometry, *Prot. Sci.* 2:183

Dong, D. L.-Y., Xu, Z-S., Chevrier, M. R., Cotter, R. J., Cleveland, D. W. and Hart, G. W. (1993) Glycosylation of mammalian neurofilaments: Localization of multiple *O*-linked *N*-acetylglucosamine moieties on neurofilament polypeptide L and M, *J. Biol. Chem.* 268:16679

Dwek, R. A., Edge, C. J., Harvey, D. J., Wormald, M. R. and Parekh, R. B. (1993) Analysis of glycoprotein-associated oligosaccharides, *Ann. Rev. Biochem.* 62:65

Gooley, A. A., Classon, B. J. Marschalek, R. and Williams K. L. (1991) Glycosylation sites identified by detection of glycosylated amino acids released from Edman degradation: The identification of Xaa-Pro-Xaa-Xaa as a motif for Thr-*O*-glycosylation, *Biochem. Biophys. Res. Comm.* 178:1194

Gooley, A. A., Packer, N. H., Pisano, A., Redmond, J. W., Alewood, P. F., Jones, A., Loughnan, M. and Williams, K. L. (1994a) Characterization of *N*-and *O*-linked glycosylation sites using Edman degradation, *Techniques in Protein Chemistry VI* In Press

Gooley, A. A., Pisano, A., Packer, N. H., Ball, M., Jones, A., Alewood, P. F., Redmond, J. W. and Williams, K. L. (1994b) Characterisation of a single glycosylated asparagine site on a glycopeptide using solid-phase Edman degradation, *Glycoconjugate J.* In Press

Gooley, A. A. and Williams, K. L. (1994) Towards characterizing *O*-glycans: the relative merits of *in vivo* and *in vitro* approaches in seeking peptide motifs specifying *O*-glycosylation sites, *Glycobiology* 4:413

Hart, G. W., Holt, G. D. and Haltiwanger, R. S. (1989) Glycosylation in the nucleaus and cytoplasm, *Annu. Rev. Biochem.* 58:841

Haynes, P. A., Batley, M. Peach, R. J. and Brennan, S. O. (1992) Characterisation of oligosaccharides from a glycoprotein variant of human serum albumin (albumin Casebrook) using high performance anion exchange chromatography and nuclear magnetic resonance spectroscopy, *J. Chromatogr.* 581:187

Jackson, P (1991) Polyacrylamide gel electrophoresis of reducing saccharide label with the fluorophore 2-aminoacridone: subpicomolar detection using an imaging system based on a cooled charge-coupled device, *Anal. Biochem.* 196:238

Ku, N-O., Omary, M. B. (1994) Expression, glycosylation, and phosphorylation of human keratins 8 and 18 in insect cells *Exp. Cell Res.*, 211:24

Laursen, R. A., Lee, T. T., Dixon, J. D. and Liang, S-P. (1991) Extending the performance of the solid-phase protein sequenater in: *Methods in Protein Sequence Analysis*, (Eds) Jornall, H., Hoog, J-O. and Gustvasson, A-M., Bikhauser Verlag, Switzerland, pp 47-54

Liang, S-P. and Laursen, R. A. (1990) Covalent immobilization of proteins and peptides for solid-phase sequencing using prepacked capillary columns. *Anal. Biochem.* 188:366

Mallett, S and Barclay, A. N. (1991) A new superfamily of cell surface proteins related to the nerve growth factor. *Immunol. Today* 12:220

Meyer, H. E., Eisermann, B., Heber, M., Hoffmann-Posorske, E., Korte, H., Weigt, C., Wegner, A., Hutton, T., Donella-Deana, A. and Perich, J. W. (1993) Strategies for nonradioactive methods in the localisation of phosphorylated amino acids in proteins, *FASEB J.* 7:776

Pisano, A., Packer, H. N., Redmond, J. W., Williams, K. L. and Gooley, A. A. (1994) Characterisation of *O*-linked glycosylation motifs in the glycopeptide domain of bovine κ-casein, *Glycobiology* In Press

Pisano, A., Redmond, J. W., Williams, K. L. and Gooley, A. A. (1993) Glycosylation sites identified by solid-phase Edman degradation: *O*-linked glycosylation motifs on human glycophorin A, *Glycobiology* 3:429

Strydom, D. J. (1994) On-line separation of phenylthiohydantoin derivatives of hydrophilic modified amino acids during sequencing, *J. Chromatogr.* 664:227

Tarr, G. E. (1986) Manual Edman sequencing system, in: *Methods of Protein Microcharacterization*, (Ed) Shively, J. E., Humana Press, Clifton, New Jersey, pp 162-163

Townsend, R. R., and Hardy, M. R. (1991) Analysis of glycoprotein oligosaccharides using high-pH anion exchange chromatography, *Glycobiology* 1:139

Townsend, R. R., Hardy, M. R. and Lee, Y. C. (1989) Separation of oligosaccharides using high performance anion exchange chromatography with pulsed amperometric detection, *Methods Enzymol.* 179:65

Williams, A. F. and Barclay, A. N. (1988) The immunoglobulin superfamily–domains for cell surface recognition, *Ann. Rev. Immunol.* 6:381

Zhou-Chou, T., Slade, M. B., Williams, K. L. and Gooley, A. A. (1994) Expression, purification and characterisation of secreted recombinant glycoprotein PsA in *Dictyostelium discoideum* , (*J Biotech.*, In Press)

8

DEACETYLATION AND INTERNAL CLEAVAGE OF POLYPEPTIDES FOR N-TERMINAL SEQUENCE ANALYSIS

Madalina T. Gheorghe and Tomas Bergman

Department of Medical Biochemistry and Biophysics
Karolinska Institutet
S-171 77 Stockholm, Sweden

INTRODUCTION

Proteins with a blocked N-terminus are common. Frequently the modification involves an acetyl-, formyl- or pyroglutamyl-moiety coupled to the α-amino group and direct sequence analysis by Edman degradation is not possible. Several enzymatic and chemical methods to remove the blocking group have been suggested (cf. Tsunasawa and Hirano, 1993), but they often suffer from poor yields and a large extent of undesirable peptide bond cleavage. Acetylation represents the most frequent N-terminal modification and is found in alcohol dehydrogenases among many other proteins. To circumvent the conventional approach to sequence analysis of blocked proteins (i.e. proteolytic cleavage, HPLC of fragments and internal sequence analysis) we have tested direct chemical deacetylation using a mixture of trifluoroacetic acid and methanol (Gheorghe *et al.*, 1995). In this manner, drawbacks as high protein consumption, long handling times and inaccessibility of the N-terminal fragment to Edman degradation, are avoided. The protocol has been applied to both a synthetic peptide corresponding to the N-terminal segment of horse liver alcohol dehydrogenase and to the intact protein.

A technique to obtain internal sequence information from N-teminally blocked proteins and from partially sequenced polypeptides, that in addition saves protein material, is treatment with cyanogen bromide directly on the sequencer filter. We have tested this approach for analysis of new structures and identification of known proteins (Bergman, 1994). Polypeptides bound to the sequencer filter are *in situ* cleaved with CNBr followed by analysis of the resulting internal sequences. Interpretation is facilitated by the varying extent to which peptide bonds are cleaved after individual methionines. Both electroblotted samples and samples applied in solution have been treated with CNBr after initial sequence analysis for a necessary number of cycles. In this manner, both unknown and known proteins available in amounts sufficient for only one sequencer application can be analyzed and identified even if they are blocked at the N-terminus.

Methods in Protein Structure Analysis, Edited by M. Z. Atassi and E. Appella
Plenum Press, New York, 1995

EXPERIMENTAL PROCEDURES

Deacetylation was tested both with an N-terminal fragment of horse liver alcohol dehydrogenase (residues 1 - 14) and with the intact protein (374 residues; cf. Jörnvall, 1970). The peptide fragment was synthetically prepared using an Applied Biosystems 430A instrument and side-chain-protected tertiary butyloxycarbonyl amino acid derivatives (cf. Kent, 1988). N-terminal acetylation was performed before cleavage from the resin and deprotection (treatment with a mixture of acetic anhydride / triethylamine / dichloromethane (9:4:87, by vol.) for 10 min at room temperature) to avoid simultaneous modification of lysine residues present. Horse liver alcohol dehydrogenase was purchased from Sigma. Deacetylation for N-terminal sequence analysis of both peptide and intact protein was performed with a mixture of trifluoroacetic acid (TFA) and methanol (MeOH). The samples were carefully lyophilized to complete dryness in small (500 μl) plastic tubes with caps after which 100 μl freshly prepared TFA / MeOH (1:1, by vol.) was added. The tubes were closed and after a brief vortex the samples were incubated at 43°C for three days. Subsequent to this treatment, the reagents were removed under vacuum and the products were analyzed by both capillary electrophoresis and Edman degradation. For capillary electrophoresis, a Beckman P/ACE 2100 system operated as described (Bergman *et al.*, 1991) was used, and the sequence analysis was performed employing an Applied Biosystems 470A instrument with reverse-phase HPLC of phenylthiohydantoin amino acids essentially as described (Kaiser *et al.*, 1988).

Cyanogen bromide cleavage of a protein immobilized on a sequencer filter (Polybrene-treated glass fiber or polyvinylidene difluoride (PVDF)) was carried out with a solution of 0.2 g CNBr/ml 70% formic acid for 22-26 h at room temperature. After a sufficient number of Edman cycles, the filter was placed in an Eppendorf tube (1.5 ml) and 30 μl CNBr solution was added. A small additional volume (60 μl) was placed in the bottom of the tube, below the filter, to maintain a CNBr-saturated atmosphere. Nitrogen gas was introduced and incubation was performed in the dark. Following this treatment, the filter was dried under vacuum and reapplied to an Applied Biosystems 470A sequencer.

RESULTS AND DISCUSSION

A combination of trifluoroacetic acid and methanol was found to cleave the N-terminal acetyl-group of polypeptides with high specificity (i.e. with a low extent of simultaneous internal peptide bond cleavage). The approach was tested on a synthetic peptide (14 residues) corresponding to the N-terminal segment of horse liver alcohol dehydrogenase and on the intact protein (374 residues). Deacetylation was monitored as a function of reaction time, temperature and ratio between trifluoroacetic acid and methanol using capillary electrophoresis and sequence analysis. The results indicate that a 1:1 (by vol.) mixture of TFA / MeOH added to the lyophilized sample followed by incubation for three days at 43°C is efficient (Gheorghe *et al.*, 1995). Both the peptide fragment and the much larger protein molecule are deblocked without predominant cleavage of internal peptide bonds. Capillary electrophoresis of the 14-residue peptide before and after deacetylation reveals that only 17% of the blocked structure remains while the deacetylated but else intact peptide corresponds to the major peak and represents 65% of the total sample (Gheorghe *et al.*, 1995). The extent of undesirable internal cleavage is low and only a minor peak corresponding to a product resulting from a cleavage after a glycine in position 4 (cf. Jörnvall, 1970) can be detected (18% of the total sample). Sequence analysis after deacetylation of the N-terminal fragment (Fig. 1) and the intact protein reveals initial yields up to 60%. Interestingly, the ratios of deblocking over unspecific cleavage of internal bonds are similar for the peptide (7:1) and the protein (8:1), despite the much larger size of the latter molecule (374 instead of 14

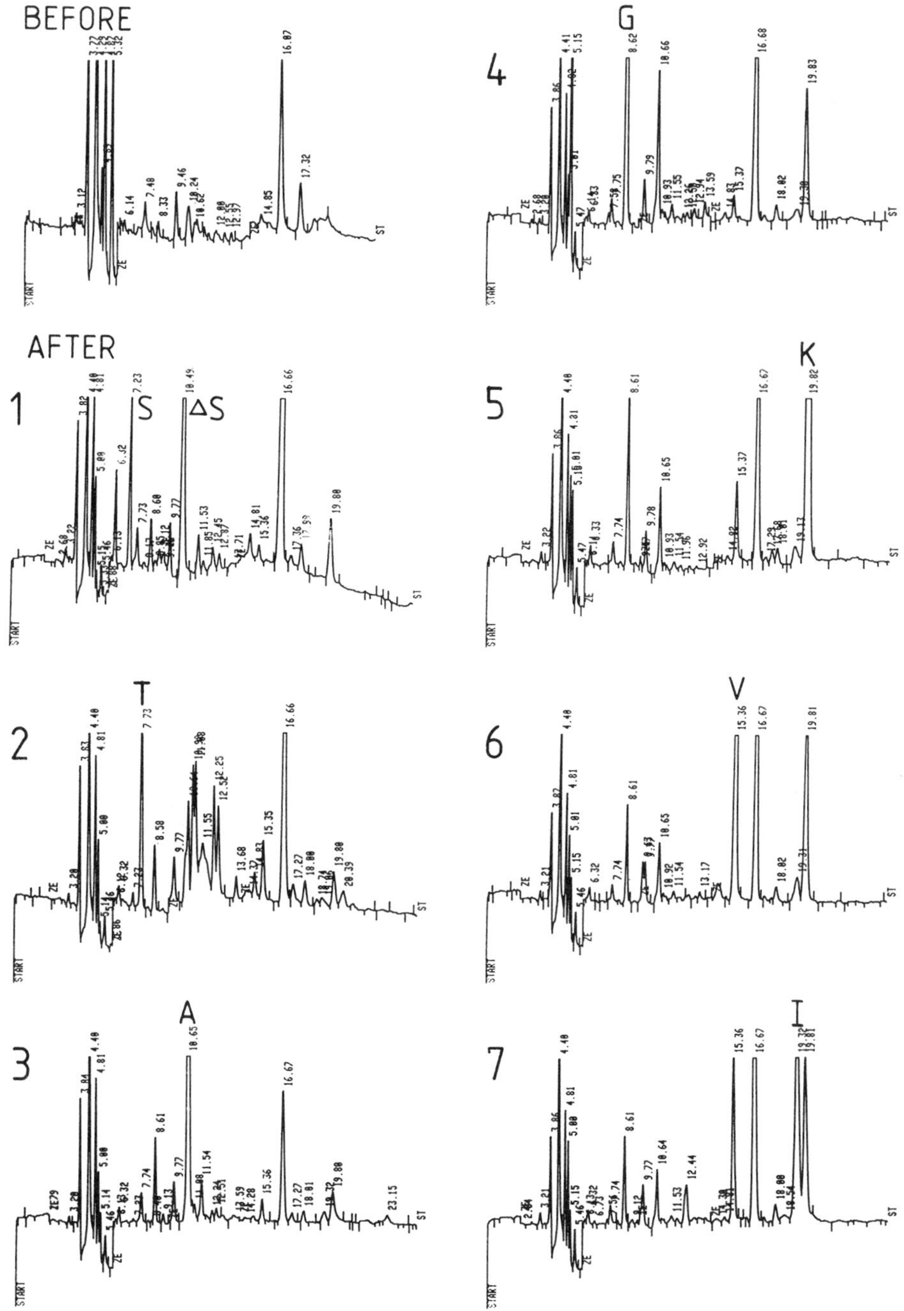

Figure 1. Sequence analysis before and after deacetylation of the N-terminal alcohol dehydrogenase fragment. The amino acid residue detected in each cycle after deblocking is indicated and in full agreement with the expected sequence (cf. Jörnvall, 1970). The symbol ΔS indicates the serine degradation product dehydroalanine.

residues) (Gheorghe *et al.*, 1995). Consequently, the results are promising for direct analysis of N-terminally acetylated proteins via Edman degradation after deblocking of the intact protein.

Electroblotting is efficient for recovery of proteins separated at the low pmol-level in SDS/polyacrylamide gels and the blotted material is easilly accessible for direct chemical cleavage with cyanogen bromide. A 42 kDa DNA-binding phosphoprotein (cf. Egyhazi *et al.*, 1991) was electroblotted onto a Polybrene-treated glass fiber filter disc as described (Bergman and Jörnvall, 1987). Although the total amount of sample available, 360 pmol, was applied to the electrophoresis gel, no significant sequence could be detected when the blotted protein was analyzed for 16 Edman cycles, establishing the absence of a free α-amino group in this 42 kDa polypeptide. The filter was subsequently removed from the sequencer and treated with CNBr. After reapplication of the filter to the sequencer, several sequences appeared and at least one major sequence could be interpreted for nine cycles (Fig. 2). The sequencer initial yield was 60 pmol or 17% of the amount applied to gel electrophoresis.

A sample of human endothelial cell proteins was separated by SDS/polyacrylamide gel electrophoresis and a 42 kDa band was isolated through electroblotting (Schuppe-Koistinen *et al.*, 1995; Bergman and Jörnvall, 1987). Edman degradation of the blotted material revealed no sequence and the protein was concluded to be blocked since approximately 700 pmol was loaded onto the electrophoresis gel. The filter disc with the blocked polypeptide was removed from the sequencer after 19 cycles and *in situ* treated with CNBr followed by reapplication. Several sequences were now detected and a major cyanogen bromide fragment was analyzed for 18 cycles at a repetitive yield of 97% which allowed identification of the 42 kDa protein as actin (Schuppe-Koistinen *et al.*, 1995).

Transthyretin (TTR) associated with amyloid deposits in the heart or in nerve tissue is known as a highly heterogeneous mixture of N-terminally blocked and truncated polypeptides with structures identical to segments of the plasma TTR sequence except for point mutations at different positions. A sample of amyloid related TTR isolated from cardiac tissue was separated by SDS/polyacrylamide gel electrophoresis and a major band at 14.5 kDa was recovered via electroblotting onto a PVDF-membrane (Hermansen *et al.*, 1995; cf. Matsudaira, 1987). Sequence analysis for 14 cycles revealed a polypeptide starting at position 49 in the plasma TTR sequence (cf. Kanda *et al.*, 1974). However, the initial yield in the Edman degradation was unexpectedly low, only 6% of the material applied to the gel, and therefore the protein was concluded to be partially blocked. After *in situ* CNBr-cleavage and reapplication of the PVDF-membrane to the sequencer, two additional sequences were detected, one starting at position 14 and the other at position 112 in the plasma TTR structure (cf. Kanda *et al.*, 1974). This result clearly shows the presence of a fraction in the cardiac amyloid TTR sample that consists of blocked polypeptides starting at positions before residue 14. Furthermore, since the sequence of plasma TTR contains only one methionine at position 13, the second fragment detected after CNBr-cleavage (starting at residue 112) indicates a point mutation to be present in amyloid related TTR isolated from heart tissue (Hermansen *et al.*, 1995).

The amino acid sequence of procarboxypeptidase A2 (PCP A2) in rat pancreas is known from the corresponding cDNA (Gardell *et al.*, 1988). It has the N-terminal structure Gln-Glu-Thr-Phe- which suggests the presence of a blocking pyroglutamic acid modification at the N-terminus. An HPLC-purified sample was analyzed for 15 Edman cycles after application of 100 pmol (the total amount available) and, as expected, no sequence data were obtained. To identify the blocked protein, *in situ* CNBr-cleavage was performed followed by reapplication of the filter to the sequencer (Oppezzo *et al.*, 1994). The resulting seven sequences detected correspond to a cleavage after each methionine present in the PCP A2 structure (cf. Gardell *et al.*, 1988). The fragment sequences could be traced up to ten cycles of Edman degradation. Interpretations were facilitated by the cleavage efficiency that varied for different methionines. The major cleavage after Met-271 resulted in a fragment giving an initial yield of 52 pmol (52%) of the starting material (Oppezzo *et al.*, 1994).

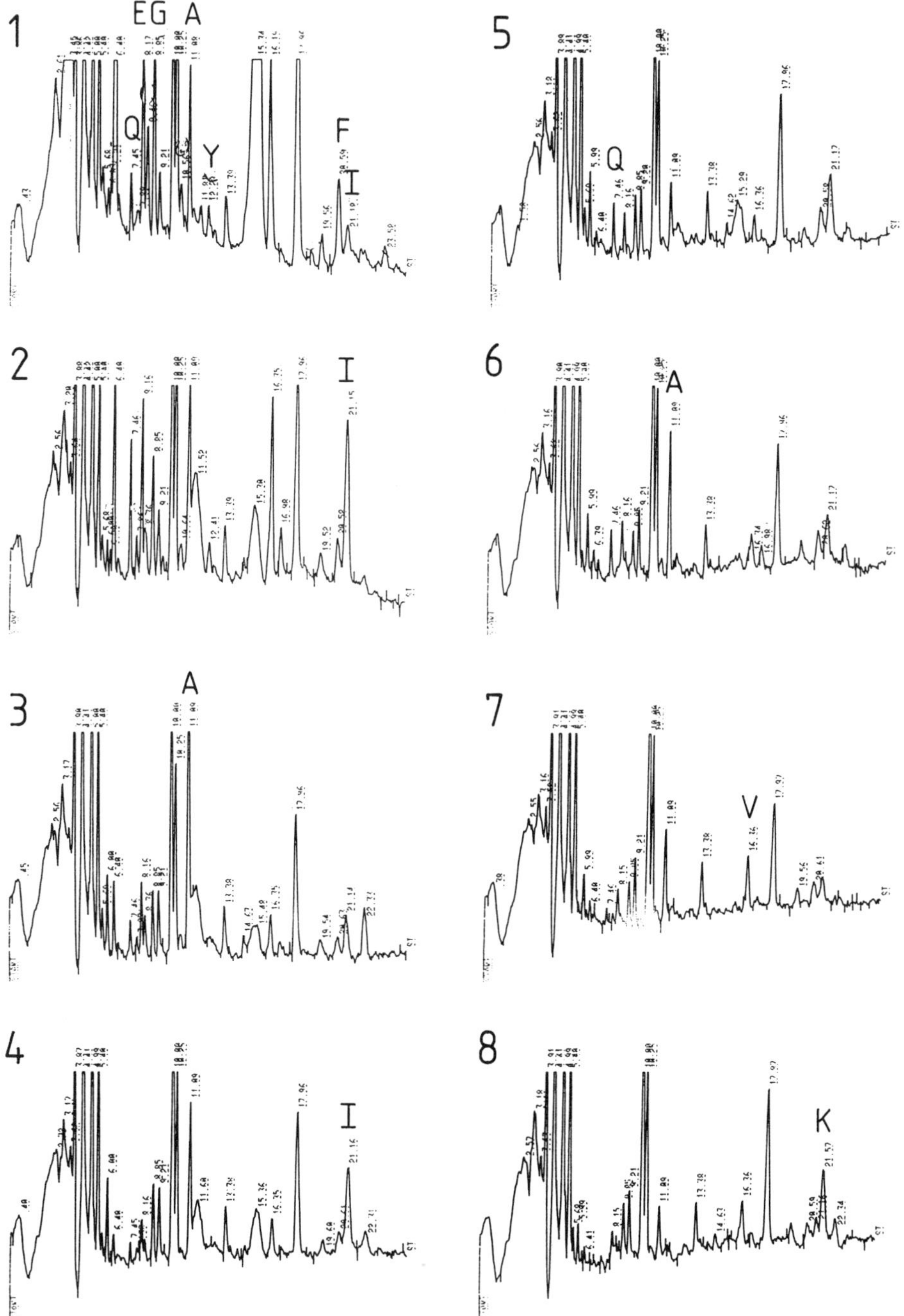

Figure 2. Sequence analysis of a 42 kDa protein after *in situ* cyanogen bromide cleavage. In the first cycle, all predominant amino acids are indicated while in cycles 2 - 8, only the residues corresponding to the major sequence are denoted.

In conclusion, efficient deblocking of N-terminally acetylated proteins using a mixture of trifluoroacetic acid and methanol reveals sequencer initial yields up to 60% and a ratio of deblocking over unspecific cleavage close to 10:1. Similarly, cyanogen bromide cleavage directly on the sequencer filter of N-terminally blocked or partially sequenced polypeptides provides an efficient approach to analysis and identification of protein structures at the pmol-level.

ACKNOWLEDGMENTS

This work was supported by grants from the Swedish Medical Research Council (projects 13X-3532 and 13X-10832), the Swedish Cancer Society (project 1806) and Stiftelsen Lars Hiertas Minne. Collaborations according to the references given are gratefully acknowledged.

REFERENCES

Bergman, T. and Jörnvall, H., 1987 Electroblotting of individual polypeptides from SDS/polyacrylamide gels for direct sequence analysis, *Eur. J. Biochem.* 169:9.

Bergman, T., Agerberth, B. and Jörnvall, H., 1991 Direct analysis of peptides and amino acids from capillary electrophoresis, *FEBS Lett.* 283:100.

Bergman, T., 1994 Internal amino acid sequences via *in situ* cyanogen bromide cleavage, *J. Prot. Chem.* 13:456.

Egyhazi, E., Stigare, J., Holst, M. and Pigon, A., 1991 Analysis of the structural relationship between the DNA-binding phosphoproteins pp42, pp43 and pp44 by *in situ* peptide mapping, *Molecular Biology Reports* 15:65.

Gardell, S.J., Craik, C.S., Clauser, E., Goldsmith, E.J., Stewart, C.-B., Graf, M. and Rutter, W.J., 1988 A novel rat carboxypeptidase, CPA2. Characterization, molecular cloning, and evolutionary implications on substrate specificity in the carboxypeptidase gene family, *J. Biol. Chem.* 263:17828.

Gheorghe, M.T., Lindh, I., Griffiths, W.J., Sjövall, J. and Bergman, T., 1995 Analytical approaches to alcohol dehydrogenase structures, *in*: Enzymology and Molecular Biology of Carbonyl Metabolism 5, Weiner, H., Holmes, R.S., and Wermuth, B., eds., Plenum Press, New York, pp 417-426.

Hermansen, L.F., Bergman, T., Jörnvall, H., Husby, G., Ranløv, I. and Sletten, K., 1995 Purification and characterization of amyloid-related transthyretin associated with familial amyloidotic cardiomyopathy, *Eur. J. Biochem.*, 227:772.

Jörnvall, H., 1970 Horse liver alcohol dehydrogenase. The primary structure of the protein chain of the ethanol-active isoenzyme, *Eur. J. Biochem.* 16:25.

Kaiser, R., Holmquist, B., Hempel, J., Vallee, B.L. and Jörnvall, H., 1988 Class III human liver alcohol dehydrogenase. A novel structural type equidistantly related to the class I and class II enzymes, *Biochemistry* 27:1132.

Kanda, Y., Goodman, D.S., Canfield, R.E. and Morgan, F.J., 1974 The amino acid sequence of human plasma prealbumin, *J. Biol. Chem.* 249:6796.

Kent, S.B.H., 1988 Chemical synthesis of peptides and proteins, *Ann. Rev. Biochem.* 57:957.

Matsudaira, P., 1987 Sequence from picomole quantities of proteins electroblotted onto polyvinylidene difluoride membranes, *J. Biol. Chem.* 262:10035.

Oppezzo, O., Ventura, S., Bergman, T., Vendrell, J., Jörnvall, H. and Avilés, F.X., 1994 Procarboxypeptidase in rat pancreas. Overall characterization and comparison of the activation processes, *Eur. J. Biochem.* 222:55.

Schuppe-Koistinen, I., Moldéus, P., Bergman, T. and Cotgreave, I.A., 1995 Reversible S-thiolation of endothelial cell actin accompanies a structural reorganisation of the cytoskeleton, Endothelium, submitted.

Tsunasawa, S. and Hirano, H., 1993 Deblocking and subsequent microsequence analysis of N-terminally blocked proteins immobilized on PVDF membrane, *in*: Methods in Protein Sequence Analysis, Imahori, K. and Sakiyama, F., eds., Plenum Press, New York, pp 45-53.

N-TERMINAL SEQUENCE ANALYSIS

9

A C-TERMINAL SEQUENCING METHOD USING PERFLUOROACYL ANHYDRIDES VAPOR

K. Takamoto, M. Kamo, K. Satake, and A. Tsugita

Research Institute for Biosciences
Science University of Tokyo
Yamazaki 2669, Noda 278, Japan

INTRODUCTION

The amino acid sequence of a protein of interest is usually one of the first pieces required in today's molecular biology, be it for gene cloning or synthesis of immunoreactive peptides. To date, amino(N)-terminal sequencing using the Edman degradation procedure has almost exclusively provided such data. Methodologies for sequencing proteins from their carboxy(C)-termini have remained relatively primitive requiring much protein in return for little sequence information. Carboxypeptidase digestion is still the most widely used method despite its intrinsic limitations of substrate specificity and endoprotease contamination. Several chemical degradation methods have been reported (Stark, 1968; Yamashita,1971; Bailey et al., 1994), and a few automated C-terminal sequencers are almost available to the public.

Recently, we observed that peptides subjected to the vapor of either 90% aqueous pentafluoropropionic acid at 90°C appeared to have amino acid residues successively cleaved from their C-termini (Tsugita et al., 1992a). By fast atom bombardment (FAB) or electrospray ionization mass spectrometry, the C-terminally successive degraded molecular ions were clearly observed and the peptide C-terminal amino acid sequence was deduced from the molecular mass differences. The predicted reaction mechanism was the formation of the oxazolone rings at the C-terminal amino acids followed by removal of the C-terminal amino acid residues. As well as the C-terminal degradation, two specific internal peptide bond cleavages were observed at the C-terminal side of internal aspartic acid residues and at the N-terminal side of serine residues.

In this paper we present an extension of C-terminal degradation method by the use of perfluoroacyl anhydride vapor instead of perfluoric acid. This method is superior to the perfluoric acid vapor method since more extensive C-terminal degradation was observed and essentially no internal peptide cleavage was observed. Preliminary reports of this method have been presented (Tsugita et al., 1992b, and two proceedings; Tsugita et al., 1992c, Tsugita

Methods in Protein Structure Analysis, Edited by M. Z. Atassi and E. Appella
Plenum Press, New York, 1995

et al., 1993). Recently we also published a preliminary report of the application to the proteins (Nabuchi et al., 1994).

FURTHER STUDIES ON REACTION METHODS

In previous experiments (Tsugita et al., 1992b, 1992c, 1993), the degradation reaction was conducted in the presence of acetonitrile.

When the same degradation reaction was performed in the dry glove box we observed that the presence of acetonitrile was not always needed from the reaction. Comparison of FAB mass spectra of the reaction products of the dodecapeptide, Ala-Arg-Gly-Ile-Lys-Gly-Ile-Arg-Gly-Phe-Ser-Gly, under the usual reagent conditions such as a vapor from 30% PFPAA in acetonitrile solution (Fig.1a), PFPAA vapor only (Fig. 1b), and by incubation with vapors from separated PFPAA and acetonitrile (Fig. 1c), clearly showed that acetonitrile was not necessary for the reaction to proceed. To clarify this seemingly contradictory data, we repeated the degradation reaction on the same dodecapeptide using a vapor from 300 μl of a PFPAA: water mixture of 10:1 (molar ratio), at -20°C for 1h either without acetonitrile (Fig. 1d) or in the presence of acetonitrile (100 μl) (Fig. 1e). Without acetonitrile, the reaction did not proceed in the presence of water, probably confirming our earlier results. However upon inclusion of acetonitrile in this reaction, the reaction proceeded as normal. We surmise that acetonitrile must absorb the moisture/water in the air ensuring that the degradation reaction can take place.

Usually operations were carried out in a glove box (1100 x 650 x 700 mm) that was continuously flushed by dry nitrogen gas. The sample peptide (2- 50 μg) was dried in a small sample tube (6 x 40 mm) in a vacuum desiccator and then transferred to the reaction tube (19 x 100 mm, Pierce , Rockford, USA). PFPAA (10% or 30%) or HFBAA (15%) in acetonitrile were used for the reagent solution. These reagents were obtained in 300 μl ampoules from Nacalai Tesque, Kyoto, Japan. The reagent solution (300~500 μl) was added to the reaction tube but outside of the sample tube(s), whilst dry nitrogen gas was continuously flushed into the sample tube(s) located in the reaction tube (Fig. 2(A) and (B)). The reaction tube was cooled with liquid nitrogen. Care must be taken to maintain dry conditions when cooling the reagent, as moisture in the air easily condenses and hydrolyses the acid anhydride. At liquid nitrogen temperature the reaction tube was evacuated (10^{-2} Torr) and sealed. The reaction tube was transferred to the reaction bath (Histo-bath, Neslab Instrument Inc., Newlngton, USA) set at -20°C when the reaction proceeded for various reaction times. After the reaction, the reaction tube was again transferred to liquid nitrogen when the reaction was stopped. The sample tube was removed from the reaction vessel, and then dried under vacuum (See Fig 2).

Partial oxidation of peptides under the present reaction conditions was observed for methionine and tryptophan residues and unmodified cysteine residues. Cysteine residues may be pyridylethylated (Amons, 1987) before the reaction. The peptides employed in this study contain almost all of the common amino acid residues and the C-terminal degradation of peptides containing these amino acids exhibited no problems. In an attempt to determine

---

Figure 1. Effect of acetonitrile as solvent and addition of water. The reactions of the dodecapeptide, ARGIK-GIRGFSG (10 μg) were carried out at -20°C for 1h with the vapor of ; (a) 30% PFPAA acetonitrile solution (300μl), (b) PFPAA (90 μl), (c) PFPAA (90 μl) and acetonitrile (210 μl) in separate tubes, (d) 300 μl of a PFPAA: water mixture of 10:1 (mol/mol) without acetonitrile and (e) the same mixture with acetonitrile (100 μl).

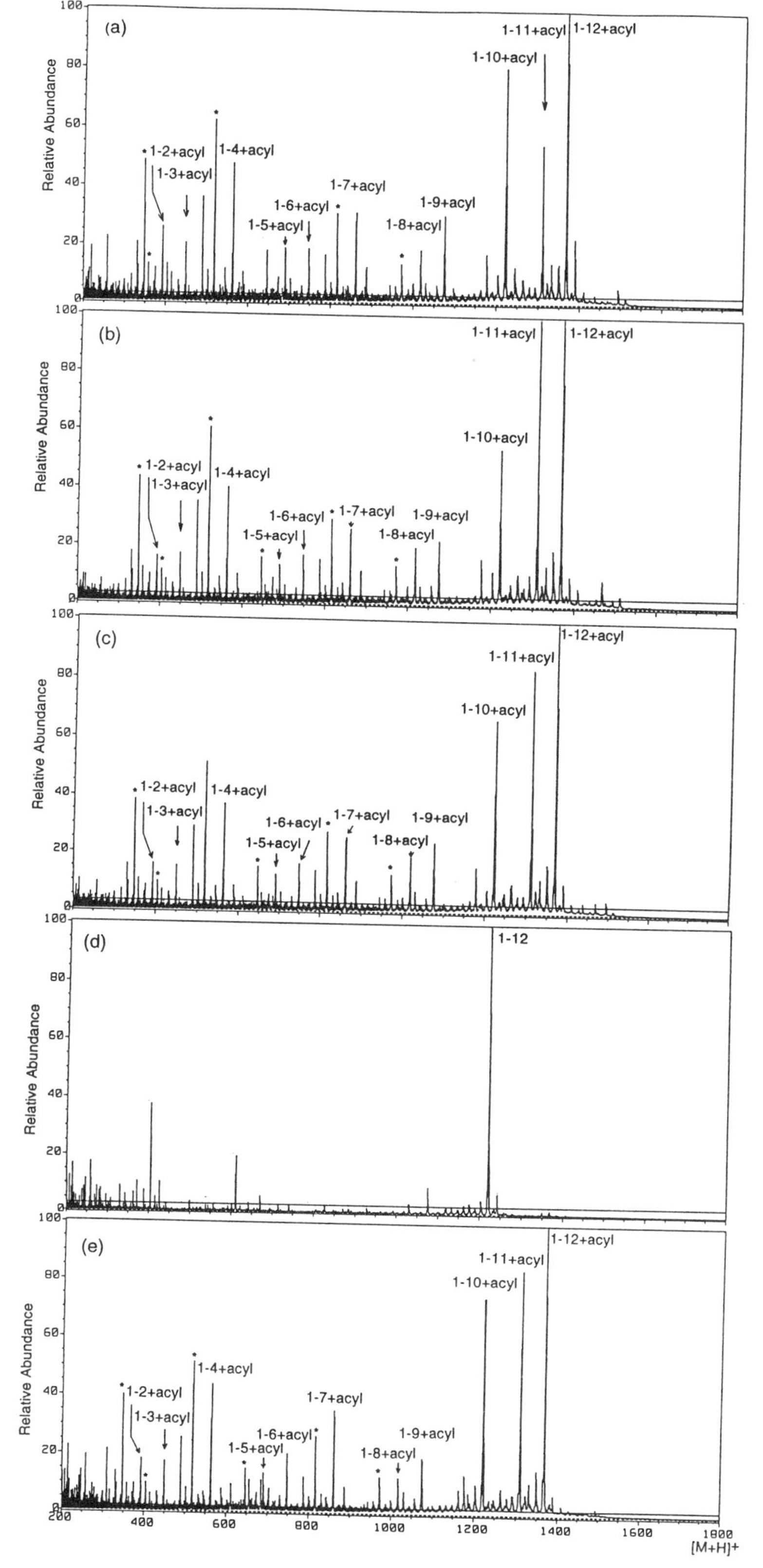
(a)
1-2+acyl
1-3+acyl
1-4+acyl
1-5+acyl
1-6+acyl
1-7+acyl
1-8+acyl
1-9+acyl
1-10+acyl
1-11+acyl
1-12+acyl
(b)
1-2+acyl
1-3+acyl
1-4+acyl
1-5+acyl
1-6+acyl
1-7+acyl
1-8+acyl
1-9+acyl
1-10+acyl
1-11+acyl
1-12+acyl
(c)
1-2+acyl
1-3+acyl
1-4+acyl
1-5+acyl
1-6+acyl
1-7+acyl
1-8+acyl
1-9+acyl
1-10+acyl
1-11+acyl
1-12+acyl
(d)
1-12
(e)
1-2+acyl
1-3+acyl
1-4+acyl
1-5+acyl
1-6+acyl
1-7+acyl
1-8+acyl
1-9+acyl
1-10+acyl
1-11+acyl
1-12+acyl
Relative Abundance
[M+H]+

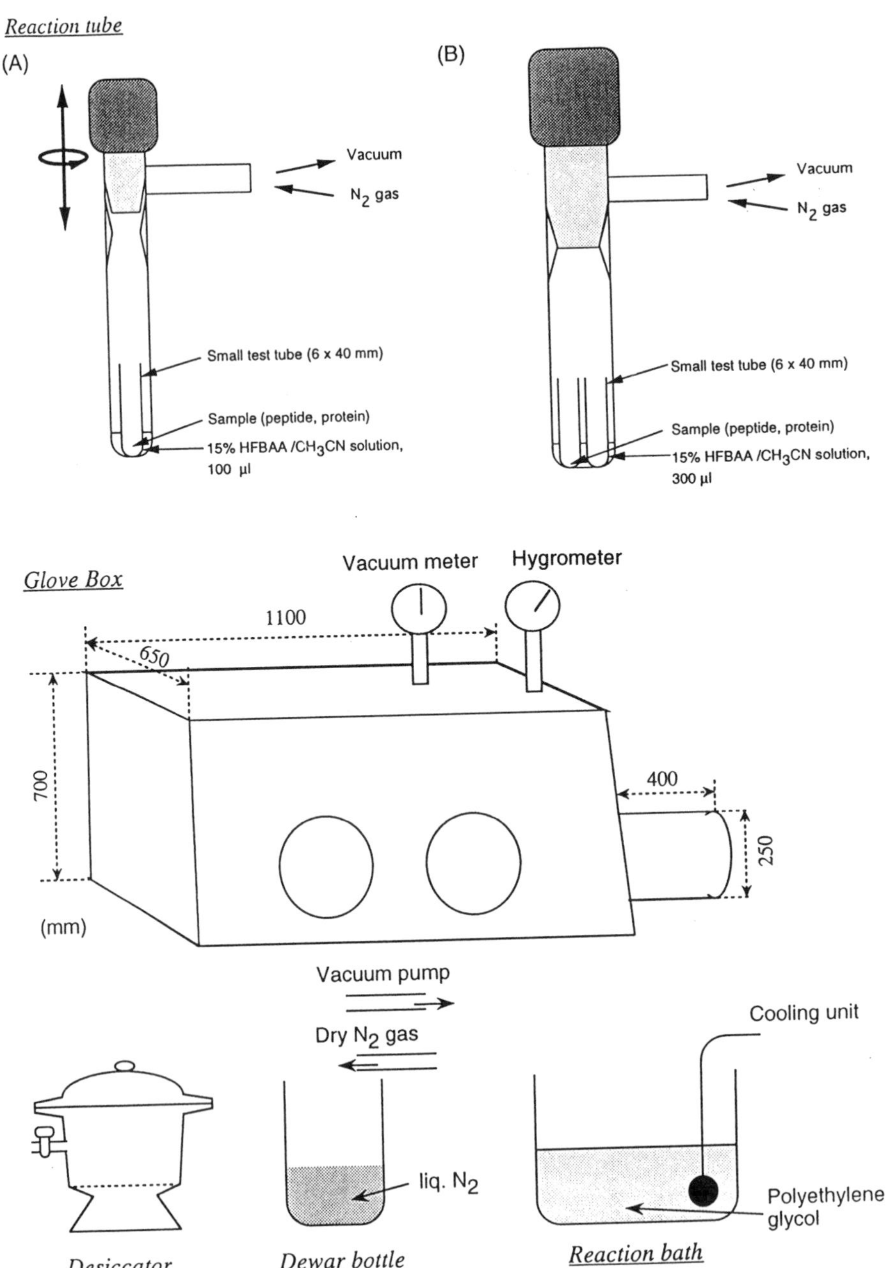

Figure 2. Reaction tube(s) and grove box; which contains desiccator, Dewar bottle and reaction bath.

the relative ease of cleavage we investigated degradation yields, as a percentage, roughly calculated from the peak height of mass spectra. Table 1 lists the truncated products observed in the present experiments from about 20 peptides including several natural C-terminal peptides and table 2 summarizes the relative cleavage ratio of peptide bonds from more than fifty experiments including the peptides listed in table 1. The vertical column is the amino

Table 1. FAB mass spectrometric identification of the C-terminal truncated fragments from the eighteen synthetic peptides and three native peptides

Sequence	Res.[†]	Acyl.[†]	Truncated main product													
MRFA	4	PFPA	1-4	(670.3)	1-3	(599.2)	1-2	(452.1)	1	(236.0)						
GRGDS	5	HFBA	1-5	(687.3)	1-4	(600.2)	1-3	(428.2)	1-2	(428.2)	Acyl-Arg	(371.2)				
YGGFM	5	PFPA	1-5	(720.1)	1-4	(589.0)										
KRTLRR	6	PFPA	1-6	(975.3)	1-5	(819.0)	1-4	(663.4)	1-3	(550.2)	1-2	(449.1)	1	(293.1)		
RYLGYLE	7	HFBA	1-7	(1109.4)	1-6	(980.3)	1-5	(867.4)	1-4	(704.3)	1-3	(647.4)	1-2	(534.1)	1	(371.3)
RRVGRPE	7	HFBA	1-7	(1065.5)	1-6	(936.3)	1-5#	(821.3)	1-4	(683.4)	1-3	(626.3)	1-2	(526.1)	1	(371.2)
HPFHLLVY	8	HFBA	1-8	(1221.7)	1-7	(1058.5)	1-6	(959.5)	1-5	(846.4)	1-4	(733.3)	1-3	(596.2)	1-2	(449.1)
LEDGPKFL	8	—	1-8#	(900.4)	1-7#	(787.3)	1-6#	(640.4)	1-5#	(512.1)						
KRNKKNNIA	9	PFPA	1-8###	(1049.1)	1-7###	(977.9)	1-6###	(865.5)	1-5##	(769.3)	1-4##	(673.4)	1-3#	(545.1)	1-2	(449.2)
RPPGFSPFR	9	PFPA	1-9	(1206.2)	1-8	(1050.8)	1-7	(903.2)	1-6	(806.1)	1-5	(719.3)	1-4	(572.2)	1-3	(515.2)
			1-2	(418.3)	1	(321.1)										
RPKPQQFFG	9	PFPA	1-9##	(1215.3)	1-8##	(1158.2)	1-7##	(1011.5)	1-6##	(863.9)	1-5#	(753.1)	1-4	(643.8)		
YGGFLRKYPK	10	PFPA	1-10	(1374.8)	1-9	(1246.5)	1-8	(1149.3)	1-7	(986.0)	1-6	(858.0)				
DRVYVHPFNL	10	PFPA	1-10#	(1389.0)	1-9#	(1275.9)	1-8	(1179.0)	1-7	(1031.7)	1-6	(934.5)	1-5	(797.3)	1-4	(697.9)
			1-3	(535.3)	1-2	(436.4)										
IKNLQSLDPSH	11	HFBA	1-11##	(1411.9)	1-9##	(1188.4)	1-8##	(1090.5)	1-7##	(975.5)	1-6##	(862.5)	1-5##	(775.4)	1-4#	(665.5)
			1-3#	(552.2)	1-2#	(438.3)										
ARGIKGIRGFSG	12	PFPA	1-12	(1364.7)	1-11	(1307.6)	1-10	(1220.6)	1-9	(1073.5)	1-8	(1016.4)	1-7	(860.4)	1-6	(747.2)
			1-5	(690.3)	1-4	(562.0)	1-3	(449.0)	1-2	(392.1)						
RRLIEDAEYAARG	13	PFPA	1-13	(1520.2)	1-12	(1462.2)	1-11	(1306.2)	1-10	(1235.1)	1-9	(1164.1)	1-8	(1001.1)	1-7	(872.4)
			1-6	(801.3)	1-5	(686.4)	1-4	(557.2)	1-3	(444.2)	1-2	(331.2)				
YGGFLRRIRPKLKWDNQ	17	PFPA	1-17##	(2258.8)	1-16##	(2130.7)	1-15#	(2033.8)	1-14+o	(1953.6)	1-13	(1750.4)	1-12	(1622.4)	1-11	(1509.3)
			1-10	(1381.2)	1-9	(1284.0)	1-8	(1127.9)	1-7	(1014.9)	1-6	(858.8)				
GIGKFLHSAGKFGKAFVGEIMKS	23	PFPA	1-22+o	(2485.0)	1-21+o	(2357.0)	1-20	(2210.0)	1-19	(2096.7)	1-18	(1967.3)	1-17	(1910.4)	1-16	(1811.0)
			1-15	(1664.2)	1-14	(1593.4)	1-13	(1465.0)	1-12	(1407.8)	1-11	(1261.1)	1-10	(1132.1)	1-9	(1076.0)
			1-8	(1004.6)	1-7	(917.7)	1-6	(780.7)	1-5	(667.7)	1-4	(519.7)				
AAEYKVLGFQG (Myoglobin - Sheep 143-153)	11	HFBA	1-11##	(1342.8)	1-10##	(1285.3)	1-9#	(1175.9)	1-8#	(1028.5)	1-7#	(971.5)	1-6#	(840.5)		
VGKVTVN (Plastocyanin 91 - 97)	7	PFPA	1-7	(862.9)	1-6	(748.6)	1-5	(649.7)	1-4	(548.3)						
IFAGIKKKTEREDLIAYLKKATNE (Cytchrome C 81 - 104)	24	PFPA	1-24##	(2890.4)	1-23##	(2761.2)	1-22#	(2665.3)	1-21#	(2564.3)	1-20#	(2492.6)	1-19#	(2364.5)	1-18#	(2237.0)
			1-17#	(2123.6)	1-16#	(1960.9)	1-15#	(1889.5)								
			1-14#*	(1730.7)	1-13#*	(1617.3)	1-12#	(1547.6)	1-11	(1437.1)	1-10*	(1234.7)				

[†]Res. = Residues; Acyl. = Acylation.
#Stands for a -18 mass, and ## for two of -18 mass.
*Stands for a -46 mass.
+o stands for a oxidized peptide mass.

Table 2. Degradation ratio of peptide bond by perfluoroacyl anhydride vapor

	D	N	T	S	E	Q	P	G	A	V	M	I	L	Y	F	H	K	R	W
D		99		99				9	96					94					
N		55				95						98	99				80		
T													22						
S		95				42	93	87	91										
E	68											76		94					
Q						87		94							50				
P	91				65	89	40	12							90		14		
G	99			97	85		29	50				65			84		95	96	
A					33			89	68						96		47	80	
V								69						75		27			
M																	99	65	
I					23			60	99		70						64	84	
L										91		38	90			81	90	90	
Y							94	21	96	75		75							
F		98		89				94	93	96	22		85		60	78		83	
H				87			56						98						
K		91		97			41	75	95				88	63	93	71	66	73	77
R		14	86				58	82		42		50	23		42		42	20	
W	98																		

Vertical column is the amino side residue of peptide bonds and horizontal column is the carboxy side residue. Degradation ratios (%) were roughly calculated from the peak heights in mass spectra.

Table 3. Peptide sequences and degraded residues

Peptide	Residue	Peptide	Residue
ARGIKGIRGFSG	R	LEDGPKFL	D
YGGFMRRVGRPE	R	RPPGFSPFR	R
YGGFLRRIRPKLKWDNQ	R	DRVYVHPFNL	R
YGGFM	G	EAKSQGGSN	A
YGGFL	G	NRVYVHPFHL	R
NRVYVHPFHL	V	YGGFLRRIRPKLKWDNQ	S
PRLIEDAEYAARG	R	WAGGDASGE	S
GIGKFLHSAGKFGKAFVGEIMKS	K	RRLIEDAEYAARG	R
HPFHLLVY	F	KRNKKNNIA	K
RFA	R	RPKQQGFFG	R
LWMRFA	M	VGKVTVN	V
MRFA	R	HSQGTFTSDYSKYLDSRRAQDFVQWLMNT	R
RPPGFSPFR	R	KKKHPDYI	K

Underline in the peptide sequence stands for identified successively degradation at positive mode. Residue stands for carboxy terminal residue in observed smallest ion peak at positive mode in FAB mass spectrometry.

side residue of peptide bond and the horizontal row is the carboxy side residue. The degradation ratio given are not quantitative and are therefore only indicative of relative ease of bond cleavage.

Most peptide bonds are readily cleaved however some residue combinations exhibited some difficulty to break. Arg residues on the amino side in peptides seemed to be somewhat resistant to degradation, as shown in Table 3. However this is not true because the negative mode of analysis showed further extension of the reaction as shown in Fig. 3.

Attention to the ion mode used in FAB mass spectrometry is important to maximize the amount of sequence information obtained in the present analysis. We degraded the peptide Tyr-Gly-Gly-Phe-Leu-Arg-Arg-Ile-Arg-Pro-Lys-Leu-Lys-Trp-Asp-Asn-Gln using a 30% PFPAA acetonitrile solution at -20°C for 30 min in vapor phase followed by aqueous pyridine vapor treatment. Mass spectrometries of the degraded peptide were carried out in both positive mode and negative mode. Positive mode analysis, as shown in Fig. 3a, suggested that degradation was only achieved until the peptide 1-6. All that has happened is that the positive charge has been lost by including acylation. Negative mode analysis

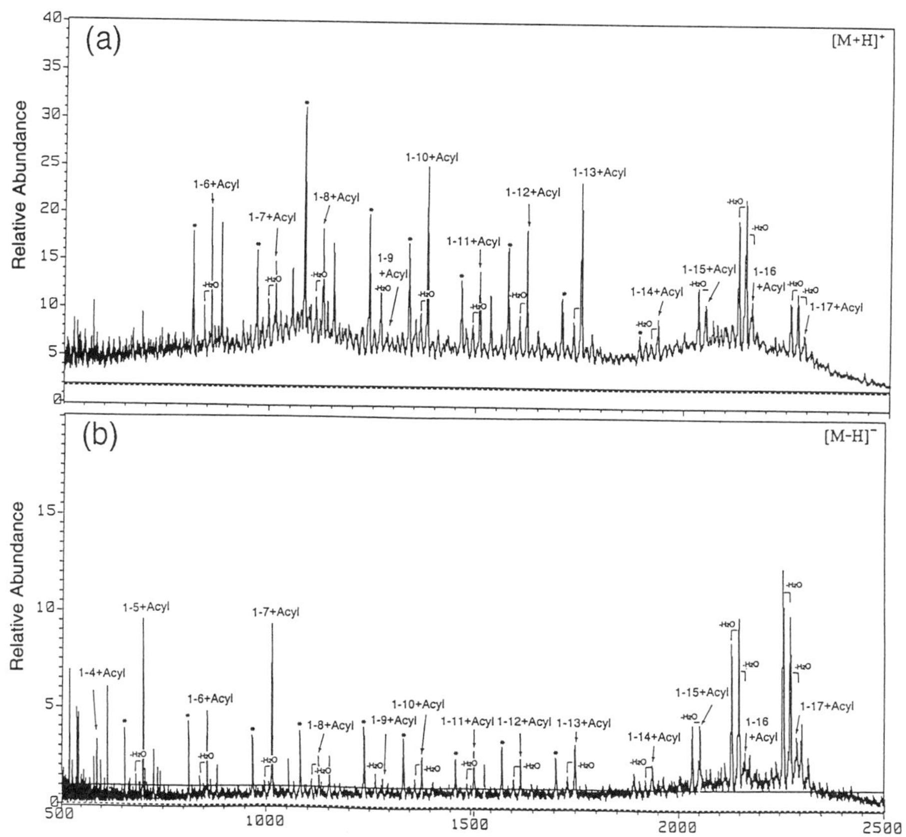

Figure 3. Positive and negative ionization of FAB mass spectra of the C-terminally degraded products of a peptide, Tyr-Gly-Gly-Phe-Leu-Arg-Arg-Ile-Arg-Pro-Lys-Leu-Lys-Trp-Asp-Asn-Gln. The peptide (5 µg) was exposed to a vapor of 10% PFPAA acetonitrile solution at -20°C for 1h. The degraded product was treated with water vapor. The product was dissolved in 2 µl of 67% acetic acid and 1 µl of solution was mixed with same volume of the matrix composed of glycerol, thioglycerol and *m* -nitrobenzyl alcohol. One µl of the solution was applied to the FAB mass spectrometer. The spectra are detected in (a) positive mode and (b) negative mode both using the same product solution and matrix.

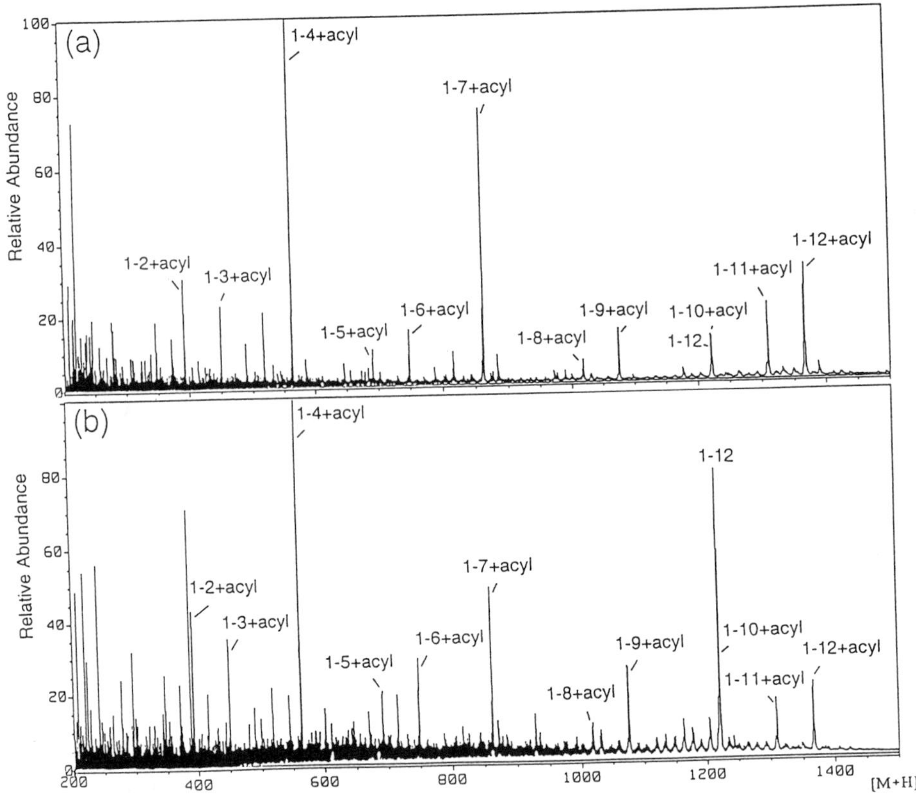

Figure 4. Matrix effect on FAB mass spectrum of C-terminally degraded product. The degradation of ARGIKGIRGFSG with vapor of 10% PFPAA acetonitrile solution followed by pyridine-water vapor and the resultant mixture treatment was dissolved in 3 μl of 67% acetic acid. Each 1 μl of solution was mixed with same volume of matrix and subjected to FAB mass spectrometer. Matrix; (a) mixture of glycerol: thioglycerol: *m* -nitrobenzyl alcohol (1:1:1 v/v), (b) glycerol.

demonstrated that the reaction has proceeded further to the peptide 1-4 (Fig. 3b). We have observed that this is a common phenomenon. Many degradations appear to stop short of what is actually achieved, such as at positive charged arginine residues. Therefore mass spectrometries using both positive and negative modes are needed to make sure all the sequence information has been retrieved, although positive mode is more sensitive than negative mode, in general.

Another feature of the analysis procedure that has a significant effect on the data obtained is the matrix used shown in Fig. 4. A mixture of glycerol, thioglycerol and *m* -nitrobenzyl alcohol in the ratio 1:1:1 (Fig. 4a), was generally found to be superior than glycerol only (fig. 4b). The mixture containing *m*-nitrobenzyl alcohol tends to stress hydrophobic peptides. When an unknown sample is analyzed then we suggest that at least two different matrices are used.

C-Terminal degradation of peptides having an α-carboxyl amide group did not occur upon exposure to perfluoroacyl anhydride vapor under the present conditions (data not shown).

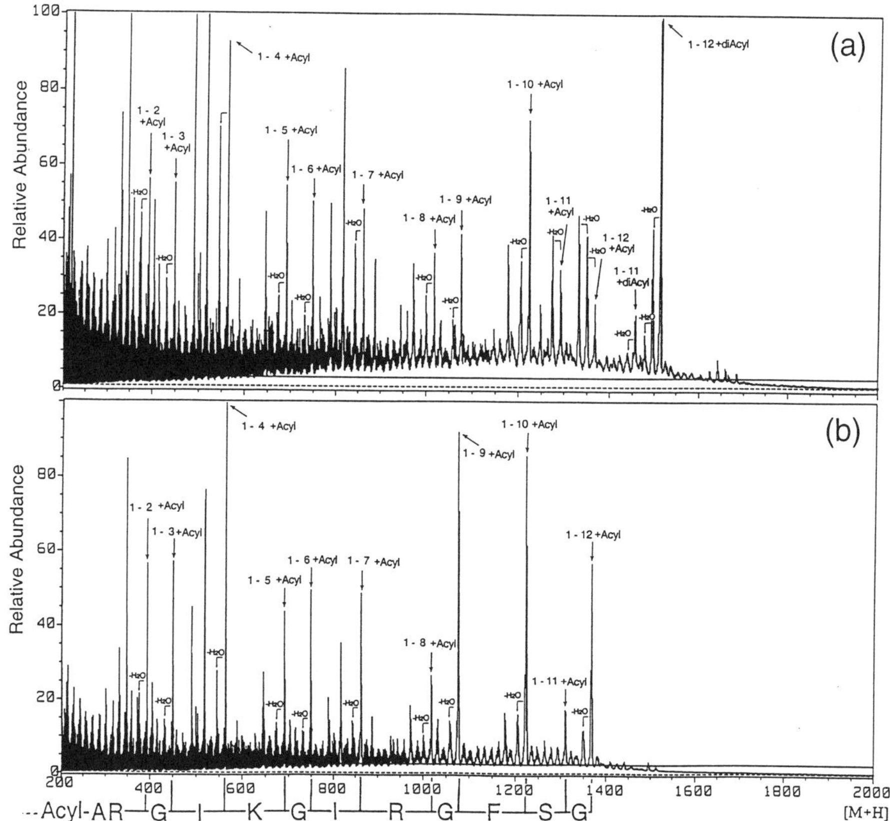

Figure 5. Water treatment simplified the degradation fragment peaks.

EXPERIMENTS TO UNDERSTAND THE REACTION MECHANISM.

The C-terminal successive degradation was carried out with the vapor of perfluoroacyl anhydrides followed by treatment of aqueous pyridine vapor. The reaction was carried out with 10% PFPAA at -18°C for 2h on ARGIKGIRGFSG.

The water vapor treatment simplify the mass spectrum as shown in Fig. 5, where the major parts of -18 molecular ions were moved to the respective C-terminal truncated molecular ions, and the part of acylated molecular ions were subjected to deacylation. The recoverable -18 molecular ions may be due to the formation of oxazolones (or mix anhydrides) at the C-terminal α–carboxyl groups while the latter unstable acylation may be *O*-acylation of the oxazolone and/or hydroxy group of Ser or Thr residue.

Evidence that the intermediate degradation products to be oxazolone was provided by converting them to the corresponding propyl esters. Successive degradation of the octapeptide His-Pro-Phe-His-Leu-Leu-Val-Tyr with a vapor of 15% HFBAA-acetonitrile solution at -20°C for 30 min was followed by esterification of the degradation products with a propanol vapor at 60°C for 15 min. The products were analyzed by FAB mass spectrometry (Fig. 6). This results show that the main degraded products were acylated intermediate compounds such as oxazolones which easily converted to their propyl esters.

The similar type of experiment was also carried out by the use of dimenthylhydrzaine where the dimethyl hydrazides were observed instead of the propyl esters (Fig. 7).

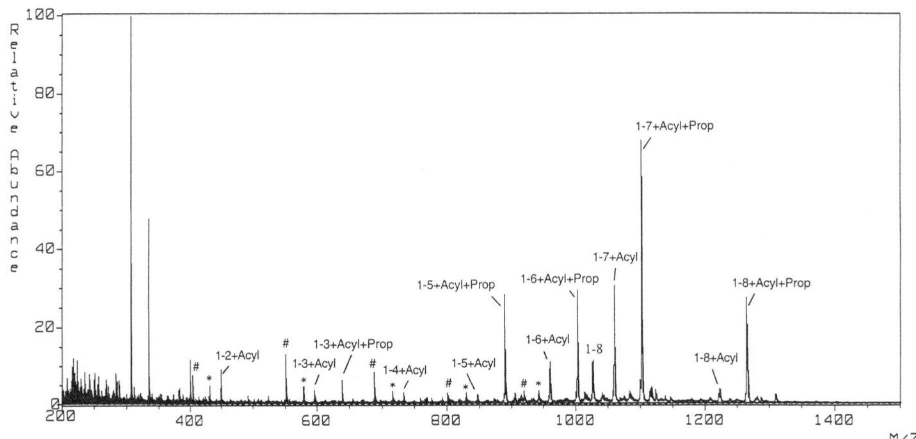

Figure 6. FAB mass spectrum of the esterified products of the C-terminally degraded His-Pro-Phe-His-Leu-Leu-Val-Tyr with HFBAA vapor. The peptide (5 μg) was exposed to a vapor of 15% HFBAA acetonitrile solution (100 μl) at -20°C for 30 min. The product was cooled by liquid nitrogen, dried in vacuo and reacted with a vapor of propanol at 60°C for 15 min. The reaction product was dried in vacuo and analyzed by FAB mass spectrometry. In this figure 'prop' stands for propylester.

The other possible -18 mass ion peak may be caused by dehydration of serine residue, acid amide residues and acidic amino acid residues converting into dehydroalanine, acid nitriles and formation of 5(6)-membered rings, respectively (Fig. 8b). These conversions are speculated by the observations of appearances and disappearances of -18 mass ion peaks in the course of C-terminal successive degradation of various peptides tested.

The accompanied molecular ions were not only -18 and acylated molecular ions but also, often seen, -1 (Fig. 8a) and -46 (Fig. 8c) molecular ions. A set of the following experiments were made to clarify these molecular ions.

A peptide Ala-Arg-Gly-Ile-Lys-Gly-Ile-Arg-Gly-Phe-Ser-Gly (20 mg) was cleaved using the vapor of 30% PFPAA-acetonitrile solution in the vapor phase at -20°C for 1h, and was exposed to the vapor of aqueous pyridine. One twentieth of the reacted peptide was analyzed by FAB mass spectrometry to confirm that the peptide was indeed degraded (Fig. 9a). The rest of the degraded peptides were fractionated by HPLC (Fig. 9b). Isolated fractions were subjected to FAB mass spectrometry and identified as the acylated products corresponding to the respective sequential degradation products. In addition to the acylated sequences, -1 mass peaks were also found (Table 4). This -1 mass peak is due to the cleaving at the amino groups resulting in the acid amides (Fig. 8a) but not to the cleaving at the peptide bond. This cleavage causes the discontinuation of the further successive degradation.

The -46 peak was thought to be due to decarboxylation of the C-termini of the degradation products. This was tested by degrading the tetrapeptide Met-Arg-Phe-Ala under the standard C-terminal degradation conditions and fractionating the degradation products by HPLC. Figure 10a shows 280 nm-profile. It is known that the λmax of a phenyl group attached to an aryl group is 280 nm, whilst that of a phenyl group alone is 254 nm with low extinction coefficient. Therefore the major peak in chromatogram at 280 nm corresponds to the degraded peptide 1-3, acyl-Met-Arg-NH-CH = CH-C_6H_5, (Fig. 8c). FAB mass spectrometry of this peak showed a molecular ion peak of 553 (Fig. 10b) which corresponded to the calculated mass of the above degraded compound (552.9). Taken together, these experiments confirm the nature of the -46 mass peak.

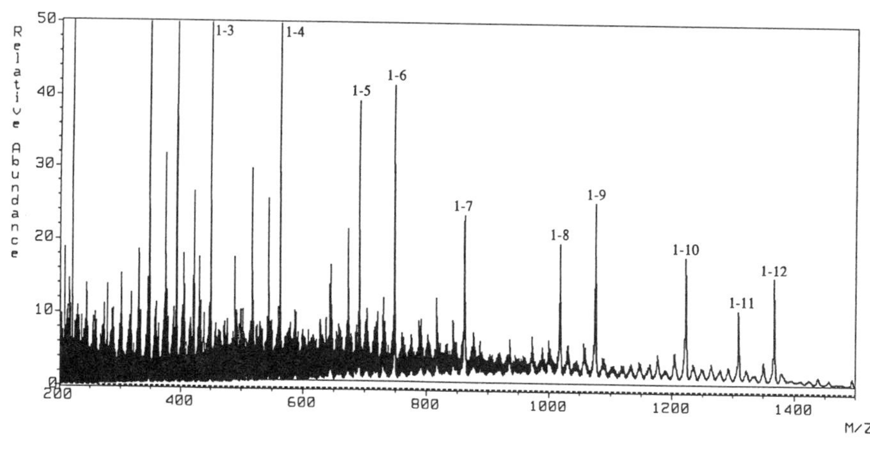

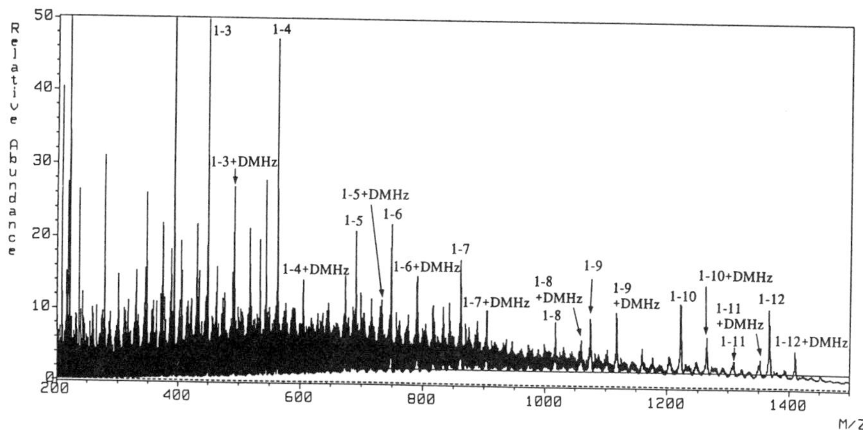

Figure 7. FAB mass spectra of degradation product derivatized with dimethylhydrazine vapor. Successive degradation was performed with 30% PFPAA acetonitrile solution at -20°C for 30 min. Degradation product was immediately exposed to vapor of dimethylhydrazine.

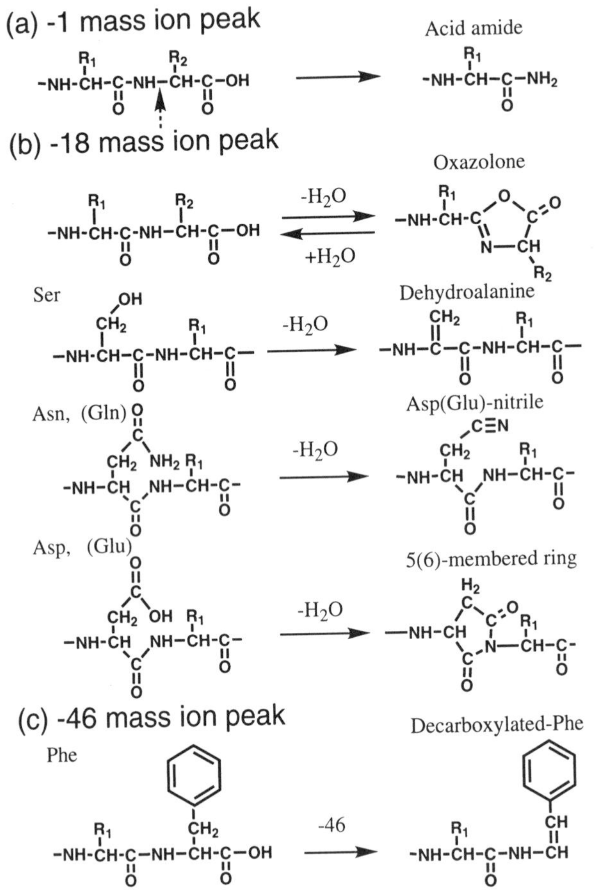

(a) -1 mass ion peak

Acid amide

(b) -18 mass ion peak

$-H_2O$ / $+H_2O$ Oxazolone

Ser $-H_2O$ Dehydroalanine

Asn, (Gln) $-H_2O$ Asp(Glu)-nitrile

Asp, (Glu) $-H_2O$ 5(6)-membered ring

(c) -46 mass ion peak

Phe -46 Decarboxylated-Phe

Figure 8. Possible schemes in the reaction of perfluoroacyl anhydride vapor on peptides.

Table 4. FAB mass spectrometric identification of the HPLC separated fragments of C-terminally degraded products.

Peak No.	Observed mass	Sequence	Calculated mass
1	391.3	1–2[†]	391.1
2	449.2	1–3	499.2
3	392.2	1–2	392.1
4	1220.3	1–10	1220.6
5	746.3	1–6[†]	746.4
6	689.3	1–5[†]	689.4
7	747.3	1–6	747.4
	1218.5	1–12	not acylated : 1218.7
8	690.4	1–5	690.4
	747.3	1–6	747.4
9	1072.1	1–9[†]	1072.6
10	561.2	1–4[†]	561.2
	746.3	1–6[†]	746.4
	1015.7	1–8[†]	1015.5
11	1016.4	1–8	1016.5
	1073.5	1–9	1073.6
12	1016.4	1–8	1016.5
	1073.3	1–9	1073.6
13	1017.5	1–8	1016.5
	1074.5	1–9	1073.6
14	562.5	1–4	562.2
	859.5	1–7[†]	859.4
15	562.2	1–4	562.2
16	562.1	1–4	562.2
17	860.5	1–7	859.4
	1364.5	1–12	1364.7
18	860.5	1–7	859.4
	1308.0	1–11	1307.7
19	1219.8	1–10[†]	1220.6
20	1219.7	1–10[†]	1220.6

The truncated degradation products were separated by reverse phase HPLC (Fig. 9b). Fractions numbered in the HPLC profile were identified by FAB mass spectrometry. Degraded product were acylated except 1–12 (Peak 7).

[†]indicates the -1 mass products.

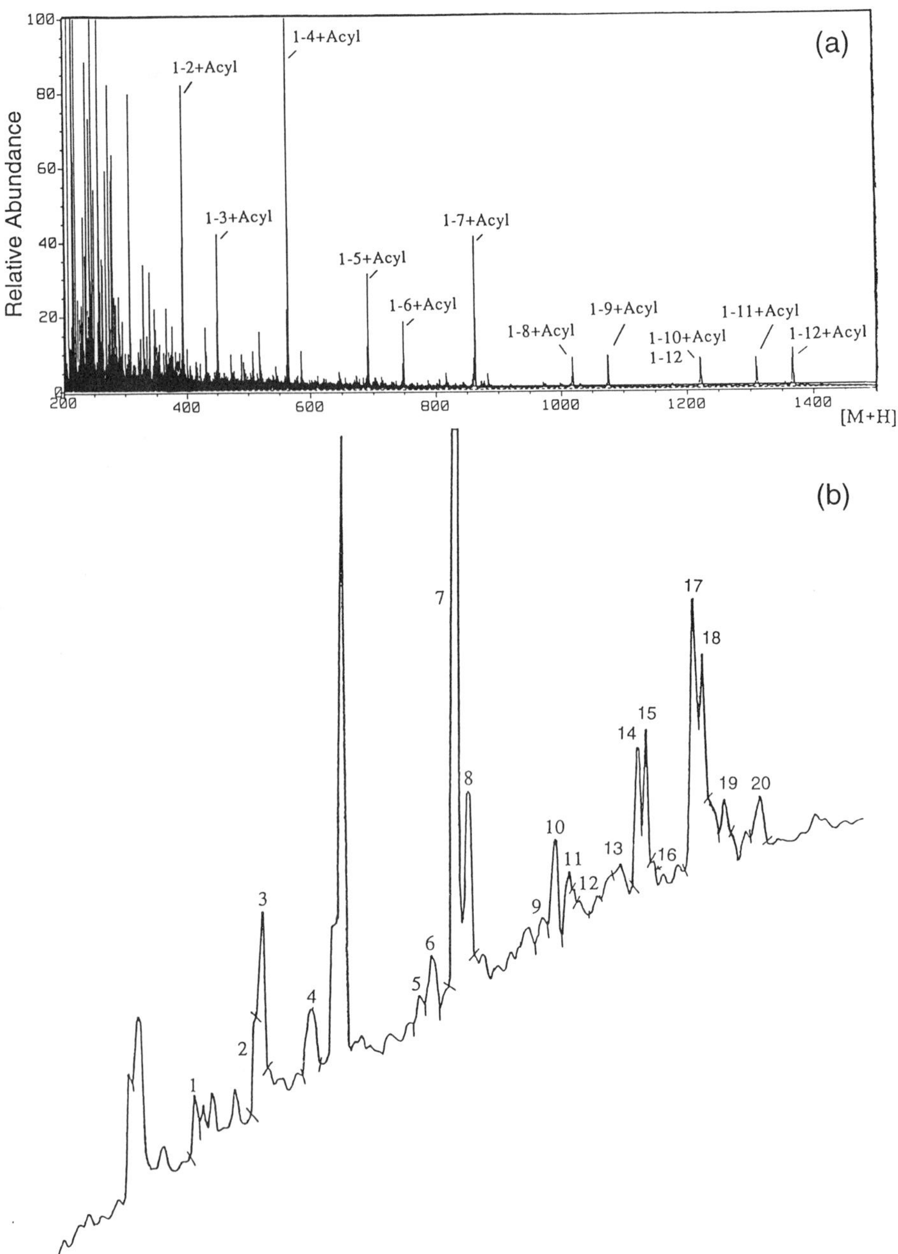

Figure 9. FAB mass spectrum and HPLC profile of the C-terminally degraded dodecapeptide. Dodecapeptide, Ala-Arg-Gly-Ile-Lys-Gly-Ile-Arg-Gly-Phe-Ser-Gly (20 μg) was degraded with 30% PFPAA acetonitrile solution at -20°C for 1h in vapor phase followed by water treatment. After reaction an aliquot (1/20) of the product was analyzed by FAB mass spectrometry. The rest of the product was separated by reverse-phase HPLC. (a) FAB mass spectrum of the degraded product mixture; (b) Elution profile of the product by reverse-phase HPLC. HPLC was made under the following conditions; System: 600E (Waters-Millipore, USA), Column: TSK-Gel (4.6x250 mm TOSOH, Japan), Flow rate: 0.8 ml/min, solvent: TFA and 0.1% TFA in 80% aqueous acetonitrile, Gradient system: a linear gradient between 0 to 48% acetonitrile for 60 min. The chromatographic peaks with numbers were analyzed by FAB mass spectrometry. The results were summarized in Table 4. The most of the peaks without numbers are amino acid acyl derivatives from the truncated C-termini.

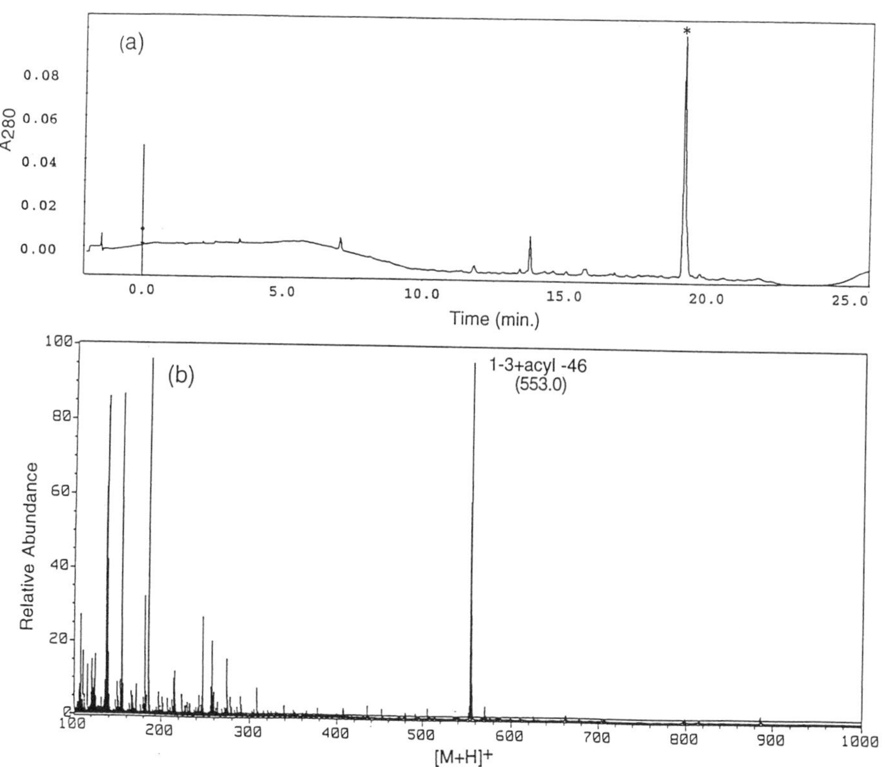

Figure 10. C-Terminal successive degradation fragments of Met-Arg-Phe-Ala were analyzed with HPLC and the major peak at 280 nm was analyzed by mass spectrometry. C-Terminal degradation of the peptide (20 μg) was carried out with a vapor of 30% PFPAA acetonitrile solution (100 μl) for 1h at -20°C. HPLC was made with the SMART HPLC system under the following conditions. Column, μRPC C2/C18 PC 3.2/3 (2.1 mm x 100 mm, Pharmacia); flow rate, 0.2ml/min; solvents, 0.1% TFA aqueous solution and 0.1% TFA acetonitrile solution. A linear gradient of acetonitrile concentration (5-60%) was made from 5 min to 17.5 min followed by an isocratic elution from 17.5 min to 20 min. The chromatogram was monitored at both wavelengths 215 nm (data not shown) and 280 nm (panel a). Peaks were fractionated and collected. After dried, the major fraction marked by * was analyzed by FAB mass spectrum shown in panel b.

APPLICATION TO BIG PROTEINS

It is not easy to apply this method to protein C-terminal sequencing because usual protein mass are too big to analyze by FAB mass directly. Electrospray ionization (ESI) mass, has been successfully applied to direct measurement of protein molecular weights, although an analysis of a sample containing a mixture of similar molecular masses like truncated molecules is not always easy.

The following three strategies were tested for protein C-terminal sequencing shown in Fig. 11. Three strategies are illustrated in the figure as follows; (1) First fragment the protein and the C-terminal fragment is isolated by somehow specific method and is degraded with perfluoroacyl anhydride vapor. (2) First fragment the protein and the fragmented mixture is degraded with the anhydride vapor, using the specific fragmentation which results in the inactive non-C-terminal fragments. (3) The protein is first degraded with the anhydride

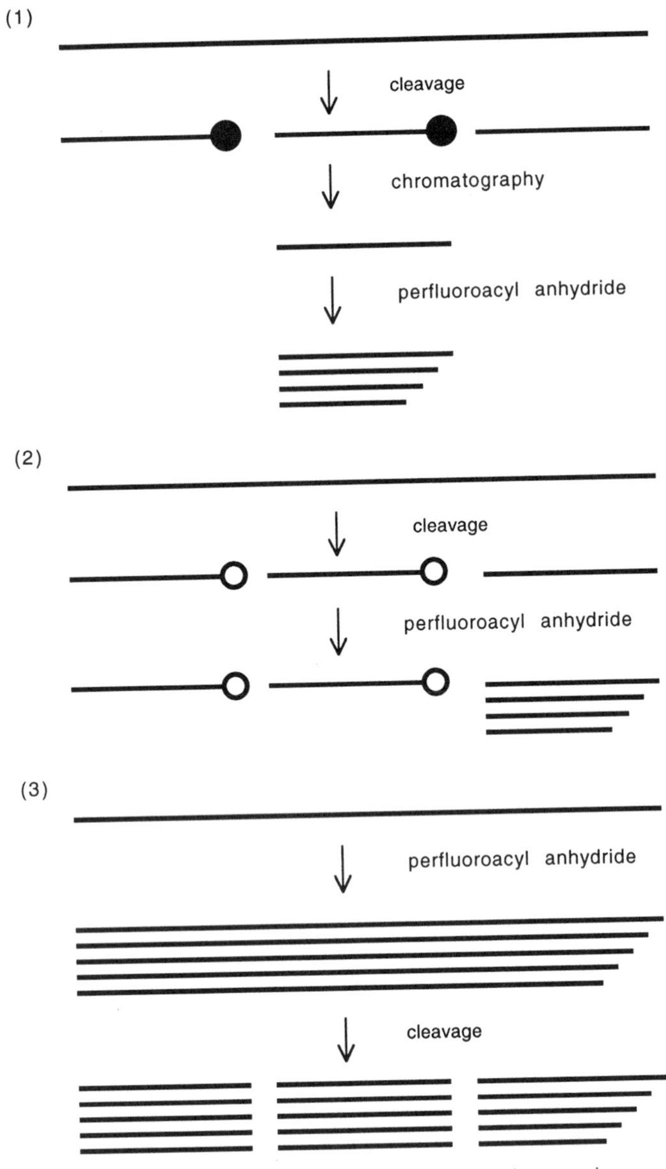

Figure 11. Three strategies for protein C-terminal sequencing.

vapor and the truncated mixture is fragmented. Then the truncated C-terminal fragments are analyzed.

According to the strategy, we tried to fragment a protein and selectively isolate the C-terminal peptide. After various trials, we selected the classical cyanogen bromide cleavage specific for methionyl peptide bond. The cleaved peptide fragments were selectively fractionated by covalent bond formation with the *N*-(2-aminoethyl)-3-aminopropyl glass (APG, LKB Biochrom Ltd.) into non-C-terminal fragments and the C-terminal peptide. The

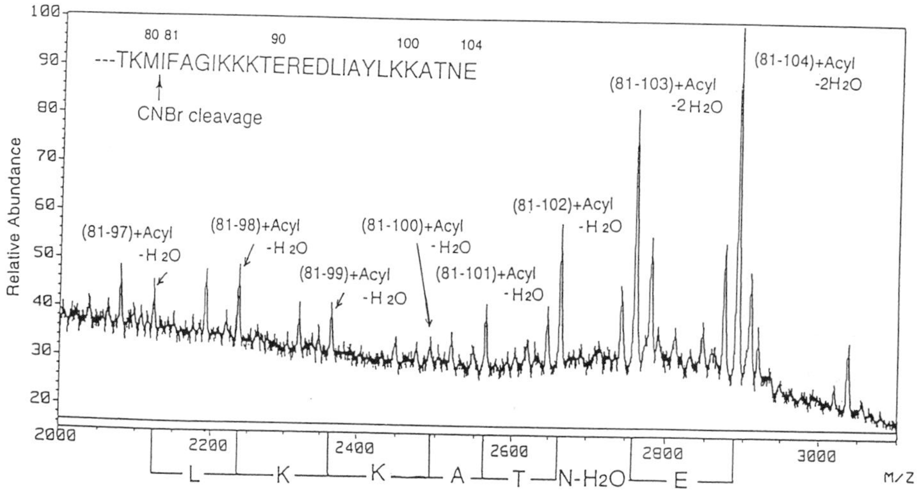

Figure 12. FAB mass spectrum of truncated C-terminal peptide of cytochrome c. Five nmoles of cytochrome c (horse heart) were cleaved by CNBr at room temperature for 16 h. The mixtures were evaporated and added anhydrous TFA for 30min. The product solution was mixed with APG, pre-equilibrated with a 2% triethylamine DMF solution and incubated for 2 h at 45°C. The products were dissolved in water and then added with DMF. The APG was washed with DMF and 0.1 M pyridine collidine buffer (pH 8.2). The unbound and washed fractions were collected and evaporated in a small test tube. The degradation was performed with 30% PFPAA acetonitrile solution at -20°C for 1h followed by aqueous pyridine vapor treatment.

C-terminal peptide was successively degraded with perfluoroacyl anhydride vapor and analyzed by FAB mass

A model dodeca spectrometry peptide, YGGFMRRVGRPE was cleaved by CNBr. The resultant N-terminal pentapeptide ending homoserine was converted to homoserine lactone with TFA at 20°C for 30 min. The APG supernatant was analyzed by FAB mass. The results showed the complete heptapeptide C-terminal sequence (Nabuchi et al. 1994).

Sheep myoglobin was subjected to the C-terminal sequencing. The C-terminal peptide (143-153) was isolated by APG treatment and subjected to HFBAA vapor. The analysis by FAB mass spectrometry showed C-terminal sequence of four amino acid residues (Nabuchi et al., 1994). The other protein, cytochrome c was sequenced after isolation the C-terminal peptide (81-104). The reaction with the vapor of 30% PFPAA acetonitrile solution at -20°C for 1h resulted in the C-terminal seven amino acid sequence as shown in Fig. 12.

As for the strategy (2) the rice Plastocyanin (10 μg) was cleaved with CNBr at 20°C for 72h and the reaction mixture was dried and treated with TFA at 20°C for 30 min. The homoserine lactones of the C-termini of non-C-terminal peptide are insensitive for the present perfluoroacyl anhydride reaction. Thus without purification of the C-terminal peptide, the mixture was subjected to the anhydride reaction. Even with high background including matrix lines, the sequence of four amino acids from the C-terminal was analyzed (Fig. 13).

As for the strategy (3) the same protein was directly degraded with the vapor of 30% PFPAA in acetonitrile at -18°C for 2h. The truncated protein was analyzed showing the C-terminal tetra peptide sequence (Fig. 14). Further more works will be carried out for the protein C-terminal sequencing.

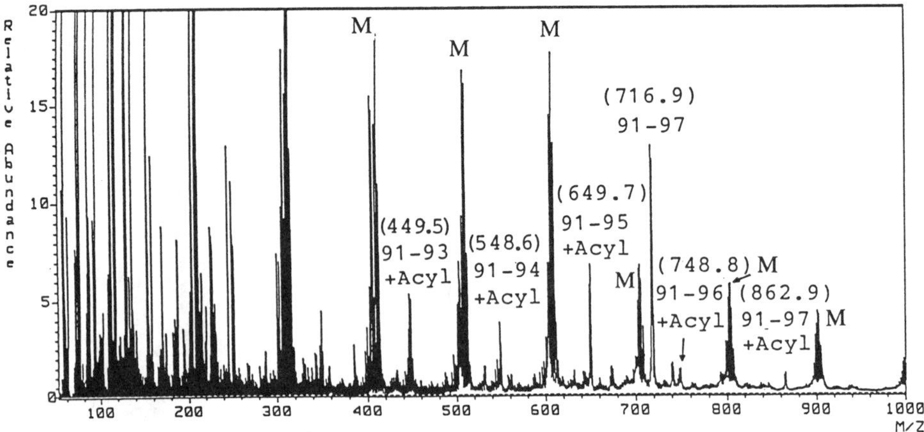

Figure 13. FAB mass spectrum of successively degraded C-terminal peptide of Plastocyanin. Rice plastocyanin, which contains one Met at 90th was cleaved with CNBr. The peptides dried and treated with TFA at 20°C for 30 min. The reaction mixture was directly subjected to the anhydride reaction. M stands for matrix line.

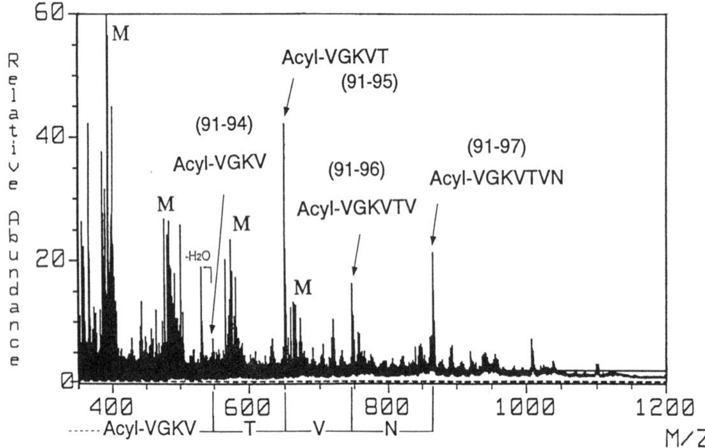

Figure 14. FAB mass spectrum of the C-terminal degradation of CNBr fragments of plastocyanin. Denatured plastocyanin (10 μg) was treated with the vapor of 30% PFPAA acetonitrile at -18°C for 2h. After aqueous pyridine vapor treatment , the product was subjected to CNBr cleavage. M stands for matrix line.

ACKNOWLEDGMENTS

This work was supported by the Ministry of Education, Science and Culture, a Grant-in-Aid for Specially Promoted Research.

REFERENCES

Amons, R., 1987, Vapor-phase modification of sulfhydryl group in proteins, *FEBS Lett.* 212:68-72.

Bailey, J.M., Tu, O., Issai, G., Ha, A., and Shively, J.E., 1994, C-terminal sequence analysis of polypeptides containg C-terminal protein, Protein Society Poster No.151-S

Biemann, K., 1987 (Pub. 1988), Contribution of mass spectrometry to peptide and protein structure, *Biomed. Environ. Mass Spectrom.* 16:99-111.

Boyd, R.K., and Beynon, J.H., 1977, Scanning of sector mass spectrometers to observe the fragmentation of metastable ions, *Org. Mass Spectrom.* 12:163-165.

Edman, P.and Begg, G., 1967, A protein sequenator, *Eur. J. Biochem.* 1:80-91.

Gaskell, S.J., Pike, A.W., and Millington, D.S., 1979, Selected metastable peak monitoring. A new, specific technique in quantitative gas chromatography- mass spectrometry, *Biomed. Mass Spectrom.* 6:78-81.

Nabuchi, Y., Yano, H., Kamo, M., Takamoto, K., Satake, K. & Tsugita, A. , 1994, C-terminal sequencing of peptdes and proteins by successive degradation with heptafluorobutyric anhydride vapor, *Chem. Lett.* 1994:757-760.

Naylor, S., Findeis, A.F., Gibson, B.W., and Williams, D.H., 1996, An approach toward the complete FAB analysis of enzymic digests of peptide and proteins, *J. Am. Chem. Soc.* 108:6359-6363.

Stark, G.R., 1968, Sequential degradation of peptides from their carboxy termini with ammonium thiocyanate and acetic anhydride, *Biochem.* 7:1796-1807.

Tsugita, A., Takamoto, K., Kamo, M., and Iwadate, H., 1992a, C-terminal sequencing of protein. A novel acid hydrolysis and analysis by mass spectrometry, *Eur. J. Biochem.* 206:691-696.

Tsugita, A., Takamoto, K., and Satake, K., 1992b, Reaction of pentafluoropropionic anhydride vapor on polypeptides as revealed by mass spectrometry. A carboxypeptidase mimetic degradation, *Chem. Lett.* 1992:235-238.

Tsugita, A., Takamoto, K., Iwadate, H., Kamo, M., and Satake, K. , 1992c, A novel protein C-terminal sequencing method using mass spectrometry, in *Biological Mass Spectrometry* (Matsuo, T., eds) pp. 242-243, San-ei pub., Kyoto.

Tsugita, A., Takamoto, K., Iwadate, H., Kamo, M., Yano, H., Miyatake, N., and Satake, K. , 1993, Development of novel C-terminal sequencing methods, in *Methods in Protein Sequence Analysis.* (Imahori, K., & Sakiyama, F., eds) pp.55-62, Plemnum Pub., New York.

Yamashita, S., 1971, Sequential degradation of polypeptides from the carboxyl ends. I. Specific cleavage of the carboxyl-end peptide bonds, *Biochem. Biophys. Acta.* 229:301-309.

10

SEQUENCING OF PROTEINS FROM THE C-TERMINUS

Victoria L. Boyd,[1] MeriLisa Bozzini,[1] Jindong Zhao,[1] Robert J. DeFranco,[1] Pau-Miau Yuan,[1] G. Marc Loudon,[2] and Duy Nguyen[2]

[1] Perkin-Elmer
Applied Biosystems Division
700 Lincoln Centre Drive, Foster City, California 94404
[2] Department of Medicinal Chemistry, School of Pharmacy
Purdue University
West Lafayette, Indiana 47907

INTRODUCTION

In 1992 we reported a new method of sequencing proteins from the carboxy-terminus (C-terminus) (Boyd *et al.*, 1992). In the past 2 years, we have continued our investigations, including the mechanism of the initial activation of the C-terminal carboxyl group and the deliberate modifications of the reactive side-chains of the amino acids with the sequencing reagents. Through the selection of the reagents and reaction conditions used for our sequencing protocol, aspartic acid, glutamic acid, serine and threonine are derivatized. The amidation of aspartic and glutamic acid, and the acetylation of serine and threonine, have led to improved yields in sequencing these residues. Aspartic and glutamic acid are now categorized, as seen in Table 1, as amino acid residues that are readily sequenced. Our criteria for determining whether a residue is reliably called is the ability to sequence through and detect that residue when it is present in one nanomole of a protein sample. On average, it is possible to sequence 5 cycles on one nanomole of protein applied to polyvinylidene difluoride (PVDF) membrane if the amino acid sequence contains those residues listed in the "reliably called" column of Table 1. Our focus for the 1994 MPSA conference is to illustrate the current utility of this C-terminal sequencing method in the sequencing of proteins immobilized onto PVDF.

SEQUENCING METHOD

Our chemical approach for sequencing proteins from the C-terminus first reported 2 years ago is presented in Scheme 1 (Boyd *et al.*, 1992). Similar to the Schlack and Kumpf

Methods in Protein Structure Analysis, Edited by M. Z. Atassi and E. Appella
Plenum Press, New York, 1995

Table 1. Alkylated thiohydantoin amino acids

Reliably called			In development	Stops sequencing
Alanine	Histidine	Glycine	Cysteine	Proline
Arginine	Isoleucine	Phenylalanine	Serine	
Asparagine	Leucine	Tryptophan	Threonine	
Aspartic acid	Lysine	Tyrosine		
Glutamic acid	Methionine	Valine		
Glutamine				

method (Schlack and Kumpf, 1926), the C-terminus is first derivatized into a thiohydantoin (TH). In the Schlack and Kumpf approach the amino acid-TH is cleaved and the truncated C-terminus is returned to a carboxylic acid. A unique feature of our sequencing method (Scheme 1) is that the C-terminal TH is alkylated. Alkylation results in an alkylated thiohydantoin (ATH) that is more readily cleaved from the C-terminus of the protein relative to the parent-TH. The ATH is cleaved by thiocyanate anion $\{NCS\}^-$ under acidic conditions. An important advantage of our method is that while $\{NCS\}^-$ cleaves the ATH, the amino acid residue adjacent to the ATH is simultaneously derivatized into a TH. The efficient and selective cleavage in addition to simultaneous derivatization of a new (n - 1) proteinyl-TH bypasses the need to return to a carboxylic acid at the C-terminus. The sequencing method shown in Scheme 1 has been successful in sequencing up to ten cycles on protein samples, noncovalently attached to PVDF. The data is presented herein.

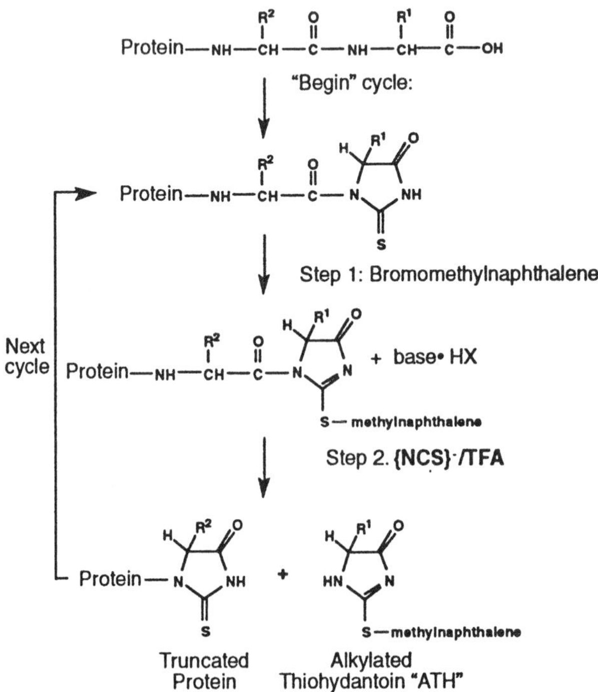

Scheme 1. The applied biosystems alkylation chemistry for automated c-terminal protein sequence analysis.

RESULTS AND DISCUSSION

Using nuclear magnetic resonance (NMR) spectroscopy, we observed that the activating reagents such as tetramethylchlorouronium chloride (Bozzini *et al.*, 1992 and Boyd *et al.*, 1992) and diphenylchlorophosphate (Guga *et al.,* 1993), under basic conditions using diisopropylethylamine (DIEA), converted the C-terminus entirely into a peptidyl-oxazolone. Using our protocol, the oxazolone is reacted with $\{NCS\}^-$ in a separate step under acidic conditions (TFA) to form a peptidyl-TH. Acidic conditions reportedly favor the cyclization into a TH. (Inglis, 1991.)

The NMR studies also revealed that the oxazolone, while under basic conditions will form an adduct with excess activating reagent. (Boyd *et al.,* manuscript in preparation) Additionally, basic conditions promote diketopiperazine formation at the C-terminus. Both of these side-reactions of the oxazolone retard or prevent proteinyl-TH formation. Using a weaker base, such as lutidine, and a less reactive activating reagent, such as acetic anhydride (Ac_2O), the oxazolone side-reactions are suppressed.

The carboxylic acid side-chains of aspartic and glutamic acid residues also react with Ac_2O, forming mixed-anhydrides. Scheme 2 portrays the formation of an oxazolone at the C-terminus, and the formation of a mixed-anhydride at a glutamic acid side-chain. The NH_3 formed from the dissociation of ammonium thiocyanate (NH_4SCN) was observed to react with the mixed anhydride, but not with the ionized oxazolone at the C-terminus while conditions were still basic. The resulting amidations of the aspartic and glutamic acid

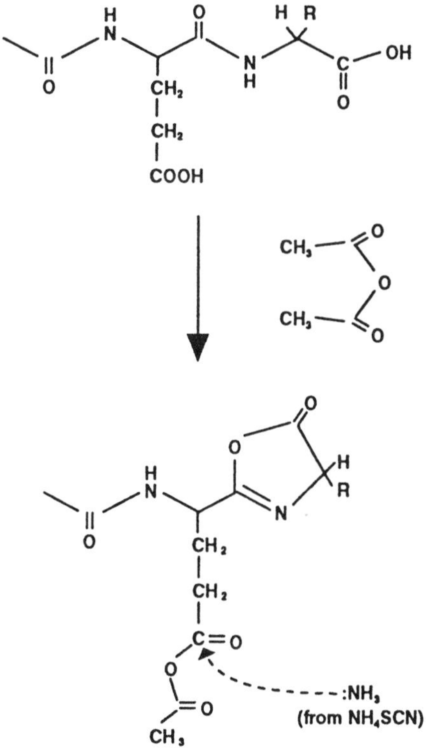

Scheme 2. Activation of side-chain carboxylic acid vs. C-terminus.

side-chains form asparagine and glutamine. If piperidine thiocyanate is used in place of NH_4SCN, aspartic and glutamic acid are converted to the piperidine amides. To avoid the possibility of the amidation at the C-terminus, tetrabutylammoniumthiocyanate (NBu_4NCS) is used as a source of $\{NCS\}^-$ for all steps in our sequencing protocol where proteinyl-TH is formed. Amidation of glutamic and aspartic acid residues is carried out preferably after proteinyl-TH formation.

Acetic anhydride will also acetylate the hydroxyl groups of serine and threonine, the phenol group of tyrosine, and the epsilon amine group of lysine. Except when serine or threonine are located at the C-terminus, the hydroxyl groups of serine and threonine interfere with the present sequencing method (Boyd *et al.*, 1992). Acetylation of the hydroxyl group prevents displacement of the alkylated sulfur atom of an adjacent ATH residue during sequencing. Therefore, "capping" the serine and threonine residues eliminates the interfering side-reaction. At present, DIEA is used with Ac_2O for the deliberate acetylation of the hydroxyl groups. Typically, a reduced yield is observed in cycles following a serine or threonine. The ATH derivative for serine and threonine, if detected, corresponds to the dehydro-analog. Whether the presence of dehydrated serine and threonine in a protein prior to sequencing interferes with the sequencing method has not yet been determined. A protein with multiple serine or threonine residues near the C-terminus remains difficult to sequence.

Acetylation of the epsilon amine group of lysine results in an ATH derivative that co-elutes with an artifact peak in our current chromatography system. At present, the lysine residues are derivatized into phenylureas prior to sequencing with phenylisocyanate (PIC). Reproducible HPLC peaks are observed for tyrosine and arginine residues when acetic anhydride is used for the initial activation. The independent synthesis of the ATH reference standards for acetylated tyrosine and arginine is in progress.

Figure 1 illustrates the aspartic acid residue in cycle 3 of enolase (....G-D-K-F) amidated to the piperidine amide during the initial cycle of our sequencing protocol. The ATH derivative for the amidated aspartic acid residue is clearly identified during sequencing. Our amidation procedure has been consistently successful on all proteins containing aspartic and glutamic acid sequenced to date.

Horse heart Cytochrome C (....L-K-K-A-T-N-E, Figure 2) has a C-terminal glutamic acid and a threonine in cycle 3. The glutamic acid is observed as both the piperidine amide and the free acid due to incomplete amidation. The hydroxyl group of threonine is acetylated during the initial cycle of sequencing. A C-terminal aspartic acid or glutamic acid residue can interfere with initial proteinyl-TH formation, and likely reduces the initial yield in this example (Stark, 1968). As described above, the ATH of threonine forms the dehydro-analog and often is not detected. However, in this example of Cytochrome C (2 nmol, PVDF) the amino acid sequence could be determined for 7 cycles.

RNase (....F-D-A-S-V, Figure 3) has both a serine and an aspartic acid residue. Prior to the recent improvements in our sequencing protocol, it was not possible for us to sequence this protein. After acetylation of the hydroxyl group of serine and amidation of the aspartic acid into a piperidine amide, the sequence could be determined for five cycles. The decline in yield in sequencing through a serine residue occurs as described above, but does not terminate the sequencing.

The final two protein sequencing examples are included to illustrate the improved sequencing performance relative to our publication 2 years ago.

One nanomole of beta-lactoglobulin (BLG) was applied to a PVDF membrane and was sequenced for 7 cycles. The fifth and sixth residues from the C-terminus of BLG are glutamic acid residues which are both detected as the piperidine amides in Figure 4. The ATH of histidine in cycle 2 is also clearly seen in this protein. The cystine in cycle 3 was

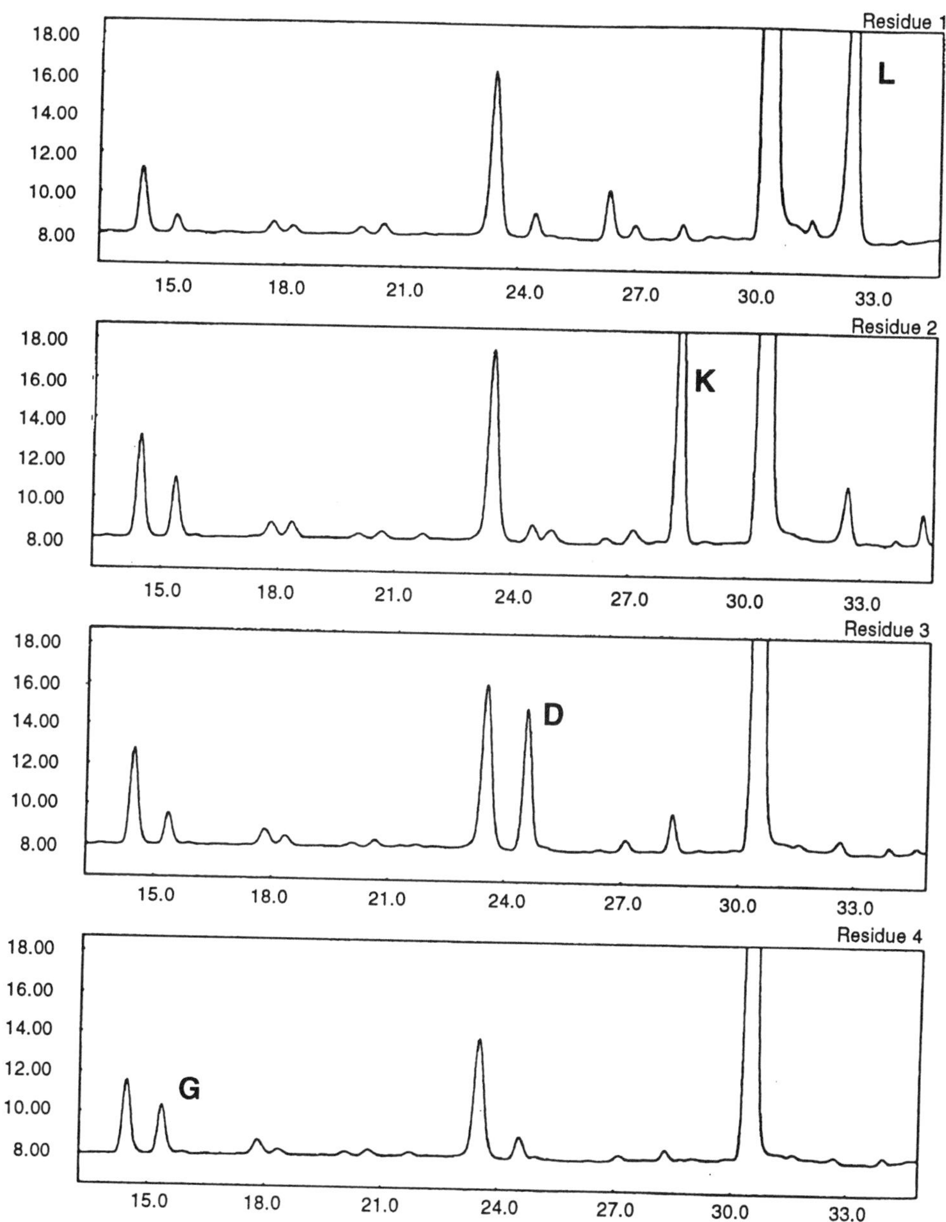

Figure 1. C-terminal sequence analysis of 2nmol of yeast enolase applied to PVDF. The lysine ATH residue in cycle 2 is the phenylurea derivative and the aspartic acid in cycle 3 is the piperidine amide derivative.

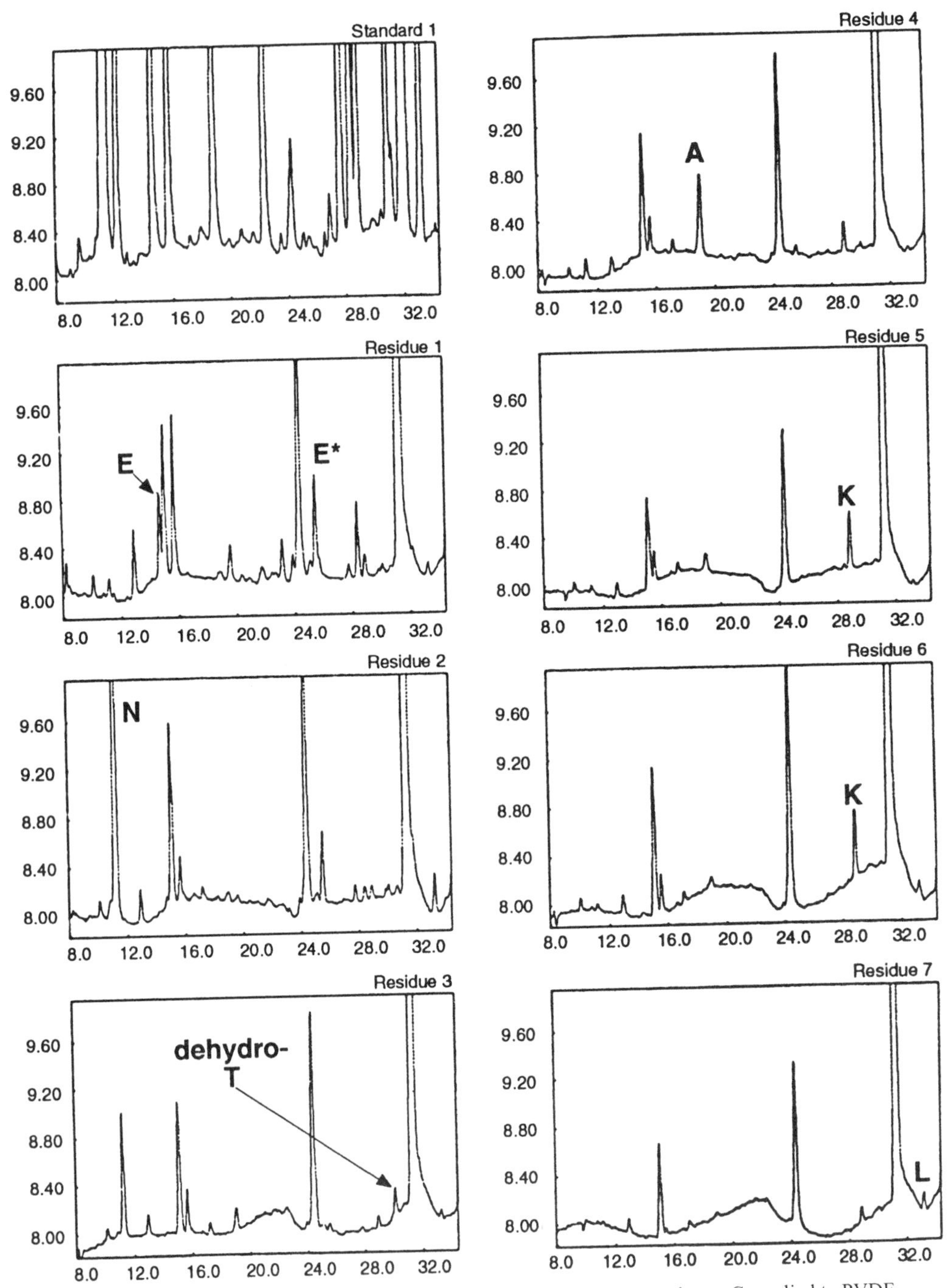

Figure 2. C-terminal sequence analysis of 2 nmol of Horse Heart CytochromeC, aaplied to PVDF.

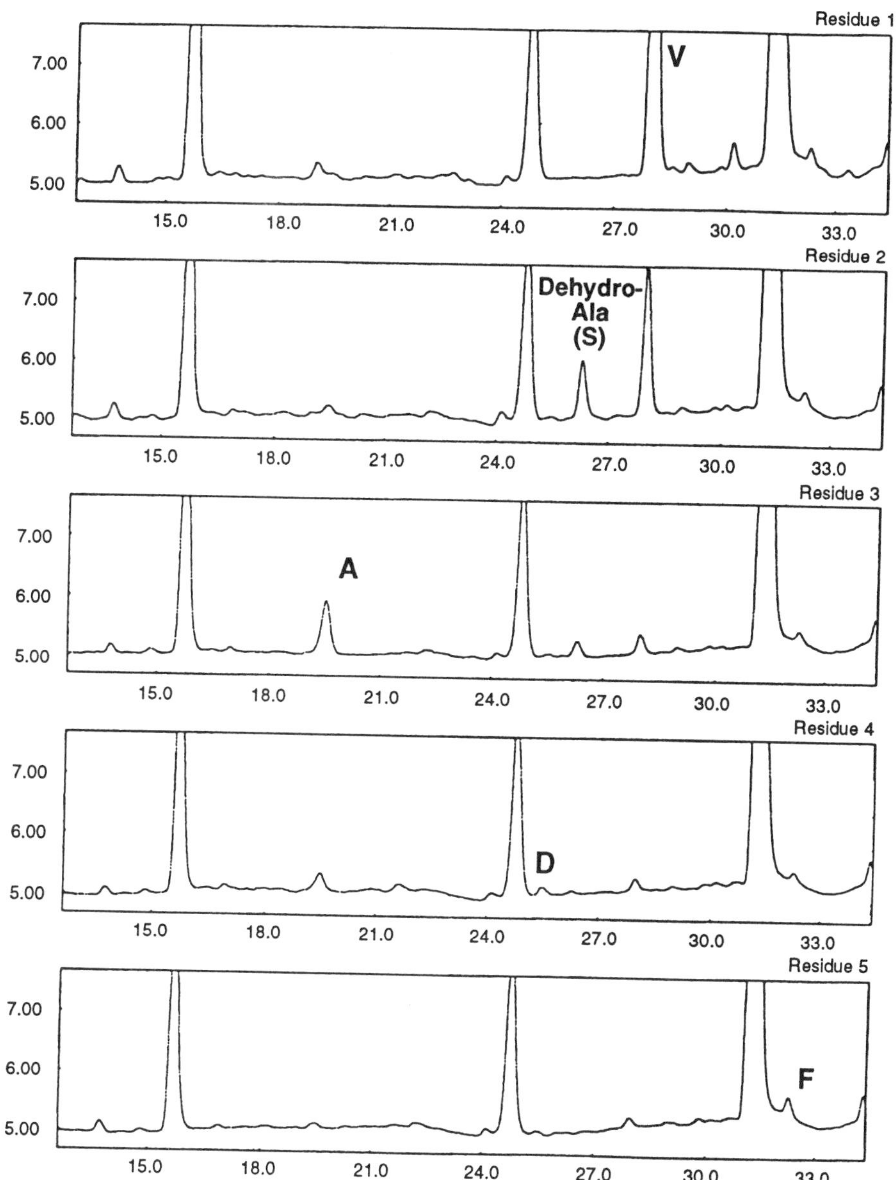

Figure 3. C-terminal sequence data for 1nmol of RNaseA. The serine ATH in cycle 2 is detected as the dehydro-analog.

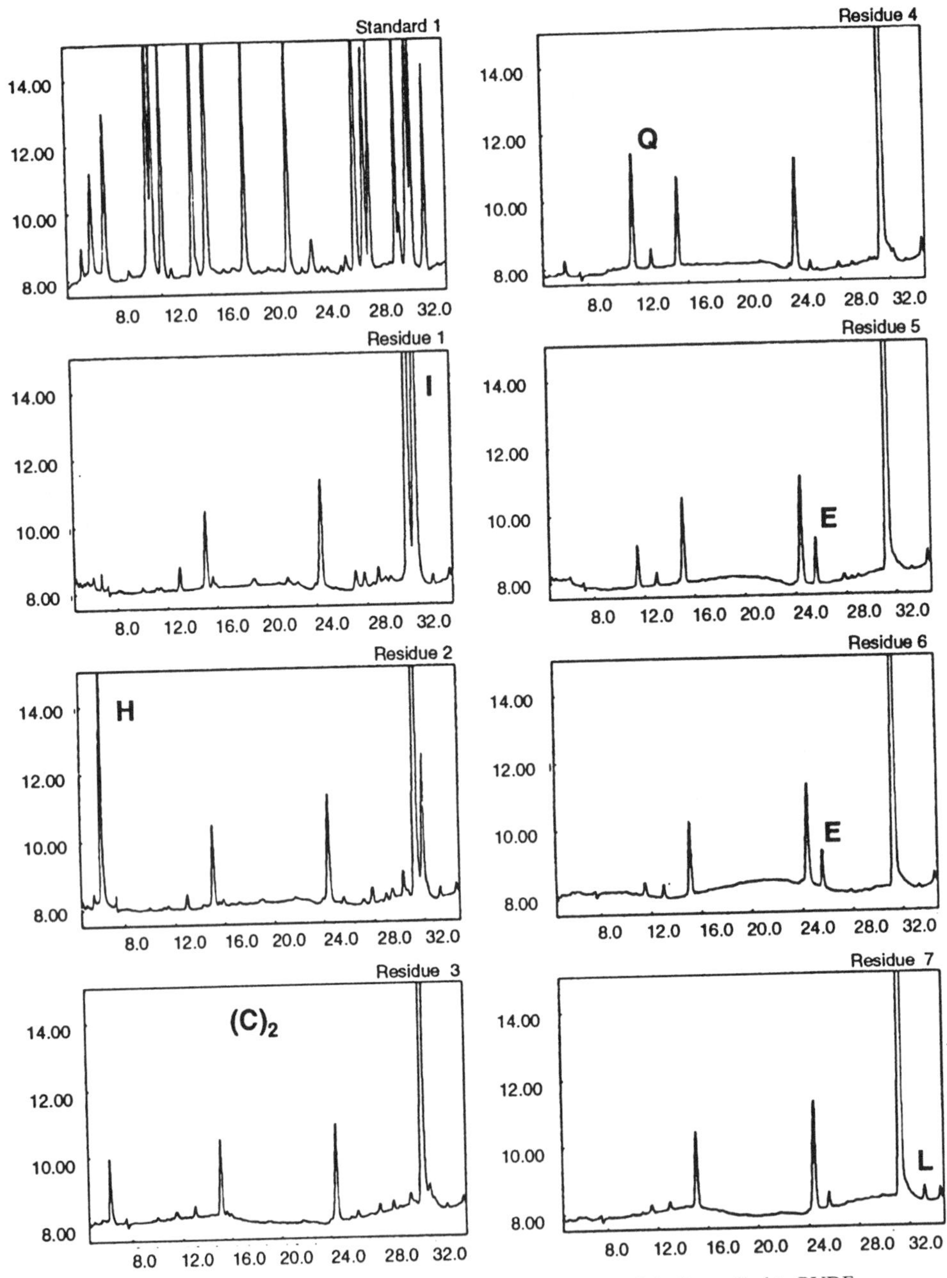

Figure 4. C-terminal sequence data of 1nmol of b-lactoglobulin applied to PVDF.

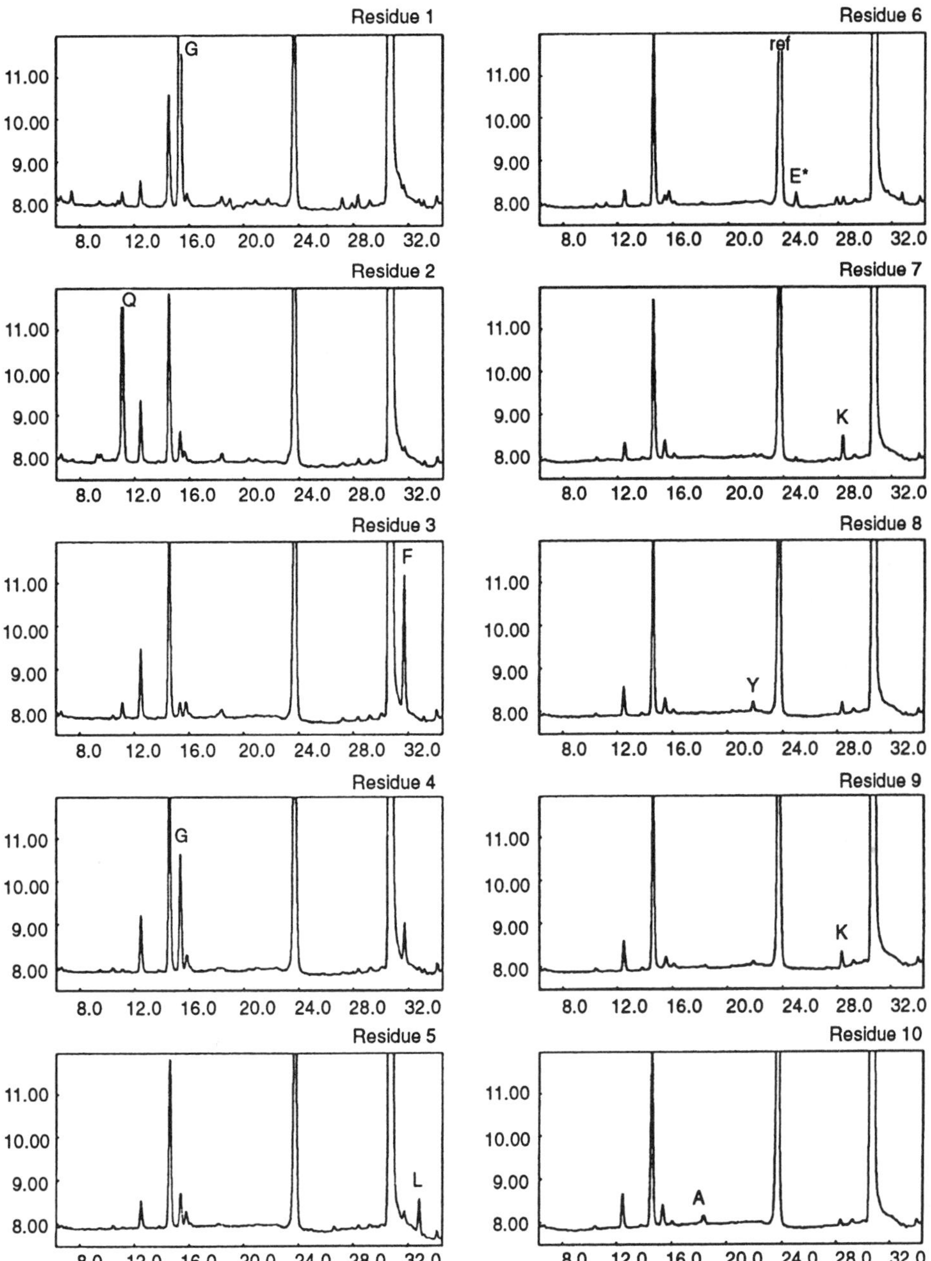

Figure 5. C-terminal sequence analysis of Horse Apomyoglobin, 500 pmol applied to PVDF.

not reduced prior to sequencing, so no signal was detected. When cysteine is present in a protein, HPLC peaks corresponding to both dehydroalanine and the S-alkylated ATH derivative have been detected (Guga *et al.,* 1993).

The C-terminal sequence data for Apomyoglobin (500 pmole) is presented in Figure 5. The amino acid sequence could be determined for ten cycles.

SUMMARY

At the seventh symposium of the Protein Society (July,1994), we demonstrated the utility of our sequencing method for the characterization of genetically engineered proteins including the sequencing of samples electroblotted onto PVDF (Bozzini *et al.,* 1994). In this article, we demonstrate the improvements in our ability to sequence amino acids with reactive side-chain groups, and in sequencing protein samples at higher sensitivity.

Acetic anhydride, a reagent with a long history of use for the activation of carboxyl groups, as well as for the acetylation of serine, threonine, lysine, and tyrosine, has been integrated into our sequencing protocol. The suitability of acetic anhydride for protein modifications including proteinyl-TH formation is well documented. Acetylation of the serine and threonine hydroxyl groups has enabled sequencing through these residues in some of the proteins we have sequenced to date. Acetylation of the epsilon amine group of lysine will make phenylisocyanate pretreatment unnecessary. The resolution of acetylated lysine ATH in the HPLC separation system is currently being optimized. The use of single reagent, acetic anhydride, for the activation of the carboxylic acid groups and for advantageous acetylation of reactive side-chain residues, reduces the number of reagents to which the protein is exposed.

The 500 pmol, 10 cycle sequencing example of apomyoglobin illustrates our progress in the development of our sequencing protocol. Our continued developement towards sequencing through all of the amino acids residues, and the enhancement in sensitivity, will expand the utility of this C-terminal sequencing method.

REFERENCES

Boyd, V. L., Bozzini, M., Zon, G., Noble, R. L., Mattaliano, R. J., 1992 Sequencing of Peptides and Proteins from the Carboxy Terminus, *Anal. Biochem.* 206:344.

Boyd, V. L., Bozzini, M., DeFranco, R. J., Guga, P. J., Yuan, P.-M., Loudon, G. M., Nguyen, D., Activation of the Carboxy Terminus of a Peptide, Manuscript in Preparation.

Bozzini, M., Boyd, V. L., Guga, P. J., Mattaliano, R. J., 1992 Alternative Chemistry Strategies for a new Protein C-terminal Sequencing Method, Presented at the Sixth Symposium of the Protein Society, San Diego CA.

Bozzini, M., Zhao, J., Yuan, P.-M., Boyd, V. L., 1994 Applications Using the Alkylation Method for Carboxy-Terminal Protein Sequencing, Presented at the Eighth Symposioum of the Protein Socieity, San Diego, CA, and accepted for publication in Techniques in Protein Chemistry VI.

Guga, P. J., Bozzini, M., DeFranco, R. J, Large, G. B., Boyd, V. L., C-Terminal Sequence Analysis of the Amino Acid Residues with Reactive Side-Chains: Ser, Thr, Glu, Asp, His, Lys, Presented at the Seventh Symposium Of the Protein Society, San Diego, CA.

Inglis, A., 1991 Chemical Procedures for C-Terminal Sequencing of Peptides and Proteins, *Anal. Biochem.* 195:183.

Schlack, P., and Kumpf, W., 1926 *Z. Physiol. Chem.* 154:125.

Stark, G. R., 1968 Sequenctial Degradation of Peptides from Their Carboxy Termini with Ammonium Thiocyanate and Acetic Anhydride *Biochem.* 7:1796.

11

C-TERMINAL PROTEIN SEQUENCE ANALYSIS USING THE HEWLETT-PACKARD G1009A C-TERMINAL PROTEIN SEQUENCING SYSTEM

Chad G. Miller, Jerome M. Bailey, David H. Hawke, Sherrell Early, and Jacqueline Tso

Protein Chemistry Systems
Hewlett-Packard Company
Palo Alto, California 94304

INTRODUCTION

An automated carboxy-terminal (C-terminal) protein sequencing technology developed by Hewlett-Packard enables the direct confirmation of the C-terminal sequence of native and expressed proteins, the detection and characterization of protein processing at the C-terminus, the identification of post-translational proteolytic cleavages, and partial sequence information on amino-terminally blocked protein samples. The approach offers sequence analysis through each of the 20 common amino acid residues including proline, which has historically been highly problematic. Additionally, the scope of typically analyzable protein samples spans a usefully broad molecular weight range and degree of structural complexity.

The automated sequencing chemistry of the HP G1009A C-terminal protein sequencing system utilizes diphenyl phosphoro-isothiocyanatidate (DPPITC) as the coupling reagent and trimethylsilanolate (KOTMS) as the cleavage reagent for the efficient generation of thiohydantoin-amino acid derivatives (TH-aa) (Bailey, J. M. et al, 1990, 1992)). The automated HPLC analyses of the sequencing cycles is accomplished using the HP 1090M liquid chromatograph with the HP specialty PTH analytical HPLC column (HP technical note, 1994).

MATERIALS AND METHODS

Protein samples were applied to precut Zitex membranes and inserted into C-terminal sequencer columns for sequence analysis using the HP G1009A C-terminal protein sequencing system (H-P, Palo Alto, CA). The system consists of the HP G1000A C-terminal protein

Methods in Protein Structure Analysis, Edited by M. Z. Atassi and E. Appella
Plenum Press, New York, 1995

sequencer, HP 1090M liquid chromatograph, HP Vectra 486/66 computer with Microsoft® MS-DOS® and Windows™ environment, and HP specialty C-terminal sequencing reagents, solvents, HPLC columns and solvents, and all related consumables.

Protein samples were obtained from Sigma Chemical Co. (St.Louis,MO). Thiohydantoin-amino acid derivatives were prepared for and quality-controlled by HP (3). The C-terminal chemical sequencing method was developed for automation using a chemistry licensed to HP from the City of Hope, Duarte, CA.

RESULTS AND DISCUSSION

Sequencing Criteria

The principal requirement for any chemical sequencing methodology is to enable efficient, rapid, and reproducible reactions that can be applied to all 20 common amino acid residues. This applies in particular to proline which has historically challenged all sequencing degradative reaction schemes resulting in no identifiable derivative and preventing sequencing chemistry. The chemical cleavage of the cyclized peptidylthiohydantoin must adequately yield the thiohydantoin-amino acid derivative as well as the shortened polypeptide, chemically suitable for the succeeding cycle of sequencing chemistry. The criteria for reliable and reproducible analyses are, in part, satisfied by the use of an inert reaction support that readily immobilizes the sample without any intervening covalent attachment protocols or pre-sequencing procedures that impart irreproducible and unpredicatable sample losses and variable yields. This sequencing methodology utilizes a Zitex membrane (a porous Teflon membrane) as a non-covalent reaction support. Protein samples are applied directly to the reaction membrane and are adsorptively immobilized for the chemical sequencing. Additionally, the protein sequence analysis relies on the implementation of a stable and reproducible chromatographic method for the analysis of the thiohydantoin-amino acid derivatives.

Sample Application on Zitex Membranes

The protein samples for C-terminal sequencing are conveniently applied directly to Zitex reaction membranes (1mm x 12mm) that have been pre-treated with alcohol (isopropanol). The liquid sample solutions, on occasion made homogeneous by the addition of a small volume (1-5ul) of dilute aqueous trifluoroacetic acid, are applied to the wetted Zitex membrane in 5-20ul volumes and allowed to dry either at room temperature (10-20 minutes) or at a controlled elevated temperature.

The dry membrane is inserted into a C-terminal sequencer column (inert Kel-F columns fitted with inert endfrits) and installed in any one of the four sample positions on the HP G1009A sequencer. The sequencer column reactions that occur on the Zitex membrane include the chemical coupling and cyclization of the C-terminal residue and the cleavage and extraction that releases the thiohydantoin-amino acid derivative. The thiohydantoin-amino acids are extracted off the Zitex membrane from the sequencer column to the sequencer transfer flask for preparation for HPLC injection.

The sample application is compatible with diverse samples recovered in various buffers (phosphate, inorganic salts) and solvents (HPLC fractions). Samples that have been subjected to amino-terminal sequence analysis using the HP G1005A N-terminal protein sequencing system and Zitex as a reaction support may be transferred to C-terminal sequencer columns and subjected to C-terminal sequence analysis with the HP G1009A C-terminal sequencing system.

C-Terminal Coupling and Cyclization Reactions

Diphenyl phosphoroisothiocyanatidate (DPPITC) in the presence of pyridine constitutes the new chemical coupling and cyclization reactions for the HP G1009A automated C-terminal sequence analysis (1). Prerequisite to the coupling reaction with DPPITC is the base activation of the protein C-terminal carboxylic acid moiety to a carboxylate species. The membrane adsorbed protein sample is treated with a suitable base such as diisopropylethylamine or trimethylsilanolate. The carboxylate salt of the C-terminal amino acid residue is highly reactive to the diphenyl phophoro-isothiocyanatidate coupling reagent, speculatively generating a reactive pentavalent species which collapses to the C-terminal acylisothiocyanate. The Zitex membrane is washed with organic solvent (acetonitrile) to eliminate the excess DPPITC.

The coupled peptidylacylisothiocyanate is treated with pyridine to induce the five-membered ring closure to the carboxy-terminal peptidylthiohydantoin. The reaction of pyridine has been found to efficiently promote the cyclization of the acylisothiocyanate to the acylthiohydantoin product. The effectiveness of pyridine in this reaction can be assigned to its nucleophilic properties in addition to its basicity. The coupled and cyclized membrane-bound peptidylthiohydantoin is washed with organic solvent to remove the excess pyridine and resultant reaction by-products.

Additional treatment of the peptidylacylisothiocyanate with liquid anhydrous trifluoroacetic acid enables the cyclization of the C-terminal prolylisothiocyanate to the corresponding prolylthiohydantoin (Bailey, J. M., et al 1995). This species is readily cleaved by treatment with 2% aqueous trifluoroacetic acid vapor (and trimethylsilanolate treatment) to yield the thiohydantoin-proline derivative. A methanolic extraction of the thiohydantoin-proline residue from the reaction column to the transfer flask is subsequently followed by the column cleavage reaction and solvent extraction to conclude the essentials of the chemical degradative cycle.

Cleavage Reaction of the Peptidylthiohydantoin

The peptidylthiohydantoin (coupled and cyclized product) is subjected to chemical cleavage to the C-terminal thiohydantoin-amino acid residue and the shortened polypeptide using an alkali salt of trimethylsilanolate (KOTMS). The cleavage reagent is a highly reactive nucleophile displacing the thiohydantoin derivative from the C-terminal acylthiohydantoin moiety. The resulting trimethylsilyl ester is rapidly cleaved to a free C-terminal carboxylate species amenable for the next cycle of chemical coupling with DPP-ITC. The released thiohydantoin-amino acid is extracted off the Zitex membrane to the transfer flask as the cleavage solution and subsequent organic solvent (acetonitrile). The extraction solvents are evaporated and the thiohydantoin-amino acid is redissolved in the HPLC transfer solvent (dilute aqueous trifluoroacetic acid) and injected into the HP 1090M HPLC for analysis.

HPLC Analysis of Thiohydantoin-Amino Acid Derivatives

The HP G1009A C-terminal protein sequencing system provides automated HPLC analysis of sequencer cycles using the HP 1090M liquid chromatograph with filter photometric detection at 269nm and the HP specialty (2.1mm x 25cm) reversed-phase PTH analytical HPLC column (3). A 39-minute binary gradient (Solvent A: phosphate buffer, pH 2.9; Solvent B: acetontrile/water) utilizes an ion-pairing reagent (alkyl sulfonate) enabling highly reproducible elution times and peak resolution. A stable thiohydantoin-amino acid standard

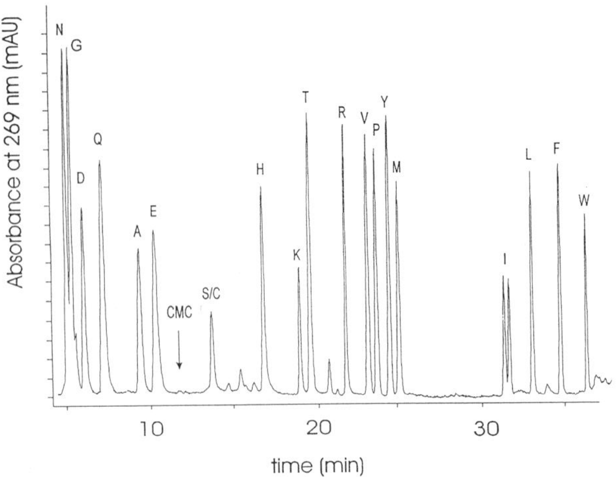

Figure 1. Thiohydantoin-amino acid standard.

mixture is incorporated on the sequencer for on-line automated peak calibration and quantitation.

The thiohydantoin-amino acid standard mixture (TH-Std) consists of the synthetic thiohydantoin derivatives corresponding to the actual sequencing products resulting from chemical sequence analysis. In particular, the serine, threonine, cysteine, and lysine thiohydantoin derivative standards correspond to their respective sequencing degradation products. The sequencing product derivatives of serine and cysteine yield the same degradation species and, without cysteine side chain modification, permit the identification of either residue for confirmatory analysis of a known sequence. The residue assignment of cysteine for unknown sequences requires the prior chemical modification of cysteine (an S-alkylation) as is routinely done with amino-terminal sequencing methods.

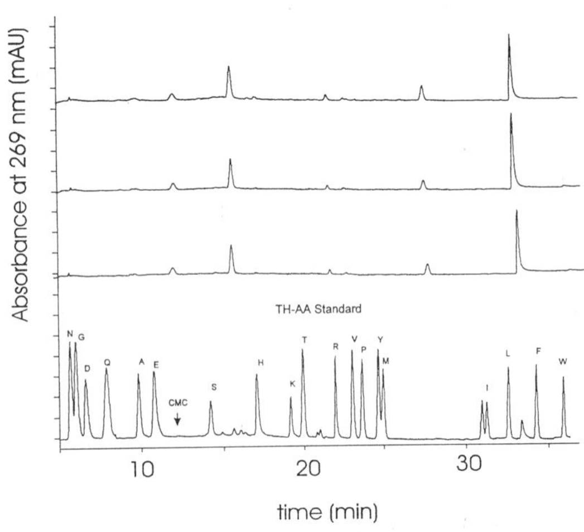

Figure 2. Chemical background of 3 blank cycles.

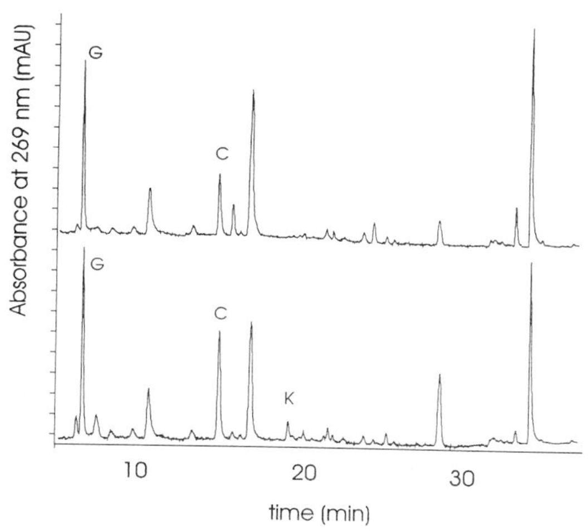

Figure 3. Cycle-1 of mouse immunoglobulin G samples.

The thiohydantoin-amino acid standard HPLC chromatogram (Figure 1) shows the elution times for each of the 20 common amino acid derivatives including thiohydantoin-Pro (P) and the common peak designated S/C, identifying Ser and Cys residues. The relative retention time for the S-carboxymethyl derivative of cysteine is indicated by the arrow. The peak identified for Lys (K) corresponds to the free-epsilon amino derivative of thiohydantoin-lysine. Thiohydantoin-Ile (I) chromatographically resolves into two components representing the structural isoforms for the amino acid residue; the Ile peaks do not coelute with any other thiohydantoin-amino acid derivative or chemical background. The thiohydantoin standard chromatogram shown represents a 50 pmol mixture of the standard thiohydantoin derivatives except for the Ser, Cys, Thr, and Lys derivatives that represent approximately 100 pmol amounts.

C-Terminal Sequence Analysis of Diverse Samples

The chemical sequencing background generated on the Zitex reaction membrane, resulting from the reaction by-products of the coupling reagent (DPPITC) appears chromatographically as three principal and two minor UV absorbing components (Figure 2). These background related species do not coelute with any of the 20 common thiohydantoin-amino acid derivatives and thus do not interfere with sequencer cycle residue assignments.

The recoveries of first-cycle residues typically result in sequencing initial yields ranging from 10%-50% of the total amount of sample applied to the Zitex membrane. As observed for amino-terminal sequencing, there is a sample dependency (and residue dependency) that contributes to the initial thiohydantoin-amino acid recoveries.

Mouse immunoglobulin G (150kDa) was applied as a phosphate buffer solution directly on a Zitex reaction membrane and subjected to C-terminal sequence analysis (Figure 3). The C-terminal cycle identified the extent of protein processing of the C-terminal heavy chains by the detection and quantitation of the heavy chain Lys (K, expected full-length sequence C-terminal residue) and Gly (G) residues. The C-terminal residue of the light chain was identified as the expected Cys (C) residue. The arrows indicate the sequencing chemical background. The results of sequence analysis of a 900 pmol and 450 pmol sample of the

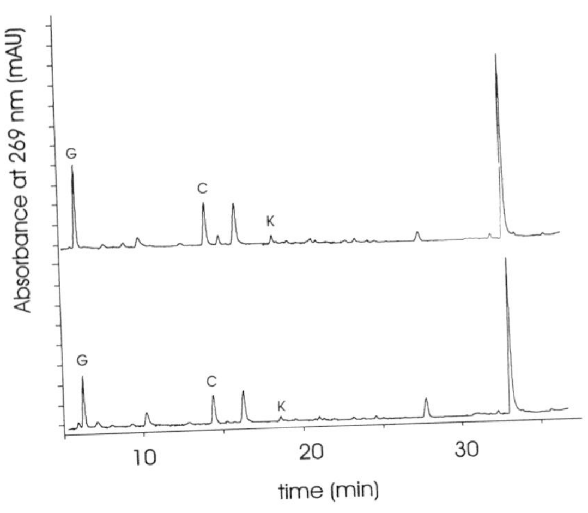

Figure 4. Cycle-1 of mouse immunoglobulin G samples (top: 900 pmols; bottom: 450 pmols).

mouse IgG show the linearity in signal response and the unambiguous residue assignments at each of these sample amounts (Figure 4).

Superoxide dismutase, an N-terminally blocked protein, was applied to a Zitex reaction membrane in the amount of approximately 1 nmol in 10ul of 1% aqueous trifluoroacetic acid. The first three cycles of C-terminal sequence analysis results in the identification of Lys (K) cycle-1, Ala (A) cycle-2, and Ile (I) cycle-3 (Figure 5). The chemical background remains relatively stable as a thiohydantoin background increases in part attributed to internal cleavages as analogously observed for amino-terminal sequencing chemistry.

The results of C-terminal sequence analysis of a chromatographically isolated hemoglobin B chain is shown for the first three sequencer cycles (Figure 6). Approximately 1 nmole of protein sample, recovered from reversed-phase HPLC separation of the hemoglobin A and B chains, was directly applied to a Zitex reaction membrane and immediately

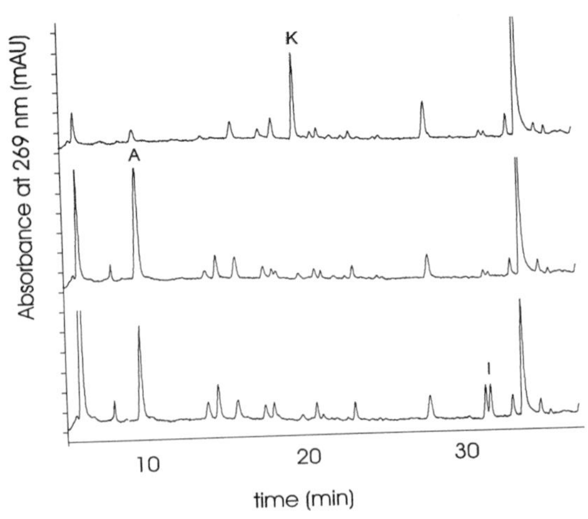

Figure 5. Cycles of superoxide dismutase.

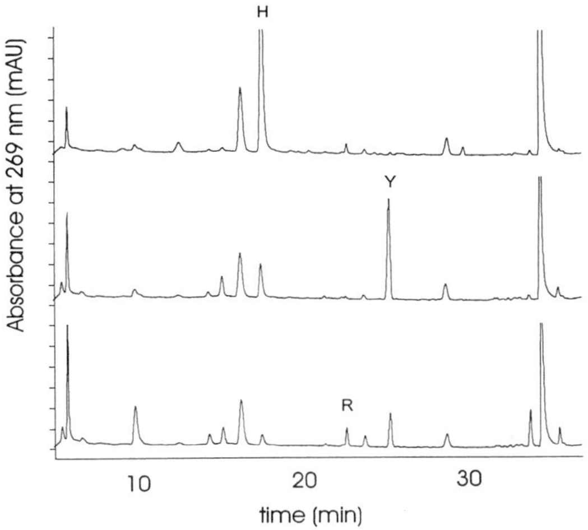

Figure 6. Cycles 1-3 of hemoglobin B chain.

subjected to C-terminal sequence analysis. The results show the unambiguous identification of the first two cycles, His-1 and Tyr-2, and the confirmatory identification of the third cycle residue, Arg-3. Sequence analysis of the hemoglobin A chain, isolated chromatographically, resulted in unambiguous residue assignments of Arg (R) cycle-1 and Tyr (Y) at cycle-2 (Figure 7).

The results of C-terminal sequence analysis on a series of proline-containing protein samples shows the identification of thiohydantoin-Pro (P) at each cycle-1 of the analyses (Figure 8) confirming the expected full-length sequences. The model polypeptide, polyproline, was applied to a Zitex reaction membrane and directly sequenced as were the additional two protein systems shown. Ovalbumin and apo-transferrin were subjected to sequence analysis and resulted in the recovery of thiohydantoin-Pro as first cycle sequencer residues. The sequencing chemistry invokes the use of neat trifluoroacetic acid (as described here and elsewhere, cf ref.3) to generate the cleavable thiohydantoin-Pro derivative which

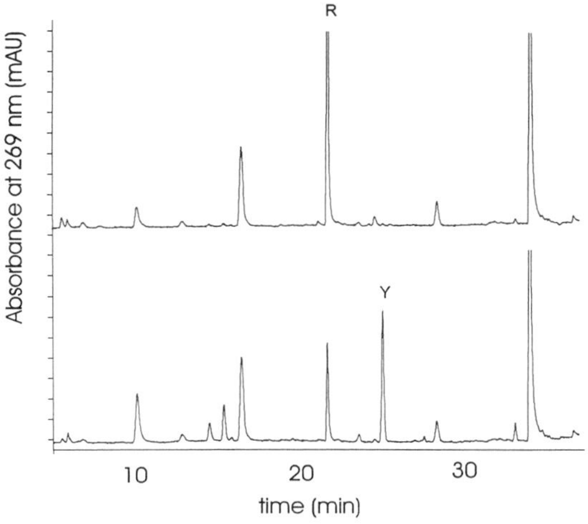

Figure 7. Cycles 1 and 2 of hemoglobin A chain.

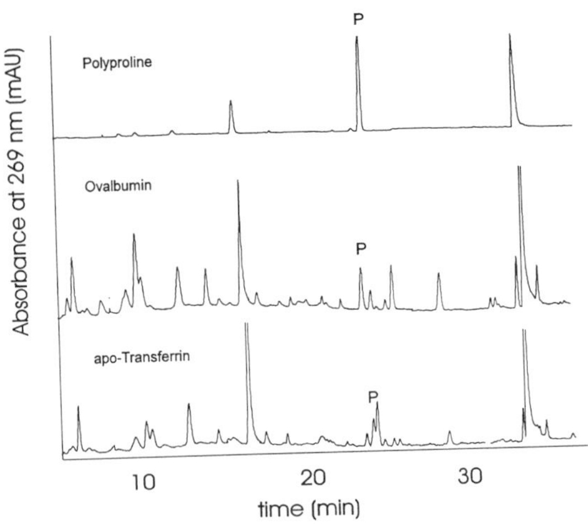

Figure 8. Cycle-1 of proline-containing proteins.

is released upon treatment with aqueous acidic vapor and the use of trimethylsilanolate. The neat trifluoroacetic acid induced cyclized prolylthiohydantoin is cleaved with aqueous trifluoroacetic acid vapor, the thiohydantoin-Pro derivative is extracted with methanol to the transfer flask, and the reaction membrane treated with trimethylsilanolate (as part of the routine cleavage steps of the chemistry cycle) that cleaves any residual thiohydantoin-Pro from the cyclized peptidylprolylthiohydantoin species.

The results of C-terminal sequence analysis on a 1 nmol sample of bovine beta-lactoglobulin A are shown in Figure 9. The first three cycles of analysis show the identification of cycle-1 Ile (I), cycle-2 His (H), and cycle-3 Cys (C) confirming the expected full-length protein sequence.

The results of C-terminal sequence analysis on a variety of low molecular weight protein samples are shown as cycle-1 chromatograms (Figure 10). Bovine insulin A and B chains (combined mol wt 5.8 kDa) yielded Asn and Ala respectively

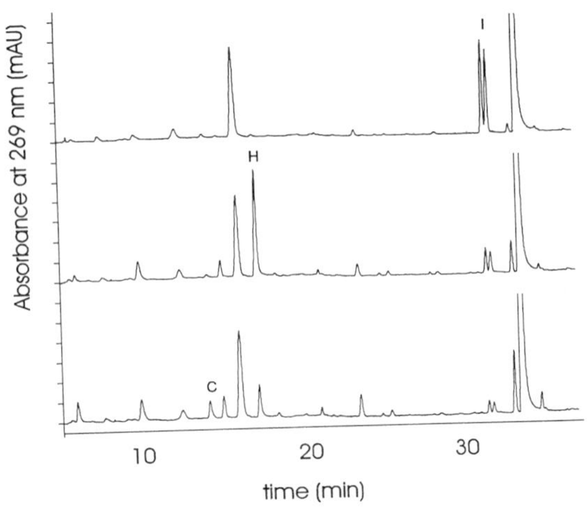

Figure 9. Cycles 1-3 of bovine beta-lactoglobulin A.

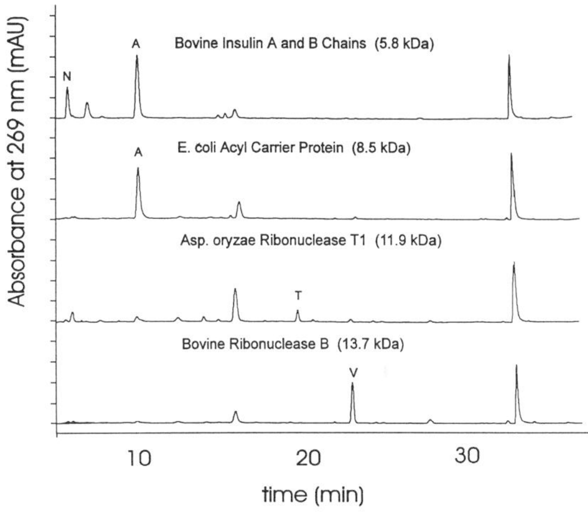

Figure 10. Cycle-1 of low molecular weight proteins.

and the 13.7 kDa protein, bovine ribonuclease B resulted in the unambiguous identification of Val at cycle-1.

Sequence analysis of an approximate 1 nmol sample of human serum albumin (68kDa) resulted in the identification of cycles 1-3 and are assigned as Leu (L) cycle-1, Gly (G) cycle-2, and Leu (L) cycle-3 (Figure 11).

A 1 nmol sample of human serum albumin was applied to a Zitex reaction membrane, inserted into an N-terminal sequencer membrane-compatible column, and installed in the HP G1005A N-terminal protein sequencer (Miller, C. G., 1995). The sample was subjected to five cycles of automated N-terminal sequence analysis (cycles-1 Asp,D and -2 Ala, A are shown) and the Zitex reaction membrane transferred to the HP G1009A C-terminal protein sequencing system (Figure 12). The first two cycles of automated C-terminal sequence analysis of the previously N-terminal sequenced sample resulted in the unambiguous

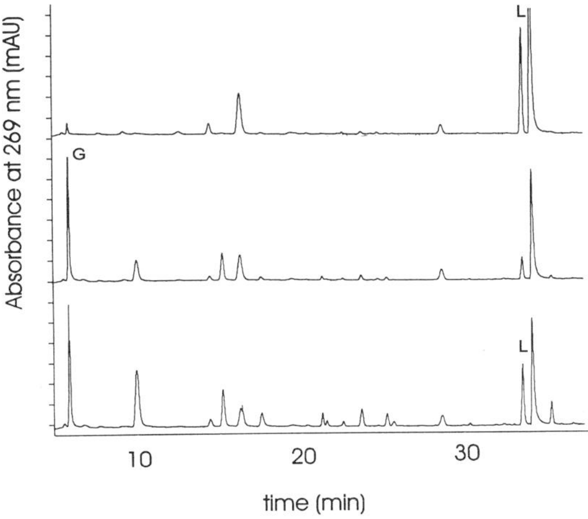

Figure 11. Cycle 1-3 of human serum albumin.

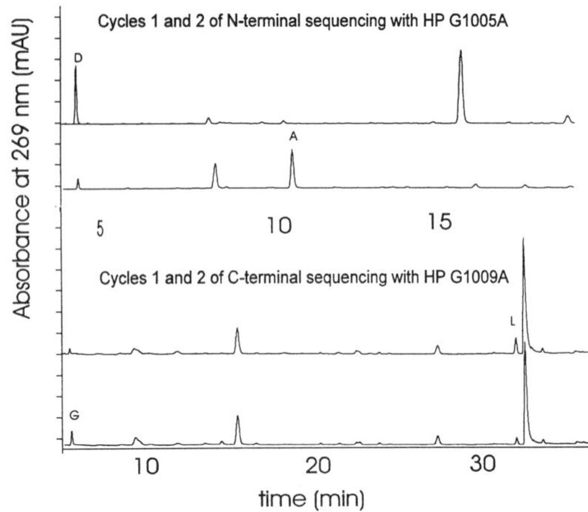

Figure 12. Human serum albumin—1 nmol on Zitex. (5 cycles N-terminal and 2 cycles C-terminal.)

C-terminal sequence residue assignments of Leu (L) at cycle-1 and Gly (G) at cycle-2. This example of the integration of amino-terminal and carboxy-terminal sequence analysis on a single sample should become an invaluable procedure for the sequence determination and structural identification of protein samples.

CONCLUSIONS

The HP G1009A C-terminal sequencing system automates an efficient, reliable, and reproducible carboxy-terminal sequencing chemistry based on the introduction of diphenyl phosphoro-isothiocyanatidate as the coupling reagent, pyridine as a cyclization reagent, and trimethylsilanolate as the cleavage reagent. The strategic incorporation of trifluoroacetic acid into an extended cyclization scheme enables the sequence analysis through all of the 20 common amino acids, including proline. The sequencing methodology does not require any pre-sequencing modifications to protect side chain residues or covalent attachment protocols to retain protein samples to the reaction support. The use of Zitex as a reaction membrane enables the adsorptive immobilization of protein samples by a facile direct sample application to the alcohol-treated membrane. A robust HPLC method for the identification of the thiohydantoin-amino acid derivatives, in addition to an on-line thiohydantoin-amino acid standard mixture, enables the reliable detection and identification of sequencer cycle residues. The sequencing system provides the flexible platform on which further developments and refinements to the chemical methodology can be accomplished. It is anticipated that the exceedingly high chemical reaction efficiencies obtained for amino-terminal sequence analysis (HP G1005A) will be approached by continued advancements in carboxy-terminal sequence analysis. The immediate needs and requirements for the identification of C-terminal sequence and the detection and quantitation of protein-processing at the C-terminus of native and expressed proteins and protein products are achievable using the current HP G1009A sequencing technology.

REFERENCES

Bailey, J. M., Nikfarjam, F., Shenoy, N. S., and Shively, J. E., 1992 Automated carboxy-terminal sequence analysis of peptides and proteins using diphenyl phophoroisothiocyanatidate, *Protein Science* 1: 1622.

HP G1009A C-Terminal Protein Sequencing System technical note (1994) **TN 94-5**.

Bailey, J. M. and Shively, J. E., 1990 Carboxy-terminal sequencing: formation and hydrolysis of C-terminal peptidylthiohydantoins, *Biochemistry* 29: 3145.

Bailey, J. M., Tu, O., Issai, G., and Shively, J. E., 1995 C-terminal sequence analysis of polypeptides containing C-terminal proline, *in*: Techniques in Protein Chemisty **VI**, Crabb, J.W. ed., Academic Press, San Diego, in press.

Miller, C.G., 1995 Adsorptive biphasic column technology for protein sequence analysis and protein chemical modification, *in*: Methods: A Companion to Methods in Enzymology, Shively, J.E. ed., Academic Press, San Diego, in press.

12

AUTOMATED C-TERMINAL SEQUENCING OF POLYPEPTIDES CONTAINING C-TERMINAL PROLINE

Jerome M. Bailey, Oanh Tu, Gilbert Issai, and John E. Shively

Beckman Research Institute of the City of Hope
Division of Immunology
Duarte, California 91010

INTRODUCTION

There has been much interest in the development of a chemical method for the sequential C-terminal sequence analysis of proteins and peptides. Such a method would be analogous and complimentary to the Edman degradation commonly used for N-terminal sequence analysis (Edman, 1950). It would also be invaluable for the sequence analysis of proteins with naturally occurring N-terminal blocking groups, for the detection of post-translational processing at the carboxy-terminus of expressed gene products, and for assistance in the design of oligonucleotide probes for gene cloning. Although a number of methods have been described, the method known as the "thiocyanate method" (Schlack and Kumpf, 1926), has been the most widely studied and appears to offer the most promise due to its similarity to current methods of N-terminal sequence analysis.

The thiocyanate method involves the reaction of a protein with an isothiocyanate reagent, in the presence of a carboxylic acid activating reagent, to form a C-terminal peptidylthiohydantoin. The C-terminal amino acid, derivatized as a thiohydantoin, is then specifically removed to yield a shortened peptide or protein and a thiohydantoin amino acid.

Many of the problems associated with the thiocyanate chemistry which have prevented its routine use in the protein chemistry lab have been addressed over the last few years. The use of sodium or potassium trimethylsilanolate for the cleavage reaction provided a method for rapid and specific hydrolysis of the derivatized C-terminal amino acid which left the shortened peptide with a free C-terminal carboxylate ready for continued rounds of sequencing (Bailey et al., 1992a). The use of diphenylphosphoroisothiocyanatidate (DPP-ITC) and pyridine combined the activation and derivatization steps and permitted the quantitative conversion of 19 of the twenty common amino acids (the exception being proline) to a thiohydantoin derivative. These improvements permitted application of the C-terminal chemistry to a wide variety of protein samples with cycle times similar to those employed for N-terminal sequence analysis (Bailey et al., 1992b). The introduction of Zitex (porous Teflon) as a support for protein sequencing permitted the C-terminal sequence

Methods in Protein Structure Analysis, Edited by M. Z. Atassi and E. Appella
Plenum Press, New York, 1995

analysis of protein samples which were non-covalently applied to the sequencing support (Bailey et al., 1992b; Bailey et al., 1993).

The inability of C-terminal proline to be derivatized to a thiohydantoin has been a major impediment to the development of a routine method for the C-terminal sequence analysis of proteins and peptides. Since the method was first described in 1926 (Schlack and Kumpf, 1926), the derivatization of C-terminal proline has been problematic. While over the years a few investigators have reported the derivatization of proline, either with the free amino acid or on a peptide, to a thiohydantoin (Kubo et al., 1971; Yamashita and Ishikawa, 1971; Inglis et al., 1989), others have been unable to obtain any experimental evidence for the formation of a thiohydantoin derivative of proline (Turner and Schmerzler, 1954; Fox et al., 1955; Stark, 1968; Bailey and Shively; 1990). Recently, utilizing a procedure similar to that described by Kubo et al. (1971), Inglis et al. (1993) have described the successful synthesis of thiohydantoin proline from N-acetylproline. This was done by the one-pot reaction of acetic anhydride, acetic acid, trifluoroacetic acid, and ammonium thiocyanate with N-acetyl proline. We have reproduced this synthesis and further developed it to a large scale synthesis of TH-Proline.

We also describe the development of chemistry based on the DPP-ITC/pyridine reaction which permits the efficient derivatization and hydrolysis of peptidyl C-terminal proline to a thiohydantoin and discuss the integration of this chemistry into an automated method for the C-terminal sequence analysis of polypeptides containing C-terminal proline.

MATERIALS AND METHODS

Materials. Diphenyl chlorophosphate, acetic anhydride, trimethylsilylisothiocyanate, anhydrous dimethylformamide (DMF), anhydrous acetonitrile, and anhydrous pyridine were from Aldrich. Water was purified on a Millipore Milli-Q system. Sodium trimethylsilanolate was obtained from Fluka. Diphenyl phosphoroisothiocyanatidate was synthesized as described (Kenner et al., 1953). All of the peptides used in this study were either obtained from Bachem or Sigma. N-Acetyl proline was from Sigma. Diisopropylethylamine (sequenal grade), trifluoroacetic acid (sequenal grade), and 1,3-dicyclohexylcarbodiimide (DCC) were obtained from Pierce. The carboxylic acid modified polyethylene membranes were from the Pall Corporation (Long Island, NY). Zitex G-110 was from Norton Performance Plastics (Wayne, NJ). The amino acid thiohydantoins used in this study were synthesized as described (Bailey and Shively, 1990). The Reliasil HPLC columns used in this study were obtained from Column Engineering (Ontario, CA).

Synthesis of Thiohydantoin Proline. Acetic anhydride (100 ml), acetic acid (20 ml), and trifluoroacetic acid (10 ml) were added to N-acetylproline (500 mg). The mixture was stirred until dissolved. Trimethylsilylisothiocyanate (3 ml) was added and mixture stirred at 60°C for 90 min. The reaction was dried to a powder by rotary evaporation and water (50 ml) added. This solution was again dried by rotary evaporation and water (20 ml) added. A white powder formed. The solution was kept on ice for approximately 30 min. The powder (thiohydantoin proline) was collected by vacuum filtration. The yield was approximately 40%. The product was characterized by UV, FAB/MS, and NMR. The UV absorption spectrum had a λ_{max} of 271 nm in methanol. FAB/MS gave the expected $MH^+ = 157$. NMR δ 4.32(H_α, m), 3.85 (H_δ, m), 3.43 (H_δ, m), 2.20 (H_γ and H_β, m), 1.70 (H_β, m).

Covalent Coupling of Peptides to Carboxylic Acid Modified Polyethylene. Peptides were covalently coupled to carboxylic acid modified polyethylene and quantitated as described (Shenoy et al., 1992).

Application of Protein Samples to Zitex. The Zitex support (2 x 10 mm) was pre-wet with isopropanol and protein samples (2-5 μl) dissolved in water were applied. The samples were allowed to dry before sequencing.

HPLC Separation of the Amino Acid Thiohydantoins. Reverse phase HPLC separation of the thiohydantoin amino acid derivatives was performed on a C-18 (3μ, 100 Å) Reliasil column (2.0 mm x 250 mm) on a Beckman 126 Pump Module with a Shimadzu (SPD-6A) detector. The column was eluted for 2 min with solvent A (0.1% trifluoroacetic acid in water) and then followed by a discontinuous gradient to solvent B (10% methanol, 10% water, 80% acetonitrile) at a flow rate of 0.15 ml/min at 35°C. The gradient used was as follows: 0% B for 2 min, 0-4% B over 3 min, 4-35% B over 35 min, 35-45% B for 5 min, and 45-0% B over 3 min. Absorbance was monitored at 265 nm.

Automation of the C-Terminal Sequencing Chemistry. The instrument used for automation of the chemistry described in this manuscript has been described previously (Bailey et al., 1993).

RESULTS AND DISCUSSION

Chemistry for Automated C-Terminal Sequence Analysis of Proline Containing Polypeptides

Application of the acetic anhydride/TMS-ITC/TFA procedure, used for the synthesis of TH-proline, to the tripeptide, N-acetyl-Ala-Phe-Pro, in our laboratory, found that thiohydantoin proline was formed in low yield (approx. 1-2% of theoretical). Recovery of the peptide products after the reaction revealed that approximately half of the starting peptide was unchanged and the remaining half had been decarboxylated at the C-terminus, thereby blocking it to C-terminal sequence analysis. This was most likely caused by the high concentration of trifluoroacetic acid, the excess of acetic anhydride present, and the high temperature (80°C) at which the reaction was performed. Substitution of TMS-ITC in place of ammonium thiocyanate and lowering the reaction to 50°C also resulted in an approximately 2% yield of TH-proline, although no decarboxylated peptide was formed.

The poor reaction with C-terminal proline most likely stems from the fact that proline cannot form the necessary oxazolinone for efficient reaction with the isothiocyanate. Previous work in our laboratory has obviated the need for oxazolinone formation by the use of diphenyl phosphoroisothiocyanatidate and pyridine. Reaction of this reagent with C-terminal proline directly forms the acylisothiocyanate. Once the acylisothiocyanate is formed, the addition of either liquid or gas phase acid followed by water was found to release proline as a thiohydantoin amino acid derivative. Unlike thiohydantoin formation with the other 19 naturally occurring amino acids, C-terminal proline thiohydantoin requires the addition of acid to provide a hydrogen ion for protonation of the thiohydantoin ring nitrogen. This step is necessary for stabilization of the proline thiohydantoin ring. The resulting quaternary amine containing thiohydantoin can then be readily hydrolyzed to a shortened peptide and thiohydantoin proline by introduction of water vapor or by the addition of sodium trimethylsilanolate (the reagent normally used for cleavage of peptidylthiohydantoins). The automation of this chemistry has allowed proline to be analyzed in a sequential fashion without affecting the chemical degradation of the other amino acids.

The chemical scheme for C-terminal sequencing is shown in Figure 1. The first step involves treatment of the peptide or protein sample with diisopropylethylamine in order to convert the C-terminal carboxylic acid into a carboxylate salt. Derivatization of the C-terminal amino acid to a thiohydantoin is accomplished with diphenylphosphorylisothiocyana-

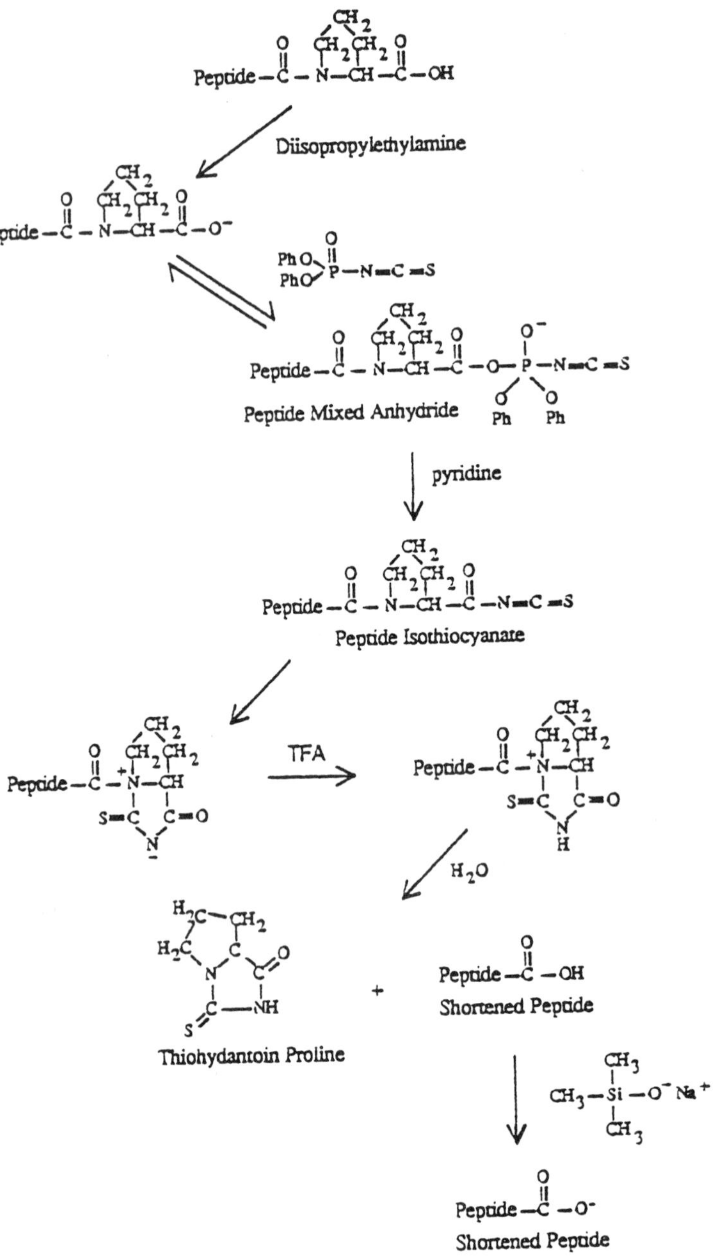

Figure 1. Reaction scheme and postulated intermediates for the sequential C-terminal degradation of polypeptides Which may contain proline.

tidate (liquid phase) and pyridine (gas phase). The peptide is then extensively washed with ethyl acetate and acetonitrile to remove reaction by-products. The peptide is then treated briefly with gas phase trifluoroacetic acid, followed by water vapor in case the C-terminal residue is a proline (this treatment has no effect on residues which are not proline). The derivatized amino acid is then specifically cleaved with sodium or potassium trimethylsilanolate to generate a shortened peptide or protein which is ready for continued sequencing. In the case of a C-terminal proline which was already removed by water vapor, the silanolate treatment merely converts the C-terminal carboxylic acid group on the shortened peptide to a carboxylate. The thiohydantoin amino acid is then quantitated and identified by reverse-phase HPLC.

The proposed role of trifluoroacetic acid (TFA) is for the protonation of the thiohydantoin proline ring. The addition of water or silanolate salt is required for cleavage of the TH-proline. If the temperature is raised to 80°C and the TFA step is prolonged (10 to 20 min) the acid alone can be used to cleave the TH-proline. TFA and water under the conditions used for the proline reaction at 50°C does not cleave peptidylthiohydantoins formed from the other 19 amino acids. This makes it possible to integrate the unique steps needed for proline as routine steps in the automated C-terminal sequencing program.

Examples of Automated Sequence Analysis

Automated C-terminal sequencing was performed on a compact protein sequencer designed and built at the City of Hope (Bailey et al., 1993). The total program run time for a cycle of C-terminal sequencing is approximately 60 min.

The performance of the automated method was evaluated by sequencing a number of peptide and protein samples. Peptide samples for C-terminal sequencing were covalently attached to carboxylic acid polyethylene (PE-COOH) prior to sequence analysis. Proteins and longer polypeptides (5 kdal and larger) were noncovalently applied to Zitex G-110 (porous Teflon). Figure 2 shows the automated C-terminal sequence analysis of the tripeptide, AFP (12 nmol). The yield of the amino acid in cycle 3 is low since this is the amino acid which is covalently attached to the solid support. This has been observed for all peptides covalently attached to PE-COOH (Bailey et al., 1992a). Figure 3 shows the automated C-terminal sequence analysis of polyproline (1 nmol) (the average molecular weight of the polyproline used was 12,000 daltons) noncovalently applied to Zitex. The reduced yields of proline in cycles two and three are consistent with the known washout of samples with molecular weights of less than 16,000 daltons. Figure 4 shows application of the sequencing chemistry to ovalbumin (approx. 5-6 nmol) noncovalently applied to Zitex. The expected sequence at the C-terminus is —Val-Ser-Pro. Although there is considerable cycle to cycle lag in this example, proline is clearly sequenced. Work is continuing toward optimizing this automated chemistry.

SUMMARY

We have described a simple procedure for the large scale (200 mg) synthesis of thiohydantoin proline from N-acetyl proline and extensively characterized this analogue. The thiohydantoin derivative of proline is conveniently obtained as a white powder which is stable to long term storage. The availability of a thiohydantoin proline standard is critical for the evaluation of automated sequencing results.

We have described automated chemistry which is capable of the C-terminal sequence analysis of polypeptides containing C-terminal proline. This chemistry has been integrated into the automated sequencing program previously used for the C-terminal sequence analysis of the other 19 amino acids without affecting performance.

We have proposed a chemical mechanism for proline sequencing via the thiohydantoin route which is consistent with the experiments performed to date.

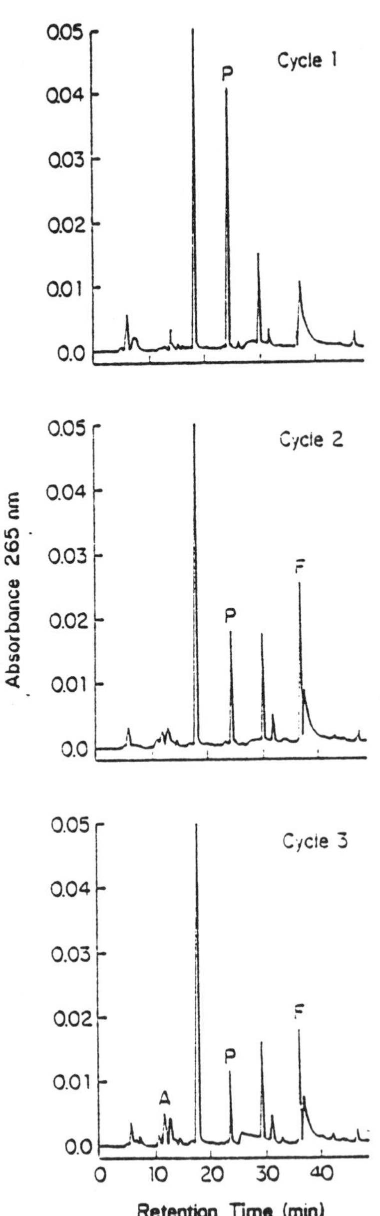

Figure 2. Automated C-terminal sequencing of the tripeptide, AFP (12 nmol), covalently attached to carboxylic acid modified polyethylene. Each thiohydantoin derivative is identified by comparison to the retention time of an authentic standard. Unlabeled peaks are background produced by reaction side products.

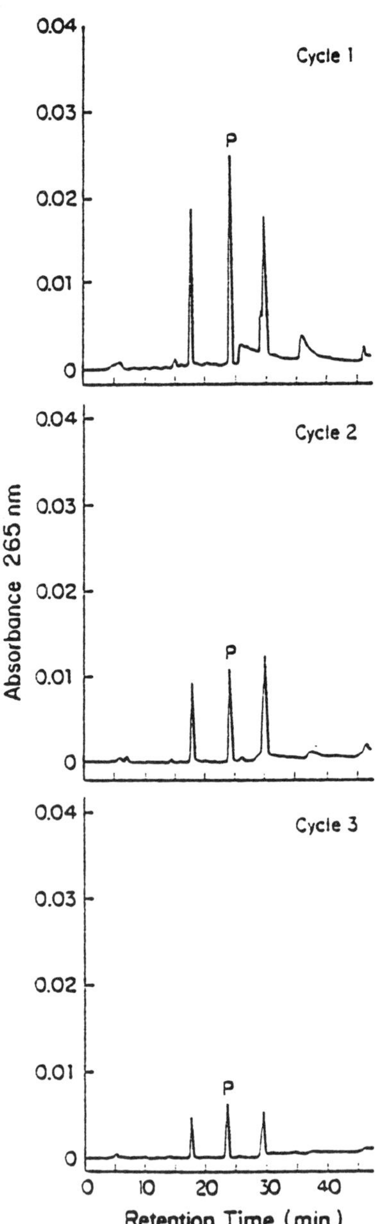

Figure 3. Automated C-terminal sequencing of polyproline (1 nmol) non-covalently applied to Zitex. The average molecular weight of polyproline used was 12,000 daltons.

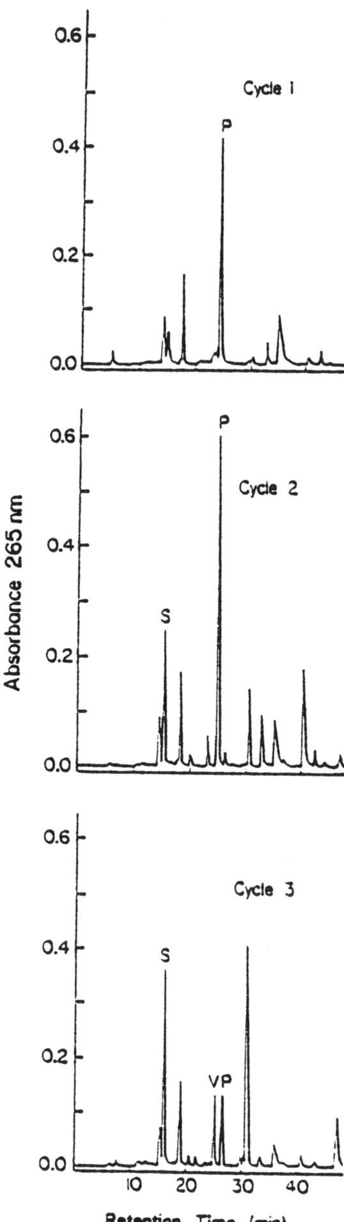

Figure 4. Automated C-terminal sequencing of ovalbumin (approx. 5-6 nmol) non-covalently applied to Zitex. The expected sequence at the C-terminus is —Val-Ser-Pro.

The failure of previous methods to derivatize C-terminal proline maybe due to the inability of proline to form an oxazolinone, a necessary step in many of the previous methods. The use of DPP-ITC/pyridine for derivatization permits the direct formation of an acylisothiocyanate at the C-terminus without the need for oxazolinone formation.

Once an acylisothiocyanate is formed it can cyclize to a quaternary amine containing thiohydantoin. This thiohydantoin, if protonated with acid, is stable. If the acid step is eliminated C-terminal proline is regenerated. The quaternary amine containing proline thiohydantoin can be readily cleaved with water vapor or alternatively with the silanolate salt normally used for the cleavage reaction.

Current Expectations for C-Terminal Sequencing

Current technology now permits 1-3 cycles of automated C-terminal sequence analysis on 200 pmol - 4 nmol of non-covalently applied protein samples which contain any of the twenty common amino acids.

Work is continuing toward the goal of extending the number of cycles of sequence information which can be obtained with this automated method.

REFERENCES

Bailey, J.M., and Shively, J.E. (1990) Carboxy-terminal sequencing: Formation and hydrolysis of C-terminal peptidylthiohydantoins, *Biochemistry* 29, 3145-3156.

Bailey, J.M, Shenoy, N.S., Ronk, M., and Shively, J.E. (1992a) Automated carboxy-terminal sequence analysis of peptides, *Protein Science* 1, 68-80.

Bailey, J.M., Nikfarjam, F., Shenoy, N.S., and Shively, J.E. (1992b) Automated Carboxy-Terminal Sequence Analysis of Peptides and Proteins Using Diphenyl Phosphoroisothio-cyanatidate, *Protein Science* 1, 1622-1633.

Bailey, J.M., Rusnak, M., and Shively, J.E. (1993) *Analytical Biochemistry* 212, 366-374.

Edman, P. (1950) Method for Determination of the Amino Acid Sequence in Peptides, *Acta Chem. Scand.* 4, 283-293.

Fox, S.W., Hurst, T.L., Griffith, J. F., and Underwood, O. (1955) A method for the quantitative determination of C-terminal amino acid residues, *J. Am. Chem. Soc.* 77, 3119-3122.

Inglis, A.S., Wilshire, J.F.K., Casagranda, F., and Laslett, R.L. (1989) C-Terminal sequencing: A new look at the Schlack-Kumpf thiocyanate degradation, *in* Methods in Protein Sequence Analysis (Wittmann-Liebold, B., Ed.) pp.137-144, Springer-Verlag.

Inglis, A.S., and De Luca, C. (1993) A new chemical approach to C-terminal microsequence analysis via the thiohydantoin, *in* Methods in Protein Sequence Analysis (Imahori, K.,Sakiyama, F., Eds.) 71-78, Plenum Publishing Corp.

Kenner, G.W., Khorana, H.G., and Stedman, R.J. (1953) Peptides. Part IV. Selective removal of the C-terminal Residue as a thiohydantoin. The use of diphenylphosphoroisothiocyanatidate, *Chem. Soc. Jour. (London),* 673-678.

Kubo, H., Nakajima, T., and Tamura, Z. (1971) Formation of thiohydantoin derivative of proline from the C-terminal of peptides, *Chem. Pharm. Bull.* 19, 210-211.

Schlack, P., and Kumpf, W. (1926) Uber eine neue methode zur Ermittlung der konstitution von peptiden, *Z. Physiol. Chem.* 154, 125-170.

Shenoy, N.S., Bailey, J.M., and Shively, J.E. (1992) Carboxylic acid modified polyethylene: A novel support for the covalent immobilization of polypeptides for C-terminal sequencing, *Protein Science* 1, 58-67.

Stark, G.R. (1968) Sequential degradation of peptides from their carboxyl termini with ammonium thiocyanate and acetic anhydride, *Biochem.* 7, 1796-1807.

Turner, R.A., and Schmerzler, G. (1954) Identification of C-terminal residues in peptides and proteins through formation of thiohydantoins, *Biochim. Biophys. Acta.* 13, 553-559.

Yamashita, S., and Ishikawa, N. (1971) Sequential degradation of polypeptides from the carboxyl ends II. Application to polypeptides, *Proc. Hoshi. Pharm.* 13, 136-138.

MASS SPECTROMETRY

13

TANDEM MASS SPECTROMETRIC CHARACTERIZATION OF MODIFIED PEPTIDES AND PROTEINS

Simon J. Gaskell, Mark S. Bolgar, and Kathleen A. Cox

Michael Barber Centre for Mass Spectrometry
Department of Chemistry
University of Manchester Institute of Science and Technology (UMIST)
Manchester M60 1QD, United Kingdom

INTRODUCTION

Mass spectrometry (MS) is now widely accepted as an analytical technique of complementary value to Edman-based approaches to peptide and protein structure determination. The value of MS derives from the accommodation of mixtures and the possibilities for characterization of modified amino acid residues. Tandem MS in particular is important in addressing both of these issues. The essential features of tandem MS are the promotion of ion fragmentation (generally following collision with a target gas) and the establishment of connectivity between precursor and product ions. Such analyses can yield structural information for individual components of mixtures. A variety of instrumental techniques have been used for tandem MS, differing in the choice of ion analyzers and the precise experimental conditions under which precursor ion activation and decomposition take place. Thus, for example, tandem MS of peptides using four-sector mass spectrometers generally involves high energy collisional activation of precursor ions (usually $[M+H]^+$) to promote fragmentations indicative of sequence and permitting the differentiation of isomeric/isobaric amino acid residues [1].

Tandem MS analyses using triple quadrupole or hybrid sector/quadrupole instruments usually include collisional activation at low energies with a correspondingly extended time period during which precursor ion decompositions can be observed. These conditions may promote fragmentation processes which differ from those observed under conditions of high energy collisional activation; aspects of this ion chemistry remain to be elucidated. Our recent work has included studies aimed at the understanding of the factors determining low energy fragmentations of protonated peptides. This work has suggested that extensive diagnostic fragmentation of protonated peptides via low energy pathways is promoted by a precursor ion population which is heterogeneous with respect to the site of charge [2]. This is consistent with the general concept that low energy decompositions are generally charge-directed and is in accord with previous observations of the unfavorable fragmentation

Methods in Protein Structure Analysis, Edited by M. Z. Atassi and E. Appella
Plenum Press, New York, 1995

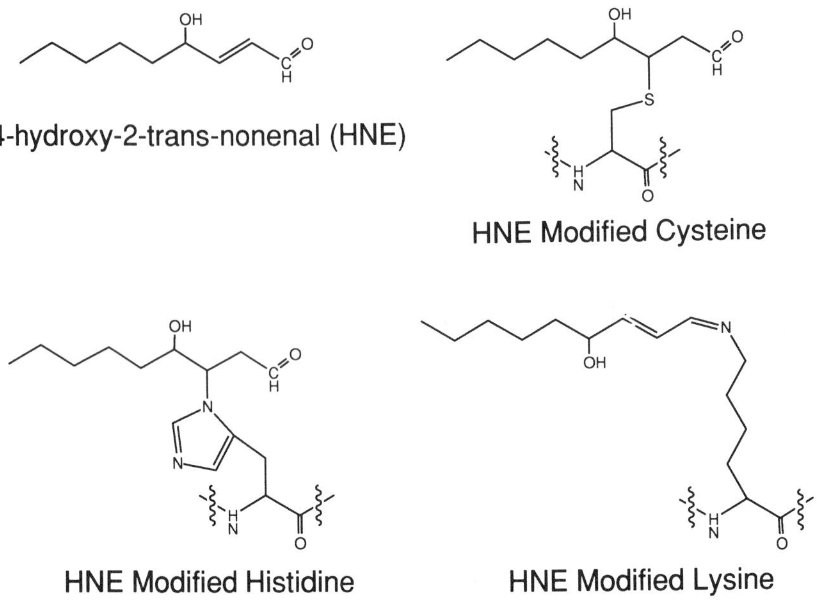

4-hydroxy-2-trans-nonenal (HNE)

HNE Modified Cysteine

HNE Modified Histidine

HNE Modified Lysine

Figure 1. Structures of 4-hydroxynonenal and putative products of reaction with amino acids.

properties of peptides which incorporate strongly basic residues (such as arginine) [3]. Definitive evidence comes from the analysis of peptides which have been converted to pre-charged derivatives; little fragmentation diagnostic of sequence is observed [2].

Electrospray ionization has proved highly compatible with tandem mass spectrometry. As anticipated in the discussion above of low energy fragmentations, the multiplicity of protonation sites promotes extensive fragmentation. Much further work is required, however, to improve our understanding of the fragmentations of multiply charged ions and facilitate the interpretation of product ion spectra derived from unknowns. Nevertheless, impressive examples have been reported of the high sensitivity characterization of peptides using tandem MS of low charge states [4].

In the present report we describe two aspects of this laboratory's recent work on the tandem MS of peptides. The first area concerns model studies on the detection and characterization of peptides modified by reaction with 4-hydroxynonenal, a common product of lipid oxidation. Figure 1 shows the structure of 4-hydroxynonenal and of putative products of reaction with cysteine and histidine (via Michael addition), and with lysine (involving formation of a Schiff's base). Secondly, we describe further studies of intra-ionic acid/base interactions in gas-phase peptide ions.

EXPERIMENTAL METHODS

Materials

Bovine insulin B-chain (in which cysteine residues were oxidized to cysteic acids) was purchased from Sigma and used as supplied. Angiotensin III (2-7) was prepared from angiotensin (Sigma) via tryptic hydrolysis followed by HPLC purification. Similarly the peptide SCFR was prepared from RLCIFSCFR (synthesized in the School of Biological

Sciences, University of Manchester) via chymotryptic digestion followed by HPLC purification. Oxidation of cysteine to cysteic acid residues was carried out using performic acid as described by Burlet et al. [5].

4-hydroxynonenal (HNE) was synthesized by a modification of the method of Esterbauer and Weger [6]; the details will be provided elsewhere (manuscript in preparation). Reaction of various peptides with HNE was carried out as follows. The peptide standards were dissolved to a concentration of 1 mg/ml in an aqueous 0.1M K_2HPO_4 buffer solution which had been adjusted to pH 7.4 by the addition of 0.1 M KH_2PO_4. A ten-fold molar excess of HNE was added and the samples were vortexed for 1 min before incubation at 37°C for 6-24 h. The incubations containing angiotensin III (2-7) or SCFR were fractionated by reverse phase HPLC, whereas the insulin B-chain incubation was subjected to rapid HPLC for de-salting purposes only.

Enzymatic Digestions

Removal of the C-terminal arginine residue of RLCIFSCFR was achieved by carboxypeptidase B hydrolysis using a modification of the procedure described by Allen [7]. The enzyme was added to a solution of the peptide in 0.1M ammonium acetate (pH 8.5) to give an enzyme:substrate molar ratio of approximately 1:100. Hydrolysis was allowed to proceed for approximatley 1 h, with monitoring of the progress of the reaction by mass spectrometry.

HPLC Separations. HPLC was performed using a Waters 600 HPLC pump and controller with a Waters 490 variable wavelength UV detector. Separation was achieved using a Waters Novapak C_{18} column (3.9 x 150 mm). A linear gradient was formed from 100% to 0% A in 50 min at 1 ml/min. Solvent A was water with 0.1% trifluoroacetic acid. The UV detector was set to monitor 217 nm and the HPLC fractions were collected into 1.5 ml polypropylene tubes and dried under reduced pressure. HPLC desalting was performed using a Vydac C_8 column (2.1 x 150 mm) with a 5 min isocratic elution at 100% solvent A followed by a step gradient elution to 20% A. The flow rate was 0.5 ml/min and detection was at 217 nm. The UV peak eluting after the step gradient was collected and dried as above.

Mass Spectrometric Analyses

FAB MS analyses were performed using a VG 7070Q instrument with the configuration, electric sector (E)/magnetic sector (B)/collision quadrupole (q)/analyzer quadrupole (Q). The FAB primary beam was xenon atoms of 8 keV energy. The liquid matrix was a 1:1 mixture of bis-(2-hydroxyethyl)-disulfide and thioglycerol. For tandem MS analyses, precursor ions were selected at 1000 resolution using EB and subjected to collisional activation in q. The pressure of argon collision gas was sufficient to decrease the transmission of the precursor ion by ca. 80%. Product ions were recorded by scanning of Q, with resolution set to achieve peak widths of of 1-2 m/z units. Alternatively, precursor ion scans were obtained by scanning of B with Q set to transmit a selected product ion. Data were recorded via a VG 11/250 data system, with acquisition in the "multi-channel analyzer" mode; 5-15 scans were accumulated.

Electrospray MS analyses were performed (through the courtesy of Dr. M. Morris, Fisons Instruments) using a VG Quattro II triple quadrupole instrument. Analytes were introduced by loop injection into a stream of acetonitrile:water (1:1) containing 0.2% formic acid. Ionization was by nebulizer-assisted electrospray; the nebulizer gas was nitrogen and the electrospray needle was held at 3.5 kV. Precursor ions were selected at a low resolution sufficient to pass all of the major isotopic variants into the collision hexapole. Selected ions were subjected to collision with argon at a pressure of 5 x 10^{-3} mbar. The collision energy was optimized for maximum fragmentation efficiency in each analysis. Product ions were scanned at unit resolution.

Electrospray MS analyses of HNE-modified angiotensin III (2-7) were performed on a Sciex API III triple quadrupole instrument. The analyte was dissolved in water/acetonitrile/acetic acid (49/49/2) and introduced into the ion source via constant infusion at 10 μl/min using a Harvard Apparatus syringe driver. The nebulizer gas was nitrogen and the electrospray needle was held at 4.5 kV. Precursor ions were selected at low resolution and introduced into the collision quadrupole which was maintained with an argon collision gas thickness of 6.4×10^{14} atoms/cm^2. Product ions were scanned with resolution set to achieve peak widths of 3 m/z units.

Matrix-assisted laser desorption/ionization (MALDI) analyses were performed (by courtesy of Fisons Instruments) on a VG TofSpec instrument equipped with a nitrogen laser. Samples were prepared in aqueous 0.1% trifluoroacetic acid at a concentration of ca. 10 picomole/μl per component. A 2 μl aliquot was mixed with 2 μl of a freshly prepared aqueous solution of 0.1% trifluoroacetic acid saturated with 2,5-dihydroxybenzoic acid. The mass resolution was approximately 500 (FWHM).

RESULTS AND DISCUSSION

As an initial assessment of the reactivity of different amino acid residues with 4-hydroxynonenal, we have investigated the sites of modification of a model oligopeptide, the B-chain of insulin incorporating oxidized cysteine residues (ie. cysteic acids). Figure 2 compares the MALDI/TOF MS analyses of the chymotryptic digest of the unmodified peptide and the product of reaction with 4-hydroxynonenal. The peptide sequence and the expected chymotryptic cleavage sites are indicated in Figure 3. Reaction with HNE resulted in the incorporation of one or two HNE moieties in fragment 1-16, consistent with modification of the histidine residues. None of the peaks corresponding to fragments 17-24, 17-25, 25-30 or 26-30 was shifted in mass, indicating that Lys^{29} and other residues were not modified or that the modifications were labile under the conditions of analysis.

Tandem MS provides a powerful approach to confirmation of the structure of lipid-modified peptides and proteins. We have therefore initiated a study of the fragmentation behavior of HNE-modified peptides. Figure 4 shows the product ion spectrum obtained by collisional activation of $[M+2H]^{2+}$ ions formed from HNE-adducted angiotensin III (2-7) during electrospray MS analysis. The decomposition is highly efficient and yields a variety of diagnostic fragment ions corresponding to the products of one or two cleavages within the peptide chain. The first category includes N-terminal fragments (a- and b-series) and C-terminal fragments (y-series), where the nomenclature used is Biemann's modification [8] of the suggestion of Roepstorff and Fohlman [9]. Product ions resulting from two chain cleavages include "internal" fragments (b_my_n) and immonium ions representing single amino acid residues. Immonium ions corresponding to the tyrosine and modified histidine residues are particularly prominent. The latter appears at m/z 266, consistent with Michael addition of HNE to the imidazole ring.

The same modified peptide was analyzed by FAB/tandem MS with CAD of the singly protonated molecule, MH^+. The product ion spectrum (Figure 5) shows the modified histidine immonium ion (m/z 266) as the most prominent product ion. The apparent propensity to fragment in this way suggests a means for screening for the presence of HNE-modified histidine-containing peptides in mixtures such as protein digests. Figure 6 illustrates this principle with the analysis of a simple binary mixture of angiotensin III (2-7) and the HNE-adducted analogue. Tandem MS analysis involved scanning of the first mass analyzer with the second set to transmit a single product ion species (in this case m/z 266). The resulting precursor ion spectrum reveals the mixture component (and its ion source-formed fragments) which incorporates a modified histidine residue.

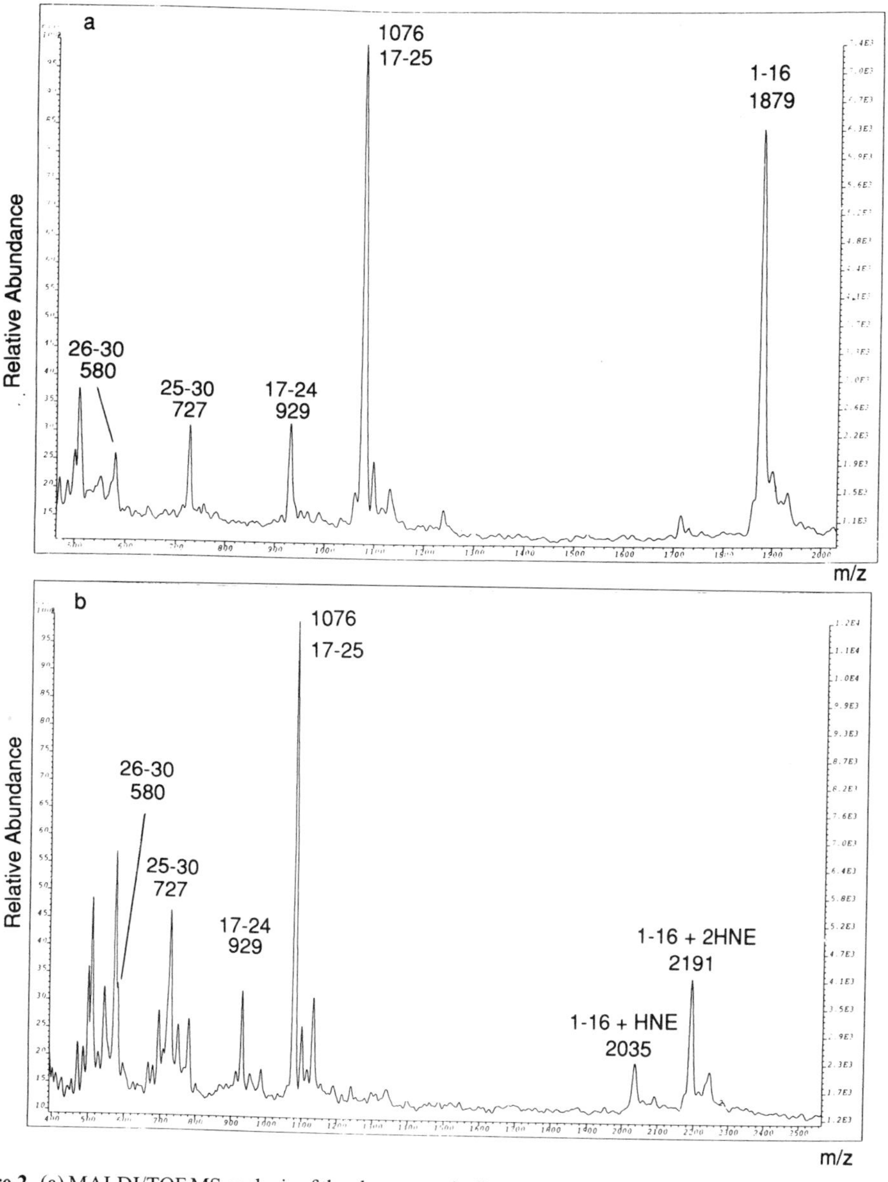

Figure 2. (a) MALDI/TOF MS analysis of the chymotryptic digest of oxidized (cysteine to cysteic acid) insulin B-chain; (b) equivalent analysis of the oligopeptide modified by reaction with 4-hydroxynonenal.

Figure 3. The sequence of the oxidized B-chain of bovine insulin. C(ox) represents cysteic acid residues. The arrows indicate the observed points of chymotryptic cleavage.

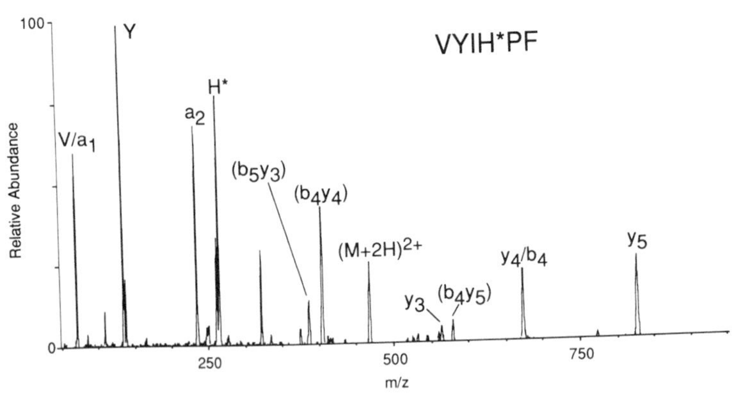

Figure 4. Product ion spectrum obtained following collisional activation of $[M+2H]^{2+}$ ions obtained by electrospray of VYIH*PF (where H* represents the modification of the histidine residue by reaction with 4-hydroxynonenal).

The analyses reported here represent the initial phase in the development of a screening and characterization strategy for the study of lipid-modified peptides and proteins. Tandem MS plays a central role in this strategy; its most effective use will be ensured by improved understanding of the relationship between peptide sequence and the propensities to fragment via a number of pathways. We have therefore pursued investigations of the low energy fragmentations of protonated peptides, with particular reference to the influence on fragmentation of the site of charge.

Figure 7 shows the product ion spectrum obtained following low energy CAD of singly protonated RLCIFSCFR (generated by FAB). A C-terminal rearrangement ion

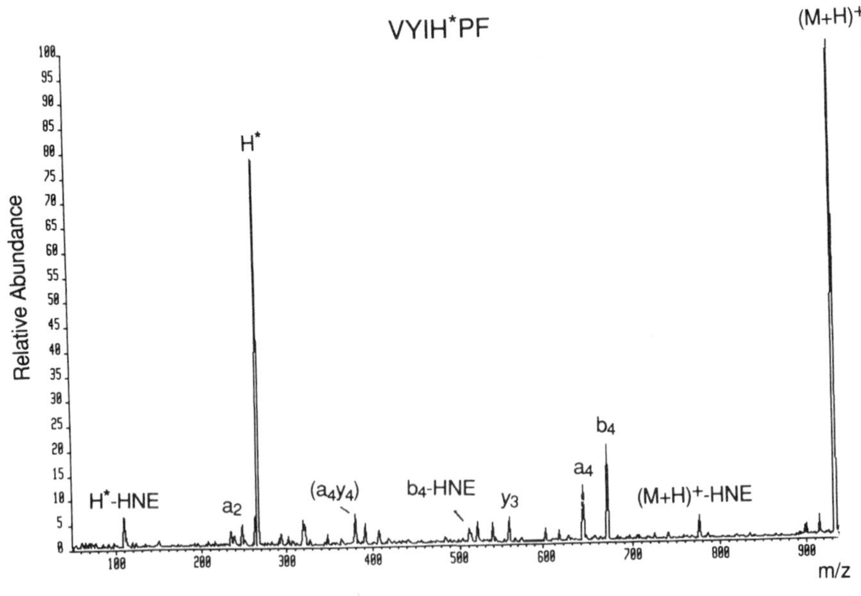

Figure 5. Product ion spectrum obtained following collisional activation of MH^+ ions obtained by FAB of VYIH*PF (where H* represents the modification of the histidine residue by reaction with 4-hydroxynonenal).

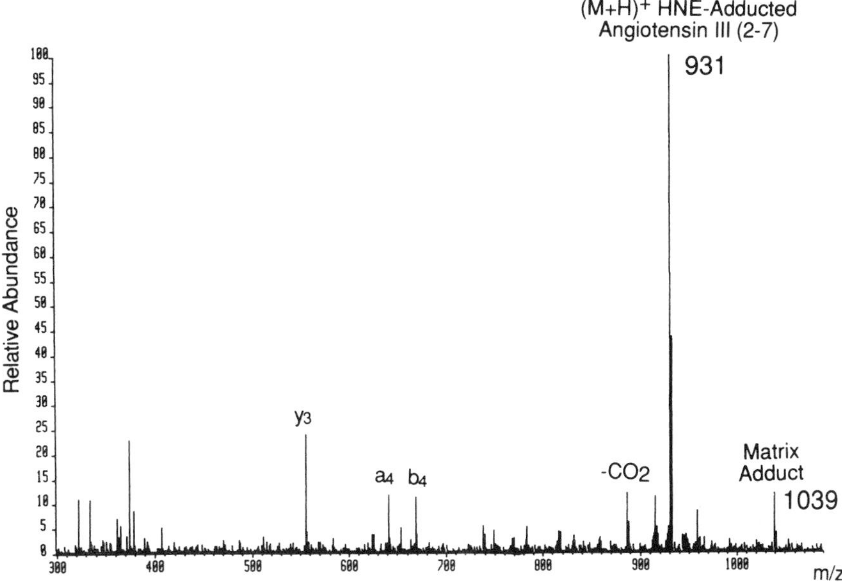

Figure 6. Spectrum of precursors of m/z 266 recorded during FAB tandem MS of a mixture of angiotensin III (2-7) and the HNE-adducted analogue. Only the modified peptide is detected, with signals orresponding to the MH^+ ion, a matrix adduct and minor fragment ions formed in the ion source.

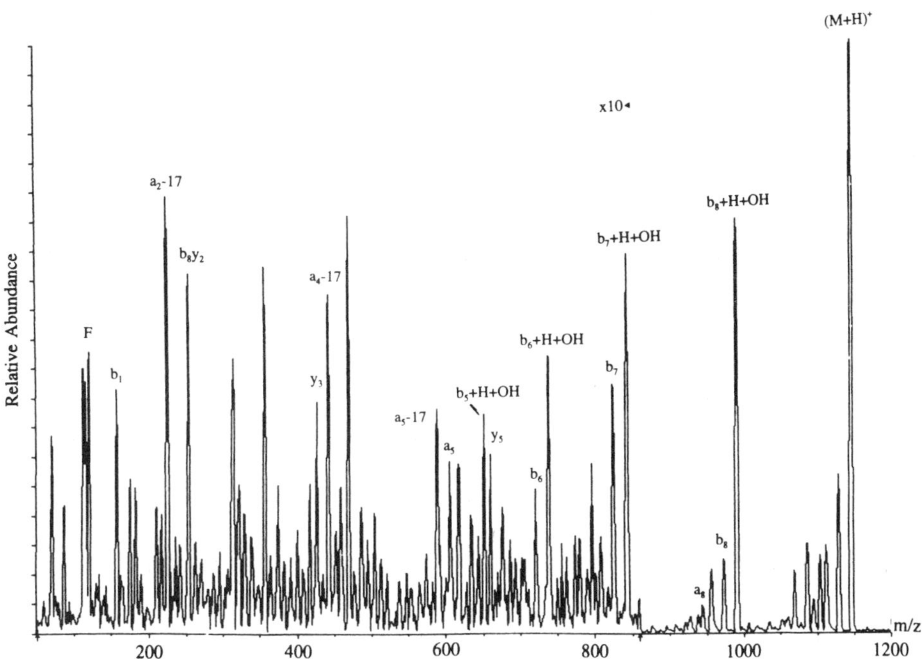

Figure 7. Product ion spectrum obtained following collisional activation of MH^+ ions obtained by FAB of RLCIFSCFR.

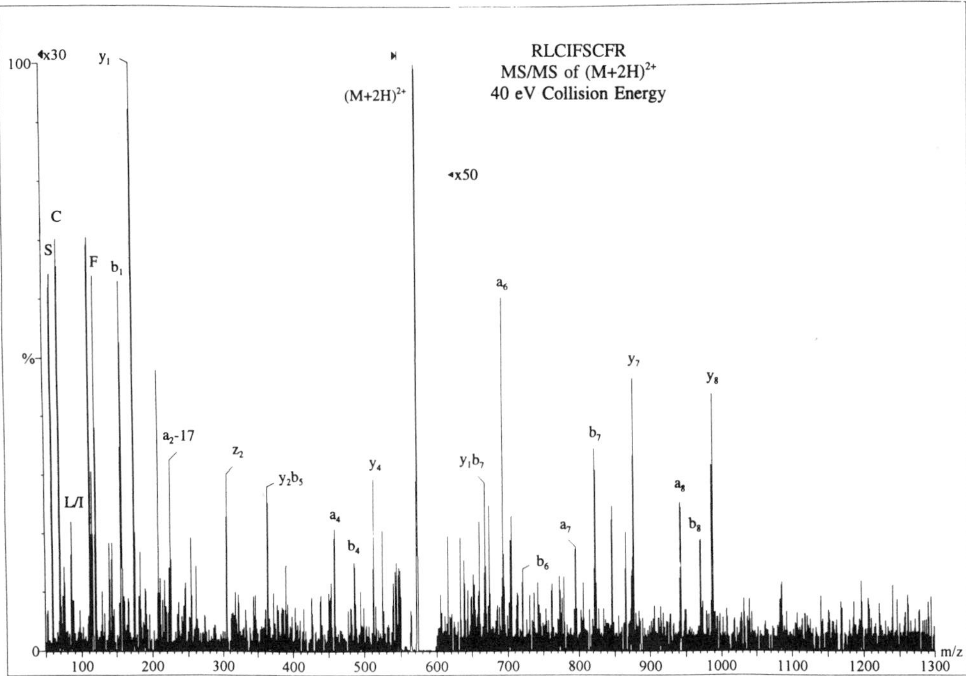

Figure 8. Product ion spectrum obtained following collisional activation of $[M+2H]^{2+}$ ions obtained by electrospray of RLCIFSCFR.

(representing loss of the C-terminal residue but retention of one carboxylic oxygen [10]) is observed as a highly favored fragmentation. Lower members of the rearrangement ion series (b_n+H+OH) are also observed but these and all other fragment ions are observed with low abundance. Low energy CAD of the doubly protonated analogue (generated by electrospray) shows a low overall efficiency of fragmentation (Figure 8; note the magnification factors), though a large number of diagnostic fragment ions are discernible. Interestingly, there is no evidence for the occurrence of the C-terminal rearrangement process. The high gas phase basicity of arginine suggests that the principal structure represented in the precursor population of doubly charged ions incorporates the two protons on the guanidino groups of the arginine residues, resulting in little charge-directed fragmentation of the peptide chain.

This hypothesis was evaluated by electrospray tandem MS analysis of the same peptide following oxidation of the cysteine residues to cysteic acid. Previous studies [5] suggested that intraionic acid-base interactions between cysteic acid and arginine residues reduce the propensity for charge location on the arginine residues, resulting in increased yields of diagnostic ions associated with cleavages of the peptide chain. Figure 9 shows the product ion spectrum obtained by CAD of doubly protonated RLC(SO_3H)IFSC(SO_3H)FR, where C(SO_3H) represents cysteic acid. In marked contrast to the equivalent data for the unoxidized peptide, a high decomposition efficiency is observed with the production of multiple C-terminal and N-terminal fragment ions. These findings are consistent with a precursor ion population heterogeneous with respect to the site of charge, a situation facilitated by the putative cysteic acid/arginine interactions.

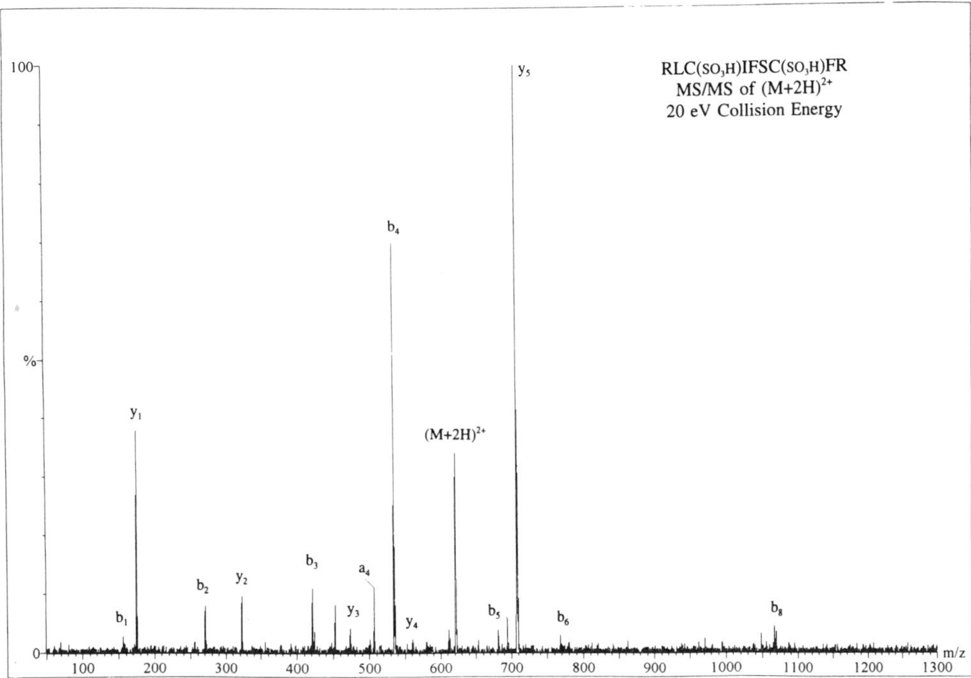

Figure 9. Product ion spectrum recorded following collisional activation of $[M+2H]^{2+}$ ions obtained by electrospray of $RLC(SO_3H)IFSC(SO_3H)FR$, where $C(SO_3H)$ represents cysteic acid.

ACKNOWLEDGMENTS

We are gratful to Fisons Instruments for providing access to the Quattro II and TofSpec instruments and to Drs. M. Morris and A. Maisey for assistance in their operation. This work was supported in part by the U.K. Engineering and Physical Sciences Research Council (GR/K18658). Funds for the purchase of synthetic peptides were provided by the British Mass Spectrometry Society. MSB thanks Bristol-Myers Squibb for financial support.

REFERENCES

1. Biemann, K., 1990, Sequencing of peptides by tandem mass spectrometry and high-energy collision-induced dissociation. In: *Mass Spectrometry (Methods in Enzymology, volume 193)*, edited by McCloskey, J.A., New York: Academic Press, p. 455.
2. Burlet, O., Orkiszewski, R.S., Ballard, K.D., and Gaskell, S.J., 1992, Charge promotion of low energy fragmentations of peptide ions, *Rapid Commun. Mass Spectrom.* 6: 658-662.
3. Poulter, L., and Taylor, L.C.E., 1989, A comparison of low and high energy collisionally activated decomposition MS-MS for peptide sequencing, *Int.J. Mass Spectrom. Ion Processes* 91: 183-197.
4. Hunt, D.F., Henderson, R.A., Shabanowitz, J., Sakaguchi, K., Michel, H., Sevilir, N., Cox, , Appella, E., and Engelhard, V.H., 1992, Characterization of peptides bound to the class I MHC molecule HLA-A2.1 by mass spectrometry, *Science* 255: 1261-1263.
5. Burlet, O., Yang, C.-Y., and Gaskell, S.J., 1992, The influence of cysteine to cysteic acid oxidation on the collisionally activated decomposition of protonated peptides: evidence for intra-ionic interactions, *J. Am. Soc. Mass Spectrom.* 3: 337-344.

6. Esterbauer, H., and Weger, W., 1967, Uber die Wirkungen von Aldehyden auf gesunde und maligne Zellen, *Monatschrift fur Chemie* 98: 1994-2000.
7. Allen, G., 1989, Sequencing of proteins and peptides. In: *Laboratory Techniques in Biochemistry and Molecular Biology* (Volume 9), edited by Burdon, R.H. and van Knippenberg, P.H., Amsterdam: Elsevier, p. 70.
8. Biemann, K., 1988, Contributions of mass spectrometry to peptide and protein structure, *Biomed. Environ.Mass Spectrom.* 16: 99-111.
9. Roepstorff, P., and Fohlman, J., 1984, Proposal for a common nomenclature for sequence ions in mass spectra of peptides, *Biomed. Mass Spectrom.* 11:601.
10. Thorne, G.C., Ballard, K.D., and Gaskell, S.J. Metastable decomposition of peptide $[M+H]^+$ ions via rearrangement involving loss of the C-terminal amino acid residue, *J. Am. Soc. Mass Spectrom.* 1: 249-257.

14

DIRECT MASS SPECTROMETRIC ANALYSES FOR PROTEIN CHEMISTRY STUDIES

Scott D. Buckel, Tracy I. Stevenson, and Joseph A. Loo

Parke-Davis Pharmaceutical Research
Division of Warner-Lambert Company
Ann Arbor, Michigan 48105

INTRODUCTION

For structural studies of various proteins, a combination of traditional sequence analysis and mass spectrometry (MS) has been effectively used in our laboratory. Protein sequencing methods and mass spectrometry are often plagued with the same problems in that samples are frequently contaminated with other materials. Many buffer components, salts, and solubilizing detergents can interfere or prevent the successful application of both methodologies. Membranes, made of materials such as polyvinyldifluoride (PVDF), have been invaluable in this regard for protein sequencing. These membranes have allowed for direct analysis of proteins from complex mixtures by allowing for separation by gel electrophoresis and subsequent electroblotting that immobilizes the protein and removes potentially interfering small molecules (Matsudaira, 1987). PVDF membranes are also stable to most organic solvents. Mass spectrometry is a sensitive bioanalytical method (McCloskey, 1990), but it is often difficult for the method to selectively discriminate against *most* species found in a sample, in search of the few components the scientist is truly interested. MS analysis of a sample containing a small amount of peptide in a great molar excess of buffer salts typically results in a mass spectrum mostly composed of buffer ions. Common desalting or chromatographic methods are often necessary prior to analysis by mass spectrometry, but this adds an additional step of complexity and increases the total analysis time.

To increase the throughput of our bioanalytical laboratory, we have been investigating simple, rapid methods of sample preparation for sequence analysis and mass determination. This report presents some of the observations we have made towards this goal and their applications. Electrospray ionization (ESI) (Fenn, Mann et al., 1989; Smith, Loo et al., 1990) and matrix-assisted laser desorption/ionization (MALDI) (Karas, Bahr et al., 1989; Hillenkamp, Karas et al., 1991) have advanced the applicability of mass spectrometry to large biomolecule analyses. We present results demonstrating the unique characteristics of an array detector for ESI detection. Its ability to discriminate against ions based on charge allows for direct detection of proteins to the low attomole level (Loo and Pesch, 1994). Detection of

Methods in Protein Structure Analysis, Edited by M. Z. Atassi and E. Appella
Plenum Press, New York, 1995

higher-charged protein molecules in the presence of higher concentration, lower molecular weight contaminants will be demonstrated.

To monitor the degree of truncation by carboxypeptidase for C-terminal sequence determination of a small protein, the molecular weights from the mixture of digest products were determined without additional sample preparation (*i.e.*, without removal of extraneous buffer materials) by ESI-MS and MALDI-MS. The number of disulfide bonds present in small tightly-bridged peptides can be determined by measuring the difference in mass of the oxidized material and the peptide reduced by tris-(2-carboxyethyl)phosphine without additional derivatization of the resulting free cysteines. We have also been able to determine the protein molecular weights from spots from two-dimensional gels that had been blotted onto PVDF membranes, stained with Coomassie blue, and extracted with hexafluoroisopropanol, and also determined the sequence of peptides via chemical digestion of the material not used for mass determination.

EXPERIMENTAL

Mass Spectrometry

ESI-MS analyses were performed with a Finnigan MAT 900Q forward geometry hybrid mass spectrometer (Bremen, Germany) equipped with a 20 kV conversion dynode/electron multiplier point detector and a focal plane array detector (position-and-time resolved ion counting, or PATRIC) (Loo, Ogorzalek Loo et al., 1993; Loo and Pesch, 1994). For operation with the microchannel plate array detector, an 8% *m/z* range of the *m/z* centered on the detector was used. The voltage across the front and back of the microchannel plates is designated as V_{MCP}. An electrospray ionization interface based on a heated glass capillary inlet was used (Fenn, Mann et al., 1989). Sample solutions for ESI-MS analysis were infused into the ESI source at a flow rate of 0.5-1.5 μl/min. The typical solution composition was 1:1 MeOH/H_2O with 1-2.5% acetic acid (v/v).

MALDI mass spectra were acquired with a PerSeptive Biosystems-Vestec (Houston, TX) LaserTec Research time-of-flight mass spectrometer operating in the linear mode. Samples were prepared by placing 1 μl of a 1-10 pmol μl^{-1} solution (0.1% trifluoroacetic acid in H_2O) of the peptide on the sample target and adding 1 μl of a solution of α-cyano-4-hydroxycinnamic acid (4-HCCA, 5 μg μl^{-1} in 1:2 acetonitrile/0.1% trifluoroacetic acid (aq.)).

Extraction from Blots

Spots from two blots of two dimensional gels were cut into about 1 mm x 1 mm pieces, placed in a microcentrifuge tube and 200 μl of hexafluoroisopropanol (HFIP) was added. This mixture was incubated in an Eppendorf Thermomixer with shaking at 25°C for 30 minutes. The solution was removed and an additional 200 μl of HFIP was added and the extraction was repeated. The second extract was pooled with the first and lyophylized. The dried sample was dissolved in 50 μl of 5% acetic acid.

Materials

Melittin, substance P, gramicidin S, ACTH, ATX II toxin, and carbonic anhydrase were purchased from Sigma Chemical Co. (St. Louis, MO). Bovine pancreatic trypsin inhibitor (BPTI or aprotinin) was obtained from BiosPacific (Emeryville, CA). Tris-(2-carboxyethyl)phosphine (TCEP) was purchased from Molecular Probes (Eugene, OR).

RESULTS AND DISCUSSION

ESI-MS of Salt-Containing Samples

The ability to directly mass analyze protein materials without additional sample cleanup results from the discriminating nature of the focal-plane array detector (Loo and Pesch, 1994). The use of a focal plane detector on a magnetic sector mass spectrometer with ESI has allowed low-to-sub femtomole detection limits for large proteins (Cody, Tamura et al., 1994; Loo and Pesch, 1994). Full scan spectra (*m/z* 500-3000) can be collected from less than 5 femtomoles consumed of bovine albumin (66 kDa), and less than 500 attomoles from a 2 fmol/μL solution of porcine pepsin (34 kDa) (Loo and Pesch, 1994). The array detector can be "tuned" for higher charged, low level species in a complex mixture and/or in the presence of low molecular weight material.

The presence of salts and buffers can often be disabling for ESI-MS experiments. However, the advantages of low molecular weight (low charge) discrimination can be realized for high levels of interfering buffers and other additives used in protein chemistry. Compared to singly charged ions, highly charged ions generate many more secondary electrons upon hitting the microchannel plate of the array detector due to the very high kinetic energy of these ions. The nature of the PATRIC array detection electronics allows only signal levels between the minimum and maximum thresholds to be counted. The array detector can discriminate against highly charged ions by changing the voltage applied to the channelplates (V_{MCP}). In order to selectively detect only highly charged ions, V_{MCP} is reduced to decrease the number of secondary electrons and place the signal within the "acceptable" window.

Triton X-100, a nonionic polyoxyethylene detergent, is often used to solubilize proteins. ESI mass spectra for a 3.4 pmol μL^{-1} solution of bovine carbonic anhydrase (29 kDa) in the presence of 0.02% (w/v) Triton X-100 are shown in Figure 1. As expected, the

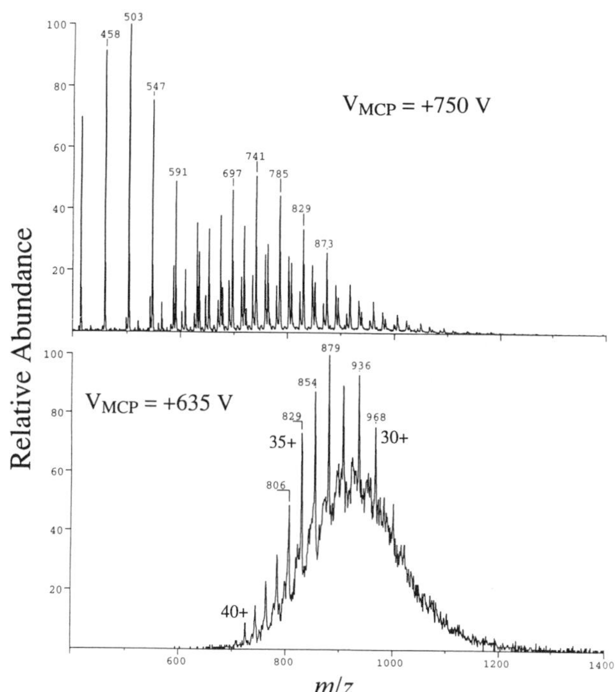

Figure 1. Electrospray ionization mass spectra of 3.4 pmol μL^{-1} of bovine carbonic anhydrase (29 kDa) with 0.02% Triton X-100 (reduced) in 2:1 MeOH:H_2O and 2.5% acetic acid with V_{MCP} at (a) +750 V and (b) +635 V. Approximately 7.9 pmol of protein was consumed during acquisition of the spectrum in Figure (b).

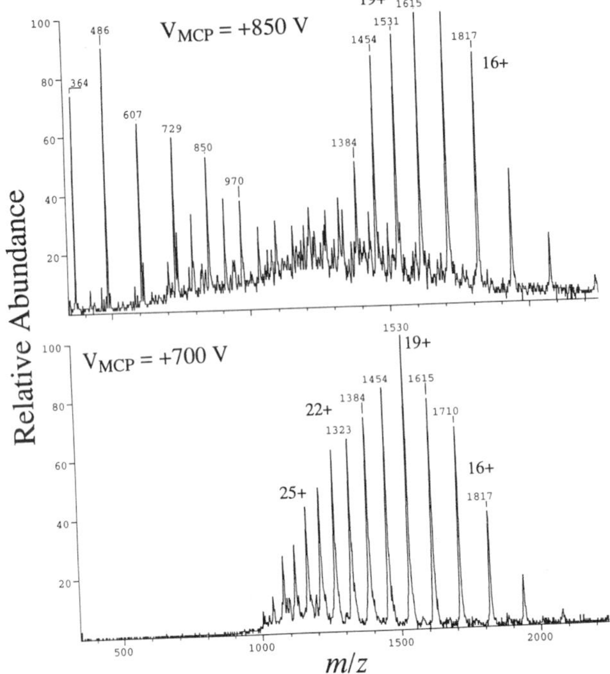

Figure 2. ESI mass spectra of 7.9 pmol µL[-1] of bovine carbonic anhydrase (29 kDa) with 50 mM Tris in 2:1 MeOH:H_2O and 2.5% acetic acid with V_{MCP} at (a) +850 V and (b) +700 V.

mass spectrum with V_{MCP} at +750 V shows only singly-charged ions for the Triton X-100 oligomers, whereas reducing V_{MCP} to +635 V allows detection of the carbonic anhydrase multiply charged molecules. Similarly, a spectrum of carbonic anhydrase in the presence of 50 mM TRIS can be obtained at the lower V_{MCP} voltages (Figure 2).

C-Terminal Sequence Analysis with MS Detection

Determination of protein sequence from the C-terminus can be directly obtained by combining traditional proteolytic chemistry and mass spectrometry. The products from enzyme reactions can be monitored at various stages of the reaction by MS. For example, a sample of Aga IVA, a highly bridged (by disulfide bonds) toxin from spider venom, was dissolved in 50 mM sodium citrate pH 4.0 with carboxypeptidase Y and allowed to incubated for 24 hours at 28°C. The sample was then submitted for mass determination by ESI-MS. A portion of the original solution was diluted by a factor of 5 with a solution of H_2O/methanol containing 2.5% acetic acid and directly infused into the ESI source. The results are shown in Figure 3. The mass of the observed peptide corresponds to the first 39 residues of the original polypeptide (i.e., 9 residues from the C-terminus were liberated from the peptide by carboxypeptidase Y). The voltage on the microchannel plates of the array detector was reduced to "selectively" detect the multiply charged peptide ions and discriminate against the much more abundant ions from the buffer components.

Similarly, carboxypeptidase Y was used to cleave C-terminal residues from epidermal growth factor (EGF 1-48). The sample, in 50 mM sodium citrate at pH 4.0, was analyzed by MALDI-MS at various time points of the reaction. The mass spectra acquired after 1 minute and 1 hour are shown in Figure 4. Interference from citrate buffer salts was reduced by diluting the sample by a factor of 10 prior to MS analysis. Protein products with masses

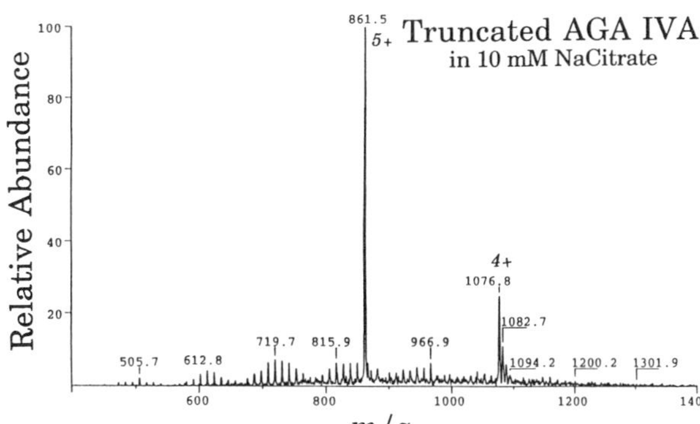

Figure 3. ESI mass spectrum of carboxypeptidase Y-digested Aga IVA in the presence of 10 mM sodium citrate buffer.

5445, 5317, and 5204 Da can be easily measured from the MALDI-MS data, representing the intact protein and the loss of Lys and Leu, respectively, from the C-terminus.

ESI-MS of Proteins Isolated from Two-Dimensional Gels

Samples thought to be Histones H2a and H2b were submitted as streaks on blots from two dimensional gels. The H2b sample yielded the expected sequence, but the H2a sample was N-terminally blocked. To further characterize these proteins, the blots were extracted with hexafluoroisopropanol and analyzed by electrospray ionization mass spectrometry. ESI

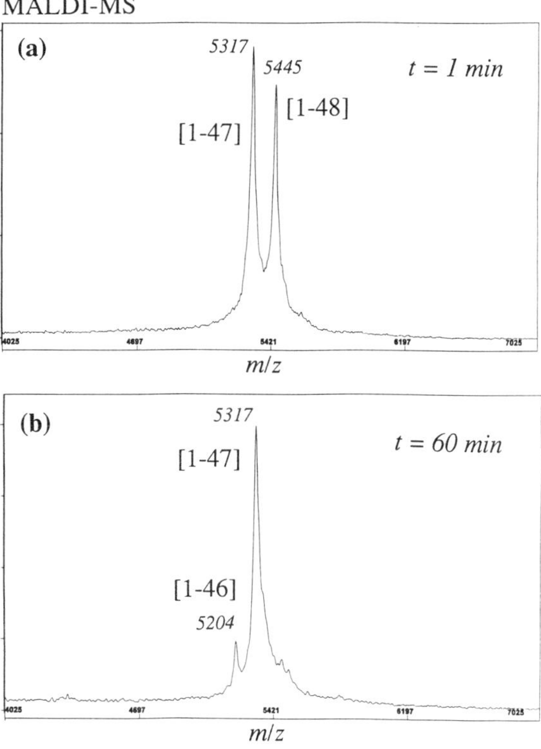

Figure 4. MALDI mass spectra of carboxypeptidase Y-digested EGF 1-48 after (a) 1 min. and (b) 60 min.

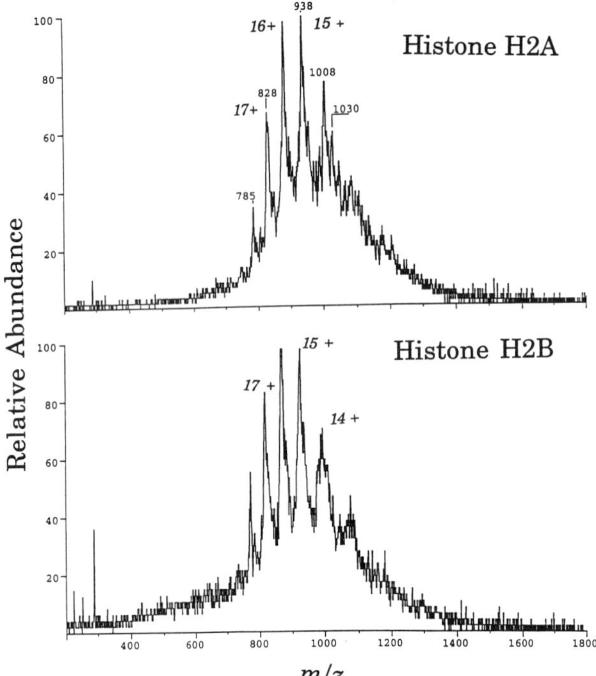

Figure 5. ESI mass spectra of (a) Histone H2A and (b) Histone H2B isolated from 2-D gels and extracted from blots in the presence of excess Coomassie blue stain.

mass spectra were acquired by using the PATRIC array detector at low microchannel plate voltages to discriminate against low molecular weight background ions from the excess Coomassie blue stain. Measured molecular weights corresponded to the expected 14 kDa proteins with multiple acetylation and multiple amino acid substitutions. (Figure 5) The material not consumed in the ESI-MS experiment was digested by the addition of cyanogen bromide and two sequences obtained corresponded to the sequence of the amino terminus of H2a and the sequence after Met51 in the sequence.

ESI-MS of Disulfide-Containing Proteins

The number of disulfide bonds in a protein can be rapidly determined by ESI mass spectrometry. As has been shown by a number of groups (Feng, Bell et al., 1990; Loo, Edmonds et al., 1990), the measured mass difference between the oxidized form and the disulfide-reduced form allows one to determine the number of disulfides. For example, a measured mass difference of 8 daltons corresponds to 4 disulfide bonds (2 daltons for each disulfide bond). The use of cysteine blocking agents (e.g., iodoacetamide or 4 vinylpyridine) would certainly lessen the mass accuracy requirements of the experiment. However, we wished to test the accuracy of the method with as few complications as possible.

Table 1 lists the ESI-MS results for a number of disulfide-containing peptides before and after reduction. These samples were dissolved in 70% acetonitrile, 0.1% trifluoroacetic acid and the oxidized form of the peptide was analyzed without further workup. To reduce the peptide, tris-(2-carboxyethyl)phosphine (TCEP) was added to a concentration of 20 mM and heated to 60°C for 2 to 4 minutes. The sample was analyzed without further work up. ESI-MS mass measurements were obtained at a resolving power of greater than 5000. Peptides, such as melittin (2845 Da), substance P (1347 Da), or gramicidin S (1141 Da) were added to the solution to provide reference peaks (internal standards) for more accurate mass

Table 1. Application of high resolution ESI-MS for disulfide determination

Peptide	MW (Sequence)	MW (Measured)	Δ MW (meas.-seq.)	MW (S-S red.)	Δ MW (red. - ox.)
Conotoxin	2800.10	2800.09	-5.4 ppm	2806.16	6.07
Apr15C	?	3674.56	—	3680.62	6.07
Apr15A	?	4389.27	—	4395.35	6.08
ATX II	4945.26	4945.25	-2.2 ppm	4951.31	6.06
BPTI	6507.04	6507.03	-2.1 ppm	6513.06	6.03

Note: Molecular weight values are monoisotopic mass. The theoretical MW difference of the disulfide reduced form versus the oxidized species containing 3 disulfides is 6.05 Da.

determination. Data was acquired by scanning the accelerating voltage at a constant magnetic field strength in the mass-to-charge region around the multiply-charged ion (and reference peak) of interest (Cody, Tamura et al., 1992).

Figure 6 shows the ESI mass spectrum of anemonia sulcata toxin II, Ile-isotoxin (ATX II), a 47 residue polypeptide. Melittin and ACTH (18-39, 2466 Da) were added to the ATX II-containing solution as internal calibrants for more accurate mass determination. A portion of the high resolution mass spectrum is shown in Figure 7. In general, mass accuracy by this high resolution method is around 5-10 parts-per-million (ppm). This accuracy is certainly sufficient to determine a mass difference of 2 daltons (1 disulfide bond). The high resolution capabilities of a magnetic sector mass spectrometer is useful for detecting the small amount of unreduced protein observed in the spectrum in Figure 8.

CONCLUSION

Protein analytical chemistry has progressed rapidly in the last several years. The development of electrospray ionization and matrix-assisted laser desorption/ionization has opened numerous research opportunities in biochemistry. Measuring the molecular weight of a 50 kDa protein to an accuracy of better than 0.05% and detecting the presence of the protein at the sub-picomole level by mass spectrometry was unimaginable 10 years ago. It

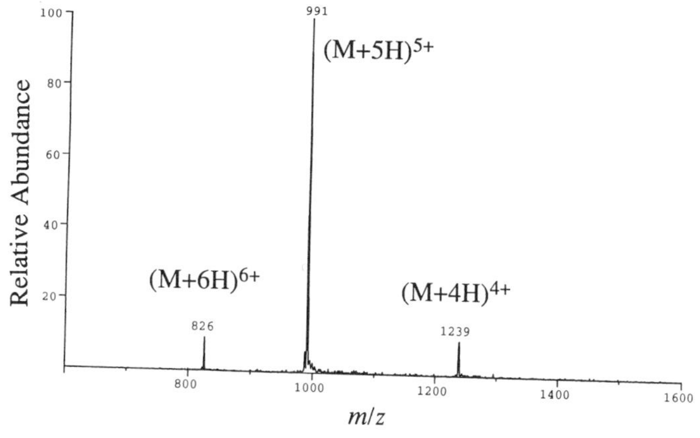

Figure 6. Electrospray ionization mass spectrum of ATX II toxin.

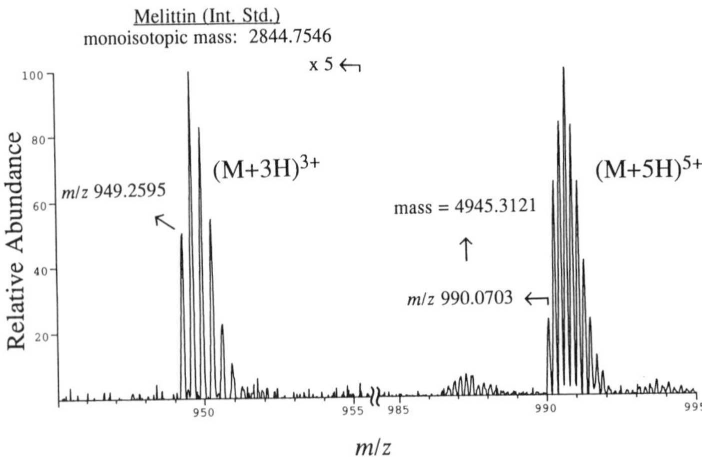

Figure 7. Portion of the high resolution (resolving power > 5000) mass spectrum of the ATX II/melittin mixture.

is becoming more frequent for a protein chemist to isolate picomole amounts of a protein and either directly obtain a mass spectrum or walk down the hallway and hand the sample off to the "mass spectrometrist" for mass analysis. With the development of improved instrumentation and methodologies based on ESI and MALDI in the near future, this interaction between the biochemist and the analytical chemist will be a common practice.

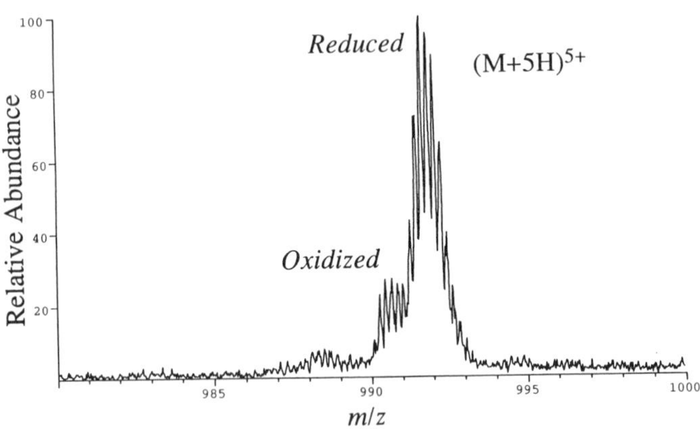

Figure 8. Mass spectrum of disulfide-reduced ATX II, showing the presence of the oxidized protein.

REFERENCES

Cody, R. B., Tamura, J., Finch, J. W., and Musselman, B. D., 1994, Improved detection limits for electrospray ionization on a magnetics sector mass spectrometer by using an array detector, *J. Am. Soc. Mass Spectrom.* 5:194-200.

Cody, R. B., Tamura, J., and Musselman, B. D., 1992, Electrospray ionization/magnetic sector mass spectrometry: calibration, resolution, and accurate mass measurements, *Anal. Chem.* 64:1561-1570.

Feng, R., Bell, A., Dumas, F., and Konishi, Y., 1990, Reduction/alkylation plus ionspray mass spectrometry: a fast and simple method for accurate counting of cysteines, disulfide bridges and free SH groups in proteins, *Proc. 38th ASMS Conf. on Mass Spectrom. Allied Topics*, Tucson, AZ: American Society for Mass Spectrometry: East Lansing, MI, pp 273-274.

Fenn, J. B., Mann, M., Meng, C. K., Wong, S. F., and Whitehouse, C. M., 1989, Electrospray ionization for mass spectrometry of large biomolecules, *Science* 246:64-71.

Hillenkamp, F., Karas, M., Beavis, R. C., and Chait, B. T., 1991, Matrix-sssisted laser desorption/ionization mass spectrometry of biopolymers, *Anal. Chem.* 63:1193A-1203A.

Karas, M., Bahr, U., Ingendoh, A., and Hillenkamp, F., 1989, Laser desorption/ionization mass spectrometry of proteins of mass 100 000 to 250 000 dalton, *Angew. Chem. Int. Ed. Engl.* 28:760-761.

Loo, J. A., Edmonds, C. G., Udseth, H. R., and Smith, R. D., 1990, Effect of reducing disulfide-containing proteins on electrospray ionization mass spectra, *Anal. Chem.* 62:693-698.

Loo, J. A., Ogorzalek Loo, R. R., and Andrews, P. C., 1993, Primary to quaternary protein structure determination with electrospray ionization and magnetic sector mass spectrometry, *Org. Mass Spectrom.* 28:1640-1649.

Loo, J. A. and Pesch, R., 1994, Sensitive and selective determination of proteins with electrospray ionization magnetic sector mass spectrometry and array detection, *Anal. Chem.*, 66: 3659-3663.

Matsudaira, P., 1987, Sequence from picomole quantities of proteins electroblotted onto polyvinylidene difluoride membranes, *J. Biol. Chem.* 262:10035-10038.

McCloskey, J. A. (Ed.)., 1990, *Mass Spectrometry*. San Diego, CA: Academic Press.

Smith, R. D., Loo, J. A., Edmonds, C. G., Barinaga, C. J., and Udseth, H. R., 1990, New developments in biochemical mass spectrometry: electrospray ionization, *Anal. Chem.* 62:882-899.

15

PEPTIDE-MASS FINGERPRINTING AS A TOOL FOR THE RAPID IDENTIFICATION AND MAPPING OF CELLULAR PROTEINS

D.J.C. Pappin,[1*] D. Rahman,[1] H.F. Hansen,[1] W. Jeffery,[1] and C.W. Sutton[2]

[1] Imperial Cancer Research Fund
44 Lincoln's Inn Fields, London WC2A 3PX, UK.
[2] Finnigan MAT Ltd.
Paradise, Hemel Hempstead, Herts HP2 4TG, UK.

INTRODUCTION

For more than 25 years protein identification has largely depended on automated Edman chemistry (Hewick et al., 1981) or western blotting with an appropriate monoclonal antibody. Several limitations, however, have never been overcome. The Edman procedure is inherently slow (generally one or two peptide or protein samples per day) and does not allow direct identification of many post-translational modifications. In addition, current detection limits are in the low-picomole to upper-femtomole range (Totty et al., 1992). Protein identification by western blotting can be extremely rapid, but requires the ready availability of an extensive library of suitable antibody probes. Large-format 2D-electrophoresis systems now make it possible to resolve several thousand cellular proteins from whole-cell lysates in the low- to upper-femtomole concentration range (Patton et al., 1990), presenting significant analytical challenges. The recent introduction of matrix-assisted laser-desorption (MALD) time-of-flight mass spectrometers (Karas and Hillenkamp, 1988) has led to the rapid analysis (at high sensitivity) of peptide mixtures. New strategies have been developed using a combination of protease digestion, MALD mass spectrometry and searching of peptide-mass databases that promise rapid acceleration in the identification of proteins (Henzel et al., 1993; Pappin et al., 1993; Mann et al., 1993; James et al., 1993; Yates et al., 1993).

Microsequence analysis of proteins electroblotted onto PVDF membranes following purification by SDS PAGE has become an essential tool for the protein chemist, and several procedures have been described for enzymatic cleavage of proteins bound to nitrocellulose or PVDF transfer membranes (Aebersold et al., 1987; Bauw et al., 1989; Fernandez et al., 1992). All these procedures require pre-treatment of membranes with polymers such as PVP-40 to prevent adsorption and denaturation of the proteolytic enzyme. This approach

[*] Author to whom correspondence should be addressed.

considerably increases both the level of background contamination during subsequent reverse-phase purification of peptides and the time required per sample for extensive washing and extraction of residual polymer. A significant improvement was made in the later work of Fernandez et al. (1994) where the use of hydrogenated Triton X-100 made blocking with PVP-40 redundant. One drawback to the use of heteropolymeric detergents, such as Triton, is that residual detergent interferes significantly when analysing peptides by MALD mass spectrometry. We report here on the development of simplified digestion methods using octyl glucoside that allow for the rapid, single step digestion of electroblotted proteins in a form suitable for both analysis by MALD mass spectroscopy or conventional Edman microsequencing. We also report here on the application of the procedure to the analysis of proteins resolved by large-format 2D electrophoresis of cellular proteins.

MATERIALS AND METHODS

2D SDS Polyacrylamide Gel Electrophoresis (PAGE)

Human Myocardial Proteins. Samples of human ventricular myocardium were taken from explanted hearts at the time of cardiac transplantation and frozen in liquid nitrogen. Frozen tissue specimens were then crushed between two cooled metal blocks. The resulting powder was homogenised in 1% w/v SDS, spun at 10,000 g for 5 min, rehomogenised and recentrifuged before harvesting the supernatant. Protein concentration was determined by the Bradford dye-binding assay and the samples stored frozen at -80°C.

Preparative 2D PAGE was performed using the Millipore Investigator system essentially as described by Patton et al. (1990). First-dimensional isoelectric focusing (IEF) was carried out in preparative rod gels (210 mm x 3 mm) containing 2.6% w/v acrylamide, 9M urea, 4%w/v CHAPS, 1% w/v DTT, 2% v/v Resolyte pH4-8 and 0.05% w/v Bromophenol blue. Samples containing up to 1 mg of total protein were applied to each gel. Gels were focused at 800V for 35,000 volt-hours (Vh), extruded onto parafilm strips and stored at -80°C. Second dimensional electrophoresis was performed on 230 x 200 x 1.5 mm SDS-PAGE gels (12% w/v acrylamide) with 2 cm stacking gels overnight at 3000 mW/gel (cooled to 10°C).

Following electrophoresis, the 2D gels were equilibrated for 30 min in 50 mM Tris/boric acid buffer, pH 8.5 (Baker et al., 1991). Proteins were then electroblotted onto FluoroTrans membranes (Pall Corp.) for 6 hr at 500 mA (10°C) using the ISO-DALT system (Hoeffer) and proteins visualised by staining with Coomassie brilliant blue or sulforhodamine B as described by Coull and Pappin (1990).

Mouse Brain Proteins. Brain tissue was prepared from parental, F1 and back-cross progeny by rinsing in physiological NaCl solution, freed from any blood vessels, then wiped dry and frozen in liquid nitrogen. Each frozen brain was pulverised in the presence of protease inhibitors in a minimum volume (v) of buffer (v μl=brain weight in mg x 0.03) and the ground tissue centrifuged at 145,000 g for 30 min. Pulverised material was then prepared and analysed by 2D gel electrophoresis essentially according to the method of Klose (1975), electroblotted onto Immobilon-P PVDF membrane (Millipore) and stained with sulforhodamine B as above.

1D SDS PAGE

Proteins were subjected to SDS polyacrylamide gel electrophoresis essentially according to Laemmli (1970). Gels were allowed to polymerise overnight to fully quench polymer free-radicals. Gel and running buffers contained only 0.02% w/v SDS, with 0.02% v/v thioglycolic acid added to the running buffer as a scavenger. Proteins were electroblotted onto Immobilon-P (PVDF) transfer membranes using the MilliBlot-SDE semi-dry electroblotting system (Millipore) and the low-ionic discontinuous 6-aminohexanoic acid buffer system. Transfer was typically accomplished by electroblotting at 1-1.5 mA/cm^2 of gel surface area for 45-60 minutes at constant current. The PVDF transfer membrane was then washed with deionised water to remove buffer salts (two changes of 200 ml, 10 minutes each with mild agitation), blotted dry with Whatman 3MM filter paper and thoroughly dried in vacuo for at least 20 minutes. Blotted proteins were visualised by staining with sulforhodamine as described above.

Digestion Procedure and Mass Spectrometric (MS) Analysis

Dried, stained spots (typically a few square mm in area) were placed in 0.5 ml eppendorf tubes and wet with 2-4 µl of 50 mM ammonium bicarbonate solution containing 1% w/v octyl glucoside and 40 ng trypsin/µl (Promega, modified). Enzyme solutions should be prepared immediately prior to use to minimise autodigestion. If using the sulforhodamine stain, it was not usually necessary to pre-wet the membrane pieces, destain, or block with polymers such as PVP-40. Digestion was performed overnight at 30°C. The following day, 10-20 µl of formic acid:ethanol (1:1 v/v, freshly prepared) was added to each sample, and the solution allowed to stand for 30-60 minutes to allow peptides to diffuse from the surface. Small aliquots (typically 0.5 µl) were sampled directly from the supernatant, applied to sample slides or strips, and dried under high vacuum for at least 30 minutes to remove residual ammonium salts. Each dry sample was then re-wet with 0.5 µl matrix solution (1% w/v alpha cyano-4-hydroxycinnamic acid in 50% aq. acetonitrile containing 0.1% TFA and 200 femtomoles/µl oxidised insulin B chain), allowed to air dry then analysed by MALD time-of-flight mass spectrometry using a Finnigan MAT LaserMat 2000 mass spectrometer (Mock et al., 1992). Spectra were calibrated using the insulin B chain as an internal standard. Observed proteolytic fragment masses were screened against the MOWSE peptide-mass database as described by Pappin et al. (1993).

The main practical difficulty encountered with the laser instruments is in finding the optimum concentration of peptides (particularly to overcome quenching). Typically, this was achieved by dilution of the digest supernatant. Thus, having sampled an initial 0.5 µl from the digest as above, additional 10 µl aliquots of the formic:ethanol mix were added, allowed to stand 30 min, and 0.5 µl sampled again for MS analysis as above. This was repeated as many times as necessary to obtain the optimum spectrum.

Purification of Peptides by Reverse-Phase HPLC

In cases where MS fingerprint analysis failed to identify the protein, the remaining volume of digest supernatant was collected, dried in vacuo, and re-dissolved in 10-30 µl 0.5% v/v heptafluorobutyric acid (HFBA) in water. The samples were then injected onto narrow-bore C8 or C18 reverse-phase columns (e.g. Aquapore RP-300, 2.1 mm x 10 cm) equilibrated with 0.025% w/v HFBA. Peptides were eluted with linear gradients (2-80%) of acetonitrile containing 0.05% v/v TFA at 0.1-0.2 ml/min over 60-80 minutes, monitored at 220 nm.

Peptide Microsequencing

HPLC purified peptides were collected, dried on to 8 mm arylamine-substituted PVDF membranes (Millipore) and covalently immobilised via carboxyl groups as described by Coull et al. (1991). Solid-phase microsequence analysis was performed on a MilliGen 6600 sequencer as described by Pappin et al. (1990).

RESULTS AND DISCUSSION

Digestion Procedure

The digestion procedure reported here was developed for the rapid screening of proteins by peptide-mass fingerprinting, primarily derived from large-format 2D gel electrophoresis of whole cell lysates. A brief comparison of the process with more typical on-membrane digestion procedures is shown in Figure 1. Key differences are the use of octyl glucoside, very low digestion volumes (typically only a few μl) and the use of formic acid and ethanol solvents to achieve efficient elution of digested peptides. The minimal time and manipulation required per sample allows the simultaneous processing of many dozens of samples in parallel.

Proteins are transferred to PVDF membranes, stained, and digested in the presence of 1% w/v octyl glucoside. Using this treatment, it is not necessary to pre-wet the membrane pieces or block with polymers such as PVP-40. If proteins are stained with sulforhodamine B or Ponceau S there is also no requirement to destain. The ability to use underivatised PVDF membranes and conventional stains side-steps more complicated staining and digestion procedures required when using positively-charged (cationic) blotting membranes such as Immobilon CD (Patterson et al., 1992). The low digestion volumes, typically just sufficient to wet the membrane pieces, significantly improve digestion kinetics for a given concentration of enzyme. For reasons that are not yet clear, we have found that the use of lower digestion temperatures (room temperature up to 30°C) give more complete digestion than identical experiments performed at 37°C.

Following overnight digestion, peptides are passively eluted from the membrane surface with formic acid and ethanol (1:1 v/v). Small aliquots can be sampled directly from the digest supernatant and analysed by matrix-assisted laser-desorption (MALD) mass spectrometry. In contrast to the reported use of heteropolymeric detergents (e.g. hydrogenated Triton X-100, Nonidet, Emulphogene) the presence of up to 1% w/v octyl glucoside has no apparent effect on the ionisation efficiency of digested peptide mixtures. In addition, the low molecular weight of this detergent does not interfere with peptide mass determination in the low-mass range (700-3000 Da). The described procedure thus has clear advantages over the use of hydrogenated Triton X-100 as reported by Fernandez et al. (1994). Residual buffer salts and stains such as sulforhodamine B or Ponceau S also have minimal suppression on the laser-desorption process. This is in very sharp contrast to the use of Coomassie blue which has a very significant suppressive effect.

The digest procedure was tested over a period of several months on proteins recovered following 1D and 2D gel electrophoresis and identified by peptide-mass fingerprinting using the MOWSE database (Pappin et al., 1993). A selection of proteins successfully identified using this process is shown in Table I. Identities of all the samples shown in Table I were confirmed by microsequencing, western blotting with an appropriate antibody or (in two cases) by genetic mapping to a precise genetic locus. Samples included human myocardial proteins, recombinant proteins, viral capsid and regulatory proteins and proteins resolved by 2D gel electrophoresis of whole mouse brain. Sample molecular weights ranged from 14.5 kDa to over 376 kDa.

The efficiency of digestion and recovery of peptides was measured by several parameters, summarised in Table II. The data was derived entirely from those proteins

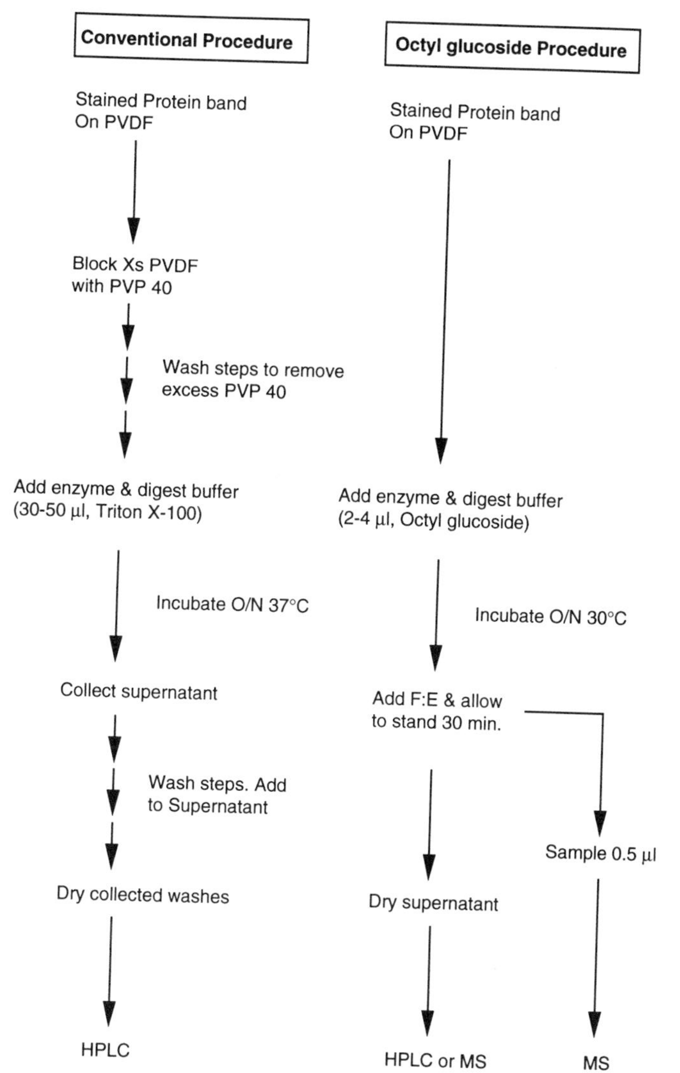

Figure 1. Comparison of enzyme digestion schemes. The conventional procedure is an amalgamation of several described methods but is principally derived from Aebersold et al. (1987) and Fernandez et al. (1992). F:E; formic acid: ethanol (1:1 v/v).

correctly identified by peptide-mass fingerprinting and whose identities were confirmed by sequence analysis, western blotting or genetic mapping.

In 42 proteins analysed by peptide fingerprinting, a total of 281 peptides were recorded where observed peptide masses could be matched with expected peptide sequences. Detailed analysis of the mass spectra recorded for several proteins showed that between 60 and 80% of all expected peptides between 700 and 3000 Da were present in the digest supernatant (passively eluted from the membranes by the formic acid:ethanol solution). In a few cases, such as the 161 kDa nitric oxide synthetase (Table I), almost 90% of all possible peptides in this size range were present in the mass spectrum. Matched peptide masses were further sorted into 'perfect' or 'partial' cleavage products. For trypsin, the MOWSE database

Table I. Protein samples identified by peptide-mass fingerprinting after enzymatic digestion using the PVDF/octyl glucoside procedure

Protein	Source	Size (kD)	Method	Confirm	Peptides used	Total hits	Partial cleavages	Perfect cleavages
Fatty acid binding protein	Human heart	14.9	2D	S	9	6	2	4
Apha crystallin B chain	Human heart	20.2	2D	S	14	8	4	4
Superoxide dismutase	Human heart	24.7	2D	S	10	7	2	5
Actin, aortic smooth muscle	Human heart	42.0	2D	S	5	4	0	4
Desmin	Human heart	53.5	2D	S	9	6	2	4
Heat shock protein 70	Human heart	69.9	2D	S	7	5	0	5
Glucose regulated protein	Human heart	73.8	2D	S	16	11	1	10
NADH ubiquitone reductase	Human heart	79.6	2D	S	8	6	0	6
Tumour necrosis factor	Recombinant	16.8	1D	S	9	6	3	3
Beta lactoglobulin	Bovine	18.3	2D	S	12	7	1	6
Bm3R1 repressor protein	Recombinant	21.9	1D	S	12	5	0	5
Carbonic anhydrase II	Bovine	28.9	1D	S	10	7	0	7
UL49 tegument protein	Herpes simplex	32.3	1D	S	6	4	0	4
Urate oxidase (Uricase)	Rat liver	34.8	1D	S	15	10	3	7
MAD3 (MHC enhancer protein)	Human	35.6	1D	A	10	7	1	6
Actin (cytoplasmic)	Rat liver	41.8	1D	S	10	8	0	8
Glycoprotein D	Herpes simplex	43.3	1D	S	5	4	0	4
P53 cellular tumour antigen	Recombinant	43.7	1D	S	16	12	2	10
Tubulin B-chain	Human T-cell	49.8	1D	S	8	5	1	4
TPL-2 homologue	Human T-cell	52.8	1D	A	13	6	0	6
Keratin, type I cytoskeletal	Rat Liver	65.5	1D	S	19	12	0	12
Heat shock protein 70-C	Human	70.8	1D	A	9	7	1	6
78 kD glucose regulated protein	Human	72.1	1D	A	10	8	1	7
Mortalin (HSP 70 family)	Mouse	73.8	1D	S	25	22	4	18

Nucleolin (C23 protein)	Human T-cell	76.2	1D	S	12	9	1	8
Heat shock protein 90-beta	Human	83.2	1D	A	11	8	1	7
Heat shock protein 90	Rat liver	83.3	1D	S	16	10	0	10
Antigenic structural protein (P100)	Herpes simplex	97.1	1D	S	7	5	0	5
Alpha-3 integrin (VLA-3)	Human keratinocyte	113.5	1D	A	26	13	0	13
Major capsid protein (VP5)	Herpes simplex	149.1	1D	S	8	6	2	4
Nitric oxide synthase	Rat brain	160.6	1D	S	28	28	2	26
376kD Golgi complex protein	Rat liver	376.1	1D	A	28	19	5	14
Synuclein SYN-2	Mouse brain	14.5	2D	S	6	6	1	5
Calcineurin B regulatory subunit	Mouse brain	24.9	2D	S	8	4	1	3
PKC inhibitor protein (14-3-3)	Mouse brain	28.0	2D	S	9	5	0	5
Lactate dehydrogenase H-chain	Mouse brain	36.5	2D	GM	7	6	0	6
Creatine kinase B-chain	Mouse brain	42.7	2D	S	8	5	0	5
Gamma enolase	Mouse brain	44.2	2D	GM	12	8	0	8
Alpha tubulin	Mouse brain	50.1	2D	S	12	6	0	6
Heat shock protein 70	Mouse brain	70.9	2D	S	11	8	2	6

Protein Identity confirmed by microsequencing (S), western blotting with appropriate monoclonal antibody (A) or genetic mapping (GM).

Table II. Assessment of digestion efficiency and recovery of peptides by passive elution with formic acid:ethanol

Trypsin	Solution digest	PVDF/octyl glucoside
Total protein samples	23	42
Total matched peptides	168	281
% Observed (700–3000 Da)	65–85	60–80
% Perfect cleavage	85.1	85.8
% Partial cleavage (nnp†)	14.9	14.2

†nnp: Nearest-neighbour pair

classifies a 'perfect' cleavage where cleavage has occurred directly C-terminal to a lysine or arginine residue except where the adjacent residue is proline (Pappin et al., 1993). The database also matches all cases of nearest-neighbour partial cleavages where cleavage has failed to occur. Examination of the 281 matched peptide masses showed that the large majority (85.8%) were derived from 'perfect' enzyme cleavages with only 14.2% resulting from nearest-neighbour partial cleavages - a ratio of almost 6:1 in favour of complete cleavage. By comparison, Table II also includes the same data obtained from 23 different protein samples (reported in the earlier study of Pappin et al., 1993) where the samples had been digested in solution. There is almost exact correspondence in all categories, including the percentages of possible peptides observed in the mass spectra and ratio of 'perfect' to 'partial' fragments. Using these criteria, on-membrane PVDF/octyl glucoside digestion and recovery of peptides by passive elution using formic acid and ethanol gives results that are indistinguishable from those obtained by digestion in solution. The validity of this result is enhanced by the fact that the data were obtained on large numbers of a wide variety of proteins derived from very different biological sources.

Influence of Adjacent Residues on Proteolytic Cleavage

Analysis of all observed nearest-neighbour pair (nnp) partial-cleavage fragments revealed significant information relating to the influence of residues C-terminal to the potential cleavage site (Figure 2).

In the case of trypsin, almost 50% of observed partial cleavages occurred where the enzyme was presented with adjacent lysine or arginine residues (RR, RK, KK, and KR pairs or longer repeats). This can result in peptides with pairs of basic residues at the C-terminus or the presence of single uncleaved lysine or arginine residues at the N-terminus (observed in almost equal numbers). This reflects the well documented poor exopeptidase efficiency of trypsin (Allen, 1989). The second largest group (approx. 25%) arises from the inhibition of cleavage where the cleavage site is followed by one of the large, hydrophobic residues L, I, F or V, where steric hindrance slows the absolute rate of cleavage. Trypsin also seems to be adversely influenced by the close proximity of acidic residues (E or D). The eight residues (E, D, L, I, F, V, K, and R) are thus associated with 88% of all observed incomplete cleavages, with the remaining 13 common amino acids present in only 12% of cases.

HPLC Purification of Recovered Peptides

The digest method was evolved principally to allow rapid screening of multiple samples by peptide-mass fingerprinting using MALD mass spectroscopy. In cases where analysis of small aliquots of the digest supernatant failed to identify the protein, the

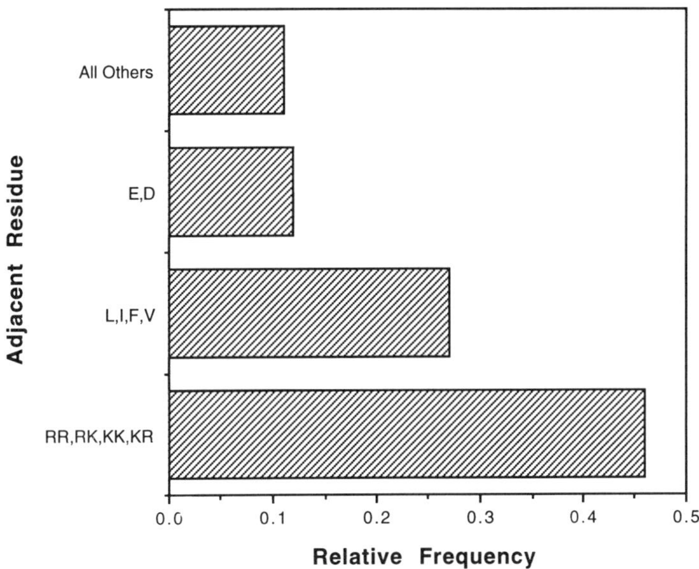

Figure 2. Observed frequencies of residues adjacent to potential cleavage sites that result in incomplete (partial) cleavage.

remaining bulk of the peptides can be purified by narrow- or microbore reverse-phase HPLC for conventional Edman microsequencing .

An example of a typical HPLC purification is shown in Figure 3A.

Approximately 15 pmol of a protein with apparent Mw of 80,000 Da (SDS PAGE) was blotted, stained with sulforhodamine B and digested on the PVDF membrane as described. Peptide fingerprint analysis of the tryptic fragments (Figure 4) identified the protein as human C23 nucleolin. This analysis was performed using only 2% of the total digest sample (0.5 μl sampled from 25 μl total volume).

The remaining bulk of the digest supernatant was collected (approx. 23 μl), dried in vacuo, redissolved in 20 μl 0.5 % v/v HFBA and injected onto an Aquapore RP-300 (C8) column (2.1 mm x 10 cm) equilibrated with 0.025% v/v HFBA. No attempts were made to re-extract peptides from the residual PVDF membrane pieces by repeated solvent washes. Peptides were eluted with a gradient of acetonitrile containing 0.05% v/v TFA (2-80% over 80 minutes at 0.2 ml/min). The presence of the residual 1% octyl glucoside minimises non-specific adsorption of peptide onto the walls of the tube and aids solubilisation. On injection, the detergent does not retain significantly and elutes at the column void (V). The presence of HFBA is very important at this stage to retain small peptides by acting as a hydrophobic ion-pairing agent. Without HFBA, small peptides are eluted by the residual detergent. As the concentration of acetonitrile rises the HFBA counterion is replaced by TFA, reducing the overall hydrophobicity of bound peptides. The concentration of TFA in the eluting acetonitrile (at 0.05% v/v) is sufficient to balance the UV absorbance at 214-220 nm. As noted by Fernandez et al. (1994) the HPLC background profiles are significantly improved by the absence of PVP-40, particularly when working at high sensitivity. The only major contaminating peak, eluting at approximately 40% ACN, is the residual sulforhodamine B dye (SR). A number of peptides were collected and subjected to solid-phase microsequence analysis as described in methods (sequences shown in Fig. 3A) with initial

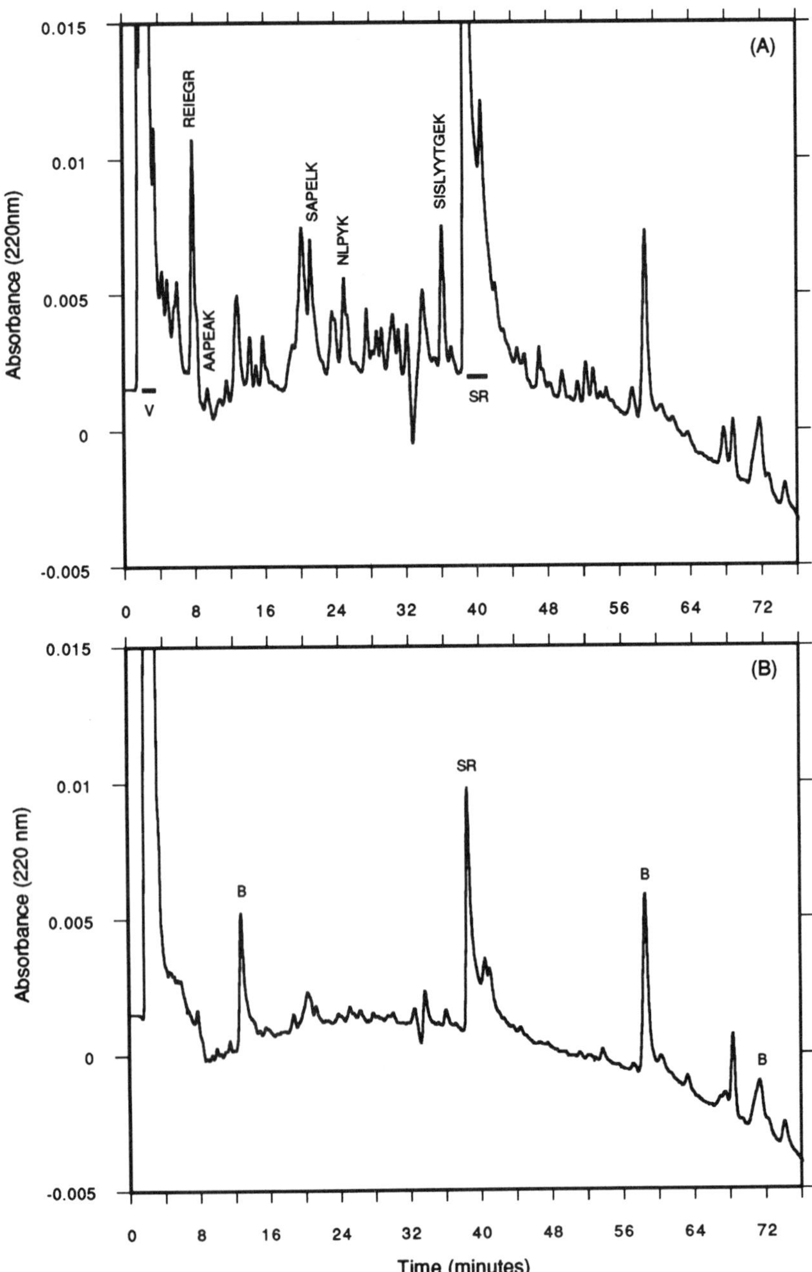

Figure 3. HPLC profile of peptides derived by on-membrane digest of 15 pmol human C23 nucleolin (experimental conditions described in the text). (3A) peptides recovered by passive elution into formic acid and ethanol added post-digest. (3B) repeat elution gradient of washings of the residual membrane pieces with aq. propanol and TFA. (V) represents the column void, (SR) the excess sulforhodamine B stain and (B) denotes those peaks present in blank gradients. 5 peaks were collected and sequenced by solid-phase microsequencing (see text).

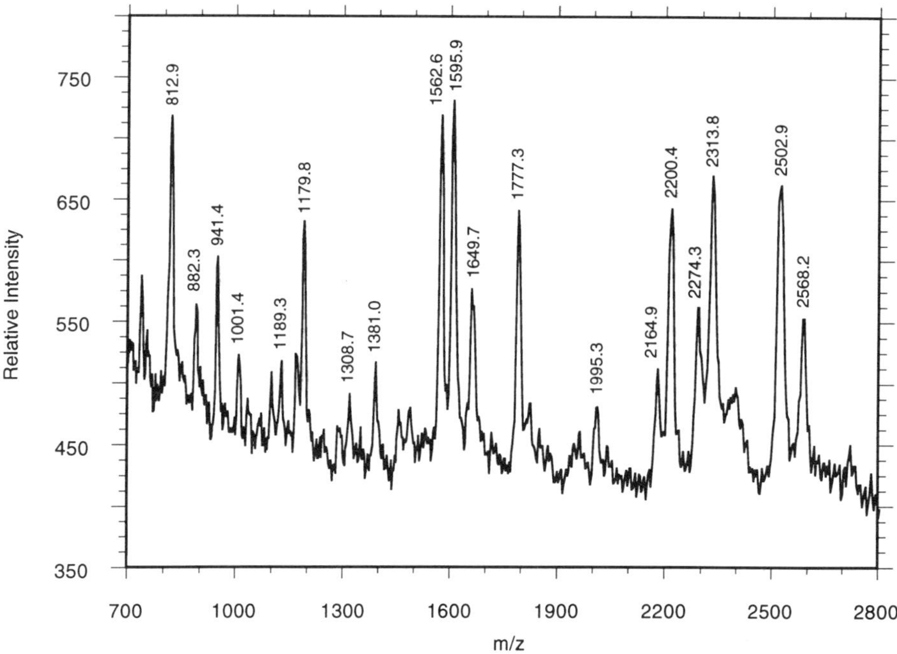

Figure 4. MALD time-of-flight mass spectrum of the tryptic peptides derived by on-membrane digest of 15 pmol human C23 nucleolin. The spectrum was obtained using 0.5 µl of the digest supernatant, corresponding to approximately 2% of the total sample.

yields of between 3 and 6 pmol. All peptide sequences corresponded exactly to sequences reported for human nucleolin C23 (Srivastava et al., 1990).

The remaining membrane pieces were then washed with 100 µl aliquots of 70% v/v isopropanol/0.1% v/v TFA (twice) and 100 µl aliquots of TFA (twice). The pooled washings were dried, redissolved in 20 µl 1% v/v HFBA and chromatographed under identical conditions (Figure 3B). Apart from a small number of background peaks present in control blank HPLC gradients (labelled B) and a small amount of residual stain (SR), the profile shows that 85-90% of the peptides had already been recovered from the membrane by passive elution into the added formic acid:ethanol solution. When handling large numbers of samples (several hundred) it is thus not necessary to exhaustively wash the membrane pieces by repeated solvent extraction. The results clearly demonstrate that the PVDF/octyl glucoside digestion can be used as a one-step procedure for the preparation of peptides suitable for analysis by MALD mass spectroscopy as well as for high-sensitivity Edman microsequencing.

Peptide Mapping of Cellular Proteins

Over the previous 12 months we have been using the described digestion procedure to analyse cellular proteins resolved by large-format 2D gel electrophoresis. Davison and Davison (1994) have used the method to identify proteins by SDS PAGE of channel catfish virus (a member of the herpesvirus family). The study has led to successful identification of 16 principal virion or capsid proteins encoded by 12 genes (11 viral and one from the host cell). The identification of viral proteins in this manner is particularly apt in that complete genome information (and hence all possible protein coding regions) are available for many viruses under study. Corbett et al. (1994) have used the procedure to identify more than 50 human myocardial proteins resolved by 2D gel electrophoresis of human heart tissue (a few examples of which are shown in Table I). In our own laboratory, we have been analysing proteins resolved by 2D PAGE of whole mouse brain as part of a large-scale genetic mapping study (Vogel and Klose, 1992). Some of the identified proteins are shown in Table I, all identified using single spots from only one gel experiment. Our study is continuing, with more than 240 individual protein spots analysed to date. Several features of the preliminary data are worth discussion.

Estimation of protein amount is difficult. From the observed intensity of the sulforhodamine stain and signal intensities recorded during MS analysis, sample amounts have ranged from tens of picomoles to less than 100 femtomoles before enzyme digestion. Rough estimates can be made by comparing the observed peptide signal intensities relative to the standard loading of insulin B chain used as internal calibrant (see Methods). From our work to date we are reasonably confident that useful spectra can be obtained from most of the proteins revealed by staining with the sulforhodamine dye. In some cases, the spots were only visible when illuminated with short-wave UV light to fully exploit the fluorescent properties of the stain (Coull and Pappin, 1990). This is well below the level at which proteins can be visualised by staining with Coomassie blue. Proteins successfully identified included creatine kinase B-chain, gamma enolase, and lactate dehydrogenase H-chain (all brain isoforms). This latter identification was of particular interest in that identity was confirmed by subsequent mapping of the spot to a precise genetic locus (J. Klose, personal communication). One surprising feature that emerged was the number of identified proteins that existed in more than one isoform. For example, the gamma enolase protein was identified in at least six separate forms over a Mw range of 35-40 kDa and approximately 1 pH unit in the IEF dimension. This was later confirmed by the finding that all the relevant protein spots again mapped to the same genetic locus. More intriguingly, several forms of the creatine kinase B-chain were observed ranging in apparent Mw from 28-45 kDa and nearly 2 units of pH. More detailed analysis is required to fully determine the cause of such heterogeneity.

Other features noted are as follows. The use of Promega modified trypsin is particularly advantageous in that autolytic digestion products are minor and do not generally interfere with the peptide spectra. Methionine containing peptides have been frequently observed in the oxidised form as the sulphoxide (+16 Da), the ratio of native/oxidised forms being variable and generally dependent on the age and/or mistreatment of blotted samples. In many instances we have observed cysteine-containing peptides as the beta-propionamide adduct (+71 Da) following reaction with unpolymerised acrylamide (Brune, 1992). Such modifications can be essentially quantitative after 2D gel electrophoresis if no effort is made to maintain reducing conditions. We have observed, however, that inclusion of thiol scavengers in both dimensions can maintain cysteine residues as the free thiol (see Methods). One final observation is that arginine containing peptides are more likely to be observed at extreme sensitivity, presumably reflecting the much greater basicity of the guanidine side-chain relative to a primary amine, with significant enhancement of ionisation efficiency. We are currently experimenting with modification of peptides with quaternary ammonium groups to further improve effective sensitivity in the presence of residual buffer, detergent and stain (Bartlet-Jones et al., 1994).

CONCLUSIONS

We have developed the PVDF/octyl glucoside procedure for the rapid analysis of cellular proteins by peptide-mass fingerprinting. The method is sensitive enough to work in the sub-picomole range, and sufficient material can be derived from a single 2D gel separation. The procedure is now in routine use in a number of large-scale analytical projects. These studies have confirmed that the analysis of cellular proteins by peptide-mass fingerprinting can provide a rapid, direct link between protein and DNA information and may supersede the use of Edman sequencing for these and related projects.

ACKNOWLEDGMENTS

This work was supported by the Imperial Cancer Research Fund

REFERENCES

Aebersold, R.H., Leavitt, J., Saavedra, R.A., Hood, L.E. and Kent, S.B. (1987) Proc. Natnl. Acad. Sci. U.S.A. 84, 6970-6974.

Allen, G. in Sequencing of Proteins and Peptides 2nd. Edition, 73-104 (Elsevier, Amsterdam, 1989).

Baker, C.S., Dunn, M.J. and Yacoub, M.H. (1991) Electrophoresis 12, 342-348.

Bartlet-Jones, M., Jeffery, W., Hansen, H.F. and Pappin, D.J.C. (1994) Rapid Commun. Mass Spectrom., in press.

Bauw, G., Van Damme, J., Puype, M., Vandekerckhove, J., Gesser, B., Ratz, G.P., Lauridsen, J.B. and Celis, J.E. (1989) Proc. Natnl. Acad. Sci. U.S.A. 86, 7701-7705.

Brune, D.C. (1992) Anal. Biochem. 207, 285-290.

Corbett, J.M., Wheeler, C., Dunn, M.J., Pappin, D.J.C., Pemberton, K.S., Sutton, C.W. and Cottrell, J.S. (1994) Poster presentation at the 9th Symposium of the Protein Society, San Diego.

Coull, J.M. and Pappin, D.J.C. (1990) J. Protein Chem. 9, 259-260.

Coull, J.M., Pappin, D.J.C., Mark, J., Aebersold, R.H. and Koster, H. (1991) Anal. Biochem. 194, 110-120.

Davison, A.J. and Davison, M.D. (1994) Virology, in press.

Fernandez, J., DeMott, M., Atherton, D. and Mische, S.M. (1992) Anal. Biochem. 201, 255-264.

Fernandez, J., Andrews, L. and Mische, S.M. (1994) Anal. Biochem. 218, 112-117.

Henzel, W.J., Billeci, T.M., Stults, J.T., Wong, S.C., Grimely, C. and Watanabe, C. (1993) Proc. Natnl. Acad. Sci. U.S.A. 90, 5011-5015.

Hewick, R.M., Hunkapiller, M.W., Hood, L.E. and Dreyer, W.J. (1981) J. Biol. Chem. 256, 7990-7997.

James, P., Quadroni, M., Carafoli, E. and Gonnet, G. (1993) Biochem. Biophys. Res. Commun. 195, 58-64.

Karas, M. and Hillenkamp, F. (1988) Anal. Chem. 60, 2299-2301.

Klose, J. (1975) Human Genetics 26, 231-243.

Laemmli, U.K. (1970) Nature 227, 680-685.

Mann, M., Hojrup, P. and Roepstorff, P. (1993) Biol. Mass Spectrom. 22, 338-345.

Mock, K,K., Sutton, C.W. and Cottrell, J.S. (1992) Rapid Commun. Mass Spectrom. 6, 233-238.

Pappin, D.J.C., Coull, J.M. and Koster, H. (1990) Anal. Biochem. 187, 10-19.

Pappin, D.J.C., Hojrup, P. and Bleasby, A.J. (1993) Current Biology 3, 327-332.

Patterson, S.D., Hess, D., Yungwirth, T. and Aebersold, R. (1992) Anal. Biochem. 202, 193-203.

Patton, W.F., Pluskal, M.G., Skea, W.M., Buecker, J.L., Lopez, M.F., Zimmermann, R., Belanger, L.M and Hatch P.D (1990) Biotechniques 8, 518-527.

Srivastava, M., Mcbride, O.W., Fleming, P.J., Pollard, H.B. and Burns, A.L. (1990) J. Biol. Chem. 265, 14922-14931.

Totty, N.F., Waterfield, M.D. and Hsuan, J.J. (1992) protein Sci. 1, 1215-1224.

Yates, J.R., Speicher, S., Griffin, P.R. and Hunkapiller, T. (1993) Anal. Biochem. 214, 397-408.

Vogel, T., and Klose, J. (1992) Biochemical Genetics 30, 649-662.

16

THE USE OF MALDITOF MASS SPECTROMETRY AND AMINO ACID SEQUENCE ANALYSIS IN CHARACTERISING SMALL AMOUNTS OF N-BLOCKED PROTEIN

Leonard C Packman,[1] Carl Webster,[2] and John Gray[2]

[1] Department of Biochemistry
[2] Department of Plant Sciences
Cambridge Centre for Molecular Recognition
University of Cambridge
Tennis Court Road, Cambridge, CB2 1QW, United Kingdom

SUMMARY

One hundred picomoles of a high mobility group (HMG) protein were isolated by reverse phase hplc from an extract of pea nuclei. Amino acid sequence and composition analysis of half the sample showed the protein to be blocked at the N-terminus. Of the remaining material, 40pmol was subjected to digestion with proteinases and the peptides from the tryptic digest were separated on reverse phase hplc. Each eluted peak was examined by matrix-assisted laser desorption time-of-flight (MALDITOF) mass spectrometry and amino acid sequence analysis. From the resulting information, it was clear that the target protein sequence correlated with the inferred sequence of a previously isolated pea leaf cDNA encoding an HMG-I-like protein. The expressed protein was smaller than the DNA sequence suggested. From a series of further digests on 1-2pmol of protein, the likely identity of the N-terminal block was established as well as several sites of C-terminal processing. This works illustrates how extensive amounts of data can be derived from a small amount of protein by the combined use of sequence analysis and MALDITOF mass spectrometry.

INTRODUCTION

High Mobility Group (HMG) proteins are abundant non-histone proteins associated with eucaryotic chromatin (Johns, 1982; Bustin et al., 1990). Their precise function is unclear but they probably play a role in regulating gene expression by affecting the conformation of

Methods in Protein Structure Analysis, Edited by M. Z. Atassi and E. Appella
Plenum Press, New York, 1995

chromatin. HMG proteins are small (<30kDa), acid-soluble, with strongly acidic and basic domains. As a group, they are subject to a variety of post-translational modifications (glycosylation, phosphorylation). In this study, structural features of an HMG protein from pea leaf nuclei are elucidated through mass and sequence analysis.

METHODS

Preparation of HMG Proteins

HMG protein was prepared from pea shoots. Chromatin was isolated from homogenised tissue by differential centrifugation and the bound HMG proteins were removed by a salt wash with 350mM NaCl (Pwee et al., 1994). Unwanted proteins were removed by precipitation with 2% TCA; the soluble (HMG) proteins were recovered by acetone precipitation. Final purification of the HMG proteins was by reverse phase hplc (Aquapore C4, 2.1x30mm) in 0.1%TFA with a gradient of acetonitrile/0.1%TFA. The position of HMG proteins in the eluate was determined by gel-retardation assays against an isolated fragment of DNA (268bp region of the pea plastocyanin promoter) (Pwee et al, 1994).

MALDITOF Mass Spectrometry

Mass spectrometer - MALDI III (Kratos) operated in linear mode.

Matrix - α-cyano,4-hydroxycinnamic acid (Sigma), 10mg/ml in 50% ethanol, 0.1%TFA, made fresh each day.

Sample preparation - 0.2μl sample was mixed on-slide with 0.5μl matrix and air dried. Each sample spot was then washed briefly with water to remove salts which may suppress the ionisation, and air dried again.

Calibration (external) of the appropriate mass ranges was done with a series of custom-made synthetic peptides 755.9 - 5961.9Da, and horse heart myoglobin (16952Da).

Proteolysis

Sequence-grade proteinases (Trypsin, Lys-C, Glu-C) were from Boehringer. Digestion was at 37°C for 0.5-3h, 100mM NH_4HCO_3;using a substrate:proteinase ratio of approx. 50:1. Samples were incubated in a closed microcentrifuge tube (0.4ml) or in a microdigestion reactor (see below).

Microdigestion Reactor

As part of this work, a quick and easy method was developed to allow digestions to occur in less than 1μl of solution. A sheet of laboratory sealing film (e.g. Nescofilm), which has a non-wettable but mouldable surface, was placed on a metal plate. Protein solution, buffer and proteinases were mixed together on the surface in sub-μl amounts using a pipette tip. The centre of a screw-cap top from a 1.4ml tube was filled with paper tissue and dampened with water. Excess water was blotted away and the cap was centred over the digestion mixture and pressed firmly into the film surface, creating a small humidity chamber (Fig. 1). The metal plate was placed into an oven or hot room at 37°C room for up to 3 hours. (A hot-block *cannot* be used because distillation of the sample solvent into the cooler tissue occurs.)

The sample was allowed to cool to room temperature; when the cap was removed, the digest was sampled for mass spectrometry. The same protocol can be done directly on

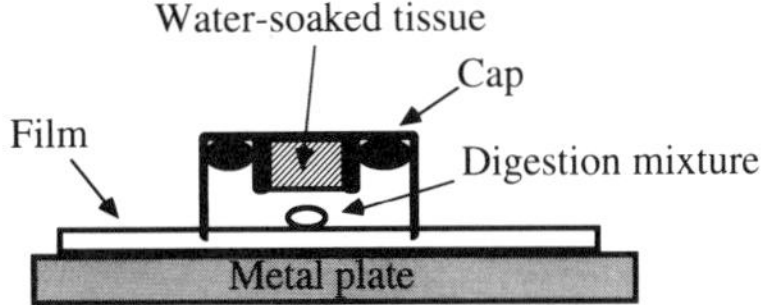

Figure 1. Microdigestion reactor design for the digestion of sub-microlitre quantities of protein.

the sample slide using a gasket of film but this is more manipulative and prevents the use of adjacent sample positions.

RESULTS

A purification profile of HMG proteins by reverse phase hplc is shown in Fig 2.

Analysis of the arrowed protein (single band on SDS gels, 100pmol by amino acid analysis) by MALDITOF mass spectrometry revealed a heterogeneous population (Fig. 3) with masses ranging from 16750-17150Da. Mass analysis was unsuccessful on the neat fraction; the sample had to be concentrated 10x under vacuum to attain a satisfactory signal and 2-3pmol of protein was required. Mass analysis of a similarly sized standard protein (horse heart myoglobin) gave a homogeneous sharp peak, suggesting that the observed heterogeneity in the HMG protein was real, and not an artefact of the analysis procedure. Such heterogeneity could result from ragged N- or C-terminal processing (if present) and/or incomplete post-translational modification at one or more sites.

Amino Acid Sequence Analysis

Sequence analysis of 50pmol of the sample gave no N-terminal sequence. Quantitation of the material on the sequencer disc confirmed the protein to be N-blocked. An attempt to release a free N-terminus through "on-disc" cyanogen bromide digestion of the sequenced

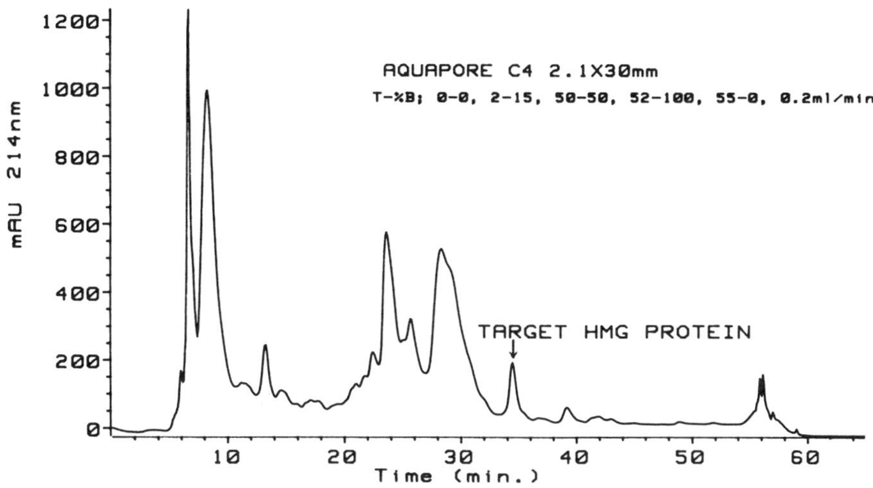

Figure 2. Reverse phase hplc purification of HMG protein from pea nuclei.

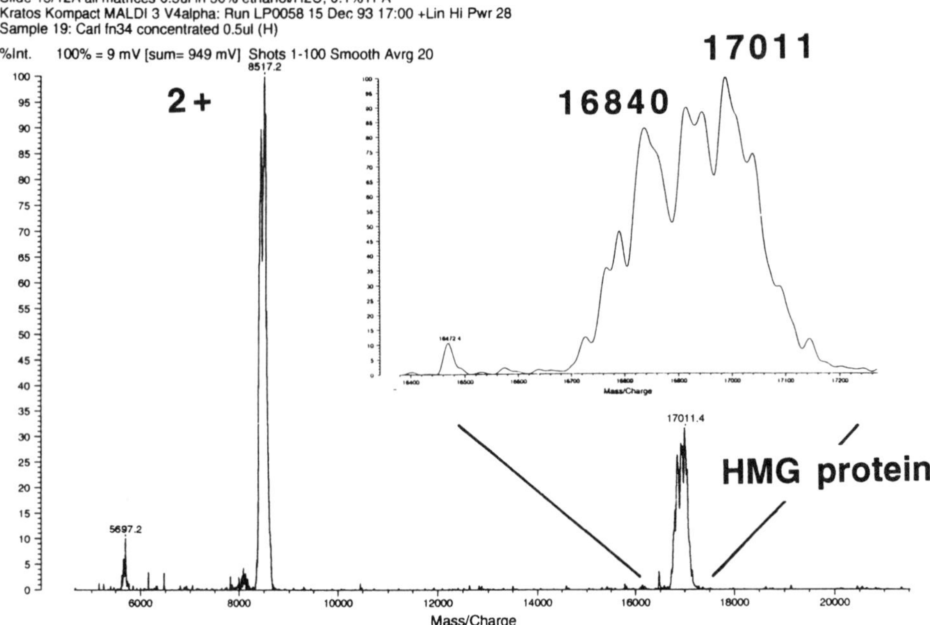

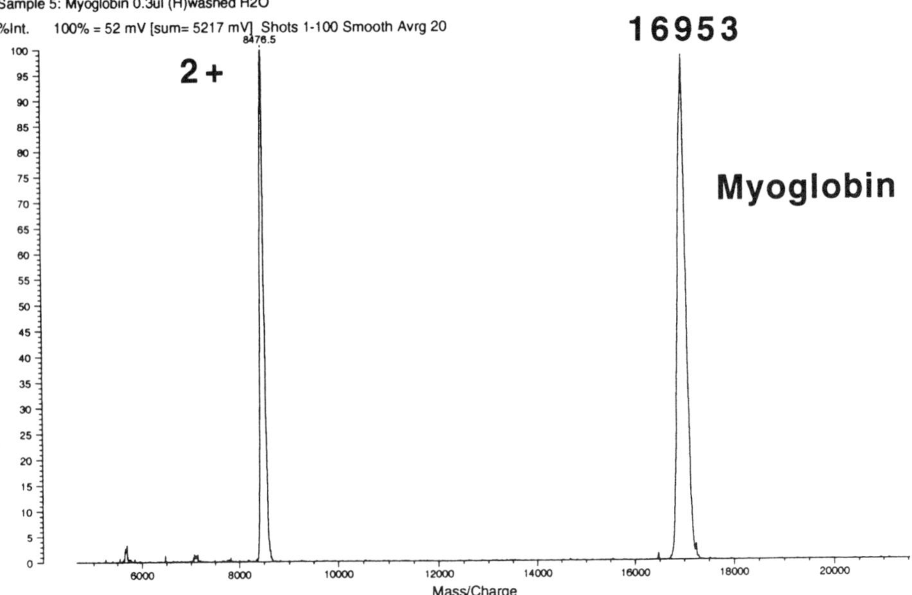

Figure 3. Mass analysis of intact purified HMG protein, demonstrating the heterogeneity of molecular size. Calibration was performed using myoglobin (lower panel).

material was unsuccessful. Trypsin digestion of a small (1-2pmol) sample of material and scanning of the masses of the fragments through the OWL database using the MOWSE program (Pappin et al., 1993) showed the HMG protein sequence to be absent. It was therefore necessary to commit a substantial amount of the remaining material to fragmentation for internal sequence analysis.

Proteolysis

40pmol of HMG protein was digested in 10μl of buffer with trypsin and the peptides separated by reverse phase hplc on Spherisorb ODS2 C18, 5μm, 1x100mm, 0.075ml/min in 0.1%TFA with acetonitrile gradient (data not shown). Fractions (70μl) were analysed directly by MALDITOF mass spectrometry (0.2μl) and sequence analysis (50-60μl). The amino acid sequencing clearly identified the protein as being the product of a pea leaf cDNA encoding an HMG-I-like protein (C. Webster and J. Gray, unpublished data). The mass analysis matched the expected length of each peptide identified from the sequence data, inferred from the DNA (Fig. 4) but no N-terminal fragment was evident. The mass of the protein from the DNA sequence (20470Da) exceeded considerably that measured for the intact protein (approx. 17kDa), showing this protein to be C-terminally processed.

Further digests were performed on small amounts of material with a view to identifying, by mass-mapping, other segments of the protein sequence. Approx. 2pmol of HMG protein was subjected to digestion with Lys-C proteinase in 2μl of buffer. A sample (0.2μl) was examined by MALDITOF mass spectrometry and the masses were scanned

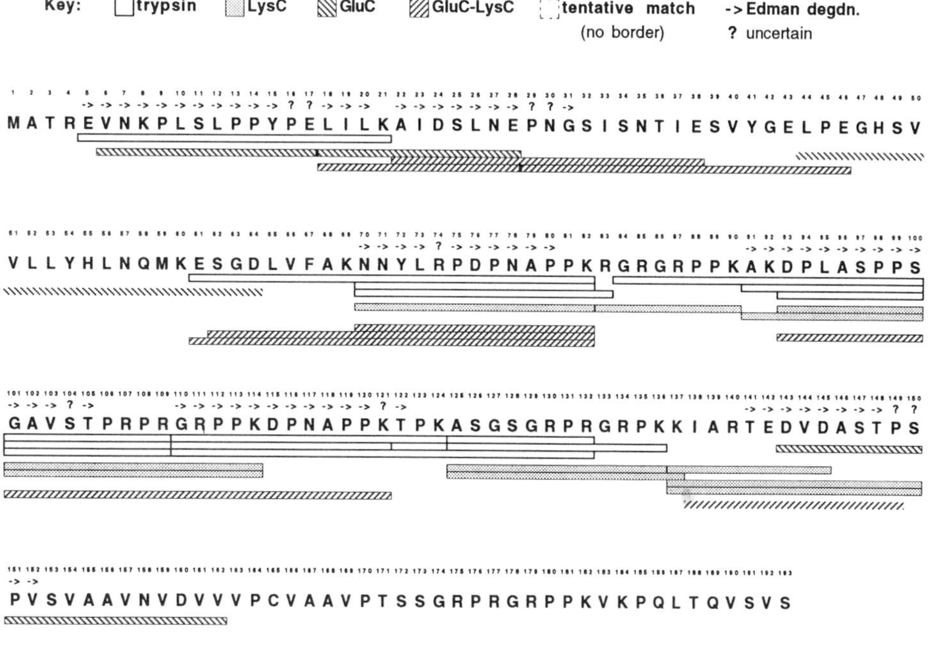

Figure 4. Mass-mapping of peptides against the DNA-derived sequence of HMG protein. Peptides were identified from their masses (error tolerance ±2Da). Edman sequencing was performed on the tryptic peptides only; in two cases (residues 22→ and 141→), partial sequence data were obtained but no mass information was forthcoming.

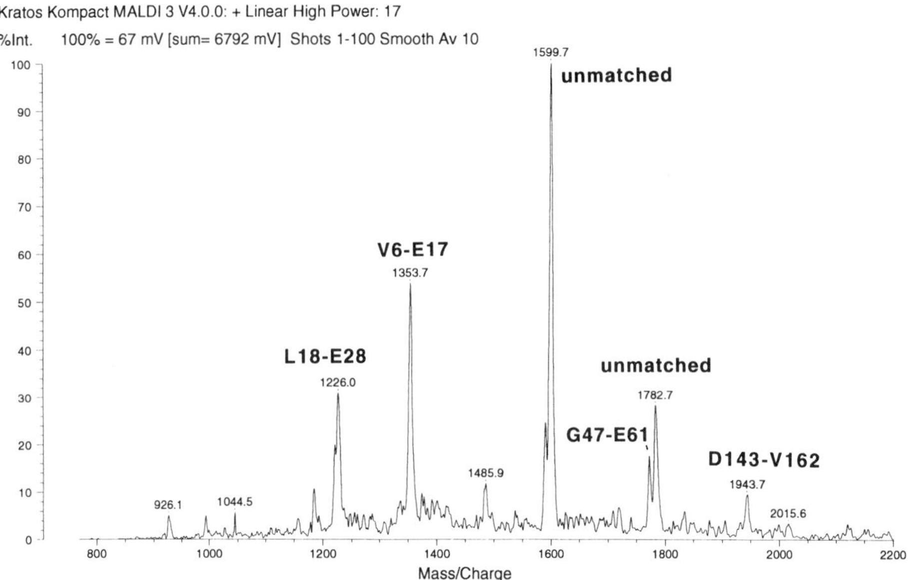

Figure 5. Mass analysis of a GluC digest of HMG protein.

against the expected protein sequence. All significant peaks could be matched and several derived from the C-terminal area of the protein. While these data were not supported by sequence analysis, the fact that they matched overlapping areas of the sequence indicated that these assignments were probably correct (Fig. 4).

The experiment was then repeated, digesting 2pmol of HMG protein with Glu-C proteinase; it was clear that not all the fragments could be matched with the protein sequence (Fig. 5). Two major peaks (1781.7±0.9Da and 1598.7±0.8Da) could not be identified and, since a fragment corresponding to Val6-Glu17 - close to the N-terminus - was also present, there was the hope that perhaps the Glu5-Val6 cleavage had been incomplete and that one of the unidentified fragments represented the full length N-terminal peptide. If this was so, the 1598.7Da peptide would have to correspond to Glu5-Glu17 with an unknown adduct of 117Da attached to an N-terminus. A more reasonable hypothesis emerged from the 1781.7Da peak. This could be explained by a peptide Thr3-Glu17, leaving an extra mass of 42.5Da to be accounted for - this is consistent with N-terminal acetylation (+42Da) of the Thr. It was therefore likely that the 1598.7Da peak derived from another segment of the sequence, but as it was unmatched, it could represent a side-chain post-translationally modified peptide. Further work will be done to establish its identity.

Further evidence for an acetyl-Thr N-terminus came from treatment of the Glu-C digest with Lys-C proteinase. This caused a loss of the unmatched fragments and the appearance of a new peptide of 786.95±0.2Da (Fig. 6). The mass difference between this and the 1781.7Da peptide (994.75±0.9) should generate a peptide of mass 1012.75±0.9Da. Given the protease specificities, the only candidate stretch of sequence with a similar mass *and* which bridges a Lys and a Glu residue is Pro9 - Glu17, mass 1012.9Da; a peak for this does not appear but its sequence (PLSLPPYPE) suggests it may be insoluble. The 786.95Da fragment must therefore be N-terminal to Pro9 and represent some altered form of the sequence. Again this can only be explained by acetyl-TREVNK (expected mass 787.8Da). Further work will be done to confirm this assignment by enzymological and chemical means.

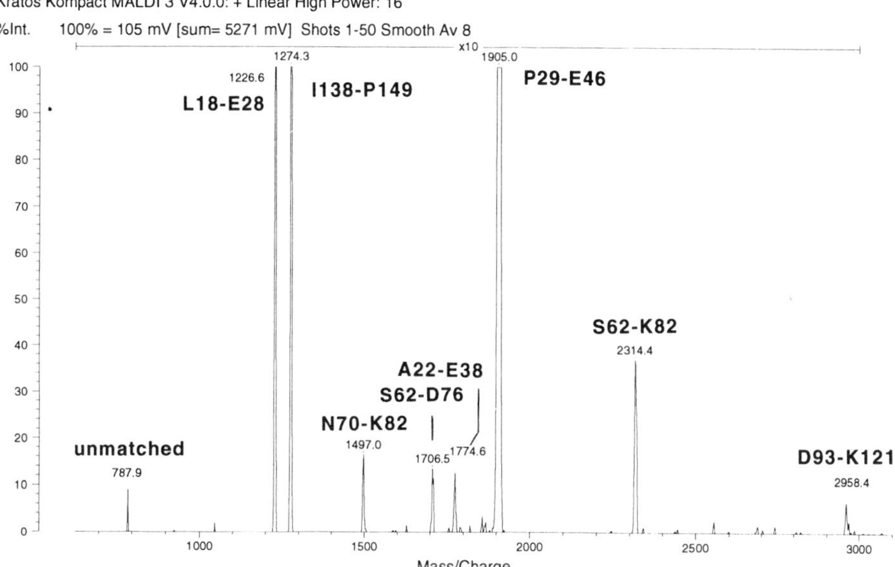

Figure 6. Mass analysis the GluC digest of HMG protein after further digestion with LysC proteinase.

DISCUSSION

This work has revealed a number of important characteristics of the HMG protein. The limited sensitivity of amino acid sequence analysis required the use of the bulk of the material for digestion and isolation of internal peptides. However, a significant amount of information has been derived from microdigestion of only a few pmoles of material. In particular, the likely nature of the N-terminal block has been established as acetyl on Thr3, and the heterogeneity seen in the mass analysis of the whole protein is supported by identification of the masses of several candidate C-terminal peptides. However, matching the observed protein mass to the peptide data, it is clear that further modifications are likely as at least 2 C-terminally matched peptides fall short of the length anticipated from the whole mass (arrowed, Fig. 7). Two internal peptides have a tentative assignment (Fig. 4) and may represent modified forms from elsewhere in the sequence , but even for identified peptides, there may be undetected modified forms of the same sequence arising from incomplete post-translational modification. Covering the entire sequence by the mass-mapping approach should not be taken as an indication that **no** part of the sequence is post-translationally modified. So far, no such modifications have been identified conclusively, but different conditions of analysis have yet to be investigated; for example, any highly phosphorylated peptides may not ionise under the positive ion mode use in this work and therefore remain undetected.

In this work it has become apparent that the use of several proteinases has given valuable information. The matching of a particular stretch of sequence from several different digests increases the confidence not only that the assignment is correct, but also that the sequence is unlikely to be present in a modified form. All the expected fragments will not be recovered from any one proteinase digest so the judicious use of several proteinases, either singly or in sequence, can significantly enhance the success rate of identifying all regions of the protein. Recent advances in MALDITOF mass spectrometry, such as high sensitivity

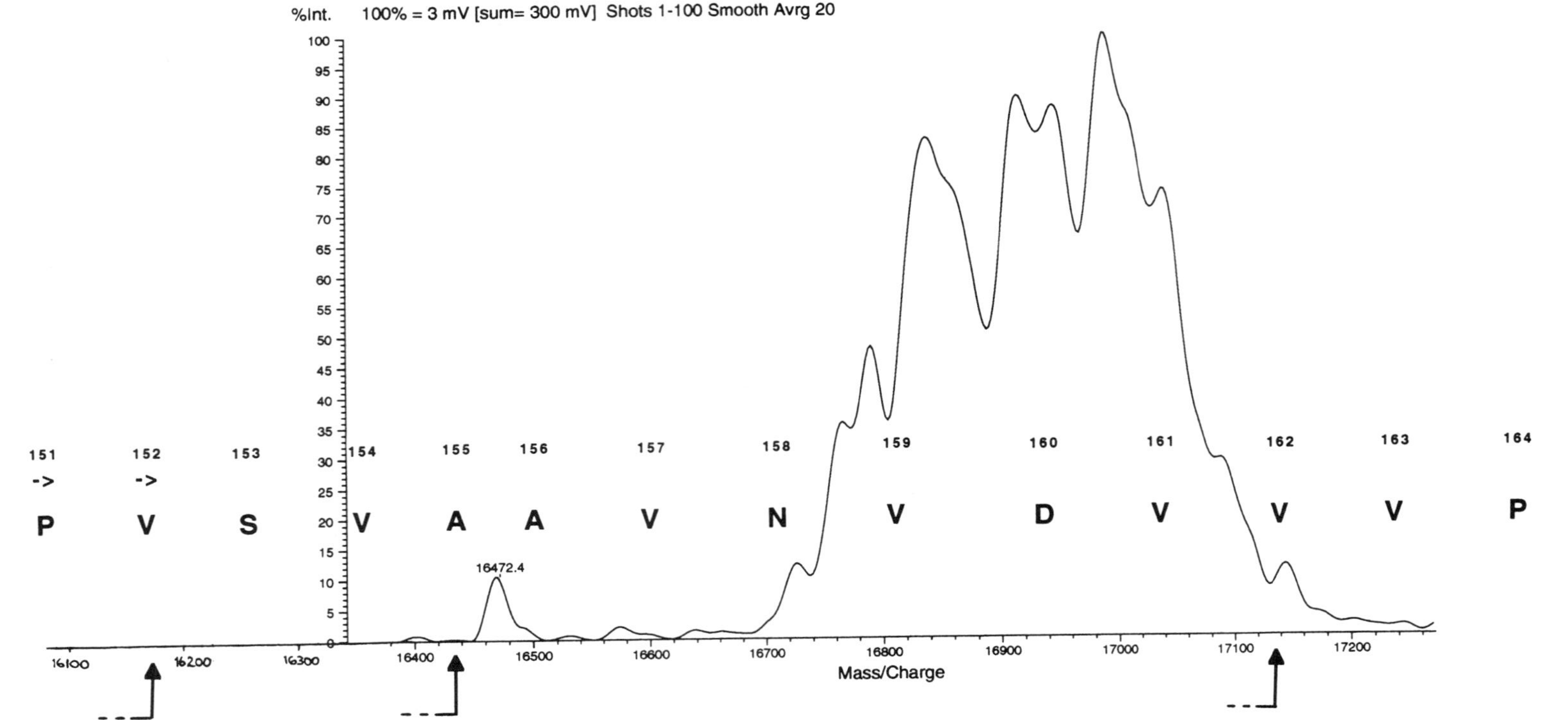

Figure 7. Alignment of the putative C-terminal peptides of HMG protein with the mass of the intact protein. The limits of the C-terminal peptides identified from mass-mapping are arrowed.

ladder sequencing (Bartlet-Jones et al., 1995) and post-source decay work (Talbo & Mann, 1994) will greatly improve the confidence of assignments made through the mass-mapping approach and make easier the detection of post-translational modifications.

CONCLUSIONS

MALDITOF mass spectrometry coupled to the approach of microdigestion enables a large amount of information to be derived from a few pmol of protein. Where a gene sequence is available, the ability to match masses with the protein sequence, coupled with the known specificity of proteinase digestion, obviates the need to support all data with sequence analysis, which requires much more material. However, the identification of modifications is not so straightforward and the advent of sequence analysis on reflector instruments will be a valuable aid in this type of analysis.

ACKNOWLEDGMENTS

We thank the AFRC and SERC for financial support, Kratos Analytical for loan of the MALDI III mass spectrometer and Mr M. Weldon for assistance with Edman sequence analysis.

REFERENCES

Bartlett-Jones, W.J., Hansen, H.F. & Pappin, D.J.C., 1995, The use of volatile N-terminal degradation reagents for rapid, high-sensitivity sequence analysis of peptides by generation of sequence ladders, *this volume*

Bustin, M., Lehn, D.A. & Landsman, D., 1990, Structural features of the HMG chromosomal proteins and their genes, *Biochim. Biophys. Acta.* 1049:231-243

Johns, E.W., 1982, The HMG Chromosomal Proteins, Academic Press, London

Pappin, D.J.C., Højrup, P & Bleasby, A.J., 1993, Rapid identification of proteins by peptide-mass fingerprinting, *Current Biology* 3:327-332

Pwee, K-H, Webster, C.I. & Gray, J.C., 1994, HMG protein binding to an A/T-rich positive regulatory region of pea plastocyanin gene promoter, *Plant Mol. Biol.*, in press

Talbo, G & Mann, M.,1994, Distinction between phosphorylated and sulfated peptides by matrix assisted laser desorption ionization reflector mass spectrometry at the sub-picomole level, in *Techniques in Protein Chemistry V*, J Crabb (ed), Academic Press, New York. pp105-113

NEW STRATEGIES FOR PROTEIN AND PEPTIDE CHARACTERIZATION

17

TRACING CELL SIGNALLING PATHWAYS USING A COMBINATION OF 2D GEL ELECTROPHORESIS AND MASS SPECTROMETRY

Manfredo Quadroni, Chantal Corti, Werner Staudenmann, Ernesto Carafoli, and Peter James

Protein Chemistry and Biochemistry III Laboratories
Swiss Federal Institute of Technology
8092 Zurich, Switzerland

INTRODUCTION

Many complex biological processes such as cell growth and differentiation are under the control of hormones, some of which act through membrane bound receptors altering the phosphorylation state of the proteins involved in the transformation, regulating their activity. The arrival of only a few hormone molecules to a specific plasma membrane receptor produces, after several steps of amplification, a phosphorylation of a few billion target molecules. In order to study the phosphorylation states of the target proteins and to follow the steps leading to their phosphorylation, an extremely high resolution separation method coupled with an equally sensitive detection method is needed.

Two dimensional gel electrophoresis as described by O'Farrel in 1975 (1) is currently the most highly resolving and simple method for protein separation, allowing up to 3,000 proteins from a cell extract to be visualised on a single gel by silver staining. The tremendous resolving power and sensitivity of the technique and the ability to electroblot proteins to an inert support for antibody detection or Edman sequencing for protein identification, makes it ideal for studying cell signalling in which the phosphorylation state of tens of proteins must be followed simultaneously. Scanned images of stained 2D-gels (or autoradiograms) of cells in different states can be analysed by computer and the changes in protein expression and phosphorylation state quantitated (2).

The main drawback in the construction of the 2D gel maps has been the sensitivity of the methods used for identifying proteins. This has improved recently with the development of peptide mass fingerprinting, the identification of a protein in a database using a set of molecular masses of peptides generated by a specific digestion. This method was independently developed by ourselves and four other groups and published simultaneously in 1993 (3, 4, 5, 6, 7) and was stressed as a means of linking two dimensional gel databases

Methods in Protein Structure Analysis, Edited by M. Z. Atassi and E. Appella
Plenum Press, New York, 1995

to protein databases. The acquisition of a mass fingerprint takes ca. 5 minutes for a matrix assisted laser desorption / ionisation time of flight (MALDI-TOF) mass spectrometer and 30 min for a capillary HPLC run on a quadrupole instrument, and requires only tens of femtomoles for detection. Since upwards of 200 or more proteins can be isolated in sufficient quantities for digestion (ca. 10 pmol) from a single experiment (running multiple gels and then digestions in parallel), mass fingerprinting is an excellent complement to the 2D analysis, providing a rapid identification of known proteins and unique tags for unknowns. Thus 2D electrophoresis in combination with mass spectrometry offers a systematic approach to the study of kinase cascades through the construction of 'cell maps'.

PROTEIN IDENTIFICATION BY MASS MAPPING

Proteins can be identified using a set of peptide fragment weights produced by a specific digestion to search a protein database in which sequences have been replaced by a list of the theoretical masses of the fragments produced by that cleavage method. Methods have been described by several groups for searching peptide mass databases derived from protein databases. The search methods described are robust for digests which yield accurate masses (+/- 0.5 amu) for five or more peptides and usually yields unequivocal data. Digests which produce only a few peptides, or where the amount of material is so low that mass accuracy suffers, can produce inconclusive results, as can proteins which are not in the database. However the major drawback using protein sequences, or protein sequences obtained by autotranslation from the cDNA sequence is that the vast amounts of data being generated by the genome and cDNA sequencing projects is being left untapped. Computerised extraction of the correct reading frame of genomic DNA sequences is possible but the extraction of sequences is not always complete due to difficulties such as; predicting boundaries for small exons/introns, reading frame shifts, the occurrence of sequences within introns of one protein which code for another protein, amongst others. Potentially the most useful source of sequence information, which is inaccessible to autotranslation, is the rapidly increasing number of Expressed Sequence Tags (EST), small cDNA sequences obtained from random primed cDNA libraries (8). In release 37 of the EMBL database there are over 4,000 such sequences present, coding on average for approximately 100-150 amino acids. Here we present methods using multi-dimensional searches which greatly increase the confidence level for identification, allowing DNA sequence databases to be examined.

One established MS technique which can greatly increase the confidence levels in database searching is Hydrogen-Deuterium exchange. This has already been used for simplifying the interpretation of MS/MS spectra for peptide sequencing (9). The number of exchangeable hydrogens in a peptide is sequence dependant, so peptides with similar masses may be distinguished after exchange (A, F, G, I, L, M, and V all have 1 exchangeable hydrogen; C, D, E, H, S, T, W, and Y have 2; K, N, and Q have 3 and R has 5). A second method for generating orthogonal data is to combine the results from two digestions using enzymes or chemicals with different cleavage specificities. The effect of using dual digestions and deuterium exchange on the certainty of the data obtained from a search are shown in table 1.

CELL SIGNALLING BY CALCIUM

In the resting state the cytoplasmic calcium concentration is of the order of 100 nM, several orders of magnitude less than the external milieu (ca. 3 mM, see figure 1). Upon stimulation by an effector (such as a hormone or an electrical signal), a calcium spike is generated as Ca^{2+} is released from the endo/sarcoplasmic reticulum (ER/SR) and let in from

Table 1. Orthogonal data increases the confidence levels of searching. Digestions were carried out using 50 pmol of phenylalanine ammonium lyase and the peptide masses were accumulated by on-line HPLC-MS. The data was used to search the EMBL database using the program MassSearch. The results give a score, the number of peptides occurring in that protein between the lowest and highest masses used for the search (n), the number of experimentally determined weights which matched (k), the accession number (AC) and a description of the entry in the database (DE)

1. Deuterated tryptic digestion of phenylalanine ammonia lyase

Score	n	k	AC	DE
80.9	18	4	X03237;	Sheep mRNA for alpha-S1-casein. Ovis aries
80.3	39	6	X16772;	P.crispum PAL-1 gene for phenylalanine ammonia-lyase exon 2. Petroselinum crispum
78.9	12	3	K03355;	HSV1 (KOS) gene for 2.7 kb spliced mRNA, splice acceptor region. Herpes simplex virus type 1
77.4	22	4	X59836;	C.hircus mRNA for as1-casein Capra hircus (goat)
74.9	22	3	J02895;	Pig non-histone chromosomal protein (HMG2) mRNA, Sus scrofa (domestic pig)
74.1	34	3	X17480;	Chicken mRNA for arylamine N-acetyltransferase (NAT-3) Gallus domesticus (chicken)

2. Combined search using deuterated and non-deuterated digests

Score	n	k	n	k	AC	DE
162.0	39	5	39	6	X16772;	P.crispum PAL-1 gene for phenylalanine ammonia-lyase exon 2 Petroselinum crispum
146.9	49	6	49	5	X17462;	P.crispum RNA for PAL4, phenylalanine ammonia-lyase Petroselinum crispum
134.7	10	2	10	3	L14214;	Human chromosome 4 (clone p4-1630) STS4-1220. Homo sapiens (human)
124.6	39	4	39	4	L11747;	Populus tricocarpa X Populus deltoides (hybrid) phenylalanine ammonia lyase (PAL)
114.8	13	3	13	2	V00567	Human messenger RNA fragment for the beta-2 microglobulin. Homo sapiens (human)
112.5	4	2	4	2	M63179;	Human HLA-DR beta x gene, exon 2. Homo sapiens (human)

3. Combined search using tryptic and AspN digestions

Score	n	k	n	k	AC	DE
133.1	49	6	27	3	X17462;	P.crispum RNA for PAL4, phenylalanine ammonia-lyase. Petroselinum crispum
133.0	39	5	17	2	X16772;	P.crispum PAL-1 gene for phenylalanine ammonia-lyase exon 2 Petroselinum crispum
106.7	30	3	11	4	M32778;	S.frugiperda insertion element IFP1.6 DNA, clone lambda 889. Spodoptera frugiperda
104.4	32	3	12	4	M32777;	S.frugiperda insertion element IFP1.6 DNA, clone lambda 883. Spodoptera frugiperda
102.5	31	5	6	4	D00511;	Rat mitochondrial acetoacetyl-CoA thiolase mRNA. Rattus norvegicus (rat)
101.8	32	3	14	4	M32775;	AcNPV mutant with an S.frugiperda insertion element IFP1.6.

outside of the cell by Ca^{2+} specific channels, raising the concentration around 100 fold. The main transducer of this signal is calmodulin (CaM) which binds Ca^{2+} with an affinity (Kd) of 10^{-6}M. The Ca^{2+} signal is self terminating, the rise in concentration activates the plasma membrane and ER Ca^{2+} pumps, simultaneously the Ca^{2+} channels close, the net effect being a rapid return to the resting Ca^{2+} levels in the cytosol.

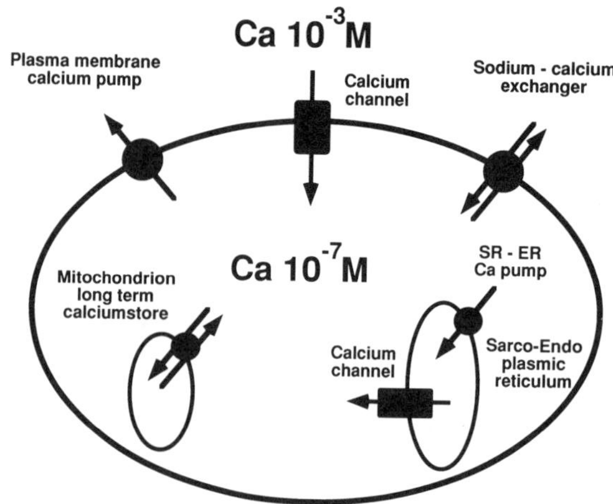

Figure 1. A simplified scheme of calcium homeostasis in the cell.

The SR/ER pump transfers Ca^{2+} out of the cytosol into the ER/SR. It is regulated by a small pentameric 5 kDa membrane protein termed phospholamban which at low Ca^{2+} concentrations binds to the pump inhibiting it by physically blocking access to the active site. Raising the Ca^{2+} concentration above 0.5 μM and / or phosphorylating phospholamban with cAMP dependent protein kinase or CaM dependent protein kinase causes a dissociation of the complex and the inhibition is released (10). The plasma membrane pump contains a 30 residue stretch of amino acids near the C-terminal of the pump which interacts with a region close to the active site inhibiting the pump. The binding of CaM in response to increased Ca^{2+} levels relieves the inhibition (11). This internal inhibition can also be removed in other ways such as by phosphorylation of the binding domain with protein kinase C, or as a final last ditch resort, by proteolytic removal of the CaM binding domain by the Ca^{2+} activated protease calpain (as is the case in ageing cells).

APPLICATION OF MS METHODS TO CELL SIGNALLING

Stimulation of cells by bombesin has been shown to produce a long calcium transient in cultured liver cells. An initial rapid release of Ca^{2+} from internal stores occurs followed by a sustained second phase >2 min where Ca^{2+} efflux is greatly reduced. Bombesin is known to produce a increase in the activity of several kinase including Casein Kinase II (CKII) and non-receptor tyrosine kinases (12). These observations are puzzling; if the cell calcium concentration is high, why is CaM not binding to the plasma membrane ATPase and lowering the calcium concentration? A clue may come from our recent finding that calmodulin can be found in a phosphorylated state in vivo. Figure 2 shows the raw and deconvoluted spectra of rat liver CaM analysed by HPLC-MS. The different phosphorylation states can be seen as peaks separated by 80 mass units. The purified phosphoCaM behaved in an entirely different manner than normal CaM. It shows a much lower affinity for the plasma membrane ATPase, only partially activating it at calcium concentrations seen in bombesin stimulated cells. PhosphoCaM is also an extremely weak activator of CaM dependant protein kinase, which is unable to activate the SR calcium ATPase by phosphorylating phospholamban. This suggests a pivotal role of CaM phosphorylation in elongation of the calcium signal.

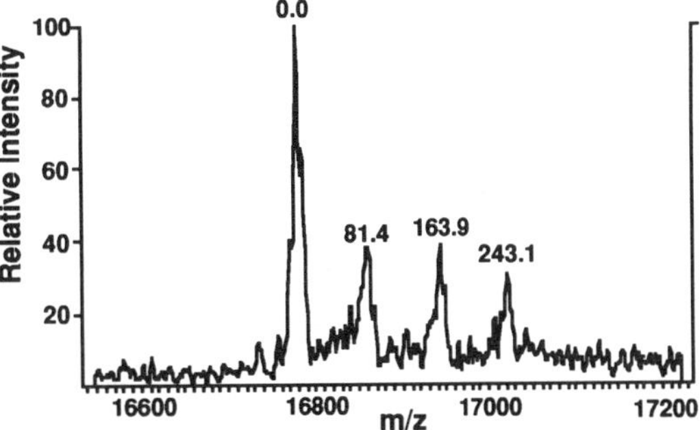

Figure 2. Deconvoluted mass spectrum of Phosphocalmodulin. 50 pmol of HPLC purified phosphocalmodulin was infused at 3 μL/min with a sheath liquid of methoxyethanol into a Finnigan MAT TSQ700 mass spectrometer. The spectrum was accumulated for 1 minute, scanning from 1000 to 2000 m/z in 3 seconds and deconvoluted with software supplied by the manufacturer. The peaks are labelled relative to the mass of non phosphorylated calmodulin and show clearly the presence of 0, 1, 2, 3 and 4 four phosphates (mass increment 80).

The next question was obvious; which kinase was responsible for this phosphorylation? In order to get an idea, we decided to determine the sites of phosphorylation. The phospho-protein was digested with CNBr and subsequently trypsin. (trypsin alone was insufficient since CaM becomes refractory to digestion after phosphorylation). The resultant peptide mixture was separated by HPLC and the effluent analysed by automated on-line MS/MS. The spectrum of the singly phosphorylated tryptic fragment T8 and also the relative ion current are shown in figure 3 . The automated MS/MS HPLC run identified the three sites of modification as being Ser79, Thr 81 and Ser101. These sites are canonical targets of

	y9	y8	y7	y6	y5	y4	y3	y2	y1
y ions	1172	1057	956	841	674	545	416	287	174
-H2PO4	1074	959	858						

LYS - ASP - THR - ASP - S(P) - GLU - GLU - GLU - ILE - ARG

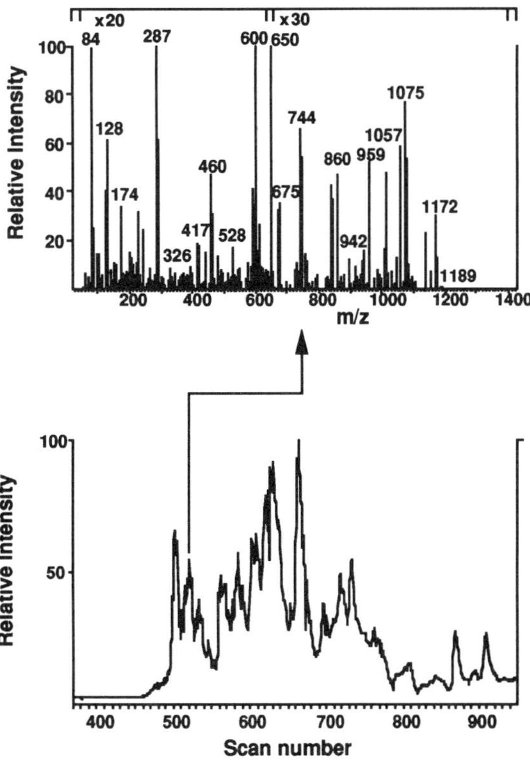

Figure 3. Automated MS/MS of a CNBr/ Tryptic digestion of Phosphocalmodulin. 100 pmol of phosphocalmodulin was digested in the dark under argon with 200 mM CNBr. The solution was diluted with water and lyophilised before digestion with trypsin overnight at 37^0 C. The digest was acidified to pH 2.0 and injected onto a self packed C-18 reverse phase column (0.375 mm X 10 cm) and eluted with a gradient of 0.1% TFA in water to 0.08% TFA in 70% acetonitrile over 30 mins. We have written an instrument control language program, Rubber, which scans the third quadrupole (in the presence of colliosion gas but no collision offset voltage) and selects ions intense enough for sequencing. The instrument then switches to MS/MS mode and sets the collision energy according to the parent mass and accumulates 6 scans. The instrument then reverts to normal MS mode to identify further candidates for MS/MS. The figure shows the ion current in the lower panel and the MS/MS accumulated for the phosphopeptide indicated by the arrow. The sequence ions are shown above the spectra.

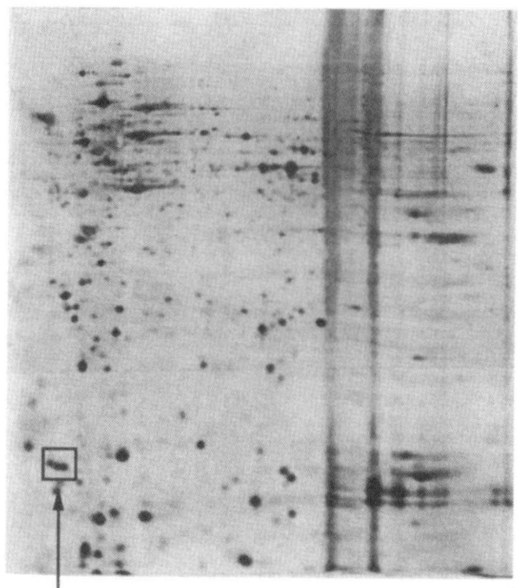

Figure 4. Two dimensional gel of baby hamster kidney cells grown in culture. The box indicates the position of non, mono and di phosphorylated calmodulin.

casein kinase II. We could reproduce the same pattern of phosphorylation using CaM phosphorylated in vitro by casein kinase II from rat liver.

In order to trace the signalling pathway backwards to try and confirm casein kinase as the kinase acting in vivo, we started analysing various cell types in culture, stimulating them with a variety of hormones and running two dimensional gels to see if CaM was becoming phosphorylated (see figure 4). If CaM was being phosphorylated by casein kinase II, then a second spot should be visible in the autoradiograms, that of the beta subunit of casein kinase, since this kinase undergoes autophosphorylation when active. By taking a series of 'snap shots', - two dimensional gels at various times after stimulation with an effector we hope to be able to trace the signalling pathway backwards to see what is stimulating casein kinase II in the cell. Critical to these experiments is the rapidity and sensitivity of mass mapping for the identification of proteins from two dimensional gels.

Database searching from a mass profile is offered as a free service by an automatic server at the ETH, Zurich. For information, send an electronic message to the address, cbrg©inf.ethz.ch with the line: help mass search, or help all. An experimental World Wide Web server has been set up at the address http://cbrg.inf.eth.ch which requires a client with forms capability.

REFERENCES

1. O'Farrell, P. H. (1975) J. Biol. Chem. 250, 4007-4021.
2. Taylor, J., Anderson, N.L., Scandora, A.E., Villard, K.E. and Anderson, N.G. (1982). *Clin. Chem. 28.* 861-66
3. Henzel, W., Billeci, T., Stults, J., Wong, S., Grimley, C. and Watanabe, C. (1993). *Proc Natl Acad Sci U S A. 90.* 5011-5
4. James, P., Quadroni, M., Carafoli, E. and Gonnet, G. (1993). *Biochem. Biophys . Res. Comm. 195.* 58-64
5. Mann, M., Hojrup, P. and Roepstorff, P. (1993). *Biol. Mass Spec. 22.* 338-345
6. Pappin, D., Hojrup, P., and Bleasby, A. (1993). *Current Biology. 3.* 327-332
7. Yates, J.R., Speicher S., Griffin P.R., and Hunkapillar T. (1993). *Anal. Biochem. 214.* 397-408

8. Adams, M.D. et al. (1991). *Science. 252.* 1651-56
9. Sepetov, N. F., Issakova, O. L., Lebl, M., Swiderek, K., Stahl, D. C. and Lee, T. D. (1993). *Rapid Comm. in Mass Spec. 7.* 58-62
10. James, P.H., Inui, M., Tada, M., Chiesi M. and Carafoli, E.(1989) *Nature,* 342, 90-92.
11. James, P.H., Maeda, M., Fischer, R., Verma, A., Krebs,J., Penniston, J.T., and Carafoli, E. (1988) *J.Biol.Chem.*, 263, 2905-2910.
12. Agostinis, P,. Van Lint, J., Sarno, S., De Witte, P., Vandenheede, J.R., Merlevede, W. (1992) *J. Biol Chem.* 267, 9732-7

18

NOVEL FLUORESCENT METHODS FOR QUANTITATIVE DETERMINATION OF MONOSACCHARIDES AND SIALIC ACIDS IN GLYCOPROTEINS BY REVERESED PHASE HIGH PERFORMANCE LIQUID CHROMATOGRAPHY

Kalyan Rao Anumula

Analytical Sciences Department, UW 2960
SmithKline Beecham Pharmaceuticals
P. O. Box 1539, 709 Swedeland Rd, King of Prussia, Pennsylvania 19406

INTRODUCTION

Carbohydrate composition analysis of glycoproteins is analogous to the amino acid analysis of proteins and, therefore, is a fundamental part of glycobiology. An accurate composition analysis can give some insights into the type and extent of glycosylation and probable structures of oligosaccharides present in an unknown glycoprotein, based on current knowledge. Obviously, identification and quantitation of the individual monosaccharides present in the purified oligosaccharides is a pre-requisite for structural determination. A review of some useful methods for carbohydrate composition analysis has been published recently (Townsend, 1993). Analysis of monosaccharides and oligosaccharides using high performance anion exchange chromatography with pulsed amperometric detection (HPAE-PAD) has gained some popularity in recent years, since this technique does not involve any derivatization prior to analysis. But, it requires its own hardware components, such as the system for carbohydrate analysis commercially available from Dionex. In addition, PAD is not a truly specific detector for carbohydrates and the glycoprotein hydrolysate containing amino acids, peptides and thiol groups interfere with the monosaccharide analysis.

Two novel reversed phase high performance liquid chromatographic (HPLC) methods with fluorescence detection are described in this report for the complete carbohydrate (monosaccharides and sialic acids) composition analysis. The monosaccharides and the sialic acids were labeled using very simple derivatization techniques to yield highly stable fluorescent derivatives. These methods have been validated for accuracy and reproducibility using both standards and the glycoproteins. The composition analysis of glycoproteins by these methods is comparable to HPAE-PAD and are without any known complications of interference in monosaccharide analysis unlike HPAE-PAD.

METHODS

Hydrolysis of Glycoproteins

Glycoproteins 5-50 μg were hydrolyzed in 0.25-0.5 ml of 20% trifluoroacetic acid in 1.6-ml conical screw cap freeze vials (polypropylene with 'O' ring seals, Sigma) at 100 °C for 6-7 hours (Anumula, 1994). The caps on the vials were further secured by applying 4-5 layers of Teflon tape in order to prevent any accidental evaporation of the sample during hydrolysis. Samples were dried overnight using a vacuum centrifuge evaporator (Savant) without heat. For hexosamine analysis specifically, the glycoproteins were hydrolyzed in 0.05 - 0.1 ml of 4N HCl at 100 °C for 16 hours and dried on a vacuum centrifuge evaporator.

Derivatization of Monosaccharides with Anthranilic Acid (ABA)

A solution of 4% sodium acetate -$3H_2O$ and 2% boric acid in methanol was prepared by shaking vigorously in a graduated cylinder with a glass stopper. This solution may be stored at room temperature for several months. The derivatization reagent was prepared by dissolving 30 mg of anthranilic acid (Aldrich) and 20 mg of sodium cyanoborohydride (Aldrich) in 1.0 ml of the methanol-sodium acetate-borate solution. Dry glycoprotein hydrolysates were dissolved in 1% fresh sodium acetate·$3H_2O$ (0.1-0.2 ml) and an aliquot (20-100 μl) was transferred to a new screw cap freeze vial. Samples were mixed with 0.1 ml of the anthranilic acid reagent solution and capped tightly. The vials were heated at 80 °C in an oven or heating block (Reacti-Therm, Pierce) for 30-45 minutes (Anumula, 1994). After cooling the vials to ambient temperature, the volume of the samples was made up to 1.0 ml with HPLC solvent A and mixed vigorously on Vortex in order to expel the hydrogen evolved during the reaction. Duplicate injections of 50 μl were made from each vial for analysis. Similarly, the monosaccharide standards were derivatized to contain 20-25 pmol each per injection and were derivatized each time for the unknown sample analysis.

HPLC Analysis of ABA-Monosaccharide Derivatives

ABA-monosaccharide derivatives of the monosaccharides were separated on a C-18 reversed phase HPLC column (Bakerbond, 5 μm, 0.46 x 25 cm, analytical, J.T Baker or Beckman, Ultrasphere-ODS, 0.46 x 25 cm) using a 1-butylamine-phosphoric acid-tetrahydrofuran mobile phase. All the separations were carried out at ambient temperature using a flow rate of 1 ml/min. Solvent A consisted of 0.15-0.3% 1-butylamine, 0.5% phosphoric acid and 1% tetrahydrofuran (0.025% BHT inhibited, Aldrich) in water and solvent B consisted of equal parts of solvent A and acetonitrile. The solvent A contained 0.2% butylamine for the Bakerbond column and the gradient program was 5% B isocratic for 25 min followed by a linear increase to 15% B at 50 min. The solvent A contained 0.15% butylamine for the Beckman column and the gradient program was 5% B isocratic for 30 min followed by a linear increase to 15% B at 50 min. The column was washed for 15 minutes with 100% B and equilibrated for 15 min with the initial conditions to ensure reproducibility from run to run. ABA derivatives were detected with a HP 1046A HPLC fluorescence detector using 230 nm excitation and 425 nm emission (Anumula, 1994).

Determination of Sialic Acids

Fifty μL of the sample containing 0.01 to 0.25 mg of protein was mixed with 50 μL of 0.5 M sodium bisulfate in a 1.6-ml conical screw cap freeze vial. After placing the caps tightly on the vials, they were incubated at 80 °C (Reacti-Therm heating module, Pierce) for 20 minutes. These mild acid hydrolysis conditions were satisfactory for the release of sialic acids from the glycoproteins.

The mild acid released sialic acids were then derivatized with o-phenylenediamine·2HCl (OPD, Aldrich). Sialic acid standard solution (1-2 nmol in 0.1 ml) in separate 1.6-mL screw cap freeze vials and the unknown samples from mild acid hydrolysis were mixed with 0.1 mL of OPD solution (20 mg/ml in 0.5 M $NaHSO_4$). After placing the caps tightly on the vials, they were incubated at 80 °C (Reacti-Therm) for 40 minutes. After cooling the tubes to room temperature, they were diluted to 1.0 mL with solvent A and mixed vigorously on a Vortex mixer. The vials were centrifuged at maximum speed for 3-5 minutes in a micro centrifuge in order to clarify the solution and the clear supernatant was used for the analysis. Sialic acids were also derivatized with 4,5 dimethyl 1,2-phenylene diamine as described earlier (Anumula, 1994a).

HPLC Analysis of OPD-Sialic Acid Derivatives

The derivatized sialic acid standards and the samples were transferred into separate auto injector vials and two injections of 0.1 mL were made from each vial. The sialic acid derivative was separated on a C18 reversed phase column (Ultrasphere-ODS, Beckman) at ambient temperature using a flow rate of 1.0 mL per min. The HPLC solvents were prepared as described for the monosaccharide analysis. Solvent A for this column contained 0.15% butylamine. The sialic acid derivatives were eluted with 13% solvent B isocratic hold for 15 min followed by 10 min wash with 95% solvent B and equilibration with 13% solvent B for 10 min. The fluorescence detector conditions were the same as described earlier for the monosaccharide analysis (Anumula, 1994) and were quantitated using an 230 nm excitation and a 425 nm emission wavelengths.

CHARACTERIZATION OF GLYCOPROTEINS

As shown in Figure 1, the characterization of glycoproteins is a rather involved process and requires a considerable amount of time and effort to determine the oligosaccharide structures specifically associated with the individual sites of glycosylation. The structure determination involves identification and quantitation of individual monosaccharides in addition to the use of various physico-chemical techniques (see mini reviews Kobata, 1992, Lee, 1992 and Lee, *et al*, 1990). Common steps involved in the determination of oligosaccharide structures using various techniques are shown in Table 1.

It is not the scope of this paper to review all the methods in detail and experimental protocols for some of these techniques may be found elsewhere (Hounsell, 1993). Notably, the analysis of carbohydrates by capillary electrophoresis has been reviewed recently (Novotny and Sudor 1993, and Oefner and Chiesa 1994). Complete release of oligosaccharides (N-linked and O-linked) from glycoproteins by hydrazinolysis is the method of choice (Patel, *et al.* 1993), but it imposes tremendous challenges in the fractionation of various oligosaccharides. For example, a mixture of high mannose, complex and O-linked oligosaccharides can not be resolved easily. Therefore, O-linked and high mannose oligosaccharides are released separately using specific methods prior to the release of complex oligosaccharides. On the other hand, it is rare to find a glycoprotein with high mannose, complex and

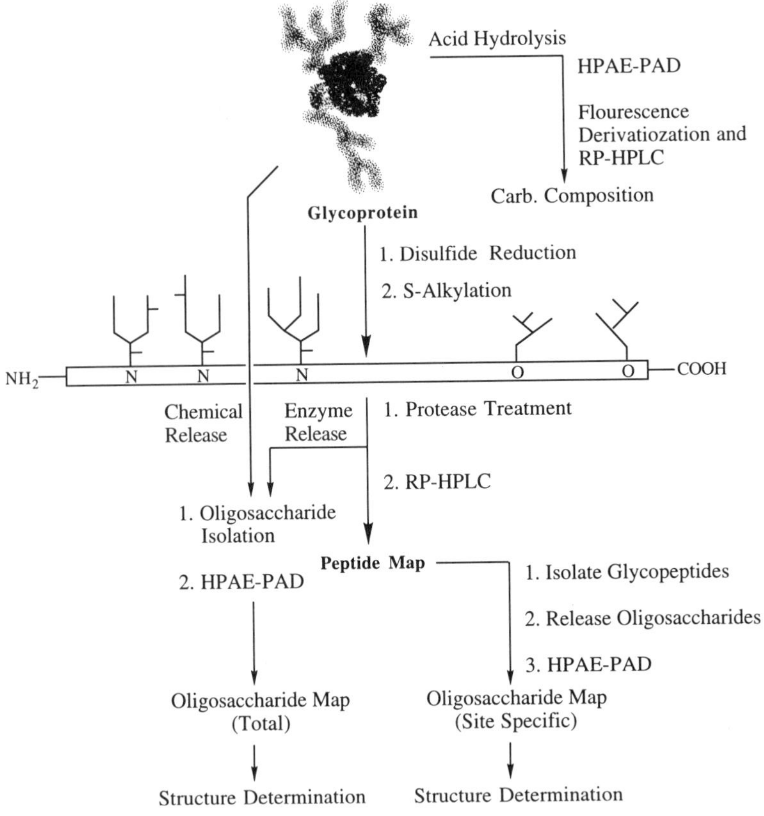

Figure 1. Characterization of glycoproteins.

O-linked oligosaccharides. At this time, HPAE-PAD remains as the method of choice for the fractionation of oligosaccharides and the analysis of monosaccharides, once the system operates in a satisfactory mode (Anumula and Taylor, 1991). Even then, the monosaccharide analysis of certain glycoproteins, for example IgG, can be problematic due to severe interference of amino acids and peptides in various monosaccharides (Anumula, 1994). However, for determining the oligosaccharide map of glycoproteins, HPAE-PAD is perhaps the best method at this time. A tentative identification of the oligosaccharide structures can be made by comparison of the various retention times with those of the reference standard oligosaccharides and based on the behavior of oligosaccharides on the HPAE-PAD column (Anumula and Taylor, 1991; Lee, 1990; and Townsend and Hardy, 1990). Obviously, the confirmation of the implied structures must be carried out by independent methods as described in the Table 1.

MONOSACCHARIDE ANALYSIS

With the intent to develop simple methods with ease of operation using common equipment to determine the carbohydrate composition of glycoproteins, two new methods with fluorescence detection for high sensitivity are developed. The monosaccharides are

Table 1. Various steps involved in the structure determination of carbohydrates

Steps involved	Comments
1. Release of sugar chains from glycoproteins	
a. Enzymatic	
PNGase F, PNGase A, Endo Fs, Endo H	Have specificities for peptides and oligosaccharides
O-glycanase	Cleaves only the neutral O-linked disaccharides
b. Chemical, e.g., hydrazine, $NaOH/NaBH_4$	Degradation products may be formed
2. Isolation of oligosaccharides	
a. HPLC (normal and reversed phase and ion exchange, etc.)	
b. Gel permeation chromatography	
c. Lectin chromatography	
d. TLC	Simple and inexpensive technique
3. Fractionation/purification of oligosaccharides	
a. HPAE-PAD	No derivatization
b. Bio-Gel P-4 chromatography	Derivatization for sensitivity
c. HPLC (normal and reversed phase and ion exchange, etc.)	
4. Determination of sugar composition	
a. HPAE-PAD	Interference with monosacchardies in hydrolysates
b. GC/CE	
c. HPLC-fluorescence detection	Requires derivatization
d. HPLC-UV detection	Not as sensitive as fluorescence
e. MS	Does not identify monosaccharides
5. Inter sugar linkage determination	
a. Methylation analysis by GC-MS	Single best method
b. NMR	Structural and conformational information with large amounts
c. MS (where applicable)	Not widely used
6. Sequential exo/endo glycosidase digestions and product analysis for anomeric sequence determination	
a. HPAE-PAD	Change in retention time
b. Bio-Gel P-4 chromatography	Change in elution volume
c. HPLC methods	Change in retention time
d. MS	Mass change in the oligosaccharide
7. Specific chemical degradations (where applicable)	
a. Periodate oxidation	Cleaves between vicinal hydroxyls
b. Acetolysis	Cleaves at 1-6 glycosidic bond
c. Nitrous acid deamination	Cleaves at sugar amines

derivatized with 2-aminobenzoic acid (ABA, anthranilic acid) and the sialic acids are derivatized with o-phenylenediamine·2HCL (OPD) as shown in Schemes 1 and 2.

Detection of the sugar derivatives by fluorescence is the most sensitive method of quantitation and with these fluorescent tags ~0.1 pmol of the monosaccharides and <2 pmol of the sialic acids can be easily determined. Furthermore, these methods use the same solvent systems and the detector conditions for convenience. Although the same column can be used for both the monosaccharide and the sialic acid determinations, it is advisable to use a Brownlee Spheri-5 RP-18 cartridge (0.46 x 10 cm) for common sialic acid estimation in order to save time in cleaning the column (4-5 blank runs) previously used for the monosaccharide analysis.

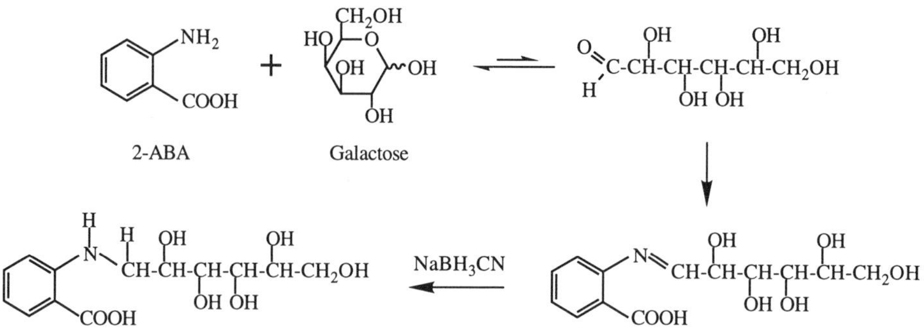

2-ABA Galactose $NaBH_3CN$

Scheme 1. Derivatization of carbohydrates with 2-aminobenzoic acid (ABA, anthranilic acid) using galactose as an example.

Analysis of Monosaccharides as ABA Derivatives

Both neutral and amino sugar residues can be derivatized with 2-amino benzoic acid (ABA) in the presence of cyanoborohydride to yield highly fluorescent stable derivatives for quantitative determination. A mixture of standard monosaccharides consisting of glucosamine, galactosamine, galactose, mannose and fucose was derivatized to give 20-25 pmols per injection. The monosaccharide derivatives were completely separated on the C-18 Bakerbond column using the butylamine-phosphoric acid system (Anumula, 1994). Tetrahydrofuran (inhibited) was used to improve the resolution of the sugar derivatives. Separation of the amino sugars from the excess reagent depends on the starting mobile phase composition and in particular the ratio of butylamine to phosphoric acid. Typical chromatograms obtained with C18-Bakerbond and Ultrasphere-ODS columns are shown in Figure 2.

Only the Ultrasphere-ODS column was suitable for analyzing both monosaccharides and sialic acids using the same solvent systems. Optimum conditions for derivatization and HPLC separation were the same as described earlier (Anumula, 1994). The results obtained using these methods for the complete carbohydrate analyses of the recombinant IgG expressed in CHO cells is shown in Table 2.

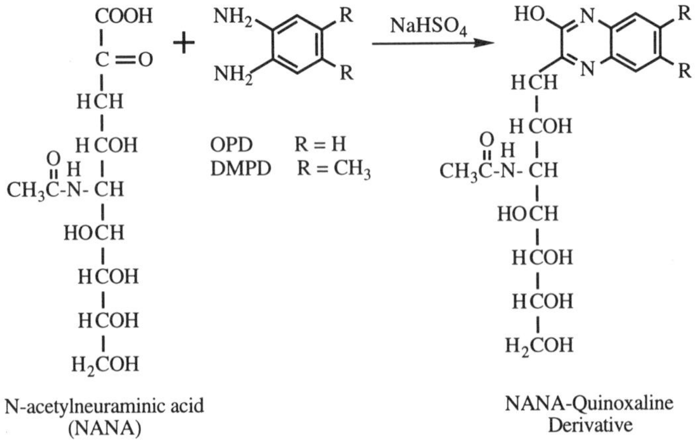

$NaHSO_4$

OPD R = H

DMPD R = CH_3

N-acetylneuraminic acid (NANA) NANA-Quinoxaline Derivative

Scheme 2. Derivatization of sialic acids with o-phenylenediamine (OPD) using N-acetylneuraminic acid as an example.

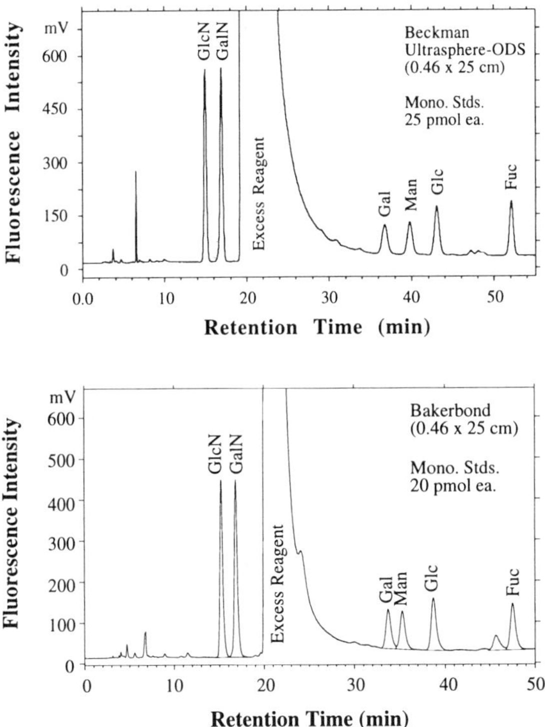

Figure 2. Typical chromatograms obtained with ultrasphere-ODS and C-18 Bakerbond columns.

Table 2. Complete carbohydrate composition of a rIgG determined by the fluorescence methods

Monosaccharide	rIgG HPLC[a] Mol/Mol[c]	rIgG Corrected[b] Mol/Mol[c]
Glucosamine	6.64	7.95 (8.0)[d]
Galactose	1.20	1.44 (1.47)
Mannose	5.02	6.01 (6.0)
Fucose	1.85	2.20 (2.0)
Sialic acid	0.10[e]	0.07[f]

[a]Values obtained by HPLC analysis.
[b]Values corrected for the recovery of monosaccharides from 20% TFA at 100°C for 7 hours of hydrolysis. Recovery of the monosaccharides was 83.5% of the expected values for well characterized glycoproteins.
[c]Calculated based on polypeptide molecular weight of 146,273 Da.
[d]Expected values based on the oligosaccharide map for rIgG.
[e]Determined by a modified thirobarbituric acid assay procedure.
[f]Determined by the OPD HPLC fluorescence method.

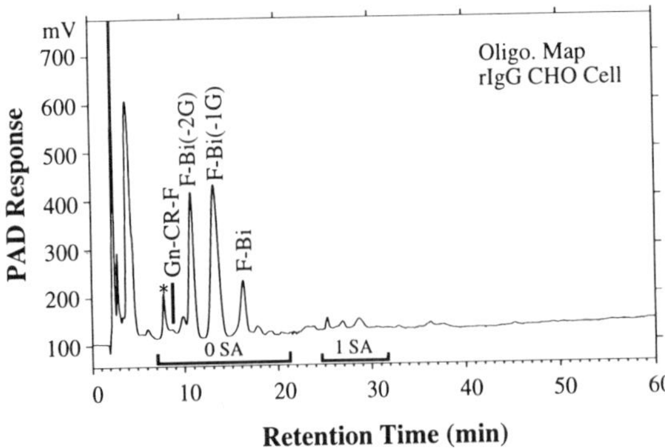

Figure 3. HPAE-PAD oligosaccharide map of a CHO cell derived recombinant IgG. The map of hydrazine released oligosaccharides was determined as described earlier (Anumula, 1994). 1SA and 0SA indicate where the oligosaccharides with one sialic acid and without any sialic acid typically elute. Hydrazinolysis artifact peak is indicated by an asterisk. For assigned structures see Figure 4.

As indicated in Table1, the composition was in good agreement with that of what is expected from the oligosaccharide map and the associated structures, following the correction for the degradation of sugar residues during hydrolysis. The oligosaccharide map of the rIgG and the associated structures are shown in Figures 3 and 4.

The expected carbohydrate composition for the rIgG was calculated from the relative ratio of oligosaccharide peaks and their structures. The oligosaccharide map of the rIgG consisted >95% of neutral complex bi-antennary oligosaccharides with peripheral heterogeneity. Oligosaccharide structures were determined by HPAE-PAD following digestions with exoglycosidases and the specificity of endo-β-N-acetylglucosaminidases Fs (Anumula, 1993). HPAE-PAD responses for these oligosaccharides were assumed equal in calculating the expected carbohydrate composition. A typical recovery of monosaccharides was between 83-85% for a number of glycoproteins examined so far and was used in determining the actual composition of the glycoproteins. For the first time an accurate estimation for the loss in recovery of the monosaccharides was determined using the predicted composition from the oligosaccharide map of a recombinant IgG produced in CHO cell line. The data for reproducibility with the monosaccharide standards and the rIgG are shown in Tables 3 and 4.

The procedures were highly reproducible with relative standard deviation of less than 3.5%. The method for monosaccharide analysis reported here is superior compared to the methods using either 2-amino pyridine (Hase, 1993) or 4-amino benzoic acid ethyl ester (Kwon and Kim, 1993) for derivatizations, since the current procedure does not involve the separation of the excess reagent from the derivatives either by extraction or by gel filtration.

Table 3. Reproducibility of monosaccharide standards (25 pmol ea.) derivatized with anthranilic acid

	Peak area $\times 10^{-6}$			
HPLC	Glucosamine	Galactose	Mannose	Fucose
Injection 1	11.15	3.954	4.177	5.474
Injection 2	11.11	3.913	4.140	5.474
Injection 3	11.28	3.995	4.211	5.575
Injection 4	11.24	3.967	4.225	5.529
Injection 5	11.16	3.971	4.195	5.481
Injection 6	11.21	3.996	4.180	5.500
Mean area	11.19	3.966	4.188	5.505
%RSD	0.55	0.77	0.71	0.72

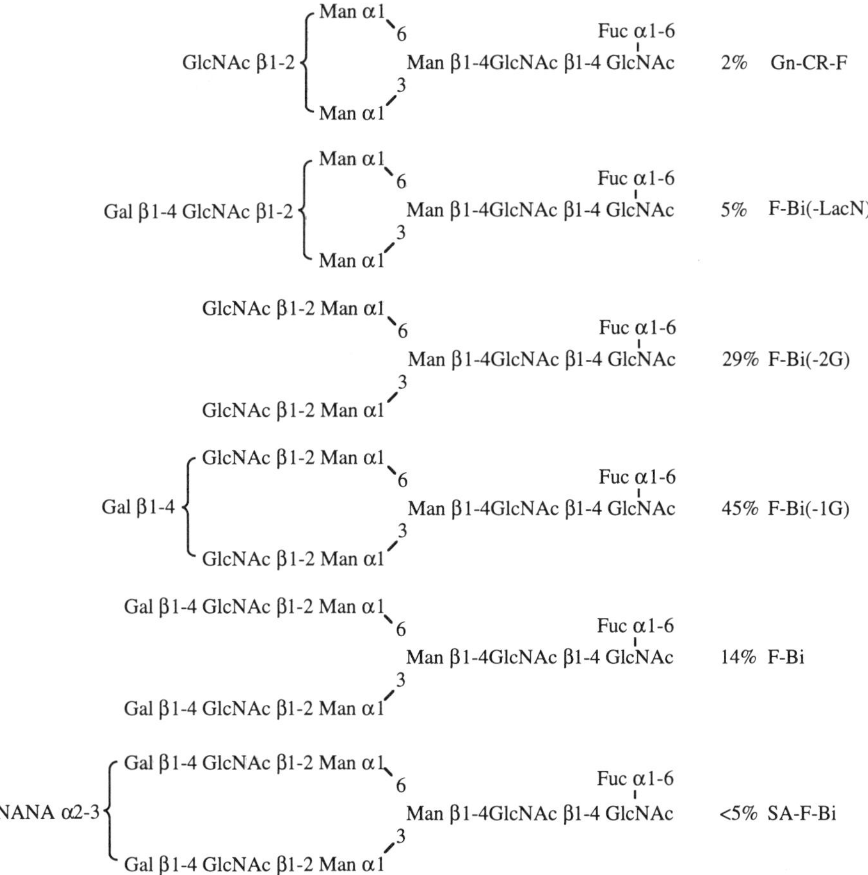

Figure 4. Oligosaccharide structures assigned to the oligosaccharide map. SA, sialic acid; F, fucose; Bi, bi-antennary complex type; Gn, N-acetylglucosamine; G, galactose and CR, tri-mannosyl chitobiose core.

Table 4. Reproducibility of rIgG monosaccharides derivatized with anthranilic acid

	Peak area $\times 10^{-6}$			
HPLC	Glucosamine 46 pmol	Galactose 6.3 pmol	Mannose 40.8 pmol	Fucose 11.5 pmol
Injection 1	20.75	0.975	6.100	2.413
Injection 2	20.80	0.963	6.143	2.407
Injection 3	20.39	0.978	5.774	2.211
Injection 4	20.60	0.948	5.878	2.257
Injection 5	20.34	0.927	5.893	2.371
Injection 6	20.71	0.940	6.068	2.439
Mean area	20.60	0.955	5.976	2.350
%RSD	0.95	2.13	2.47	3.25

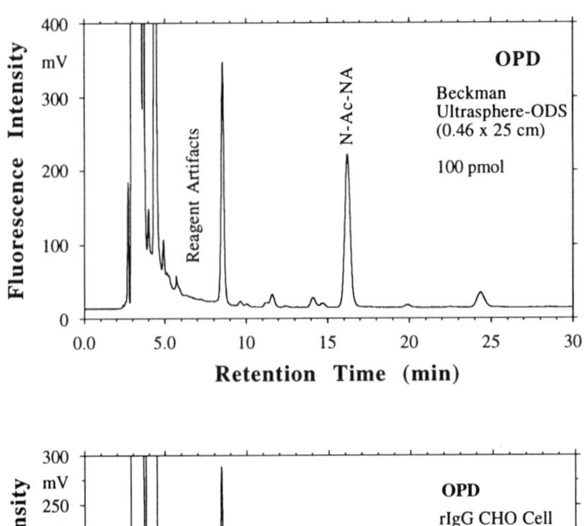

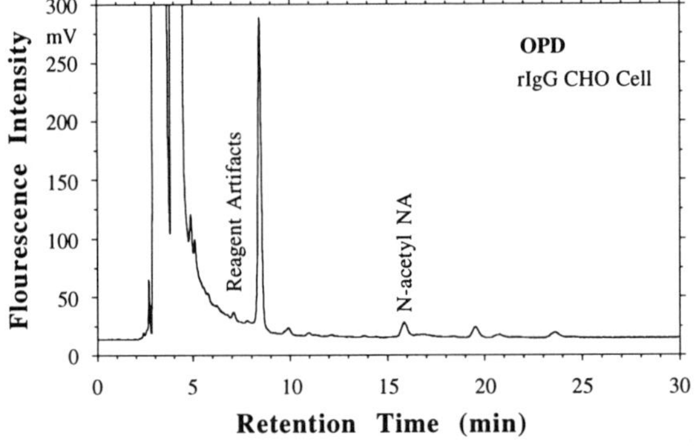

Figure 5. RP-HPLC chromatograms obtained with OPD-N-acetylneuraminic acid and the sialic acid from rIgG.

SIALIC ACID ANALYSIS

Determination of Sialic Acids as Quinoxaline Derivatives

Since the sialic acids could not be derivatized with ABA, they were derivatized with either DMPD (Anumula, 1994) or OPD to yield stable fluorescent quinoxaline derivatives with similar excitation and emission maximums to that of the monosaccharides. Experience with these derivatives indicated that the OPD derivatives are more amenable to the chromatographic procedures than the DMPD derivatives (Anumula, 1994a). The sialic acid derivatives were easily separated on the same column using the same eluents and quantitated using the same detector settings. Optimum condition for mild hydrolysis and derivatization at 80 °C were 20 and 40 min, respectively. Typical chromatograms of the N-acetyl neuraminic acid-OPD and sialic acid derivatives obtained from the rIgG are shown in Figure 5.

Less than 2 pmols of sialic acid can be easily determined by these two different tags. It should be noted that in the case of the rIgG with ~2.0% carbohydrate and ~95% neutral oligosaccharides, the sialic acid content can be easily determined by this method. Data for reproducibility of the standard and the rIgG sample are shown in Table 5.

The relative standard deviation for the N-acetylneuraminic acid from the rIgG was less than 3% at ~6 pmol level.

Table 5. Reproducibility of sialic acid standard and rIgG sample derivatized with OPD

HPLC	Peak area $\times 10^{-5}$ Standard 100 pmol	rIgG 6.32 pmol
Injection 1	47.72	2.949
Injection 2	47.98	2.989
Injection 3	47.54	3.019
Injection 4	48.10	3.117
Injection 5	47.94	3.149
Injection 6	48.51	2.981
Mean area	47.96	3.034
%RSD	0.70	2.65

SUMMARY

Monosaccharide composition analysis is the fundamental part of glycobiology. In this regard, two novel and simple high performance liquid chromatographic (HPLC) methods with fluorescence detection, for the determination of neutral and amino sugar residues and for the determination of sialic acids in glycoproteins are described.

Following acid hydrolysis of the glycoproteins in 20% trifluoroacetic acid, the released monosaccharides were labeled by reductive amination with anthranilic acid (2-amino benzoic acid) in the presence of sodium cyanoborohydride and the derivatives were separated from each other and from the excess reagent. The method is suitable for quantitative determination of less than 100 fmols of monosaccharides.

Sialic acids were released from glycoproteins by mild acid ($NaHSO_4$) hydrolysis and were derivatized with commercial o-phenylenediamine.2HCL (OPD) to obtain stable fluorescent quinoxaline derivatives. OPD-silaic acid derivatives were separated on C18 reversed phase column using the same solvent systems used for monosaccharide analysis. Fluorescence properties of these derivatives were similar to that of anthranilic acid derivatives and they were quantitated using the same excitation and emission wavelengths.

These two methods use the same C-18 column, mobile phase and detector wavelengths of 230 nm and 430 nm for excitation and emission, respectively, but changing the column for sialic acid analysis is recommended in order to save time. Both methods are highly reproducible with relatively low limits (3-4%) of relative standard deviations, since the procedures do not involve separation of excess reagents from the derivatives. Therefore, the fluorescence methods reported here provide a convenient and efficient means of determining both monosaccharides and sialic acids in glycoproteins with high sensitivity. For the first time the recovery of monosaccharides from the hydrolysates was determined using the oligosaccharide map and the recovery was from 83-85% for most of the glycoproteins examined.

REFERENCES

Anumula, K.R., 1994 Quantitative determination of monosaccharides in glycoproteins by high performance liquid chromatography with highly sensitive fluorescence detection, *Anal. Biochem.* 220:275-283.

Anumula, K.R., 1994a Novel fluorescent methods for quantitative determination of monosaccharides in glycoproteins, *J. Protein. Chem.* 13:496-497.

Anumula, K. R., 1993 Endo β-N-acetylglucosaminidase F cleavage specificity with peptide free oligosaccharides, *J. Mol. Recog.* 6:139:145.

Anumula, K. R and Taylor, P. B., 1991 Rapid characterization of asparagine -linked oligosaccharides isolated from glycoproteins using a carbohydrate analyzer, *Eur. J. Biochem.* 195:269-280.

Hase, S. (1993) *Methods in Molecular Biology:* Analysis of sugar chains by pyridylamination, Hounsell, E. F, ed., Vol. 14, pp 69-80.

Hounsell, E. F., ed., 1993 *Methods in Molecular Biology:* Glycoprotein Analysis in Biomedicine. Vol. 14.

Kobata, A., 1992 Structures and functions of the sugar chains of glycoproteins, *Eur. J. Biochem.* 209:483-501.

Kwon, H. and Kim, J., 1993 Determination of monosaccharides in glycoproteins by reversed phase high performance liquid chromatography, *Anal. Biochem.* 215:243-252.

Lee, Y. C., 1992 Perspective of glycotechnology: Carbohydrate recognition for better or worse, *Trends. Glycosci. Glycotech.* 17:251-261.

Lee, K. B, Loganathan, D, Merchant, Z. M and Linhardt, R.J, 1990 Carbohydrate analysis of glycoproteins, *Appl. Biochem. Biotech.* 23:53-80.

Lee, Y. C., 1990 High Performance Anion-Exchange Chromatography for Carbohydrate *Analysis, Anal. Biochem.* 189:151-162.

Novotny, M. V. and Sudor, J., 1993 High performance capillary electrophoresis of glycoconjugates, *Electrophoresis.* 14:373-389.

Oefner, P. J. and Chiesa. C., 1994 Capillary electrophoresis of carbohydrates, *Glycobiology*. 4:397-412.

Patel, T., Bruce, J., Merry, A., Bigge, C., Wormald, M., Jaques, A. and Parekh, R., 1993 Use of hydrazine to release in intact and unreduced form both N- and O-linked oligosaccharides from glycoproteins, *Biochemistry*. 32:679-693

Townsend, R.R., 1993 Quantitative monosaccharide analysis of glycoproteins, *Am. Chem. Soc. Sym.* pp 86-101.

Townsend, R.R. and Hardy, M.R., 1991 Analysis of glycoprotein oligosaccharides using high-pH anion exchange chromatography, *Glycobiology.* 1:139-147.

19

IDENTIFICATION OF THE DISULFIDE BONDS OF HUMAN COMPLEMENT COMPONENT C9

Comparison with the Other Components of the Membrane Attack Complex

Stephan Lengweiler, Johann Schaller, and Egon E. Rickli

Institute of Biochemistry
University of Bern
Freiestr. 3, CH-3012 Bern, Switzerland

INTRODUCTION

C9 is the last acting protein in the complement cascade. Upon interaction with the membrane bound precursor complex, composed of C5b, C6, C7, and C8, C9 undergoes a conformational change from a monomeric globular plasma protein into an oligomeric integral membrane protein, thus generating an amphiphilic channel across the lipid bilayer of the target cell.

Mature C9 is a single chain glycoprotein of an apparent molecular mass of about 71 kDa. The complete sequence has been determined by cDNA analysis (DiScipio *et al.*, 1984; Stanley *et al.*, 1985; Marazziti *et al.*, 1988). C9 contains 538 amino acids, 24 of which are half-cystines. Most of them occur in clusters that have been designated as thrombospondin type I repeat (TSP I), low density lipoprotein receptor class A (LDL A) module and low density lipoprotein receptor class B (LDL B) or epidermal growth factor-like domain.

C9 shows partial sequence similarites to the other terminal complement proteins (Haefliger *et al.*, 1989; DiScipio and Hugli, 1989). All the structural motifs found in C9 are also present in variable number in C6, C7, C8α, and C8β.

EXPERIMENTAL PROCEDURES

Isolation

C9 was isolated from human plasma according to published procedures (Biesecker and Mueller-Eberhard, 1980; Dankert *et al.*, 1985), which involve precipitation with $BaCl_2$

Methods in Protein Structure Analysis, Edited by M. Z. Atassi and E. Appella
Plenum Press, New York, 1995

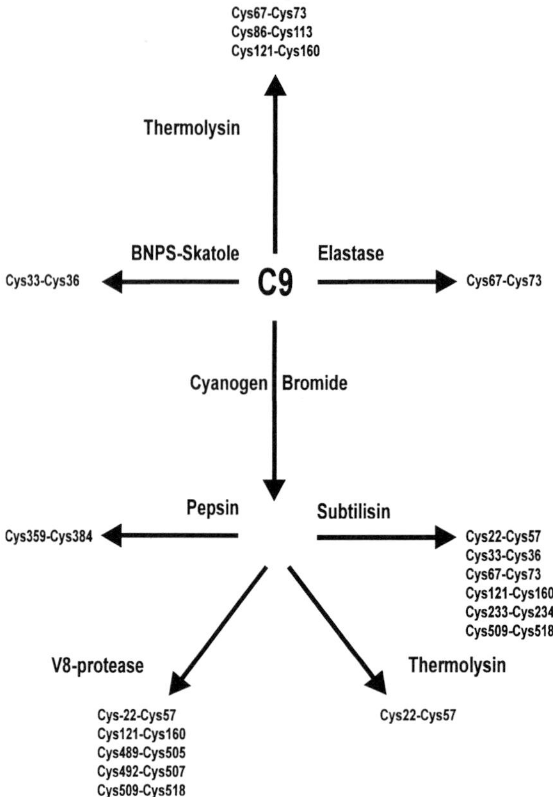

Figure 1. Schematic representation of the cleavage strategy applied for the identification of the disulfide bonds of human C9.

and polyethylene glycol, plasminogen depletion on lysine-Bio-Gel, and chromatography on DEAE-Sephacel and hydroxlapatite. The purity was checked by SDS-PAGE, RP-HPLC, and amino acid analysis.

Fragmentation

Cleavage with cyanogen bromide was carried out for 48 h in 70% trifluoroacetic acid in the dark at a reagent:substrate ratio of 5:1 (w/w). BNPS-skatole was used in a 5-fold excess (w/w) for 72 h in 60% acetic acid, containing 0.2 M phenol.

Enzymatic digestions were generally performed at an enzyme:substrate ratio of 1:50, in the presence of 1 mM iodoacetamide to avoid disulfide interchange. Cleavage with elastase was achieved in 0.2 M Tris-HCl, pH 8.8, for 96 h. Thermolysinolysis was performed in 10 mM borate buffer, pH 6.5, containing 50 mM NaCl and 2 mM $CaCl_2$, for 96 h. Subdigestion with V8-protease was carried out in 50 mM sodium phosphate buffer, pH 7.8, at 37 °C during 74 h. Subdigestion with subtilisin was done in 10 mM NH_4HCO_3, pH 8.0, at 37 °C for 30 h. Subdigestion with pepsin was performed in 10 mM HCl at 37 °C for 40 h.

Preparation of Fragments

Fragments generated by cleavage with BNPS-skatole and cyanogen bromide were separated by gel filtration on a Sephadex G-50 superfine column (1.8 x 90 cm) in 0.13 M formic acid. Enzymatic digests were fractionated by RP-HPLC using a Bakerbond Butyl column (4.6 x 250 mm, wide pore 33 nm, 5 μm; J.T. Baker, Chemicals, Deventer, the Netherlands) in a Hewlett-Packard 1090 liquid chromatograph (Hewlett Packard, Waldbronn, FRG). The acetonitrile gradient system used is described in Fig. 3. Final purification was achieved on Aquapore Butyl, Phenyl, or Octadecyl columns (2.1 x 100 mm, 7 μm, Brownlee Columns, Foster City, USA) or on an Aquapore RP-300 column (1.0 x 100 mm, 7 μm).

Identification of Cystine-Containing Peptides

Aliquots of each HPLC-fraction were reduced with tri-n-butylphosphine and the generated thiol groups specifically labeled with ammonium 7-fluorobenz-oxa-1,3-diazole-4-sulfonate (SBD-F) (Sueyoshi *et al.*, 1985). The resulting fluorescence intensities were measured with excitation at 385 nm and emission at 515 nm.

Alternatively the cystine content was monitored by means of amino acid analysis, using gas phase hydrolysis with 6 M HCl containing 0.1 % (v/v) phenol for 24 h at 115 °C under vacuum (Chang and Knecht, 1991). The liberated amino acids were reacted with PITC and the PTC-derivatives analyzed by RP-HPLC on a Nova Pak Octadecyl column (3.9 x 150 mm; 4 μm; Waters, Milford, USA) (Bidlingmeyer *et al.*, 1984).

Sequence analysis was carried out in a pulsed-liquid-phase sequenator 477A from Applied Biosystems. PTH-amino acids were analyzed on-line according to instructions of the manufacturer. Thereby it is essential to use only DTT-free reagents and solvents. Di-PTH-cystine was detected as a double peak in the vicinity of PTH-tyrosine.

RESULTS AND DISCUSSION

Generation of Cystine-Containing Peptides

Native C9 was cleaved with BNPS-skatole, thermolysin, elastase, or cyanogen bromide. Selected cyanogen bromide fragment pools were subdigested with pepsin, subtilisin, V8-protease, or thermolysin according to Fig. 1.

The generated peptides were separated either by gel filtration or RP-HPLC and all fractions examined for cystine content as described above. Positive fractions were usually rechromatographed once or twice on different columns.

The chromatograms in Fig. 2 and 3 illustrate the strategy that led to the identification of several disulfide bonds primarily in the amino terminal half of C9:

20 mg of native C9 were cleaved with cyanogen bromide. Upon gelfiltration on a Sephadex G-50 superfine column the cyanogen bromide digest yielded 60 fractions. SDS-PAGE, amino acid analysis, and sequence analysis indicated that the fractions 23 to 26 consisted mainly of a fragment ranging from Gln_1 to Met_{272}, whereas fraction 30 essentially contained a chain extending from Lys_{464} to Lys_{538} (Fig. 2). Fractions 23 to 26 and fraction 30 were subsequently treated with subtilisin as described above. The digests were then separated by RP-HPLC on a Bakerbond Butyl column and the fractions checked for cystine

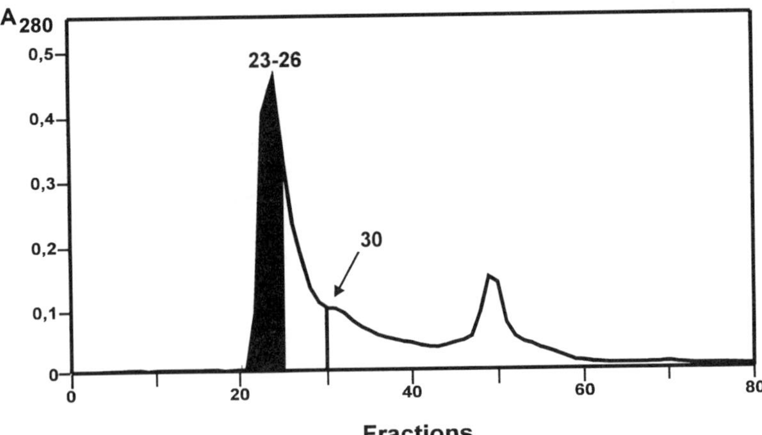

Figure 2. Separation of cyanogen bromide fragments of native C9. Gel filtration on Sephadex G-50 superfine (1.8 x 90 cm) using 0.13 M formic acid as eluant. Flow rate: 10 ml/h; fraction size: 5 ml; detection: absorbance at 280 nm. Pools corresponding to fractions 23 to 26 and to fraction 30 are shaded.

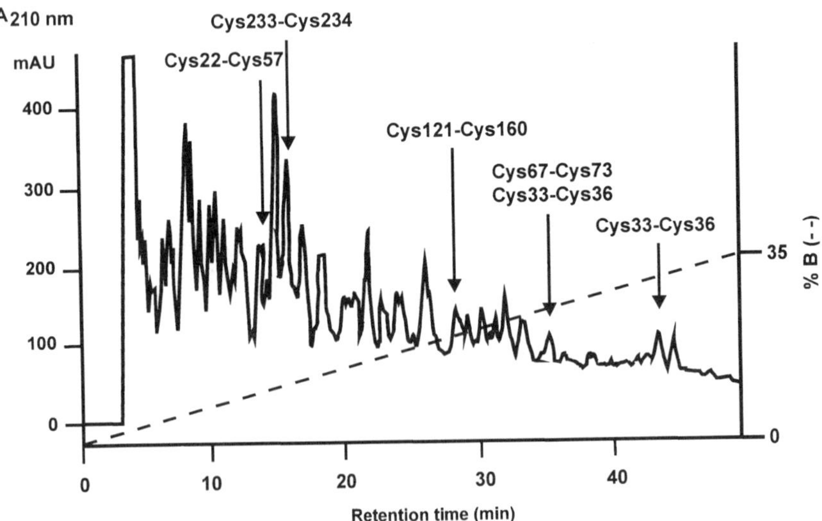

Figure 3. Partial separation of the subtilisin digest of the pooled fractions 23 to 26. RP-HPLC on Bakerbond Butyl (wide pore, 33 nm; 5 µm; 4.6 x 250 mm) using a linear acetonitrile gradient (0 - 35 % B in 50 min). Solution A: 0.1 % TFA in H_20; solution B: 0.1 % TFA in 80 % acetonitrile. Flow rate: 1 ml/min; detection: absorbance at 210 nm. Pools containing disulfide-paired peptides are marked with arrows and the assigned disulfide bond indicated.

content (Fig. 3). Further purification of the Cys-containing peptides was achieved on Aquapore columns (chromatograms not shown).

Identification of the Cystine-Containing Peptides

The Cys-containing peptides were first characterized by amino acid analysis, thus allowing a preliminary localization of the fragments within the known polypeptide chain of C9.

Final identification of the disulfide bond was achieved by amino acid sequence analysis. The corresponding results are summarized in Table 1.

Assignment of the Individual Disulfide Bonds in C9

On the basis of the Edman degradation data the disulfide bridges in C9 were assigned as follows (Fig. 4):

The terminal thrombospondin type I repeat (TSP I) contains six half-cystine residues, which are joined in a 1 to 4, 2 to 3, and 5 to 6 pattern (Cys_{22}-Cys_{57}, Cys_{33}-Cys_{36}, Cys_{67}-Cys_{73}).

The pairing of the six cysteines in the adjacent low density lipoprotein receptor class A module (LDL A) is not yet fully secured. On the basis of sequence comparisons with complement factor I, however, a link between Cys_{80} and Cys_{91} was postulated (Sim *et al.*, 1993). Moreover compositional analyses suggest that Cys_{86} is bonded with Cys_{113}. These findings taken together would then indicate a linkage between Cys_{98} and Cys_{104}, thus implying an overall 1 to 3, 2 to 6, and 4 to 5 arrangement.

The comparatively heterologous central part of the molecule, containing six cysteine residues, has already been investigated earlier. Protease digestion experiments indicated that

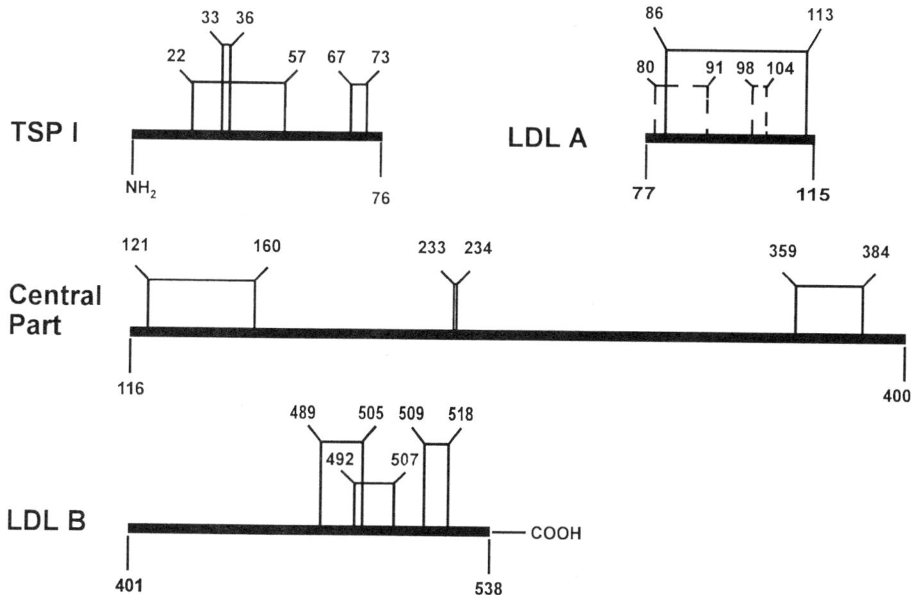

Figure 4. Disulfide bond pairing of human C9. TSP I: thrombospondin type I module; LDL A: low density lipoprotein receptor class A module; LDL B: low density lipoprotein receptor class B module (epidermal growth factor precursor module). Definitively assigned disulfide bridges are indicated by solid lines; tentatively assigned disulfide bridges are indicated by dashed lines.

Table 1. Edman degradation data and identification of disulfide bonds in human C9

Fragmentation	Subfragmentation	Structural data	Disulfide bond
Thermolysin		60AVGDRRQCVPTEPCEDA	67-73
Thermolysin		115SEPRPPCR 159LC	121-160
Elastase		67CVPTEPCEDA	67-73
BNPS-Skatole		31SQCDPCLR*	33-36
Cyanogen Bromide	Pepsin	356IKRCL 380NKDDCV	359-384
Cyanogen Bromide	V8-Protease	13SSGSASHIDCR 48VFGQFNGKRCT	22-57
Cyanogen Bromide	V8-Protease	115SEPRPPCRDR 155FYNGLCNRDR	121-160
Cyanogen Bromide	V8-Protease	484FSVRKCHTCQ 503GKC L CACPF 515GIACE	489-505 492-507 509-518
Cyanogen Bromide	Thermolysin	20IDCR 52FNGKRCTDA	22-57
Cyanogen Bromide	Subtilisin	33CDPCLR	33-36
Cyanogen Bromide	Subtilisin	231EQCCEET	233-234
Cyanogen Bromide	Subtilisin	19HIDC 57CT	22-57

Table 1. *Continued*

Cyanogen Bromide	Subtilisin	115SEPRPPC 160CNRDRDG	121-160
Cyanogen Bromide	Subtilisin	67CVPTEPCED	67-73
Cyanogen Bromide	Subtilisin	33CDPCL	33-36
Cyanogen Bromide	Subtilisin	115SEPRPPCRD 158GLC	121-160
Cyanogen Bromide	Subtilisin	C 507CACPF 516IAC	509-518

Only amino acids that were unambigously identified are shown. Asteriks indicate chain ends that have not been rigorously identified.

Cys_{121} is linked to Cys_{160}, Cys_{233} to Cys_{234}, and Cys_{359} to Cys_{384} (Stanley *et al.*, 1985). These results are fully confirmed by the sequence analysis data shown above. Thereby the very unusual vicinal pairing of Cys_{233} and Cys_{234} is noteworthy.

The six cysteine residues of the low density lipoprotein receptor class B module (LDL B) at the carboxyl terminus of C9 are connected in a 1 to 3, 2 to 4, and 5 to 6 pattern (Cys_{489}-Cys_{505}, Cys_{492}-Cys_{507}, Cys_{509}-Cys_{518}). These findings are entirely compatible with the arrangement inferred by analogy to the known disulfide linkages in the epidermal growth factor (Savage *et al.*, 1973) and complement component C1s (Hess *et al.*, 1991).

Comparison with the other Components of the Membrane Attack Complex

Complement component C9 is partly similiar to C6, C7, C8α, and C8β, with cysteine as the most conserved of all amino acids (Haefliger *et al.*, 1989; DiScipio and Hugli, 1989). This is reflected by a similiar molecular architecture of the terminal complement proteins. As depictet in Fig. 5 all the late acting complement components are based on the same set of modules. C9 in particular contains three types of modules that are omnipresent in the other compounds.

The aim of this study was to establish the cystine pattern of human complement component C9. Due to common structural features the disulfide-bonding pattern of C9, however, can help to predict cystine linkages in the other terminal complement proteins.

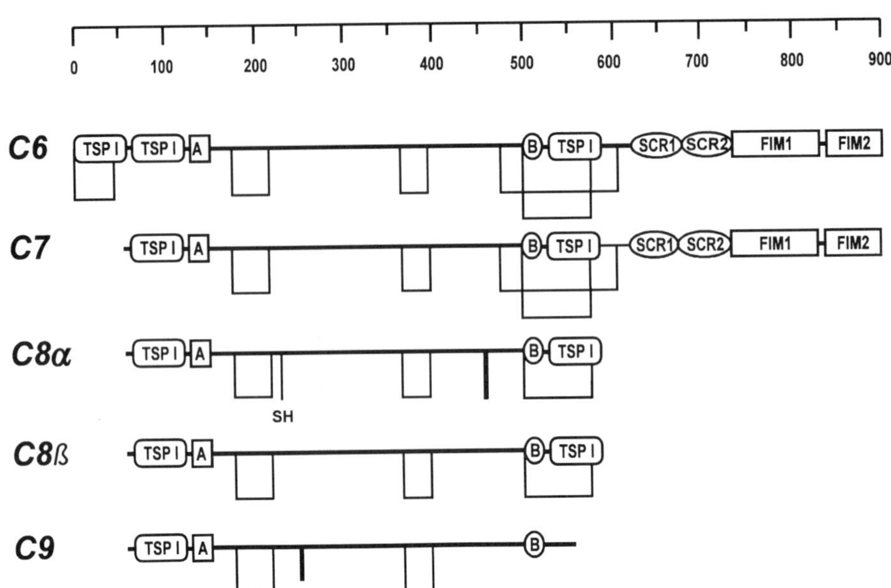

Figure 5. Structural organization of the terminal complement proteins. The scheme shows the distribution of the cysteine-rich modules in human C6, C7, C8α, C8β, and C9 (Haefliger *et al.*, 1989; DiScipio and Hugli, 1989). TSP I: thrombospondin type I module; A: low density lipoprotein receptor class A module; B: low density lipoprotein receptor class B module (epidermal growth factor precursor module); SCR 1, SCR 2: short consensus repeat (sushi domain); FIM 1, FIM 2: factor I module. Cystines found outside modules are indicated by solid lines.

ACKNOWLEDGMENTS

We are particularly indebted to Dr. Juerg Tschopp, University of Lausanne, for kindly providing us with a generous gift of complement component C9 at the very beginning of the project and for valuable advice concerning the C9 isolation in our laboratory. We express our gratitude to Urs Kaempfer and Daniel Wymann for skillful technical assistance. We are grateful to the Central Laboratory of the Blood Transfusion Service, Swiss Red Cross, for the generous supply of human plasma.

REFERENCES

Bidlingmeyer, B. A., Cohen, S. A., and Tarvin, T. L., 1984, Rapid analysis of amino acids using pre-column derivatization, *J. Chromatogr.* 336:93.

Biesecker, G., and Mueller-Eberhard, H. J., 1980, The ninth component of human complement: purification and physiochemical characterization, *J. Immunol.* 124:1291.

Chang, J.-Y., and Knecht, R., 1991, Direct analysis of the disulfide content of proteins: methods for monitoring the stability and refolding process of cystine-containing proteins, *Anal. Biochem.* 197:52.

Dankert, J. R., Shiver, J. W., and Esser, A. F., 1985, Ninth component of complement: selfaggregation and interaction with lipids, *Biochemistry* 24:2754.

DiScipio, R. G., Gehring, M. R., Podack, E. R., Kan, C. C., Hugli, T. E., and Fey, G. H., 1984, Nucleotide sequence of cDNA and derived amino acid sequence of human complement C9, *Proc. Natl. Acad. Sci. U.S.A.* 81:7298.

DiScipio, R. G., and Hugli, T. E., 1989, The molecular architecture of human complement component C6, *J. Biol. Chem.* 264:16197.

Haefliger, J.-A., Tschopp, J., Vial, N., and Jenne, D. E., 1989, Complete primary structure and functional characterization of the sixth component of the human complement system, *J. Biol. Chem.* 264:18041.

Hess, D., Schaller, J., and Rickli, E. E., 1991, Identification of the disulfide bonds of human complement C1s, *Biochemistry* 30:2827.

Marazziti, D., Eggertsen, G., Fey, G. H., and Stanley, K. K., 1988, Relationships between the gene and protein structure in human complement component C9, *Biochemistry* 27:6529.

Savage, C. R., Hash, J H., and Cohen, S., 1973, Epidermal growth factor, *J. Biol. Chem.* 248:7669.

Sim, R. B., Day, A. J., Moffat, B. E., and Fontaine, M., 1993, Complement factor I and cofactors in control of complement system convertase enzymes, *Methods Enzymol.* 223:13-35.

Stanley, K. K., Kocher, H. P., Luzio, J. P., Jackson, P., and Tschopp, J., 1985, The sequence and topology of human complement C9, *EMBO J.* 4:375.

Sueyoshi, T., Miyata, T., Iwanaga, S., Toko'oka, T., and Imai, K., 1985, Application of a fluorogenic reagent, ammonium 7-fluorobenzo-2-oxa-1,3-diazole-4-sulfonate for detection of cystine-containing peptides, *J. Biochem. (Tokyo)* 97:1811.

20

LOCALIZATION OF *IN VIVO* PHOSPHORYLATION SITES IN MULTIPHOSPHORYLATED PROTEINS

Application of S-Ethylcysteine Derivatization and Mass Spectrometric Analysis

Esben S. Sørensen,[1] Lone K. Rasmussen,[1] Peter Højrup,[2] and Torben E. Petersen[1]

[1] Protein Chemistry Laboratory
University of Aarhus
Science Park, Gustav Wieds Vej 10, DK-8000 Aarhus C, Denmark
[2] Department of Molecular Biology
University of Odense
DK-5230 Odense M, Denmark

INTRODUCTION

Phosphorylation of proteins is one of the most frequent forms of posttranslational modification in eukaryotic cells and is linked to the control of a multitude of cellular functions. Proteins involved in this type of regulation are typically only phosphorylated at a single or a few sites. Another type of phosphoproteins are those containing multiple phosphorylations. In these proteins the phosphorylations usually possess different functions than in proteins phosphorylated at single sites. In the case of multisite phosphorylated proteins (for review see Roach, 1991), the phosphorylations are often important as physical interactors with divalent metal ions, especially Ca^{2+}.

Multiphosphorylated proteins are important factors in all biologically regulated mineralization processes. They are hypothesized to be involved in the nucleation of mineral crystals within and upon the organic matrix of tissues such as bone and dentin. When crystal growth has been induced, the multiphosphorylated proteins are thought to regulate the rate of mineralization by adsorption to the surface of the crystals.

Besides calcifying tissue, multiphosphorylated proteins are especially abundant in physiological fluids containing high Ca^{2+} concentrations such as saliva, urine and milk. In saliva and urine, the phosphoproteins are speculated to form complexes with developing calcium salt crystals, thereby inhibiting the formation of dentin plaques (Holt and van Kemenade, 1989) and urinary stones (Shiraga et al., 1992), respectively.

Methods in Protein Structure Analysis, Edited by M. Z. Atassi and E. Appella
Plenum Press, New York, 1995

The significance of the position of the phosphorylations in these multiphosphorylated proteins is unknown. It is likely that the effect of having phosphoserine residues at specific locations in the protein is important in the regulation of crystallization.

The inability of the standard Edman degradation procedure to deal with phosphoamino acids has made the localization of *in vivo* phosphorylation sites a laborious procedure. Subfragmentation, purification and amino acid analysis of smaller peptides are necessary if a phosphopeptide contains more than one serine residue and this is an even more tedious challenge if two or more serines are in series.

A method to overcome these difficulties has been reported. Meyer et al. (1986) have described a one-step micro batch reaction in which peptide-bound phosphoserine is quantitatively converted into S-ethylcysteine, while serines without O-linked phosphate are left unaffected. Subjected to Edman degradation chemistry, S-ethylcysteine yields a stable PTH-derivative which elutes between valine and DPTU in the system used (Applied Biosystems 477A/120A), thereby making it possible to assign phosphoserines directly by automated sequencing. This technique has been employed in the localization of *in vivo* phosphorylation sites in several proteins containing one or few phosphorylations (for review see Meyer et al., 1993). Here we demonstrate that the method is also applicable to the localization of phosphorylation sites in multiphosphorylated proteins.

Strategy

The strategy used for the localization of phosphorylation sites in a multiphosphorylated protein is outlined in figure 1. Osteopontin, a heavily phosphorylated glycoprotein, was isolated from bovine milk (Sørensen and Petersen, 1993a). The protein was digested with proteases and the digests separated by reverse-phase HPLC. Fractions containing phosphoamino acids were identified by a amino acid analysis and further purified. Aliquots of the phosphopeptides were subjected to mass spectrometric analysis. Peptides containing phosphoserine were subjected to the ethanethiol treatment, resulting in the conversion of phosphoserine to S-ethylcysteine, and subsequently sequenced. The S-ethylcysteine derivatization technique is discussed on the basis of selected phosphopeptides from osteopontin.

Modification of Phosphoserine with Ethanethiol

The S-ethylcysteine derivatization of peptide-bound phosphoserine is performed as a β-elimination of the phosphate group followed by addition of ethanethiol to the double bond of the dehydroalanine intermediate (Figure 2). The experimental procedure is performed essentially as described (Meyer et al., 1991). The dried peptide was incubated for 1 h at 50°C under nitrogen with 50 μL of a freshly prepared derivatization mixture (80 μL ethanol, 60 μL ethanethiol, 65 μL 5M NaOH, 400 μL H_2O). The sample was then cooled to room temperature and neutralized by addition of 10 μL glacial acetic acid. Derivatized samples were vacuum-dried and frozen. Prior to sequencing the samples were dissolved with 50 μL 0.1% trifluoroacetic acid and applied to Biobrene-treated glass fiber filters.

RESULTS AND DISCUSSION

Amino acid analysis of the peptide LPVKPTSSGSSEEK, corresponding to residues 1-14 in bovine osteopontin, showed that the peptide contained phosphoserine. The phosphoserine(s) could potentially be located at any of the four serines in the peptide. Yields of

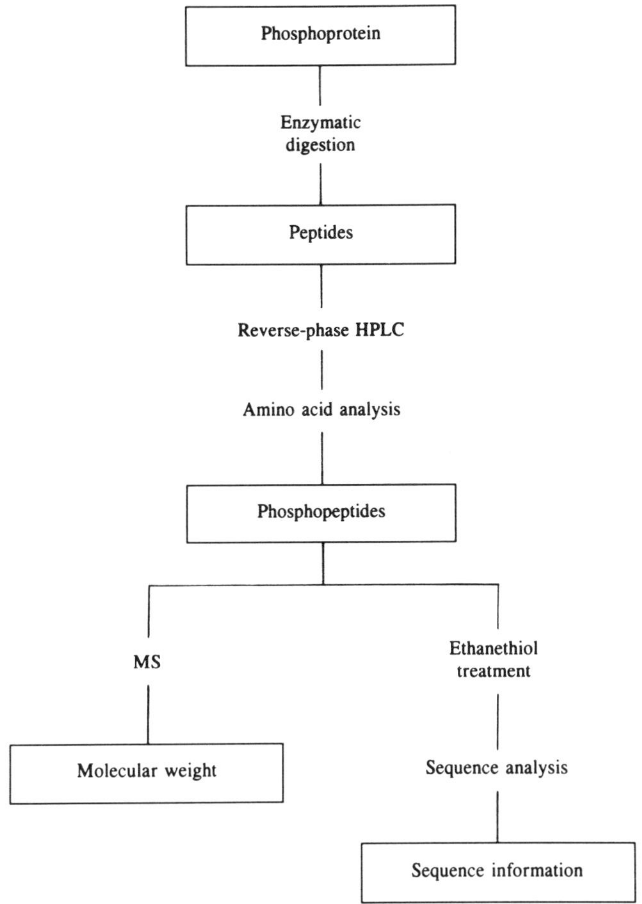

Figure 1. Strategy for localization of phosphorylation sites in multiphosphorylated proteins.

the PTH-amino acids generated by sequencing of the native as well as the ethanethiol-treated peptide are shown in Table 1 (the yields in the two experiments are not comparable, due to different amounts of sample applied to the sequencer).

Sequencing of the native phosphopeptide shows that the yield of PTH-serine is zero or very low in all cycles predicted to contain a serine. This indicates that the four serines in

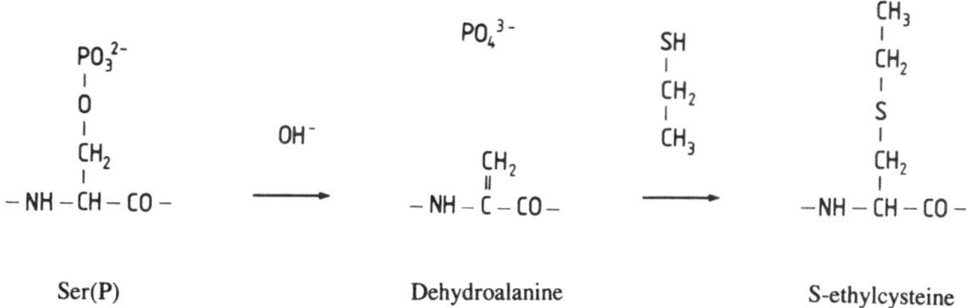

Figure 2. S-ethylcysteine-derivatization of peptide-bound phosphoserine. Experimental procedures are described in the text.

Table 1. Sequence analysis of the peptide LPVKPTSSGSSEEK, corresponding to residues 1-14 in bovine osteopontin

Amino acid sequence[a]	PTH-AA yields native peptide[b]	PTH-AA yields EtSH peptide[c]
L^1	419	215 (1)
P^2	154	81 (17)
V^3	207	100 (20)
K^4	87	46 (7)
P^5	59	63 (5)
T^6	20	11 (4)
S^7	3	0 (43)*
S^8	0	0 (58)*
G^9	37	28 (32)
S^{10}	6	0 (41)*
S^{11}	5	0 (55)*
E^{12}	12	9 (30)
E^{13}	12	15 (17)
K^{14}	d	d

[a]Amino acid sequence according to the cDNA (Kerr et al., 1991).
[b]PTH-amino acid yields obtained by sequencing of the native underivatized peptide.
[c]PTH-amino acid yields obtained by sequencing of the ethanethiol-treated peptide. PTH-S-ethylcysteine yields, given in parenthesis in each sequence step, are quantified by use of the response factor for PTH-methionine. Sequence cycles in which PTH-S-ethylcysteine were assigned are marked by asterisks.
[d]The yields of PTH-lysine were not determined.

the peptide are modified, but direct evidence for the phosphorylation is missing. Treatment of the phosphopeptide with ethanethiol as described, followed by sequencing, gave PTH-S-ethylcysteine at all four serine residues (PTH-S-ethylcysteine yields are shown in parenthesis), thereby showing that all four serines in this peptide are actually phosphorylated. Mass spectrometric analysis of the underivatized peptide (data not shown) revealed an excess mass of 320.5 Da compared to the calculated peptide mass, which corresponds to four phosphate groups. In this example, the phosphorylations could have been assigned simply by sequencing of the native peptide combined with the mass spectrometric data, but this is not the case when the phosphorylation pattern becomes more complicated.

By amino acid analysis the peptide TSQLTDHSKETNSSELSK, corresponding to residues 182-199 in bovine osteopontin, was shown to contain phosphoamino acids. Mass spectrometric analysis of the peptide showed a MH^+ ion at m/z 2153.7 (Figure 3). This mass exceeds the calculated peptide mass by 160.6 Da, corresponding to two phosphorylations. The phosphopeptide contains five serines and three threonines, constituting a total of eight potential phosphorylation sites. In this case, where the phosphopeptide contains both phosphorylated and unphosphorylated hydroxy amino acids, sequence analysis is necessary for the localization of the phosphorylations. The phosphopeptide was subjected to the ethanethiol treatment and subsequently sequenced (Table 2). Sequence analysis revealed PTH-S-ethylcysteine in cycles corresponding to Ser189 and Ser194 in the native protein. Examination of the PTH-serine and PTH-S-ethylcysteine yields shows that the conversion of these serines was essentially complete while the other three serines in the sequence were

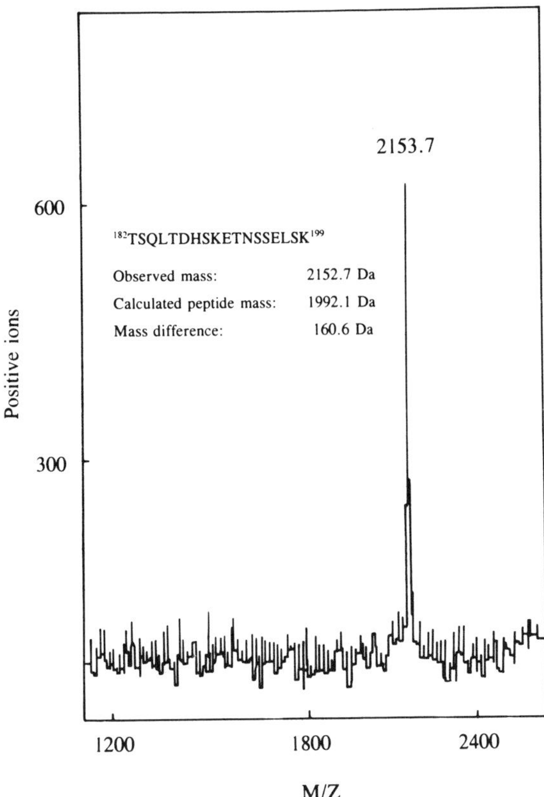

Figure 3. PDMS spectra of the peptide TSQLTDHSKETNSSELSK, corresponding to residues 182-199 in bovine osteopontin.

not affected by the treatment. In cases like this, the S-ethylcysteine derivatization is essential for the localization of the phosphorylation sites.

Limitations of the Technique

It has been observed that certain cysteines gave false positive results for S-ethylcysteine when subjected to the ethanethiol treatment. This problem has been solved by performic acid oxidation of cysteines to cysteic acid before subjecting the peptide or protein to the ethanethiol treatment (Meyer et al., 1989).

O-glycosylated serine residues have also been hypothesized to yield S-ethylcysteine by use of this method, but in combination with a second analysis such as a carbohydrate or mass spectrometric analysis, peptides containing O-glycosylations can easily be identified. Moreover, a selective elimination procedure for the elimination of phosphate groups has been described (Byford, 1991).

The position of the phosphoserine in the peptide subjected to the ethanethiol treatment is also of importance. If the phosphoserine residue is positioned as the N-terminal amino acid in the phosphopeptide, S-ethylcysteine is not formed during the ethanethiol treatment. Instead, elimination of the phosphate group is followed by rearrangement of the double bond in the dehydroalanine intermediate, resulting in pyruvate formation (Meyer et

Table 2. Sequence analysis of the peptide TSQLTDHSKETNSSELSK, corresponding to residues 182-199 in bovine osteopontin

Amino acid sequence[a]	PTH-AA yields EtSH peptide[b]	Amino acid sequence[a]	PTH-AA yields EtSH peptide[b]
T^{182}	166 (1)	E^{191}	33 (15)
S^{183}	63 (4)	T^{192}	26 (8)
Q^{184}	75 (1)	N^{193}	16 (5)
L^{185}	146 (1)	S^{194}	2 (22)*
T^{186}	55 (1)	S^{195}	7 (11)
D^{187}	26 (1)	E^{196}	14 (6)
H^{188}	9 (7)	L^{197}	19 (3)
S^{189}	2 (53)*	S^{198}	7 (1)
K^{190}	52 (34)	K^{199}	9 (2)

[a]Amino acid sequence according to the cDNA (Kerr et al., 1991).
[b]PTH-amino acid yields obtained by sequencing of the ethanethiol-treated peptide. PTH-S-ethylcysteine yields, given in parenthesis in each sequence step, are quantified by use of the response factor for PTH-methionine. Sequence cycles in which PTH-S-ethylcysteine were assigned are marked by asterisks.

al., 1991). Likewise, ethylamine is formed during the elimination process of phosphoserines located as the C-terminal amino acid in the phosphopeptide.

In our work with osteopontin, we have encountered another sequence position of phosphoserine which seems to hinder the formation of S-ethylcysteine, namely the presence of a proline residue as the C-terminal neighbour. Mass spectrometric analysis of a peptide, TLPSKSNESPEQ, comprising residues 57-68 in bovine osteopontin showed a MH^+ ion at m/z 1559.3, corresponding to the calculated peptide mass plus three phosphorylations (Figure 4). The peptide was subjected to the ethanethiol treatment and subsequently sequenced (Table 3). The yield of PTH-threonine in cycle one is at an expected level, thereby indicating that the phosphorylations must be located at the three serines in the sequence. As anticipated, PTH-S-ethylcysteines were observed at positions corresponding to Ser60 and Ser62. However, at the position corresponding to Ser65 virtually no PTH-amino acids were identified. Similar to this, two phosphoserines in κ-casein both followed by prolines failed to yield S-ethylcysteine when subjected to the ethanethiol treatment (Rasmussen et al., unpublished data). These results strongly indicate that addition of ethanethiol to the double

Table 3. Sequence analysis of the peptide TLPSKSNESPEQ, corresponding to residues 57-68 in bovine osteopontin

Amino acid sequence[a]	PTH-AA yields EtSH peptide[b]	Amino acid sequence[a]	PTH-AA yields EtSH peptide[b]
T^{57}	146 (1)	N^{63}	47 (25)
L^{58}	213 (0)	E^{64}	38 (6)
P^{59}	100 (1)	S^{65}	5 (4)
S^{60}	2 (89)*	P^{66}	40 (2)
K^{61}	90 (19)	E^{67}	26 (1)
S^{62}	2 (91)*	Q^{68}	29 (1)

[a]Amino acid sequence according to the cDNA (Kerr et al., 1991).
[b]PTH-amino acid yields obtained by sequencing of the ethanethiol-treated peptide. PTH-S-ethylcysteine yields, given in parenthesis in each sequence step, are quantified by use of the response factor for PTH-methionine. Sequence cycles in which PTH-S-ethylcysteine were assigned are marked by asterisks.

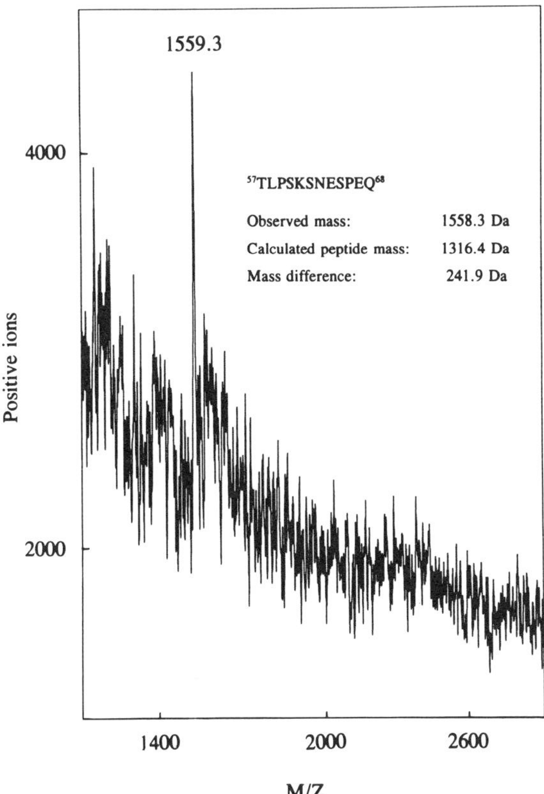

Figure 4. MALDI-TOF spectra of the peptide TLPSKSNESPEQ, corresponding to residues 57-68 in bovine osteopontin.

bond of the dehydroalanine intermediate is impaired when neighboured by a proline residue at the C-terminal side.

As seen in Table 3, the position of a proline residue as the N-terminal neighbour to a phosphoserine (Ser60), does not affect the formation of S-ethylcysteine. We are not aware of any phosphoserine located in a -Ser(P)-Pro- sequence motif that has been identified as S-ethylcysteine after ethanethiol treatment.

As summarized here, the technique has certain limitations but these can easily be managed by taking the appropriate precautions in the individual cases, or simply confirming phosphorylations by mass spectrometric analysis.

CONCLUSION

By the described combination of amino acid analysis, sequencing of S-ethylcysteine-derivatized phosphopeptides and mass spectrometric analysis we have identified a total of 27 phosphoserines and one phosphothreonine in bovine osteopontin (Sørensen and Petersen, 1994; Sørensen et al.,1995). Osteopontin contains 41 serines, 17 threonines and 2 tyrosines, which make up 60 potential phosphorylation sites, corresponding to 23% of the amino acids in the protein. A considerable number of these serines and threonines occur in series of 4-5 potential phosphorylations sites. This clustering of potential phosphorylation sites makes the localization of phosphoamino acids by the subfragmentation method impossible. Even application of mass spectrometric analysis of series of extensive digest of the protein would not result in unambiguous identification of the phosphorylation sites in this multiphosphorylated protein.

Likewise, we have used the S-ethylcysteine derivatization method to identify five phosphoserines in component PP3 (Sørensen and Petersen, 1993b,c), a major non-casein phosphoprotein from milk, which has recently been shown to be homologous with mouse and rat adhesion molecule GlyCAM-1 (Johnsen et al., 1995). In PP3 the five phosphorylations were all located within a stretch of 18 amino acids, which would have made localization by the subfragmentation method very difficult. By the S-ethylcysteine derivatization procedure all five phosphorylations were assigned in a single sequence run.

The examples discussed here emphasize the importance of both mass spectrometric and sequence data in the exact localization of phosphorylated amino acids.

ACKNOWLEDGMENTS

We thank Lise Møller for technical assistance. This work is part of the FØTEK programme supported by the Danish Dairy Board and the Danish Government.

REFERENCES

Byford, M. S., 1991 Rapid and selective modification of phosphoserine residues catalysed by Ba^{2+} ions for their detection during peptide microsequencing, *Biochem. J.* 280:261

Holt, C., van Kemenade, M. J. J. M., 1989 The interaction of phosphoproteins with calcium phosphate, *in*: Calcified tissue, Hukins, D. W., ed., MacMillan, London, pp 175-213.

Johnsen, L., Sørensen, E. S., Petersen, T. E., Berglund, L., 1995 Characterization of a bovine mammary gland PP3 cDNA reveals homology with mouse and rat adhesion molecule GlyCAM-1, *Biochim. Biophys. Acta* 1260:116.

Kerr, J. M., Fisher, L. W., Termine, J. D., Young, M. F., 1991 The cDNA cloning and RNA distribution of bovine osteopontin, *Gene* 108:237.

Meyer, H. E., Hoffmann-Posorske, E., Korte, H., Heilmeyer jr., M. G., 1986 Sequence analysis of phosphoserine-containing peptides. Modification for picomolar sensitivity, *FEBS Lett.* 204:61.

Meyer, H. E., Kuhn, C. C., Meyer, G. F., Swiderek, K., Weber, C., Hoffmann-Posorske, E., Heilmeyer jr., L. M. G., 1989 Localization of endogenous phosphoserine residues in the primary structure of proteins, *in:* Methods in Protein Sequence Analysis, Wittmann-Liebold, B., ed., Springer-Verlag, Berlin, pp 19-22.

Meyer, H. E., Hoffmann-Posorske, E., Heilmeyer jr., M .G., 1991 Determination and location of phosphoserine in proteins and peptides by conversion to S-ethylcysteine, *Methods in Enzymology* 201:169-185.

Meyer, H. E., Eisermann, B., Heber, M., Hoffmann-Posorske, E., Korte, H., Weigt, C., Wegner, A., Hutton, T., Donella-Deana, A., Perich, J. W., 1993 Strategies for nonradioactive methods in the localization of phosphorylated amino acids in proteins, *FASEB J.* 7:776.

Roach, P. J., 1991 Multisite and hierarchal protein phosphorylation, *J. Biol. Chem.* 266:14139.

Shiraga, H., Min, W., VanDusen, W. J., Clayman, M. D., Miner, D., Terrell, C. H., Sherbotie, J. R., Forman, J. W., Przysiecki, C., Neilson, E. G., Hoyer, J. R., 1992 Inhibition of calcium oxalate crystal growth *in vitro* by uropontin: Another member of the aspartic acid-rich protein superfamily. *Proc. Natl. Acad. Sci. USA* 89:426.

Sørensen E. S., Petersen T. E., 1993a Purification and characterization of three proteins isolated from the proteose peptone fraction of bovine milk, *J. Dairy Res.* 60:189.

Sørensen E. S., Petersen T. E., 1993b Phosphoproteins, *Protein Sci.* 2 (Suppl.1):111.

Sørensen E. S., Petersen T. E., 1993c Phosphorylation, glycosylation and amino acid sequence of component PP3 from the proteose peptone fraction of bovine milk, *J. Dairy Res.* 60:535.

Sørensen E. S., Petersen T. E., 1994 Identification of two phosphorylation motifs in bovine osteopontin, *Biochem. Biophys. Res. Commun.* 198:200.

Sørensen E. S., Højrup, P., Petersen, T. E., 1995 Posttranslational modifications of bovine osteopontin: Identification of twenty-eight phosphorylation and three O-glycosylation sites, *Protein Sci.* 4, in press.

21

IDENTIFICATION AND CHARACTERIZATION OF TRANSDUCIN FUNCTIONAL CYSTEINES, LYSINES, AND ACIDIC RESIDUES BY GROUP-SPECIFIC LABELING AND CHEMICAL CROSS-LINKING

José Bubis,[1] Julio O. Ortiz,[1] Carolina Möller,[2] and Enrique J. Millán[1]

[1] Departamento de Biología Celular
[2] Departamento de Química
Universidad Simón Bolívar
Apartado 89.000, Caracas 1081-A, Venezuela

INTRODUCTION

Guanine nucleotide binding proteins or G-proteins function as molecular switches in a diverse set of signaling pathways by coupling seven-helix transmembrane receptors to specific intracellular effectors (Kaziro et al., 1991; Dohlman et al., 1991). G-proteins are heterotrimers composed of α-, β-, and γ-subunits. Activation of the appropiate receptor causes a GDP molecule bound to the resting form of a G-protein to be exchanged for GTP. As a consequence, the G-protein dissociates to form the α-subunit complexed to GTP, and the βγ-dimer. The GTP-bound conformation of the α-subunit is capable of activating or inhibiting a variety of downstream effectors including enzymes as well as ion channels (Birnbaumer, 1992; Hepler & Gilman, 1992; Simon et al., 1991). The released βγ-complex can itself activate or modulate some effectors (Logothetis et al., 1987; Tang et al., 1991; Katz et al., 1992). A GTPase-controlled timing mechanism inherent in all α-subunits and, in some cases, modulated by other proteins (Berstein et al., 1992; Arshavsky & Bownds, 1992), returns the GTP-activated α-subunit to the inactive GDP-bound conformation. The α-subunit complexed to GDP reassociates with the βγ-complex and forms again the hetero-trimer in its resting state. Conklin & Bourne (1993) proposed a structural model for a general G-protein α-subunit, on the basis of biochemical, immunologic, and molecular genetic observations. This model provided a blurred but revealing view of the orientation of membrane-bound G_{α} with regard to $G_{\beta\gamma}$, receptors, and effectors.

One of the best characterized heterotrimeric G-protein-coupled systems is the visual cascade of retinal rod outer segments (Lagnado & Baylor, 1992; Hargrave & McDowell, 1992). Here, the visual G-protein, transducin, serves as an intermediary between the

Methods in Protein Structure Analysis, Edited by M. Z. Atassi and E. Appella
Plenum Press, New York, 1995

photoreceptor rhodopsin (R), and the effector protein cGMP phosphodiesterase (cGMP PDE) during signaling. Visual signal transduction begins with the absorption of a photon by the 11-cis-retinal chromophore of R. Rapid photoisomerization to all-trans-retinal triggers a series of structural conformational changes that lead to the formation of the activated intermediate metarhodopsin II (R^*). R^* binds the heterotrimeric GDP-bound form of transducin ($T_{\alpha\beta\gamma}$) and catalyses the exchange of GTP for GDP, resulting in the dissociation of $T_{\alpha\beta\gamma}$ into T_α-GTP and $T_{\beta\gamma}$. T_α-GTP, in turn, activates a potent cGMP PDE by binding and displacing its inhibitory subunits. The resulting decrease in second messenger cGMP concentration causes cation-specific cGMP-gated channels to close, leading to hyperpolarization of the rod cell membrane and to the generation of the nerve impulse. As a result of its intrinsic GTPase activity, T_α is inactivated by hydrolysis of GTP to GDP, returning the system to its resting state.

Recently, the three-dimensional structures of a 325-amino acid fragment of T_α bound to either $GTP_\gamma S$ (Noel et al., 1993), or to GDP (Lambright et al., 1994) have been solved. Together, the two T_α structures furnish contrasting freeze-frame pictures of two key intermediates in the G-protein cycle. Although both structures are quite similar, there are differences induced by nucleotide exchange that are localized to three adjacent regions on one face of the protein, which have been implicated in effector activation (Lambright et al., 1994). However, little information is available concerning the contacts among transducin subunits. Furthermore, it is not known which residues are directly involved in transducin-rhodopsin and transducin-cGMP PDE interactions. Bubis & Khorana (1990) found that Cys-25 of T_β is in close proximity to Cys-36 and/or Cys-37 of T_γ, by using cupric phenanthroline, a reagent known to catalyse the formation of disulfide bonds between suitably placed sulfhydryl groups. To continue the studies on the structure-function of transducin, we have used group-specific labeling and chemical cross-linking techniques to identify some of the functionally important amino acid residues of the protein.

EXPERIMENTAL PROCEDURES

Materials

Bovine eyes were obtained from the nearest slaughterhouse (Matadero Caracas, C.A., Venezuela). Retinae were extracted in the dark, under red light, and were mantained frozen at -70°C. Chemical reagents were obtained from the following suppliers: [2-^{3}H] iodoacetic acid ([^{3}H] IAA, 131 mCi/mmol), β, γ-imido-[^{3}H] guanosine 5′-triphosphate ([^{3}H] GMP-PNP, 12.8 Ci/mmol), and [γ-^{32}P] GTP (30 Ci/mmol), Amersham; [8,5-^{3}H] GTP (15 Ci/mmol), American Radiolabeled Chemicals Inc.; 4-acetamido-4′-maleimidyl-stilbene-2,2′ disulfonic acid (AMDA), and 2,5-dimethoxystilbene-4′-maleimide (DM), Molecular Probes, Inc.; 2-nitro 5-thiocyano benzoic acid (NTCBA), N,N′-dicyclohexyl- carbodiimide (DCCD), phenyl isothiocyanate (PITC), o-phtalaldehyde (OPA), acetic anhydride (AC), and dansyl chloride (DnsCl), Sigma; iodoacetic acid (IAA), Kodak; 4-vinyl pyridine (VP), Fluka; 1-ethyl 3-(3-dimethylaminopropyl) carbodiimide (EDC), and citraconic anhydride (CA), Pierce; N,N′-1,2-phenylenedimaleimide (o-PDM), and N,N′-1,4-phenylenedimaleimide (p-PDM), Aldrich. All other reagents were analytical grade.

Rod Outer Segments and Washed Membranes

Rod outer segments (ROS) were isolated from frozen bovine retinae, by flotation and subsequent centrifugation on discontinuos sucrose gradients, as described previously (Bubis & Khorana, 1990; Bubis et al., 1993). ROS membranes were washed with 2 mM EDTA

(Baehr et al., 1979), or 5 M urea (Shichi & Somers, 1978), to remove ROS peripheric proteins. Washed ROS were used as the source of rhodopsin in guanine nucleotide binding and GTPase assays.

Transducin Isolation

Transducin was isolated from ROS membranes prepared under room light, at 4°C, following the affinity binding procedure carried out by Kühn (1980). GTP (100 μM) was used to elute transducin from the washed illuminated ROS membranes (Baehr et al., 1982), and transducin was further purified to homogeneity by ion-exchange chromatography on DE 52, as described elsewhere (Bubis & Khorana, 1990; Bubis et al., 1993).

Isolation of T_{α} and $T_{\beta\gamma}$ by Chromatography *in Tandem* through Blue Agarose followed by ω-Amino Octyl Agarose

GTP-extracted transducin was supplemented with 100 μM EDTA and 10% glycerol and chromatographed on a blue agarose column followed by an ω-amino octylagarose column (Bubis & Khorana, 1990). The α-subunit, which was bound to the blue agarose column, was eluted following the procedure of Shichi et al. (1984). The βγ-complex was eluted from the ω-amino octylagarose as described by Fung (1983).

Binding of [^{3}H] GMP-PNP or [^{3}H] GTP to Transducin

Guanine nucleotide binding to native or reconstituted transducin was measured by Millipore filtration, as described previously for cyclic nucleotide binding to cAMP-dependent protein kinase (Bubis & Taylor, 1985; 1987). The binding reaction was carried out in Buffer I [10 mM HEPES (pH 7.4), 100 mM NaCl, 5 mM magnesium acetate, 5 mM β-mercaptoethanol] containing 0.1 μM rhodopsin (as urea-washed ROS membranes), and a fixed concentration of [^{3}H] GMP-PNP or [^{3}H] GTP (0.2 μM).

Modification of Transducin Cysteyl, Acidic and Lysyl Residues

For cysteine modification, transducin (1-2 μM) was incubated either with DM (5 mM), AMDA (5 mM), NTCBA (5 mM), [^{3}H] IAA (4 mM), or VP (74.9 mM), in 20 mM Tris-HCl (pH 8.0), 1 mM dithiothreitol (DTT), 100 mM NaCl, and 5 mM magnesium acetate. To label acidic residues, transducin (0.2 μM) was incubated with 5 mM EDC or DCCD, in 50 mM PIPES (pH 6.2), and magnesium acetate (20 or 30 mM). For lysine labeling, transducin (0.2 μM) was incubated with either 5 mM of PITC, OPA, AA, CA, or DnsCl, in 0.1 M Tris-HCl (pH 8.0). At designated time intervals (0-60 min), the reaction mixtures for the cysteines, acidic residue, and lysines modifications were terminated with 5 mM DTT, 12 mM acetic acid (pH 6.3), or 15 mM β-mercaptoethanol and 30 mM magnesium acetate, respectively. Then, the kinetics of inactivation of transducin was assayed measuring the rhodopsin-dependent [^{3}H] GMP-PNP or [^{3}H] GTP binding. Similar procedures were used to modify T_{α} and $T_{\beta\gamma}$. T_{α} and $T_{\beta\gamma}$ (0.15 μM) were incubated with the chemical reagents, and after terminating the reactions, they were reconstituted with 0.15 μM of the complementary unit, to reform the holoenzyme.

Transducin (0.2 - 0.4 mg) also was modified with [^{3}H] IAA, AMDA, and VP, under native conditions. AMDA-modified transducin was chromatographed on a Sephadex G-25 gel filtration column. The fractions containing protein were pooled, dialyzed extensively against 50 mM NH_4HCO_3 (pH 8.3), and concentrated. IAA- and VP-modified transducin also were dialyzed against 50 mM NH_4HCO_3 (pH 8.3). Labeled-transducin samples were then digested with TPCK-treated trypsin in a molar ratio of 50:1 protein to protease for 24 h, at 37°C, lyophilized, and the resulting peptides were separated by high performance liquid chromatography (HPLC).

Interaction Assay between Modified Transducin and Photoexcited Rhodopsin

Transducin was incubated either with 5 mM DM, AMDA, NTCBA, IAA, or with 74.9 mM VP. After 30 min of incubation, at 4°C, the reactions were terminated with 5 mM DTT. These samples were denominated T(Cys-X), (X = H or labeling group). EDTA-washed ROS (3.5 μM rhodopsin) were mixed with T(Cys-X) (2.4 μM), in an isotonic solution [5 mM Tris-HCl (pH 7.5), 0.1 M NaCl, 5 mM magnesium acetate, 5 mM β-mercaptoethanol], and incubated for 1 h, at 4°C, under light. To assess whether a functional T-R^* complex was formed, succesive cycles of centrifugation (40,000 rpm, 30 min) and extraction were carried out. Initial supernatants (S-ISO) were separated and the pellets were resuspended in a hypotonic solution [5 mM Tris-HCl (pH 7.5), 5 mM magnesium acetate, 5 mM β-mercaptoethanol] (P_r). After centrifugation, we obtained S-HYPO's and the resulting pellets were extracted in the hypotonic buffer containing 150 μM GTP (P_r-GTP). P_r-GTP's were centrifuged as before, and the supernatants (S-GTP) were separated from the final pellets (P_f). During the course of these experiments, aliquots were taken from T(Cys-X), S-ISO, S-HYPO, P_r-GTP, S-GTP, and P_f, for analysis by SDS-polyacrylamide gel electrphoresis and Western blot.

Assay of Transducin Functionality in the T-R^* Complex Incubated with Sulfhydryl Group-Specific Reagents

Transducin (1.4 μM) was incubated with EDTA-washed ROS (rhodopsin = 5.6 μM) for 30 min, at 4°C, under light. The mixture was centrifuged at 40,000 rpm for 30 min, and T-R^* was obtained in the sedimented pellets. The pellets were resuspended in Buffer II (hypotonic buffer, pH 8.0) and incubated with either 5 mM DM, AMDA, NTCBA, IAA, or 74.9 mM VP. After a 30 min incubation on ice, the reactions were stopped with 5 mM DTT. These samples were denominated P_r(Cys-X), (X = H or labeling group). P_r(Cys-X)'s were centrifuged, and the supernatants were designated S-1. The resulting pellets were resuspended in Buffer II (P_r), centrifuged to obtain second supernatants (S-2). Then, the pellets were extracted in the hypotonic buffer containing 150 μM GTP and were named P_r-GTP. P_r-GTP's were centrifuged as above, and the supernatants (S-GTP) were separated from the final pellets which were resuspended in Buffer II and denominated P_f. During the course of these experiments, aliquots were taken from P_r(Cys-X), S-1, P_r, S-2, S-GTP, and P_f, for analysis by SDS-polyacrylamide gel electrphoresis and Western blot.

Cross-Linking with o-PDM and p-PDM

Transducin (0.2 μM) was incubated with 2 mM o-PDM or p-PDM for 30 min, at room temperature, in 100 mM HEPES (pH 8.0). The function of the modified enzyme was assessed determining its light-dependent [^{3}H] GMP-PNP binding and GTPase activities. Both transducin functional units also were reacted with 2 mM o-PDM or p-PDM in a similar fashion, and at designated time intervals (0-60 min) two aliquots of the mixture were removed and terminated with 17 mM DTT. One aliquot was analyzed by SDS-polyacrylamide gel electrophoresis, and the second aliquot was assayed for [^{3}H] GMP-PNP binding after holoenzyme reconstitution with the complementary native functional unit.

Spontaneous Cross-Linking of T_α

T_α was dialyzed either against 25 mM sodium phosphate (pH 6.8) or 50 mM Tris-HCl (pH 8.0), containing 5 mM magnesium acetate. Parallel experiments were performed in the same buffers containing 5 mM β-mercaptoethanol. T_α was then incubated with $T_{\beta\gamma}$ to reform the holoenzyme, and the light-dependent GTP-hydrolytic activity of the reconstituted protein was determined in the absence or presence of 2 mM DTT. To isolate the peptides containing the residues involved in the spontaneous formation of disulfide linkages in the α-subunit, T_α (0.3 mg) was dialyzed at pH 6.8, modified with [^{3}H] IAA after treatment with DTT, digested with trypsin, and separated by HPLC. A similar procedure was carried out in the absence of DTT.

GTPase Assay

GTP hydrolysis assays were performed as described by Franke et al. (1992), using 0.2 μM of native or reconstituted transducin, 0.1 μM rhodopsin (as EDTA-washed ROS membranes), and 20 μM [^{32}P] GTP in Buffer I.

Electrophoresis on Polyacrylamide Gels with SDS

Electrophoresis on polyacrylamide slab gels (10%, 1.5 mm thick) was carried out in the presence of SDS according to Laemmli (1970).

Western Blot Analyses

Following SDS-polyacrylamide gel electrophoresis, the proteins were electrotransferred to nitrocellulose. The filters were processed as described by Towbin et al. (1979) using polyclonal antibodies directed against transducin, raised in mice (Bubis et al., 1993). These antibodies recognize T_α very specifically, and also cross-react with rhodopsin.

HPLC Separations

HPLC was carried out on either a Hewlett-Packard HP 1090 Liquid Chromatograph instrument or a Waters 625 LC System using a Merck LiChrospher 100 RP-8 (5 μm) column. The solvents employed were: 10 mM sodium phosphate (pH 6.8) in one vessel and acetonitrile in the second vessel. Further purification of peptides modified with [^{3}H] IAA, AMDA and VP, was achieved by rechromatographing the peptides using a Hibar LiChrosorb RP-18 (5 μm) column (0.40 x 25 cm) and a different buffer system: 0.1% trifluoroacetic acid in one vessel and 0.1% trifluoroacetic acid in acetonitrile in the second vessel. Absorbance was monitored specifically at 210 nm. For VP-modified peptides, we used a photodiode array detector (Waters, 990 Series) to monitor absorbance between 200 and 300 nm. For AMDA-modified peptides, fluorescence was monitored with a Bio-Rad model 1700 flowthrough Fluorimeter (excitation source: 350 nm; emission filter cut off: 440 nm). [^{3}H] IAA-modified peptides were identified by scintillation counting.

Sequencing

Gas-phase sequencing was carried out on an Applied Biosystems protein sequenator. Phenylthiohydantoin amino acids were identified by HPLC, as described by Matsudaira (1987).

RESULTS

Transducin Labeling with Sulfhydryl Group-Specific Reagents

The role of transducin sulfhydryl groups was examined by chemical modification with five reagents: AMDA, DM, VP, NTCBA, and IAA, which possess different sizes and ionic properties. AMDA, NTCBA and IAA are hydrophilic compounds, while DM and VP are hydrophobic. Furthermore, AMDA and DM are more bulky than VP, IAA and NTCBA. In the case of NTCBA, a very small group (-CN) will be incorporated onto the reactive cysteine(s) of the protein.

All these reagents inhibited rhodopsin-dependent guanine nucleotide binding activity of transducin. Figure 1 shows the kinetics of inactivation of transducin [^{3}H] GMP-PNP binding activity by modification with NTCBA. Transducin modification with NTCBA was carried out in the presence of different concentrations of β-mercaptoethanol (1.4, 2, or 3 mM) (Fig. 1). The reducing agent was shown to protect against the inactivation in a concentration dependent manner, demonstrating the specificity of the modification reaction. Similar results were obtained with the other four reagents (Data not shown). Figure 1 also illustrates that the solvent (5% DMF) did not have any effect on transducin guanine nucleotide binding. In the case of transducin modification with IAA, we used [^{3}H] IAA in the reaction mixture and were able to measure the stoichiometry of incorporation of the reagent to the protein, which was 1.3 mol of [^{3}H] IAA per mol of transducin (Data not shown).

Interaction between T(Cys-X) and Photoexcited Rhodopsin

Using the sedimentation assay described under Experimental Procedures, we were able to determine whether modified transducin was capable of interacting either with R* or GTP. Sedimentation experiments followed by SDS-polyacrylamide gel electrophoresis and Western blot analyses, showed that modification with AMDA or VP hindered the binding of transducin to R*. As illustrated in figure 2, VP-modified transducin lost its rhodopsin binding capacity bacause it was completely recovered in the supernatant after the first centrifugation (S-ISO), and the resuspended pellets (P_r-GTP and P_f) only showed rhodopsin. Similar results were observed with AMDA-modified transducin (Data not shown). On the other hand, the modification of transducin with NTCBA allowed its interaction with R*, as shown in figure 3 by SDS-polyacrylamide gel electrophoresis (Panel A) and Western blot (Panel B). NTCBA-modified transducin was not extracted with isotonic (S-ISO) or hypotonic washes (S-HYPO), and transducin was recovered in the resuspended pellet obtained after both of these washes

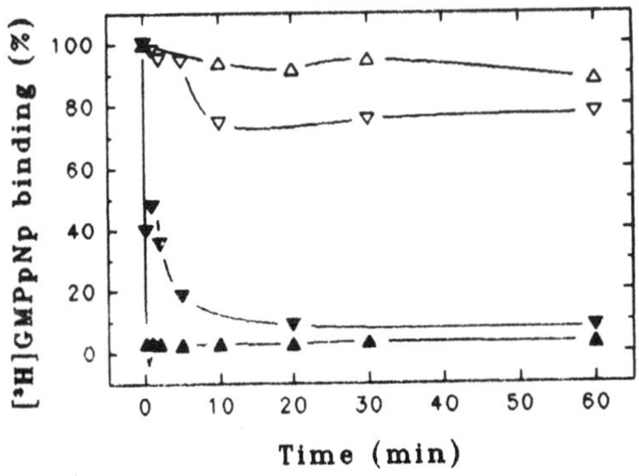

Figure 1. Inactivation of transducin GMP-PNP binding activity by modification with NTCBA. Transducin was incubated with 5% DMF (Δ) as a control, or with 5 mM NTCBA in the presence of either 1.4 (▲), 2 (▼), or 3 mM β-mercaptoethanol (▽). At the indicated time intervals, the reactions were terminated with 5 mM DTT and assayed for [^{3}H] GMP-PNP binding.

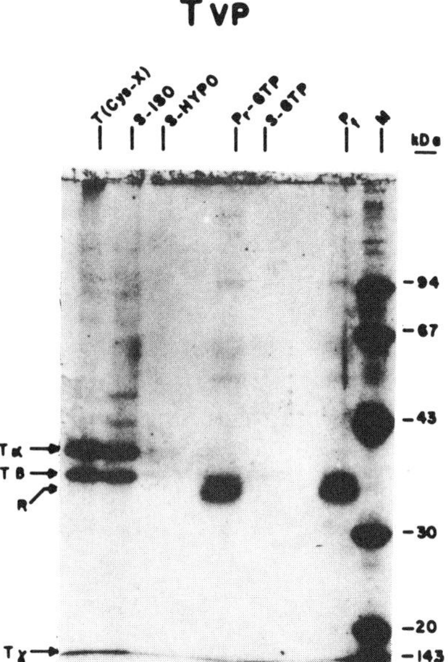

Figure 2. Interaction of photoexcited rhodopsin with VP-modified transducin. Transducin labeled with VP was incubated with illuminated washed ROS membranes. The supernatants and pellets produced by the sedimentation assay described under Experimental Procedures, were analysed by SDS-polyacrylamide gel electrophoresis. The abbreviations used in the figure are explained in the text. M = molecular weight markers.

(P_r-GTP). However, this interaction was maintained even in the presence of GTP; no transducin was observed in S-GTP, and all the transducin was recovered in the final pellet (P_f). DM- and IAA-modified transducin showed identical gel and Western blot patterns as NTCBA-labeled transducin (Data not shown).

The results observed for the samples of transducin modified with NTCBA, IAA, and DM, could also be explained by a non-specific precipitation of the protein due to denatura-

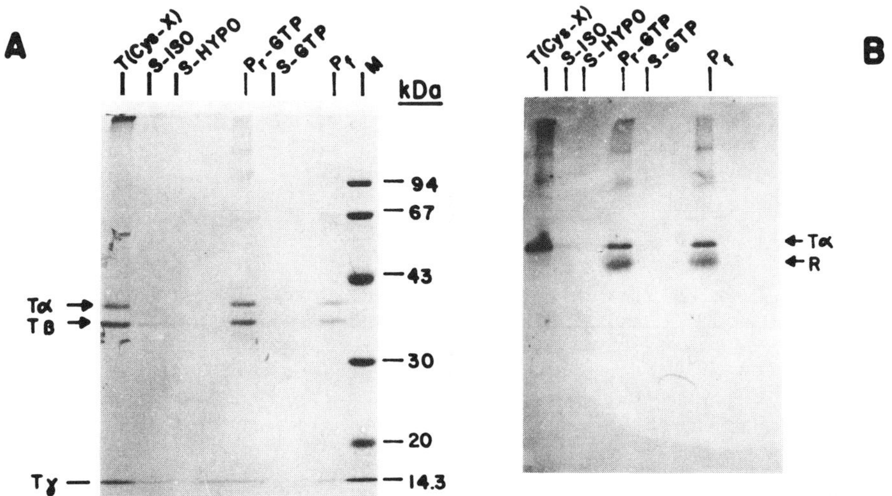

Figure 3. Interaction of photoexcited rhodopsin with NTCBA-modified transducin. Transducin labeled with NTCBA was incubated with illuminated washed ROS membranes. The supernatants and pellets produced using the sedimentation assay described under Experimental Procedures, were analysed by SDS-polyacrilamide gel electrophoresis (Panel A), and Western blot (Panel B). The abbreviations used in the figure are explained in the text. M = molecular weight markers.

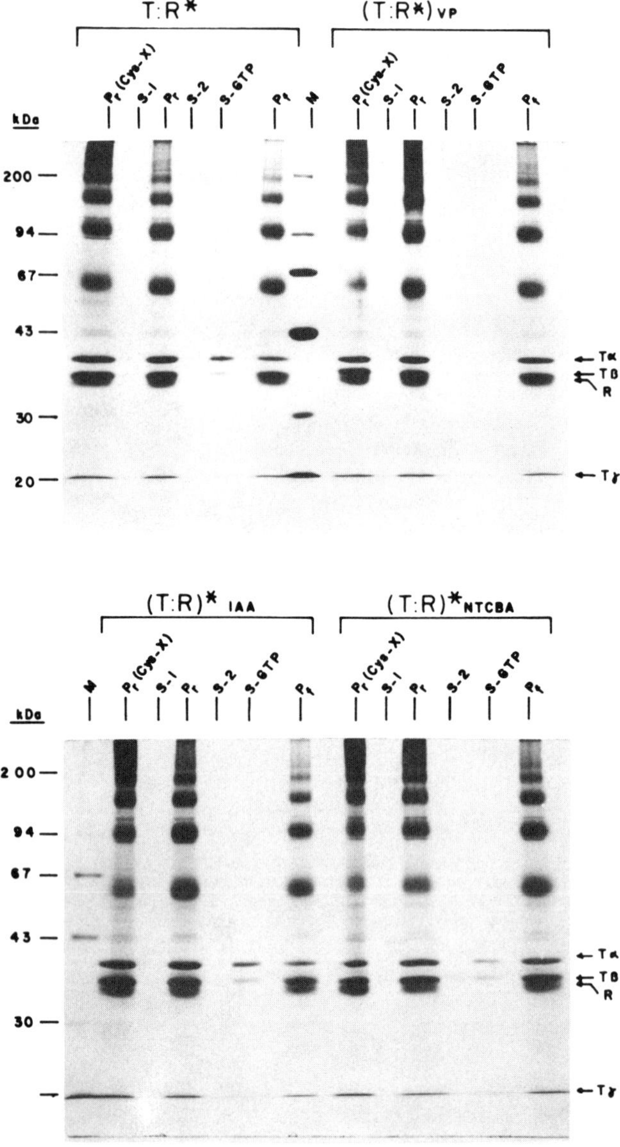

Figure 4. Transducin functionality in transducin-illuminated rhodopsin complexes treated with the sulfhydryl group-specific labels. T:R*, control; (T:R*)$_{VP}$, modification with VP; (T:R*)$_{IAA}$, modification with IAA; (T:R*)$_{NTCBA}$, modification with NTCBA. Transducin-photoexcited rhodopsin complexes were incubated with the different cysteine modification reagents. The supernatants and pellets produced by the sedimentation assay described under Experimental Procedures, were analysed by SDS-polyacrylamide gel electrophoresis. The abbreviations used in the figure are explained in the text. M = molecular weight markers.

tion. To examine this possibility, parallel experiments were performed with transducin incubated with NTCBA, IAA, or DM, but the sedimentation assays were carried out without washed ROS membranes. We observed that DM precipitated the protein (Data not shown). DM-modified transducin was observed in the sedimented pellets (P_r-GTP and P_f) even in the absence of rhodopsin. However, NTCBA- and IAA-modified transducin behaved as unmodified transducin in the absence of washed ROS membranes. These results proved that NTCBA- and IAA-modified transducin specifically interact with R^*.

Transducin Functionality in Transducin-Illuminated Rhodopsin Complexes Treated with the Sulfhydryl Group-Specific Labels

$T\text{-}R^*$ complexes were incubated with the sulfhydryl group-specific reagents, and the dissociation of these complexes in the presence of GTP was evaluated. As shown by SDS-polyacrylamide gel electrophoresis in figure 4, $T\text{-}R^*$ complexes modified with NTCBA and IAA, behaved as unmodified $T\text{-}R^*$, allowing the dissociation of transducin in the presence of GTP. Transducin was recovered in the supernatants after treatment with GTP (S-GTP). These results indicate that rhodopsin was able to protect against the inactivation previously observed for NTCBA- and IAA-modified transducin (Fig. 1 and 3). AMDA-treated $T\text{-}R^*$ showed identical results as IAA- and NTCBA-treated $T\text{-}R^*$ by SDS-polyacrylamide gel electrophoresis (Data not shown). On the other hand, VP-modified $T\text{-}R^*$ was not able to interact with GTP. Transducin was present in the sedimented pellets (P_r and P_f), for VP-treated $T\text{-}R^*$ (Fig. 4). As DM precipitated transducin, we decided to exclude DM from this set of experiments.

HPLC Separations of Tryptic Peptides from IAA-, AMDA-, and VP-Labeled Transducin and Identification of the Cysteine Residues Involved in the Modification

Transducin was modified with [^{3}H] IAA, dialyzed against 50 mM NH_4HCO_3 (pH 8.3), digested with trypsin, and the products separated by HPLC, as described under Experimental Procedures. The HPLC profile showed one major radioactive peak (Data not shown). This peak was rechromatographed using a different gradient system (Fig. 5) and a unique [^{3}H] containing peptide was observed. The purified radioactive tryptic peptide was subjected to gas-phase sequencing, and as seen in figure 5, it corresponded to the carboxy-terminal tryptic peptide of T_α (residues 342-350). The radioactivity was released at sequencing cycle 6 of the peptide. This result showed that Cys-347 of T_α is the residue derivatized by IAA in native transducin.

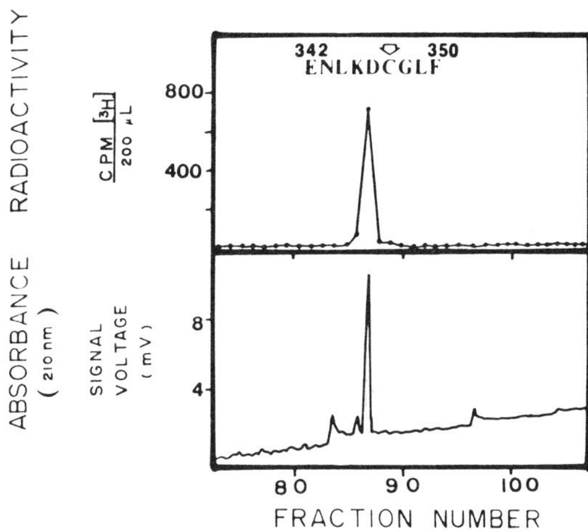

Figure 5. Rechromatogram of the major [^{3}H] IAA-labeled tryptic peptide of transducin. The carboxymethylated [^{3}H]-peptide was rechromatographed on a Hibar LiChrosorb RP-18 column. The buffers employed were: A) 0.1% trifluoroacetic acid and B) 0.1% trifluoroacetic in CH_3CN. The peptides were eluted isocratically in buffer A for 10 min, and then with a 120-min linear gradient from 0-40% B. Top Panel, radioactivity. Bottom Panel, absorbance at 210 nm. The purified peptide was subjected to gas-phase sequencing, and the sequence obtained is shown in the Top Panel.

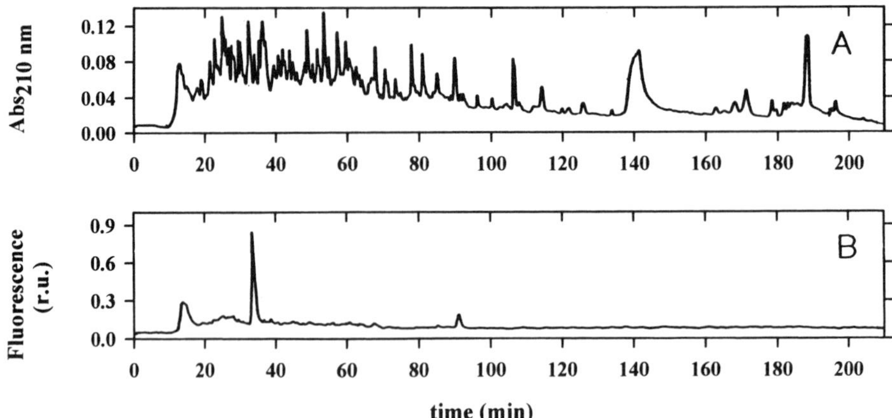

Figure 6. HPLC separation of trypsin digest of AMDA-modified transducin. Transducin was incubated with AMDA and digested with trypsin. The resulting peptides were separated by HPLC on a LiChrospher 100 RP-8 column. The buffers employed were: A) 10 mM sodium phosphate (pH 6.8) and B) CH_3CN. The peptides were eluted with a 180-min linear gradient from 0-40% B, and then with a 30-min linear gradient from 40-60% B. Panel A, absorbance at 210 nm. Panel B, fluorescence.

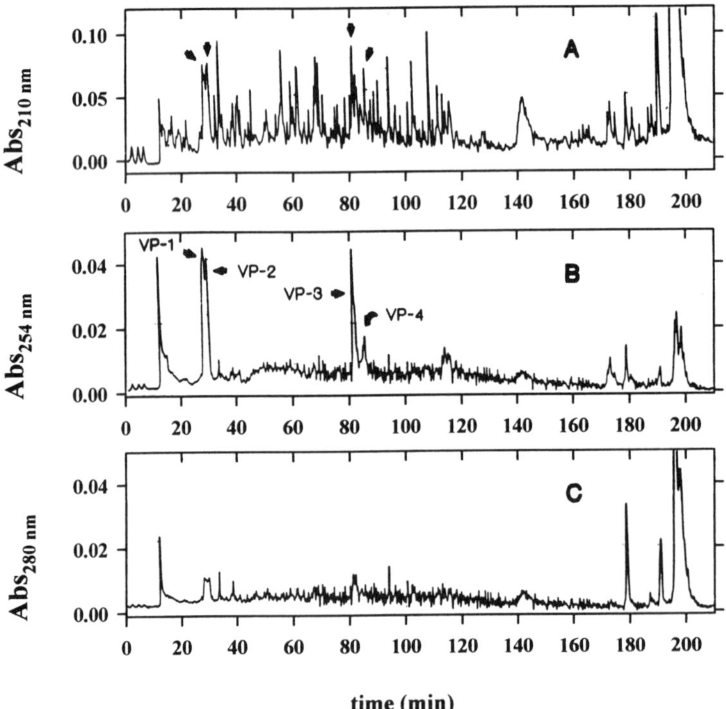

Figure 7. HPLC elution profile of tryptic peptides of VP-modified transducin. Transducin labeled with VP was digested with trypsin. The peptides were separated by HPLC as described in fig. 6, and their elution was monitored at 210 (Panel A), 254 (Panel B), and 280 nm (Panel C).

Transducin also was modified with AMDA and VP on a preparative scale, and digested as described for IAA-labeled transducin. Figure 6 illustrates the HPLC separation of trypsin digest of AMDA-modified transducin. The fluorescent profile obtained for the sample treated with AMDA was similar to the radioactive profile obtained for [^{3}H] IAA-modified transducin. A major fluorescent peak appeared in the AMDA-treated transducin (Fig. 6, Panel B). On the other hand, the HPLC separation of the resulting peptides from VP-modified transducin showed the existence of pyridylethyl cysteine in at least four different transducin tryptic peptides, as determined by their absorbance at λ: 254 nm (Fig. 7, Panel B). We also monitored the absorbance at λ: 280 nm, to discriminate peptides containing aromatic amino acid residues from the peptides containing the derivatized cysteines (Fig. 7, Panel C). These results showed clear differences in labeling of transducin sulfhydryl groups with AMDA and VP.

Derivatization of Transducin Carboxyl Groups

Chemical modification also was used to examine transducin functional acidic amino acids. We used two different carbodiimides, EDC, a water-soluble compound that was used to determine the role of solvent-accesible carboxyl groups in the protein, and DCCD, a non-polar carbodiimide that will partition into the hydrophobic environments of proteins. As seen in figure 8, EDC inhibited the [^{3}H] GTP binding activity of holotransducin and its isolated subunits. Transducin, T_α, and $T_{\beta\gamma}$ showed almost complete inhibition (more than 80%) of their light-dependent guanine nucleotide binding, in the presence of EDC. In contrast, the holoenzyme and the βγ-complex were only slightly affected by treatment with DCCD. We observed only 10% inhibition of their activity (less in the case of $T_{\beta\gamma}$), when incubated with DCCD (Fig. 8). DCCD-modified T_α showed 40% inactivation, more than

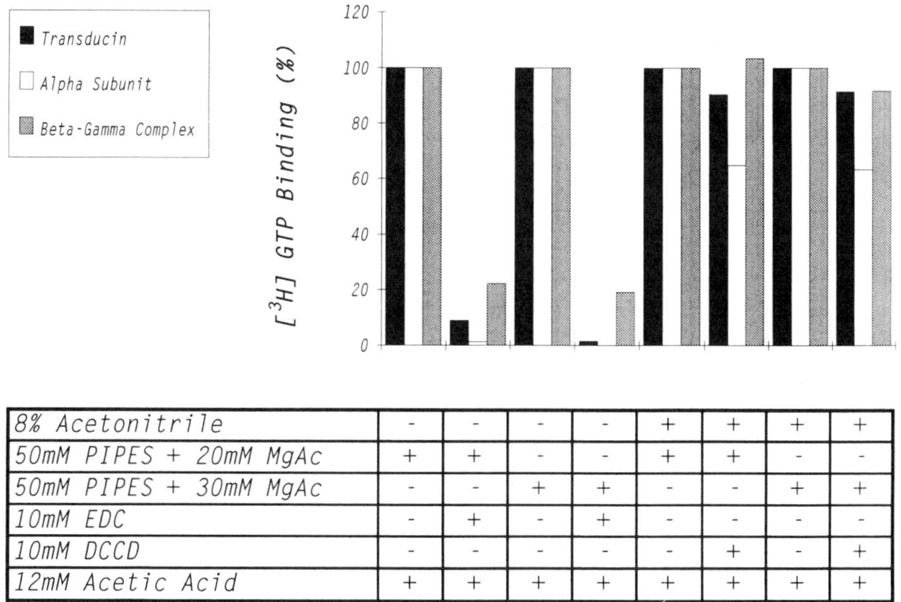

8% Acetonitrile	-	-	-	-	+	+	+	+
50mM PIPES + 20mM MgAc	+	+	-	-	+	+	-	-
50mM PIPES + 30mM MgAc	-	-	+	+	-	-	+	+
10mM EDC	-	+	-	+	-	-	-	-
10mM DCCD	-	-	-	-	-	+	-	+
12mM Acetic Acid	+	+	+	+	+	+	+	+

Figure 8. Effect of EDC and DCCD treatment of transducin, T_α, and $T_{\beta\gamma}$ on the [^{3}H] GTP binding activity of native and reconstituted holoenzyme. The bottom Panel shows the reagents and buffers used for each experiment. MgAc = magnesium acetate.

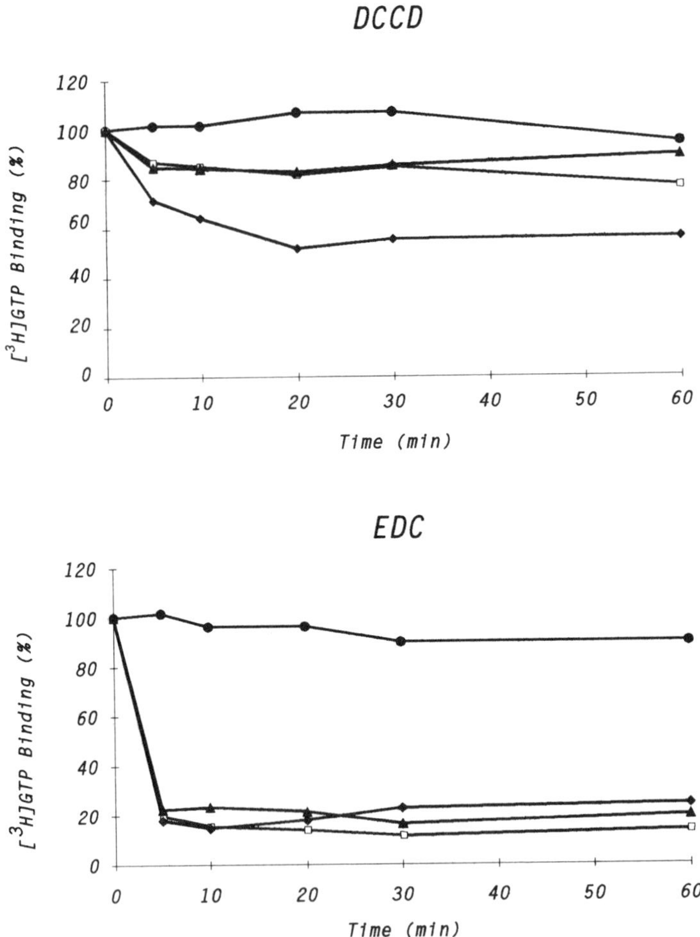

Figure 9. Time course of the modification of transducin, T_{α}, and $T_{\beta\gamma}$ with EDC and DCCD. Control experiments contained either 8% acetonitrile or 50 mM Pipes and 30 mM magnesium acetate, for DCCD and EDC labeling experiments, respectively. Control (●); Transducin (□); T_{α} (♦); $T_{\beta\gamma}$ (▲).

four times the inhibition observed for transducin and $T_{\beta\gamma}$, treated with DCCD. Similar results were obtained whether the experiments were performed in the presence of 20 or 30 mM magnesium acetate. The kinetics of the modification of transducin, T_{α}, and $T_{\beta\gamma}$ with EDC and DCCD, and the effect on the guanine nucleotide binding activity of the protein, are shown in figure 9.

Derivatization of Transducin Lysyl Residues

Transducin functional lysines were examined by chemical modification with five different reagents: PITC, OPA, AA, CA, and DnsCl. As illustrated in figure 10, with the exception of PITC, all lysine modification reagents produced the functional inactivation of transducin (more than 60% inhibition in GTP binding activity). Incubation of T_{α} or $T_{\beta\gamma}$ with PITC or OPA, also resulted in more than 60% inactivation of the reconstituted holoenzyme. On the other hand; AA, CA, and DnsCl caused inactivation of the reconstituted enzyme with

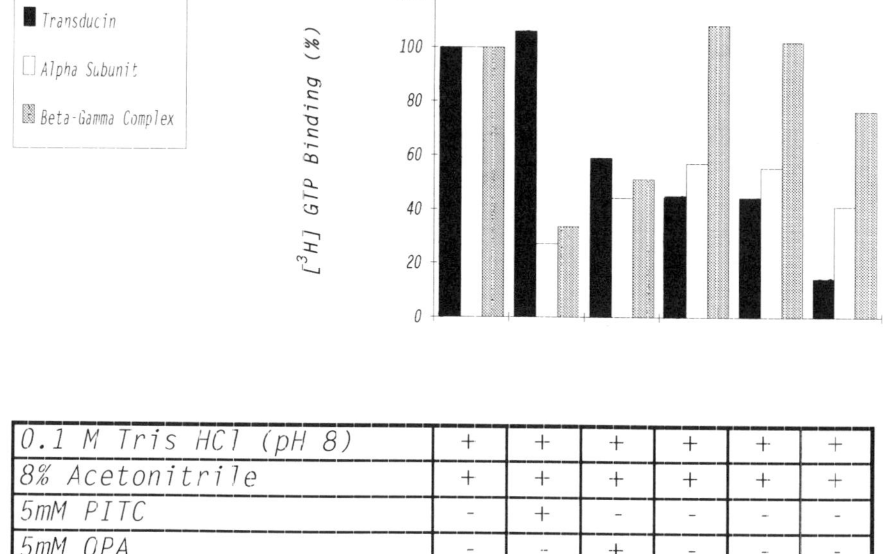

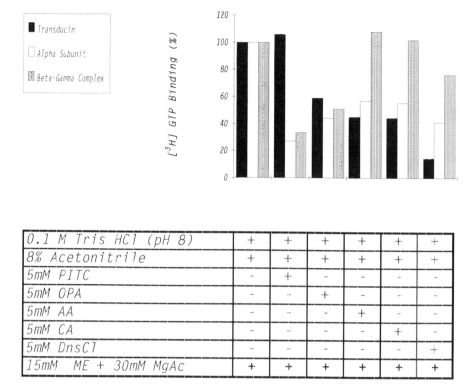

0.1 M Tris HCl (pH 8)	+	+	+	+	+	+
8% Acetonitrile	+	+	+	+	+	+
5mM PITC	-	+	-	-	-	-
5mM OPA	-	-	+	-	-	-
5mM AA	-	-	-	+	-	-
5mM CA	-	-	-	-	+	-
5mM DnsCl	-	-	-	-	-	+
15mM ME + 30mM MgAc	+	+	+	+	+	+

Figure 10. Effect on the [^{3}H] GTP binding activity of native and reconstituted holoenzyme after treatment of transducin, T_α, and $T_{\beta\gamma}$ with lysyl group-specific reagents. The bottom Panel shows the reagents and buffers used for each experiment. ME = β-mercaptoethanol. MgAc = magnesium acetate.

modified T_α, but not with modified $T_{\beta\gamma}$. Figure 11 shows the time course of inactivation of transducin by OPA (Panel A), AA (Panel B), CA (Panel C), and DnsCl (Panel D), with DnsCl displaying the fastest inhibition.

Cross-Linking of Transducin by Sulfhydryl Group-Specific Bifunctional Reagents

Two sulfhydryl group-specific bifunctional labels: o-PDM and p-PDM, were used as cross-linking agents for transducin and transducin subunits. As seen in figure 12 (Panel A), incubation of T_α with o-PDM or p-PDM resulted in the formation of high molecular weight oligomers, as well as bands that migrated with apparent molecular masses of 37 and 35 kDa. Incubation of $T_{\beta\gamma}$ with both reagents produced a new major species, 46 KDa, which resulted from the cross-linking between T_β and T_γ (Fig. 12, Panel B). Transducin modified with o-PDM or p-PDM showed a complete inactivation of its [^{3}H] GMP-PNP binding and GTPase activities (Data not shown). We carried out holoenzyme reconstitution experiments combining modified with native transducin functional units, to discriminate which unit was responsible for the observed inhibition in transducin function. The combination of intact T_α and o-PDM-treated $T_{\beta\gamma}$ reconstituted native transducin GMP-PNP binding activity, indicating that the formation of the 46 KDa species did not affect the function of the reconstituted protein. However, o-PDM-modified T_α incubated with intact $T_{\beta\gamma}$, exhibited

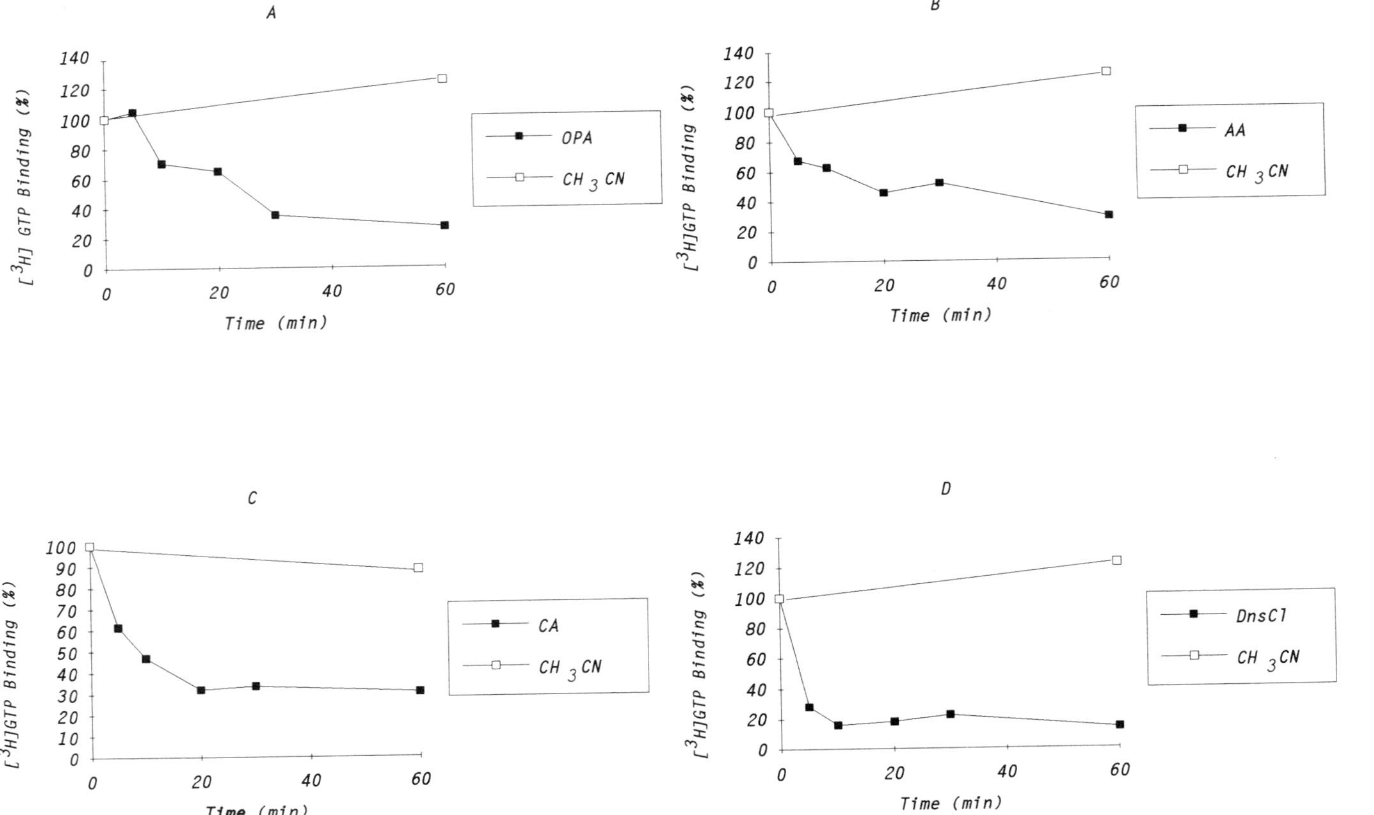

Figure 11. Kinetics of inactivation of transducin by lysine modification reagents. Panel A, modification with OPA; Panel B, modification with AA; Panel C, modification with CA; Panel D, modification with DnsCl. Control experiments contained 8% acetonitrile.

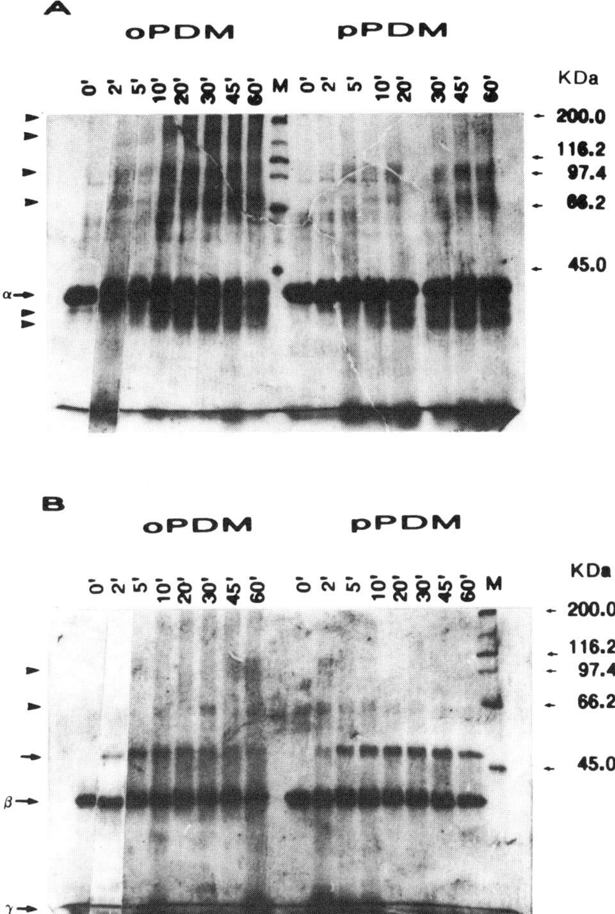

Figure 12. Time course of the cross-linking of T_α and $T_{\beta\gamma}$ with o-PDM and p-PDM. T_α and $T_{\beta\gamma}$ were incubated with 2 mM o-PDM and p-PDM. At the indicated time intervals, an aliquot of the reaction mixture was terminated with 17 mM DTT and analyzed by SDS-polyacrylamide gel electrophoresis. Panel A, T_α; Panel B, $T_{\beta\gamma}$. M = molecular weight markers.

more than 90% inhibition in transducin native enzymatic function (Data not shown). These results implied that the formation of intra and/or intermolecular cross-links in T_α were responsible for the observed inactivation. Similar results were obtained when the individual functional units of transducin were treated with p-PDM (Data not shown).

Spontaneous Formation of Disulfide Bonds in T_α

Spontaneous oxidation of sulfhydryl groups, and formation of disulfide bonds in T_α, were examined at different pH values and under non-reducing conditions. Dialysis of T_α at pH 8.0, followed by reconstitution with native $T_{\beta\gamma}$, produced the total inactivation of its GTPase activity (Fig. 13). However, native GTP-hydrolytic activity was obtained after incubation of the reconstituted holoenzyme with 2 mM DTT (Fig. 13). Dialysis of T_α at pH 6.8, followed by reconstitution with native $T_{\beta\gamma}$, produced only a reduction in GTPase activity of about 30-40% (Fig 14), and as before, the addition of 2 mM DTT restored native transducin GTPase functionality. As seen in figure 14, T_α dialyzed in the presence of 5 mM β-mercaptoethanol and reconstituted with intact βγ-complex, exhibited native GTP-hydrolytic capacity, in the presence or absence of 2 mM DTT.

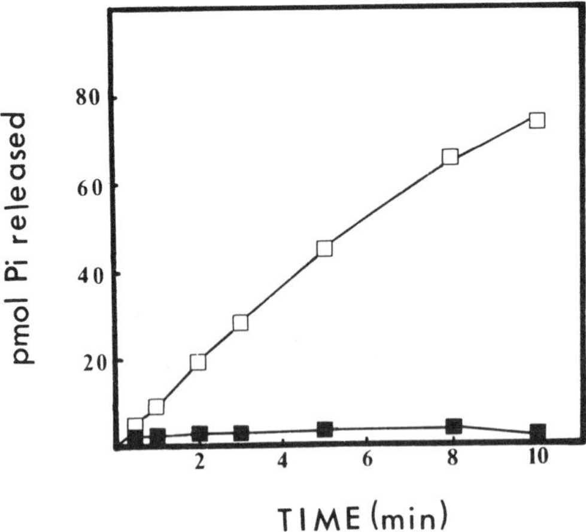

Figure 13. GTPase activity of T_α dialyzed at pH 8.0 and reconstituted with $T_{\beta\gamma}$. Shown is the light-dependent GTP-hydrolytic activity of the reconstituted enzyme in the absence (■) or presence (□) of 2 mM DTT.

In order to localize the cysteines involved in the spontaneous formation of disulfide linkages, T_α was dialyzed at pH 6.8, and two aliquots containing 0.3 mg of protein were taken. DTT (2 mM) was added to one of the samples. Then, both samples were incubated with [^{3}H] IAA, digested with trypsin, and the products separated by HPLC. Figure 15 shows the resulting absorbance and radioactivity HPLC profiles. One peptide was alkylated by IAA in both cases (peptide II, Panel A and C). However, the sample containing DTT showed three new radioactively labeled peptides (peptides I, III, and IV). Gas-phase sequencing of peptides IIa and IIc (Fig. 16) identified Cys-135 of T_α, as the amino acid residue carboxymethylated by IAA in the presence or absence of DTT. Sequencing of peptides I and III showed that both contained Cys-347 of T_α, being peptide III an incomplete proteolytic

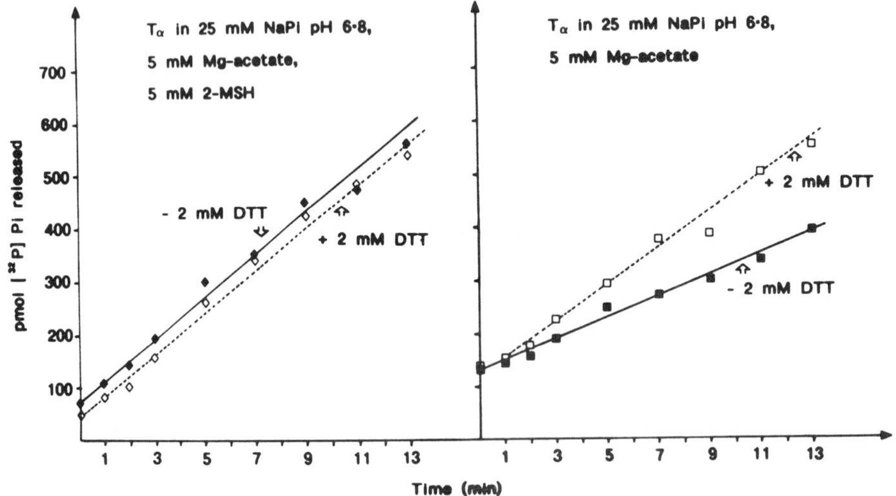

Figure 14. GTPase activity of T_α dialyzed at pH 6.8 and reconstituted with $T_{\beta\gamma}$. T_α was dialyzed in 25 mM sodium phosphate (NaPi), 5 mM magnesium acetate (Mg-acetate), pH 6.8 (Right Panel), or in the same buffer containing 2 mM β-mercaptoethanol (2-MSH) (Left Panel). Shown is the light-dependent GTP-hydrolytic activity measured in the absence (■, ♦) or presence (□, ◊) of 2 mM DTT, after reconstitution with the βγ-complex.

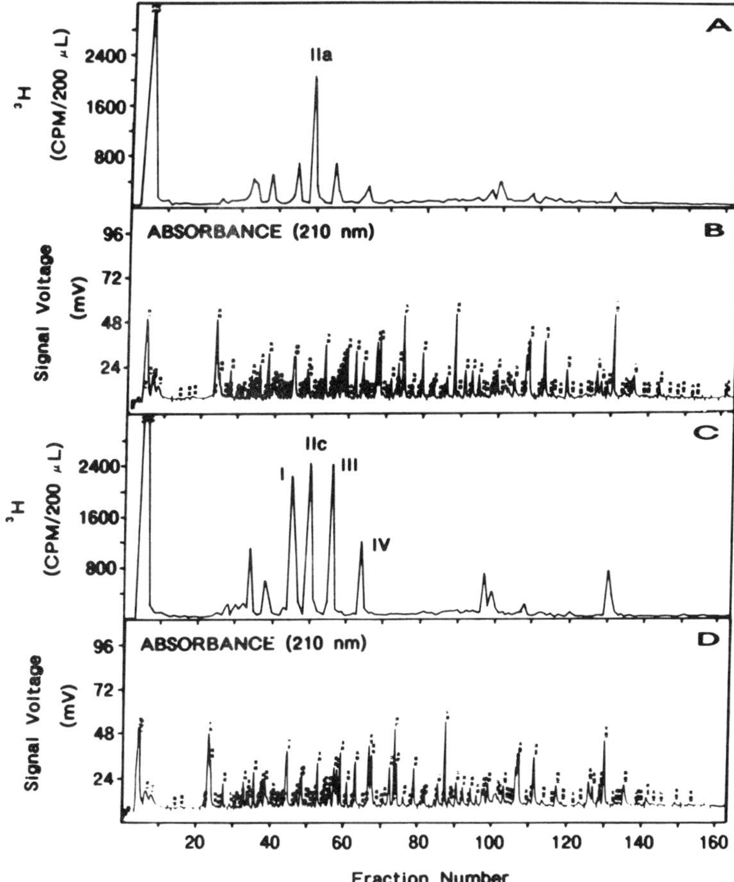

Figure 15. HPLC separation of tryptic peptides from T_α dialyzed at pH 6.8, and modified with [^{3}H] IAA. The peptides were separated by HPLC as described in Fig. 6. Panel A and B, radioactivity and absorbance at 210 nm, respectively, of T_α dialyzed in the absence of DTT. Panel C and D, radioactivity and absorbance at 210 nm, respectively, of T_α dialyzed in the presence of DTT. The major radioactive tryptic peptides were designated with roman numerals.

product of peptide I. These results identified Cys-347 of T_α as one of the cysteines involved in the spontaneous formation of disulfide cross-links in the α-subunit of transducin. Sequence analysis of peptide IV identified Cys-321 of T_α as the carboxymethylated residue. This cysteine also may be participating in T_α disulfide formation, however the amount of [^{3}H] IAA incorporated onto peptide IV is lower than the level of alkylation observed for peptides I and III (Fig.15). As seen in figure 15, there are other minor [^{3}H]-containing tryptic peptides in T_α dialyzed in the absence of DTT.

DISCUSSION

Five different sulfhydryl group-specific labels were used for the chemical modification of transducin, and in all cases the protein was completely inactivated. The kinetics of

PEPTIDE	SEQUENCE	CYSTEINE
I	346 350 D-C*-G-L-F	Cys 347
IIa	129 137 D-S-G-I-Q-A-C*-F-D	Cys 135
IIc	129 137 D-S-G-I-Q-A-C*-F-D	Cys 135
III	342 350 E-N-L-K-D-C*-G-L-F	Cys 347
IV	314 322 E-I-Y-S-H-M-T-C*-A	Cys 321

Figure 16. Sequencing of carboxymethylated tryptic peptides from T_α dialyzed at pH 6.8, in the presence or absence of DTT.

modification of transducin cysteines with AMDA, DM, VP, NTCBA and IAA showed that the loss in GMP-PNP binding activity occurred inmediately after the reaction.

With the exception of DM which caused the precipitation of transducin, the other sulfhydryl group-specific reagents were used to examine whether R^* was capable to interact in a productive manner with labeled transducin. The modification of transducin with AMDA or VP hindered the binding of transducin to R^*. The inactivation in guanine nucleotide binding activity observed for AMDA- and VP-transducin must be attributed to this blockade. In contrast, NTCBA- and IAA-modified transducin behaved as untreated transducin, and were capable of interacting with illuminated rhodopsin. These results can be explained by the small size of the cysteine alkylating groups, -CN and carboxymethyl groups, from NTCBA and IAA, respectively. However, GTP was not able to dissociate the complex between NTCBA- or IAA-modified transducin and R^*. Probably, transducin GDP/GTP exchange reaction, induced by illuminated rhodopsin, was sterically prevented as a result of transducin modification with either NTCBA or IAA. Then, with NTCBA- and IAA-labeled transducin, we were able to stabilize and freeze a transducin-rhodopsin complex intermediate, which could not dissociate even after incubation with GTP.

AMDA, NTCBA and IAA are similar in nature, the three compounds are charged and polar at our working pH. Due to their hydrophilic characteristics, these reagents are capable of modifying cysteine(s) localized on the external surface of the protein. In particular, sulfhydryl groups located in close proximity to the receptor binding site of transducin constitute excellent targets for these compounds. The three reagents also may modify the same amino acid residue(s) in transducin. Nevertheless, a different behaviour was observed between AMDA- and NTCBA- or IAA-modified transducin in the sedimentation assays with R^*. As explained above, the cyanide and carboxymethyl groups derived from NTCBA and IAA, will allow transducin interaction with R^*, due to their small volume. However, the labeling group incorporated onto transducin after modification with AMDA is very bulky and will produce steric hindrance preventing rhodopsin binding.

Chemical modifications performed on T-R^* complexes were used to explore the ability of rhodopsin to protect against the inactivation observed in transducin function.

Rhodopsin protected when AMDA, NTCBA and IAA were used as labeling reagents, but not when VP was employed. The more hydrophilic compounds, AMDA, NTCBA and IAA did not affect transducin function when the T-R* complex was formed previous to the modification reactions. However, the non-polar reagent VP, inactivated guanine nucleotide binding by transducin even in the presence of photoexcited rhodopsin. These results will favor the hypothesis that AMDA, NTCBA and IAA are modifying the same cysteine(s) in transducin.

Transducin carboxymethylated with [^{3}H] IAA under native conditions showed an incorporation of approximately one mole of the compound per mole of protein. HPLC separation of [^{3}H] IAA-labeled tryptic peptides of transducin also revealed that a unique peptide was alkylated. The labeled amino acid residue corresponded to Cys-347 of T_α. Then, the modification of this cysteine must be responsible for the complete inhibition observed in transducin function after IAA alkylation. HPLC separation of AMDA-labeled tryptic peptides of transducin also showed a major modified peptide, as with IAA. Although the cysteines modified by AMDA and NTCBA were not mapped in this study, Cys-347 of T_α represents a good candidate, due to the similar polar properties shared by AMDA, NTCBA, and IAA. In contrast, several VP-labeled tryptic peptides (at least four) were identified in the HPLC profile of VP-modified transducin proteolized with trypsin. Even if Cys-347 of T_α is one of the residues modified by VP, it is clear that this compound is labeling more than one sulfhydryl group in transducin. As discussed above, R* did not protect against transducin inactivation by VP. Then, some of the cysteines modified with the pyridylethyl group in transducin may be located in regions different than the receptor binding site; for example, in the protein guanine nucleotide binding pocket. VP also may be labeling cysteine(s) located in regions that sterically will hinder the GDP/GTP exchange induced by rhodopsin, preventing either the exit of GDP or the entrance of GTP. These results proved that transducin sulfhydryl groups are differentially labeled depending on the hydrophilicity or hydrophobicity of the reagent.

Ho & Fung (1984) examined the role of the sulfhydryl groups of transducin by 5, 5′-dithiobis-(2-nitrobenzoic acid) (DTNB) titration and N-ethylmaleimide (NEM) modification. They, as well as Reichert & Hofmann (1984) showed that derivatization of a reactive sulfhydryl by NEM or DTNB inhibited rhodopsin-catalised GDP/GTP exchange. In addition, Hofmann & Reichert (1985) showed that replacement of the thionitrobenzoate derivative, with the less bulky cyanide, reversed the hindrance of transducin binding to photoexcited rhodopsin, similar to our observations when NTCBA was used as the alkylating reagent. Neither the studies of Ho & Fung (1984) nor those of Reichert & Hofmann (1984) and Hofmann & Reichert (1985) mapped the reactive cysteine derivatized by NEM or DTNB in the primary sequence of transducin.

Dhanasekaran et al. (1988) used [^{125}I] N-(3-iodo-4-azidophenyl propionamido)-S-(2-thiopyridyl) cysteine ([^{125}I] ACTP) to derivatize reduced sulfhydryls of transducin, and showed the incorporation of 1-1.3 mol of the compound into T_α. They found that both Cys-347 and Cys-210 in T_α were derivatized by [^{125}I] ACTP in a ratio of approximately 70 and 30%, respectively. The modification of these two reactive cysteines inhibited rhodopsin-catalised activation of transducin. Van Dop et al. (1984) showed that the ADP-ribosylation of transducin by pertussis toxin blocked the light-stimulated hydrolysis of GTP. Subsequently, West et al. (1985) identified Cys-347 as the site of pertussis toxin-catalysed ADP-ribosylation in transducin. These studies strengthen our results that identify Cys-347 of T_α as one of the functionally important residues in the protein.

Cys-347 is the fourth residue upstream from the carboxyl terminus of T_α (Phe 350). Various reports (Conklin et al., 1993; Weingarten et al., 1990) have shown that the carboxyl terminal region of several α subunits are particularly important for interaction with the receptor. Hamm et al. (1988) showed that a synthetic peptide corresponding to two regions near the carboxyl terminus of T_α, Glu^{311}-Val^{328} and Ile^{340}-Phe^{350} (which contained Cys-347), competed with transducin for interaction with rhodopsin. Furthermore the 11-amino acid peptide from the COOH terminus of T_α (Ile^{340}-Phe^{350}) mimics transducin effects in stabilizing the rhodopsin active form, metarhodopsin II. Dratz et al. (1993) studied the structure of the interface between excited and unexcited rhodopsin and the carboxyl terminus peptide of T_α (Ile^{340}-Phe^{350}), using NMR. They observed conformational differences between the two bound forms and suggested a mechanism for activation of G proteins by agonist-stimulated receptors. Among the changes, the Cys side chain of residue 347 appears to pivot from pointing outside the dark rhodopsin-bound peptide to being tucked inwards in the metarhodopsin II bound structure (Dratz, et al., 1993). Then, the insertion of alkylating groups, or an ADP-ribose group by pertussis toxin, at Cys-347 would be predicted to change the conformation of the COOH terminus of T_α, and may affect transducin-rhodopsin interactions.

Chemical modification also was used to examine transducin functional acidic amino acids. EDC, hydrophilic in nature, completely abolished the light-dependent [^{3}H] GTP binding activity of transducin. When, T_α or $T_{\beta\gamma}$ were individually treated with EDC, and then incubated with the native complementary unit, the GTP binding activity of the reconstituted holoenzyme also was inhibited. Binding to rhodopsin probably is hindered in EDC-modified native and reconstituted transducin, as EDC will target accesible carboxyl groups located on the surface of the protein. EDC-labeling of Asp(s) and/or Glu(s) located on the individual transducin units also may be preventing the reconstitution of the holoenzyme. In contrast, transducin and the $\beta\gamma$-complex were only slightly affected by treatment with the hydrophobic carbodiimide, DCCD. DCCD-modified T_α showed 40% inactivation, more than quadruple the amount of inhibition observed for transducin and $T_{\beta\gamma}$, treated both with DCCD. We believe that the Asp and/or Glu residues located in or near the metal and nucleotide interaction sites in transducin, are the best targets for DCCD-labeling. Guanine nucleotide binding sites contain a hydrophobic pocket that constitutes the primary recognition site for the guanine ring. Since Mg^{+2}-GDP is strongly bound to purified transducin, the metal-nucleotide complex may protect against DCCD inactivation. On the other hand, T_α is purified free of nucleotide from the blue agarose column (Shichi et al., 1984). In this case, the acidic residues involved in the GTP binding pocket will be more susceptible to DCCD-labeling. However, T_α was always stored and maintained in buffers containing Mg^{+2}, and the metal may be protecting some against the modification by DCCD. Furthermore, the labeling of T_α with DCCD also may be affecting its interaction with the $\beta\gamma$-complex, and viceversa, hindering the formation of the native holoenzyme.

There are several possible reaction pathways for the interaction of carbodiimides with carboxyl groups on proteins. The reaction of a carbodiimide with an acidic residue may produce a stable N-acylurea adduct, after the initial formation of an O-acylurea intermediate. Alternatively, the O-acylurea that is formed may interact with a nucleophile, for example water. If, however, the nucleophile is a nearby amino group of an amino acid side chain, an inter or intramolecular "zero-length" cross-link may be formed (Toner-Webb & Taylor, 1987). As described above, the COOH terminus peptide of T_α have been shown to be important for interaction with the receptor. The NMR structure of rhodopsin and metharhodopsin bound to the 11-amino acid peptide from the C-terminus of T_α (Ile^{340}-Phe^{350}) showed that residue 345, which is a lysine, is capable of forming different salt bridges with carboxyl groups from the same peptide, depending on the state of the receptor (Dratz et al., 1993). Lys-345 which is close to Glu-342 in the dark rhodopsin-bound peptide, appears to be closer to the free carboxyl at the C-terminus in the activated receptor-bound peptide. The hydrophylic carbodiimide EDC may be cross-linking Glu-342 with Lys 345 in T_α. The formation of this covalent cross-link will impede the breakage and replacement of salt bridges involving Lys 345. EDC also may be modifying either Glu-342 or the C-terminal carboxyl group, hindering the light-induced transducin conformational change mediated by interaction with activated rhodopsin. The crystal structures of a 325-amino acid fragment of T_α bound either to GDP (Lambright et al., 1994) or to $GTP_\gamma S$ (Noel et al., 1993) have been solved. A view of both three-dimensional structures of T_α showed that several acidic residues were involved directly either in the coordination of Mg^{+2}, in guanine nucleotide binding, or in the mechanism for GTP hydrolysis. These amino acids were: Glu-39, Asp-146, Asp-196, Glu-203, and Asp-268. The labeling by DCCD of any of these target residues will result in the inactivation observed in DCCD-modified T_α.

Five different modification reagents were used to evaluate the existence of functional lysines in transducin. One of the most interesting compounds was PITC, which did not affect the guanine nucleotide binding properties of the modified holoenzyme. Hingorani & Ho (1987) studied the effect of fluorescein 5′-isothiocyanate labeling on transducin function, and they found no effect on the transducin-rhodopsin interaction or on the binding of GMP-PNP in the presence of R^*, similar to our observations with PITC. However, When T_α or $T_{\beta\gamma}$ was incubated with PITC, we observed more than 60% inhibition in the function of the reconstituted protein. These results suggested that PITC was modifying Lys residues located in or near the region of intersubunit contact, hindering the interactions among transducin subunits and impeding the formation of the holoenzyme. The other Lys modification compounds produced inactivation of the native modified transducin. Incubation of T_α or $T_{\beta\gamma}$ with OPA, also resulted in more than 60% inactivation in the function of the reconstituted holoenzyme. On the other hand; AA, CA, and DnsCl caused inactivation of the reconstituted enzyme with modified T_α, but not with modified $T_{\beta\gamma}$. For OPA, AA, CA, and DnsCl, we were not able to discriminate the functionality(ies) affected in transducin: the recognition site for photoexcited rhodopsin, the guanine nucleotide binding pocket, or the site of interaction between the α-subunit and the βγ-heterodimer. Furthermore, these labels also may be modifying Lys located in regions that sterically hinder either the exit or the entrance of guanine nucleotides, producing an indirect inhibitory effect.

Lys-345 of T_α is an excellent candidate for modification by the amino group-specific reagents used in this work. As described above, Lys-345 is located in the C-terminal region of T_α, and the interaction of transducin with photoexcited rhodopsin mediates conformational changes in the position of this residue. This is also true for Lys 341 which is located in the surroundings of Lys 345. In particular, OPA also may cross-link suitably placed ε-amino groups from lysine side chains, with sulfhydryls from cysteine residues. Lys 345 and Cys 347 are neighbouring functional amino acids that may be cross-linked in OPA-modified transducin and/or T_α. The three-dimensional view of T_α guanine nucleotide binding pocket, also showed the direct involvement of lysyl residues in the binding site. The key lysines were: Lys-42, and Lys 266. Modification of either of these lysine residues in transducin and/or T_α will produce the functional inactivation of the protein.

Although the crystal structures of T_α bound to $GTP_\gamma S$, and bound to GDP are virtually identical for 86% of the positions examined in both (Lambright et al., 1994), there are changes in a small surface area located on one face of T_α. The structural differences are localized to three adjacent regions referred as switch I (Ser 173-Thr 183), switch II (Phe 195-Thr 215), and switch III (Asp 227-Arg 238). These three regions contain residues of the type we characterized in this manuscript: Lys-176, and Glu-182 in switch I; Asp-196, Glu-203, Lys-205, Lys-206, Cys-210, and Glu-212 in switch II; Asp-227, Glu-232, Asp-233, Asp-234, and Glu-235 in switch III. The chemical modification of any of these amino acids probably will hinder the conformational change induced in T_α by nucleotide exchange, and will cause the inactivation of transducin.

Cross-linking studies with o-PDM and p-PDM demonstrated the formation of intra and intermolecular species in T_α, which were responsible for the inactivation observed in the function of the reconstituted holotransducin. Formation of a 46 KDa cross-linking species between T_β and T_γ, when the $\beta\gamma$-heterodimer was incubated with o-PDM or p-PDM, did not affect the function of the reassociated enzyme. Hingorani et al. (1988) reported the formation of similar cross-linking products in transducin, when p-PDM was used. In their case, the cross-linked products were identified by Western immunoblotting using antisera against purified subunits of transducin (T_α and $T_{\beta\gamma}$). However, the studies by Hingorani et al. (1988) failed to measure the effect of p-PDM chemical cross-linking on transducin function. Bubis & Khorana (1990) have shown that cupric phenanthroline catalises a single interchain disulfide bond formation between the β- and γ-subunits of transducin. The same disulfide bond was formed when holotransducin or the complex of βγ-subunits were treated with the reagent. The residues participating in the disulfide bond were identified as Cys-25 in T_β and Cys-36 and/or Cys-37 in T_γ. Cupric phenanthroline induced the formation of a new species with an apparent molecular mass of 43 KDa (Bubis and Khorana, 1990), very similar to the 46 Kda cross-link observed in o-PDM- and p-PDM-treated $T_{\beta\gamma}$. These results suggest that the same cysteine residues of transducin β- and γ-subunits may be involved in o-PDM- and p-PDM-cross-linked $T_{\beta\gamma}$.

We also studied the spontaneous formation of disulfide bonds in T_α. Sequence analysis of the radioactive peptides obtained by tryptic digestion of [^{3}H] IAA-modified T_α, identified Cys-347 as one of the cysteines involved in the cross-links. The formation of disulfide linkages in transducin α-subunit inhibited the light-dependent GTPase activity of the reconstituted holoenzyme. Finding that Cys-347 of T_α participated in the formation of these disulfide bonds, demonstrated again the important role of this residue in the function of transducin. Wessling-Resnick & Johnson (1989) also reported the formation of intermolecular disulfide linkages between the α-subunits of transducin molecules when the purified protein was placed in a non-reducing buffer system. The specific oligomeric association of α-subunits provides a physical basis for the cooperative activation kinetics described for the rhodopsin transduction system.

The work reported here identifies different functionally important cysteines, lysines, and acidic residues in transducin that are located either in the domains of intersubunit contact, in the proximity of the interaction site with rhodopsin, or near the guanine nucleotide binding pocket.

ACKNOWLEDGEMENTS

This work was supported in part by CONICIT (Consejo Venezolano de Investigaciones Científicas y Tecnológicas), under grant # S1-2171. We also would like to thank Dr. Rafael Rangel-Aldao from Empresas Polar and his laboratory (Laboratorio de Biotecnología, Universidad Simón Bolívar) for supplying equipment and reagents throughout the course of this work.

REFERENCES

Arshavsky, V. Y., and Bownds, M. D., 1992, Regulation of deactivation of photoreceptor G protein by its target enzyme and by cGMP, *Nature* 357: 416-417.

Baehr, W., Devlin, M. J., and Applebury, M. L., 1979, Isolation and characterization of cGMP phosphodiesterase from bovine rod outer segments, *J. Biol. Chem.* 254: 11669-11677.

Baehr, W., Morita, E. A., Swanson, R. J., and Applebury, M. L., 1982, Characterization of bovine rod outer segment G-protein, *J. Biol. Chem.* 257: 6452-6460.

Berstein, G., Blank, J. L., Jhon, D.-Y., Exton, J. H., Rhee, S. G., and Ross, E. M., 1992, Phospholipase C-β1 is a GTPase-activating protein for $G_{q/11}$, its physiologic regulator, *Cell* 70: 411-418.

Birnbaumer, L., 1992, Receptor-to-effector signaling through G proteins: Roles for $\beta\gamma$ dimers as well as α subunits, *Cell* 71: 1069-1072.

Bubis, J., and Khorana, H. G., 1990, Sites of interaction in the complex between β- and γ-subunits of transducin, *J. Biol. Chem.* 265: 12995-12999.

Bubis, J., Millán, E. J., and Martínez, R., 1993, Identification of guanine nucleotide binding proteins from *Trypanosoma cruzi*, *Biol. Res.* 26: 177-188.

Bubis, J., and Taylor, S. S., 1985, Covalent modification of both cAMP binding sites in cAMP-dependent protein kinase I by 8-azidoadenosine 3′,5′-monophosphate, *Biochemistry* 24: 2163-2170.

Bubis, J., and Taylor, S. S., 1987, Correlation of photolabeling with occupancy of cAMP binding sites in the regulatory subunit of cAMP-dependent protein kinase I, *Biochemistry* 26: 3478-3486.

Conklin, B. R., and Bourne, H. R., 1993, Structural elements of G_α subunits that interact with $G_{\beta\gamma}$, receptors, and effectors, *Cell* 73: 631-641.

Conklin, B. R., Farfel, Z., Lustig, K. D., Julius, D., and Bourne, H. R., 1993, Substitution of three amino acids switches receptor specificity of $G_q\alpha$ to that of $G_i\alpha$, *Nature* 363: 274-276.

Dhanasekaran, N., Wessling-Resnick, M., Kelleher, D. J., Johnson, G. L., and Ruoho, A. E., 1988, Mapping of the carboxyl terminus within the tertiary structure of transducin's α subunit using the heterobifunctional cross-linking reagent, ^{125}I-N-(3-iodo-4-azidophenylpropionamido)-S-(2-thiopyridyl) cysteine, *J. Biol. Chem.* 263: 17942-17950.

Dohlman, H. G., Thorner, J., Caron, M. C., and Lefkowitz, R. L., 1991, Model systems for the study of seven-transmembrane-segment receptors, *Annu. Rev. Biochem.* 60: 653-688.

Dratz, E. A., Furstenau, J. E., Lambert, C. G., Thireault, D. L., Rarick, H., Schepers, T., Pakhlevaniants, S., and Hamm, H. E., 1993, NMR structure of a receptor-bound G-protein peptide, *Nature* 363: 276-281.

Franke, R. R., Sakmar, T. P., Graham, R. M., and Khorana, H. G., 1992, Structure and Function in Rhodopsin. Studies of the interaction between the rhodopsin cytoplasmic domain and transducin, *J. Biol. Chem.* 267: 14767-14774.

Fung, B. K.-K., 1983, Characterization of transducin from bovine retinal rod outer segments. I. Separation and reconstitution of the subunits, *J. Biol. Chem.* 258: 10495-10502.

Hamm, H. E., Deretic, D., Arendt, A., Hargrave, P. A., Koenig, B., and Hofmann, K. P., 1988, Site of G protein binding to rhodopsin mapped with synthetic peptides from the α subunit, *Science* 241: 832-835.

Hargrave, P. A., and McDowell, J. H., 1992, Rhodopsin and phototransduction: A model system for G protein-linked receptors, *FASEB J.* 6: 2323-2331.

Hepler, J. R., and Gilman, A. G., 1992, G Proteins, *Trends. Biochem. Sci.* 17: 383-387.

Hingorani, V. N., and Ho, Y.-K., 1987, Chemical modification of bovine transducin: Effect of fluorescein 5′-ispthiocyanate labeling on activities of the transducin α subunit, *Biochemistry* 26: 1633-1639.

Hingorani, V. N., Tobias, D. T., Henderson, J., T., and Ho, Y.-K., 1988, Chemical cross-linking of bovine retinal transducin and cGMP phosphodiesterase, *J. Biol. Chem.* 263: 6916-6926.

Ho, Y.-K., and Fung, B. K.-K., 1984, Characterization of transducin from bovine retinal rod outer segments. The role of sulfhydryl groups, *J. Biol. Chem.* 259: 6694-6699.

Hofmann, K. P., and Reichert, J., 1985, Chemical probing of the light-induced interaction between rhodopsin and G-protein. Near-infrared light-scattering and sulfhydryl modifications, *J. Biol. Chem.* 260: 7990-7995.

Kaziro, Y., Itoh, H., Kozasa, T., Nakafuku, M., and Satoh, T., 1991, Structure and function of signal-transducing GTP-binding proteins, *Annu. Rev. Biochem.* 60: 349-400.

Katz, A., Wu, D., and Simon, M. I., 1992, Subunits $\beta\gamma$ of heterotrimeric G protein activate β2 isoform of phospholipase C, *Nature* 360: 686-689.

Kühn, H., 1980, Light- and GTP-regulated interaction of GTPase and other proteins with bovine photoreceptor membranes, *Nature* 283: 587-589.

Laemmli, U. K., 1970, Cleavage of structural proteins during the assembly of head bacteriophage T4, *Nature* 227: 680-685.

Lagnado, L., and Baylor, D., 1992, Signal flow in visual transduction, *Neuron* 8: 995-1002.

Lambright, D. G., Noel, J. P., Hamm. H. E., and Sigler, P. B., 1994, Structural determinants for activation of the α-subunit of a heterotrimeric G protein, *Nature* 369: 621-628.

Logothetis, D. E., Kurachi, Y., Galper, J., Neer, E. J., and Clapham, D. E., 1987, The $\beta\gamma$ subunits of GTP-binding proteins activate the muscarinic K^+ channel, *Nature* 325: 321-326.

Matsudaira, P., 1987, Sequence from picomole quantities of proteins electroblotted onto polyvinylidene difluoride membranes, *J. Biol. Chem.* 262: 10035-10038.

Noel, J. P., Hamm, H. E., and Sigler, P. B., 1993, The 2.2 A crystal structure of transducin-α complexed with $GTP_\gamma S$, *Nature* 366: 654-663.

Reichert, J., and Hofmann, K. P., 1984, Sulfhydryl group modification of photoreceptor G-protein prevents its light-induced binding to rhodopsin, *FEBS Lett.* 168: 121-124.

Shichi, H., and Somers, R. L., 1978, Light-dependent phosphorylation of rhodopsin. Purification and properties of rhodopsin kinase, *J. Biol. Chem.* 253: 7040-7046.

Shichi, H., Yamamoto, K., and Somers, R. L., 1984, GTP binding protein: Properties and lack of activation by phosphorylated rhodopsin, *Vision Res.* 24: 1523-1531.

Simon, M. I., Strathmann, M. P., and Gautam, N., 1991, Diversity of G proteins in signal transduction, *Science* 252: 802-808.

Tang, W. J., Krupinski, J., and Gilman, A. G., 1991, Expression and characterization of calmodulin-activated (type I) adenylylcyclase, *J. Biol. Chem.* 266: 8595-8603.

Toner-Webb, J., and Taylor, S. S., 1987, Inhibition of the catalytic subunit of cAMP-dependent protein kinase by dicyclohexylcarbodiimide, *Biochemistry* 26: 7371-7378.

Towbin, H., Staehlin, T., and Gordon, J., 1979, Electrophoretic transfer of proteins from polyacrylamide gels to nitrocellulose sheets: Procedure and some applications, *Proc. Natl. Acad. Sci. U.S.A.* 76: 4350-4354.

Van Dop, C., Yamanaka, G., Steinberg, F., Sekura, R. D., Manclark, C. R., Stryer, L., and Bourne, H. R., 1984, ADP-ribosylation of transducin by pertussis toxin blocks the light-stimulated hydrolysis of GTP and cGMP in retinal photoreceptors, *J. Biol. Chem.* 259: 23-26.

Weingarten, R., Ransnäs, L., Mueller, H., Sklar, L. A., and Bokoch, G. M., 1990, Mastoparan interacts with the carboxyl terminus of the α subunit of G_i, *J. Biol. Chem.* 265: 11044-11049.

Wessling-Resnick, M., and Johnson, G. L., 1989, Evidence for oligomeric forms of transducin alpha subunit: Formation of intermolecular alpha-alpha disulfide linkages, *Biochem. Biophys. Res. Comm.* 159: 651-657.

West, R. E., Moss, J., Vaughan, M., Liu, T., and Liu, T.-Y., 1985, Pertussis toxin-catalyzed ADP-ribosylation of transducin. Cysteine 347 is the ADP-ribose acceptor site, *J. Biol. Chem.* 260: 14428-14430.

X-RAY PHOTOELECTRON SPECTROSCOPY OF AMINO ACIDS, POLYPEPTIDES AND SIMPLE CARBOHYDRATES

Kenneth E. Dombrowski,[1] Stephen E. Wright,[1] Jannine C. Birkbeck,[2] and William E. Moddeman[2]

[1] Department of Veterans Affairs Medical Center and Department of Internal Medicine
Texas Tech University Health Sciences Center
Amarillo, Texas 79106
[2] Mason & Hanger-Silas Mason Co., Inc.
Pantex Plant Amarillo, Texas 79177

ABSTRACT

X-ray photoelectron spectroscopy (XPS) is a surface sensitive analytical technique which measures the binding energy of electrons in atoms and molecules. The binding energy can be related to the molecular bonding or oxidation state of an element in the outermost layer of a material, that is < 100 Å. Thus, XPS is able to identify chemical species present on the surface of a molecule. In this paper XPS is briefly described. Spectra demonstrating its potential use for probing the surface properties of amino acids, polypeptides, proteins, carbohydrates and glycoproteins are discussed.

INTRODUCTION

XPS has also been referred to as electron spectroscopy for chemical analysis (ESCA). The basis for XPS is the photoelectric effect (1). Irradiation of a material with monochromatic x-rays results in the expulsion of photoelectrons from electron orbitals (e.g., s-orbitals) of the sample. The energy of an incident x-ray is transformed into the kinetic energy of a photo-emitted electron (Figure 1). By measuring the kinetic energy (E_k) of the ejected photoelectron and known x-ray photon energy ($h\nu$), the binding energy (E_b) of that electron can be deduced using the following equation:

$$E_b = h\nu - E_k - w$$

where w is the experimentally determined work function of the spectrometer.

Methods in Protein Structure Analysis, Edited by M. Z. Atassi and E. Appella
Plenum Press, New York, 1995

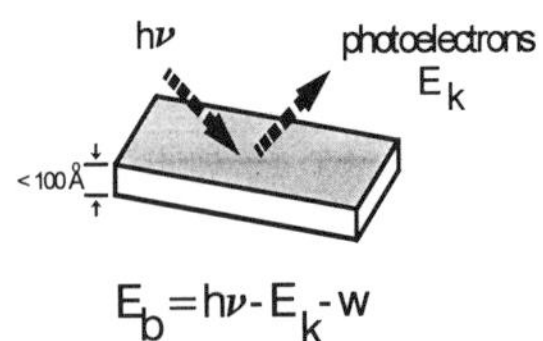

Figure 1. Schematic representation of the XPS process.

XPS generates its information from two modes of analyses: low resolution, or survey, spectra and high resolution spectra. Qualitative information is normally obtained and atomic composition can be obtained from a survey spectrum of the sample surface. Detailed chemical bonding information (e.g., oxidation state) is acquired from high resolution scans on each element.

The binding energy of electrons in an element is unique to the element as well as unique to its chemical environment. E_b can change several eV due to changes in oxidation state. When examining inner-shell electrons, the binding energies of these electrons in any element, X, can be directly related to the oxidation state of that element in a molecule in the following progression: $X^- < X^{\delta-} < X^0 < X^{\delta+} < X^+$; i.e., the larger the positive charge on the element, the greater the affinity of the nucleus for the remaining electrons, and hence, the larger the E_b.

XPS has routinely been used to examine the chemical structure of various organic and inorganic materials (2). In an article on the microencapsulation of an explosive, which contains the elements of carbon, nitrogen and oxygen (similar to the biological compounds to be discussed), equations were written that allowed for the determination of the thickness of the coating and the mechanism of polymer bonding to the explosive (3).

XPS has not been extensively used in the study of biological systems. A few XPS papers have been published which examined the role of metals in biological systems. Chiu et al. (4) studied the bonding of oxygen to selenium in a glutathione peroxidase model system, Meisenheimer et al. (5) determined monovalent cation compositions in erythrocyte membranes, and Pickart et al. (6) studied Ca^{2+} flux in hepatoma cells during DNA synthesis. In the latter study, an intramembrane Ca^{2+} gradient was established with the highest levels of Ca^{2+} being at the cytoplasmic side and *not* towards the extracellular space. A few studies have been published that have made use of XPS for the characterization of surfaces of bacterial cells (7), and for estimating the protein content in seeds (8).

Only a few XPS studies have been performed to study in detail the surface chemistries of biological macromolecules such as proteins (9-11). These papers showed the zwitterionic nitrogen to be in a less positive state following amide formation. In this paper, low and high resolution XPS spectra for several amino acids related to the core protein of a glycoprotein are reported. These results show the potential usefulness of this technique in characterizing carbohydrate coatings on polypeptides and proteins.

EXPERIMENTATION

Samples are applied to a 25 mm^2 Au metal surface as a multilayer in one of the following methods: 1) as a powder, 2) as a solution in water (doubly distilled), 3) as a slurry in methanol, or 4) as a slurry in a mixture of water and methanol. In the latter cases, the solvents were allowed to evaporate before being placed in the analysis chamber.

The instrument used in this study was a Kratos AXIS spectrometer which uses a hemispherical electrostatic analyzer to determine electron kinetic energies. Following sample introduction, the chamber was evacuated to < 10^{-5} torr. The x-ray source was an aluminum monochromator that emits an x-ray beam of 1486.67 Pa. The x-ray power was 300 watts (20 mA and 15 kV). During the photoelectron process, the surface acquires a positive charge. In order to minimize sample decomposition and charging, the sample was bathed in low energy electrons of less than 1 eV by a charge neutralizer. In this work, copious amounts of electrons were generated by the neutralizer which were sufficient to minimize differential sample charging. Samples were irradiated until sufficient data were collected. The amount of x-ray degradation previously determined on glycine was found to be < 1% over two hours of irradiation. No detectable x-ray damage was noted on the samples analyzed in this paper. Data collection required about 90 min per sample.

XPS spectra were deconvoluted to a best fit using peak shapes of 70 % Gaussian and 30 % Lawrencian character to account for tailing toward the high E_b side. Atomic % compositions of each elemental species present were calculated by dividing the area under each XPS elemental peak by an instrumental sensitivity factor. The sensitivity factors were theoretically calculated (12) from photoionization cross-sectional data. It is very difficult for the XPS technique to quantify hydrogen since the H 1s electron is part of the valence level. These valence electrons are often associated with, or shared between, two or more elements and thus can not be easily used for elemental quantification. Therefore, hydrogen is not included in the atomic % determinations.

RESULTS AND DISCUSSION

As discussed earlier, XPS data are often accumulated in two ways: either low resolution or high resolution. In the first case, the electron binding energies of a protein or peptide are measured to approximately ± 1 eV. From this determination, qualitative and semi-quantitative information about the protein surface is acquired. That is, the elemental constituents are determined and the atomic % composition is calculated. In the second case, E_b are determined to within ± 0.1 eV. In our work, E_b are determined to < 0.1 eV. From these measurements, the electron distributions about each atom (or the charge distributions) in the molecule are determined, and oxidation state and chemical bonding information are inferred.

Low Resolution Spectra

Figure 2 illustrates a survey scan for the simple amino acid glycine. The spectrum shows the three major constituents found in most amino acids: C at 285 eV, N at 401 eV and O at 531 eV. After correcting for differences in sensitivity factors, the area under each photoelectron peak is proportional to the atomic concentration. For this amino acid, the experimentally determined composition (in atomic %) from the survey scan is 42 % C, 20 % N and 38% O. These values are in good agreement with the theoretically calculated atomic compositions of 40% C, 20% N and 40% O.

The atomic concentrations obtained from the XPS spectra of the nine different amino acids found in the human mucin tandem repeat sequence MUC1 (12, 13) are given in Table 1. MUC1 is a 20 amino acid polypeptide with the sequence GSTAPPAHGVTSAPDTRPAP. An amino acid mixture with the composition of $G_2S_2T_3A_4P_5HVDR$ corresponding to the MUC1

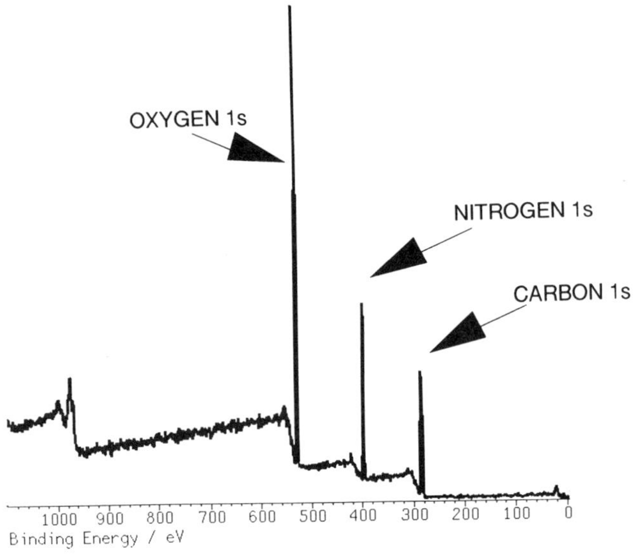

Figure 2. Representative low resolution XPS spectrum of glycine showing the presence of carbon, nitrogen and oxygen characteristic of amino acids.

peptide was prepared and analyzed by XPS. Again, good agreement between experimental and theoretical values was noted. These amino acids have a range of carbon from 43% for Gly to 64% for Val. The nitrogen composition is much lower: from 12% for Asp to 32% for Arg. The usual range for % N in proteins is about 15 to 20%. The % atomic composition of oxygen varies much like that of nitrogen. Ranges for this element are from 16% in Arg to 41% in Ser.

Table 1. Summary of quantitative XPS results on mucin-related materials

	Atomic % Compositions		
Material	%C	%N	%O
Ala	52[1] (50)[2]	17 (17)	32(33)
Arg	53 (50)	32 (36)	16 (17)
Asp	46 (44)	12 (11)	40 (44)
Gly	43 (40)	21 (20)	36 (40)
His	52 (50)	17 (17)	32 (33)
Pro	63 (62)	17 (16)	30 (31)
Ser	45 (43)	14 (14)	41 (43)
Thr	53 (50)	13 (12)	37 (38)
Val	64 (62)	13 (12)	23 (31)
MUC1 peptide [3]	52 (53)	17 (16)	30 (31)

[1]Values given are the experimentally determined atomic % composition for each element in the corresponding amino acid.
[2]Values in parentheses are the theoretically derived atomic % compositions for each element in the corresponding amino acid.
[3]Amino acids present in the MUC1 peptide were combined in the molar ratios of $G_2S_2T_3A_4P_5$HVDR.

Table 2. Summary of quantitative XPS results on mucins and related materials

	Atomic % compositions[1]		
Material	%C	%N	%O
Carbohydrate, normal [2]	55 (50)	0.3 (0.4)	45 (49)
Carbohydrate, cancer [3]	55 (52)	3.4 (3.6)	42 (44)
Porcine mucin [4]	77	4.9	18
Porcine mucin, partially deglycosylated [5]	65	8.5	27
Bovine mucin [4]	72	3.2	25
Bovine mucin, partially deglycosylated [5]	67	9.7	28

[1]Values given are the experimentally determined atomic % composition for each element in the corresponding amino acid. Values in parentheses are the theoretically derived atomic % compositions for each element in the corresponding amino acid.

[2]Carbohydrates common to normal mucins (13, 14) were combined in the molar ratios of $Gal_{12}GlcNAc_{11}GalNAc_1$ where Gal = galactose, GlcNAc = N-acetylglucosamine and GalNAc = N-acetylgalactosamine.

[3]Carbohydrate common to cancer-associated mucins (13) were combined in the molar ratios of $NANA_1Gal_1GalNAc_1$, where NANA = N-acetylneuraminic acid.

[4]Mucins (from porcine stomach and bovine submaxillary gland) were purchased from Sigma Chemical Co (St. Louis, MO).

[5]Mucins were partially deglycosylated by periodate oxidation (15).

The full range of sensitivity of this technique for identifying specific amino acid substitutions in simple polypeptides and proteins is not fully established. Using a human mucin MUC1 tandem repeat peptide, we have preliminary evidence suggesting that XPS is capable of easily identifying a point mutation in a mutant peptide containing the $T^{16} \rightarrow N^{16}$ mutation. This mutant peptide represents a 20% decrease in the number of hydroxyl groups present on this peptide.

A survey of carbohydrate structures found in human mucins (13, 14) is beginning to reveal some similarities and differences between protein and carbohydrate in an XPS spectrum (Table 2). Carbohydrates are not too dissimilar from amino acids in carbon content (~ 55 % C in carbohydrates as compared to ~ 50% in amino acids). The % O is slightly higher in carbohydrates (~ 45-49 % O) than in protein (~ 35 % O). However, the major difference between amino acids and carbohydrates is their nitrogen content. In human mucins, an average of about 0.3 atomic % nitrogen can be observed in samples representing normal mucin oligosaccharide side chains to 3 atomic % N in the oligosaccharide side chains of breast cancer-associated mucin (13, 14).

The atomic % of nitrogen in carbohydrates are clearly distinguishable from the nitrogen composition of protein. For example, fully glycosylated porcine mucin showed a composition of 77 % C, 4.9 % N and 18 % O (Table 2). The measured composition of periodate-oxidized porcine mucin (Table 2) showed a decrease in the atomic % of carbon and an increase in the atomic % of both nitrogen and oxygen. This indicates that the mucin core protein is being exposed by the removal of carbohydrate during periodate oxidation. Since the core protein of a related porcine mucin has a theoretical composition of 51% C, 16 % N and 33 % O (16), it can be concluded that the periodate treatment did not fully deglycosylate this mucin and that it still bears a high degree of oligosaccharide side chains branched on the GalNAc-O-Thr which are not susceptible to periodate oxidation (15). Similar to the porcine mucin, bovine mucin showed an increase in atomic % N content from 3.2% to 9.7% after periodate treatment. Again, indicating that bovine

Table 3. Summary of binding energies for glycine, polyglycine and glucose: High resolution data

	E_b, eV		
	Glycine (12 runs)	Polyglycine (8 runs)	Glucose (4 runs)
C 1s			
α-C	286.19 ± 0.07	286.42 ± 0.07	
carboxylate	288.38 ± 0.07		
amide		288.42 ± 0.07	
alcohol			286.60 ± 0.03
anomeric			287.94 ± 0.05
N 1s			
zwitterion	401.45 ± 0.05		
amide		400.19 ± 0.05	
O 1s			
carboxylate	531.15 ± 0.06		
amide		531.78 ± 0.06	
alcohol			532.88 ± 0.04
pyranose ring			533.62 ± 0.07

E_b given are the average energies of the indicated number of runs ± the standard deviation.

mucin is not fully susceptible to periodate oxidation to remove oligosaccharide side chains.

High Resolution Spectra

Figures 3 through 5 illustrate the XPS high resolution C 1s, N 1s and O 1s spectra, respectively, of one amino acid (glycine), one polypeptide (polyglycine with average M_r ~ 4,500) and one carbohydrate (glucose). The analyses of these data are given in Table 3. As can be seen from the data (also see Fig. 3a), glycine has two types of carbon atoms: one methylene and one carboxylate. (A small third carbon peak is also found to be present. Almost all materials contain a carboneous contaminant that is sometimes referred to as ubiquitous, or residual, carbon. The source is often not readily identifiable. The C 1s electrons from this carbon source is often found at an E_b of ~ 285 eV and has not been assigned to any specific carbon species of the amino acid.) The higher binding energy peak at 288.38 ± 0.07 eV corresponds to the carbon atom of the carboxylate group with the methylene carbon appearing at a lower binding energy of 286. 19 ± 0.07 eV. This is consistent with the molecular bonding and the electron charge distribution where the carbon of the carboxylate is at a higher positive oxidation state than the methylene carbon. Also, the intensities of the peaks are in about a 1:1 ratio, consistent with their abundance in this amino acid. Fig. 3B illustrates the C 1s data from polyglycine. Again, two peaks are seen, one of which represents the α-carbon at 286.19 ± 0.07 eV and the other which represents the amide carbon at 288.42 ± 0.07 eV. The glucose C 1s spectrum (Fig. 3c) is completely different from either of the above two glycine compounds. This carbohydrate contains two carbon atoms in different chemical environments: one at 286.60 ± 0.03 eV characteristic of an alcohol and the other at 287.94 ± 0.05 eV, characteristic of the anomeric carbon. Thus, the E_b's are distinguishable between the different oxidation states of each carbon species (i.e., a carboxylate of a zwitterion is identifiable from an anomeric carbon, or an alcohol or an aliphatic carbon). The binding energy data reported above is referenced to the C 1s level of the two methyl groups on Leu which was set at 285.00 eV.

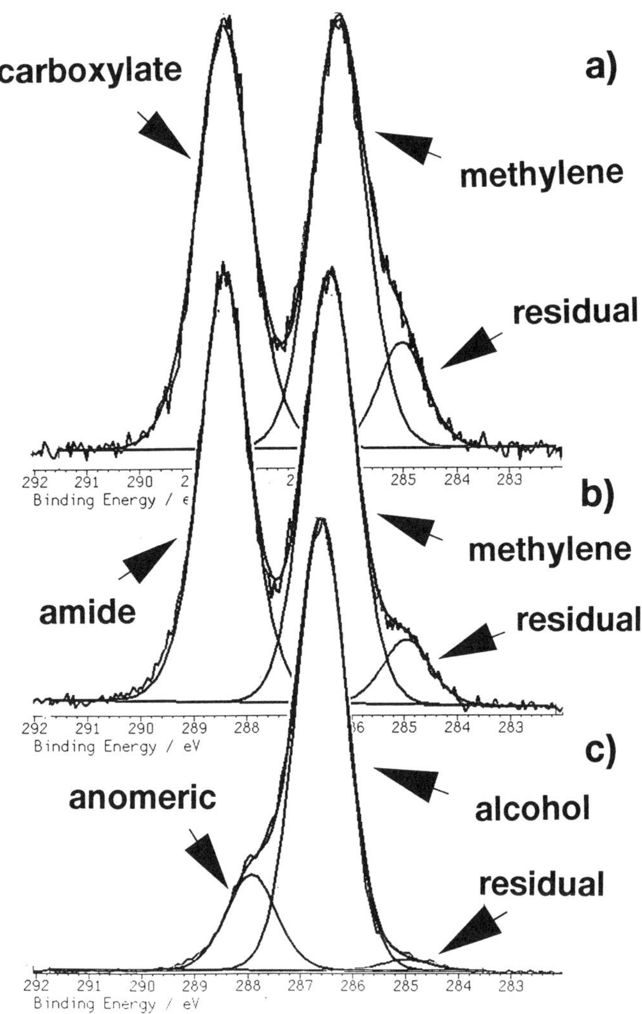

Figure 3. High resolution XPS C 1s spectra of (a) glycine, (b) polyglycine and (c) glucose showing unique patterns for the different carbon atoms in each compound. Jagged curves are the raw data and the smooth curves are the best fit approximations as described in the text.

Fig. 4 exhibits the high resolution N 1s spectra of glycine, polyglycine and glucose. A single N 1s peak is observed in both the glycine (Fig. 4a) and polyglycine (Fig. 4b) spectra, but no nitrogen is observed in the glucose spectrum (Fig. 4c). The N 1s peaks appear at different positions in the first two spectra: the nitrogen atoms in the zwitterionic form of glycine are observed at $E_b = 401.45 \pm 0.05$ eV which is characteristic of nitrogen in a +1 oxidation state and the amide nitrogen at $E_b = 400.19 \pm 0.05$ eV where nitrogen has a lone pair of electrons. These nitrogen results are similar to those reported previously (8-10).

Glycine has two oxygen atoms. However, these electrons are equivalent due to the resonance structures of the carboxylate anion of the acid. Fig. 5a depicts a single O 1s peak with a binding energy of 531.15 ± 0.06 eV. Upon polymerization to form the peptide, again, only one O 1s peak is observed (Fig. 5b), but the binding energy is

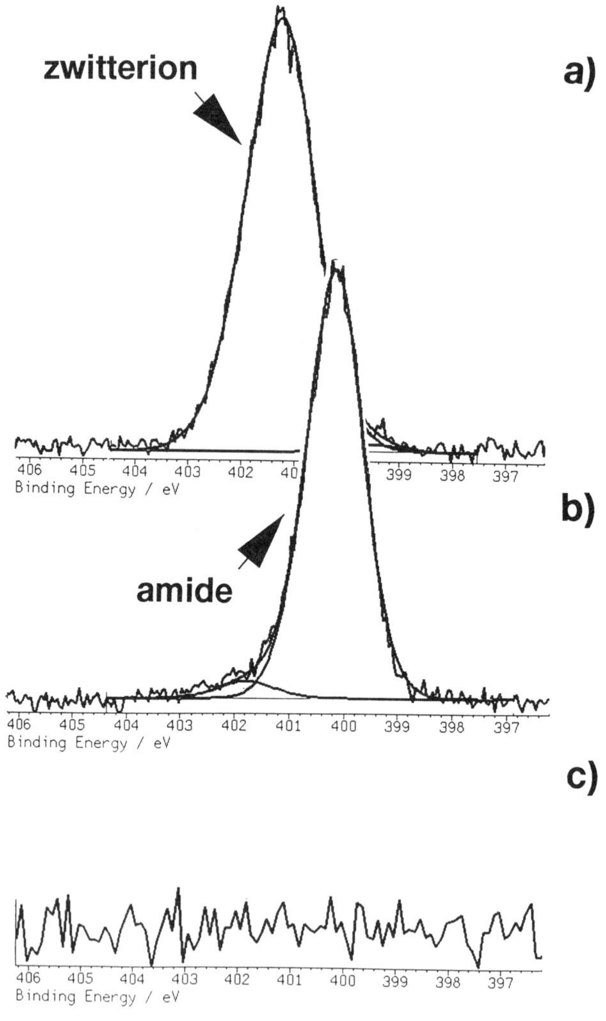

Figure 4. High resolution XPS N 1s spectra of (a) glycine, (b) polyglycine and (c) glucose showing nitrogen binding energy differences between zwitterion and amide structures. The data also shows the absence of nitrogen in the carbohydrate glucose. Jagged curves are the raw data and the smooth curves are the best fit approximations.

increased to 531.78 ± 0.06 eV. The carbohydrate O 1s spectrum shows a broad peak that has been deconvoluted into two structures: the primary one at 532.88 ± 0.04 eV and a smaller one with approximately one-fifth the area at 533.62 ± 0.07 eV. These two peaks are characteristic of oxygen atoms of the alcohol and pyranose ring environments, respectively.

SUMMARY

XPS is a surface sensitive technique capable of distinguishing between core protein and carbohydrate coatings on glycoproteins. Deglycosylation of periodate-oxidized porcine

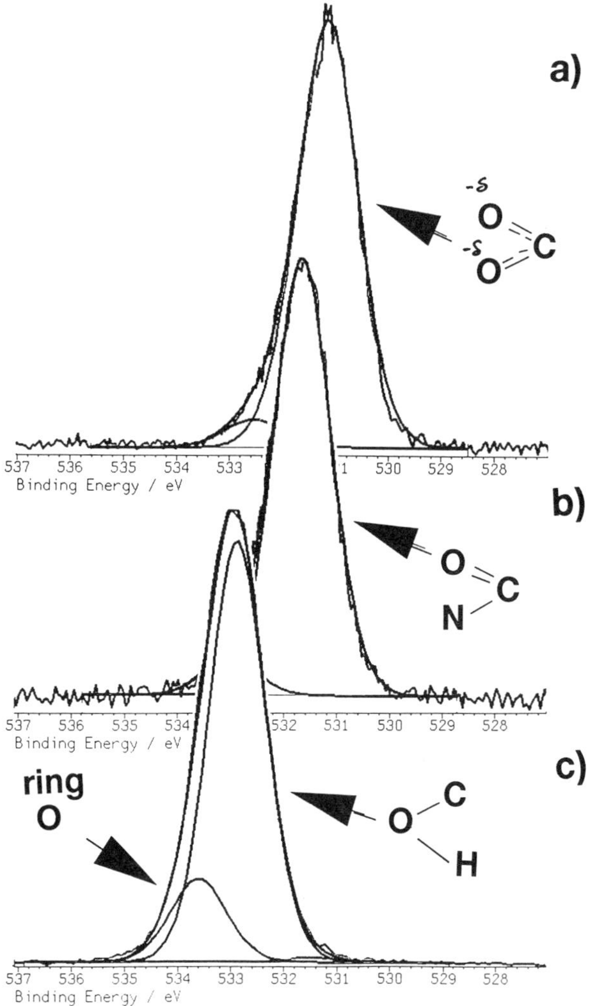

Figure 5. High resolution XPS O 1s spectra of (a) glycine, (b) polyglycine and (c) glucose showing oxygen binding energy differences between carboxylate, amide and carbohydrate (alcohol). Jagged curves are the raw data and the smooth curves are the best fit approximations.

and bovine mucin were not complete. There is a difference in the composition between coatings of normal and breast cancer-associated oligosaccharide side chains. Further examination of XPS in terms of its sensitivities for sample amount and limits of detection is an exciting area of protein structure analysis.

ACKNOWLEDGMENTS

This work was supported in part from the following sources: the Elsa U. Pardee Foundation (K. E. D. & S. E. W.), U.S. Army Medical Research and Materiel Command Grant #DAMD17-94-J-4272 (K. E. D. & S. E. W.), U.S. Army Medical Research and

Materiel Command Career Development Award #DAMD17-94-J-4161 (K. E. D.), Department of Veterans Affairs Medical Research Funds (S. E. W.) and Department of Energy contract #DE-AC04-91AL-65030 (W. E. M., J.C.B.),

REFERENCES

1. Carlson, T. A., 1975 in Photoelectron and Auger Spectroscopy, Plenum Press, NY.
2. Robinson, J. W., 1991 in Practical Handbook of Spectroscopy, CRC Press, Boca Raton, FL.
3. Worley, C. M., Vannet, M. D., Ball, G. L. and Moddeman, W. E., 1987 Surface chemistry of a microcoated energetic material, pentaerythritoltetranitrate (PETN). Surface and Interface Analysis, 10:273.
4. Chiu, D., Tappel, A.L. and Millard, M.M., 1977 Improved procedure for x-ray photoelectron spectroscopy of selenium-glutathione peroxidase and application to the rat liver enzyme. Arch. Biochem. Biophys. 184:209.
5. Meisenheimer, R. G., Fisher, J. W. and Rehfeld, S. J., 1976 Thallium in human erythrocyte membranes: an x-ray photoelectron spectroscopy study. Biochem. Biophys. Res. Commun. 68, 994.
6. Pickart, L., Millard, M. M., Beiderman, B. and Thaler, M. M., 1978 Surface analysis and depth profiles of calcium in hepatoma cells during pyruvate-induced synthesis. Biochim. Biophys. Acta, 544:138.
7. Millard, M. M., Scherrer, R. and Thomas, R. S. 1976 Surface analysis and depth profile composition of bacterial cells by x-ray photoelectron spectroscopy and oxygen plasma etching. Biochem. Biophys. Res. Commun. 72:1209.
8. Peeling, J., Clark, D. T., Evans, M. and Boulter, D. 1976 Evaluation of the ESCA technique as a screening method for the estimation of protein content and quality in seed meals. J. Sci. Fd. Agric. 27:331.
9. Siegbahn, K., Mordling, C., Fahlman, A., Mordbert, R., Hedman, J., Johnsson, G., Bergmark, T., Karlsson, S. E., Lindgren, I. and Linberg, B., 1967 ESCA-Atomic, Molecular, and Solid State Structure Studied by Means of Electron spectroscopy, Nova Acta Regiae Soc. Sci. Upsaliensis Ser. IV Vol. 20.
10. Bumben, K. D. and Dev, S. B. 1988 Investigation of poly(L-amino acids) by x-ray photoelectron spectroscopy. Anal. Chem. 60:1393.
11. Clark, D. T., Peeling, J. and Colling, L., 1976 An experimental and theoretical investigation of the core level spectra of a series of amino acids, dipeptides and polypeptides. Biochim. Biophys. Acta 453:533.
12. Scofield, J. H., 1973 in Theoretical cross-sections from 1-1500 KeV, Lawrence Livermore Laboratory Report UCRL-51326.
13. Gendler, S. J., Spicer, A. P., Lalani, E. -N, Duhig, T., Peat, N., Burchell, J., Pemberton, L., Boshell, M. and Taylor-Papadimitriou, J., 1991 Structure and biology of a carcinoma-associated mucin, MUC1. Amer. Rev. Resp. Dis. 144:S42.
14. Hanisch, F. -G., Uhlenbruck, G., Peter-Katalinic, J., Egge, H., Dabrowski, J. and Dabrowski, U., 1989 Structures of neutral O-linked polylactosaminoglycans on human skim milk mucins. J. Biol. Chem. 264:872.
15. Gerken, T. A., Gupta, R and Jentoft, N., 1992 A novel approach for chemically deglycosylating O-linked glycoproteins. The deglycosylation of submaxillary and respiratory mucins. Biochemistry 31:639.
16. Gupta, R and Jentoft, N., 1989 Subunit structure of porcine submaxillary mucin. Biochemistry 28:6114.

23

MEASUREMENT OF ASP/ASN DAMAGE IN AGING PROTEINS, CHEMICAL INTERCONVERSION OF ASPARTYL ISOMERS, ^{18}O TAGGING OF ENZYMATICALLY REPAIRED ASPARTYL SITES, AND ENZYME AUTOMETHYLATION AT SITES OF ASP/ASN DAMAGE

Jonathan A. Lindquist and Philip N. McFadden

Department of Biochemistry and Biophysics
Oregon State University
Corvallis, Oregon 97331

INTRODUCTION

Asp/Asn damage in aging proteins, resulting from the propensity of L-Asn and L-Asp residues to spontaneously convert to a mixture of α-epimerized and β-isomerized aspartyl products via succinimide intermediates (Figure 1), is a practical problem from the standpoint of researchers seeking to isolate and study pure proteins. In particular, the advent of protein overexpression systems and the convenience of working with large quantities of protein has made it increasingly common for investigators with no prior intention of studying spontaneous protein damage to find that a protein of interest has undergone a transformation that is ultimately found to be due to Asp/Asn damage. The first indication of this problem is generally the detection of isoforms of a polypeptide with altered chromatographic or electrophoretic properties, often as a function of a heat-step involved in the purification of the protein. Other times the formation of these spontaneously formed isoforms has been traced to a prolonged fermentor run or a lengthy storage period during the production of the protein. Though the isoforms of the protein may make up only a few percent of the total material, their presence is troubling since the purity of the protein is compromised.

Successful efforts have been made in the chemical and physical characterization of such spontaneously altered proteins, with recent examples pertaining to characterization of Asp/Asn damage in overexpressed forms of hirudin, deoxyribonuclease I, calbindin, interleukin-1β, epidermal growth factor, human growth hormone, anti-p185^{HER2}, tissue plasminogen activator, phosphocarrier protein, CD4, somatotropin, and interleukin-1α (Bischoff,

Methods in Protein Structure Analysis, Edited by M. Z. Atassi and E. Appella
Plenum Press, New York, 1995

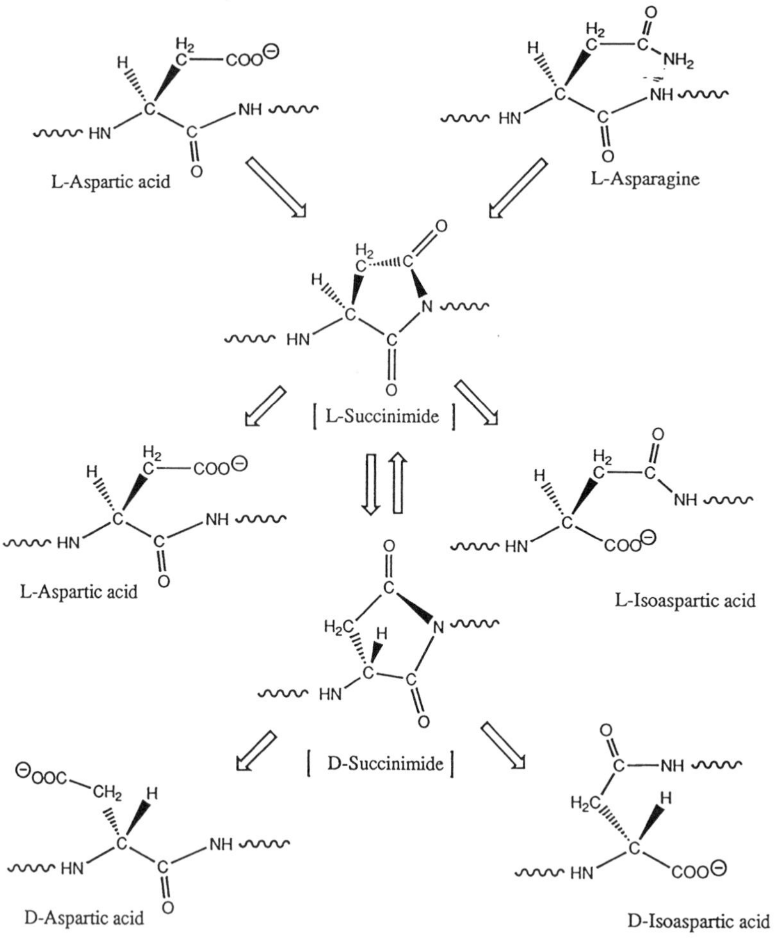

L-Aspartic acid

L-Asparagine

[L-Succinimide]

L-Aspartic acid

L-Isoaspartic acid

[D-Succinimide]

D-Aspartic acid

D-Isoaspartic acid

Figure 1. Asp/Asn damage in proteins. L-Aspartic acid and L-asparagine cyclize to an L-succinimide, losing water and ammonia, respectively. Epimerization at the α-carbon results in a mixture of L-succinimide and D-succinimide. Hydrolysis of the succinimides at either the α- or β-carbonyl groups results ultimately in a mixture of L-aspartic acid, D-aspartric acid, L-isoaspartic acid, and D-isoaspartic acid.

et al., 1993, Cacia, et al., 1993, Chazin, et al., 1989, Daumy, et al., 1991, George-Nascimento, et al., 1990, Johnson, et al., 1989, Kwong and Harris, 1994, Paranandi, et al., 1994, Sharma, et al., 1993, Teshima, et al., 1991, Violand, et al., 1990, Wingfield, et al., 1987). Such efforts have made use of a varied combination of methods to pinpoint sites of Asp/Asn damage in the respective proteins, the most general of which have included a) peptide mapping and detection of chromatographically altered protein fragments obtained through enzymatic and/or chemical cleavage reactions, b) determination of asparagine deamidation by mass spectral analysis, c) the failure of the Edman cleavage reaction at β-isomerized aspartyl residues, and d) diagnostic enzymatic methylation of peptide fragments by protein (D-aspartyl/L-isoaspartyl) carboxyl methyltransferase (PCM). The widespread occurrence of protein Asp/Asn damage in many sequence contexts and in many different classes of proteins indicate that this type of protein damage will continue to plague researchers and companies who are interested in producing large amounts of pure protein.

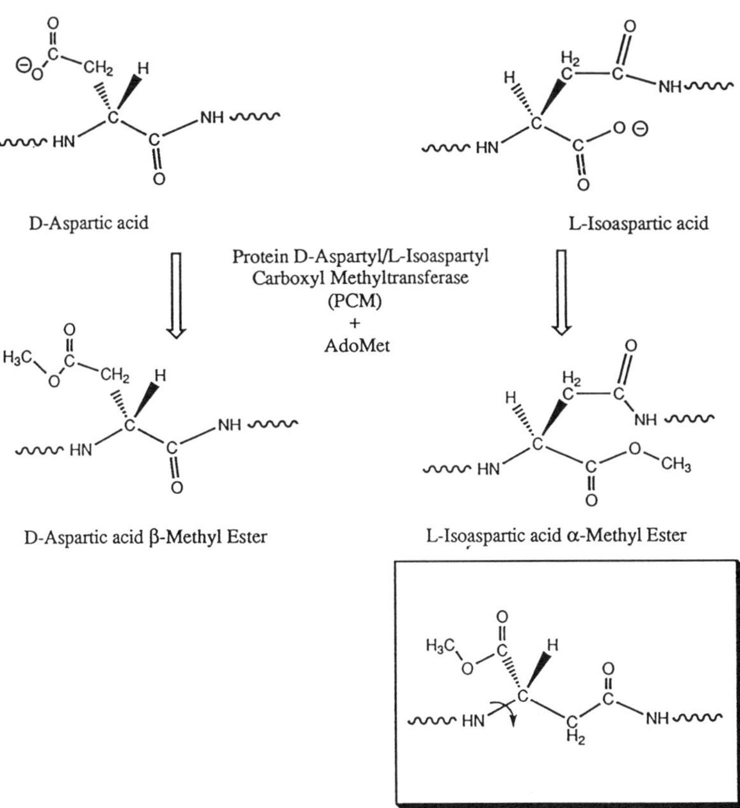

Figure 2. Enzymatic methylation of damaged protein. Protein (D-aspartyl/L-isoaspartyl) carboxyl methyltransferase (PCM) incorporates the methyl group from S-adenosylmethionine into ester linkage with the β-carboxyl of D-aspartyl residues and the α-carboxyl of L-isoaspartyl residues. The same enzyme active site is capable of methylating both forms of Asp/Asn damage, possibly because, as shown in the box, free rotation about the N-C bond can yield similar configurations of D-Asp and L-Isoasp in which the esterified carboxyl group is in approximately the same position in space relative to the α-carbon and the rest of the protein backbone.

Asp/Asn damage of aging proteins is also a physiological problem for organisms dependent on the integrity of their proteins. Many tissue and cell proteins have been found to contain such forms of damage, including, for example, a large proportion of the β-amyloid protein associated with Alzheimer's dementia (Roher, et al., 1993). Perhaps an even stronger indication of the physiological relevance of Asp/Asn damage is the presence in most cells of the above-mentioned enzyme, PCM, whose function is the methylation and processing of D-aspartyl and L-isoaspartyl residues that form as the result of intracellular Asp/Asn damage (Figure 2). The methylation of these sites and their rapid demethylation by a spontaneous mechanism at physiological pH has been shown in model studies to convert these abnormal aspartyl isomers to normal L-aspartyl residues (Figure 3). Hence, the function of this enzyme evidently relates to the repair of sites of Asp/Asn damage, which could either restore a protein's function or could allow complete proteolytic degradation of a protein that might otherwise resist degradation because of the presence of abnormal amino acid isomers.

This paper covers four aspects related to Asp/Asn damage and the D-Asp/L-Isoasp enzymatic methylation pathway. First, some suggestions are made for improvements in the use of chemical reduction assays for the products of Asp/Asn damage (Carter and McFadden,

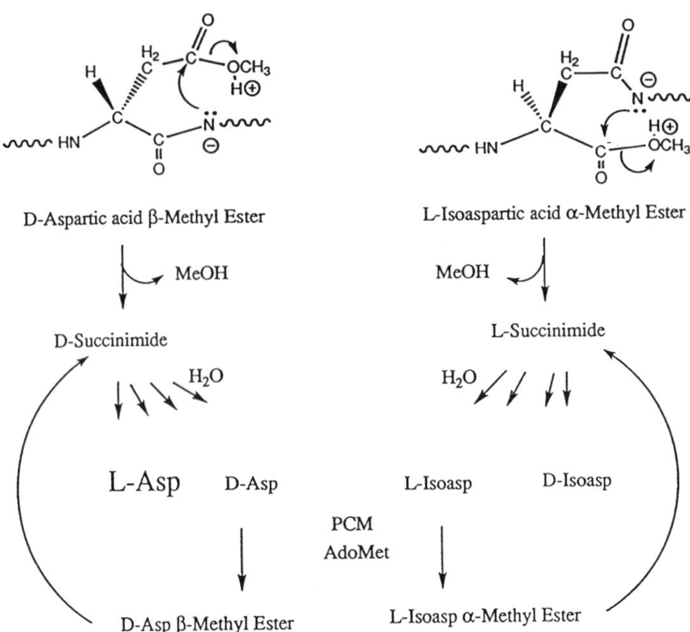

Figure 3. Methylation-dependent protein repair. Enzymatically formed methyl esters of D-aspartyl and L-isoaspartyl residues are subject to rapid nonenzymatic displacement by nucleophilic attack by the amide nitrogen of the adjacent amino acid residue toward the C-terminus. This results in formation of succinimides, which are in turn subject to spontaneous epimerization and hydrolysis to yield four products (L-Asp, D-Asp, L-Isoasp and D-Isoasp). Two of these (D-Asp, L-Isoasp) can be methylated again by PCM, affording another opportunity for demethylation, succinimide formation, and succinimide hydrolysis. Eventually, the products of the process are highly enriched in L-Asp and D-Isoasp. L-Asp, being a normal α-amino acid, may restore function to the damaged protein, or may at least permit the aging protein to be fully destroyed by proteolytic mechanisms. While D-Isoasp is not a normal α-amino acid, it is possible that D-Isoasp may be a functional equivalent for L-Asp since L-Asp and D-Isoasp might adopt similar configurations as might D-Asp and L-Isoasp as shown in Figure 2.

1994a, Carter and McFadden, 1994b). Second, a technique is described that can chemically interconvert normal aspartyl residues and isoaspartyl residues for purposes of synthesizing model damaged peptides, and conversely, for enabling the Edman sequencing through an isoaspartyl linkage. Third, recent isotope-labeling evidence in support of methylation-dependent repair of Asp/Asn damage is reviewed. Fourth, automethylation involving an Asp/Asn-damaged subpopulation of PCM (termed the αPCM fraction) is described (Lindquist and McFadden, 1994a), and a model is presented for the unusual kinetics of PCM automethylation.

MEASUREMENT OF ISOASPARTIC ACID AND PROTEIN SUCCINIMIDES BY CHEMICAL REDUCTION

The hydrolysis of a protein to free amino acids results in the loss of any information as to how an aspartyl residue was linked within the protein, and so it has not been possible to measure succinimidyl and isoaspartyl residues in conjunction with conventional amino acid analysis. Recently we investigated whether two related approaches of chemical reduction might show promise in converting succinimides and isoaspartic acid to derivatives that

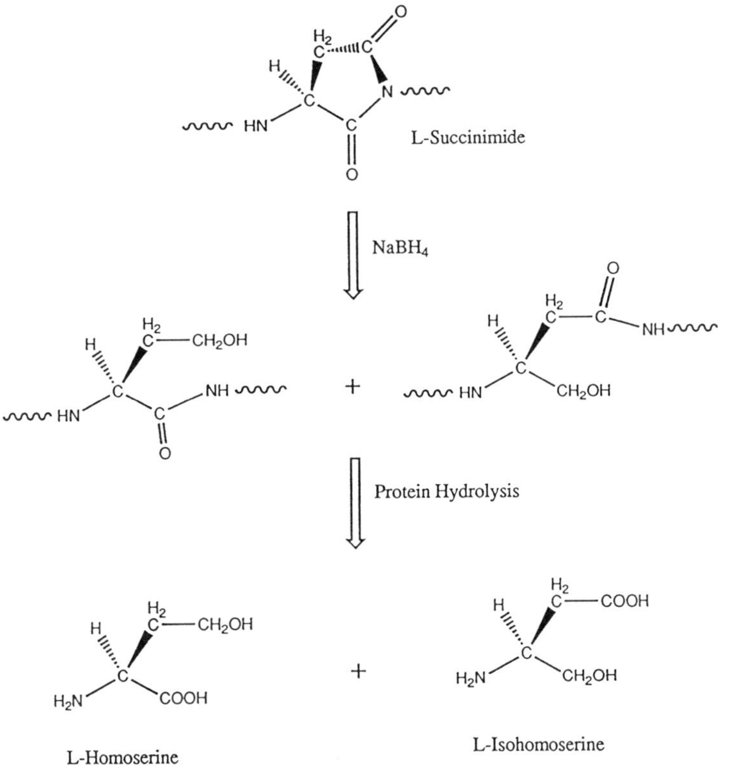

L-Succinimide

$NaBH_4$

Protein Hydrolysis

L-Homoserine

L-Isohomoserine

Figure 4. Trapping succinimides by chemical reduction. The reductive ring-opening of succinimides (Kondo and Witkop, 1968) as applied to protein succinimides, results in formation of a mixture of homoserine and isohomoserine residues. These are stable derivatives that can be analyzed following protein hydrolysis.

are stable to protein hydrolysis. The first method involves reductive ring-opening of protein succinimides by sodium borohydride to yield homoserine and isohomoserine upon protein hydrolysis (Figure 4). This technique was validated for model compounds containing known quantities of succinimides (Carter, et al., 1994b). The second method uses borane (BH_3) reduction to convert the free α-carboxyl group of isoaspartyl residues to an alcohol, resulting in isohomoserine that can be detected by protein hydrolysis and amino acid analysis (Figure 5). This method has also been validated with model polypeptides (Carter, et al., 1994a).

An improvement in borane reduction of polypeptides can be made by using as a reducing agent the commercially available borane dimethylsulfide complex in place of borane tetrahydrofurn complex that was the reagent used previously (Carter, et al., 1994a) This change is to be recommended partly because borane dimethylsulfide is a more stable and less hazardous reagent than borane tetrahydrofuran, and because borane dimethylsulfide excelled in a direct comparison of the effectiveness of the two reagents in the reduction of N-carbobenzoxy-L-aspartic acid-β-benzyl ester to N-carbobenzoxy-L-isohomoserine-β-benzyl ester.

A general difficulty in applying reduction methods to the assay of Asp/Asn damage in large proteins is in detection of the substoichiometric content of isohomoserine that typically is expected to result from borohydride or borane reduction. For example, less than one percent of the total aspartic acid and asparagine present may be present as succinimide and/or isoaspartic acid, and so only a small amount of isohomoserine could possibly be

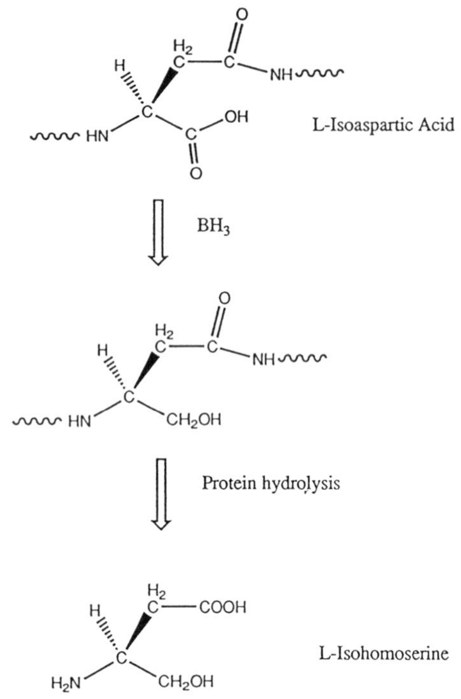

Figure 5. Detection of isoaspartic acid as the corresponding alcohol. The reduction of protein carboxyl groups by borane treatment (Atassi and Rosenthal, 1969, Rosenthal and Atassi, 1967) as applied to protein isoaspartyl groups results in formation of isohomoserine.

expected. A useful adjunct in such cases, then, is to degrade most of the α-amino acids in an amino acid hydrolyzate by snake venom L-amino acid oxidase, leaving the β-amino acid isohomoserine as a stronger signal above the background. However, a further difficulty in the analysis of small amounts of isohomoserine is the low ninhydrin color constant exhibited by this amino acid. The reddish color intensity following ninhydrin spraying of thin-layer separated isohomoserine is less than 1/10th that of equivalent amounts of aspartic acid, and ion-exchange separation of isohomoserine with post-column ninhydrin detection indicates a ninhydrin color yield (570 nm+460nm) that is as little as 1/50th that of equivalent amounts of aspartic acid. Depending on the amount of starting material, then, a means other than ninhydrin may be necessary for detection of isohomoserine. Derivatization of isohomoserine with either phenylisothiocynate or dabsyl chloride, and chromatographic separation of the conjugates by reversed phase HPLC are promising routes to isohomoserine analysis since the isohomoserine derivatives absorb in the ultraviolet (PTC-isohomoserine) and at 460 nm (dabsyl-Ihser) approximately as well on a molar basis as derivatives of α-amino acids.

INTERCONVERSION OF ASP AND ISOASP RESIDUES BY CHEMICAL ESTERIFICATION AND DE-ESTERIFICATION

There is a convenient procedure for converting normal L-Asp peptides to peptide mixtures containing the several aspartyl isomers (McFadden and Clarke, 1986). Here the methyl ester of the L-aspartyl side chain is first formed by acidic methanol tretament. The ester is then displaced via a succinimide intermediate upon mild alkaline treatment. Hydrolysis of the succinimide finally gives rise to the mixture. The main product in the mixture is the L-isoaspartyl derivative, which can be purified to serve as a stoichiometric substrate in various studies of protein (D-aspartyl/ L-isoaspartyl) carboxyl methyltransferase. For example, L-Trp-L-Met-L-Isoasp-L-Phe-NH2 was prepared in this manner for studies described below, and recently an esterified/de-esterified preparation of bovine serum albumin has been prepared and found in enzymatic assays to be extensively methylated by PCM. Similar chemistry could be used to convert isoaspartyl linkages to normal aspartyl linkages, which could permit Edman sequencing beyond an otherwise sequence-terminating isoaspartyl site.

The major limitation of this procedure is that glutamic acid and C-terminal carboxyls are also methyl esterified and are not extensively de-esterified under mild alkaline conditions. This results in additional complexity of the mixture with the varied presence of these other methylated carboxyl groups. We have recently explored a general solution to this problem by chemically forming the benzyl esters of peptide carboxyl groups. Here, the mild alkaline treatment of the peptide ester is again expected to facilitate the formation of succinimide rings that are then hydrolyzable to mixtures of aspartyl isomers. The remaining benzyl esters of glutamic acid residues and of the C-termini can then be removed by catalytic reduction, most conveniently by palladium catalyst with a hydrogen donor such as formate. This approach has met limited success thus far in our hands because the initial formation of the benzyl ester is not as facile as the formation of the methyl ester. Even so, it will be worthwhile to develop improved esterification conditions since this approach can enable the interconversion of aspartyl isomers without modifying the other carboxyls.

METHYLATION-DEPENDENT PROTEIN REPAIR VIA REPEATED PASSAGE THROUGH A SUCCINIMIDE INTERMEDIATE

A rather inefficient aspect of the pathway for methylation-dependent repair of L-isoaspartyl sites in peptides is the requirement for repeated formation and hydrolysis of the succinimide intermediate. While kinetic evidence suggested that indeed on the order of 5 cycles of methylation, demethylation, and succinimide hydrolysis are necessary to effect the complete repair of a peptide, this assumption had not been directly tested. Recently, the expectation that ^{18}O from [^{18}O]water would be incorporated into the peptide upon succinimide hydrolysis was used to test for multiple passages through the succinimide intermediate during peptide repair (Figure 6). Here, the model peptide used was L-Trp-L-Met-L-Isoasp-L-Phe-NH2, which itself had been prepared by chemical esterification/de-esterification as described above. [^{18}O]Water (43%) was included in the reaction medium, which, in addition to the isopeptide, consisted of bovine erythrocyte protein carboxyl methyltransferase, S-adenosylmethionine (AdoMet), and pH 7.8 phosphate buffer. Following a 48 hour reaction, the peptide products were purified by reversed phase HPLC, and peptide masses were measured by fast-atom bombardment mass spectrometry. The identification and quantification of doubly-^{18}O labeled normal aspartyl peptide as a repaired product fit closely with quantitative predictions of the extent of methylation/ demethylation/ succinimide hydrolysis that would occur in a period of about 27 hours, rather than the actual 48 hours of the reaction (Lindquist and McFadden, 1994b) . Thus, the theoretical kinetics of repair closely matched the experimental work, verifying that multiple succinimide hydrolyses take place, but with the caveat that during the final ~21hours of the reaction little peptide repair took place. We originally ascribed the failure of the repair reaction to go to completion to either enzyme denaturation during the lengthy incubation, or to the buildup of S-adenosylhomocysteine (AdoHcy), the end-product inhibitor of methyltransferases. Recently, however, we have found an additional factor that may block the completion of the repair reaction in that S-methylthioadenosine (MTA) has been detected in repair reaction mixtures as a substantial spontaneous breakdown product of AdoMet. Since MTA may be equal to or more potent

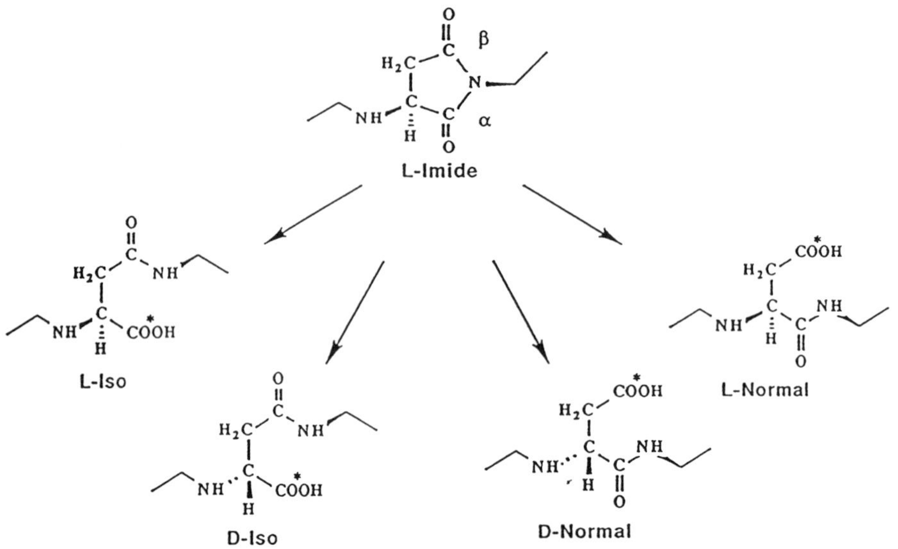

Figure 6. L-Succinimide (L-imide) postulated as the central intermediate during peptide repair, showing the sites (asterisks) of potential incorporation of ^{18}O during hydrolytic ring opening in the presence of [^{18}O]water.

than AdoHcy as a methyltransferase inhibitor, this spontaneous side-reaction may be a major cause of the slowing of the rate of peptide repair.

A DAMAGED SUBPOPULATION OF PROTEIN (D-ASPARTYL/ L-ISOASPARTYL) CARBOXYL METHYLTRANSFERASE IS METHYLATED BY A HIGH AFFINITY, LOW-TURNOVER REACTION

Interesting possibilities for feedback can be predicted to occur if enzymes that detect damage in other aging proteins are themselves damaged with age. As an example, bovine erythrocyte PCM was recently found to methylate itself on a subpopulation of enzyme molecules (Lindquist, et al., 1994a). The subpopulation of presumably damaged L-Isoasp- and D-Asp-containing enzyme molecules has been termed the αPCM fraction. From the known specific activity of [^{3}H-*methyl*]S-adenosylmethionine used in the radioactive automethylation assay it is calculated that αPCM molecules make up approximately 1% of the total PCM population in the cell. Such a low stoichiometry of methylation is expected, given that over the lifetime of the cell there is only a partial spontaneous conversion of amino acid sites to D-aspartyl and L-isoaspartyl residues, and given the hypothesis that part of this spontaneous damage is repaired by the enzymatic methylation pathway. αPCM can be partly enriched by anion-exchange chromatography and then quantified by HPLC (Figure 7), probably on the basis that deamidation of an Asn yields the slightly more negatively charged αPCM. Preparations with up to about 10% αPCM have been obtained in this manner.

To investigate the mechanism of PCM automethylation, assays were performed at several different enzyme dilutions. It was found that the specific rate of αPCM methylation increases with PCM concentration. This shows that the automethylation reaction involves more than a single PCM molecule since an intrapeptide methylation reaction would not have its rate affected by enzyme concentration. Most likely, enzyme automethylation involves the incorporation of a methyl group into an αPCM molecule by the activity of a second PCM molecule.

The specific rate of αPCM plateaus at high concentrations of total PCM (Figure 8), indicative of a saturating reaction. The assumption of a rapid equilibrium in the interaction between αPCM and active PCM has allowed the derivation of a rate equation that lends a good theoretical fit to our dilution experiments (McFadden and Lindquist, 1994). In this equation,

$$v' = v/[PCM]_{tot} = \alpha k_p [PCM]_{tot} / (K_s + [PCM]_{tot}),$$

the specific rate, v′, of αPCM methylation is given as a function of the total PCM concentration, $[PCM]_{tot}$; α is the fractional population of αPCM (e.g.2%); k_p is the turnover number for the αPCM methylation reaction; and K_s is the dissociation constant between PCM and αPCM. By applying the above equation to the experimental dilution studies, values for the kinetic constants were calculated to be k_p, 0.0095min^{-1}, and K_s, 0.5 μM. These values were constant in experiments with different PCM preparations, including those containing different percentages of αPCM (Figure 8). These values for K_s and k_p are interesting and somewhat surprising. The turnover number for αPCM methylation, K_s, is lower than for the methylation of most other polypeptides by PCM (Lowenson and Clarke, 1991), which could indicate that PCM has a high affinity for its damaged "brethren". While this could be considered a logical adaptation to preserve the integrity of the repair system, the turnover number for the automethylation reaction, k_p, is so low as to suggest that the enzyme-substrate complex, PCM*αPCM, is nearly a dead-end since the complex decays to methylated product at a rate of less than once every hundred minutes. This combination of high affinity and low turnover suggests that as more αPCM is formed by spontaneous aging, the enzyme could conceivably become self-occupied by its slow self-methylation reaction, interfering with the methylation and further metabolic processing of other age-damaged proteins.

CONCLUSIONS

Asp/Asn damage* affects numerous if not all proteins both in vitro and in vivo. Combined approaches, including the chemical reduction methods described here, can be used to measure Asp/Asn damage in a given protein. While protein engineering can enable the elimination of particularly troublesome sites that are prone to Asp/Asn damage, this may not be a complete answer to the practical problem of Asp/Asn damage since it is not uncommon for proteins to develop multiple sites of Asp/Asn damage, and in the long run, essentially every Asp or Asn residue could develop some degree of damage. Nature has evidently taken an active approach to solving the problem of Asp/Asn damage by selecting for the presence in most living cells of a PCM activity that specifically methylates and metabolizes Asp/Asn protein damage. PCM is being exploited increasingly as an in vitro tool to diagnose sites of Asp/Asn damage, and given the present good understanding of methylation-dependent repair of Asp/Asn damage it is now conceivable that a similarly active approach can be used to repair damaged proteins of pharmaceutical or industrial importance and to help ensure the preservation of activity of proteins during in vitro reactions of various kinds. Untangling the full complexity of the metabolism of intracellular Asp/Asn

* Asp/Asn damage is the subject of a recent excellent book, *Deamidation and Isoaspartate Formation in Peptides and Proteins*, Aswad, D.W., Ed., CRC Press, Boca Raton, FL 1995.

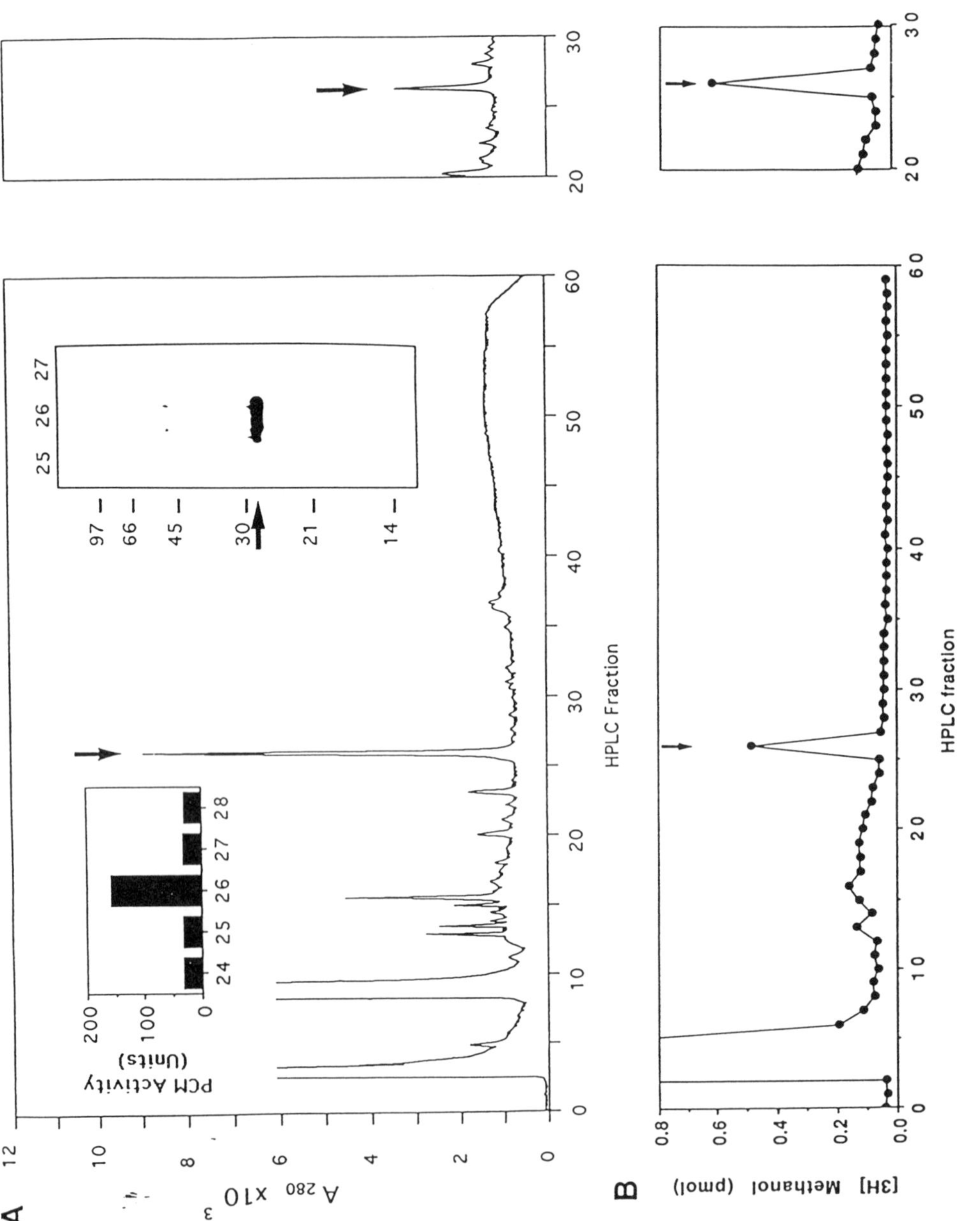
A
A 280 x10 3
PCM Activity (Units)
HPLC Fraction
B
[3H] Methanol (pmol)
HPLC fraction

Figure 7. High-pressure liquid chromatography of [^{3}H]automethylated PCM, showing that the acidic shoulder of PCM as purified by anion exchange chromatography is enriched in automethylatable αPCM. A) HPLC elution profile of UV-absorbing species after [^{3}H]automethylation of PCM. *Left chromatogram*: An aliquot (60 μl) from an anion-exchange fraction containing the peak of PCM activity was reacted with [^{3}H-*methyl*] S-adenosylmethionine for 18 hours and then injected onto a reversed phase (C_4) HPLC column. Elution by acetonitrile gradient yielded the chromatogram shown here of reaction components absorbing at 280 nm. The arrow marks the position of the sharp peak attributable to the PCM polypeptide, the other peaks being evident as background species in similarly analyzed reaction mixtures in which PCM was not present (data not shown). The *left inset* shows the units of PCM activity surviving the denaturing HPLC conditions in fractions 24-28 in a parallel HPLC analysis of nonradiolabeled PCM. The *right inset* shows Western blot detection of the PCM polypeptide (horizontal arrow; M_r=27,000) in HPLC fractions 25-27 by specific PCM antiserum and colorimetrically detectable alkaline-phosphatase-linked secondary antibody. A small amount of dimerized PCM (M_r=54,000) was also immunodetectable, with control experiments showing that it was formed after the enzyme incubation, probably during sample lyophilation. *Right chromatogram:* An aliquot (60 μl) from an anion-exchange fraction immediately following the peak of PCM activity, and hence enriched in the slightly more acidic (and thus presumably deamidated) αPCM, was similarly reacted with [^{3}H-*methyl*]S-adenosylmethionine. Only a ten-fraction window of the right-hand chromatogram is shown, as the rest of the chromatogram was essentially identical to the 60-fraction chromatogram on the left. *B) HPLC elution of [^{3}H]radioactivity.* HPLC fractions from the two runs in "A" were assayed for their content of [^{3}H] volatile radioactivity in a vapor-phase assay for protein methyl esters (Chelsky, et al., 1984) (●). From Lindquist and McFadden (1994a).

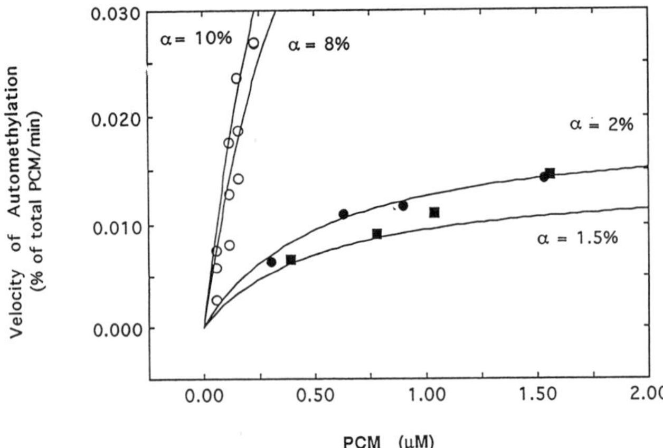

Figure 8. The effect of dilution on the rate of PCM automethylation. The initial rate of PCM automethylation (percent of the total PCM methylated per minute) was measured as a function of enzyme dilution. Since the reactions contained differing volumes of purified PCM, the additional volumes were replaced with buffer containing identical salts as in the purified PCM solution. The experiment was repeated using different samples of anion-exchange fractionated PCM, containing estimated percentages of αPCM as follows: ○ , 8-10% content of αPCM; ● , 2% αPCM; ■ , 1.5% αPCM. These estimates of αPCM content were made by measuring the final level of automethylation in extended time courses, and taking into account a small competing rate of ester hydrolysis that occurs in such time courses. Methyl ester formation in the PCM polypeptide chain was quantified following either HPLC separation (●) or following acidic gel electrophoresis (○, ■). The theoretical saturation curves were generated by the equation in the text, using $k_p = 0.0095\ min^{-1}$, $K_s = 0.50\ \mu M\ PCM_{tot}$, and the values for α shown in this figure.

damage is still in the future, though, and factors such as Asp/Asn damage in PCM itself and the resulting automethylation of PCM may have considerable importance in determining the effectiveness of the metabolic systems that process damaged protein.

REFERENCES

Atassi, M. Z., and Rosenthal, A. F., 1969 Specific reduction of carboxyl groups in peptides and proteins by diborane, *Biochem. J.*, 111: 593.

Bischoff, R., Lepage, P., Jaquinod, M., Cauet, G., Acker-Klein, M., Clesse, D., Laporte, M., Bayol, A., Van Dorsselaer, A., and Roitsch, C., 1993 Sequence-specific deamidation: Isolation and biochemical characterization of succinimide intermediates of recombinant hirudin, *Biochemistry*, 32: 725.

Cacia, J., Quan, C. P., Vasser, M., Sliwkowski, M. B., and Frenz, J., 1993 Protein sorting by high-performance liquid chromatography. I. Biomimetic interaction chromatography of recombinant human deoxyribonuclease I on polyionic stationary phases, *J. Chromatog.*, 634: 229.

Carter, D. A., and McFadden, P. N., 1994a Determination of β-isomerized aspartic acid as the corresponding alcohol, *J. Protein Chem.*, 13: 97.

Carter, D. A., and McFadden, P. N., 1994b Trapping succinimides in aged polypeptides by chemical reduction, *J. Protein Chem.*, 13: 89.

Chazin, W. J., Kordel, J., Thulin, E., Hofmann, T., Drakenberg, T., and Forsen, S., 1989 Identification of an isoaspartyl linkage formed upon deamidation of bovine calbindin D-9k and structural characterization by 2D 1-H NMR, *Biochemistry*, 28: 8646.

Daumy, G. O., Wilder, C. L., Merenda, J. M., McCall, A. S., Geoghegan, K. F., and Otterness, I. G., 1991 Reduction of biological activity of murine recombinant interleukin-1β by selective deamidation at asparagine-149, *FEBS Lett*, 278: 98.

George-Nascimento, C., Lowenson, J., Borissenko, M., Calderon, M., Medina-Selby, A., Kuo, J., Clarke, S., and Randolph, A., 1990 Replacement of a labile aspartyl residue increases the stability of human epidermal growth factor, *Biochemistry*, 29: 9584.

Johnson, B. A., Shirokawa, J. M., Hancock, W. S., Spellman, M. W., Basa, L. J., and Aswad, D. W., 1989 Formation of isoaspartate at two distinct sites during in vitro aging of human growth hormone, *J Biol Chem*, 264: 14262.

Kondo, Y., and Witkop, B., 1968 Reductive ring openings of glutarimides and barbiturates with sodium borohydride, *J. Org. Chem.*, 33: 206.

Kwong, M. Y., and Harris, R. J., 1994 Identification of succinimide sites in proteins by N-terminal sequence analysis after alkaline hydroxylamine cleavage, *Protein Sci.*, 3: 147.

Lindquist, J. A., and McFadden, P. N., 1994a Automethylation of protein (D-aspartyl/ L-isoaspartyl) carboxyl methyltransferase: A response to enzyme aging, *J. Protein Chem.*, 13: 23.

Lindquist, J. A., and McFadden, P. N., 1994b Incorporation of two ^{18}O atoms into a peptide during isoaspartyl repair reveals repeated passage through a succinimide intermediate., *J. Protein Chem.*, 13: 553.

McFadden, P. N., and Clarke, S., 1986 Chemical conversion of aspartyl peptides to isoaspartyl peptides: A method for generating new methyl-accepting substrates for the erythrocyte D-aspartyl/L-isoaspartyl protein methyltransferase, *J. Biol. Chem.*, 261: 11503.

McFadden, P. N., and Lindquist, J. A., 1994 A damaged subpopulation of protein (D-aspartyl/ L-isoaspartyl) carboxyl methyltransferase is methylated by a high-affinity, low-turnover reaction., *J. Protein. Chem.*, 13: 453.

Paranandi, M. V., Guzzetta, A. W., Hancock, W. S., and Aswad, D. W., 1994 Deamidation and isoaspartate formation during in vitro aging of recombinant tissue plasminogen activator, *J. Biol. Chem.*, 269: 243.

Roher, A. E., Lowenson, J. D., Clarke, S., Wolkow, C., Wang, R., Cotter, R. J., Reardon, I. M., Zurcher-Neely, H. A., Heinrikson, R. L., Ball, M. J., and Greenberg, B. D., 1993 Structural alterations in the peptide backbone of beta-amyloid core protein may account for its deposition and stability in Alzheimer's disease, *J. Biol. Chem.*, 268: 3072.

Rosenthal, A. F., and Atassi, M. Z., 1967 Specific reduction of carboxyl groups in peptides, *Biochem. Biophys. Acta*, 147: 410.

Sharma, S., Hammen, P. K., Anderson, J. W., Leung, A., Georges, F., Hengstenberg, W., Klevit, R. E., and Waygood, E. B., 1993 Deamidation of HPr, a phosphocarrier protein of the phosphoenolpyruvate: sugar phosphotransferase system, involves asparagine 38 (HPr-1) and asparagine 12 (HPr-2) in isoaspartyl acid formation, *J. Biol. Chem.*, 268: 17695.

Teshima, G., Porter, J., Yim, K., Ling, V., and Guzzetta, A., 1991 Deamidation of soluble CD4 at asparagine-52 results in reduced binding capacity for the HIV-1 envelope glycoprotein gp120, *Biochemistry*, 30: 3916.

Violand, B. N., Schlittler, M. R., Toren, P. C., and Siegel, N. R., 1990 Formation of isoaspartate-99 in bovine and porcine somatotropins, *J. Prot. Chem.*, 9: 109.

Wingfield, P. T., Mattaliano, R. J., MacDonald, H. R., Craig, S., Clore, G. M., Gronenborn, A. M., and Schmeissner, U., 1987 Recombinant-derived interleukin-1α stabilized against specific deamidation, *Protein Eng.*, 1: 413.

24

PURIFICATION SCHEME FOR ISOLATION AND IDENTIFICATION OF PEPTIDES CROSS-LINKED TO THE rRNA IN RIBOSOMES

Henning Urlaub, Volker Kruft,* and Brigitte Wittmann-Liebold

Max-Delbrück-Centrum für Molekulare Medizin, Proteinchemie
Robert-Rössle-Straße 10
D-13125 Berlin-Buch, Germany

INTRODUCTION

Ribosomes are complex organelles which contain 50 to 80 different ribosomal proteins (r-proteins) and several RNA molecules (rRNAs). Bacteria sediment at 70S with 30S and 50S subunits whereas eukaryotes have a sedimentation coefficient of 80S with 40S and 60S subunits. Accordingly, the ribosomes from different organisms vary considerably although their main function, namely to perform the translation of the genetic message into proteins is generally maintained. Without detailed knowledge of the molecular structure of the components and their topography within the organelle the translational processes in which more than 200 molecules are involved cannot be understood. Therefore, information on the nearest neighborhoods of the RNAs and the r-proteins and their individual domains is the basis for understanding the translational machinery on a molecular level.

Several approaches have led to general models on the topography in *E. coli* ribosomes (Brimacombe et al, 1990; Walleczek et al., 1988). These were assembled by: i, immuno-electron microscopy data obtained by binding studies of antibodies directed against individual purified ribosomal proteins to the bacterial ribosome and its subunits (Stöffler-Meilicke and Stöffler, 1990); ii, reconstitution assays employing individual components whereby the so-called assembly maps of the ribosomal constituents were established for the 30s and 50S subunit, respectively (Nomura and Held, 1976; Nierhaus and Dohme, 1974); iii, neutron scattering resulting in distances between various r-proteins within the subunit (Moore et al., 1986; Nowotny et al., 1986); iv, X-ray analysis on crystals of intact bacterial ribosomal particles yielding informations about the shape of the ribosome and the arrangement of structured elements within the subunits (Yonath et al, 1990; Wittmann et al., 1982); v,

* Present address: Applied Biosystems GmbH, Brunnenweg 13, D-64331 Weiterstadt, Germany.

Methods in Protein Structure Analysis, Edited by M. Z. Atassi and E. Appella
Plenum Press, New York, 1995

cross-linking experiments resulting in informations about protein-protein-, RNA-RNA-, and protein-RNA interactions.

Cross-linking of ribosomal RNA to ribosomal proteins is a straightforward approach to gain insight into the structural arrangement of these components within the complex. Many cross-link sites between ribosomal proteins and rRNA were analyzed at the rRNA level of *E. coli* (for review see Brimacombe, 1991) whereas only for ribosomal proteins S7 and L4 of *E. coli* the exact cross-link position to the rRNA has also been identified at the amino acid level (Ehresmann et al., 1976, Möller et al., 1978; Maly et al., 1980). On the other hand, more information about protein neighborhoods was gained by cross-linking proteins within the ribosome (Traut et al., 1986) and several cross-linked protein-protein positions were identified at the amino acid level (Allen et al., 1979; Pohl and Wittmann-Liebold, 1988; Brockmöller and Kamp, 1988; Herwig et al., 1993; Bergmann and Wittmann-Liebold, 1993).

Only by precisely determining the direct contact points between the constituents can the three dimensional structure of the ribosome be solved at the molecular basis. Without this knowledge the mechanism of the tranlational machinery cannot be understood.

Here we present a method for the analysis of rRNA-protein cross-links induced in ribosomal subunits from *E. coli* and *Bacillus stearothermophilus* at the amino acid and nucleotide level and we discuss the results obtained with various peptides cross-linked to the 16S or 23S RNAs in *E. coli* or *B. stearothermophilus* ribosomes.

EXPERIMENTS

Cross-Linking

Cross-linking of the ribosomal subunits was done with the heterobifunctional reagent 2-iminothiolane (Traut et al., 1973; Wower et al., 1981) or by mild UV irridiation for 10-15 min at a concentration of 5 A_{260} units/ml in a buffer consisting of 5mM mangnesium acetate, 50 mM KCL, 6mM β-mercaptoethanol, 10 mM Tris-HCl (pH 7.8) as described by Möller et al., 1978. After cross-linking ribosomal subunits were redissolved in 25mM Tris-buffer pH 7,8 containing 0.1 % SDS, 2 mM EDTA and 6 mM β-mercaptoethanol.

Isolation of Cross-Linked RNA-Proteins

The general strategy for the isolation of the proteins cross-linked to rRNA is given in Fig. 1.

Cross-linked proteins were separated from the non cross-linked moiety either by sucrose gradient centrifugation or by size exclusion chromatography on a S300 column (Pharmacia LKB Biotechnologie, Uppsala, Sweden) in the same buffer, see Figure 2. The proteins cross-linked to the rRNA eluted together with the non cross-linked rRNA moiety in the chromatogram.

Isolation of Cross-Linked Oligonucleotide-Peptide Heteromers

For digestion of the cross-linked ribosomal proteins the rRNA containing fractions were precipitated and redissolved in an appropriate volume of buffer. Proteins cross-linked to the rRNA were digested with endoproteases Lys-C, Glu-C or chymotrypsin. The remaining cross-linked peptides were separated from released peptides by size exclusion chromatography as described above. For identification of the cross-linked peptides in the fractions the rRNA was fully digested with ribonucleases A and T1 or partially treated with ribonuclease

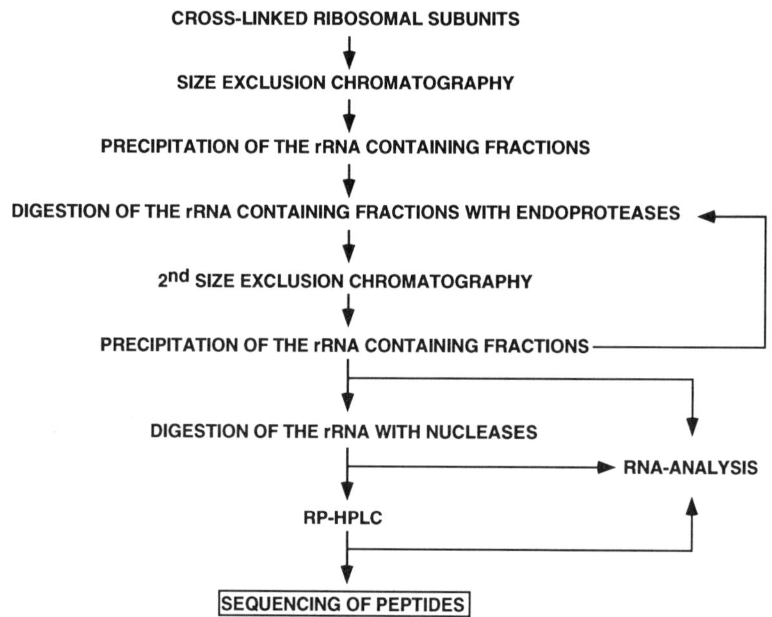

Figure 1. General strategy for the isolation of cross-linked peptides to the ribosomal rRNA.

T1 and injected directly onto a HPLC 100 RP-18 LiChrospher® endcapped column (250 x 4 mm, 5 µm, E. Merck, Darmstadt, Germany), see Figure 3.

Solvent A was water with 0.1% TFA, solvent B acetonitrile with 0.1% TFA. Fractions which showed absorption at 220 and 260 nm (corresponding to the cross-linked peptide-nucleotide portion) were sequenced directly in a Model 477A pulsed-liquid-gas-phase sequencer equipped with a model 120A amino acid analyzer (Applied Biosystems, Inc., Foster City, U. S. A.). Sequences were identified by comparison with known ribosomal sequences (Wittmann-Liebold et al., 1990) in the NBRF databank (National Biomedical Research Foundation, Washington DC, USA). A typical Edman degradation run is shown in Figure 4 for an individual peptide from protein BstL6 starting at position 149 with PTH-alanine cross-linked to an oligonucleotide of the 23S RNA. In the 8th degradation step the expected tyrosine-156 could not be detected due to its covalent attachment to the rRNA and is designed as X in Figure 4.

RESULTS

Cross-Linking Sites

Using the strategy above we were able to localize and sequence peptides of the ribosomal proteins S7 and S17 from the small ribosomal subunit and of L4, L6 and L14 from the large ribosomal subunit cross-linked to the rRNA as shown in Figure 5.

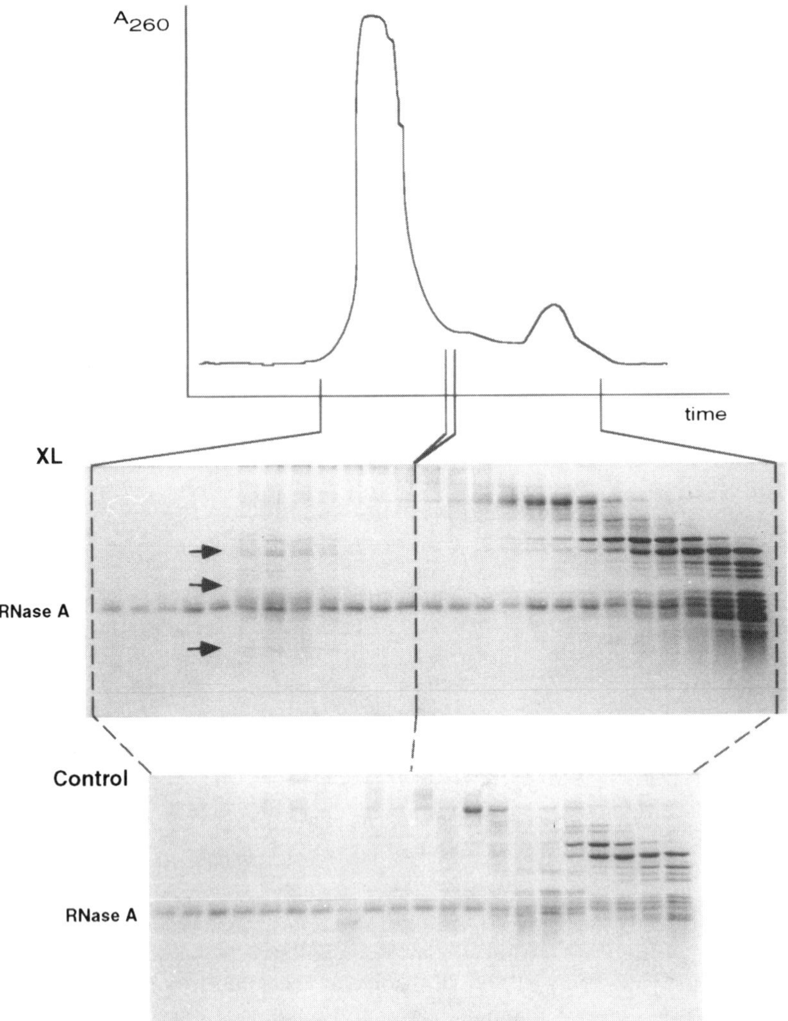

Figure 2. Separation of cross-linked ribosomal proteins of the large subunit from *B. stearothermophilus* from the non-cross-linked moiety by size exclusion. Cross-linked proteins were verified by SDS-PAGE. **XL:** SDS-PAGE of cross-linked ribosomal subunits after size exclusion chromatography, cross-linked proteins are indicated by arrows; **Control:** SDS-PAGE of non cross-linked ribosomal subunits after size exclusion chromatography. For SDS gel electrophoresis all aliquots from size exclusion chromatography were digested with 1 μg RNase A (present as one band in the lower part of the gels).

The Binding Domain of Protein S7 and S17 to the 16S RNA

From the 30S ribosomal subunit we isolated peptides of protein S7 (fragment positions 109-129 from *E. coli* and fragment positions 1-21 and 114-135 from *B. stearothermophilus*). During Edman degradation of the cross-linked fragments no PTH-methionine in position 114 of *E. coli* protein S7 and in position 115 of *B. stearothermophilus* protein S7 was detected, whereas the following amino acid residues in the sequence could be positively identified, indicating that these methionines are the cross-link sites to the rRNA-oligonucleotide. These data are in agreement with those of Möller et al. (1978), who found the

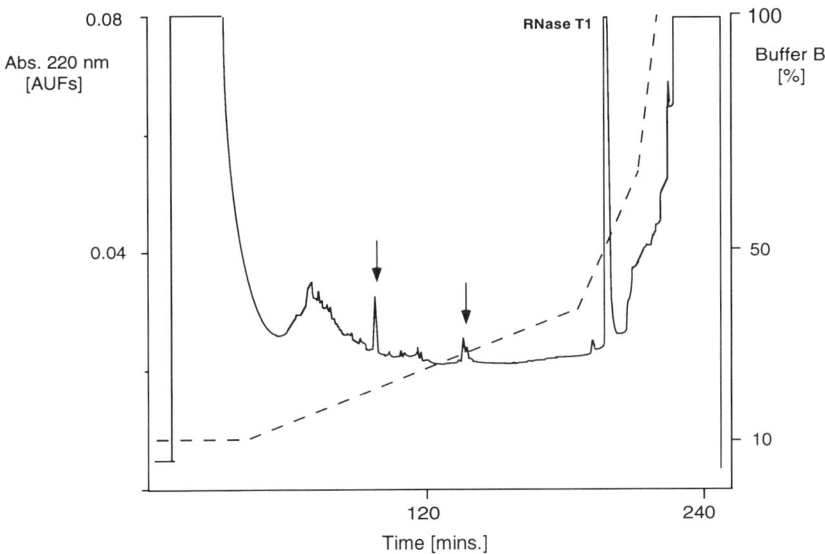

Figure 3. RP-HPLC chromatogram of cross-linked ribosomal peptides of the large ribosomal subunit from *B. stearothermophilus* after digestion with RNase T1.

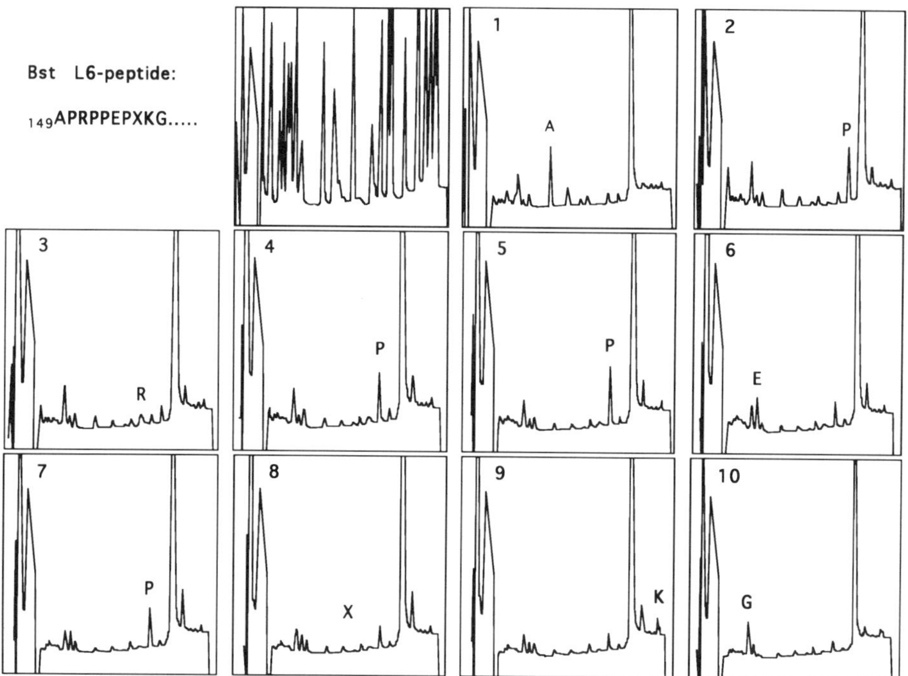

Figure 4. Individual Edman-degradation steps of an isolated cross-linked peptide of the ribosomal protein L6 from *B. stearothermophilus*. PTH-amino acids are given in the one letter code, X symbolize no detectable PTH-amino acids, due to its covalent attachment to the rRNA

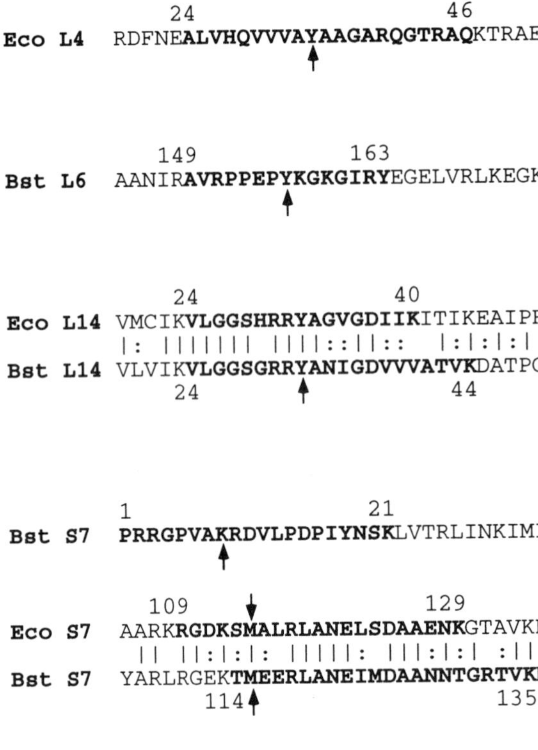

Figure 5. Identified homologions peptides (bold) with to the rRNA cross-linked amino acids (arrows) from ribosomal subunits of *E. coli* (Eco) and *B. stearothermophilus* (Bst).

cross-link between methionine-114 and uridine-1240 in the 16S rRNA of *E. coli*. In addition we determined the lysine in position 8 of *B. stearothermophilus* protein S7 as a second cross-link site. We also identified a fragment of protein S17 in both bacteria (at positions 19-35 in *E. coli* and at positions 21-37 in *B. stearothermophilus*) cross-linked at lysine 29 (*E. coli*) and 31 (*B. stearothermophilus*) to the 16S rRNA. This domain of S17 was assumed to be in close contact to the rRNA by ^{3}H-NMR and ^{15}N-NMR structure analysis (Golden et al., 1993b). Our data allow for the first time to resolve the S17 contact site to the 16S rRNA at the amino acid level.

Peptide-rRNA Cross-Links within the Large Ribosomal Subunit

Within the 50S ribosomal subunit of both bacteria we sequenced peptides of protein L14 from position 24 to position 44 in *B. stearothermophilus* and to position 39 in *E. coli*. Both peptides contain a tyrosine in position 32 but only in *B. stearothermophilus* this amino acid could be identified as the cross-link site to the 23S rRNA. In the same way we detected tyrosine-35 in protein L4 cross-linked to the 23S rRNA within the 50S ribosomal subunit of *E. coli*, which confirms the data of Maly et al. (1980) for the cross-link position of tyrosine-35 to uridine-615 in the 23S rRNA. As mentioned above we found tyrosine-156 in protein L6 from *B. stearothermophilus* cross-linked to the 23S rRNA. These data again agree well with the three dimensional structure of the protein. The recently published three dimensional structure of protein L6 shows tyrosine-156 within the putative RNA binding domain (Golden et al., 1993a).

CONCLUSIONS

The new approach directly determines amino acid residues of ribosomal proteins involved in interaction with the rRNA and thereby generates data on the structural organisation of the ribosome at the molecular level. By screening ribosomal subunits from both bacteria 30 other peptides cross-linked to the rRNA could be established on the molecular level (H. Urlaub, and B. Wittmann-Liebold, manuscript in preparation). Many of the peptides identified to be cross-linked to the rRNA are homologous in both organisms as shown for peptides derived from ribosomal proteins S7, S17 and L14 revealing a nearly identical topography of the contact sites in these organisms. Cross-linked amino acids found were basic residues particulary lysines after chemical reaction with 2-iminothiolane, and tyrosines and methionine after cross-linking via mild UV irridation.

The detected peptide sequences determined so far show no significant sequence similarities to other structural sequence elements found in RNA-complexes like the common RNP-motif (for review see Mattaj, 1989), although the isolated peptides are rich in basic residues (lysine, arginine), aromatic residues (especially tyrosines) and small hydrophobic amino acids (glycine, valine, leucine, and isoleucines). Furthermore, the precise analysis of the cross-link site at the amino acid level allows to substantiate whether one or more domains of a ribosomal protein are in direct contact to the rRNA and which secondary structure elements are involved. Experiments to analyze the oligonucleotide part of the various sequenced cross-linked peptides are in progress in order to determine the corresponding nucleotide sites on the rRNAs.

ACKNOWLEDGMENTS

This work was supported by a grant to Dr. B. W.-L. from the Deutsche Forschungsgemeinschaft (SFB 344, YE6).

REFERENCES

Allen, G., Capasso, R., Gualerzi, C., 1979 Identification of the amino acid residue of proteins S5 and S8 adjacent to each other in the 30S ribosomal subunit of *Escherichia coli*, *J. Biol. Chem.* 254:9800-9806.

Bergmann, U., and Wittmann-Liebold, B., 1993 Localization of proteins HL29 and HL31 from *Haloarcula marismortui* within the 50S ribosomal subunit by chemical crosslinking, *J. Mol. Biol.* 232:693-700.

Brimacombe, R., 1991 RNA-protein interaction in the *E. coli* ribosome, *Biochimie* 73:927-936.

Brimacombe, R., Greuer, B., Mitchell, P., Osswald, M., Rinke-Appel, J., Schüler, D., and Stade, K., 1990 Three-dimensional structure and function of *Escherichia coli* 16S and 23S rRNA as studied by cross-link techniques, in: The Ribosome Structure Function and Evolution, Hill, W. E., Dahlberg, A., Garrett, R. A., Moore, P. B., Schlessinger, D., and Warner, J. E., eds., Am. Soc. Microbiol., Washington, D.C., pp.93-106.

Brockmöller, J., and Kamp,. R. M., 1988 Cross-linked amino acids in the protein pair S13-S19 and sequence analysis of protein S13 of *Bacillus steraothermophilus* ribosomes, *Biochemistry* 27:3372-3381.

Ehresmann, B., Reinbolt, J., Backendorf, C., Tritsch, D., and Ebel, J. P., 1976 Studies of the binding of *Escherichia coli* ribosomal protein S7 with the 16S RNA by ultraviolet irridation, *FEBS Lett.* 67:316-319.

Golden, B. L., Ramankrishnan, V., and White, S. W., 1993a Ribosomal protein L6: structural evidence of gene duplication from a putative RNA binding protein, *EMBO J.*, 12:4901-4908.

Golden, B. L., Hoffman, D. W., Ramankrishnan, V., and White, S. W., 1993b Ribosomal protein S17: Characterization of the three-dimensional structure by ^{1}H and ^{15}N NMR, *Biochemistry* 32:12812-12820.

Herwig, S., Kruft, V., Eckart, K., and Wittmann-Liebold, B., 1993 Crosslinked amino acids in the protein pairs L3-L9 and L23-L29 of *Bacillus stearothermophilus* ribosomes after treatment with diepoxybutane, *J. Biol. Mol.* 268:4643-4650.

Maly, P., Rinke, J., Ulmer, E., Zwieb, C., and Brimacombe, R., 1980 Precise location of the site of crosslinking between protein L4 and 23S RNA induced by mild ultraviolet irridation of *Escherichia coli* 50S ribosomal subunits, *Biochemistry* 19:4179-4188.

Mattaj, I. W., 1989 A binding consensus: RNA-protein interactions in splicing, snPNPs, and sex, *Cell* 57:1-3.

Möller, K., Zwieb, C., and Brimacombe, R., 1978 Identification of the oligonucleotide and oligopeptide involved in an RNA-protein crosslink induced by UV irridation of *E. coli* 30S, *J. Mol. Biol.* 126:489-506.

Moore, P. B., Capel, M., Kjeldgaard, M and Engelmann, D. M., 1986 A 19 protein map of the 30S ribosomal subunit of *Escherichia coli*, in: Structure Function and Genetics of Ribosomes, Hardesty, B. and Kramer, G., eds., Springer Verlag, Heidelberg, New York, pp.87-100.

Nowotny, V., May, P. R., and Nierhaus, K. H., 1986 Neutron scattering analysis of structural and functional aspects of the ribosome: The strategy of the glassy ribosome, in: Structure Function and Genetics of Ribosomes, Hardesty, B. and Kramer, G., eds., Springer Verlag, Heidelberg, New York, pp.101-111.

Nierhaus, K. H., and Dohme, F., 1974 Total reconstitution of functionally active 50S ribosomal subunits from *Escherichia coli*, *Proc. Nat. Acad. Sci. USA* 71:4713-4717.

Nomura, M., and Held, W. A., 1976 Reconstitution of ribosomes: Studies of ribosome structure, function, and functional aspects of the ribosome, in: Ribosomes, Nomura, M., Tissières, A., and Lengyel, P., eds., Cold Spring Habour Labratory, New York, pp.193-124.

Pohl, T., and Wittmann-Liebold, B., 1988 Identification of a crosslink in *Escherichia coli* ribosomal protein pair S13-S19 at the amino acid level, *J. Biol. Chem.* 263:4293-4301.

Stöffler-Meilicke, M., and Stöffler, G., 1990 Topography of the ribosomal proteins from *Escherichia coli* within the intact subunits as determined by immunoelectron mikroscopy and protein-protein cross-linking, in: The Ribosome Structure Function and Evolution, Hill, W. E., Dahlberg, A., Garrett, R. A., Moore, P. B., Schlessinger, D., and Warner, J. E., eds., Am. Soc. Microbiol., Washington, D.C., pp.123-133.

Traut, R. R., Bollen, A., Sun, T. T., Hershey, J. W. B., Sundberg, J., and Pierce, L. R., 1973 Methyl 4-mercaptobutyrimidate as a cleavable cross-linking reagent and its application to the *Escherichia coli* 30S ribosome, *Biochemistry* 12:3266-3272.

Traut, R. R., Tewari, D. S., Sommer, A., Gavino, G. R., Olson, H. M., and Glitz, D. G., 1986 Protein topography of ribosomal functional domains: Effects of monoclonal antibodies to different epitopes in *Escherichia coli* protein L7/L12 on ribosome function and structure, in: Structure Function and Genetics of Ribosomes , Hardesty, B. and Kramer, G., eds., Springer Verlag, Heidelberg, New York, pp.286-308.

Urlaub, H., and Wittmann-Liebold, B.; manuscript in preparation

Walleczek, J., Schüler, D., Stöffler-Meilicke, M., Brimacombe, R., and Stöffler, G. 1988 A model for the spatial arrangement of the proteins in the large ribosomal subunit of the *Escherichia coli* ribosome, *EMBO J.* 7:3571-3576.

Wittmann, H. G., Müssig, J., Gewitz, H. S., Rheinberger H. J., and Yonath, A., 1982 Crystallisation of *Escherichia coli* ribosomes, *FEBS Lett.* 146:217-220.

Wittmann-Liebold, B., Köpke, A. K. E., Arndt, E., Krömer, W., Hatakeyama, T., and Wittmann, H. G., 1990 Sequence comparison and evolution of ribosomal proteins and their genes, in: The Ribosome Structure Function and Evolution, Hill, W. E., Dahlberg, A., Garrett, R. A., Moore, P. B., Schlessinger, D., and Warner, J. E., eds., Am. Soc. Microbiol., Washington, D.C., pp.598-616.

Wower, I., Wower, J., Meinke, M., and Brimacombe, R., 1981 The use of 2-iminothiolane as an RNA-protein cross-link agent in *Escherichia coli* ribosomes, and the location on 23S RNA of sites cross-linked to proteins L4, L6, L21, L23, L27 and L29, *Nucleic Acids Res.* 9:4285-4302.

Yonath, A., Bennett, W., Weinstein, S., and Wittmann, H. G., 1990 Crystallography and image reconstructions of ribosomes, in: The Ribosome Structure Function and Evolution, Hill, W. E., Dahlberg, A., Garrett, R. A., Moore, P. B., Schlessinger, D., and Warner, J. E., eds., Am. Soc. Microbiol., Washington, D.C., pp.134-147.

25

INVESTIGATION OF PROTEIN-PROTEIN INTERACTIONS IN MITOCHONDRIAL STEROID HYDROXYLASE SYSTEMS USING SITE-DIRECTED MUTAGENESIS

Rita Bernhardt,[1,2] Regine Kraft,[1,3] Heike Uhlmann,[1] and Vita Beckert[1]

[1] Max-Delbrück-Centrum für Molekulare Medizin
Robert-Rössle-Str. 10, D-13125 Berlin
[2] Freie Universität Berlin
FB Chemie, Institut für Biochemie
Thielallee 63, D-14195 Berlin
[3] Humboldt-Universität Berlin
FB Chemie
Hessische Str. 1-2, D-10115 Berlin

INTRODUCTION

Adrenodoxin belongs to the family of /2Fe-2S/ type ferredoxins being widely distributed in bacteria, plants and animals. Although adrenodoxin is a small (~14 kDa) and soluble protein, its three-dimensional structure has not been elucidated as yet. It functions as an electron carrier from the FAD-containing NADPH-dependent ferredoxin reductase to the cytochromes P450scc (CYP11A1), which catalyzes the side-chain cleavage of cholesterol, the initial step in adrenal steroidogenesis, and $P450_{11\beta}$ (CYP11B1), being involved in the formation of cortisol and aldosterone (Fig. 1).

Different models of electron transfer via adrenodoxin have been proposed, the "shuttle" model (Hanukoglu and Jefcoate, 1980), a ternary complex formation of adrenodoxin reductase, adrenodoxin, and the cytochrome P450 (Kido and Kimura, 1979), and a model suggesting the occurrence of two adrenodoxin molecules in the electron transport chain (Hara et al., 1994).

Recognition and interaction of adrenodoxin with adrenodoxin reductase and CYP11A1 was shown to be mainly of an electrostatic nature. Replacement of acidic amino acids 76 and 79, which have been proposed to be involved in protein interaction on the basis of chemical modification studies (Geren et al., 1984), with neutral amino acids resulted in a considerable decrease in activity and in a decreased affinity of mutant adrenodoxins for their reaction partners (Coghlan and Vickery, 1991). This result is supported by mutation of the conserved lysine residues 377 and 381 of CYP11A1 to either neutral or negative amino

Methods in Protein Structure Analysis, Edited by M. Z. Atassi and E. Appella
Plenum Press, New York, 1995

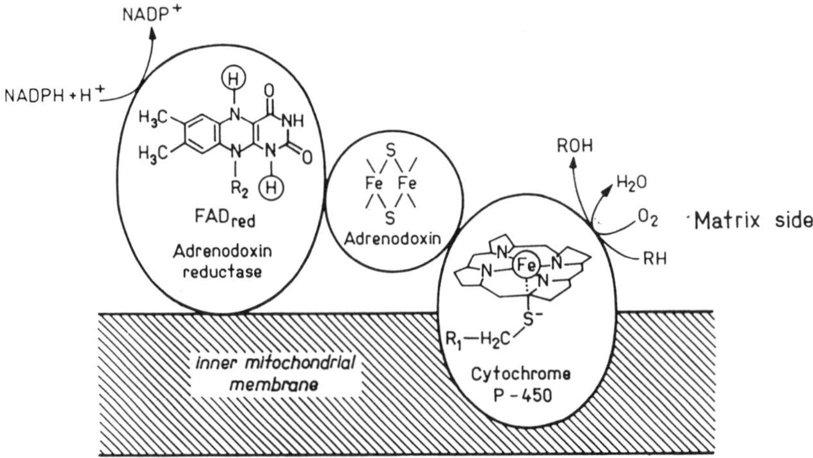

Figure 1. Mitochondrial steroid hydroxylase system.

acids. The mutants have been shown to cause greatly increased K_d values for adrenodoxin binding, indicating that these lysines are the key sites in binding of bovine adrenodoxin by CYP11A1 (Wada and Waterman, 1992). There are only a few reports on the involvement of other amino acid residues into recognition and interaction site of adrenodoxin with the reductase and P450s.

Adrenodoxin is synthesized in the cytoplasm as a large precursor molecule. The 58 amino acids of the N-terminal leader peptide are processed upon mitochondrial uptake (Nabi et al., 1983, Sagara et al., 1984, Matocha and Waterman, 1984). The primary structure of mature adrenodoxin was first elucidated by amino acid sequencing (Tanaka et al., 1973) and shown to consist of 114 residues. Later, a 14 amino acid C-terminal extension peptide was found in the nucleotide sequence of adrenodoxin cDNA (Okamura et al., 1985), so the mature full-length adrenodoxin contains 128 amino acids. Western blotting using an antibody against a peptide consisting of C-terminal amino acids 115-128 of adrenodoxin revealed the presence of an adrenodoxin longer than 114 amino acids in adrenocortical mitochondria (Bhasker et al., 1987).

When isolating adrenodoxin from bovine adrenals (Driscoll and Omdahl, 1986, Hiwatashi et al., 1986, Sagara et al., 1984, Sagara et al., 1992, Sakihama et al., 1988), multiple forms of the protein have been observed. Proteinchemical analysis revealed different sizes of the C-termini, varying in length from 114 amino acids (Tanaka et al., 1973), 121, 124, and 125 amino acid residues (Hiwatashi et al., 1986) up to 127 amino acids (Sakihama, et al., 1988). However, when bovine adrenodoxin was purified in the presence of protease inhibitors, a protein consisting of 128 amino acids could be obtained as determined by carboxypeptidase Y digestion (Cupp and Vickery, 1989).

In-vitro studies have shown that adrenodoxin, from which amino acids 116-128 were removed by trypsin cleavage, revealed an identical UV/vis spectrum in comparison to that of native adrenodoxin (Cupp and Vickery, 1989). Furthermore, it has been demonstrated that adrenodoxin lacking residues 116-128 exhibited higher biological activity towards CYP11A1 and CYP11B1 and higher affinity to CYP11A1 as compared to the full-length molecule, but interaction with adrenodoxin reductase was not significantly affected (Cupp and Vickery, 1989).

To systematically study the role of the C-terminal region of adrenodoxin and of different amino acid residues in interaction with the electron donor, adrenodoxin reductase, and the electron acceptors, CYP11A1 and CYP11B1, mutants of adrenodoxin have been prepared by site-directed mutagenesis, expressed in *E. coli* as described (Uhlmann et al., 1992), and their structural and functional properties have been characterized in detail.

MATERIALS AND METHODS

E. coli strains HB101 and BL21 were used as host strains. Site-directed mutagenesis and synthesis of deletion mutants were performed using PCR as described recently (Uhlmann et al., 1992, Beckert et al., 1994). Mutants of adrenodoxin were expressed in a high-level expression system using the expression vector pKKAdx (Uhlmann et al. 1992). Bacteria were grown and recombinant adrenodoxin was purified as previously described (Uhlmann et al., 1992). Adrenodoxin reductase, CYP11A1 and CYP11B1 were isolated from bovine adrenals according to Akhrem et al. (1979). Proteins were analyzed as described (Uhlmann et al., 1994). The amino acid compositions of the recombinant proteins were determined by vaporphase hydrolysis in a SYKAM analyzer. N-termini were analyzed using a 477A gas phase sequenator (Applied Biosystems). Mass spectrometry measurements were carried out on a FINNIGAN MAT triplestage quadrupole TSQ 700 instrument. EPR spectroscopy was carried out at -163°C on a Varian E3 spectrometer using whole *E. coli* cells with expressed proteins which were reduced by dithionite. CD spectra were recorded on a Jasco J720 spectropolarimeter at room temperature in the ultraviolet and visible region. Fluorescence spectra were taken on an RF-5001 PC spectrofluorometer at room temperature. The exciting wavelength was 270 nm. Redox potentials of adrenodoxin mutants were measured by the dye photoreduction method with Safranin T as indicator and mediator (Sligar et al., 1979). Data were analyzed according to the Nernst equation. Cytochrome c reduction was assayed in 50 mM potassium phosphate buffer, pH 7.4, containing 0.1 % Tween 20 at room temperature. Reaction mixtures with the deletion mutants of adrenodoxin contained 0.05 μM adrenodoxin reductase, 65 μM horse heart cytochrome c, various amounts of the respective adrenodoxin and 140 μM NADPH. The mixtures with the substitution mutants contained 0.2μM adrenodoxin reductase, 100 μM cytochrome c, variable amounts of recombinant adrenodoxin and 100 μM NADPH. The reduction of cytochrome c was monitored at 550 nm. Calculations based on a molar extinction coefficient ε of 20 $(mM \cdot cm)^{-1}$ for cytochrome c.

The cholesterol side-chain cleavage activity was measured in the reconstituted assay system according to Sugano et al. (1989). The incubation mixtures contained 0.5 μM CYP11A1, 0.2 μM adrenodoxin reductase, 100 μM cholesterol, and a NADPH-regenerating system for the substitution mutants of adrenodoxin (Beckert et al., 1994). Reaction mixtures of the deletion mutants were composed in the same way except that they contained 0.5 μM adrenodoxin reductase (Uhlmann et al., 1994). NADPH was added to start the reaction. After a 10 minute-incubation at 37°C, the reaction was stopped by boiling for 5 minutes. The steroids were converted into their corresponding 3-one-4-ene forms by adding cholesterol oxidase to the reaction mixtures. The steroids were extracted with dichloromethane and analyzed by reverse-phase HPLC.

CYP11B1 dependent 11β-hydroxylation of deoxycorticosterone was performed as described (Beckert et al., 1994, Uhlmann et al., 1994) with 0.4 μM adrenodoxin reductase, 0.4 μM CYP11B1, 100 μM deoxycorticosterone, various amounts of adrenodoxin, and a NADPH regenerating system. The reaction was started by addition of NADPH and carried out for 10 minutes at 37°C. Dichloromethane was used to stop the reaction and to extract the steroids, which were then analyzed by HPLC.

Binding affinity of adrenodoxin to CYP11A1 was measured by affinity chromatography on biotin-labelled CYP11A1 bound to an avidin-sepharose column. 10 nmol of the appropriate adrenodoxin sample were loaded onto the column, followed by washing to remove unbound adrenodoxin. A KCl gradient (0 - 0.7 M; flow rate, 0.3 ml/min) was used to elute bound adrenodoxin. The bound amount of the respective adrenodoxin mutant was determined assuming the binding capacity of wild type adrenodoxin as 100 %. Differential spectral titration was performed according to Kido & Kimura (1979). Binding of cholesterol to CYP11A1, facilitated by the binding of adrenodoxin, causes absorbance changes in the Soret region (393 - 417 nm) of the cytochrome due to conversion of CYP11A1 from its low to its high spin form.

RESULTS

The role of the C-terminal region of adrenodoxin was studied by analyzing deletion mutants 4-128, 4-114, 4-108, and 4-107, lacking amino acids 1-3, 1-3 and 115-128, 1-3 and 109-128, or 1-3 and 108-128, respectively. In addition to the deletion of the C-terminal peptides, amino acids 1-3 have been removed to avoid proteolytic cleavage at the N-terminus and to study the influence of these residues on adrenodoxin structure and function. To check whether partial proteolytic digestion of the mutants occurs, all mutant proteins were analyzed with respect to amino acid composition, mass spectrometry as well as N- and C- terminal microsequencing. The mutants were shown to be of the expected composition (Table 1), but contained an additional methionine at the first position resulting from an uncleaved start codon.

Table 1. Amino acid analysis of wild type adrenodoxin and adrenodoxin mutants

Amino acid	Adrenodoxin wild type th*	Adrenodoxin wild type obs†	Adrenodoxin mutants Y82F obs	Y82S obs	Y82L obs	4-114 obs	4-108 obs
Asx	20	20.0	19.8	19.1	18.8	17.1	18.6
Thr	10	9.5	9.0	8.8	8.4	9.8	10.8
Ser	9	8.7	7.4	6.7	5.5	4.2	2.4
Glx	13	13.1	12.0	11.5	11.0	10.7	11.9
Gly	10	10.3	10.3	9.7	9.4	8.2	9.0
Ala	7	7.0	7.0	7.0	7.0	7.0	5.0
Val	7	7.7	6.3	6.3	5.7	6.5	6.0
Met	5	5.5	5.4	4.4	4.7	4.0	4.6
Ile	10	9.3	8.8	7.6	7.6	7.5	8.3
Leu	12	11.9	12.4	12.2	13.4	12.2	13.9
Tyr	1	1.2	—	—	—	1.1	1.0
Phe	4	4.0	5.1	4.1	3.9	4.1	4.3
His	3	3.1	3.1	3.1	2.9	3.1	3.3
Lys	6	5.7	6.1	5.8	5.3	4.9	4.9
Arg	5	4.7	5.0	5.2	4.9	4.0	3.8
Pro	1	n. d.	n. d.	n. d.	n. d.	n. d.	n. d.
Cys	5	n. d.	n. d.	n. d.	n. d.	n. d.	n. d.
Total	128		128	128	128	112	106

*th = theoretical number.
†obs = observed number of the respective amino acid.

Table 2. Mass spectrometry of adrenodoxin mutants

Adx	Theoretical mass (Da)	Observed mass (Da)
Wild type	14017.8	14015.5 ± 2.2
Y82F	14001.8	14001.1 ± 1.6
4-128	13917.7	13916.4 ± 2.5
4-114	12338.9	12338.0 ± 1.5
4-108	11780.4	11777.7 ± 1.5

A highly purified sample (A414/A276>0.9) of each adrenodoxin mutant was desalinated and dissolved in methanol/water (1:1), 1 % acetic acid to a final concentration of 10 pmol/µl. Mass spectra were recorded on a FINNIGAN MAT triplestage quadrupole TSQ 700 instrument

No proteolytic digestion has been observed, even in the case of mutant 4-128 (Table 2) when freshly purified proteins were analyzed. In contrast, native adrenodoxin was shown to undergo proteolytic digestion (Driscoll and Omdahl, 1986).

Deletion of amino acids 1-3 did not lead to any significant changes of the structure and function of adrenodoxin. The absorption spectra of all mutants studied were identical to that of the wild type. However, EPR, CD, and redox potential measurements of mutants 4-114 and 4-108 revealed that the structure of these mutants differs from that of wild type adrenodoxin. EPR spectra of adrenodoxin are characterized by two g-values: $g_\perp = 1.94$ and $g_\parallel = 2.03$ (Uhlmann et al., 1992). The deletion mutants 4-114 and 4-108 showed signals, where the position of $g_\perp$ was identical to that of native adrenodoxin, but broadened, while $g_\parallel$ was shifted to a smaller value. The CD signals of these mutants were increased in all three wavelenght ranges measured (absorption of the peptide region, aromatic residues, and iron-sulfur cluster). The molar ellipticity increases from ± 0 (wild type) to 2.600 $deg \cdot cm^2 \cdot dmol^{-1}$ (mutant 4-114) and 5.500 $deg \cdot cm^2 \cdot dmol^{-1}$ (mutant 4-108) at 195 nm. In addition, the redox potentials of these mutants were lower than that of wild type adrenodoxin. Furthermore, mutant 4-107, lacking the single proline residue contained in adrenodoxin, P108, did not show EPR signals indicating that P108 plays an essential role for the assembly of the /2Fe-2S/ cluster. Deletion of residues 115-128 or 109-128 did not essentially affect the interaction with the electron donor adrenodoxin reductase as shown by nearly unchanged cytochrome c reduction activity. Although this reaction does not occur physiologically, it is a widely used model for the electron transfer from reduced adrenodoxin reductase to adrenodoxin since the flavin-to-iron electron transfer appears to be the rate-limiting step in cytochrome c reduction (Lambeth and Kamin, 1979). In contrast, interaction with the electron acceptors, CYP11A1 and CYP11B1, was influenced.

In CYP11A1-dependent cholesterol conversion, mutants 4-108 and 4-114 exhibited 3-fold and 5-fold decreased K_m values (Fig. 2A, Table 3), respectively, while the binding affinity for CYP11A1 raised nearly 3-fold and 2-fold, respectively (Fig. 3).

The V_{max} values did not change upon deletion of the C-terminal region. When measuring the CYP11B1-dependent conversion of deoxycorticosterone to corticosterone, mutants 4-108 and 4-114 again showed decreased K_m values (6-fold and 3-fold, respectively, Fig. 2B, Table 3). In this reaction, however, also the V_{max} values increased, being 5.5 nmol product/min/nmol CYP11B1 for wild type adrenodoxin, 11.8 nmol product/min/nmol CYP11B1 for mutant 4-114, and 19.7 nmol product/min/nmol CYP11B1 for mutant 4-108.

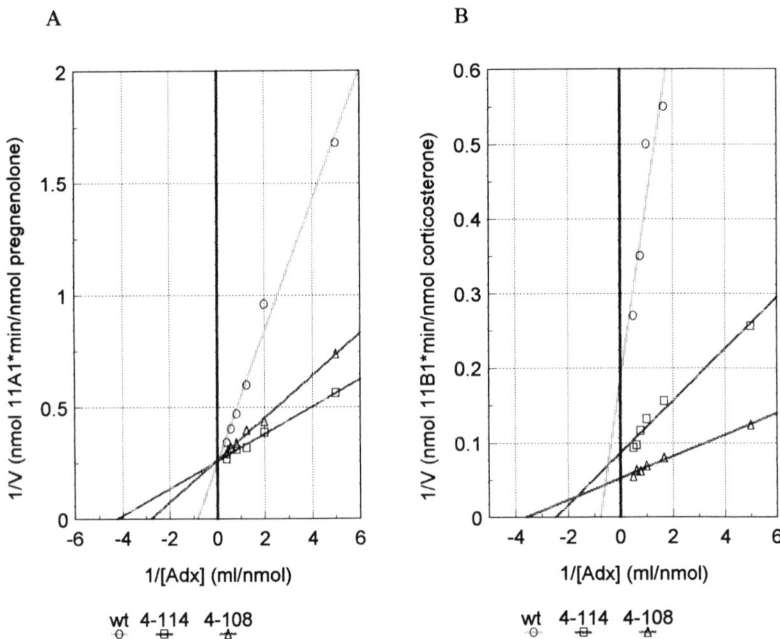

Figure 2. CYP11A1 and CYP11B1 dependent hydroxylation assays of the deletion mutants of adrenodoxin. (A) CYP11A1 dependent cholesterol side-chain cleavage, producing pregnenolone. (B) CYP11B1 dependent 11β-hydroxylation of deoxycorticosterone. The reactions were carried out in a reconstituted system, containing a NADPH regenerating system, adrenodoxin reductase, variable amounts of adrenodoxin, the cytochrome P450 (CYP11A1 or CYP11B1), and the respective substrate.

Table 3. Michaelis constants of the adrenodoxin mutants

Adrenodoxin mutant	Cytochrome c reduction assay K_m (nM)	CYP11A1 assay K_m (μM)	CYP11B1 assay K_m (μM)
Wild type	5.9 ± 0.9	1.1 ± 0.1	1.2 ± 0.2
4-128	5.6 ± 0.5	1.1 ± 0.1	1.3 ± 0.1
4-114	3.4 ± 0.2	0.2 ± 0.02	0.4 ± 0.04
4-108	7.3 ± 1.0	0.4 ± 0.04	0.2 ± 0.02
Wild type	6.4 ± 0.1	1.8 ± 0.05	1.7 ± 0.24
Y82F	4.7 ± 0.2	1.2 ± 0.04	0.4 ± 0.03
Y82S	5.8 ± 0.2	2.5 ± 0.07	4.6 ± 0.15
Y82L	6.1 ± 0.2	2.1 ± 0.05	0.7 ± 0.09
Wild type	19.1 ± 0.2	2.5 ± 0.07	2.0 ± 0.15
H56Q	40.2 ± 0.5	3.2 ± 0.04	12.2 ± 0.62
H56T	54.2 ± 0.4	3.4 ± 0.06	8.1 ± 0.25

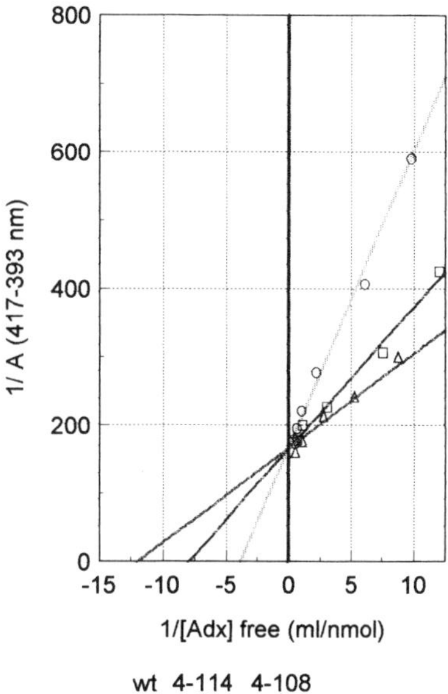

Figure 3. Binding affinity of the adrenodoxin deletion mutants to CYP11A1, determined by optical titration according to Kido & Kimura (1979). The absorption change of CYP11A1 in the Soret region (393 - 417 nm), which was induced by binding of cholesterol facilitated by binding of adrenodoxin, was monitored.

The data suggest that the electron transfer-coupled interaction of adrenodoxin with CYP11A1 and CYP11B1 is determined at least in part by different features of the cytochromes. This observation is further supported by site-directed mutagenesis studies of amino acid residues Y82 and H56.

As shown in Table 1, adrenodoxin does not contain tryptophan and contains only one tyrosine residue, which is in position 82. This unique tyrosine of adrenodoxin, which previously had been proposed to be involved in reductase binding and/or electron transfer by chemical modification studies (Taniguchi and Kimura, 1975, 1976) was replaced by phenylalanine, leucine or serine. Again the mutants were tested for amino acid composition (Table 1), mass spectrometry (Table 2), the structural integrity of the iron-sulfur cluster, the function in enzymatic assays (with cytochromes c, CYP11A1 and CYP11B1 as electron acceptors) and the binding to CYP11A1. As demonstrated in Table 1, the mutations can be clearly identified on the protein level when analyzing the amino acid composition of wild type and mutant proteins. Amino acid composition and mass spectrometry of mutant Y82F (Tables 1, 2) revealed that there was no proteoloytic digestion of the mutant protein. N-terminal microsequencing revealed residues GSSEDK, corresponding to amino acids 2-6 of native adrenodoxin with a substitution of the N-terminal serine to glycine, which was engineered to improve the proteolytic stability of the protein. Unchanged absorption, CD and EPR spectra as well as redox potential measurements indicate that the environment of the /2Fe-2S/ cluster was not affected by the mutations (Beckert et al., 1994). Replacement of Y82 also did not affect adrenodoxin reductase binding as shown by unchanged cytochrome c activity. Determination of the hydroxylating activities of CYP11A1 and CYP11B1 reconstituted with adrenodoxin mutants, however, indicated marked changes in the K_m values up to 4-fold (Table 2) at an unchanged V_{max} value. These changes again differ in dependence on the P450 (CYP11A1 or CYP11B1) used, exerting a more pronounced effect of tyrosine replacement on interaction with CYP11B1 as compared to CYP11A1.

Since H56 has been supposed to be proximal to the domain between E74 and D86, being involved in ferredoxin reductase and P450 binding by adrenodoxin (Miura and Ichikawa, 1991), it was checked whether replacement of H56 by other amino acids would

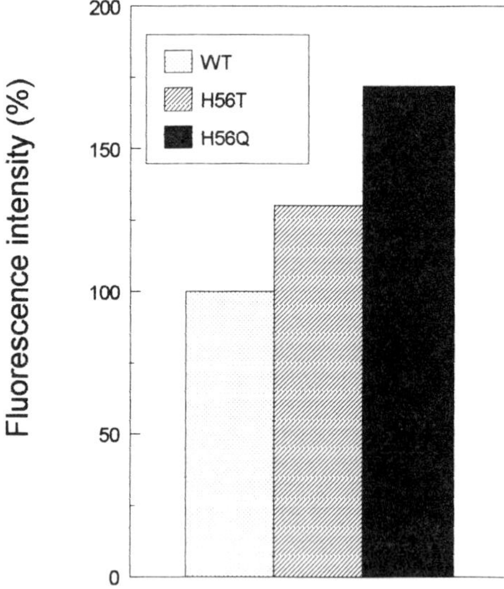

Figure 4. Intensity of Y82 fluorescence of wild type adrenodoxin and H56 mutants. Fluorescence measurements of 80 μM wild type adrenodoxin and H56 mutants were carried out in 10 mM potassium phosphate buffer, pH 7.4 at room temperature by exciting at 270 nm. The fluorescence intensity of wild type adrenodoxin is assumed as 100%.

change interaction with adrenodoxin reductase, CYP11A1 and CYP11B1. At first it was investigated whether replacements in position 56 would affect the fluorescence of Y82. In fact, it could be shown that replacement of H56 by glutamine and threonine causes changes in the intensity of adrenodoxin tyrosine fluorescence (Fig. 4). Thus, H56 appears to be in the immediate vicinity of Y82 and therefore the intermolecular interface of adrenodoxin with its redox partners.

Furthermore, it could be demonstrated that replacement of H56 by the indicated amino acids lead to changes in adrenodoxin reductase binding as indicated by an increased K_m value for cytochrome c reduction, but also to changes in the K_m values in CYP11A1 and CYP11B1-dependent substrate conversions (Table 3).

DISCUSSION

The importance of various amino acid residues as well as the N- and C-terminal regions of adrenodoxin for its structure and function as electron mediator in mitochondrial steroid hydroxylases has been investigated using site-directed and deletion mutants. The effect of the N- and C-terminal regions on the interaction of adrenodoxin with various redox partners, the role of the unique proline residue 108 of adrenodoxin, which is conserved among vertebrate ferredoxins (Usanov et al., 1990) for protein structure and function, and the importance of tyrosine 82 and histidine 56 in protein-protein interaction have especially been considered. It could be clearly demonstrated that analysis of functional and structural properties of proteins changed by site-directed mutagenesis requires a thorough analysis of the desired changes.

The mutants were characterized by nucleotide sequencing as well as by N-terminal microsequencing, estimation of the amino acid composition and mass spectrometry. As seen in Tables 1 and 2, the C-terminal region of wild type adrenodoxin and of the mutants as well did not undergo proteolytical degradation under our experimental conditions. Since the deletion mutants, in addition to the C-terminal deletion, lack the N-terminal three serines to avoid partial cleavage, which has been observed in previous studies on native adrenodoxin (Coghlan et al., 1988, Mittal et al., 1988), the effects of deletion of amino acids 1-3 and introduction of a methionine on the properties of this mutant (4-128) were investigated. It has been demonstrated that mutant 4-128 behaves similarly to the wild type with regard to spectral properties, redox potential, kinetics in cytochrome c reduction, CYP11A1 and CYP11B1 dependent conversion assays and reduction rates of cytochromes P450. The binding affinity of mutant 4-128 to CYP11A1 was also identical to the affinity of wild type adrenodoxin.

Considering the C-terminal domain of adrenodoxin, P108 was revealed to be critical for the formation of a biologically active protein. Mutant 4-108, which contains P108, exhibits unchanged absorption characteristics of adrenodoxin and functions as an electron shuttle to cytochrome c, CYP11A1 and CYP11B1, whereas mutant 4-107 did not show an incorporation of the /2Fe-2S/ cluster. Thus, it was concluded that P108 promotes a correct folding of adrenodoxin as a necessary prerequisite for the assembly of the /2Fe-2S/ cluster. When studying the influence of deletions on protein-protein interaction, it was shown that removal of residues 115-128 or 109-128 did not essentially affect adrenodoxin reductase binding as shown by nearly unchanged cytochrome c reduction activity. In a CYP11A1 assay, mutants 4-108 and 4-114 exhibited 3.2-fold and 5-fold decreased K_m values, respectively, whilst the K_d values for CYP11A1 decreased 3-fold and 1.9-fold, respectively (Table 3, Fig. 3). Additionally, in a CYP11B1 assay, mutants 4-108 and 4-114 showed decreased K_m values. The results suggest that electron donation from adrenodoxin to CYP11A1 and CYP11B1 is determined at least in part by different features of the cytochromes.

Similar differences have been observed when analyzing mutants where Y82 has been replaced by serine, leucine or phenylalanine. While cytochrome c reduction was not influenced by these replacements, the K_m values changed up to 4-fold when measuring the enzymatic activities of mutant adrenodoxins with CYP11B1 and CYP11B2 (Table 3). Again, changes differ in dependence on the P450 used. Finally, the K_m values for some redox partners with H56 mutants are increased (Table 3). The K_m value for mutant H56Q in the cytochrome c assay raised twofold. Marked changes in K_m values were also found in the CYP11B1-dependent conversion of deoxycorticosterone to corticosterone. The K_m values for CYP11B1 mediated by H56Q and H56T mutant increase 6-fold and 4-fold, respectively (Table 3). The effect of replacement in position 56 is less pronounced when hydroxylating activities of CYP11A1 were studied. It can been seen from the Table 3 that both H56Q and H56T mutants exhibit only slightly increased K_m values when compared to wild type adrenodoxin. Furthermore, the intensity of Y82 fluorescence is strongly affected by replacement of H56.

From this data it can be concluded, that H56 is located in the immediate environment of Y82 and directly or indirectly involved in binding to the redox partners of adrenodoxin, adrenodoxin reductase and cytochromes P450. Furthermore, the results obtained support our conclusion that although the major determinant of adrenodoxin for binding to the redox partners is its highly acidic region, these proteins (adrenodoxin reductase, CYP11A1, and CYP11B1) have slightly different binding requirements. This is not unlikely if the different physico-chemical properties of CYP11A1 and CYP11B1 are taken into consideration, such as amino acid composition (Watanuki et al., 1978) in conjunction with the net charge of the proteins, their solubility and stability (Takemori et al., 1975) as well as their immunological (Suhara et al., 1978) and EPR properties (Kominami et al., 1979). These different features of the cytochromes P450 obviously require differences in the site and/or mechanism of interaction with the electron donor adrenodoxin.

These differences may also be important for allowing the proteins to discriminate between the oxidized and reduced forms of adrenodoxin to promote the production of productive electron transfer complexes.

ACKNOWLEDGMENTS

The authors thank Dr. D. Schwarz for support in EPR measurements, Dr. P. Franke for carrying out mass spectrometry and Dr. K. Gast for aid in CD measurements. We appreciate the commitment of R. Dettmer in affinity chromatography and of Ch. Jaeger in isolation of adrenodoxin. This work was supported by a grant from the Deutsche Forschungsgemeinschaft (Be 1343/1-2). H.U. is sponsored by the Boehringer Ingelheim Fonds.

REFERENCES

Akhrem, A. A., Lapko, V.N., Lapko, A.G., Shkumatov, V.M., and Chashchin, V.L. (1979) Isolation, structural organization and mechanism of action of mitochondrial steroid hydroxylating systems. *Acta biol. med. ger.* **38**, 257-273

Beckert, V., Dettmer, R., and Bernhardt, R. (1994) Mutations of Tyrosine 82 in Bovine Adrenodoxin That Affect Binding to Cytochromes P45011A1 and P45011B1 but Not Electron Transfer. *J. Biol. Chem.* **269**, 2568-2573

Bhasker, C.R., Okamura, T., Simpson, E.R., and Waterman, M.R. (1987) Mature bovine adrenodoxin contains a 14-amino-acid COOH-terminal extension originally detected by cDNA sequencing. *Eur. J. Biochem.* **164**, 21-25

Coghlan, V.M., Cupp, J. R., and Vickery, L.E. (1988) Purification and Characterization of Human Placental Ferredoxin. *Arch. Biochem. Biophys.* **264**, 376-382

Coghlan, V.M., and Vickery, L.E. (1991) Site-specific Mutations in Human Ferredoxin That Affect Binding to Ferredoxin Reductase and Cytochrome P450scc. *J. Biol. Chem.* **266**, 18606-18612

Cupp, J.R., and Vickery, L.E. (1989) Adrenodoxin with a COOH-terminal Deletion (*des* 116-128) Exhibits Enhanced Activity.*J. Biol. Chem.* **264**, 1602-1607

Driscoll, W.J., and Omdahl, J.L. (1986) Kidney and Adrenal Mitochondria Contain two Forms of NADPH-adrenodoxin Reductase-dependent Iron-Sulfur Proteins. *J. Biol. Chem.* **261**, 4122-4125

Geren, L.M., O'Brien, P., Stonehuerner, J., and Millett, F. (1984) Identification of specific Carboxylate Groups on Adrenodoxin That Are Involved in the Interaction with Adrenodoxin Reductase. *J. Biol. Chem.* **259**, 2155-2160

Hanukoglu, I., and Jefcoate, C. R. (1980) Mitochondrial Cytochrome P-450scc. Mechanism of electron transport by adrenodoxin. *J. Biol. Chem.* **255**, 3057-3061

Hara, T., and Takeshima, M. (1994) Conclusive evidence of a quaternary cluster model for cholesterol side chain cleavage reaction catalyzed by cytochrome P-450scc. In: Lechner, M.C. (ed) *Cytochrome P450. 8th International Conference*. John Libbey Eurotext, Paris. pp. 417-420

Hiwatashi, A., Sakihama, N., Shin, M., and Ichikawa, Y. (1986) Heterogeneity of adrenocortical ferredoxin. *FEBS Lett.* **209**, 311-315

Kido, T., and Kimura, T. (1979) The Formation of Binary and Tertiary Complexes of Cytochrome P-450scc with Adrenodoxin and Adrenodoxin Reductase·Adrenodoxin Complex. *J. Biol. Chem.* **254**, 11806-11815

Kominami, S., Ochi, H., and Takemori, S. (1979) Electron paramagnetic resonance studies of the purified cytochrome P-450scc and P-$450_{11\beta}$ from bovine adrenocortical mitochondria. *Biochim. Biophys. Acta* **577**, 170-176

Lambeth, J.D., and Kamin, H. (1979) Adrenodoxin Reductase·Adrenodoxin Complex. Flavin to iron-sulfur electron transfer as the rate-limiting step in the NADPH-cytochrome c reductase reaction. *J. Biol. Chem.* **254**, 2766-2774

Matocha, M.F., and Waterman, M. (1984) Discriminatory Processing of the Precursor Forms of Cytochrome P-450scc and Adrenodoxin by Adrenocortical and Heart Mitochondria. *J. Biol. Chem.* **259**, 8672-8678

Matocha, M.F., and Waterman, M.R. (1985) Synthesis and Processing of Mitochondrial Steroid Hydroxylases. *J. Biol. Chem.* **260**, 12259-12265

Mittal, S., Zhu, Y.-Z., and Vickery, L.E. (1988) Molecular Cloning and Sequence Analysis of Human Placental Ferredoxin. *Arch. Biochem. Biophys.* **264**, 383-391

Miura, S., and Ichikawa, Y. (1991) Conformational Change of Adrenodoxin Induced by Reduction of Iron-Sulfur Cluster. *J. Biol. Chem.* **266**, 6252-6258

Nabi, N., Ishikawa, T., Ohashi, M., and Omura, T. (1983) Contributions of Cytoplasmatic Free and Membrane-Bound Ribosomes to the Synthesis of NADPH-Adrenodoxin-Reductase and Adrenodoxin of Bovine Adrenal Cortex Mitochondria. *J. Biochem.* **94**, 1505-1515

Okamura, T., John, M.E., Zuber, M.X., Simpson, E.R., and Waterman, M.R. (1985) Molecular cloning and amino acid sequence of the precursor form of bovine adrenodoxin: Evidence for a previously unidentified COOH-terminal peptide.*Proc. Natl. Acad. Sci. U.S.A.* **82**, 5705-5709

Sagara, Y., Ito, A., and Omura, T. (1984) Partial Purification of a Metalloprotease Catalyzing the Processing of Adrenodoxin Precursor in Bovine Adrenal Cortex Mitochondria. *J.Biochem.* **96**, 1743-1752

Sakihama, N., Hiwatashi, A., Miyatake, A., Shin, M., and Ichikawa, Y. (1988) Isolation and Purification of Mature Bovine Adrenocortical Ferredoxin with an Elongated Carboxyl End. *Arch. Biochem. Biophys.* **264**, 23-29

Sligar, S.G., Cinti, D.L., Gibson, G.G., and Schenkman, J.B. (1979) Spin State Control of the Hepatic Cytochrome P450 Redox Potential. *Biochem. Biophys. Res. Commun.* **90**, 925-932

Sugano,S., Morishima, N., Ikeda, H., and Horie, S. (1989) Sensitive Assay of Cytochrome P450scc Activity by High-Performance Liquid Chromatography. *Anal. Biochem.* **182**, 327-333

Takemori, S., Sato, H., Gomi, T., Suhara, K., and Katagiri, M. (1975) Purification and properties of cytochrome P-$450_{11\beta}$ from adrenocortical mitochondria. *Biochem. Biophys. Res. Comm.* **67**, 1151-1157

Tanaka, M., Haniu, M., Yasunobu, K.T., and Kimura, T. (1973) The Amino Acid Sequence of Bovine Adrenodoxin. *J. Biol. Chem.* **248**, 1141-1157

Taniguchi, T., and Kimura, T. (1975) Studies on NO_2-Tyr^{82} and NH_2-Tyr^{82} Derivatives of Adrenodoxin. Effects Of Chemical Modification On Electron Transferring Activity. *Biochemistry* **14**, 5573-5578

Taniguchi, T., and Kimura, T. (1976) Studies on Nitrotyrosine-82 and Aminotyrosine-82 Derivatives of Adrenodoxin. Effects of Chemical Modification on the Complex Formantion with Adrenodoxin Reductase. *Biochemistry* **15**, 2849-2853

Uhlmann, H., Beckert, V., Schwarz, D., and Bernhardt, R. (1992) Expression of Bovine Adrenodoxin in E. coli and Site-Directed Mutagenesis of /2Fe-2S/ Cluster Ligands. *Biochem. Biophys. Res. Commun.* **188**, 1131-1138

Uhlmann, H., Kraft, R., and Bernhardt, R. (1994) C-Terminal Region of Adrenodoxin Affects Its Structural Integrity and Determines Differences in Its Electron Transfer Function to Cytochrome P-450. *J. Biol. Chem.*, **269**, 22557-22564

Usanov, S.A., Chashchin, V.L., and Akhrem, A.A. (1990) Cytochrome P-450 Dependent Pathways of the Biosynthesis of Steroid Hormones. In: *Frontiers in Biotransformation* (K. Ruckpaul and H. Rein, Eds.), vol. 3, pp. 1-57, Akademie-Verlag Berlin

Wada, A., and Waterman, M.R. (1992) Identification by Site-directed Mutagenesis of Two Lysine Residues in Cholesterol Side Chain Cleavage Cytochrome P450 That Are Essential for Adrenodoxin Binding. *J. Biol. Chem.* **267**, 22877-22882

Watanuki, M., Tilley, B.E., and Hall, P.F. (1978) Cytochrome P-450 for 11β- and 18-Hydroxylase Activities of Bovine Adrenocortical Mitochondria: One Enzyme or Two? *Biochemistry* **17**, 127-130

26

MODIFICATION OF CHROMATIN STRUCTURE FOLLOWING EXPOSURE OF MOLT4 CELLS TO THE CARCINOGENIC CHROMIUM(VI)

Subhendra N. Mattagajasingh[1] and Hara P. Misra[2]

[1] John Hopkins University, School of Medicine
1033 Ross Research Building
720 Rotland Drive, Baltimore, MD 21205
[2] Department of Biomedical Scicences and Pathobiology
Virginia-Maryland Regional College of Veterinary Medicine
Virginia Polytechnic Institute and State University
Blacksburg, VA 24061-0442

ABSTRACT

DNA-protein complexes-induced by potassium chromate in human leukemic T-lymphocyte MOLT4 cells were isolated by ultracentrifugal sedimentation in the presene of 2% sodium dodecyl sulfate (SDS) and 5 M urea. The complexes were analyzed by two-dimensional SDS-polyacrylamide gel electrophoresis (PAGE). Three acidic proteins of 74, 44 and 42 kD, and a basic protein of 51 kD were primarily complexed to DNA following 25 μM chromate treatment indicating selectivity in chromate-induced DNA-protein complexes. Higher concentrations of chromate cross-linked many other proteins to DNA. A 43 kD protein predominantly localized in the cytoplasmic fraction was found to be cross-linked to DNA upon chromate treatment. Partial N-terminal amino acid sequencing of p43 showed that it could be a human lectin. Treatment of the complexes with DNase I, RNase and EDTA revealed that sedimentation of the proteins was not due to formation of protein aggregates, but due to their association with DNA. The complexes were disrupted, to some extent, by EDTA indicating the involvement of a chelatable form of chromium in the complex. Because chromate-induced DNA-protein complexes are resistant to treatments such as 2% SDS and 5 M urea, but disrupted under gel electrophoretic conditions, it is possible that chromium could be used as a cross-linking agent for identification of other proteins such as transcription factors, that interact with DNA.

INTRODUCTION

A number of physical and chemical carcinogenic agents have been shown to induce DNA-protein cross-linking (Oleinick et al., 1987). Hexavalent chromium [Cr(VI)] compounds have been considered potent carcinogens and have been shown to cause various types of DNA damage including DNA-protein cross-linking in various cells and tissues (see Cohen et al. 1990, for a review). Cr(VI) does not bind to DNA in cell free systems (Fornace et al., 1981; Koster et al., 1985), however it readily enters into the cell through the sulfate anion transport system (Arslan et al., 1987; Jannette et al., 1985) and is reduced by the cells' redox system to chromium (III) [Cr(III)], which in turn binds to DNA in cell free systems (Tsapakos, et al., 1983). Cr(III) is poorly taken up into the cell, possibly explaining why this form of chromium has not been shown to be carcinogenic (De Flora et al., 1989).

Although chromate-induced DNA-protein complexes are implicated in chromate carcinogenicity, the mechanisms of their formation, composition, and biological significance are not well understood. It has been postulated that cross-linking of proteins to DNA could disrupt chromatin structure and the normal regulation of gene expression. This, in turn could play a role in carcinogenesis in that deletion of DNA bases may result when portions of replicating DNA are buried under DNA-protein complexes (Bedinger et al, 1983; Briggs and Briggs, 1988). Such deletions may also give rise to loss or inactivation of tumor suppressor genes (Bouck and Benjamin, 1989). During normal regulation of gene expression, protein(s) reversibly interact with specific DNA sequences (Stein and Kleinsmith, 1979). Cross-linking of DNA with inappropriate proteins could disrupt the normal regulation of DNA-protein interactions, an event that may be anticipated to have serious genetic consequences, including disruption in or alteration of gene expression. Therefore, it is necessary to determine the identity of proteins which are cross-linked to DNA upon chromate exposure as well as the nature of the interaction of such proteins with DNA. Identification of proteins cross-linked to DNA may further contribute to a better understanding of chromatin, protein interactions, including the three-dimensional orientation of proteins around DNA.

In the present study, we have analyzed the changes in the protein constituent of chromatin following chromate treatment of MOLT4 cells and have attempted to identify a 43 kD protein complexed to DNA upon chromate treatment, by partial sequencing. Inappropriate complexing of proteins of structural or functional importance to DNA, rather than the DNA-protein complexes themselves, may have important role in chromate-carcinogenicity. Thus, identification of proteins participating in chromate-induced DNA-protein complexes will be required in order to better understand the potential consequences of this lesion.

Chemicals

Potassium chromate was purchased from J.T. Baker (Phillipsburg, NJ). Acrylamide, N', N', N', N'-tetramethylenediamine (TEMED), ammonium persulfate, protein determination kit, coomassiee brilliant blue R-250, ampholines, urea and sodium dodecyl sulfate (SDS) were purchased from Bio-Rad Laboratories (Richmond, CA). Polyvinylidene difluoride (PVDF) membrane, ^{3}H-thymidine, ^{35}S-methionine and Aquassure LSC cocktail were purshased from New England Nuclear (Boston, MA). DNase free RNase was purchased from Boehringer Manheim (Indianapolis, IN). All other chemicals and enzymes were purchased from Sigma Chemical Co. (St. Louis, MO)

Cell Culture and Treatment

Human leukemic T-lymphocyte MOLT4 cells were purchased from American Type Culture Collection (Bethesda, MD) and were maintained in suspension at exponential growth phase in RPMI 1640 [N-2-hydroxyethylpiperazine-N'-2-ethanesulfonic acid (HEPES) modified] medium supplemented with 10% heat-inactivated fetal bovine serum, 10 U penicillin and 10 μg/ml streptomycin solution at 37°C in a humidified atmosphere of 5% CO_2 and 95% air. Cells, in exponential growth phase, were radiolabeled with ^{3}H-thymidine and ^{35}S-methionine (0.02 μCi/ml each) for ~36 hr, in methionine free RPMI 1640 medium. Radiolabeled cells were collected by centrifugation, washed three times in cold Saline A (5 mM $NaHCO_3$, 6 mM dextrose, 5 mM KCl and 140 mM NaCl, pH 7.2) and resuspended in salts-glucose medium [SGM: 50 mM HEPES, 100 mM NaCl, 5 mM KCl, 2 mM $CaCl_2$, 5 mM dextrose, pH 7.2] at a concentration of $1x10^6$ cells/ml. Potassium chloride (control) or potassium chromate were added to the cell suspensions at different concentration for incubations periods of 2 hr or 16 hr. Following treatment, cells were collected and cytotoxicity was determined by exclusion of trypan blue.

Isolation and Quantitation of DNA-Protein Complexes

Potassium chromate-treated and control cells were collected by centrifugation at 500xg for 10 min, after which cells were washed three times in Saline A and incubated for 15 min on ice in cold hypotonic buffer (10 mM Tris-HCl, pH 7.5 containing 10 mM NaCl, 1.5 mM $MgCl_2$). Cells were collected by centrifugation at 300xg for 5 min, and were lysed in the above solution supplemented with 0.5% Nonidet P-40 and 1mM phenylmethylsulfonyl fluoride (PMSF), using a loose-fitting glass homogenizer. The nuclei were sedimented at 700xg for 5 min at 4°C, resuspended in 10 mM Tris-HCl containing 250 mM sucrose, 5 mM $MgCl_2$ and 1 mM PMSF (pH 7.5), and were layered over a similar solution but containing 880 mM sucrose. Nuclei were subsequently collected by centrifugation for 10 min at 1000xg at 4°C and used for isolation of DNA-protein complexes.

The DNA-protein complexes were isolated from the nuclei of control or chromate treated cells by modification of a previously described method (Miller and Costa, 1989). Briefly, the purified nuclei were solubilized in 35 ml of 10 mM Tris-HCl containing 2% SDS, 1 mM PMSF (pH 7.5) by shaking on a platform shaker for 6 hr at room temperature. The samples were then homogenized using a tight-fitting homogenizer and then sedimented at 100,000xg for 16 hr at 18°C, using a Beckman SW 27 rotor (Beckman Instruments, Fullerton, CA). The pellets were placed in a solution of 5 M urea, and 1 mM PMSF, shaken at 4°C for 6 hr, and homogenized again. SDS was added to 2% final concentration and the DNA-protein complexes were isolated by ultracentrifugation as above. The pellets were resuspended in 1 ml of 10 mM Tris-HCl containing 1mM PMSF (pH 7.5) by gentle sonication using a micro probe (Model W 225 R, Ultrasonics, Inc., Plainview, New York), and were precipitated in 70% acetone at -20°C. The different steps in isolation of DNA-protein complexes are summarized in Scheme I.

The DNA-protein complexes were collected by centrifugation at 12,500xg for 15 min at 4°C using a Beckman microfuge and were resuspended in 1 ml of 10 mM Tris-HCl containing 1mM PMSF (pH 7.5) by gentle sonication or by shaking in a Nutator for about 16 hr at 4°C. The DNA content was determined by measuring the absorbance at 260 nm (Maniatis et al., 1982). Both ^{3}H and ^{35}S activity were determined by dissolving the samples in Aquassure Cocktail (NEN, Boston, MA) and counting in a Beckman LS 5800 Liquid Scintillation counter (Beckman Instruments, Inc., Irvine, CA). The protein content of the Nonidet P-40 homogenate and ^{35}S specific activity were determined by using Bio-Rad dye (Bradford, 1976) and by measuring the ^{35}S activity in the acid-insoluble material, respectively.

Determination of Stability of DNA-Protein Complexes

DNA-protein complexes containing 100 μg DNA were taken in siliconized micro-centrifuge tubes. $MgCl_2$ was added to 5 mM in samples treated with DNase 1 and RNase. DNase 1 (200 μg/ml), RNase (40 μg/ml) or EDTA (10-50 mM) were added, mixed and the tubes were incubated at room temperature for 2 hr. SDS was then added to a final

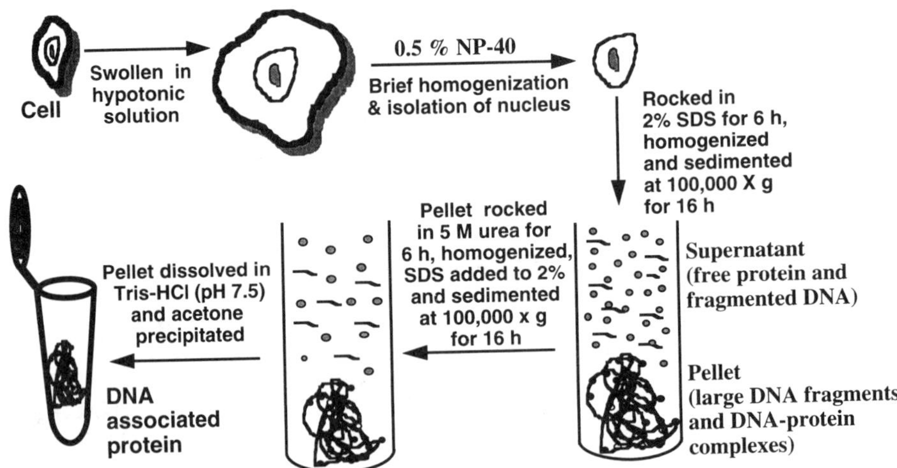

Scheme I. Isolation of DNA-protein cross-links by SDS/urea extraction method

concentration of 0.5% and samples were centrifuged at 100,000xg for 16 hr at 18°C. The supernatants were carefully removed and the pellets were resuspended in 10 mM Tris-HCl (pH 7.5) by brief sonication. DNA and protein contents were determined from the ^{3}H and ^{35}S activity, respectively, by liquid scintillation counting as described above.

Analysis of Proteins by Two-Dimensional Gel Electrophoresis

Proteins were analyzed by the nonequilibrium method of two-dimensional gel electrophoresis as described by O'Farrell et al. (1977). DNA-protein complexes containing 150 μg of DNA were acetone precipitated or lyophilized (FTS Systems, Inc., Stone Ridge, NY) and solubilized in 30 μl solubilizing buffer [9 M urea, 4% Nonidet P-40, 2% β-mercaptoethanol and 2% ampholines (Bio-Rad, pH range 3-10 and 8-10 (4:1)]. Isoelectric focusing was carried out in 200 μl capillary tubes (1.5 mm diameter, Fisher Scientific, NJ) cointaining 4% polyacrylamide and 2% ampholines (pH range 3-10). Cytochrome c, a colored protein, was used to indicate the mobility of basic proteins. Second dimensional separation was carried out on 12% SDS-polyacrylamide gels. The gels were subjected to silver staining by following the method of Sammons et al. (1981).

Cytoplasmic (Nonidet P-40 soluble cytoplasmic material) and nuclear (SDS soluble material of the isolated nuclei) protein fractions were saved and their protein contents were determined as described above. Thirty μg of protein from each fraction was acetone precipitated and solubilized in 20 μl of solubilizing buffer. Nonequilibrium focusing, separation in the second dimention, and silver-staining of samples were carried out as described above.

Electroblotting and Amino Acid Sequencing

DNA-protein complexes containing 250 μg of DNA were analyzed by two-dimensional gel electrophoresis as previously described (Mattagajasingh and Misra, 1994). The second dimensional gel was pre-run in presence of 1 mM sodium thioglycolate to protect the N-terminal from reactive compounds. Proteins were electroblotted onto a PVDF membrane in a Bio-Rad Transblot appartaus (Bio-Rad, Richmond, CA) using 10 mM 3-cyclohexylamino-1-propanesulfonic acid (CAPS) and 10% HPLC grade methanol (pH 11) as the electroblotting buffer, at 50 V for 1 hr at room temperature. Coomassie brilliant blue R-250 (Bio-Rad, 0.025% in 40% methnol) was used to visualize the proteins. Acetic acid was omitted from the staining and destaining solution as it may cause N-terminal blocking (Bio-Rad technical bulletin # 240). The protein band of interest was excised and automated Edman degradation was performed using an Applied Biosystems 477A protein sequencer equipped with a 120 A analyzer (Applied Biosystems, Inc., Foster City, CA).

RESULTS

Cytotoxicity

Exposure of MOLT4 cells to 0 $\rightarrow$ 200 μM potassium chromate in SGM for 2 hr was found to have little cytotoxic effects, as assessed by trypan blue exclusion (viability was within 98 ± 2% of the control). The viability of cells treated with 200 μM chromate for 16 hr in SGM was decreased to 72 ± 3% of control.

Effect of Potassium Chromate on DNA-Protein Crosslinking in MOLT4 Cells

Cell exposure to 0 $\rightarrow$ 200 μM potassium chromate for 2 hr resulted in a dose-dependent increase in the formation of DNA-protein complexes in MOLT4 cells. Cells treated with 200μM chromate for 2 hr had about two-fold increased DNA-protein complex formation as compared to the control (data not shown). When cells were treated with 200 μM chromate for 16 hr, an 8-10 fold increase in the formation of DNA-protein complexes was observed as compared to the control cells.

Stability of DNA-Protein Complexes

The stability of DNA-protein complexes were tested by monitoring the recovery of DNA and protein in the pellet following treatment of DNase I, RNase and EDTA. The control samples (without DNase I or EDTA) had almost 100% recovery of both DNA and protein in the pellet following ultracentrifugation, as determined by ^{3}H- and ^{35}S-radioactivity, respectively. Treatment of DNA-protein complexes, isolated from both control and chromate treated cells, with DNase I significantly reduced the recovery of ^{3}H and ^{35}S in the pellet (Table 1). RNase treatment of DNA-protein complexes did not interfere with recovery of DNA or protein (Table 1). These data indicate that chromate treatment induces the cross-linking of proteins to DNA and does not cause sedimentable protein aggregates, and are consistent with the findings of Miller and Costa (1989) for chromate-induced DNA-protein complexes in cultured chinese hamster ovary (CHO) cells.

In order to test whether chromium is directly participating in the DNA-protein complexes, excess amounts of EDTA was used to disrupt the complex by chelating chromium. To test whether EDTA chelation of chromium is complete, 0—>100 mM EDTA was employed to disrupt the DNA-protein comoplexes isolated from both the ^{35}S-methionine as well as ^{51}Cr-labeled cells. Dissociation of ^{35}S or ^{51}Cr did not increase above 50 mM EDTA

Table 1. Stability of potassium chromate-induced DNA-protein complexes in intact MOLT4 cells

	Chromate-induced DNA-protein complexes		Control DNA-protein complexes	
Treatment	DNA recovered (%)	Protein recovered (%)	DNA recovered (%)	Protein recovered (%)
Control	100	100	100	100
DNase I	3.7 ± 2.4	7.4 ± 3.7	1.8 ± 3.1	2.3 ± 2.4
EDTA	98.3 ± 3.5	81.3 ± 4.3	98.8 ± 4.9	96.7 ± 2.8
RNase	96.4 ± 5.1	98.2 ± 6.3	96.2 ± 5.8	97.3 ± 5.7

^{35}S- and ^{3}H-labeled DNA-protein complexes were treated either with DNase I (200 μg/ml), RNase (40 μg/ml) or EDTA (50 mM) and incubated for 2h at room temperature and sedimented by ultracentrifugation. The recovery of protein and DNA from the complexes was determined by recovery of ^{35}S and ^{3}H cpm in the pellet, respectively. Each value is a mean ± SD of three normalized values.

(data not shown). As shown in Table 1, EDTA treatment of DNA-protein complexes isolated from control cells did not affect the recovery of ^{3}H or ^{35}S-radioactivity in the pellet. When chromate-induced DNA-protein complexes were treated with EDTA, recovery of ^{35}S-radioactivity in the pellet decreased without affecting the recovery of ^{3}H-activity. These results indicate that the decrease in ^{35}S activity was not due to fragmentation of DNA. The maximum decrease in ^{35}S recovery after EDTA (50 mM) treatment was found to be approximately 18% of the control (Table 1).

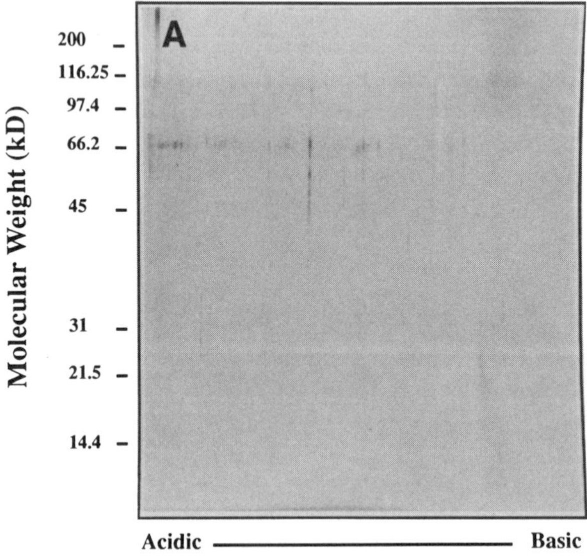

Figure 1. Non equilibrium two-dimentional electrophoresis of DNA-protein complexes isolated from nuclei of control or chromate-treated cells. DNA-protein complexes were isolated from potassium chloride (control) or potassium chromate treated cells. (A) Two-dimensional gel of the control DNA-protein complexes. (B) and (C) Two-dimensional gel of proteins dissociated from DNA-protein complexes generated by treatment of MOLT4 cells with 25 and 200 μM chromate for 16 hr, respectively.

Two-Dimensional Gel Electrophoretic Analysis of Proteins Complexed to DNA

The proteins complexed to DNA in both the control (potassium chloride) or potassium chromate treated cells were analyzed by two-dimensional gel electrophoresis. DNA-protein complexes were loaded on the acidic end of the gel, in order to avoid the entry of nucleic acids into the first dimensional focusing gels. Silver-staining of two-dimensional gels of DNA-protein complexes isolated from control cells did not show any protein in the gel, indicating that the SDS/urea method used for isolation of DNA-protein complexes effectively dissociates the DNA-protein complexes in the control cells (Fig 1 A).

The proteins complexed to DNA, in cells exposed to 25 μM chromate for 16 hr, but did not dissociate from DNA by 2% SDS and 5 M urea treatments, are shown in Fig

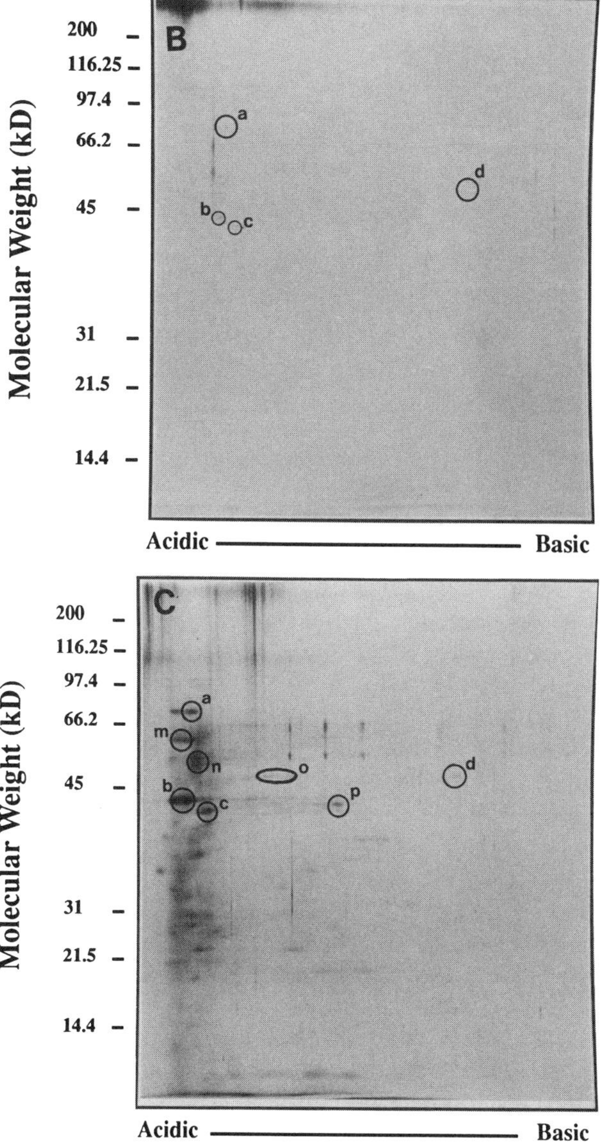

Table 2. Molecular weight and pI of major proteins complexed to DNA upon chromate treatment of intact MOLT4 cells

Molecular weight (x10^3)		pI
74	(a*)	5.2-5.6
63		5.2-5.4
53		5.2
51	(d*)	8.8-9.2
44	(b*)	5.3
43	(p43)	6.0-6.5
42	(c*)	5.8
40		4.8-5.0
36		5.0-5.2
36-38	(CNP)	5.5-7.2
29		6.8
25-28	(CNP)	7.0-8.5
19		6.4-6.8
16	(CNP)	5.6-6.8

CNP: Cluster of nuclear proteins.
*Proteins marked in Figure 1 B.

1 B. As shown in this figure, there were primarily three acidic proteins ('a', 'b' and 'c') and a basic protein, 'd', complexed to DNA upon chromate treatment. Analysis of the molecular weight and pI of these proteins showed that the protein 'a' has a pI of 5.2-5.6 and a molecular weight of 74 kD, the protein 'b' has a pI of 5.2-5.4 and a molecular weight of 44 kD, and the protein 'c' has a pI of ~5.8 and molecular weight of 42 kD, respectively. The protein 'd' on the other hand has a pI of 8.8-9.2 and a molecular weight of 51 kD. When cells were treated with 200 μM of chromate for 16 hr many other proteins, in addition to the above proteins, were found to be complexed to DNA (Fig 1 C), indicating that the number of proteins cross-linked to DNA is dependent upon the dose of chromate. Molecular weight and pI of the major proteins complexed to DNA upon 200 μM chromate treatment are listed in Table 2.

To determine the subcellular localization of the four proteins primarily complexed to DNA, cytoplasmic and nuclear protein fractions were analyzed by two-demensional gel electrophoresis (Fig 2). Proteins 'b', 'c', and 'd' were visualized and were found to correspond to proteins in the cytoplasmic fraction. Proteins 'a', and 'd' were predominantly present in the nuclear fraction. Additional proteins ('m', 'n', 'o', and 'p') which were predominantly present in the cytoplasmic fraction were also found complexed to DNA upon treatment of cells with 200 μM chromate for 16 hr.

Analysis of Amino Acid Sequence

The protein 'p' (p43, pI 6.0-6.5) was predominantly detected in the cytoplasmic fraction but was found to be abundantly cross-linked to DNA. Therefore, attempts were made to identify this protein by partial N-terminal sequencing and homology comparision to proteins listed in the Swiss protein DataBank. The N-terminal sequencing of p43 revealed six consecutive amino acids as listed in Fig 3. This sequence was found to have absolute homology with amino acid residues 24-29 of lectin bra-3. This sequence is also partially homologous to many glycoproteins and the human multidrug-resistance protein 1.

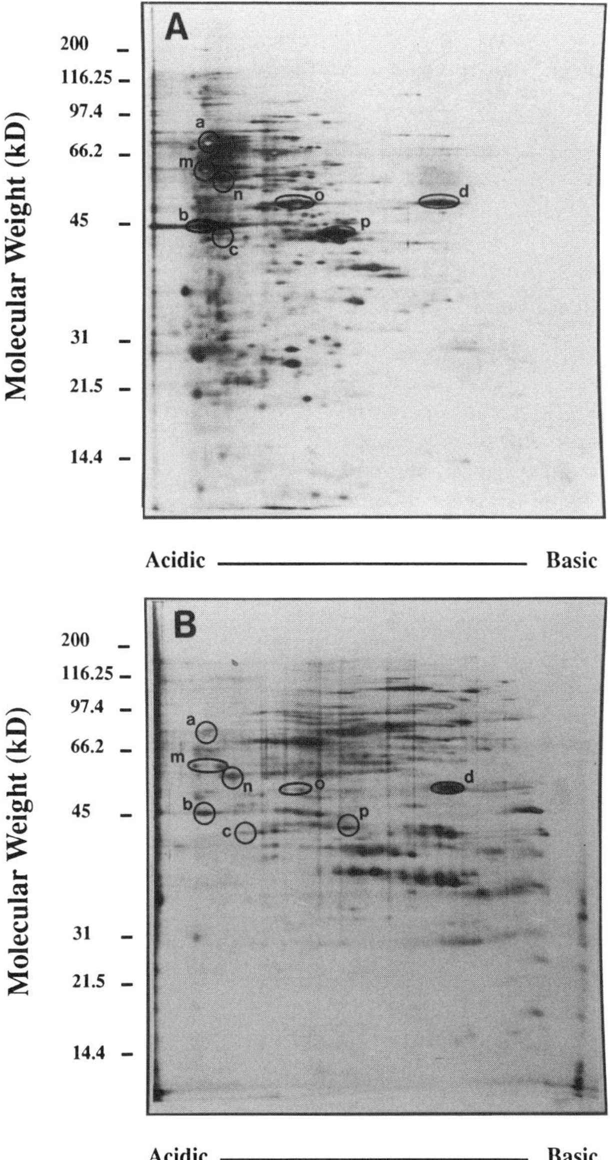

Figure 2. Localization of major proteins crosslinking to DNA upon chromate treatment in cytoplasmic and nuclear protein fractions. Thirty μg of cytoplasmic proteins (A) and nuclear proteins (B) were analyzed by nonequilibrium two-dimensional gel electrophoresis and silver stained.

DISCUSSION

In the present report we have shown that treatment of cells with 0 → 200 μM chromate increases the association of proteins with DNA in a dose-dependent manner without causing immediate lethality to the cell. This indicates that formation of DNA-protein complexes was not related to cell killing at least at the early stages of chromate interaction. The proteins

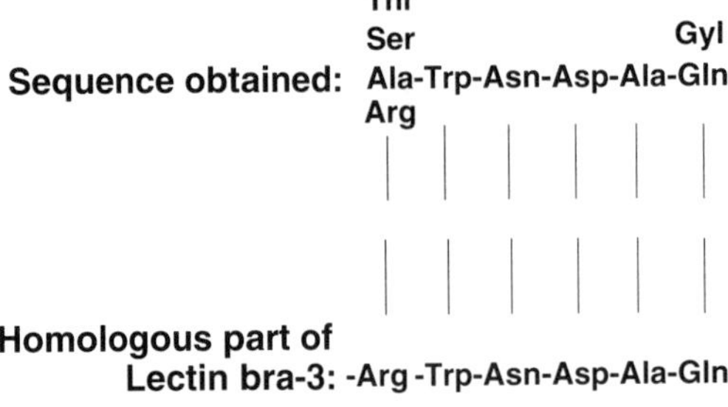

Figure 3. Microsequencing of p43. Following two-dimensional gel electrophoresis, proteins were electroblotted to PVDF membrane. The protein band if interest was trimmed and Edman degradation was performed to in a Applied Biosystems 477 A protein sequencer.

associated with DNA upon chromate exposure are expected to be nuclear proteins because, in this study, DNA-protein complexes were isolated from nuclei of chromate-exposed cells, and therefore free from cytoplasmic proteins.

Treatment of DNA-protein complexes isolated from control or chromate-treated cell nuclei with DNase I dissociated most of the proteins associated with DNA (Table 1), indicating that the sedimentable nature of the proteins is due to the association of proteins with the genomic DNA and not due to protein aggregation following chromate treatment. The small amount of DNA-protein complexes that were sedimented after DNase I digestion appears to mostly be in the form of stable chromium-nucleoprotein complexes. This is consistent with the findings of other investigators who have demonstrated the resistance of chromium-bound nucleoli to nuclease digestion (Ono et al., 1981), and the cross-linking of nuclear matrix proteins to DNA by heavy metals and UV irradiation (Wedrychowski et al., 1986; Bouliakas, 1986). The resistance of the DNA-protein complexes to RNase digestion indicates that chromate treatment does not induce the formation of RNA-protein complexes. The stability of the DNA-protein complexes was further assessed by monitoring the resistance of the complexes to EDTA treatment. Treatment of EDTA dissociated only 18% of ^{35}S activity from the DNA-protein complexes. Because EDTA effectively chelates Cr(III) but poorly binds with oxyanion of chromate, EDTA-dissociable proteins from DNA-protein complexes could have been mediated by a chelatable form of chromium such as Cr(III). However, the majority of the chromate-induced DNA-protein complexes were resistant to EDTA treatment. These data suggest that the predominant form of chromium in the DNA-protein complexes is not Cr(III). It is, however, plausible that there may be some direct interaction of proteins with DNA due to generation of free radicals during the intracellular reduction of chromate.

Intracellular Cr(III) is predominately generated by the reduction of Cr(VI), a process shown to generate oxygen free radicals (Kawashini et al., 1986; Shi and Dalal, 1989). Although, the role of free radicals in the chromate-induced DNA-protein cross-linking is uncertain (Standeven and Wetterhahn, 1991), free radical generating systems such as ionizing radiation as well as Fenton type reactions have also been shown to cause DNA-protein cross-linking (Chiu et al., 1985; Lesko et al., 1982). Furthermore, we have shown that treatment of cells with antioxidants prior to chromate treatment inhibited chromate-induced DNA-protein cross-linking (Mattagajasingh and Misra, 1993). Collectively, these results suggest that free radicals may be, at least in part, involved in chromate-induced DNA-protein cross-linking. However, the present results suggest that free radical independent mechanisms may also play a role in some chromate-induced DNA-protein crosslinking, because the the electrophoretic conditions would not disrupt the radical-induced covalent DNA-protein cross-links, and we were able to visualize the proteins cross-linked to DNA in two-dimensional gels without digesting DNA. Visualization of proteins in two-dimensional gels may, at least in part, be due to the reduction of sulfhydryl groups of proteins by 2-mercaptoethanol, used in SDS-PAGE, leading to disruption of the complex. Such mechanisms have been shown to be the leading cause of chromate and cisplatin-induced DNA-protein complexes (Wedrychowski et al. (1986).

In order to be cross-linked to DNA by any form of cross-linking agent, a protein must reside in close proximity to DNA and its reactive groups should be oriented such that they are able to interact with reactive groups of DNA. Present studies show that only four proteins ('a','b','c' and 'd') were found to be primarily complexed to DNA, although several other proteins were seen in the nuclear protein fraction (Fig 2B). Since chromate was required for the crosslinking of proteins with DNA, it is conceivable that not only a close proximity of these proteins to DNA but also their selective interaction with chromium could be important factors necessary for the crosslinking of these proteins to DNA. Other investigators have reported the association of a 45 kD protein (similar in mol. wt. and pI to protein 'b' we have detected) to DNA by chromium (Wedrychowski et al., 1985; 1986) and ionizing radiation (Chiu et al., 1985). This protein has been identified as nuclear actin (Miller and costa, 1989) in CHO cells exposed to chromate. The identity of proteins a, c, and d remains to be ascertained. Although histones constitute a substantial part of the chromatin, these basic proteins were not complexed to DNA upon chromate treatment. Similar results were reported by Miller and Costa (1989). Because Cr(III) has high affinity for sulfur-containing ligands, and there is scarcity of cystiene residues among histones, it appears plausible that histones may not complex to DNA by chromate due to unavailability of appropriate ligands.

In the present study, homology of p43 microsequence with amino acids 24-29 of lectin bra-3 indicates that it could be a human lectin. Lectins have not been previously shown as DNA-binding proteins. It has been shown that lectins are located in a wide variety of cells and cell membranes. Although lectin receptors have been found on the cytoplasmic surface of intracellular membranes such as the nuclear envelope and mitochondrial outer membrane, recent evidences indicate that lectin binding takes place on the noncytoplasmic surface of these organelles (Lis and Sharon, 1986 a). Alteration in lectin levels upon malignant transformation (Gabius et al., 1986) and their involvement in developmental processes (Kolb-Bochofen, 1986) has been reported. Lectins have also been shown to function as receptors (Kolb-Bochofen, 1986) and mitogenic regulators (Lis and Sharon, 1986 b). Because chromate-induced DNA-protein complexes predominantly occur in transcriptionally active DNA (Hamilton and Wetterhahn, 1989), it is tempting to speculate that lectins may be involved in the transcription process. Nonetheless, the cross-linking of lectins to DNA could lead to serious physiological and genetic consequences.

In summary, our results indicate that chromate treatment modifies the chromatin structure through complex formation with a selected group of non-histone proteins. Lectin has been suspected as one of the proteins involved in chromate-induced DNA-protein complexes. The exact nature of the interaction between the DNA and protein remains obscure. However, our results suggest both the participation of a chelatable form of chromium such as Cr(III) as well as the involvement of oxidative mechanisms in the process of chromate-induced DNA-protein cross-linking. Although chromate-induced DNA-protein complexes are found to be resistant to treatments such as 2% SDS and 5 M urea, their dissociation in gel electrophoretic conditions indicates their association in the form of non-covalent interactions. These characteristics of chromate-induced DNA-protein complexes suggest that it is possible to use chromium in studies involving chromatin structure as well as indentification of proteins participating in DNA-protein interactions, specifically those that undergo transient interaction with DNA, such as transcription factors.

ACKNOWLEDGMENTS

We are thankful to Dr. C. Rutherford and Ms. L. Sporakowski of the Department of Biochemistry and Anaerobic Microbiology at Virginia Tech for their assistance in amino acid sequencing and to the Ministry of Human Resources Development, Govt. of India for financial assistance in terms of a National Fellowship to S.N.M. This work was supported in part by grant HL 42009 from NIH.

REFERENCES

Arslan, P., Beltrame, M., and Tomasi, A., 1987, Intracellular chromium reduction, Biochem. Biophys. Acta, 931:10-15.

Bedinger, P., Hochstrasser, M., Jongenel, C.V., and Alberts, B.M., 1983, Properties of the T4 bacteriophage DNA replication apparatus: the T4 dda DNA helicase is required to pass a bound RNA polymerase molecule, Cell , 34:115-123.

Bouck, N.P., and Benjamin, B.K., 1989, Loss of cancer suppressors, a driving force in carcinogenesis, Chem. Res. Toxicol, 2:1-11.

Bouliakas, T., 1986, Protein-protein and protein-DNA interactions in calf thymus nuclear matrix using cross-linking by ultraviolet irradiation, Cell Biol, 64:474-484.

Briggs, J.A., and Briggs, R.C., 1988, Characterization of chromium effects on a rat liver epithelial cell line and their relevance to *in vitro* transformation, Cancer Res, 48:6484-6490.

Chiu, S., Friedman, L., Sokany, N., Xue, L., Oleinick, N., 1986, Nuclear matrix proteins are crosslinked to transcriptionally active gene sequences by ionizing radiation, Radiat. Res, 107:24-38.

Cohen, M., Latta, D., Coogan, T., and Costa, M., 1990, Mechanisms of Metal carcinogenesis: The reactions of metals with nucleic acids, In: E.C. Foulkes (ed.) The Biological effects of heavy metals, Vol. II, CRC Press, Boca Raton, Fl,.pp 19-76.

De Flora, S., and Wetterhahn, K.E., 1989, Mechanisms of chromium metabolism and genotoxicity, Life. Chem. Rep, 7:169-277.

Fornace, A.J., Jr., Seres, D.S., Lechner, J.F., and Harris, C.C., 1981, DNA-protein cross-linking by chromium salts, Chem.-Biol. Interact, 36:345-354.

Gabius, H., Engelhardt, R., Graupner, G., and Cramer, F., 1986, Lectins in carcinoma cells: Level reduction as possible regulatory event in tumor growth and colonization, In: Lectins: Biology, Biochemistry, Clinical Biochemistry. Bog-Hansen, T.C., and van Driessche, E. (eds) vol. 5, Walter de Gruyter & Co., Berlin, pp. 237-242.

Hamilton, J.W., and Wetterhahn, K.E., 1989, Differential effects of chromium(VI) on constitutive and inducible gene expression in chick embryo liver in vivo and correlation with chromium(VI)-induced DNA damage, Mol. Carcinog. 2:274-286.

Kawanishi, S., Inoue, S., Sano, S., 1986, Mechanism of DNA cleavage induced by sodium chromate(VI) in the presence of hydrogen peroxide, J. Biol. Chem. 261:5952-5989.

Kolb-Bachofen, V., 1986, Mammalian lectins and their function - A Review, In: Bog-Hansen, T.C., and van Driessche, E. (eds), Lectins: Biology, Biochemistry, Clinical Biochemistry (de Gruyter, Berlin), pp. 197-206.

Koster. A., and Beyersmann, D., 1985, Chromium binding by calf thymus nuclei and effects on chromatin, Toxicol. Environ. Chem. 10:307-313.

Lesko, S.A., Drocourt, J., Yang, S., 1982, Deoxyribonucleic acid-protein and deoxyribonucleic acid interstrand cross-links induced in isolated chromatin by hydrogen peroxide and ferrous ethylenediaminetetraacetate chelates, Biochemistry 21:5010-5015.

Lis, H., and Nathan, S., 1986(a), Application of lectins, In: The lectins: Properties, function, and applications in biology and medicine, Liener, I.E., Sharon, N., and Goldstein, I.J. (eds) Academic press, Inc., New York, pp.293-370.

Lis, H., and Nathan, S., 1986(b), Biological properties of lectins, In: The lectins: Properties, function, and applications in biology and medicine, Liener, I.E., Sharon, N., and Goldstein, I.J. (eds) Academic press, Inc., New York, pp.265-291.

Maniatis, T., Fritsch, E.F., and Sambrook, J. (eds.) 1982, Molecular Cloning: A Laboratory Manual, Cold Spring Harbor Laboratory Pubications, New York, New York.

Mattagajasingh, S.N., and Misra, H.P., 1994, Partial Sequencing of a Protein Cross-linking to DNA upon treatment of Cultured Intact Human Cells (MOLT4) with the Carcinogen Chromium(VI), J. Protein Chem. 13:449-450.

Mattagajasingh, S.N., and Misra, H.P., 1993, Vitamin E suppresses the potassium chromate induced crosslinking of proteins to DNA in MOLT4 cells, FASEB J. 7:469.

Miller, C.A., III, and Costa, M., 1989, Analysis of proteins cross-linked to DNA after treatment of cells with formaldehyde, chromate, and cis-diamminedichroloplatinum(II), Mol. Toxicol. 2:11-26.

O'Farrell, P.Z., Goodman, H.M., and O'Farrell, P.H., 1977, High resolution two dimensional electrophoresis of basic as well as acidic proteins, Cell 12:1133-1142.

Oleinick, N.L., Chiu, S., Ramakrishnan, N., and Xue, L., 1987, The formation, identification, and significance of DNA-protein cross-links in mammalian cells, Br. J. Cancer, (suppl. 8) 55:135-140.

Ono, H., Wada, O., and Ono, T., 1981, Distribution of trace metals in nucleoli of normal and regeneratin rat liver with special reference to the different behavior of nickel and chromium, J. Toxicol. Environ. Health, 8:947-957.

Sammons, D.W., Adams, L.D., and Nishiziwa, E.E., 1981, Ultracensitive silver based color staining of polypeptides in polyacrylamide gels, Electrophoresis. 2:135-141.

Shi, X., and Dalal, N.S., 1989, Chromium(V) and hydrogen radical formation during the glutathione reductase-catalyzed reduction of chromium(VI), Biochem. Biophys. Res. Commun, 163:627-634.

Standeven, A.M., and Wetterhahn, K., 1991, Is There a Role for Reactive Oxygen Species in the Mechanism of Chromium(VI) Carcinogenesis?, Chem. Res. Toxicol, 4:616-625.

Stein, G.S., and Kleinsmith, L.J., eds., 1979, Chromosal Proteins and Their Role in Regulation of Gene Expression, Academic Press, New York.

Sugiyama, M., Patierno, S.,R., Catoni, O., and Costa, M., 1986, Characterization of DNA lesions induced by CaCrO4 in synchronous and asynchronous cultured mammalian cells, Mol. Pharmacol, 29:606-613.

Tsapakos, M.J., and Wetterhahn, K.E., 1983, The interaction of chromium with nucleic acids, Chem.-Biol. Interact, 46:265-277.

Wedrychowski, A., Schmidt, W.N., and Hnilica, L.S., 1986, The in vivo crosslinking of proteins and DNA by heavy metals, J. Biol. Chem, 261:3370-3376.

Wedrychowski, A., Ward, W.S., Schmidt, W.N., and Hnilica, L.S., 1985, Chromium-induced cross-linking of nuclear proteins and DNA, J. Biol. Chem, 260:7150-7155.

IMMUNOLOGICAL RECOGNITION, PHAGE AND SYNTHETIC LIBRARIES

27

REGIONS OF INTERACTION BETWEEN NICOTINIC ACETYLCHOLINE RECEPTOR AND α-NEUROTOXINS AND DEVELOPMENT OF A SYNTHETIC VACCINE AGAINST TOXIN POISONING

M. Zouhair Atassi and Behzod Z. Dolimbek

Department of Biochemistry
Baylor College of Medicine
Houston, Texas 77030

INTRODUCTION

The nicotinic acetylcholine receptor, (AChR) is a membrane protein on the postsynaptic neuromuscular junction. It has a principal role in postsynaptic neuromuscular transmission because it mediates ion flux across the membrane (1-2). The receptor is a pentamer composed of four subunits $\alpha_2\beta\gamma\delta$. Functional studies have focused mostly on the α-subunit because it is responsible for binding acetycholine (3-5) and α-neurotoxins (6). Snake venom postsynaptic neurotoxins form a large family of related proteins of which two subgroups, the long and short neurotoxins, are major constituents. Both long and short neurotoxins are known to bind specifically to the α-chain of AChR in a competitive manner with cholinergic ligands (7-8), but display differences in their association and dissociation kinetics.

The primary structures of the four AChR subunits of *Torpedo californica* (*t*) (9-12) and mouse (*m*) (13-16) and the α-subunits of human (*h*) and bovine (17) have been deduced from the respective cDNA or mRNA sequences. From the primary structure of each AChR subunit, it was possible to identify transmembrane hydrophobic regions and the extracellular part of the chain (11,18,19). We carried out immunological and toxin-binding studies on inter-transmembrane synthetic peptides which confirmed (20) the model postulating five transmembrane regions (18,19). These investigations afforded an outline for the transmembrane organization of the AChR subunits and a working 3-D model for the α-neurotoxin and the AChR binding cavity on AChR.

In recent work, applying a comprehensive synthetic strategy previously introduced and developed by this laboratory (21,22), we mapped (23) the extracellular surface of the α-chain of *t*AChR for regions that are accessible to binding with antibodies against a panel of synthetic overlapping peptides which encompassed the entire extracellular

Methods in Protein Structure Analysis, Edited by M. Z. Atassi and E. Appella
Plenum Press, New York, 1995

parts of the chain. The binding of the anti-peptide antibodies to membrane-bound *t*AChR and to isolated, soluble *t*AChR was determined. These binding studies, which enabled a comparison of the accessible regions in membrane-bound AChR and free AChR, revealed that the receptor undergoes considerable changes in conformation upon removal from the cell membrane (23). Also, we determined the structural organization of the main extracellular domain of *m*AChR α-subunit on ***live*** mouse muscle cells in culture (24). A comparison of this binding profile and the profile obtained with membrane-bound *t*AChR in isolated membrane fractions showed some similarities as well as significant differences between the subunit organization in the isolated membrane fraction and in the membrane on ***live*** muscle cells. The exposed regions defined by this study (24) may be the primary targets for the initial autoimmune attack on the receptors in experimental autoimmune myasthenia gravis.

Recent studies from this laboratory mapped the full profile of binding regions for long α-neurotoxins [α-bungarotoxin (BTX) and cobratoxin (Cbt)] on the α-subunits of the *t*AChR (25-27) and *h*AChR (28) as well as the binding regions for short neurotoxins [erabutoxin (Eb) and cobrotoxin (Cot)] on the α-subunits of *t*AChR and *h*AChR (29). Conversely, the binding regions for AChR on BTX were mapped by synthetic peptides representing each of the toxin loops (30,31). Identification of the binding regions on AChR for short and long neurotoxins has provided a molecular explanation for the observed differences between the two toxin groups in their actions on AChR (29).

Interaction of acetylcholine receptor with acetylcholine and α-neurotoxins

Localization and synthesis of the acetylcholine-binding site

On the basis of sequence analysis and structural topology of the α-subunit, it has been proposed that the invariant cysteine residues 128 and 142 form a disulfide bridge, the integrity of which is essential for the binding of ACh to the receptor (1,9-11,32). We have localized, by peptide synthesis, the acetylcholine-binding site in both human and *Torpedo* receptors (5). A peptide containing this loop region (residues α125-147) was synthesized and solid-phase radiometric binding assays demonstrated that it had a high binding of ^{125}I-labeled BTX (5, 25). It was further shown that the free peptide bound well to [^{3}H] acetylcholine (5). Pretreatment of peptide α-125-147 with 2-mercaptoethanol destroyed its binding activity, clearly showing that the integrity of the disulfide bonded loop structure was essential for binding. Unlabeled ACh also inhibited the binding of labeled ACh to the synthetic peptide. The region α125-147, therefore, contains essential elements of the ACh-binding site of AChR (5). It is not surprising, therefore, that immune responses to this peptide are involved in the pathogenesis of experimental autoimmune myasthenia gravis (33,34). It has been noted (5), however, that the results do not preclude the possibility that additional residues, residing outside the region α125-147, are involved in the binding of acetylcholine to AChR.

The α-Neurotoxin Binding Regions on Human and *Torpedo* AChR

A comprehensive synthetic-peptide strategy we had originated (21,22) was applied to *t*AChR and *h*AChR (Figure 1) and enabled us to map the full profile of the continuous binding regions for long and short α-neurotoxins on the extracellular part (residues

Peptide Position	Species	Structure
α1-16	Human	S E H E T R L V A K L **F K D Y S**
	Torpedo	- - - - - - - - - - N - **L E N - N**
α12-27	Human	**F K D Y S** S V V R P V **E D H R Q**
	Torpedo	**L E N - N** K - I - - - **- H - T H**
α23-38	Human	**E D H R Q** V V E V T V **G L Q L I**
	Torpedo	**- H - T H** F - D I - - **- - - - -**
α34-49	Human	**G L Q L I** Q L I N V D **E V N Q I**
	Torpedo	**- - - - -** - - - S - - **- - - - -**
α45-60	Human	**E V N Q I** V T T N V R **L K Q Q W**
	Torpedo	**- - - - -** - E - - - - **- R - - -**
α56-71	Human	**L K Q Q W** V D Y N L K **W N P D D**
	Torpedo	**- R - - -** I - V R - R **- - - A -**
α67-82	Human	**W N P D D** Y G G V K K **I H I P S**
	Torpedo	**- - - A -** - - - I - - **- R L - -**
α78-93	Human	**I H I P S** E K I W R P **D L V L Y**
	Torpedo	**- R L - -** D D V - L - **- - - - -**
α89-104	Human	**D L V L Y** N N A D G D **F A I V K**
	Torpedo	**- - - - -** - - - - - - **- - - - H**
α100-115	Human	**F A I V K** F T K V L L **Q Y T G H**
	Torpedo	**- - - - H** M - - L - - **D - - - K**
α111-126	Human	**Q Y T G H** I T W T P P **A I F K S**
	Torpedo	**D - - - K** - M - - - - **- - - - -**
α122-138	Human	**A I F K S** Y G E I I V T **H F P F D**
	Torpedo	**- - - - -** - - - - - - - **- - - - -**
α134-150	Human	**H F P F D** E Q N G S M K **L G T W T**
	Torpedo	**- - - - -** Q- - - -T - - **- - I - -**
α146-162	Human	**L G T W T** Y D G S V V A **I N P E S**
	Torpedo	**- - I - -** - - - T K - S **- S - - -**
α158-174	Human	**I N P E S** D Q P D L S N **F M E S G**
	Torpedo	**- S - - -** - R - - - - T **- - - - -**
α170-186	Human	**F M E S G** E W V I K E S **R G W K H**
	Torpedo	**- - - - -** - - - M - D Y **- - - - -**
α182-198	Human	**R G W K H** S V T Y S G G **P D T P Y**
	Torpedo	**- - - - -** W - Y - T - - **- - - - -**
α194-210	Human	**P D T P Y** L D I T Y H F V M Q R L
	Torpedo	**- - - - -** - - - - - - - I - - - I
α262-276	Human	E L I P S T S S A V P L I G K
	Torpedo	- - - - - - - - - - - - - - -

Figure 1. Covalent structures of the synthetic overlapping peptides representing the extracellular part of each of the α-chains of human and *Torpedo californica* AChRs. The upper sequences of each pair of peptides give the full primary structures of the human AChR peptides and, under these, only the residues that are different in the corresponding *Torpedo* peptides are given. Segments in **bold type** represent the five-residue overlaps between consecutive peptides.

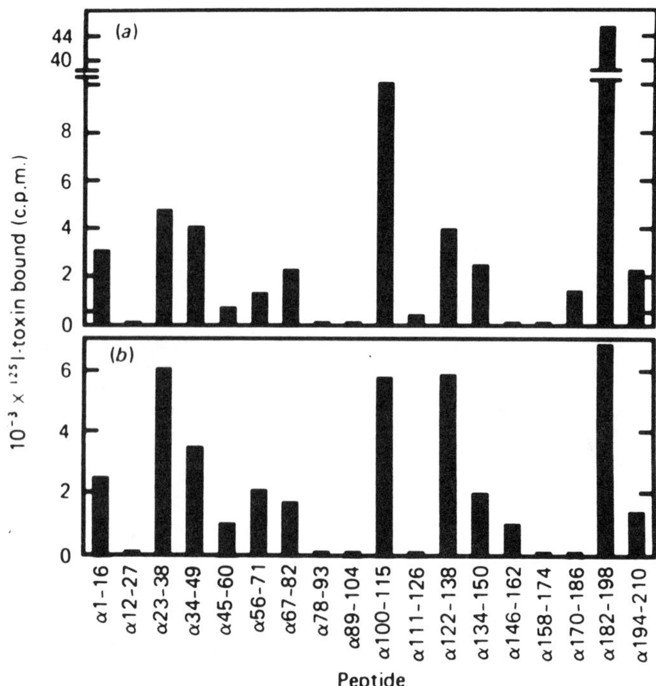

Figure 2. Summary of the binding profiles of (a) BTX and (b) cobratoxin to the synthetic overlapping peptides of *t*AChR. The bars represent the maximum binding values to 25μl of a 1:1 (v/v) suspension of each peptide adsorbent. (From Mulac-Jericevic and Atassi, refs. 26, 27.)

α1-210) of the α-chains of *t*AChR and *h*AChR. In *t*AChR, the binding regions for long neurotoxins were found (25-27) to reside within (but may not include all of) residues *t*α1-10, *t*α32-49, *t*α100-115, *t*α122-138 and *t*α182-198 (Figure 2). In human AChR, long neurotoxins bind to regions *h*α32-49, *h*α100-115, *h*α122-138 and *h*α194-210 (28,35). For short-neurotoxin binding on the α-chain of *t*AChR, five Cot-binding regions (Figure 3) were found to reside within peptides *t*α1-16, *t*α23-38/*t*α34-49 overlap, *t*α100-115, *t*α122-138 and *t*α194-210. The Eb-binding regions were localized (Figure 3) within peptides *t*α23-38/*t*α34-49/*t*α45-60 overlap, *t*α100-115 and *t*α122-138. The main binding activity for both toxins resided within region *t*α122-138. The binding of long α-neurotoxins [BTX and Cbt] involved the same regions of *t*AChR as well as an additional region within the residues α182-198. Thus, region α182-198, which is the strongest binding region for long neurotoxins on *t*AChR, was not a binding region for short neurotoxins (29). On *h*AChR, peptide *h*α122-138 possessed the highest activity with both toxins (Figure 4), and lower activity was found in the overlap *h*α23-38/*h*α34-49/*h*α45-60 and in peptide *h*α194-210. In addition, peptides *h*α100-115 and *h*α56-71 showed strong and medium binding activities to Eb, but low activity to Cot, whereas peptide *h*α1-16 exhibited low binding to Cot and no binding to Eb. Comparison with the aforementioned studies (28,35) indicated that, for *h*AChR, the binding regions of short and long neurotoxins were essentially the same (29). The finding that the region within residues α122-138 of both human and *Torpedo* AChR possessed the highest binding activity with short neurotoxins indicated that this region constitutes a universal binding region for long and short neurotoxins on AChR from various species (29).

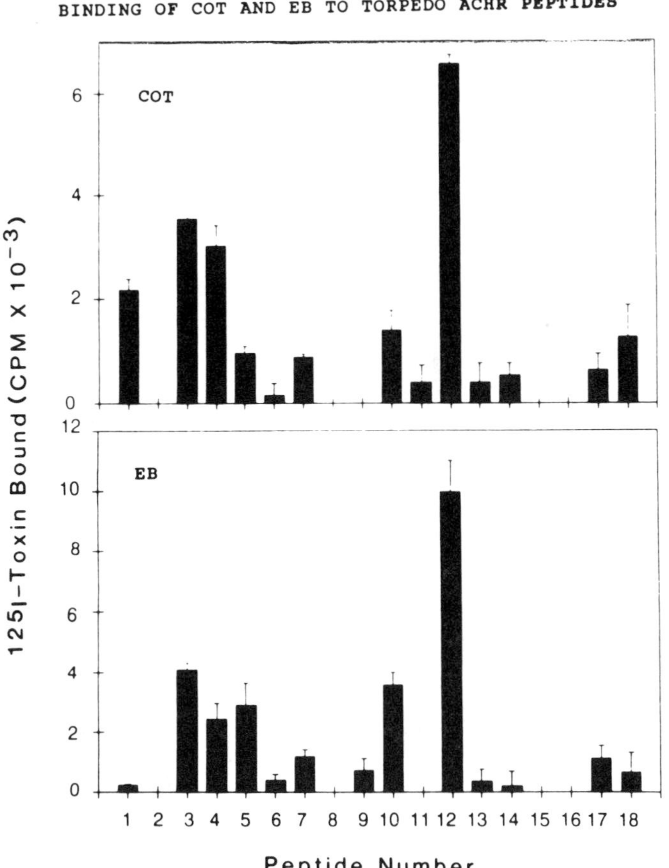

Figure 3. Summary of the profiles of (a) cobrotoxin (Cot) and (b) erabutoxin (Eb) binding to the synthetic overlapping peptides of the extracellular part of the α-chain of *t*AChR. The peptide numbers refer to the sequences given in Fig. 1. The binding values of ^{125}I- labeled Cot and Eb to *t*AChR were 55140 ± 1350 and 68550 ± 1520 cpm, respectively. Binding to unrelated proteins (bovine serum albumin, horse myoglobin) and peptides ((sperm-whale myoglobin synthetic peptides 1-17, 25-41 and 121-137 (ref. 41)) (negative controls) was 650 ± 220 cpm. (From Ruan *et al.*, ref. 29.)

Mapping of the Acetylcholine Receptor-Binding Sites on α-Bungarotoxin

The amino acid sequences of numerous snake venom toxins have been determined (36). These sequences tend to fall under three classification: short neurotoxins, long neurotoxins, and cytotoxins (reviewed in ref. 36). Pharmacologically active peptides, with effects ranging from those of the neurotoxins (i.e., muscle paralysis and respiratory failure) to those of the cytotoxins (i.e., hemolysis, cytolysis, cardiotoxic effects, and muscle depolarization), can be designed and synthesized based on the structure of short neurotoxins, long neurotoxins, and cytotoxins (37). Therefore, we adopted a synthetic approach to dissect the activities of these toxins. The approach has, thus far, been applied to localize the distinct AChR-binding regions on BTX. The entire toxin molecule was essentially subdivided into unique, potentially active regions, and the peptides were designed to mimic as closely as possible the native regional structure (31).

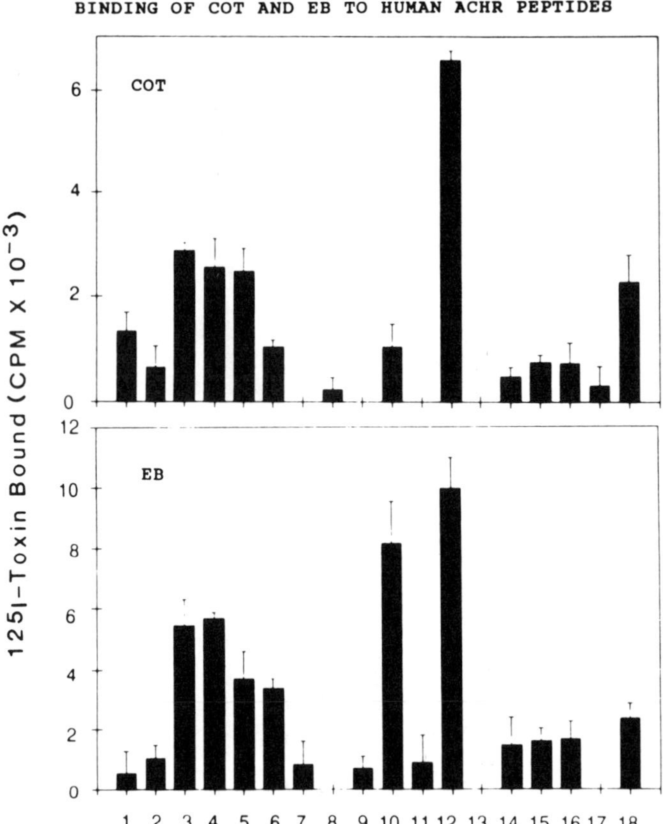

Figure 4. Summary of the binding profiles of Cot and Eb to the synthetic overlapping peptides of the extracellular part of the α-chain of human AChR. The peptide numbers refer to the sequences given in Fig. 1. (From Ruan *et al.*, ref. 29.)

Accordingly the following panel of BTX peptides were constructed (31) (Figure 5)

Loop 1 Peptide (L1). residues 3-16, with an artificial disulfide between two terminal cysteines.

NH_2-Terminal Extension of the Loop 1 Peptide (L1/N-Tail). residues 1-16, constructed as in L1 but further extended with the hydrophobic potentially interactive NH_2-terminal residues 1 and 2.

Loop 2 Peptide (L2). residues 26-41, with cysteine substitutions at both ends of the peptide providing an artificial intramolecular disulfide linkage between these two residues; alanine and threonine substitutions at residues 29 and 33, respectively, to eliminate the disulfide of BTX loop 5 and, thereby, avoid the formation of disulfide-linked polymers (30).

Loop 3 Peptide Corresponding to Loop 3 (L3). residues 48-59, clasped at naturally-occurring cysteines 48 and 59 of BTX on the terminals of the peptide.

A. Covalent structure of the synthetic BgTX peptides

```
                     3┌──────────────────────────┐16
L1                    C-H-T-T-A-T-I-P-S-S-A-V-T-C-(G)

                     3┌──────────────────────────┐16
L1/N-tail         I-V-C-H-T-T-A-T-I-P-S-S-A-V-T-C-(G)

                      ┌26  28                          41┐
L2                    C-K-M-W-A-D-A-F-T-S-S-R-G-K-V-V-E-C-G

                      ┌26                              41┐
L2 (G)                C-K-M-G-A-D-A-F-T-S-S-R-G-K-V-V-E-C-G

                    48┌──────────────────────┐59
L3                    C-P-S-K-K-P-Y-E-E-V-T-C-(G)

              45      ┌──────────────────────┐59
L3/Ext        A-A-T-C-P-S-K-K-P-Y-E-E-V-T-C-(G)

                    60┌─────────┐ 66            74
L4/C-tail             C-S-T-D-K-C-N-H-P-P-K-R-Q-P-G

                                  66            74
C-tail                            N-H-P-P-K-R-Q-P-G
```

B. Covalent structure of the randomized sequence analogs of the BgTX peptides.

```
R.L1-N-tail     T.H.C.I.T.V.A.S.T.P.I.T.S.V.A.C.G.

R.L2            C.W.V.R.D.T.A.M.F.K.G.A.K.S.E.V.S.C.G.

R.L3/Ext        K.S.P.C.A.Y.K.E.P.E.T.T.V.A.C.G.
```

Figure 5. Structure of the synthetic peptides representing: (a) the loops and exposed regions of BTX; and (b) three peptide analogs which had the same amino acid composition as the respective peptides L1/N-tail, L2 and L3/Ext, but whose sequences were randomized. (Figure is from Atassi *et al*, ref. 39.)

Loop 3 Extended toward the NH_2 terminal by the Three Residues (L3/Ext). Ala-45, Ala-46, and Thr-47 added at the N-terminal of loop 3 (i.e. residues 45-59).

An Extension of the COOH-Terminated Tail (L4/C-Tail). residues 60-74 which included the fourth loop of BTX between Cys-60 and Cys-65.

A COOH-Terminal Linear Peptide (C-Tail). residues 66-74.

In all experiments, the peptides were purified and, when appropriate, the monomeric cyclic structures were prepared.

The ability of these peptides to bind *t*AChR was studied (31) by radiometric absorbent titrations. Three regions, represented by peptides 1-16, 26-41, and 45-59, were able to bind ^{125}I-labeled *t*AChR and, conversely, ^{125}I-labeled peptides were bound by *t*AChR. In these regions, residues Ile-1, Val-2, Trp-28, Lys-26 and/or Lys-38, and one or all of the three residues Ala-45, Ala-46, and Thr-47, are essential contact residues in the binding BTX to receptor. Other synthetic regions of BTX showed little or no *t*AChR-binding activity. The specificity of *t*AChR binding to peptides 1-16, 26-41, and 45-49 was confirmed by inhibition with unlabeled BTX.

Other parts of the BTX molecule make little or no contribution to its binding to AChR. The region within peptide L2 makes a higher contribution to the binding activity of BTX than do the regions within the peptides L1/N-tail and L3/Ext The radioiodination of peptide L1/N-tail (most likely at His-4) and L3/Ext (most likely at Tyr-54), appear to have some adverse effects on the binding of the respective peptide to *t*AChR. Thus, peptide L1/N-tail and L3/Ext exhibited lower affinity than peptide L2 when the binding of the labeled peptides to *t*AChR adsorbent was inhibited by unlabeled BTX (IC_{50} values: L2, 8.4 x 10^{-8}M; L1/N-tail, 8.2 x 10^{-7}M; L3/Ext, 4.4 x10^{-7}M). By contrast, the three peptides had comparable affinities (IC_{50} values: L2,1.5 x 10^{-7}M; L1/N-tail, 4.2 x 10^{-7}M; L3/Ext 5.1 x 10^{-7}M) when binding of ^{125}I-labeled *t*AChR to peptide adsorbents was inhibited by unlabeled BTX, giving almost superimposable inhibition curves (31).

It was concluded (31) that BTX has three main AChR-binding regions (loop 1 with NH_2-terminal extension, loop 2, and loop 3 extended toward the NH_2-terminal by residues 45-47).

The α-Neurotoxin Binding Cavity of Human AChR

We have recently described an approach for studying the details of protein-protein recognition (35). Each of the active peptides of one protein is allowed to interact with each of the active peptides of the other protein. Based on the relative binding affinities of peptide-peptide interactions, the disposition of two protein molecules in a complex can be described if the 3-D structure of one of the two molecules is known. The peptides of the binding site of one protein (whose 3-D structure is not known) are docked onto the appropriate regions of the other whose 3-D structure is known, by computer graphics and energy minimization thus allowing a 3-D model to be constructed of the unknown binding-site cavity. The validity of this approach was first established with peptides corresponding to regions on the β chain of human hemoglobin involved in binding to the α chain (35). As mentioned above, the regions on *h*AChR and *t*AChR which bind BTX have been localized. Also, the binding regions for *t*AChR on BTX were mapped by synthetic peptides representing each of the BTX loops (30,31). In recent work, (35), peptides representing the active regions of one molecule were allowed to bind to each of the active-region peptides of the other molecule. Thus, the interaction of three-BTX synthetic loop peptides with four synthetic peptides representing the toxin-binding regions on *h*AChR permitted the determination of the region-region interactions between BTX and the human receptor. Based on the known 3-D structure of BTX (38), the active peptides of *h*AChR were then assembled to their appropriate toxin-contact regions by computer model building and energy minimization. This allowed the three-dimensional construction of the toxin-binding cavity on *h*AChR (Figure 6). The cavity appears to be conical, 30.5Å in depth, involving several AChR regions that make contact with the BTX loop regions. One AChR region (within residues α125-136) involved in the binding to BTX also resides in the aforementioned ACh binding site (5). This demonstrates in three dimensions a critical site involved in both ACh activation and BTX blocking. Thus, studying the interaction between peptides representing the binding regions of two protein molecules may provide an approach in molecular recognition by which the binding site on one protein can be described if the 3-D structure of the other protein is known (35)

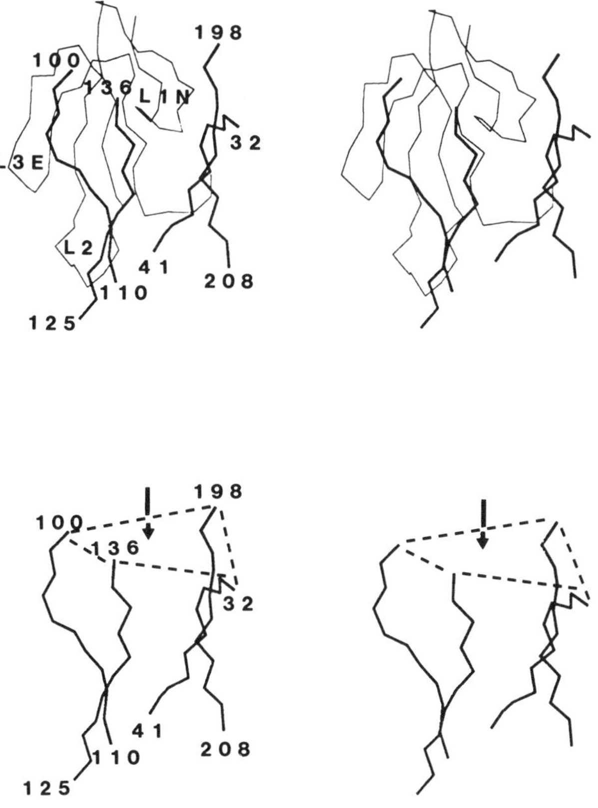

Figure 6. A stereo drawing of a 3-D construction of the toxin-binding cavity in AChR with the BTX molecule (backbone only) bound in the cavity *(Upper)* and without the BTX molecule *(Lower)*. The somewhat conical cavity has the following dimensions: residues 100-136, 21.3Å; residues 136-32, 35.0Å; residues 32-198, 16.06Å; and residues 198-100, 22.13Å. The depth of the cavity is 30.48Å. (From Ruan *et al.*, ref. 35.)

Antibody and T-cell Recognition of α-Bungarotoxin and Its Synthetic Peptides

The aforementioned peptides representing the loops and surface regions of BTX, as well as control peptide analogs in which these sequences were randomized were used to map the recognition profiles of the antibodies and T-cells obtained after BTX immunization (39). Also, the abilities of anti-peptide antibodies and T-cells to recognize the immunizing peptide and BTX were determined (39).

Responses to Immunization with BTX

Three regions of BTX were immunodominant by both rabbit and mouse anti-BTX antibodies (Table 1). These regions resided within loops L1 (residues 3-16), L2 (residues 26-41) and the C-terminal tail (residues 66-74) of the toxin. The regions recognized by BTX-primed T-lymphocytes were mapped in five mouse strains: C57BL/6 (H-2^b), Balb/c (H-2^d), CBA (H-2^k), C3H/He (H-2^k) and SJL (H-2^s). The H-2^b and H-2^d haplotypes were high

Table 1. Binding of anti-BTX antibodies to synthetic BTX peptides.

Peptides	^{125}I-labeled antibodies bound (cpm)$^{+}$			
	Mouse antiserum #233	Mouse antiserum #253	Mouse antiserum #236	Rabbits antiserum
BTX	**36,480**	**34,746**	**38,015**	**44,120**
L1	**14,675**	**15,220**	**18,715**	**16,768**
L1/N-tail	13,040	14,706	16,420	14,487
L2	**5,204**	**6,500**	**8,590**	**10,952**
L2(G)	1,879	2,361	3,340	4,953
L3	2,419	3,380	2,570	3,682
L3/Ext	1,796	1,830	2,340	4,952
L4/C-tail	5,105	7,256	7,374	9,122
C-tail	**16,481**	**11,588**	**12,256**	**17,468**
Controls				
R.L2	875	1,092	1,324	1,108
R.L3/Ext	962	867	1,011	1,487
Nonsense peptide	103	683	764	684
Bovine serum albumin	895	1,121	1,157	985
Myoglobin	725	985	1,270	nd

$^{+}$Results were obtained by radioimmunoadsorbent titrations and represent the average plateau values of three replicate analyses which varied ± 1.3% or less. Values have not been corrected for non-specific binding, but values of negative controls are shown. (Table is from ref. 39.)

responders to BTX, while the H-2^k and H-2^s were intermediate responders. The T-cell recognition profile of the peptides varied with the haplotype (Figure 7), consistent with Ir-gene control of the responses to the individual regions. The submolecular specificities of antibodies and T-cells were compared in three of the mouse strains (C57BL/6, Balb/c and SJL) (Table 2). In a given mouse strain, there were regions that were strongly recognized by both antibodies and T-cells as well as regions that wee predominantly recognized either by antibodies or by T-cells.

Table 2. Comparison of the specificities of antibody and T-cell responses against BTX in three independent mouse haplotypes

BTX or peptide	Response levels following BgTX immunization*					
	C57BL/6 (H-2^b)		Balb/c (H-2^d)		SJL (H-2^s)	
	Antibody	T-cell	Antibody	T-cell	Antibody	T-cell
BTX	5+	4+	5+	5+	5+	3+
L1	3+	2+	4+	2+	4+	2+
L1/N-tail	3+	3+	4+	3+	3+	2+
L2	2+	2+	2+	4+	2+	2+
L3	1+	2+	2+	3+	—	2+
L3/Ext	1+	3+	2+	3+	2+	2+
L4/C-tail	2+	3+	3+	4+	2+	2+
C-tail	3+	3+	4+	4+	4+	2+

*The number of + signs denote the following levels of antibody and T-cell responses (in net cpm values): 1+, 3000-5000, 2+, 5,100-10,000; 3+, 10,000-20,000; 4+, 20,100-30,000; 5+ > 30,000. (Table is from ref. 39.)

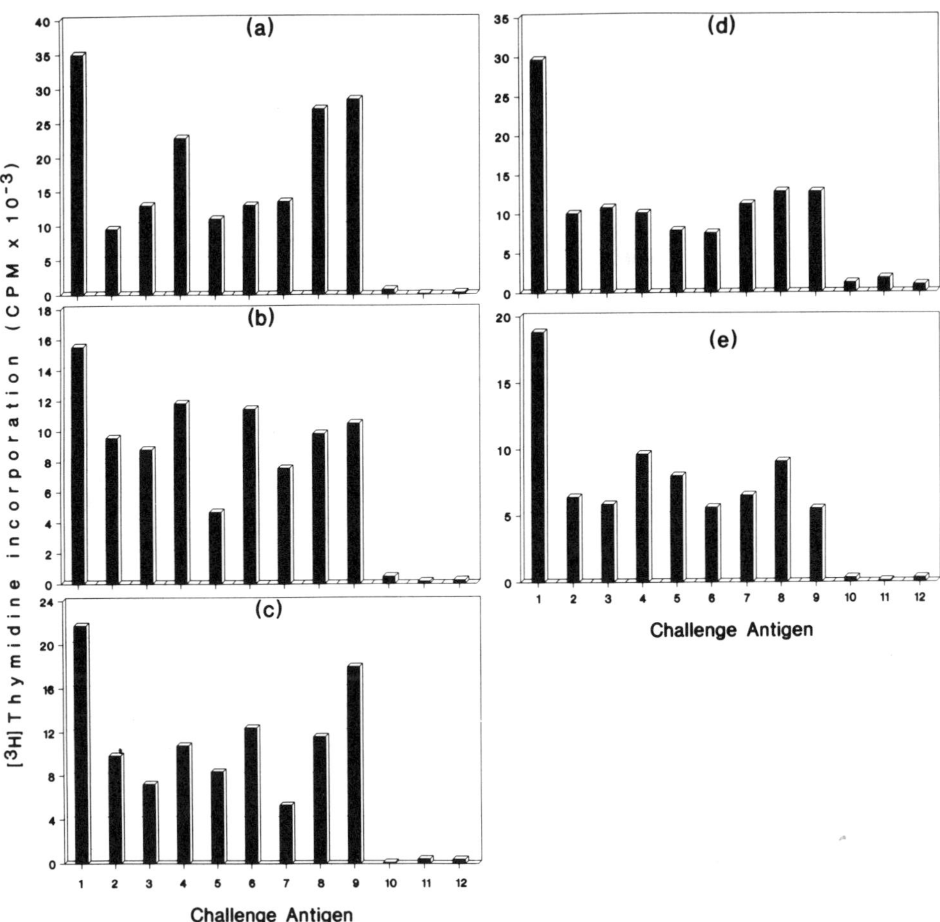

Figure 7. A bar diagram showing the maximum proliferative responses (at optimum doses) *in vitro* to BTX, its peptides, or negative controls of T cells from BTX-primed mice of: (a) Balb/c; (b) CBA/JCr; (c) C3H/He; (d) C57BL/6; (e) SJL. The challenge antigens were: (1) BTX; (2) L1; (3) L1/N-tail; (4) L2; (5) L2 (G); (6) L3; (7) L3/Ext; (8) L4/Ctail; (9) C-tail; (10) bovine serum albumin; (11) myoglobin: (12) nonsense peptide. (Figure is from Atassi *et al.*, ref. 39.)

Reponses to Immunization with BTX Peptides

The peptides were used as immunogens in their free form (i.e. without coupling to any carrier) in two of the mouse strains, Balb/c and SJL (39). In both mouse strains, the peptides gave strong antibody responses (Table 3). Antibodies against peptide L2 showed the highest binding to intact BTX. Antibodies against the other peptides exhibited lower binding activity to the intact toxin and this activity was dependent on the peptide and the mouse strain (39). The response of peptide-primed T-cells to a given immunizing peptide was not related to whether this region was immunodominant with BTX-primed T-cells. The ability of peptide-primed T-cells to recognize the intact toxin varied with the peptide and was dependent on the host strain (Figure 8). These results indicate that anti-peptide antibody and T-cell responses are also under genetic control and that their ability to cross-react with the parent toxin is not only dependent on the conformational exposure of the correlate region in intact BTX.

Protective Immune Responses by Immunization with the Synthetic BTX Peptides

Since the differences in the abilities of the antibodies against the various peptides to bind intact BgTX were not very significant and since these activities did not necessarily correlate with the ability of anti-peptide T-cells to recognize BTX, it was decided to test each of the peptides for its capacity to generate protective immune responses (40).

In both Balb/c and SJL, peptides L1, L2 and C-tail were most protective against BTX poisoning (Table 4) (protection index, PI, defined as LD_{50} of immunized mice/LD_{50} for unimmunized controls: Balb/c, 3.2; SJL, 2.5-2.7). Protective immunity exhibited by the other peptides was also quite substantial (PI: Balb/c, 2.5-2.6; SJL, 2.2-2.4). It is noteworthy that the three most protective peptides (L1, L2, and C-tail) were also immunodominant in terms of binding of anti-toxin antibodies, suggesting perhaps that, for identification of the most protective regions, it would have been sufficient to map the immunodominant regions towards anti-BTX antibodies (40).

Since each of the peptides L1, L2, and C-tail was quite protective (increasing the LD_{50} of BTX about 3 fold relative to control mice), it was important to determine whether

Table 3. Binding of anti-peptide antibodies to the immunizing peptide and to BTX

	Antibodies bound (net cpm)					
	Balb/c antibodies			SJL antibodies		
Antigen	No. of mice	Binding to peptide	Binding to BTX	No. of mice	Binding to peptide	Binding to BTX
L1	10	58,942	16,432	8	77,639	2,128
L1/N-tail	8	95,793	8,460	9	62,183	1,964
L2	9	88,619	12,831	9	65,676	10,735
L3	9	49,601	6,200	10	38,654	1,074
L3/Ext	10	35,217	1,447	8	43,704	3,361
L4/C-tail	8	63,909	4,170	10	73,795	3,802
C-tail	8	43,306	5,149	9	44,393	6,062

Antisera were raised against each of the peptides in Balb/c and SJL mice and represent pools of the 87-day bleeds from the number of mice shown. For RIA, the antisera were pre-diluted 1:500 (v/v) with 0.15 M NaCl in 0.01 M sodium phosphate buffer, pH 7.2, containing 0.1% bovine serum albumin (details given in ref. 39). Values have been corrected for non-specific binding of each antiserum to negative controls, the levels of which were similar to those shown in Table 1. (Table is from ref. 39.)

higher protection will be achieved by immunizing mice with all three peptides simultaneously. These studies (which were done only in Balb/c) clearly showed that this was indeed the case (40). Immunization with an equimolar mixture of the peptides allowed the mice to survive BTX challenge doses which were 4.6 fold higher than control mice (Table 4). In other words, immunization with an equimolar mixture of peptides L1, L2 and C-tail was 42% more protective, in terms of survivable BTX challenge dose, than any of the three peptides by itself. Clearly, antibodies against all three regions are more efficient at neutralizing toxin poisoning than antibodies against any single region. The protective capacity of the peptide mixture was somewhat related to titer of the fraction, in anti-peptide antibodies, that binds to BTX. But the titers of these antibodies were moderate and did not increase substantially over an extended period of immunization. It was therefore decided to determine the protective ability of a peptide-carrier conjugate.

The three peptides L1, L2 and C-tail were conjugated to a single carrier, ovalbumin (40). Analysis of the conjugate showed that coupling levels of the peptides differed. This is to be expected because each peptide has different reactivity of side chains and accessibility requirements on the surface of the ovalbumin carrier. It was important to find that the conjugate generated high titer antibodies that bound to intact BTX. This immunogen (i.e. the conjugate) afforded excellent protection against BTX challenge (PI>18.1) (Table 4). In

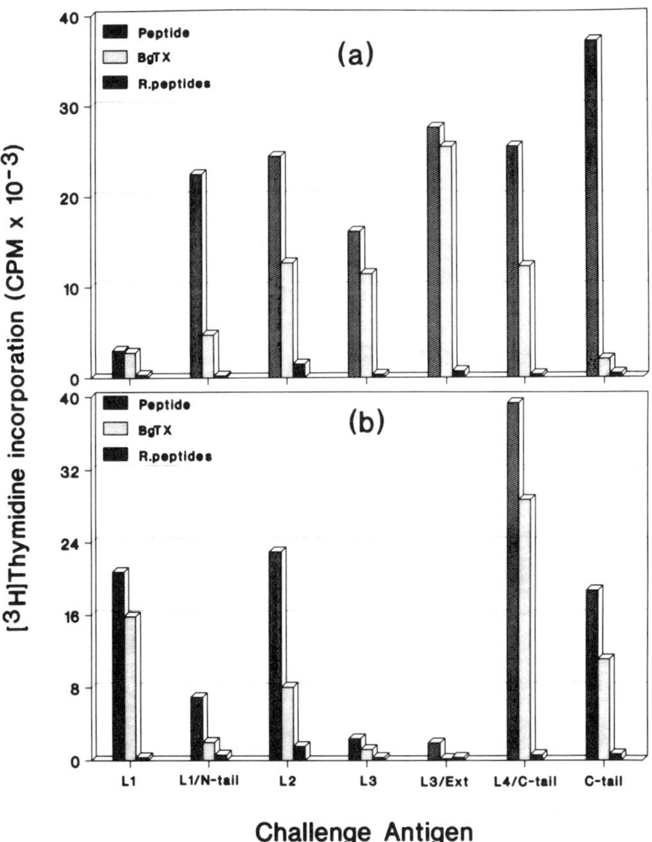

Figure 8. A bar diagram showing the responses of peptide-primed T cells to the optimum challenge dose of the respective priming peptide, BTX or random peptides. Peptide-primed T-cells were from: (a) Balb/c mice; (b) SJL mice. (Figure is from Atassi *et al,* ref. 39.)

Table 4. Protection of mice against BTX by immunization with BTX or with synthetic BTX peptides

Immunizing antigen	Protection parameters for Balb/c and SJL mice			
	Balb/c		SJL	
	LD_{50} (μg BTX/mouse)	Protection index	LD_{50} (μg BTX/mouse)	Protection index
None or random peptides[+]	3.20	1.00	3.60	1.00
L1	10.27	3.21	8.86	2.46
L1/N-tail	8.36	2.61	7.86	2.18
L2	10.27	3.21	9.76	2.71
L3	7.86	2.46	7.94	2.21
L3/Ext	8.36	2.61	8.57	2.38
L4/C-tail	8.36	2.61	8.64	2.40
C-tail	10.27	3.21	8.86	2.46
BTX	31.0	9.69	26.50	7.36
Mixture of L1, L2, C-tail	14.63	4.57	nd	nd
Multi-peptide conjugate of L1, L2, C-tail	>57.8	>18.1	nd	nd

[+]This group includes 45 unimmunized mice and mice that were immunized with randomized sequence peptides R.L1/N-tail (45 mice), R.L2 (45 mice), and R.L3/Ext (45 mice).

fact, the multi-peptide conjugate was almost twice as protective as whole toxin immunization (PI = 9.7). In addition, unlike BTX, the multi-peptide conjugate is not toxic and, therefore, there is no risk of poisoning the recipient by the immunogen in the process of vaccination. Clearly, the multi-peptide conjugate will constitute an excellent vaccine against toxin poisoning.

CONCLUSIONS

Using a comprehensive synthetic peptide strategy which originated in this laboratory (21,22), we have mapped the extracellular domain of AChR α-chain for accessibility to anti-peptide antibodies in isolated membrane fractions and in live muscle cells in culture. We also mapped on this domain in *t*AChR and in *h*AChR, the regions that are recognized by antibodies and by T-lymphocytes against the respective AChR and by long and short α-neurotoxins. Conversely, we mapped, by synthesis, the regions on BTX that bind the receptor as well as the antigenic regions of BTX that are recognized by rabbit and mouse anti-BTX antibodies. Three regions residing within peptides L1, L2 and C-tail were immunodominant. The regions recognized by BTX primed T-cell were also mapped in five mouse strains. Immunization of Balb/c and SJL mice with each of the synthetic peptides in its free form afforded considerable protection against BTX poisoning. Peptides L1, L2 and C-tail were most protective and mice immunized with these peptides survived LD_{50} values that were three times higher than non-immune control mice. Immunization with an equimolar mixture of the three peptides was even more protective and these mice survived even higher challenge doses of BgTX (4.6 fold higher than LD_{50} of controls). An ovalbumin conjugate carrying all three peptides, when used as an immunogen, displayed a high protection capability which was almost double that obtained by BTX immunization. The conjugate of the three peptides should serve as an effective vaccine against BTX poisoning.

ACKNOWLEDGMENTS

The work described in this article was supported by a grant (NS-26280) from the National Institute of Health and by a contract (DAMD 17-89-C-9061) from the Department of Defense, U.S. Army Medical Research and Development Command.

REFERENCES

1. Karlin, A. (1980) Molecular properties of nicotinic acetylcholine receptors. In *Cell Surface and Neuronal Function* (Edited by Colman, C.W., Poste, G. and Nicolson, G.L.), pp. 191-260. Elsevier/North-Holland Biomedical Press, New York.
2. Changeux, J.P., Devillers-Thiery A. and Chemouilli, P. (1984) Acetylcholine receptor: an allosteric protein. *Science* **225**, 1335-1345.
3. Sobel, A., Weber, M. and Changeux, J.P. (1977) Large-scale purification of the acetylcholine-receptor protein in its membrane-bound and detergent-extracted forms from *Torpedo marmorata* electric organ. *Eur. J. Biochem.* **80**, 215-224.
4. Tzartos, S.J. and Changeux, J.P. (1983) High affinity binding of α-bungarotoxin to the purified α-subunit and its 27,00-dalton proteolytic peptide from *Torpedo* marmorata acetylcholine receptor. Requirements for sodium dodecil sulfate. *EMBO J.* **2**, 381-387.
5. McCormick, D.J. and Atassi, M.Z. (1984) Localization and synthesis of the acetylcholine-binding site in the α-chain of the *Torpedo californica* acetylcholine receptor. *Biochem. J.* **224**, 995-1000.
6. Lee, C.Y. (1979) Recent advances and pharmacology of snake toxins. *Adv. Cytopharmacol.* **3**, 1-16.
7. Maelicke, A., Fulpius, B.W., Klett, R.P. and Reich, E. (1977) Acetylcholine receptor. Responses to drug binding. *J. Biol. Chem.* **252**, 4811-4830.
8. Haggerty, J.G. and Froehner, S.C. (1981) Restoration of ^{125}I-α-bungarotoxin binding activity to the α-subunit of *Torpedo* acetylcholine receptor isolated by gel electrophoresis in sodium dodecyl sulfate. *J. Bio Chem.* **256**, 8294-8297.
9. Noda, M., Takahashi, H., Tanabe, T., Toyosato, M., Furutani, Y., Hirose, T., Asai, M., Inayama, S., Miyata, T. and Numa, S. (1982) Primary structure of α-subunit precursor of *Torpedo californica* acetylcholine receptor deduced from cDNA sequence. *Nature (London)* **299**, 793-797.
10. Noda, M., Furutani, Y., Takahashi, H., Toyosato, M., Tanabe, T., Shimizu, S., Kikyotani, S., Kayano, T., Hirose, T., Inoyama, S. and Numa, S. (1983) Cloning and sequence analysis of calf cDNA and human genomic DNA encoding alpha-subunit precursor of muscle acetylcholine receptor. *Nature* **305**, 818-823.
11. Noda, M., Takahashi, H., Tanabe, T., Toyosato, M., Kikyotani, Miyata, T. and Numa, S. (1983) Structural homology of *Torpedo californica* acetylcholine receptor subunits. *Nature* **302**, 528-532.
12. Claudio, T., Ballivet, M., Patrick, J. and Heinemann S. (1983) Nucleotide and deduced amino acid sequences of *Torpedo californica* acetylcholine receptor subunit. *Proc. Nat. Acad. Sci. USA* **80**, 1111-1115.
13. Isenberg, K.E., Mudd, J., Shah, V. and Merlie, J.P. (1986) Nucleotide sequence of the mouse muscle nicotinic acetylcholine receptor alpha subunit. *Nucleic Acids Res.* **14**, 5111-5111; Boulter, J., Evans, K, Goldman, D., Martin, G., Treco, D., Heinemann, D. and Patrick J. (1986) Isolation of a cDNA clone coding for a possible neural nicotinic acetylcholine receptor α-subunit. *Nature (London)* **319**, 368-374.
14. Buonanno, A., Mudd, J., Shah, V. and Merlie, J.P. (1986) A universal oligonucleotide probe for acetylcholine receptor genes: Selection and sequencing of cDNA clones of the mouse muscle beta subunit. *J. Biol. Chem.* **261**, 16451-16458.
15. Yu., L, LaPolla, J. and Davidson, N. (1986) Mouse nicotinic acetylcholine receptor gamma subunit: cDNA sequence and gene expression. *Nucleic Acids Res.* **14**, 3539-3555.
16. LaPolla, R.J., Mayne, K.M. and Davidson, N. (1984) Isolation and characterization of a cDNA clone for the complete protein coding region of the delta subunit of the mouse acetylcholine receptor. *Proc. Natl. Acad. Sci. USA* **81**, 7970-7974.
17. Noda, M., Takahashi, H., Tanabe, T., Toyosato, M., Kikyotani, S., Hirose, T., Asai, M., Takashima, H., Inayama, S., Miyata, T., Numa, S. (1983) Primary structures of β- and δ-subunit precursors of *Torpedo californica* acetylcholine receptor deduced from cDNA sequences. *Nature (London)* **301**, 251-255.
18. Guy, H.R. (1983) A structural model of the acetylcholine receptor channel based on partition energy and helix packing calculations. *Biophys. J.* **45**, 249-261.

19. Finer-Moore, J. and Stroud, R.M. (1984) Amphipathic analysis and possible formation of the ion channel in an acetylcholine receptor. *Proc Natl. Acad. Sci.* **81**, 155-159.
20. Atassi, M.Z., Manshouri, T. and Yokoi, T. (1988) Recognition of inter-transmembrane regions of acetylcholine receptor α subunit by antibodies, T cells and neurotoxins. Implications for membrane subunit organization. *FEBS Lett.* **228**, 295-300.
21. Kazim, A.L. and Atassi M.Z. (1980) A novel and comprehensive synthetic approach for the elucidation of protein antigenic structures. Determination of the full antigenic profile of the α-chain of human haemoglobin. *Biochem. J.* **191,** 261-264.
22. Kazim, A.L. and Atassi, M.Z. (1982). Structurally inherent antigenic sites. Localization of the antigenic sites. Localization of the antigenic sites of the α-chain of human haemoglobin in three host species by a comprehensive synthetic approach. *Biochem. J.* **203**:201-208.
23. Atassi, M.Z. and Mulac-Jericevic, B. (1994). Mapping of the extracellular topography of the α-chain in free and in membrane-bound acetylcholine receptor by antibodies against overlapping peptides spanning the entire extracellular parts of the chain. *J. Prot. Chem.* **13**, 37-47.
24. Jinnai, K., Ashizawa, T. and Atassi, M.Z. (1994) Analysis of exposed regions on the main extracellular domain of mouse acetylcholine α-subunit in ***live*** muscle cells by binding profiles of anti peptide antibodies: *J. Prot. Chem.* **13**, 715-722
25. Mulac-Jericevic, B. and Atassi, M.Z. (1986) Segment α182-198 of *Torpedo californica* acetylcholine receptor contains a second toxin-binding region and binds anti-receptor antibodies. *FEBS Lett.* **199**, 68-74.
26. Mulac-Jericevic, B. and Atassi, M.Z. (1987) α-neurotoxin binding to acetylcholine receptor: localization of the full profile of the cobratoxin-binding regions in the α-chain of *Torpedo californica* acetylcholine receptor by a comprehensive synthetic strategy. *J. Prot. Chem.* **6**, 365-373.
27. Mulac-Jericevic, B. and Atassi, M.Z. (1987) Profile of the α-bungarotoxin binding regions on the extracellular part of the α-chain of *Torpedo californica* acetylcholine receptor. *Biochem. J.* **248**, 847-852.
28. Mulac-Jericevic, B., Manshouri, T., Yokoi, T. and Atassi, M.Z. (1988) The regions of α-neurotoxin binding on the extracellular part of the α-subunit of human acetylcholine receptor. *J. Prot. Chem.* **7**, 173-177.
29. Ruan, K.-H., Stiles, B.G. and Atassi, M.Z. (1991) The short-neurotoxin binding regions on the α-chain of human and *Torpedo californica* acetylcholine receptors. *Biochem. J.* **274**, 849-854.
30. McDaniel, C.S., Manshouri, T. and Atassi, M.Z. (1987) A novel peptide mimicking the interaction of α-neurotoxins with acetylcholine receptor. *J. Prot. Chem.* **6**, 455-461.
31. Atassi, M.Z., McDaniel, C.S. and Manshouri, T. (1988) Mapping by synthetic peptides of the binding sites for acetylcholine receptor on α-bungarotoxin. *J. Prot. Chem.* **7**, 655-666.
32. Devillers-Thiery, J.A., Giraudat, J., Bentaboulet, M. and Changeux, J.P. (1983). Complete mRNA coding sequence of the acetylcholine binding α-subunit of *Torpedo marmorata* acetylcholine receptor: A model for the transmembrane organization of the polypeptide chain. *Proc. Natl. Acad. Sci. USA,* **80**:2067-2071.
33. Lennon, V.A., McCormick, D.J., Lambert, E.H., Griesmann, G.E. and Atassi, M.Z. (1985) Region of peptide 125-147 of acetylcholine receptor α-subunit is exposed at neuromuscular junction and induces experimental autoimmune myasthenia gravis, T-cell immunity and modulating autoantibodies. *Proc. Natl. Acad. Sci. USA,* **82**, 8805-8809.
34. Atassi, M.Z., Ruan, K.H., Jinnai, K., Oshima, M. and Ashizawa, T. (1992) Epitope-specific suppression of antibody response in experimental autoimmune myasthenia gravis by an mPEG conjugate of a myasthenogenic synthetic peptide. *Proc. Natl. Sci. USA,* **89**, 5852-5856.
35. Ruan, K.-H., Spurlino, J., Quiocho, F.A. and Atassi, M.Z. (1990) Acetylcholine receptor α-bungarotoxin interactions: determination of the region-to-region contacts by peptide-peptide interactions and molecular modeling of the receptor cavity. *Proc. Natl. Acad. Sci. USA,* **87**, 6156-6160.
36. Endo, T. and Tamiya, N. (1987). Current view on the structure function relationship of post-synaptic neurotoxins from snake venom. *Pharmacol. Ther.* **34**, 403-451.
37. Atassi, M.Z. (1991). Postsynaptic-neurotoxin-acetylcholine receptor interactions and the binding sites on the two molecules. In: *Handbook of Natural Toxins,* ed. A. Tu, pp. 53-83, Marcel Dekker, New York.
38. Love, R.A. and Stroud, R.M. (1986) The crystal structure of α-bungartoxin 2.5. Å resolution related to solution structures and binding to acetylcholine receptor. *Protein Eng.* **1** 37-46.
39. Atassi, M.Z., Dolimbek, B.Z. and Manshouri, T. (1995) Antibody and T-cell recongition of α-bungarotoxin and its synthetic loop peptides. *Mol. Immunol.*, in press
40. Dolimbek, B.Z. and Atassi, M.Z. (1994) α-Bugarotoxin peptides afford a synthetic vaccine against toxin poisoning. *J. Prot. Chem.* **13**, 490-493.
41. Bixler, G.S. and Atassi, M.Z. (1983) Molecular localization of the full profile of the continuous regions recognized by myoglobin-primed T-cells using synthetic overlapping peptides encompassing the entire molecule. *Immunol. Commun.* **12**, 593-603.

28

IMMUNOLOGICAL APPROACH TO STUDY THE STRUCTURE OF OXIDIZED LOW DENSITY LIPOPROTEINS

Chao-Yuh Yang, Natalia V. Valentinova, Manlan Yang, Zi-Wei Gu, John R. Guyton, and Antonio M. Gotto, Jr.

Department of Medicine
Baylor College of Medicine and
The Methodist Hospital
6565 Fannin Street, MS/A601, Houston, Texas 77030

INTRODUCTION

Human low density lipoproteins (LDL), the major carriers of cholesterol in the bloodstream, plays the major role in supplying cells of tissues and organs with cholesterol. It is derived from the metabolism of the triglyceride-rich very low density lipoproteins (VLDL). Pathologic and epidemiologic studies have implicated that higher concentration of LDL in circulation is correlated with the development of atherosclerosis. Apolipoprotein (apo) B-100 serves as the ligand for the LDL receptor on cell surfaces. Thus, apoB-100 occupies a crucial position in the metabolic pathway of cholesterol and LDL. The complete primary structure of apoB-100 has been determined from its cDNA sequence (Chen et al., 1986; Knott,et al, 1986) and from its proteolytic peptide sequence information (Yang, et. al, 1986). ApoB-100 consists of 4536 amino acid residues with a calculated molecular mass of 513 kDa. Based on the differential trypsin releasibility of apoB-100 in LDL, apoB can be divided into 5 domains. Domain 1 contains 14 of the 25 cysteine (Cys) residues in apoB. Sixteen of the 25 Cys residues (which are numbered from 1 to 25 from the amino end to the carboxy end in apoB-100) exist in disulfide form. All 14 Cys residues in domain 1 are linked in disulfide form, and all except Cys1-Cys3 and Cys2-Cys4 are linked to neighboring Cys. Domain 4 contains 7 of the 16 N-glycosylated carbohydrates (Yang et al., 1989). Based on the published structural information (Yang et al., 1990), we proposed that the structure of apoB-100 in LDL is likely to be an elongated form that wraps around the LDL molecule as shown in Fig. 1 (Yang et al., 1992). The process of atherogenesis is believed to involve transformation of macrophages to lipid-laden foam cells. Degradation of native LDL by macrophages occurs at relatively low rates and doesn't cause any apparent accumulation of lipids in these cells (Goldstein et al.,1979). LDL modified *in vitro* by endothelial cells (Henriksen et al., 1983) or chemically (Goldstein et al.,1979) has been shown to enhance LDL interaction with macrophages and cause their *in vitro* transformation to foam cells.

Methods in Protein Structure Analysis, Edited by M. Z. Atassi and E. Appella
Plenum Press, New York, 1995

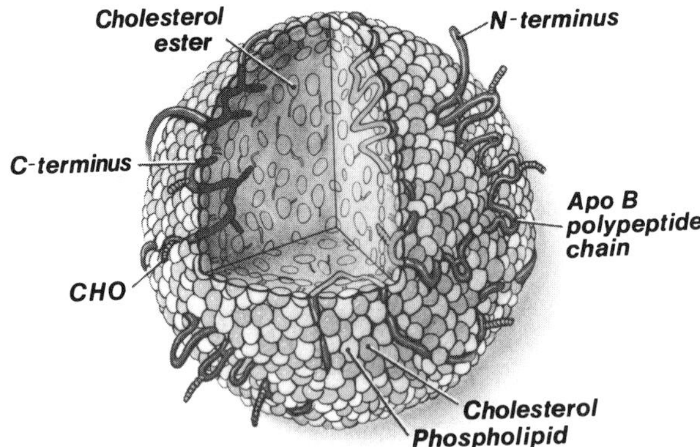

Figure 1. Structure of apoB-100 in LDL (From Yang, C.Y. and Pownall, H.J., In *Structure and Function of Plasma Apolipoproteins*, M. Rosseneu, Ed., CRC Press, Inc. 1992. With Permission).

Modification of LDL that occurs upon incubation with endothelial cells was demonstrated to be similar to LDL oxidation in the presence of oxygen and transition metal ions, causing similar changes in LDL structure that could be prevented by addition of antioxidants such as vitamin E and butylated hydroxytoluene (BHT) or chelators such as EDTA (Steinbrecher et al., 1984).

The changes of immunoreactivity of apoB upon LDL oxidation in the presence of Cu^{++} had been reported by a number of laboratories. They found that immunoreactivity of ox-LDL with different antibodies showed no changes (Gandjini et al 1991), reduced reactivity (Young et al. 1986; Zawadski et al. 1989) or increased reactivity and then decreased reactivity (Zawadski et al. 1989; Gandjini et al. 1991). We investigated 12 well characterized MAbs. Of these MAbs, 2 showed non or very little change, 6 had reduced reactivity, one (B6) had increased reactivity at the first 8 hours and then diminished upon prolonged oxidation, and 1 (4C11) had enhanced immunoreactivity as a function of time up to 24 hours and stabilized the immunoreactivity for another 16 hours (Valentinova et al., 1994).

MATERIALS AND METHODS

Antibodies

MAbs against LDL were Mb43 (Pease et al., 1990), Mb47 (Young et al., 1986), BL3, BL7 (Pease et al., 1990; Salmon et al., 1984), 4C11, 2G8, 8G4, and 5F8 (Yang et al., 1993; Yanushevskaya et al., 1993) - mouse antibodies; B1, B3, B5, B6 (Fievet et al., 1989; Pease et al., 1990) - rat antibodies. The production and specificity of each have been previously described as indicated. MAbs 4C11, 2G8, 8G4, and 5F8 were raised in Cell Engineering group of The Institute of Experimental Cardiology (Moscow, Russia), others were the generous gifts from Dr. S.G.Young (Mb43, Mb47) of the Gladstone Foundation, San Francisco; and from Dr. J.C.Fruchart (BL3, BL7, B1, B3, B5, and B6) of Institut Pasteur, Lille, France. Peroxidase conjugated goat anti-mouse and anti-rat IgG were purchased from Jackson ImmunoResearch Lab, Inc.

Immunoreactivity of ox-LDL

Competitive ELISA with different MAbs was used for comparison of LDL immunoreactivity before and after oxidation. To prepare the required dilutions of antibody or LDL samples 10 mM phosphate buffer saline, pH 7.4, containing 0.5% bovine serum albumin

(BSA, Fraction V, Boehringer Mannheim Corp.) and 0.1% Tween 20 (Sigma) (PBS-BSA-Tw) was used.

Microtitration plates (96-well, Corning) were coated with LDL (20 μg/ml in PBS, 100 μl per well) by incubating at 4°C overnight. The plates were washed with PBS + 0.5% BSA, incubated with 300 μl of this solution for 1 hour at ambient temperature and then washed once with PBS-BSA-Tw. The wells' contents were discarded and 50 μl of serially diluted LDL samples (ox-LDL) were mixed in the wells with 50 μl of MAb. Optimal concentration of each MAb was determined previously as a middle point of MAb dose-dependent binding curve with immobilized LDL and were 500 ng/ml for B1 and B6, 12μg/ml for B3, 600 ng/ml for B5, 40 ng/ml for 8G4, 160 ng/ml for 4C11 and 2G8, 15 ng/ml for BL3, 20 ng/ml for BL7. Mb43 and Mb47 were used as ascites diluted 1:60000 and 1:20000, respectively. Plates were allowed to stand at ambient temperature for 2 hours, then washed and the second antibody - HRP-conjugated goat-anti-mouse or goat-anti-rat IgG at appropriate dilution was added (100 μl per well). After incubation for 1 hour at ambient temperature and thorough washing with PBS plates were assayed for peroxidase activity. The substrate mixture (100 μl per well) contained 1 mg o-phenylenediamine in 10 ml of 20 mM citrate buffer, pH 4.7, and 15 μl H_2O_2; the reaction was stopped with 25 μl of 5 M H_2SO_4.

RESULTS

LDL Oxidation in the Presence of Cu^{++}

The physico-chemical properties of LDL changed during incubation with Cu^{++}. As expected, apo B fragmentation was observed by SDS-PAGE (Figure 2) and LDL mobility in native agarose gel increased with increase of oxidation time (Figure 3). Time-dependent increase in TBARS concentration was found in Cu^{++}-treated LDL, the effect was abolished by the concomitant presence of EDTA or BHT.

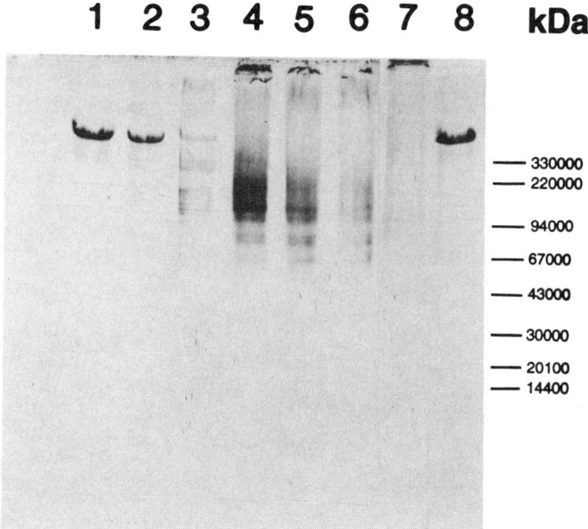

Figure 2. The 5%-25% SDS-PAGE of LDL after different periods of oxidation: 1 - 1 h; 2 - 2 h; 3 - 4 h; 4 - 8 h; 5 - 12 h; 6 -16 h; 7 - 24 h; 8 - native LDL.

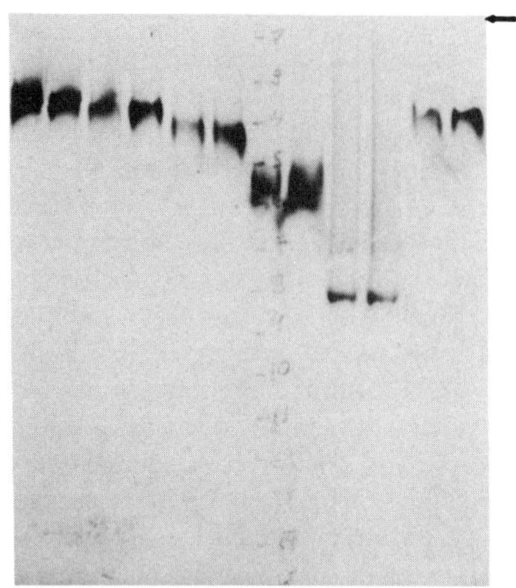

Figure 3. Non-denaturing agarose-gel electrophoresis of LDL after different periods of oxidation: 1 - 0 h (native LDL); 2 and 3 - 1 h; 4 - 2 h; 5 and 6 - 4 h; 7 and 8 - 8 h; 9 and 10 - 24 h; 11 - LDL incubated for 4 h at 37°C without Cu^{++} and oxygen flow; 12 - 24 h at 37°C without Cu^{++} and oxygen flow.

Immunoreactivity of apoB upon Cu^{++} -Mediated Oxidation of LDL

MAbs against different apoB epitopes (Figure 4) were used to test the immunoreactivity of ox-LDL. MAb 5F8-HRP was used for apoB quantification in native and oxidatively modified LDL because the expression of this epitope seemed to be relatively independent of lipid environment. The behavior of different apoB epitopes after oxidation varied widely. No changes in interaction of ox-LDL with MAb 5F8 and 8G4 (residues 1-1297) were observed. Mb47 (3441-3569), the antibody to the apoB-100 region involved in LDL-receptor interaction (Young et al., 1986) didn't display any changes in binding to ox-LDL, however, immunoreactivity of acetylated and MDA-LDL was shown to decrease significantly 4- and 20-fold, respectively. The apoB epitope interacting with MAb B6 (2239-2331) showed enhanced immunoreactivity during the first 4 hours of oxidation and then exhibited gradual decline of its immunoreactivity upon prolonged incubation with Cu^{++}. MAb B1 (405-539),

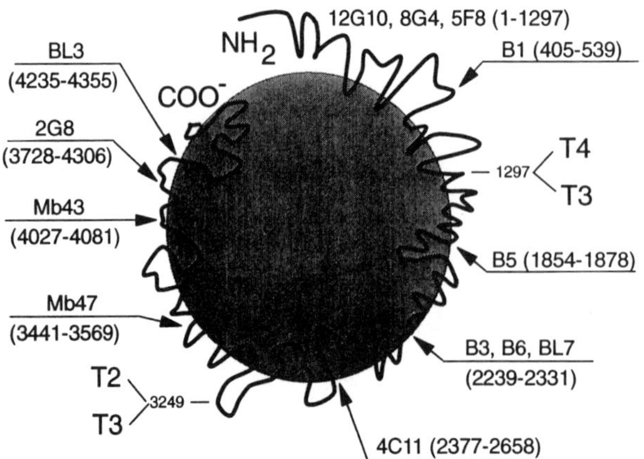

Figure 4. Schematic presentation of LDL particle and apoB-100 epitopes. Apo B-100 (solid black line) is shown to be located on the surface of LDL particle (faded dark circle) and partly buried in the lipid phase. This LDL model is based on that proposed by Yang et al. (1989). ApoB-100 fragments T2, T3, T4 generated by thrombin cleavage are shown. Approximate location of apoB-100 epitopes studied is indicated by arrow with the name of corresponding MAb. Numbers in parenthesis define the amino acid residues corresponding to the immunoreactive apoB-100 fragment. (From Valentinova, N.V.; Gu, Z.W.; Yang, M.; Yanuchevskaya, E.V.; Antonov, I.V.; Guyton, J.R.; Smith, C.V.; Gotto, A.M.,Jr.; and Yang, C.Y., Biol. Chem. Hoppe-Seyler, 375,Oct. 1994, With Permission.)

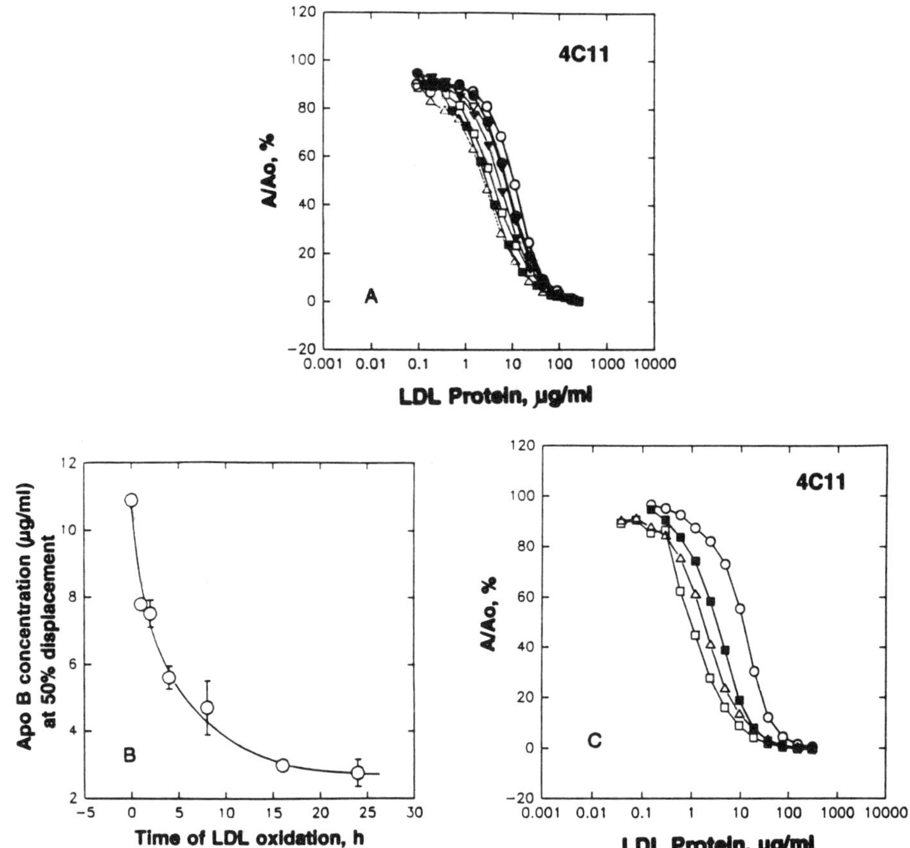

Figure 5. A) Interaction of 4C11 with ox-LDL (competitive ELISA); time of oxidation: 0 h (○), 1 h (●), 2 h (▽), 4 h (▼), 8 h (□), 16 h (■), and 24 h (△); concentration of 4C11 was 160 ng/ml. B) Concentration of ox-LDL apo B required for 50% displacement of maximal 4C11 binding in competitive ELISA. C) Interaction of 4C11 with modified LDL. (□) - MDA-LDL, (■) - acetylated LDL; (○) - native LDL, and (△) -ox-LDL, 24 h of oxidation. MDA-LDL and acetylated LDL were prepared as described elswhere. (From Valentinova, N.V.; Gu, Z.W.; Yang, M.; Yanuchevskaya, E.V.; Antonov, I.V.; Guyton, J.R.; Smith, C.V.; Gotto, A.M.,Jr.; and Yang, C.Y., Biol. Chem. Hoppe- Seyler, 375,Oct. 1994, With Permission.)

in contrast, had slightly reduced binding affinity in the first 8 hours of LDL oxidation, after which an apparent increase in affinity was observed, however, these changes were not statistically significant. Six MAbs, 2G8 (3728-4306), BL3 (4235-4355), Mb43 (4027-4081), BL7 (in the vicinity of residue 2331), B3 (2239-2331), and B5 (1854-1878), displayed considerably reduced binding affinity to ox-LDL. Consistent changes in ox-LDL immunoreactivity were observed with MAb 4C11 (Figure 5A). The immunoreactivity of the epitope recognized by 4C11 increased as a function of time. ApoB concentrations required for 50 % displacement of 4C11 maximal binding at different times of LDL oxidation are shown in Figure 5B. Progressive increase in ox-LDL immunoreactivity was observed up to 16 h of oxidation. The immunoreactivity of ox-LDL to 4C11 remained stable up to 40 h of LDL oxidation. Enhanced binding of 4C11 to MDA- and acetylated LDL was also observed (Figure 5C).

DISCUSSION

Changes in immunoreactivity of the epitope, both increases and decreases, may result from direct modification of amino acid residues such as lysine, tyrosine, arginine and histidine, and oxidative cleavage of peptide bonds. Also, conformational changes of apoB domains caused by fragmentation of polypeptide chain or by changes of lipid microenvironment are probably involved. In the present study, 12 Mab were used. 6 of them, B5, B3, BL7, 2G8, Mb43, and BL3, displayed significant decrease in binding to ox-LDL. All these epitopes are located in the middle part and carboxyterminal region of apo B-100.

It is interesting to note that MAbs B3, B6 and BL7 recognize epitopes located proximately, but patterns of their binding to ox-LDL were different. Those were originally mapped to the same apoB region, i.e. between residues 2239 and 2331 (Pease et al., 1990), but fine epitope specificity of these MAB's seems to be different. Epitopes BL7 and B3 demonstrated progressive decreases in immunoreactivity (more pronounced for B3), whereas B6 displayed a slight increase in immunoreactivity during the first 4 h of LDL oxidation, followed by a gradual decrease.

The epitope for Mb47 was expected to display marked changes upon oxidation because the affinity of the B,E-receptor to ox-LDL is much lower than to native LDL (Steinbrecher et al., 1987). However, no changes in its immunoreactivity were observed in our experiments. Our results are in agreement with those reported by Negri et al. (1993) that show the epitope Mb47 to remain unchanged in ox-LDL. In contrast, MDA-LDL and acetylated LDL displayed significant decrease in binding to Mb47, may be due to higher degree of amino groups modification. It is possible that modification of amino acid residues located in other regions of apoB influence receptor-binding properties of ox-LDL.

MAb 8G4 revealed no change in binding to ox-LDL and only minor changes were observed for 5F8-HRP and B1 (405-539). Our data suggest that the N-terminal region of apoB-100 (thrombin-digest fragment T4) is less susceptible to Cu^{++}-mediated LDL oxidation as compared to C-terminal and the middle region of apoB and may largely retain its secondary structure. Results reported earlier (Yang et al., 1989) showed that N-terminal domain of apoB-100 (between residues 1-1297) contains 15 of the 25 cysteines, and 14 of them occur in intramolecular disulfide bonds (Yang et al., 1990), which apparently stabilize the tertiary structure of this region.

In the present study, noteworthy changes in the structure of the epitope recognized by MAb 4C11 were observed. Progressive increase in 4C11 binding affinity was demonstrated to be a function of oxidation time and TBARS concentration in ox-LDL samples. The MAb 4C11 showed a sustained increase in its affinity even to severely oxidized LDL, up to 40 h of oxidation. No intact apoB was detected in this LDL by SDS-PAGE. Hence, the marked immunoreactivity seems to be attributable to apoB fragments still associated with lipid matrix (Zawadzki et al., 1989). The observed increase in binding of 4C11 to acetylated and MDA-LDL suggests that amino groups modification may be responsible for changes of 4C11 interaction with ox-LDL. Both type and degree of modification seems to influence 4C11 binding: the highest affinity of 4C11 was observed for MDA-LDL which has the highest degree of modification (about 87% of reactive amino groups), however, ox-LDL had higher immunoreactivity and lower degree of modification (about 43%) in comparison to acetylated LDL (70%). Our results demonstrated that immunological approach can be used to understand the structure of ox-LDL and the epitope for 4C11 has the potential to become a useful marker for monitoring LDL oxidation.

ACKNOWLEDGMENTS

The authors would like to express their gratitude to Drs. S.G. Young and J.C. Fruchart for their generous contribution of the anti-apoB-100 MAbs. This research was supported by Grants HL-45619 and HL-27341 from the National Institute of Health and Grants from the Methodist Hospital Foundation and The DeBakey Heart Center.

REFERENCES

Chen, S.-H., Yang, C.-Y., Chen, P.-F., Setzer, D., Tanimura, M., Li, W.-H., Gotto, A.M., Jr., and Chan, L., 1986, The complete cDNA and amino acid sequence of human apolipoprotein B-100, J. Biol. Chem., 261:12918.

Fievet, C., Durieux, C., Milne, R., Delaunay, T., Agnani, G., Bazin, H., Marcle, Y., and Fruchart, J.C., 1989, Rat monoclonal antibodies to human apolipoprotein B: advantages and applications. J. Lipid Res. 30, 1015-1024

Gandjini, H., Gambert, P., Athias, A., and Lallemant, C., 1991, Resistance to LDL oxidative modifications of an N-terminal apolipoprotein B epitope. Atherosclerosis 89, 83-93.

Goldstein, J.L., Ho, Y.K., Basu, S.K., and Brown, M.S., 1979, Binding site on macrophages that mediates uptake and degradation of acetylated low density lipoprotein, producing massive cholesterol deposition. Proc. Natl. Acad. Sci. USA 76, 333-337.

Henriksen, T., Mahoney, E.M., and Steinberg, D. 1983, Enhanced macrophage degradation of biologically modified low density lipoprotein. Arteriosclerosis 3, 149-159.

Knott, T.J., Pease, R.J., Powell, L.M., Wallis, S.C., Rall, S.C., Jr., Innerarity, T.L., Blackhart, B., Taylor, W.H., Marcel, Y.L., Milne, R.W., Johnson, D., Fuller, M., Lusis, A.J., McCarthy, B.J., Mahley, R.W., Levy-Wilson, B., and Scott, J.L., 1986, Complete protein sequence and identification of structural domains of human apolipoprotein B, Nature (London) 323:734.

Negri, S., Roma, P., Fogliatto, R., Uboldi, P., Marcovina, S., and Catapano, A.L., 1993, Immunoreactivity of apo B towards monoclonal antibodies that inhibit the LDL-receptor interaction: effects of LDL oxidation. Atherosclerosis 101, 37-41.

Pease, R.J., Milne, R.W., Jessup, W.K., Law, A., Provots, P., Fruchart, J.C., Dean, R.T., Marcel, Y.L., and Scott, J., 1990, Use of bacterial expression cloning to localize the epitopes for a series of monoclonal antibodies against apolipoprotein B100. J. Biol. Chem. 265:553-568.

Salmon, S., Goldstein, S., Pastier, D., Theron, L., Berthelier, M., Ayrault-Jarrier, M., Dubarry, M., Rebourcet, R., and Pau, B., 1984, Monoclonal antibodies to low density lipoprotein used for the study of low- and very-low-density lipoproteins, in "ELISA" and immunoprecipitation techniques. Biochim. Biophys. Res. Commun. 125:704-711.

Steinbrecher, U.P., Parthasarathy, S., Leake, D.S., Witztum, J.L., and Steinberg, S., 1984, Modification of low density lipoprotein by endothelial cells involves lipid peroxidation and degradation of low density lipoprotein phospholipids. Proc. Natl. Acad. Sci. USA 81:3883-3887.

Steinbrecher, U.P, Witztum, J.L., Parthasarathy, S., and Steinberg, D. (1987). Decrease in reactive amino groups diring oxidation or endothelial cell modification of LDL. Correlation with changes in receptor-mediated catabolism. Arteriosclerosis 7, 135-143.

Valentinova, N.V., Gu, Z.W., Yang, M., Yanushevskaya, E.V., Antonov, I.V., Guyton, J.R., Smith, C.V., Gotto, A.M. Jr., and Yang C.Y. (1994) Immunoreactivity of apolipoprotein B-100 in oxidatively modified low density lipoprotein. Biological Chemistry Hoppe-Seyler in print.

Yang, C.Y., Chen, S.-H., Gianturco, S.H., Bradley, W.A., Sparrow, J.T., Tanimura, M., Li, W.-H., Sparrow, D.A., DeLoof, H., Rosseneu, M., Lee, F.-S., Gu, Z.-W., Gotto, A.M., Jr., and Chan, L., 1986, Sequence, Structure, Receptor Binding Domains and Internal Repeats of Human Apolipoprotein B-100, Nature (London), 323:734.

Yang, C.-Y, Gu, Z.-W., Weng, S.-A., Kim, T.W., Chen, S.-H., Pownall, H.J., Sharp, P.M., Liu, S.-W., W.-H., Gotto, A.M., Jr., and Chan, L., 1989, Structure of apolipoprotein B-100 of human low density lipoproteins, Arteriosclerosis 9:96.

Yang, C.-Y., Kim, T.W., Weng, S.-E., Lee, B., Yang, M., and Gotto, A.M., Jr., 1990, Isolation and characterization of sulfhydryl and disulfide peptides of human apolipoprotein B-100, Proc. Natl. Acad. Sci. U.S.A., 87:5523.

Yang, C.-Y. and Pownall, H.J. Structure and function of apolipoprotein B. In *Structure and Function of Plasma Apolipoproteins.* (M. Rosseneu, Ed.), CRC Press, Inc. 63-84 (1992).

Yang, C.Y., Gu, Z.W., Valentinova, N., Pownall, H.J., Lee, B., Yang, M., Xie, Y.H., Guyton, J.R., Vlasik, T.N., Fruchart, J.C., and Gotto, A.M., Jr. (1993). Human very-low-density lipoprotein structure: interaction of the C apolipoproteins with apolipoprotein B-100. J. Lipid Res. 34, 1311-1321.

Yanushevskaya, E.V., Vlasik, T.N., Valentinova, N.V., Medvedeva, N.V., Fantappie, S., and Catapano, A.L. (1993). Monoclonal antibodies as a specific tool for studying apo B conformation. Abstracts for 62nd EAS Congress, Jerusalem, Israel, p. 68.

Young, S.G., Witztum, J.L., Casal, D.C., Curtiss, L.K., and Bernstein, S. (1986a). Conservation of the low density lipoprotein receptor-binding domain of apoprotein B. Demonstration by a new monoclonal antibody, MB47. Arteriosclerosis 6, 178-188.

Zawadzki, Z., Milne, R.W., Marcel, Y.L. (1989). An immunochemical marker of low density lipoprotein oxidation. J. Lipid Res 30, 885-891.

29

SYNTHETIC COMBINATORIAL LIBRARIES: A NEW TOOL FOR DRUG DESIGN

Methods for Identifying the Composition of Compounds from Peptide and/or Nonpeptide Libraries

Michal Lebl,[1] Viktor Krchňák,[1] Nikolai F. Sepetov,[1] Victor Nikolaev,[1] Magda Staňková,[1] Petr Kočiš,[1] Marcel Pátek,[1] Zuzana Flegelová,[1] Ronald Ferguson[1] and Kit S. Lam[2]

[1] Selectide Corporation
1580 E. Hanley Blvd., Oro Valley, Arizona 85737
[2] Arizona Cancer Center and
Department of Medicine, University of Arizona College of Medicine
Tucson, Arizona 85724

INTRODUCTION

Development of new leads for drug design and structure/function relationship studies were revolutionized by the introduction of combinatorial or "library" techniques (for review see e.g. (Moos et al., 1993; Gallop et al., 1994; Gordon et al., 1994)). These techniques allow for the generation and screening of millions of potentially active structures. Due to the well developed and finely tuned synthetic methodology, peptides were the first group of compounds evaluated by this new approach. However, the next logical challenge is to synthesize libraries of nonpeptidic structures. The combinatorial library approach applied at Selectide consists of three basic steps: (i) chemical synthesis based on the split synthesis method yielding a library with one test compound structure per one bead; (ii) screening of the library either using an on-bead binding assay or a multiple step release assay; and (iii) recovery of positive beads and determination of the structure of the test compound (Lam et al., 1991).

CHEMICAL LIBRARY TYPES

Each chemically synthesized combinatorial library represents a certain structural diversity and multiplicity. Libraries containing sequential repetition of amino acids (peptide libraries) are easy to synthesize and the structure of compound of interest can be easily determined by sequencing. However, such libraries do not contain very high structural diversity, since the only variable parameter is the type of side-chain connected to the C-alpha

Methods in Protein Structure Analysis, Edited by M. Z. Atassi and E. Appella
Plenum Press, New York, 1995

carbon of the peptide backbone, and those side-chains occupy only limited conformational space. Combining natural L amino acids with D amino acids brings more diversity, nevertheless, it is still quite limited. Over the last three years we have synthesized and screened approximately 400 peptide libraries. These libraries ranged from linear (with exposed N- or C-terminus), to cyclic (homo or heterodetic), to libraries with a high probability of regular structural features (alpha helix, beta turn), covering most of the conformational space which can be explored by a peptide structure with molecular weight below 1000.

The advent of non-peptide libraries increased the diversity of conformational space filled by the test compound subunits, as well as increased chemical diversity due to the nature of the subunits (see e.g. Simon et al., 1992; Cho et al., 1993; DeWitt et al., 1993; Nikolaiev et al., 1993; Bunin et al., 1994; Chen et al., 1994; Gordon et al., 1994; Lebl et al., 1994; Staňková et al., 1994). Combinatorial libraries of chemically synthesized compounds can be classified into several distinct groups in which libraries from individual groups represent certain structural types: (i) Libraries of small, compact, and relatively rigid structures (e.g. N-acyl-N-alkyl amino acids); (ii) Libraries based on a more or less rigid scaffold structure (usually multifunctional cyclic scaffold, e.g. derivatized cyclopentane or cyclohexane ring, functionalized steroid skeleton, tricarboxybenzene, diaminobenzoic acid); (iii) Libraries based on a flexible scaffold that is built during the synthesis of the library and can be randomized (branched scaffold based on diamino acids, α, β, γ, δ-library); (iv) Libraries of linear, sequential compounds (typical example is peptide library, including also N-substituted glycines — peptoids, or α, β, and γ amino acids containing library); (v) Libraries of small organic molecules (e.g. benzodiazepine type). Library types which we have explored are shown in figure 1.

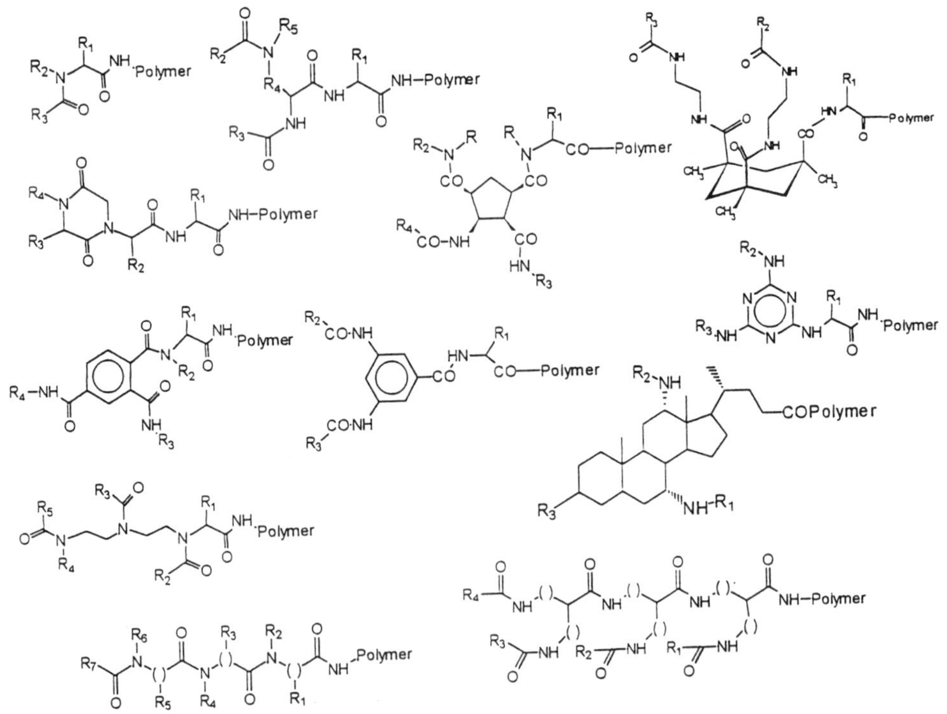

Figure 1. Structure of studied nonpeptidic library types.

DETERMINATION OF POSITIVELY REACTING STRUCTURES

Once the bead of interest is selected by the screening protocol, it is neccessary to determine the structure of the test compound responsible for the observed effect. Peptide structures can be easily determined by sequencing using automatic microsequencers. The structure determination of hits from nonpeptide libraries is complicated by the fact that the amount of positively reacting compound is limited. Standard bead of 100 μm diameter carries approximately 100 pmoles of the functional group onto which the library can be built. This amount of organic structure does not allow application of modern analytical methods for structure elucidation. The only exception is mass spectroscopy, which can be applied in cases when the library is composed of a limited number of structures or in cases where the fragmentation patterns are known and predictable. An example of mass spectroscopical structure determination is shown in figure 2. Beads expressing binding to streptavidin were selected from the small library of N-acyl-N-alkyl amino acids. The compound was cleaved from the bead and all components of the generated mixture were analyzed by MS/MS experiment. The deduced structures were resynthesized and their mass spectra matched those obtained for components cleaved from the beads. Binding to streptavidin was verified by solution assay (Staňková et al., 1994). The structures from a library based on the attachment of carboxylic acids to a modified Kemp's triacid scaffold were also analyzed by the MS/MS technique (figure 3).

In cases when mass spectroscopy cannot be used, a coding principle is applied (Brenner & Lerner, 1992; Kerr et al., 1993; Needels et al., 1993; Nielsen et al., 1993; Nikolaiev et al., 1993; Ohlmeyer et al., 1993). The various formats of coded libraries are given in figure 4. Linear coding is based on parallel synthesis of the screening and coding structure. Fractional coding is realized in two ways: (i) Simultaneous coupling of a tag together with tagged building block - e.g. coupling 0.05 equivalents of norleucine together with a D-amino acid to identify the configuration of the amino acid during sequencing, or (ii) Capping part of the growing chain by the tag which can be later cleaved and identified as such (Ohlmeyer et al., 1993), or as a tagged molecule (Sepetov, 1992; Youngquist et al., 1994). This last possibility is illustrated in figure 5, showing the tagging of a growing peptide chain by bromobenzoylation. After cleaving the mixture of full length peptide and truncated bromobenzoylated fragments, mass spectroscopic evaluation allows the elucidation of the peptide sequence (Sepetov, 1992). Binary coding utilizes a mixture of several blocks instead of a single coding block for coding building block of screening structure. Using a different set of coding blocks for coding different positions in the library allows for the construction of a coding structure in such a way that the coding blocks are cleaved and analyzed in a single step (Ohlmeyer et al., 1993).

Nature has coded proteins by nucleic acids for ages. However, coding nonpeptidic compounds by peptide structures is robust and reliable (Kerr et al., 1993; Nikolaiev et al., 1993). Each chemical individuum in the synthetic library is independently coded by a peptide whose composition can be easily resolved using an established technique (Edman degradation). The synthesis scheme of a coded library is shown in figure 6. Peptidic tags can be constructed in such a way that one cycle of Edman degradation will cleave all coding amino acids and a single HPLC run will reveal all components. The structure of a coding molecule (more appropriately a mixture of coding molecules) is shown in figure 7 together with the HPLC trace of the product of one Edman degradation cycle of this molecule.

The coding principle bears one inherent complication. If the screening process is being performed on the bead, the coding structure can interact with the target molecule. Three possibilities exist to prevent the interaction of the coding structure with the target: (i) The coding structure can be present in a very low concentration so that the interaction with the target molecule will not be seen under the conditions of the experiment; (ii) The coding and test structures can be physically separated; (iii) The test structure can be coded by a

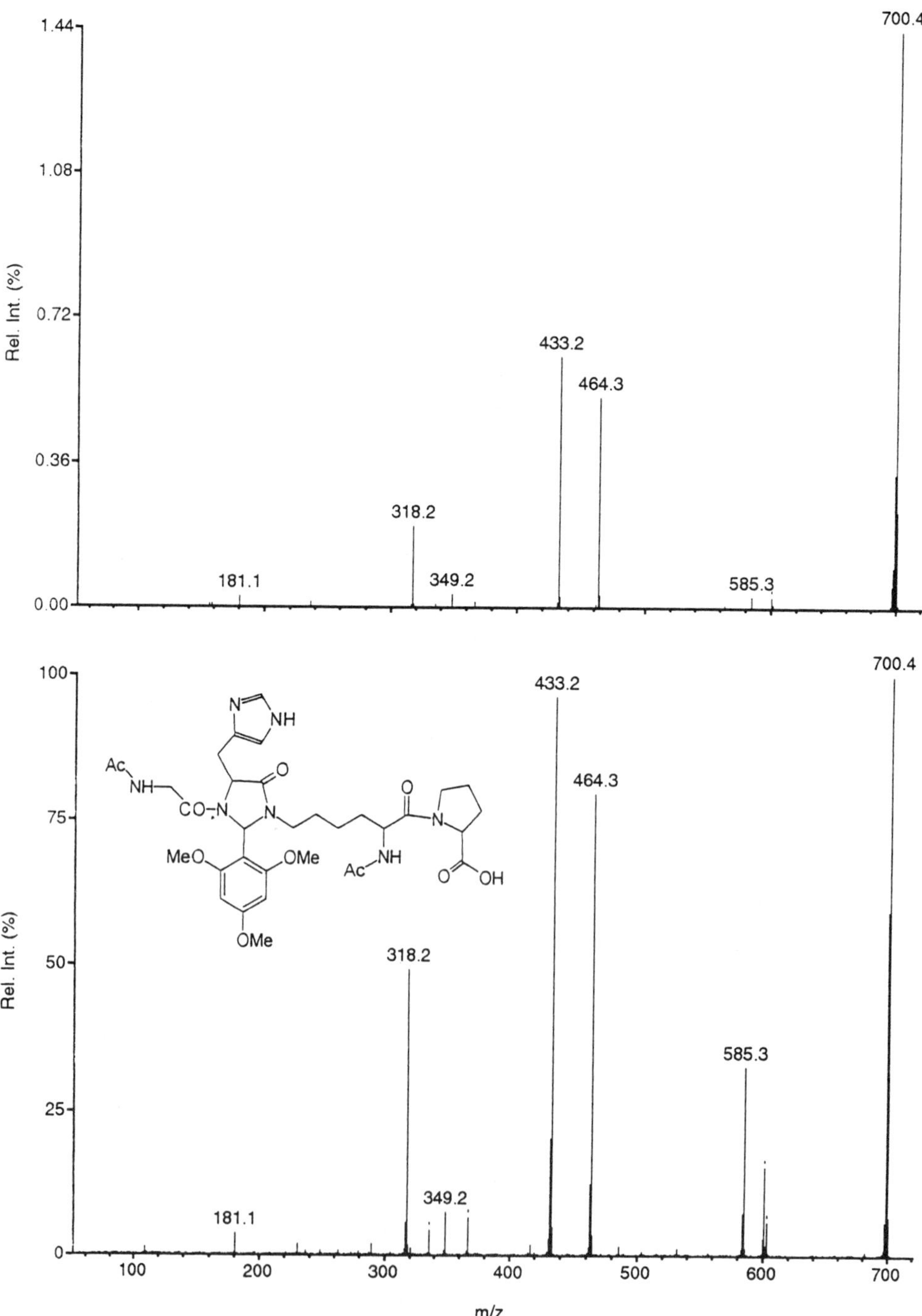

Figure 2. MS/MS spectrum of a compound released from a positively reacting bead (upper trace), and spectrum of the resynthesized compound, the structure of which was deduced from the upper spectrum (lower trace).

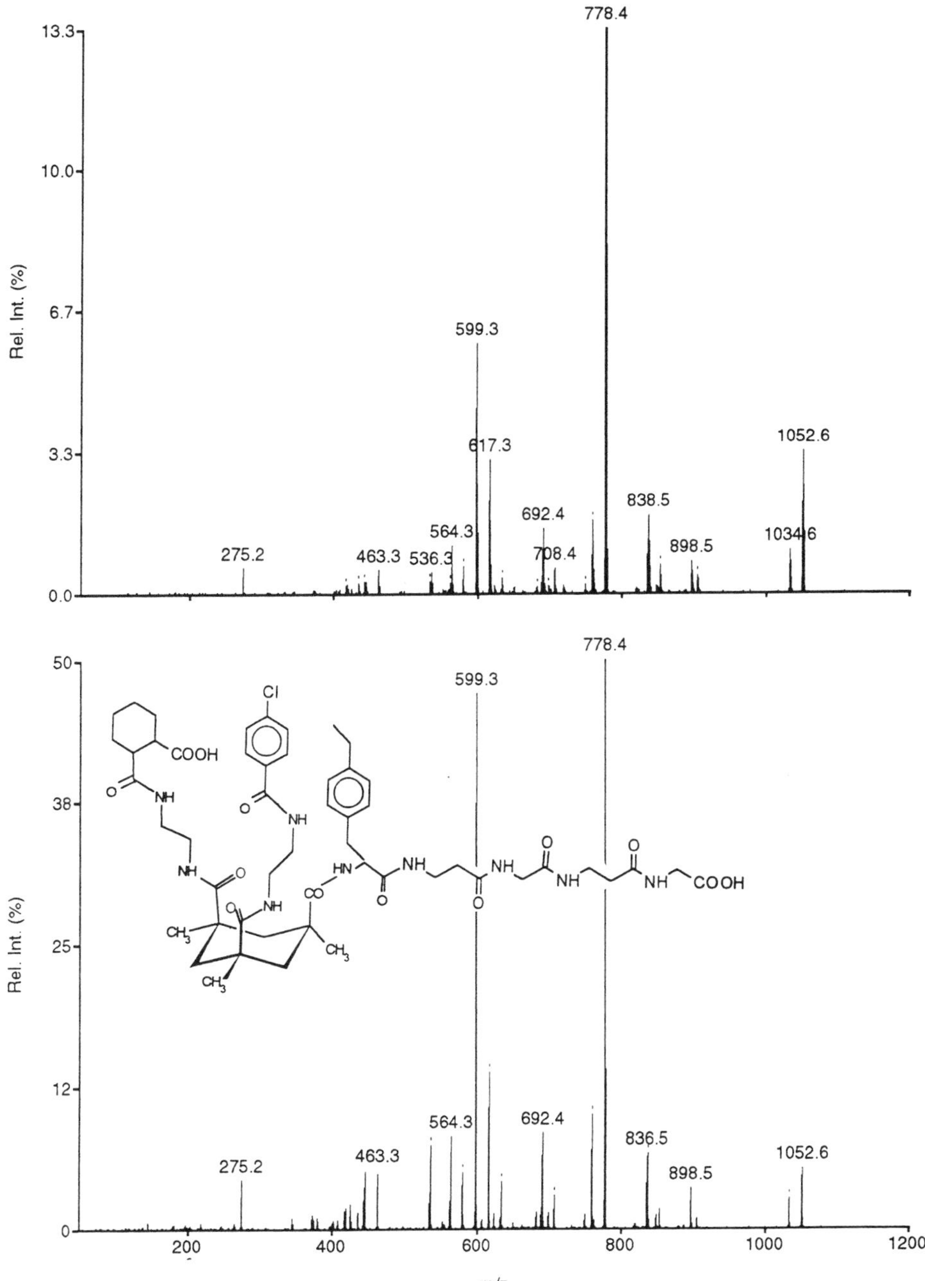

Figure 3. MS/MS spectrum of a compound from library constructed on Kemp's triacid. Spectrum of bead bound compound (upper trace) and of resynthesized compound.

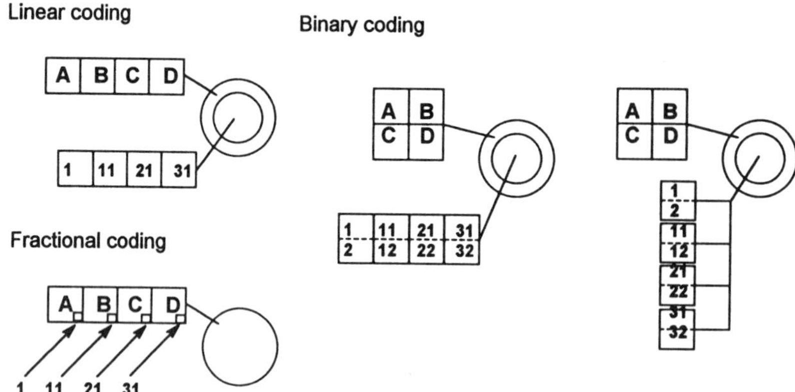

Figure 4. Coded library formats.

multiplicity of coding structures. The first possibility is not realistic in the case of peptide coding due to the limited sensitivity of peptide sequencing. However, it can be used advantageously in cases of coding by nucleic acids, where the coding structure can be conveniently amplified (Needels et al., 1993). The second option was explored by us recently (Vágner et al., 1994). Separation of the "surface" of the bead, which is available for interaction with the macromolecular target, from the "interior" of the bead, was achieved by enzymatic "shaving". To this target inaccessible "interior" was coupled the coding structure. The last possibility is based on the idea of coding using a different set of structures rather than one unique structure. This set of structures must provide unambiguous information about the chemistry performed on screening arm.

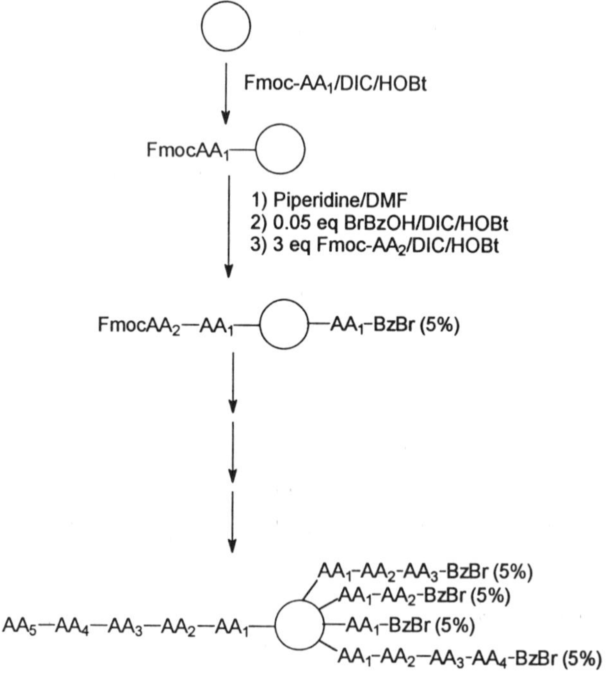

Figure 5. Coding by bromobenzoyl cap.

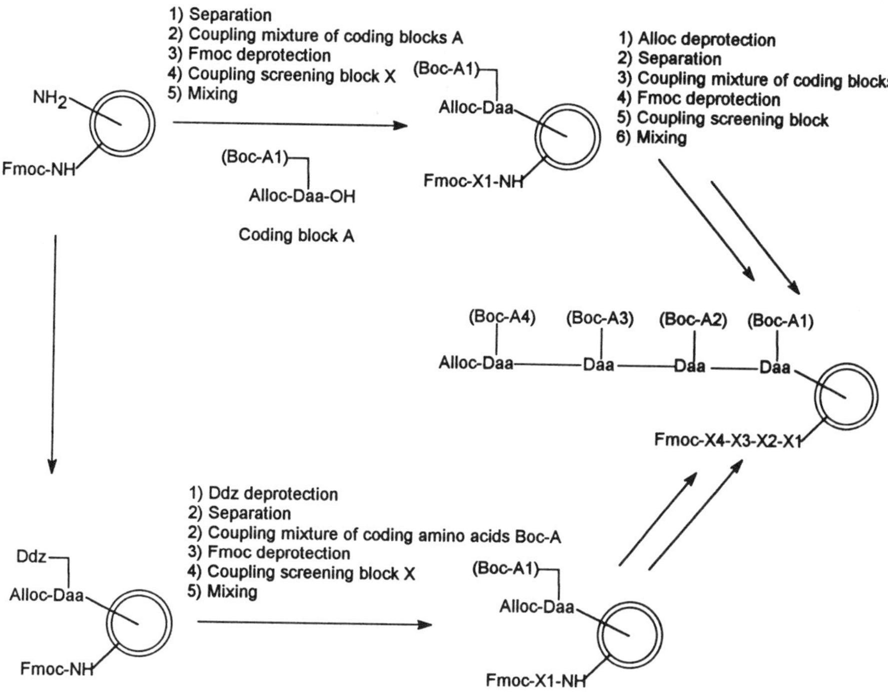

Figure 6. Synthesis scheme of a coded library.

ACKNOWLEDGMENTS

This work was supported in part by the NCDDG Cooperative Agreement (CA57723).

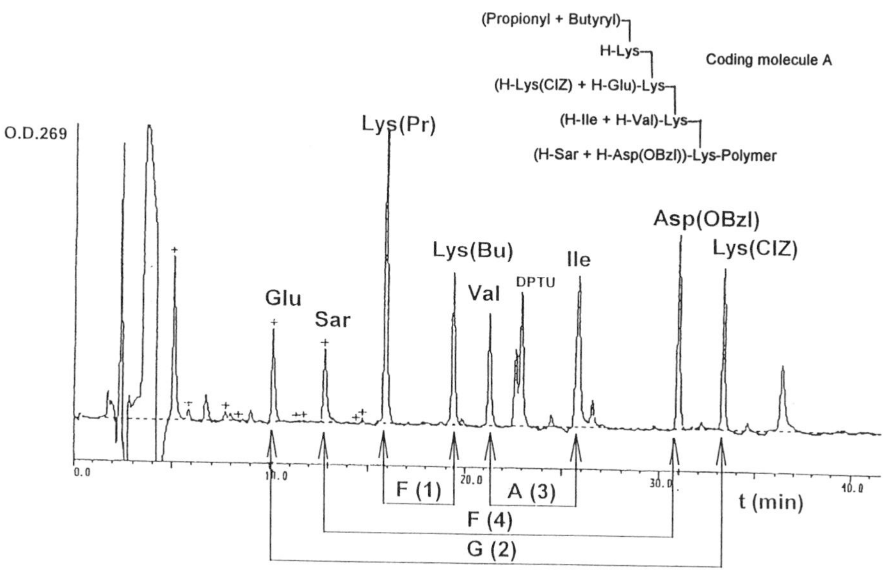

Figure 7. HPLC trace of the digital code generated by Edman degradation of the coding molecule A.

REFERENCES

Brenner, S., Lerner, R. A., 1992 Encoded combinatorial chemistry, *Proc.Natl.Acad.Sci.USA* 89(June):5381.

Bunin, B. A., Plunkett, M. J. and Ellman, J. A., 1994 The combinatorial synthesis and chemical and biological evaluation of 1,4-benzodiazepine library, *Proc.Natl.Acad.Sci.USA* 91(May):4708.

Chen, C., Randall, A. L. A., Miller, B. R., Jones, D. A. and Kurth, M. J., 1994 Analogous Organic Synthesis of Small-Compound Libraries: Validation of Combinatorial Chemistry in Small-Molecule Synthesis, *J.Am.Chem.Soc.* 116(6):2661.

Cho, C. Y., Moran, E. J., Cherry, S. R., Stephans, J. C., Fodor, S. P. A., Adams, C. L., Sundaram, A., Jacobs, J. W. and Schultz, P. G., 1993 An unnatural biopolymer, *Science* 261(3 Sep):1303.

DeWitt, S. H., Kiely, J. S., Stankovic, C. J., Schroeder, M. C., Reynolds Cody, D. M. and Pavia, M. R., 1993 "Diversomers": An approach to nonpeptide, nonoligomeric chemical diversity, *Proc.Natl.Acad.Sci.USA* 90:6909.

Gallop, M. A., Barrett, R. W., Dower, W. J., Fodor, S. P. A. and Gordon, E. M., 1994 Applications of Combinatorial Technologies to Drug Discovery .1. Background and Peptide Combinatorial Libraries, *J. Med. Chem.* 37(9):1233.

Gordon, E. M., Barrett, R. W., Dower, W. J., Fodor, S. P. A. and Gallop, M. A., 1994 Applications of combinatorial technologies to drug discovery.2. Combinatorial organic synthesis, library screening strategies, and future directions, *J. Med. Chem.* 37:1385.

Kerr, J. M., Banville, S. C. and Zuckermann, R. N., 1993 Encoded combinatorial peptide libraries containing non-natural amino acids, *J.Am.Chem.Soc.* 115:2529.

Lam, K. S., Salmon, S. E., Hersh, E. M., Hruby, V. J., Kazmierski, W. M. and Knapp, R. J., 1991 A new type of synthetic peptide library for identifying ligand-binding activity, *Nature* 354(7 November):82.

Lebl, M., Krchňák, V., Šafář, P., Stierandová, A., Sepetov, N. F., Kočiš, P. and Lam, K. S., 1994 Construction and screening of libraries of peptide and nonpeptide structures, in: Techniques in Protein Chemistry, Crabb, J. W., ed., Academic Press, San Diego, pp. 541-548.

Moos, W. H., Green, G. D. and Pavia, M. R., 1993 Recent advances in the generation of molecular diversity, in: Annual Reports in Medicinal Chemistry, Bristol, J. A., ed., Academic press. Inc., San Diego, CA, pp. 315-324.

Needels, M. C., Jones, D. G., Tate, E. H., Heinkel, G. L., Kochersperger, L. M., Dower, W. J., Barrett, R. W. and Gallop, M. A., 1993 Generation and screening of an oligonucleotide-encoded synthetic peptide library, *Proc.Natl.Acad.Sci.USA* 90(Nov):10700.

Nielsen, J., Brenner, R. A. and Janda, K. D., 1993 Synthetic methods for the implementation of encoded combinatorial chemistry, *J.Am.Chem.Soc.* 115:9812.

Nikolaiev, V., Stierandová, A., Krchňák, V., Seligman, B., Lam, K. S., Salmon, S. E. and Lebl, M., 1993 Peptide-encoding for structure determination of nonsequenceable polymers within libraries synthesized and tested on solid-phase supports, *Pept. Res.* 6:161.

Ohlmeyer, M. H., Swanson, R. N., Dillard, L. W., Reader, J. C., Asouline, G., Kobayashi, R., Wigler, M. and Still, W. C., 1993 Complex synthetic chemical libraries indexed with molecular tags, *Proc.Natl.Acad.Sci.USA* 90:10922.

Sepetov, N. F., Issakova, O. L., Krchňák, V. and Lebl, M., 1992 Peptide sequencing using mass spectrometry, U.S.A., 07/939,811.

Simon, R. J., Kaina, R. S., Zuckermann, R. N., Huebner, V. D., Jewell, D. A., Banville, S., Simon, N. G., Wang, L., Rosenberg, S., Marlowe, C. K., Spellmeyer, D. C., Tan, R., Frankel, A. D., Santi, D. V., Cohen, F. E. and Bartlett, P. A., 1992 Peptoids: A modular approach to drug discovery, *Proc.Natl.Acad.Sci.USA* 89(October):9367.

Staňková, M., Issakova, O., Sepetov, N. F., Krchňák, V., Lam, K. S. and Lebl, M., 1994 Application of one-bead-one-structure approach to identification of nonpeptidic ligands, *Drug Development Research*, 33:146.

Vágner, J., Krchňák, V., Sepetov, N. F., Štrop, P., Lam, K. S., Barany, G. and Lebl, M., 1994 Novel methodology for differentiation of "surface" and "interior" areas of poly(oxyethylene)-polystyrene (POE-PS) supports: Application to library screening procedures, in: Innovation and Perspectives in Solid Phase Synthesis, Epton, R., ed., Mayflower Worldwide Ltd., Birmingham, pp. 347-352.

Youngquist, S., R, Fuentes, G., R, Lacey, M., P and Keough, T., 1994 Matrix-assisted Laser Desorption Ionization for Rapid Determination of the Sequences of Biologically Active Peptides Isolated from Support-bound Combinatorial Peptide Libraries, *Rapid Comm.Mass Spectrom.* 8:77.

30

DNA-PROTEIN INTERACTIONS AND PROTEIN-PROTEIN INTERACTIONS IN FILAMENTOUS BACTERIOPHAGE ASSEMBLY

Implications for Epitope Display

Richard N. Perham, Donald A. Marvin, Martyn F. Symmons, Liam C. Welsh, and Tamsin D. Terry

Cambridge Centre for Molecular Recognition
Department of Biochemistry
University of Cambridge
Tennis Court Road, Cambridge CB2 1QW, England

INTRODUCTION

The virion of the filamentous bacteriophage fd (fl and M13 are very similar strains) is a flexible rod about 1 μm long and 6 nm in diameter, comprising a tubular sheath of approx. 2700 copies of the major coat protein subunit surrounding a DNA core. The DNA is a single-stranded circular molecule of 6408 nucleotides embodying 10 genes; these genes are tightly packed and, in some instances, overlapping, apart from a short region (the intergenic space) which encodes no protein component but which contains a double-stranded helical hairpin loop responsible for initiating assembly of the virion. There are a few copies (about 5) of each of two minor coat proteins at the two ends of the virion: gVIIp and gIXp at the end where assembly is initiated, and gIIIp and gVIp at the end where the process is terminated [for general reviews, see Model & Russel,1988; Russel, 1991)].

The major coat protein (gVIIIp) contains 50 amino acid residues and is largely α-helical (Glucksman et al., 1992; Marvin et al., 1994). It has a tripartite structure: an N-terminal segment that is rich in acidic and hydrophilic residues; a 19-residue stretch of apolar and hydrophobic amino acids; and a C-terminal region rich in basic residues. In the virion, these protein subunits are in a shingled, helical array, with five subunits in the 16.26 Å axial repeat. The long axes of the helices make a small angle with the axis of the virion and their N-terminal segments occupy the outside of the particle. The first four or five amino acids are conformationally mobile, as judged by NMR spectroscopy (Colnago et al., 1987), in keeping with their location at the viral surface. The apolar regions generate a hydrophobic girdle of protein-protein interactions, leaving the positively-charged C-terminal regions to

Methods in Protein Structure Analysis, Edited by M. Z. Atassi and E. Appella
Plenum Press, New York, 1995

line the inside of the tubular capsid where they can interact with the negatively-charged sugar-phosphate backbone of the DNA, a mechanism of encapsidation that obviates the need for DNA sequence specificity (Hunter et al., 1987; Glucksman et al., 1992; Marvin et al., 1994).

One of the most exciting new technologies of recent years is the display of foreign peptides and proteins on the surface of filamentous bacteriophage particles. The peptide or protein is encoded by DNA inserted at an appropriate site in the structural gene of one of the virus coat proteins; provided the modified coat protein remains compatible with virus assembly, the foreign polypeptide will appear on the surface of the progeny virion where it can act as a ligand or effector in a wide variety of biologically important systems [for recent reviews, see Scott (1992), Cesareni (1992) and Smith (1993)]. If random DNA sequences are used as inserts, the corresponding amino acid sequences form vast libraries of displayed peptides. These can be screened in various ways as a powerful means of identifying novel biologically active peptides or of testing large numbers of mutants for loss or acquisition of biological activity.

We describe here a series of experiments that has led to a deeper understanding of the unusual mode of DNA-protein interaction that underlies the process of DNA encapsidation, the parameters that govern peptide display on gVIIIp, and the immunological properties of peptides displayed in this way.

DIRECTED MUTAGENESIS OF THE MAJOR COAT PROTEIN

The C-terminal segment of the major coat protein contains 4 lysine residues (positions 40, 43, 44 and 48). The importance of their positively charged side-chains in interacting with the negatively charged sugar-phosphate backbone of the DNA has been emphatically proved by directed mutagenesis of the phage gene VIII encoding the major coat protein. If Lys48 is replaced by a neutral amino acid (Gln, K48Q; Thr, K48T; or Ala, K48A), viable mutant virions are produced but are found to be 35% longer than the wild-type particles. On the other hand, if Lys48 is replaced with arginine (K48R), which conserves the positive charge on the side-chain, the length of the virion is unchanged (Hunter et al., 1987). Further experiments have shown that the K48E mutant protein, which would further lower the positive charge density inside the capsid, is unacceptable, unless it is incorporated into hybrid virions that contain some wild-type or K48Q mutant coat proteins to help restore the positive charge, at least in part (Rowitch et al., 1988). These hybrid virions are all longer than the wild-type but their lengths are variable, depending on the ratio of K48E to wild-type or K48Q coat proteins in the capsid (Rowitch et al., 1988).

These experiments are all consistent with the proposal (Marvin & Wachtel, 1976; Marvin, 1978) that there is direct but non-specific electrostatic interaction between the DNA and coat protein in filamentous bacteriophages. A reduction in the positive charge density per unit length inside the protein sheath of the K48Q, K48T or K48A virions would require a matching fall in the negative charge density per unit length of the DNA core. This would be achieved most simply by an elongation of the encapsidated DNA, leading to a corresponding increase in the length of the virion in which the protein-protein interactions remain essentially unchanged (Hunter et al., 1987). Similarly, the existence of the hybrid virions with variable particle lengths can be explained by postulating that the number of nucleotides packaged per coat protein subunit is not restricted to any particular value and that, depending on the positive charge density generated by the protein sheath, the DNA must adopt a compatible electrostatic (and thus spatial) arrangement during the elongation phase of virus assembly (Rowitch et al., 1988; Greenwood et al., 1991a).

FIBRE DIFFRACTION ANALYSIS OF BACTERIOPHAGE FD

We have now obtained direct structural evidence in support of this interpretation by means of X-ray fibre diffraction analysis of bacteriophage fd and the K48A mutant (Symmons et al., 1995). The overall distribution of intensity and the pattern of layer lines were found to be essentially unchanged by the K48A mutation. There is an intrinsic ambiguity in fibre patterns that can make it difficult to distinguish integral differences in helical symmetry (discussed by Marvin, 1978; Nave et al., 1981). In the case of bacteriophage fd, this could obscure a change in the number of protein subunits per 16.26 Å axial repeat from the wild-type value of five, to four or six in the mutant. However, no such change was detected by comparative sedimentation analysis of wild-type fd and K48A or K48Q mutants (Molina-Garcia et al., 1992). We conclude that the structure of the coat protein subunit and the helical symmetry of its packing in the capsid are essentially unchanged in the K48A mutant.

The distribution of intensity along the equator of a fibre diffraction pattern gives direct information about the radial electron density distribution in a single virion. In hydrated gels of fd and K48A, the virions are sufficiently far apart for them to diffract as individual particles [see Wachtel et al. (1974)] and the molecular transform of the virion can be measured directly from these patterns. The first maximum on the equator at about $R = 0.025$ $Å^{-1}$ is stronger (relative to neighbouring maxima) for the K48A mutant than for wild-type (Fig. 1a and 1b). This kind of change in intensity was also noted for the equatorial intensity of the filamentous bacteriophage strain Pf1 relative to strain If1 (Wachtel et al., 1974), and is best explained by a reduction in the DNA:protein ratio. The effects on the low-resolution electron density distribution attributable to these changes in the diffraction pattern are illustrated in Fig. 1c. There is a reduction in electron density of the K48A mutant relative to wild type at a radius of about 4 Å, corresponding to the average radius of the DNA. The reduction in electron density at about 12 Å radius may correspond to the replacement of the side-chain of Lys48 by the methyl group of alanine, with a displacement of protein to occupy space vacated by DNA. There may also be parts of the DNA (phosphate?) at 12Å. However, it is abundantly clear that the electron density in the centre of the virions, attributable to the DNA core, is lower in the K48A mutant than in wild-type fd. Moreover, the difference in electron density distribution near the centre is consistent with a reduction in the mass of DNA per protein subunit in the K48A mutant to 75% of its wild-type value, as expected for the same amount of DNA packaged in a virion that is 35% longer.

THE MECHANISM OF DNA PACKAGING

Taken together the results of directed mutagenesis and X-ray fibre diffraction show that lowering the positive charge density per unit length inside the protein tube forces the DNA to increase the length it occupies by adopting a more elongated configuration, thereby lowering its negative charge density per unit length in a matching process. A longer virion is thus required to package the same amount of DNA. Moreover, given the existence of hybrid virions with varying lengths, the number of nucleotides packaged per coat protein subunit is not restricted to any particular value. The capsid can thus be regarded as a protein sheath lined with positive charges interacting electrostatically and non-specifically with a negatively-charged DNA core of matching charge density, and with the length of the virion dictated by the length that the DNA molecule is required to adopt.

It has previously been shown that the replacement of Ser47 in the major coat protein with lysine, which would raise the positive charge density lining the capsid, is unacceptable (Greenwood et al., 1991a). This suggests that it is impossible to force the DNA to adopt a

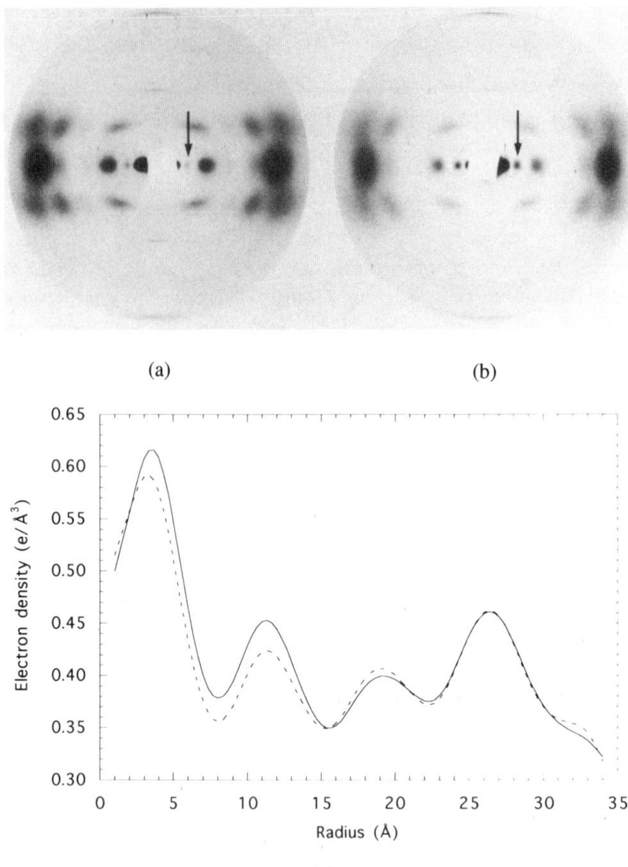

Figure 1. X-ray fibre diffraction patterns of filamentous bacteriophage fd and radial electron density distributions. a) Gel of wild-type fd, prepared at pH 6.5, showing the strong $l = 0$ and $l = 1$ layer lines for $c = 32.52$ Å; b) gel of K48A mutant, prepared at pH 6.5, layer lines as in a). The fibre axis is vertical and the sharp meridional reflexions near the top and bottom are at 8.13 Å. c) Radial electron density distribution calculated for wild-type fd (continuous curve) and K48A mutant (broken curve). For full details, see Symmons et al. (1995).

shorter length than it does in the wild-type virion, probably because the central hole it occupies is too narrow to accommodate the compression. We can thus regard the virus as having reached evolutionary perfection in packaging its DNA in a capsid containing the minimum number of coat protein subunits: one extra positive charge in the C-terminal segment is impossible, one fewer and 35% more coat protein subunits are required to package the same amount of DNA (Greenwood et al., 1991a).

SURFACE DISPLAY OF FOREIGN PEPTIDES ON BACTERIOPHAGE PARTICLES

Since the first description of the incorporation of foreign peptides into a minor coat protein of filamentous bacteriophage (Smith, 1985), numerous developments and applications of the concept have been described. Most of the experiments have relied on expression

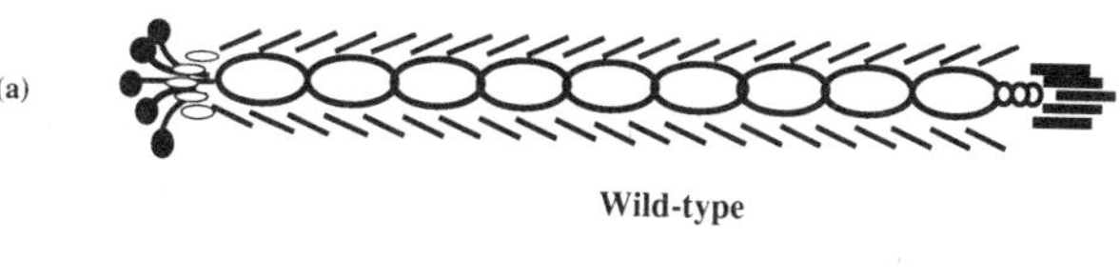

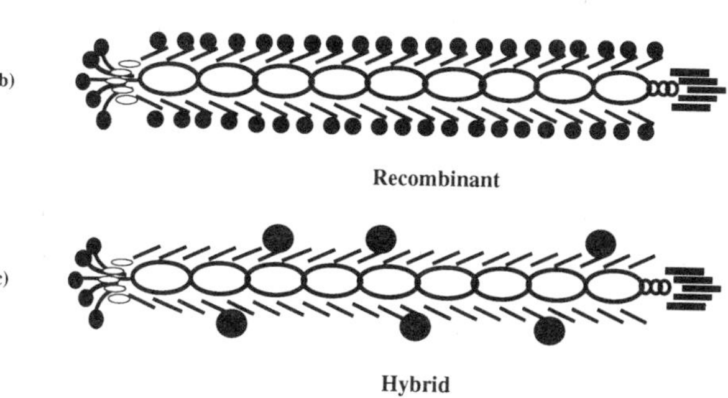

Figure 2. Schematic diagram showing (a) wild-type bacteriophage fd; (b) recombinant virions; and (c) hybrid virions. In the recombinant virions, all 2700 copies of the major coat protein are displaying a peptide insert, whereas in the hybrid virions, mutated coat proteins are interspersed with wild-type coat proteins.

systems that utilize modification of the phage gene III to allow foreign peptides or proteins to be inserted at or near the N-terminus of gIIIp (Scott, 1992; Cesareni, 1992; Smith, 1993). This limits the number of peptides displayed to a few copies at one end of the virion. However, it is also possible to modify phage gene VIII so that the foreign peptides are incorporated near the N-terminus of gVIIIp, which ensures prominent exposure of multiple copies of the peptide on the surface of the virion (Il'ichev et al., 1989; Kang et al, 1991; Greenwood et al., 1991b; Felici et al., 1991). It turns out that five (Il'ichev et al., 1989) or up to six (Greenwood et al., 1991b) amino acids can readily be inserted in this way, generating a recombinant virion in which all 2700 copies of the major coat protein are displaying the peptide (Fig. 2). However, it proved difficult or impossible to accommodate larger peptides in recombinant virions. To overcome this, hybrid phage capsids can be constructed (Kang et al, 1991; Greenwood et al., 1991b; Felici et al., 1991) in which the modified coat proteins are interspersed with copies of the wild-type protein (Fig. 2).

Depending on the peptide, up to 30-40% of the major coat protein subunits in a hybrid capsid can carry an insert of 12 or more amino acid residues (Greenwood et al., 1991b; Veronese et al., 1994). However, with appreciably larger peptides or intact proteins, the frequency of display falls substantially: only a few copies of Fab antibody fragments (Kang et al., 1991; Huse et al., 1992) or trypsin (Corey et al., 1993) can be incorporated into the phage particle. In the X-ray model of the virion, there is sufficient room on the surface of the virion for each N-terminal region of a gVIIIp subunit to accommodate peptides much larger than the six or so residues currently found acceptable (Makowski, 1993). Recent experiments in our laboratory indicate that the inability to achieve a high frequency of display of large peptides or proteins lies, at least in part, in the insertion and processing of the enlarged pro-coat molecule in the bacterial cell membrane where assembly takes place (A. Langara, L. Gowda, P. Malik and R.N.Perham, unpublished work). A deeper understanding of this problem may lead to ways to overcome it.

IMMUNOLOGICAL PROPERTIES OF DISPLAYED PEPTIDES

As expected from their surface location on the virion, peptides displayed close to the N-terminus of gVIIIp in a bacteriophage capsid are highly immunogenic, in the presence or absence of adjuvant (Greenwood et al., 1991b; Minenkova et al., 1993), much more so than peptides on gIIIp (de la Cruz et al., 1988). Anti-wild-type antibodies can be removed by passage of the antiserum through a column of immobilized wild-type phage particles, leaving an antibody preparation directed solely against the peptide insert (Greenwood et al., 1991b). Moreover, the antibody specificity is very high (Willis et al., 1993). Thus we have a new means of preparing antibodies against peptides, substantially simpler and less expensive than the conventional methods of chemical synthesis followed by chemical coupling of the peptide to a suitable carrier protein before injection.

As shown by the poor antibody titre generated in nude BALB/c mice, the immune response against hybrid bacteriophage epitopes is T-cell dependent, a conclusion supported by the observation of class-switching from IgM to IgG during the maturation of the response in heterozygous (nu+/-) but not homozygous (nu/nu) nude mice (Willis et al., 1993). The generation of a strong immune response in the absence of adjuvants suggests that helper T-cell activity is being stimulated direct, adding to the ease of the procedure as a means of raising anti-peptide antibody.

STRUCTURAL MIMICRY OF NATURAL EPITOPES

The ability of short peptides displayed on filamentous bacteriophage particles to mimic natural protein epitopes has been tested using the principal neutralizing determinant of the human immunodeficiency virus HIV. This is an intra-chain disulphide-bridged loop, designated V3, in the third hypervariable region of the variable surface glycoprotein gp120. A hybrid filamentous bacteriophage displaying the 12-residue sequence IHIGPGRAFYTT derived from the tip of the loop, turns out to be a remarkably effective structural mimic of the natural epitope, capable of eliciting neutralizing antibodies in mice against the parental strain HIV-1_{MN} and also against related but different strains of HIV such as IIIB and Rutz (Veronese et al., 1994).

The high level of structural mimicry achieved is emphasized by the fact that in ELISA assays employing natural anti-HIV antisera from HIV-infected patients, the 12-residue peptide displayed on a phage particle is at least one and perhaps two orders of magnitude more sensitive as a substrate than the same peptide covalently coupled to the microtitre plate by conventional chemical means (Fig. 3). This suggests that the N-terminal region of gVIIIp

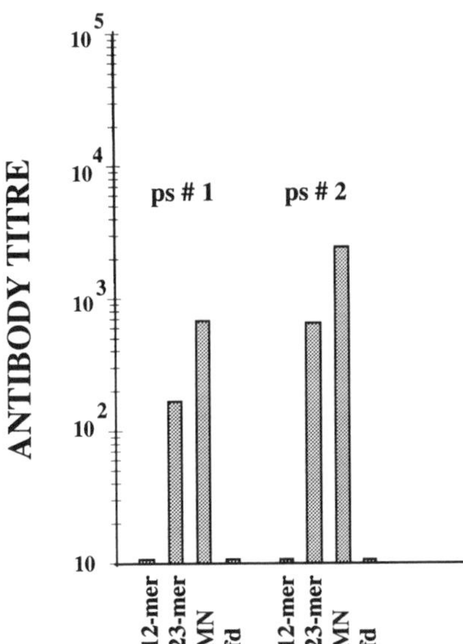

Figure 3. ELISA assays of sera from two different human HIV-1-infected individuals (ps#1, ps#2). The ELISA assays were conducted with peptides (2μg/well) or hybrid MN phage and wild-type phage fd (20μg/well). The hybrid MN phage is equivalent to 1μg peptide/well. The MN phage is displaying the peptide IHIGPGRAFYTT inserted between residues 3 and 4 of the major coat protein. The peptides were: 12-mer, IHIGPGRAFYTT; 23-mer, YNKRKRIHIGPGRAFYTTKNIIG. For full details, see Veronese et al. (1994).

on the bacteriophage surface offers a sympathetic folding milieu in which an inserted peptide can adopt a conformation close to its state in the V3 loop in the HIV virion.

It has similarly been reported that a corrupt 5-residue peptide related to the HIV-1 gag protein, when displayed in a recombinant phage virion, is capable of eliciting rabbit antibodies that cross react with the native protein (Minenkova et al., 1993). On the other hand, using a library of peptides displayed on hybrid bacteriophage particles, a 9-residue amino acid sequence can be selected as a mimic of a discontinuous epitope of *Bortadella pertussis* toxin, but antibodies raised in mice against the modified phage are unable to recognize the natural antigen (Felici et al., 1993). Nonetheless, it is clear that in the right circumstances peptides displayed on filamentous bacteriophages are capable of acting as highly effective structural mimics of natural continuous epitopes of native proteins.

PEPTIDE ACCESSIBILITY

Peptides displayed on gVIIIp are efficient mediators of immunological reactions of various kinds, but there is still little knowledge of the factors that might govern the physical accessibility of a peptide to a target receptor. X-ray fibre diffraction analysis of recombinant phage with peptides incorporated between residues 3 and 4 of the major coat protein has revealed no significant change in the helical parameters of the protein subunit packing (M.F. Symmons, L.C. Welsh, C. Nave, D.A. Marvin and R.N. Perham, unpublished work). Thus the structural model of phage fd derived from X-ray fibre diffraction (Makowski, 1993; Marvin et al., 1994) can serve as a sound basis for interpreting studies of peptide accessibility.

A straightforward test of accessibility is to analyse the susceptibility of the peptide insert to proteolysis with defined proteinases *in vitro*. Thus, when a recombinant phage particle with the sequence GPGRAF inserted between residues 3 and 4 in each copy of gVIIIp is treated with trypsin, there is rapid cleavage between the arginine and alanine residues (positions 7 and 8, respectively, measured from the N-terminus). On the other hand, when treated with chymotrypsin, cleavage after the phenylalanine residue (position 9) is much less facile (T.D. Terry and R.N. Perham, unpublished work). This strongly suggests that a protein receptor comparable in size to the serine proteinases (M_r about 25,000) should be able to come into intimate contact with peptides inserted betweeen residues 3 and 4 of gVIIIp, but that amino acid sequences inserted further from the N-terminus will be shielded by the bulk structure of the virion. Information of this kind will be valuable in the design of future display systems.

CONCLUSIONS

The viral DNA is packaged inside the filamentous bacteriophage capsid by a novel mechanism of electrostatic charge matching, in which the protein forms a sheath lined with positive charges that neutralize, without base sequence specificity, the negative charges of the sugar-phosphate backbone. If the positive charge density per unit length inside the protein tube is lowered, the DNA is forced to increase the length it occupies by adopting a more elongated configuration, thereby lowering its negative charge density per unit length to match that of the protein. The length of the virion is thus dictated by the length of the DNA molecule, but we can manipulate the scale by which it is read. This property is the basis of the widespread use of the virus (M13) as a cloning and DNA sequencing vector.

Displaying foreign peptides on the surface of the bacteriophage particle offers a powerful means of studying the immunological recognition of proteins. The specificity of the immune response, the ability to recruit helper T-cells, the lack of need for external adjuvants, the structural mimicry of defined peptide epitopes, and the accessibility of the peptide inserts to analysis by means of protein chemical and biophysical techniques, all favour it as a technique. It may also prove to be an inexpensive and simple route to the production of effective vaccines. More work now needs to be done to determine, if possible, the structure of a displayed peptide in comparison with the structure of a natural epitope, and to probe more fully the ways in which T-cell epitopes might similarly be displayed to full advantage.

ACKNOWLEDGMENTS

We thank the Science and Engineering Research Council and The Wellcome Trust for financial support and for the award of a Wellcome Prize Studentship to T.D.T. R.N.P. gratefully acknowledges the award of a Fogarty International Scholarship by the National Institutes of Health, Bethesda, MD during which time some of this work was carried out. We thank numerous colleagues, principally Dr E. Appella, Dr A.E.Willis, Dr D. Wraith and Dr F. Di Marzo Veronese, for their collaboration and Mr C. Fuller for skilled technical assistance.

REFERENCES

Cesareni, G., 1992, Peptide display on filamentous phage capsids. A new powerful tool to study protein-ligand interactions, *FEBS Lett.* 307:66-70.

Colnago, L.A., Valentine, K.G., and Opella, S.J., 1987, Dynamics of fd coat protein in the bacteriophage, *Biochemistry* 26:847-854.

Corey, D.R., Shiau, A.K., Yang, Q., Janowski, B.A., and Craik, C.S., 1993, Trypsin display on the surface of bacteriophage, *Gene* 128:129-134.

de la Cruz, V., Lal, A., and McCutchan, T., 1988, Immunogenicity and epitope mapping of foreign sequences via genetically engineered filamentous phage, *J. Biol. Chem.* 263:4318-4322.

Felici. F., Castagnoli, L., Musacchio, A., Jappelli, R., and Cesareni, G., 1991, Selection of antibody ligands from a large library of oligopeptides expressed on a multivalent exposition vector, *J.Mol.Biol.* 222:301-310.

Felici, F., Luzzago, A., Folgori, A., and Cortese, R., 1993, Mimicking of discontinuous epitopes by phage-displayed peptides, II. Selection of clones recognized by a protective monoclonal antibody against the *Bortadella pertussisi* toxin from phage peptide libraries, *Gene* 128:21-27.

Glucksman, M.J., Bhattacharjee, S., and Makowski, L., 1992, Three-dimensional structure of a cloning vector. X-ray diffraction studies of filamentous bacteriophage M13 at 7 Å resolution, *J. Mol. Biol.* 226:455-470.

Greenwood, J., Hunter, G. J., and Perham, R. N., 1991a, Regulation of filamentous bacteriophage length by modification of electrostatic interactions between coat protein and DNA, *J. Mol. Biol.* 217:223-227.

Greenwood, J., Willis, A.E., and Perham, R.N., 1991b, Multiple display of foreign peptides on a filamentous bacteriophage. Peptides from *Plasmodium falciparum* sporozoite protein as antigens, *J.Mol.Biol.* 220:821-827.

Hunter, G.J., Rowitch, D.H., and Perham, R.N., 1987, Interactions between DNA and coat protein in the structure and assembly of filamentous bacteriophage fd, *Nature* 327:252-254.

Huse, W.D., Stinchcombe, T.J., Glaser, S.M., Starr, L., MacLean, M., Hellström, K.E., Hellström, I., and Yelton, D.E., 1992, Application of a filamentous phage pVIII fusion protein system suitable for efficient production, screening, and mutagenesis of F(ab) antibody fragments, *J. Immunol.* 149:3914-3920.

Il'ichev, A.A., Minenkova, O.O., Tat'kov, S.I., Karpyshev, N.N., Eroshkin, A.M., Petrenko, V.A., and Sandakhchiev, L.S., 1989, M13 filamentous bacteriophage in protein engineering, *Doklady Akademii Nauk. SSSR.* 307:431-433.

Kang, A.S., Barbas, C.F., Janda, K.D., Benkovic, S.J., and Lerner, R.A., 1991, Linkage of recognition and replication functions by assembling combinatorial antibody Fab libraries along phage surfaces, *Proc. Natl. Acad. Sci. U.S.A.* 88:4363-4366.

Makowski, L., 1993, Structural constraints on the display of foreign peptides on filamentous bacteriophages, *Gene* 128:5-11.

Marvin, D. A., 1978, Structure of the filamentous phage virion. In *The Single-stranded DNA Phages.* (Denhardt, D. T., Dressler, D & Ray, D. S., eds), pp. 583-603, Cold Spring Harbor Press, Cold Spring Harbor, New York.

Marvin, D.A., Hale, R.D., Nave, C., and Citterich, M.H., 1994, Molecular models and structural comparisons of native and mutant class I filamentous bacteriophages, *J. Mol. Biol.* 235:260-286.

Minenkova, O.O., Ilyichev, A.A., Kishchenko, G.P., and Petrenko, V.A., 1993, Design of specific immunogens using filamentous phage as the carrier, *Gene* 128:85-88.

Model, P. and Russel, M., 1988, Filamentous bacteriophage. In *The Bacteriophages* (Calendar, R. ed.) vol. 2, pp. 375-456, Plenum Press, New York

Molina-Garcia, A. D., Harding, S. E., Diaz, F. G., de la Torre, J.-G., Rowitch, D., and Perham, R. N., 1992, Effect of coat protein mutations in bacteriophage fd studied by sedimentation analysis, *Biophys. J.* 63:1293-1298.

Nave, C., Brown, R. S., Fowler, A. G., Ladner, J. E., Marvin, D. A., Provencher, S. W., Tsugita, A., Armstrong, J., and Perham, R. N., 1981, Pf1 filamentous bacterial virus. X-ray fibre diffraction analysis of two heavy-atom derivatives, *J. Mol. Biol.* 149, 675-707.

Rowitch, D. H., Hunter, G. J., and Perham, R. N., 1988, Variable electrostatic interaction between DNA and coat protein in filamentous bacteriophage assembly. *J. Mol. Biol.* 204:663-674.

Russel, M. ,1991, Filamentous phage assembly, *Mol. Microbiol.* 5:1607-1613.

Scott, J.K., 1992, Discovering peptide ligands using epitope libraries, *Trends Biochem.Sci.* 17:241-245.

Smith, G.P., 1985, Filamentous fusion phage: novel expression vectors that display cloned antigens on the virion surface, *Science* 228:1315-1317.

Smith, G.P., 1993, Surface display and peptide libraries, *Gene* 128:1-2.

Symmons, M.F., Welsh, L.C., Nave, C., Marvin, D.A., and Perham, R.N., 1995, Matching electrostatic charge between DNA and coat protein in filamentous bacteriophage. Fibre diffraction of charge-deletion mutants, *J. Mol. Biol.* 245:86-91.

Veronese, F. Di Marzo, Willis, A.E., Boyer-Thompson, C., Appella, E., and Perham, R.N., 1994, Structural mimicry and enhanced immunogenicity of peptide epitopes displayed on filamentous bacteriophage. The V3 loop of HIV-1 gp120. *J. Mol. Biol.*, 243, 167-172.

Wachtel, E. J., Wiseman, R. L., Pigram, W. J., Marvin, D. A., and Manuelidis, L., 1974, Filamentous bacterial viruses. XIII. Molecular structure of the virion in projection. *J. Mol. Biol.* 88, 601-618.

Willis, A.E., Perham, R.N., and Wraith, D., 1993, Immunological properties of foreign peptides in multiple display on a filamentous bacteriophage, *Gene* 128:79-83.

ANALYSIS OF PROTEIN STRUCTURES OF SPECIAL INTEREST

31

POST-TRANSLATIONAL MODIFICATION BY COVALENT PHOSPHORYLATION OF HUMAN APOLIPOPROTEIN B-100

Protein Kinase C-Mediated Regulation of Secreted apo B-100 in Hep G-2 Cells

Zafarul H. Beg,[1] John A. Stonik,[2] Jeffrey M. Hoeg,[2] and H. Bryan Brewer, Jr. [2]

[2] Department of Biochemistry
J.N. Medical College, A.M.U.
Aligarh 202 002, India
[2] Molecular Disease Branch
National Institutes of Health
Bethesda, Maryland 20892

1. INTRODUCTION

Within human plasma, apolipoprotein B exists as two antigenically-related isoproteins, designated apoB- 100 (Mr 512,000) and apoB-48 (Mr 250,000). The major apoB secreted *in vitro* by normal human hepatocytes and Hep G-2 cells is apoB-100 (Edge, et al., 1985). The peripheral metabolism of VLDL and apoB in part determine the level of circulating LDL (Dolphin, 1985). The LDL, possesing apoB-100 as the principal apolipoprotein constituent, are the major cholesterol transporting lipoproteins in human plasma. Since both LDL-cholesterol (Grundy, 1986) and apoB-100 (Brunzell et al., 1984) levels are directly and positively correlated with premature coronary artery heart disease, an understanding of the control of hepatic apoB-100 synthesis and secretion is important. The human apolipoproteins have been demonstrated to undergo several co-translational and post-translational modifications including proteolytic cleavage (Gordon et al., 1983; Stoffel et al., 1983; Zannis et al, 1983; Bojanovski et al., 1984), glycosylation (Swaminathan and Aladjem, 1976; Lee and Breckenridge, 1967; Brewer et al., 1974; Zannis and Breslow, 1981), covalent phospholyration (Beg et al., 1989; Davis et al., 1984; Sparks et al., 1988 and Jackson et al., 1990), fatty acid acylation (Hoeg et al., 1986 and Hoeg et al., 1988) and deamidation (Ghisseli et al., 1985). These structural alterations may have important physiologic as well as pathologic roles.

Methods in Protein Structure Analysis, Edited by M. Z. Atassi and E. Appella
Plenum Press, New York, 1995

Davis *et al* (Davis et al., 1984) initially suggested that apoB phosphorylation in rat hepatocytes may play a role in the intracellular transport of hepatic VLDL during lipid assembly and secretion. Recently, Sparks *et al* (Sparks et al., 1988) reported that both apoB-48 and apoB-100 were secreted as a phosphoapolipoprotein by primary cultures of rat hepatocytes. Jackson et al have recently demonstrated that addition of insulin to rat hepatocytes decreased the phosphorylation of apoB-100 with only a small effect on apoB-48 (Jackson et al., 1990).

Phophorylation of apoB may be an important mechanism for the intracellular assembly and secretion of VLDL as demonstrated for vitellogenin (Wang and Williams, 1982). Since apoB-100 is the principal apolipoprotein associated with circulating plasma cholesterol (Osborne and Brewer, 1977), and recently human apolipoprotein A-1 has been shown to undergo covalent reversible phosphorylation (Beg et al., 1989), potential role of covalent phosphorylation in the newly secreted apoB-100 from Hep G-2 cells as well as circulating human plasma apoB-100 was investigated. We have also investigated whether alterations in intracellular apoB-100 turnover and phosphorylation are associated with changes in secretion of phosphorylated apoB-100 following cellular induction of protein kinase C (PKC) in Hep G-2 cells. The data from these studies support the concept that phospholipase C (PLC) mediated activation of PKC and covalent phosphorylation plays a role in the regulation of apoB-100 synthesis and/or processing, and intracellular transport, and secretion.

2. RESULTS

2.1. Phosphorylation and Dephosphorylation of Human Plasma apoB-100

Fig. 1 represents the time course of phosphorylation of human plasma circulating apoB-100 in LDL, mediated by purified protein kinase C. Increasing incorportion of radiolabeled phosphate was observed with increasing time. Within 60 min. of apoB-100:LDL (0.04 mg/ml) incubation with protein kinase C at 30°C was associated with a stoichiometry of four mol phosphate per mol of apoB-100 (Fig.1).

Dephosphorylation of maximally phosphorylated ^{32}P-apoB-100 (4 mol of phosphate/ mol) with hepatic phosphoprotein phosphate I (Brandt et al., 1975) was associated with a time-dependent loss of ^{32}P-bound radioactivity (Fig. 2A). Incubation of ^{32}P-apoB-100

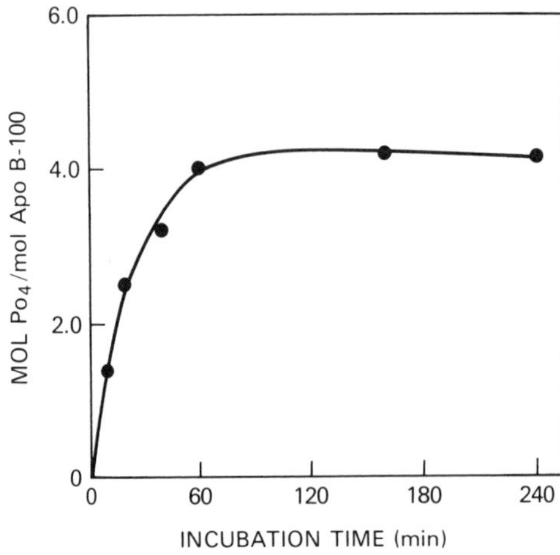

Figure 1. In vitro phosphorylation of human plasma apoB-100:LDL. Purified human plasma apoB-100:LDL (0.04 ng/m) was phosphorylated in the presence of purified rat brain portein kinase C (0.19 mg/ml). At the indicated time intervals, aliquots were analyzed by NaDodSO$_4$-PAGE, ^{32}P-apoB-100 bands were cut from gel and analyzed for ^{32}P-bound radioactivity.

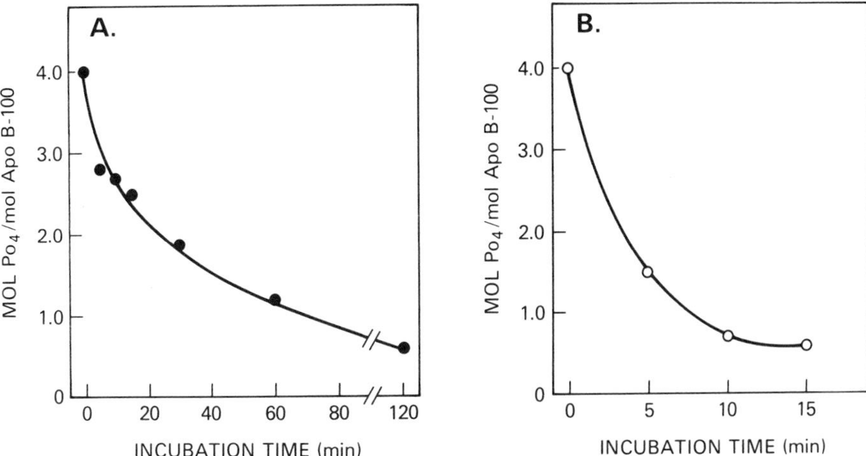

Figure 2. Dehphosphorylation of phosphorylated ^{32}P-apoB-100:LDL. A, aliquot (40μl) of maximally phosphorylated (30°C, 3 h, 4 mol/mol) apoB-100:LDL was incubated at 37°C in a total volume of 700 μl in a buffer containing 50 mM imidazole, pH 7.5, I mM EDTA, 2 mM dithiothreitol, 100 mM sucrose and 250 mM NaCl. ^{32}PapoB-100:LDL samples (3.1 μg) were incubated with either NaF-inactivated rat liver type I phosphotase or active (NaCl-treated phosphatase (0.078 mg/ml). At the indicated time intervals, 100 μl aliquots were mixed with 100 μl of buffer B and analyzed by NaDodSO$_4$-PAGE for the ^{32}P-bound apoB-100:LDL. After autoradiography of dry NaDodSO$_4$-gel, ^{32}P-apoB-100:LDL bands were cut and analyzed for radioactivity. B, 40 μl (3.1 μg) of phosphorylated ^{32}P-apoB- 100:LDL (4 mol PO$_4$/mol) was incubated at 37°C in a total volume of 700 μl in a buffer containing 25 mM glycine- HCL, pH 9.6 and 50 mM MgCl$_2$. Samples were either incubated with inactivated (boiled) or active alkaline phosphatase (1463 units). At the indicated time intervals, 100 μl aliquots were analyzed by NaDodSO$_4$-PAGE and residual radioactivity in ^{32}PapoB-100:LDL band quantified as described in a panel A.

with alkaline phosphatase revealed a similar time-dependent dephosphorylation (Fig 2B). Dephosphorylation of ^{32}P-apoB-100 with hepatic phosphatase-I and alkaline phosphatase was associated with approximately 85% loss of radioactivity in the apoB-100 band in 120 amd 15 min, respectively (Fig. 2A and B). Dephosphorylation of purified human plasma apoB-100:LDL by incubating with phosphatase and subsequent PKC-mediated phosphorylation failed to exhibit an increase in the degree of phosphorylation in comparison to control apoB-100 treated with inactivated phosphatase (data not shown). These results suggest that human plasma apoB-100 exists primarily in the dephosphorylated form.

2.2. Isolation and Analysis of Thrombolytic ^{32}P-Peptides from ^{32}P-ApoB-100:LDL

Human plasma apoB-100:LDL was maximally phosphorylated (4mol PO$_4$ /mol) by protein kinase C purified by ultracentrifugation and digested with human thrombin. The digested apoB-100 was analyzed by NaDodSO$_4$-PAGE. After 18 hr of digestion at 37°C, 80-90% of apoB-100 protein was cleaved into four major peptides- T$_1$ (M$_r$ = 385,000 Da), T$_2$ (M$_r$ = 170,000 Da), T$_3$ (M$_r$ = 238,000 Da) and T$_4$ peptide (M$_r$ = 145,000 Da) (Fig. 3A, lanes 1-3). These results are consistent with the thrombin peptides generated from unphosphorylated LDL (Fig. 3A, lanes 4-6) and with the previously published reports describing the thrombin-mediated cleavage of human apoB-100 (Cardin et al., 1984 and Knott et al., 1985). Autoradiography of the gel revealed ^{32}P-radioactivity bands corresponding to T$_1$, T$_2$, T$_3$ and T$_4$ petides (Fig. 3B). Analysis of the protein bound ^{32}P radioactivity and quantification

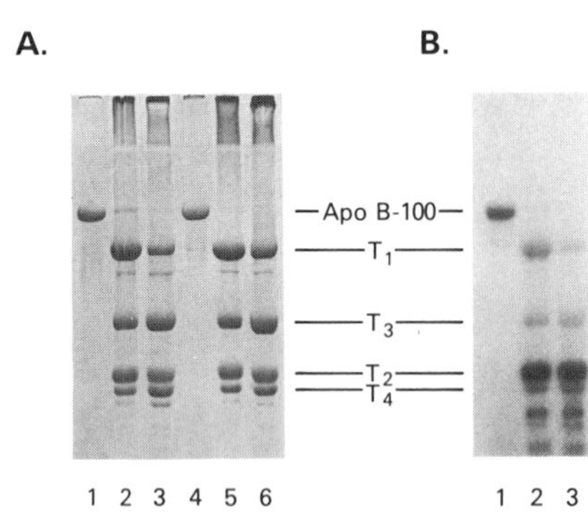

Figure 3. Analysis of ^{32}P-peptides generated by thrombin-mediated digestion of ^{32}-apoB-100:LDL. Protein kinase C-mediated phosphorylated ^{32}-apoB-100:LDL (4 mol PO_4/mol) was purified by ultracentrifugation, digested with thrombin and delipidated. The thrombolytic ^{32}P-peptides thus generated were analyzed by $NaDodSO_4$-PAGE as described under 'Methods'. The gel was dried and autoradiographed. The stained bands of T peptides and corresponding radioactive bands in the autoradiogram were quantified by scanning, after which each T peptide bands were cut and analyzed for apoB-100-bound radioactivity. A, $NaDodSO_4$gel: lane 1 is phosphorylated ^{32}P-apoB-100:LDL (8.5 μg). Lanes 2 and 3 represent the stained T peptide bands (T_1, T_2, T_3 and T_4) generated after digestion (37°C, 18 h) of ^{32}P-apoB-100:LDL (68 μg) with two levels (32 and 127 units) of thrombin, respectively. Lane 4 depicts native (8 μg) apoB-100 band. Lanes 5 and 6 represent the stained T peptides after hydrolysis of cold apoB- 100:LDL (64 μg) with two concentrations (32 and 127 units) of thrombin, respectively. B, autoradiogram of $NaDodSO_4$ gel shown in A. Lane 1 shows the ^{32}P-apoB-100 band. Lanes 2 and 3 represent the radiolabeled peptides, T_1, T_2, T_3, and T_4, after digestion of ^{32}P-apoB-100:LDL with 32 and 127 units of thrombin, respectively.

of protein content within the band in each peptide revealed that T_3 and T_4 peptides were associated with approximately one mol of ^{32}P-PO_4 each, whereas T_2 peptide was associated with 2 mol of phosphate.

Autoradiography of the thin-layer cellulose chromatographic sheet following electrophoresis of acid hydrolysate of ^{32}P-apoB- 100 revealed radioactivity only in phosphoserine band (Fig.4, lane 1) comigrating with the phosphoserine standard detected by staining with ninhydrin (Fig. 4, lane 2). These results demonstrated that PKC-mediated phosphorylation occured only on serine residues of circulating human plasma apoB-100:LDL (Fig.4).

2.3. Secretion of Phosphorylated apoB-100 by Hep G-2 Cells

Incubation of Hep G-2 cells with ortho [^{32}P]-phosphate was associated with the synthesis and secretion of radiolabeled apoB-100:LDL in the media. Immunoprecipitation of secreted ^{32}P-apoB-100 with a monospecific anti-apoB-100 IgG, and $NaDodSO_4$-PAGE of the immunoprecipitates revealed a protein band which comigrated with the apoB-100 standard. Autoradiography of the gel revealed a radioactive band which comigrated with the

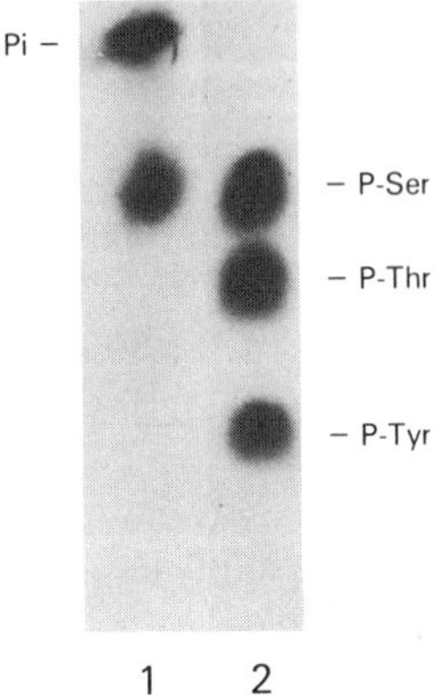

Figure 4. Autoradiogram of ^{32}P-phosphoamino acids from acid hydrolysate of phosphorylated human plasma ^{32}P-apoB-100:LDL. Purified ^{32}P-apoB-100:LDL (4 mol PO_4/mol) was hydrolyzed by HCl. The hydrolysate was analyzed for ^{32}P-phosphoamino acids and autoradiogram was prepared. Lane 1 represents the migration of ^{32}P-phosphoserine. The migrations of the standard phosphoserine (P-Ser), phosphothreonine (P-Thr), and phosphotyrosine (P-tyr) are indicated in lane 2. Pi denotes free ^{32}P-phosphate.

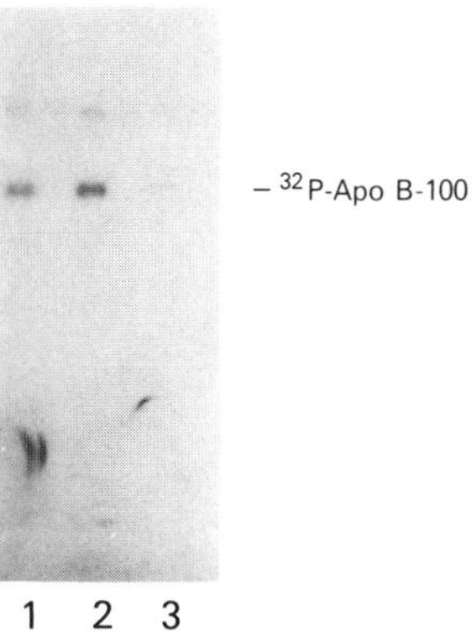

Figure 5. Analysis of media ^{32}P-apoB-100:LDL secreted by Hep G-2 cells incubated in the absence (control) and presence of PLC. After incubation of Hep G-2 cells with and without PLC for 5 h, media were collected and secreted ^{32}P-apoB-100:LDL in each group was purified by ultracentrifugation. Aliquots (2.1 μg control; 1.6 μg PLC-treated) of ^{32}P-apoB-100:LDL wre analyzed by $NaDodSO_4$- PAGE. The gel was dried, autoradiogram was prepared, after which ^{32}P-apoB-100 bands were cut and quantified for radioactivity. Lanes 1 and 2 depict the secreted radiolabeled apoB-100 bands from control and PLC-treated cells, respectively. Lane 3 represents the band of in vitro phosphorylated human plasma ^{32}P-apoB-100:LDL.

stained protein band and radioactive band of plasma phosphorylated apoB-100, similar to Fig.5, lane 1.

Purification of media ^{32}P-apoB-100:LDL secreted over a period of 5 h, by two successive ultracentrifugation and analysis by $NaDodSO_4$-PAGE demonstrated that newly secreted apoB-100 from Hep G-2 cells was phosphorylated and comigrated with human plasma *in vitro* phosphorylated ^{32}P-apoB-100. Autoradiography of the $NaDodSO_4$ gel revealed phosphorylation and co-migration of purified media ^{32}P-apoB-100 (Fig.5, lane 1) with plasma ^{32}P-apoB- 100 (Fig. 5, lane 3). Analysis of the immunoprecipitated cellular ^{32}P-apoB-100 of Hep G-2 cells by $NaDodSO_4$-PAGE and autoradiography of the gel revealed that newly synthesized intracellular apoB-100 was phosphorylated, consistent with the phosphorylation of secreted media apoB-100 (Table I).

Thrombin digestion of secreted ^{32}P-apoB-100, isolated by ultracentrifugation (above), followed by $NaDodSO_4$-PAGE and autoradiography of the gel demonstrated the presence of protein bound ^{32}P radioactivity in T_1, T_2, T_3 and T_4 peptides, similar to shown in Fig. 3 for *in vitro* phosphorylated plasma apoB-100 (data not shown).

Phosphoamino acid analysis of secreted ^{32}P-apoB-100 by Hep G-2 cells revealed phosphorylation only on the serine residue (data not shown). These results are identical to

Table I. Effect of protein kinase C activation on the phosphorylation and secretion of apo B-100:LDL by Hep G-2 cells

Incubation (min)	Concentration munits/ml	^{32}P-Apo B-100 Scanning Units (% of control)	
		Medium	Cellular
30	0	100	100
30	10	235	196
60	0	100	100
60	10	52	60
300	0	100	100
300	10	55	94

Confluent Hep G-2 cells pulsed with ortho[^{32}P]-phosphate for 3 h were incubated with (10 munits/ml) and without phospholipase C for 30, 60 and 300 minutes, after which media and cellular [^{32}P]apo B-100 were immunoprecipitated with a monospecific rabbit anti-human apo B-100 IgG. The analysis of immunoprecipitates by $NaDodSO_4$-PAGE, autoradiography of the gel, and scanning of the autoradiogram were carried out.

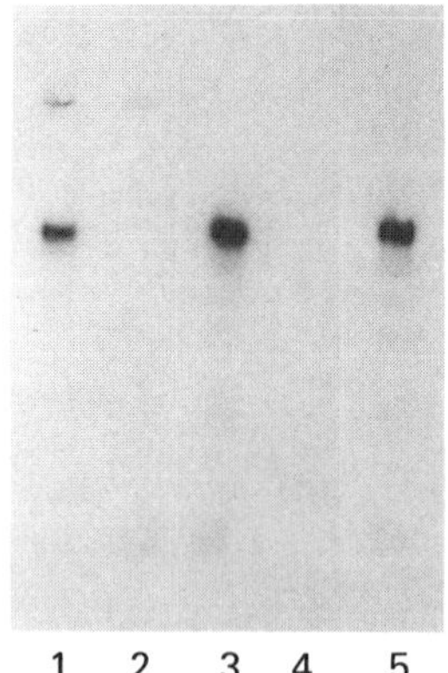

Figure 6. Hepatic phosphatase I mediated dephosphorylation of phosphorylated ^{32}P-apoB-100:LDL secreted by control and PLC- treated Hep G-2 cells. Secreted ^{32}P-apoB-100:LDL from untreated control and PLC-treated Hep G-2 cells (5 h) were purified and aliquots (2.1 µg, control; 1.2 µg, PLC-treated) were incubated at 37°C for 16 h, either with NaF-treated inactive phosphatase-I or NaCl-treated active phosphatase I (1.29 mg/ml). The samples were analyzed by NaDodSO$_4$-PAGE and autoradiogram of the gel was prepared. Lanes 1 and 2 show autoradiogram of control (inactive phosphatase) and dephosphorylated ^{32}P-apoB-100 respectively, secreted by untreated control cells whereas lanes 3 and 4 in the autoradiogram represent the inactive phosphatase-treated and active phosphatase treated (dephosphorylated) ^{32}P-apoB-100:LDL from PLC-treated cells, respectively. Lane 5 is in vitro phosphorylated human plasma ^{32}P-apoB-100 band.

and consistent with the *in vitro* phosphorylation of human plasma apoB-100 mediated by PKC as depicted in Fig. 4.

Incubation of purified secreted ^{32}P-apoB-100:LDL with hepatic phosphatase-I from Hep G-2 cells resulted in the loss of >95% of the ^{32}P-radioactivity in the apoB-100 band when compared to the controls containing inactive phosphatase (Fig. 6, lanes 1 and 2).

2.4. Phospholipase C-Mediated Induction of Protein Kinase C and the Phosphorylation of apoB-100 in Hep G-2 Cells

In order to demonstrate the physiological relevance of PKC-mediated *in vitro* phosphorylation, we initiated a series of studies to evaluate *in vivo* phospholipase C-mediated activation of PKC and the potential impact of increased phosphorylation on cellular and secreted apoB-100 in Hep G-2 cells. Cultures were exposed to control media or media containing 10 munits/ml of phospholipase C for various time intervals. At the end of each incubation, cells were harvested and seperated into membrane and cytosol fractions to determine PKC activity. Exposure of cells to phospholipase C resulted in a transient 2.5 fold increase in membrane associated PKC activity which reached a maximum after 15 min then declined (Fig. 7). In order to evaluate the effects of PLC mediated induction of PKC on phosphorylation and secretion of apoB-100, Hep G-2 cells were pulsed with ortho [^{32}P]-PO$_4$ for 3 hrs., then incubated with and without PLC for 30, 60 and 300 minutes. After 30 min incubation of Hep G-2

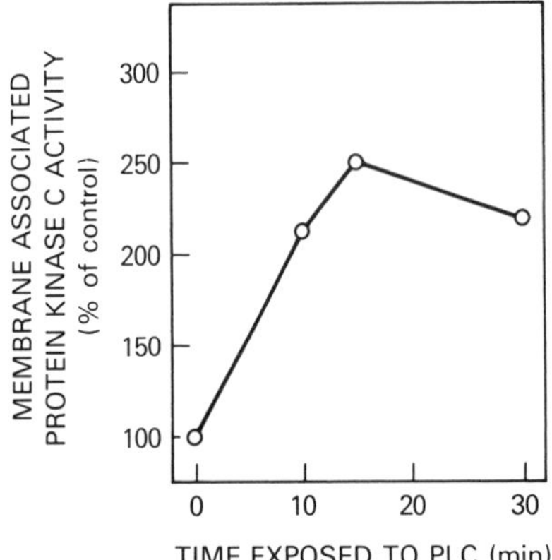

Figure 7. Activation of protein kinase C by phospholipase C. Hep G- 2 cells were incubated for 18 h in DMEM and then exposed to 10 munits/ml of PLC for indicated times at 30°C. Following incubation membrane and cytosol fractions were prepared. The solubilized membrane fractions were subjected to DEAE chromatography, and fractions eluted with 200 mM NaCl were assayed for PKC activity. Results are expressed as percent of membrane associated activity compared with controls. Each point represents the average of duplicate assays.

Table II. Specific activity of purified media [^{32}P]-labeled apo B-100:LDL secreted from control and phospholipase C treated Hep G-2 cells

Treatment	Apo B-100 specific activity (dpm/μg protein)	% of control	Ratio[a]
Control	303	100	1.9
Phospholipase C	578	191	

Hep G-2 cells were pulsed with ortho[^{32}P] phosphate for 3 h then incubated for 5 h with and without phospholipase C. After which media radiolabeled apoB-100:LDL was purified by two sequential ultracentrifugation. The purified [^{32}P]-apoB-100:LDL was dialyzed, concentrated and analyzed for protein and apo B-100 radioactivity.
[a]Ratio of the specific activity of ^{32}P-apo B-100 secreted by phospholipase C-treated to that secreted by untreated control Hep G-2 cells.

cells with PLC, a significant increase in the degree of phosphorylation of both newly synthesized cellular (196%) and secreted (235%) ^{32}P-apoB-100 was evident when compared to untreated controls (100%, Table I). After 60 min of incubation, both media and cellular phosphorylated ^{32}P-apoB-100 in PLC-treated cells were significantly declined to 48% and 40% respectively, when compared to respective controls (Table I). This decrease in the ^{32}P-apoB-100 appears to be due to activated PKC-mediated enhanced phosphorylation and increased degradation of ^{32}P-apoB-100 in PLC-treated cells. At the end of a 5 hr incubation of PLC the cellular ^{32}P-apoB-100 content was restored to control levels due to apparent loss of PLC-mediated stimulation of PKC. However, the net secretion of ^{32}P-apoB-100 during 5 h incubation of Hep G-2 cells with PLC remained 45% inhibited (Table I). Surprisingly the specific activity (dpm/μg protein) of secreted ^{32}P-apoB-100 was significantly increased (Table II). the data demonstrated a 191% increase in the specific activity of purified secreted phosphorylated apoB-100 in Hep G-2 cells treated with PLC for 5 h in comparison to untreated control (100%, Table II). These results are consistent with increased phosphorylation, increased intracellular degradation and decreased secretion of media ^{32}P-apoB-100 in PKC-stimulated Hep G-2 cells.

SDS-PAGE of thrombinized secreted ^{32}P-apoB-100:LDL from control and PLC-treated Hep G-2 cells revealed hydrolysis of ^{32}P-apoB-100:LDL into T peptides, similar to shown in Fig. 3. Thrombinization and SDS-PAGE of secreted ^{32}P-apoB-100:LDL from control and PLC-treated Hep G-2 cells revealed an identical stained peptide pattern as in human plasma apoB-100:LDL similar shown in Fig. 3A. The autoradiogram of the above gel demonstrated that both control and PLC-treated T peptides of ^{32}P-apoB-100 demonstrated a pattern similar to ^{32}P- peptides obtained from the thrombinization of plasma *in vitro* phosphorylated ^{32}P-apoB-100 (Fig.3B). Quantification of protein and ^{32}P-radioactivity in T peptides were consistent with the incorporation of one mol of phosphate each in T_3 and T_4 peptides where as two mol of phosphate in T_2 peptide as seen in T peptide mapping of plasma ^{32}P-apoB-100, except higher radioactive counts were present in T peptides derived from ^{32}P-apoB-100 of PLC-treated Hep G-2 cells (Table II).

Phosphoamino acid analysis of PLC-treated media ^{32}P-apoB-100 showed phosphorylation only on serine residues, similar to untreated controls, consistent with data presented in Fig.4.

2.5. Pulse-Chase Studies on the Effect of PLC on the Intracellular Degradation and Secretion of ^{35}S-apoB-100

In order to confirm that increased PKC-mediated phosphorylation of apoB-100 caused increased intracellular degradation and decreased secretion, Hep G-2 cells were pulsed with ^{35}S-methionine:^{35}S-cysteine and chased in the presence or absence of PLC. PKC activation in PLC incubated cells induced a significant increase in the intracellular degradation of apoB- 100, reaching a maximum (70%) begining at 60 min. into the chase in comparison to untreated cells with no decline at 30 min. (Fig. 8A). Secretion of radiolabeled apoB-100:LDL from the cells which were chased in the presence of PLC was significantly reduced. At 90, 120 and 180 min of the chase period, a decline of approximately 50% was observed (Fig.8B). The data in Fig.8A and B suggested that enhanced degradation coupled with reduced secretion of ^{35}S-apoB-100 in PLC treated cells may be due to increased phosphorylation.

3. DISCUSSION

The apolipoproteins as constituents of lipoproteins particles are known to interact noncovalently with lipids and play a pivotal role in directing the metabolism of the lipoprotein transport system (Gofman et al., 1954 and Brewer, 1981). Recent reports indicate that human plasma apoA-I (HDL) as well as cellular and secreted apoA-I:HDL of Hep G-2 cells undergo post-translational modification involving reversible phosphorylation (Beg et al., 1989). In the present report we describe the initial description of *in vitro* phosphorylation

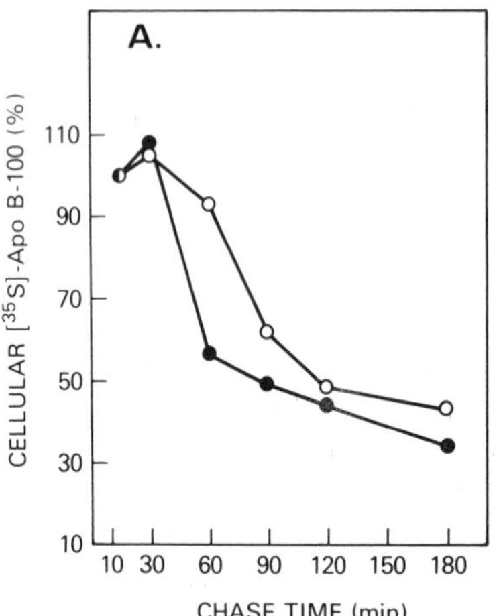

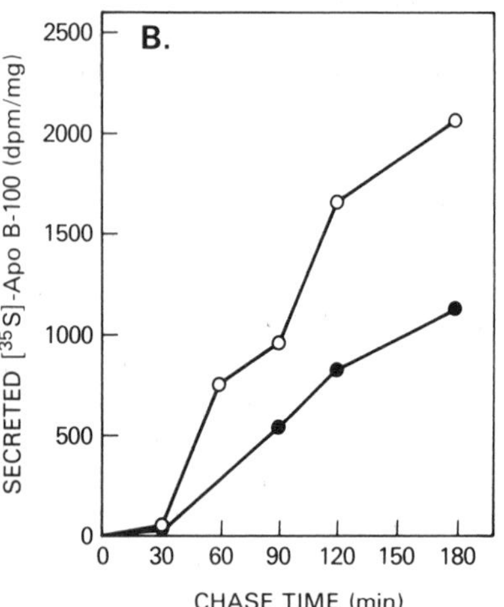

Figure 8. Effect of protein kinase C activation on intracellular degradation and secretion of ^{35}S-apoB-100:LDL. Hep G-2 cells were pulsed (15 min) with ^{35}S-meth:^{35}S-cysteine, incubated with and without PLC and chased for 15,30,60,90,120 and 180 min. After the indicated time, cellular and secreted apoB-100:LDL immunoprecipates were subjected to $NaDodSO_4$-PAGE and autoradiogram of the dry gel was prepared. The stained protein and radioactive bands corresponding to apoB-100 were quantified by densitometric scanning, after which ^{35}S-apoB-100 bands were cut and analyzed for radioactivity. Open circles, the control cells; closed circles, the PLC-treated cells. A, intracellular ^{35}S-apoB-100 bands were quantified by densitometric scanning. Area of the peak at 15 min. into the chase is taken as 100% B, apoB-100:LDL secreted in the medium. Ordinate, ^{35}S-apoB-100 content expressed as dpm per miligram scanning units of protein. Abscissa represents the time after initiation of the cold methionine chase. Each point represents the average of duplicate assays.

and dephosphorylation of circulating human plasma apoB-100:LDL. *In vitro* phosphorylation of human apoB-100:LDL was mediated by a purified rat brain protein kinase C with a stoichiometry of four mol of phosphate per mol of apoB-100, assuming a molecular mass of 512 KDa for apoB. Thrombolytic digestion of maximally phosphorylated ^{32}P-apoB-100:LDL demonstrated that T_1 (T_3 and T_4) and T_2 fractions are phosphorylated. T_2 fraction has been shown to be generated by thrombin cleavage at lysine residue 3,249 from the amino-terminal end of apoB-100:LDL (Cardin et al., 1984 and Knott et al., 1985). Analysis of ^{32}P radioactivity in each thrombolytic peptides revealed that T_1 and T_2 peptides were associated with two mol of phosphate each, whereas T_3 and T_4 peptides generated by cleavage of T_1 peptide (Cardin et al., 1984 and Knott et al., 1985) contained one mol of phosphate each. Phosphoamino acid analysis of phosphorylated ^{32}P-apoB-100 demonstrated that all sites of phosphorylation are associated with serine residues only. These results established that PKC-mediated phosphorylation of human plasma apoB-100 is specific to serine residues.

Dephosphorylation of ^{32}P-apoB-100 (4 mol phosphate per mol) with either hepatic phosphatase-I or alkaline phosphatase resulted into virtually complete loss of esterified ^{32}P-phosphate. These results suggested the possible *in vivo* role of phosphorylation and dephosphorylation in the regulation of apoB-100:LDL synthesis, processing and secretion. The degree of phosphorylation of purified apoB-100 after *in vitro* dephosphorylation was similar to the phosphorylation of non-dephosphorylated apoB-100, consistent with the concept that circulating plasma apoB-100 is dephosphorylated.

The physiological relevance of *in vitro* protein kinase C-mediated phosphorylation and phosphoprotein phosphatase- dependent dephosphorylation of human plasma apoB-100:LDL was confirmed *in vivo* in human hepatoma cells, Hep G-2. ApoB-100 is shown to be secreted as a phosphoapolipoprotein. Incorporation of radiolabeled phosphate; removal of $^{32}PO_4$ upon dephosphorylation with phosphatases; immunoprecipitation of phosphorylated ^{32}P- apoB-100 with a monospecific human apoB-100 antibody; isolation of secreted ^{32}P-apoB-100:LDL by ultracentrifugation and immunoblotting established that both cellular and secreted apoB- 100 of Hep G-2 cells were modified post-translationally by covalent phosphorylation. Thrombolytic digestion of secreted ^{32}P-apoB-100:LDL revealed the phosphorylation of T_1, (T_3 and T_4) and T_2 peptides. Although it is not possible to quantify the stoichiometry of extracellular phosphorylated ^{32}P-apoB-100 from Hep G-2 cells, however, based on the level of incorporation of $^{32}PO_4$ and protein mass in each peptide suggest that *in vivo* phosphorylation of apoB-100 appears to be similar to the protein kinase C- mediated *in vitro* phosphorylation of circulating human plasma apoB-100. Phosphoamino acid analysis of the secreted ^{32}P-apo-B- 100 demonstrated the phosphorylation of only serine residues similar to the *in vitro* phosphorylated apoB-100. These results established that phosphorylation of human apoB-100 *in vitro* and *in vivo* (Hep G-2 cells), is associated with specfic serine residues. These results are consistent with previous reports showing that in rat hepatocytes both apoB-48 (Davis et al., 1984 and Sparks et al., 1988) and apoB-100 (Sparks et al., 1988) are secreted as a phosphoprotein with the radiolabel on serine residues (Davis et al., 1984 and Sparks et al., 1988). Currently, the sites of apoB-100 phosphorylation remained undefined, largely because of problems in working with apoB-100, such as self association and unique hydrophobic nature of this large (512 KDa) glycosylated apolipoprotein. However, the data presented in this manuscript indicate the possibility of four sites of phosphorylation in apoB-100 molecule.

We investigated the possibility of a PKC-mediated phosphorylation of apoB-100 in Hep G-2 cells by stimulating the cellular PKC activity with phospholipase C. We have also examined the regulatory function of enhanced apoB-100 phosphorylation on apoB synthesis, degradation and secretion. PLC has also been shown to stimulate PKC by raising intracellular concentrations of diacylglycerol (Allan et al., 1978 and Kaibuchi et al., 1983), which is generated as a result of hydrolysis of phosphatidylinositol. The results presented in this manuscript established that the increase in the protein kinase C activity (2.5 fold) observed within 15 minutes of PLC challenge to Hep G-2 cells, caused two-fold increase in the level of phosphorylation of both cellular and secreted apoB- 100 within 30 min of incubation. With increasing time (60 min) of incubation with PLC, both cellular and secreted phosphorylated ^{32}P-apoB-100:LDL showed a decrease of 40% and 48%, respectively, in comparison to untreated control cells (Table I). After 5 h of incubation the net secreted ^{32}P-apoB-100 remain inhibited (45%) but with a two-fold higher specfic activity (dpm/μg apoB-100 protein, Table II), whereas cellular ^{32}P-apoB-100 returned to basal level similar to untreated control, apparently because of the concomitant decline in PKC activity and level of cellular apoB-100 phosphorylation (Table I). These results are consistent with increased degradation and decreased secretion because of increased phosphorylation of apoB-100:LDL in PKC-stimulated Hep G-2 cells when compared to control cells. Phosphatase I-mediated dephosphorylation of secreted ^{32}P-apoB-100 from control and PLC- treated Hep G-2 cells revealed a virtual loss of ^{32}P radioactivity in apoB-100 band in comparison to control samples treated with inactivated phosphatase I. In order to confirm the enhanced degradation and reduced secretion of phosphorylated apoB-100:LDL, we conducted pulse-chase experiments to monitor the intracellular ^{35}S-labeled apoB-100 turnover and secretion. The data is consistent with increased intracellular degradation of newly synthesized apoB-100, because of increased phosphorylation in PKC stimulated Hep G-2 cells. Enhanced phosphorylation and intracellular degradation of apoB-100 were coupled with significantly reduced secretion of newly synthesized apoB-100 during chase period of cells exposed to PLC.

In this report we have demonstrated the phosphorylation- dephosphorylation of circulating plasma apoB-100 as well as secreted apoB-100 from control and PLC-treated Hep G-2 cells. Thus, apoB-100 is one of several proteins whose phosphorylation/dephosphorylation state is affected by protein kinases and phosphatases. Recently, Capasso *et al* (Capasso et al.,1989) have demonstrated that intact rat liver Golgi vesicles translocate ATP into their cisternal space and used it to phosphorylate a set of secretory proteins. Enhanced degradation of phosphorylated apoB-100 in Hep G-2 cells is consistent with several other phosphorylated enzymes and proteins known to be degraded at a faster rate than the dephosphorylated form (Engstrom et al., 1982; Muller and Holzer, 1981 and Pontremoli et al., 1987) including HMG-CoA reductase (Parker et al., 1984 and Parker et al., 1989), which is a transmembrane protein bound to endoplasmic reticulum. The mechanism for protease activation and it's role in subsequent enhanced degradation of phosphorylated apoB-100 in Hep G-2 cells remains to be eluciated. However, it is intriguing to postulate that Ca^{2+}-signaled protein kinases such as protein kinase C and Ca^{2+}/calmodulin-dependent kinase, which phosphorylate apoB-100 and also activate Ca^{2+}-dependent endoproteases, may act in concert in the differential degradation of phosphorylated and less phosphorylated forms of apoB-100. In a recent report a role for a Ca^{2+}-evoked or Ca^{2+}-dependent hormones in the regulation of apoB secretion in Caco-2 cells has been suggested (Hughes et al., 1988). Ionophore-induced increased calcium ion availability, may be hypothesized to influence the phosphorylation state of apoB-100:LDL by affecting the activities of Ca^{2+} /calmodulin-dependent kinase (Hughes et al., 1988 and Cohen, 1985). Indeed, we have demonstrated at least *in vitro*, a Ca^{2+}-calmodulin-dependent protein kinase-mediated phosphorylation of human plasma apoB-100, as well as *in vitro* and *in vivo* phosphorylation of human plasma apoA-I (Beg et al., 1989).

A potential important physiological role for PKC-regulated changes in the phosphorylation of apoB-100:LDL could be to regulate VLDL particle size. Powell and Glenrey (Powell and Glenrey, 1987) have demonstrated an increased affinity of dephosphorylated calpactin (lipocortin 1) for phosphatidylserine liposomes compared to the phosphorylated form. Thus, increased phosphorylation of intracellular apoB-100:LDL could result in decreased lipid association and VLDL particle size. Since human liver secretes primarily apoB-100:LDL, our demonstration of the PKC effects on apoB-100:LDL phosphorylation and turnover may be relevant to human apoB metabolism. Recently, it has been demonstrated that in cultured human fibroblasts, activation of PKC following binding of HDL (apoA-I and apoA-II) to the HDL receptor is involved in the HDL-mediated translocation and efflux of intracellular cholesterol (Mendez et al., 1991). In summary, our results establishes a role of signal transduction through protein kinase C-mediated enhanced phosphorylation of cellular apoB-100:LDL, which is degraded at a faster rate, thus causing reduced secretion and in turn potentially a reduced apoB-100:LDL level in plasma. The site and mechanism of phosphorylated apoB- 100:LDL degradation remains to be established. It has been well established that elevated apoB-100 and LDL, the major cholesterol carrying lipoprotein in human plasma has been directly and positively correlated with premature coronary heart disease. Therefore, mechanism(s) which may reduce plasma apoB-100:LDL would be beneficial in the prevention of premature coronary heart disease.

REFERENCES

Allan, D., P . Thomas, and R. H. Michell. 1978. Rapid transbilayer diffusion of 1,2-diacylglycerol and its relevence to control of membrane curvature. Nature (london) **276:** 289-290.

Beg, Z. H., J. A. Stonik, J. M. Hoeg, S. J. Demosky, Jr., T. Fairwell, and H. B. Brewer, Jr. 1989. Human apolipoprotein A-I: Post-translational modification by covalent phosphorylation, J. Biol. Chem. **264:** 6913-6921.

Beg, Z. H., J. A. Stonik, and H. B. Brewer, Jr. 1987. Phosphorylation and modulation of enzyme activity of nature and 2rotease cleaved purified hepatic 3-hydroxy-3methylglutaryl coenzyme A reductase by a calcium/calmodulin-dependent kinase. J. Biol. Chem. **262:** 13228-13240.

Bojanovski, D., R. E. Gregg, and H. B. Brewer , Jr. 1984. Tangier disease: in vitro conversion of proapoA-I Tangier to mature apoA-I Tangier. J. Biol. Chem. **259:** 6049-6051.

Bradford, M. M. 1976. A rapid and sensitive methods for the quantitation of microgram quantities of protein utilizing thr principle of protein-dye binding.Anal.Biochem. **72:** 248-254.

Brandt, H. Z. L. Capulong, and E.Y.C. Lee. 1975. Purification and properties of rabbit liver phosphorylase phosphatase. **250:** 8038-8044.

Brewer, H. B., Jr., R.Shulman, P.Herbert, R.Ronan, and K. Wehrly.1974. The complete amino acid sequence of alanine apolipoprotein (apoC-III), an apolipoprotein from human plasma very low density lipoproteins.J.Biol. Chem. **249:** 4975-4984.

Brewer,H. B. Jr.1981.Current concepts of molecular structure and metabolism of human apolipoproteins and lipoproteins. Klin. Wochschr. **59:** 1023-1035.

Brunzell, J.D.A.D. Sniderman, J.J. Albers, and P.O. Kwiterovich, Jr.1984.Apoprotein B and A1 and coronary artery disease in humans.Arteriosclerosis **4:** 79-83.

Cardin, A.D., K. R. Witt, J. Chao, H. S. Margolius, V. H. Donaldson, and R. L. Jackson. 1984. Degradation of apolipoprotein B-100 of human plasma low density lipoproteins by tissue and plasma kallikreins. J.Biol. Chem. **259:** 8522-8528.

Capasso, J. M., T. W. Keenan, C. Aberjon, and C. B. Hirschberg. 1989. Mechanism of phosphorylation in the lumen of the Golgi apparatus: translocation of adenosine 5'-triphosphate in to Golgi vesicles from rat and mammary gland. J. Biol. Chem. **264:** 5233-5240.

Cohen,P. 1985. Thr role of protein phosphorylation in the hormonal control of enzyme activity. Eur.J. Biochem. **151:** 439- 448.

Davis, R.A., G.M.Clinton, R.A.Borchardt, M.MaloneMcNeal, T.Tan, and G.R. Lattier.1984. Intrahepatic assembly of very low density lipopoproteins: phosphorylation of small molecular weight apolipoprotein B.J.Chem. **259:** 3383-3386.

Dolphin, P.J. 1985. Lipoprotein metabolism and the role of apolipoproteins as metabolic programmers.Can.J. Biochem.Cell Biol. 63: **850-869**.

Edge, S.B., J.M.Hoeg, P.D.Schneider, and H.B.Brewer, Jr.1985. Apolipoprotein B synthesis in man : liver synthesizes only apolipoprotein B-100. Metabolism. **34:** 726-730.

Engstrom, L., U. Zetterqvist, U. Ragnorson, P. Ekman, and U.Dahlqvist- Edberg. 1982. Cell function and diffrentiation, pp. 203-212. Alan R. Liss, New York.

Ghisseli, G., M. F. Rohde, S. Tanenbaum, S. Krishnan, and A. M.Gotto. 1985. Origin of apolipoprotein A-I polymorphism in plasma. J. Biol. Chem. **29:** 15662-15668.

Gordon, J. I., H. F. Sims, S. R. Lentz, C. Edelstein, A. M. Scanu, and A. W. Strauss. 1983. Proteolytic processing of human preproapolipoprotein A-I: a proposed defect in the conversion of proA-I to A-I in Tangier's disease. J. Biol. Chem. **258:** 4037-4044.

Grundy, S. M. 1986. Cholestrol and coronary heart disease. JAMA (J. Am. Med. Assoc.) **256:** 2849-2858.

Gofman, J. W., F. Glazier, A. Tomplin, B. Strisower, and O. De Lalla. 1954. Lipoproteins, coronary heart disease, and atherosclerosis. Physiol. Rev. **34:** 589-607.

Havel, R. J., H. A. Eder, and J. H. Bragdon. 1955. The distribution and chemical composition of ultracentrifugally seperated lipoproteins in human serum. J. Clin. Invest. **34:** 1345-1353.

Hoeg, J. M., M. S. Meng, R. Ronan, T. Fairwell, and H. B. Brewer, Jr. 1986. Human apolipoprotein A-I: post-translational modification by atty acid acylation. J. Biol. Chem. **261:** 3911- 3914.

Hoeg, J. M., M. S. Meng, R. Ronan, S. J. demosky, Jr., T. Fairwell, and H. B. Brewer, Jr. 1988. apolipoprotein B synthesized by Hep G-2 cells undergoes fatty acid acylation. J. Lipid. Res. **29:** 1215-1220

Hughes, T. E., J. M. Ordovas, and E. J. Schaefer. 1988. Regulation of apoliproteinB synthesis and secretion by caco-2 cells: lack of fatty acid effects and control by intracellular calcium ion. J. Biol. Chem. **263:** 3425-3431.

Hunter, T., and B. M. Sefton. 1980. Transforming gene product of Rous sarcoma virus phosphorylates tyrosine. Proc. Natl. Acad. Sci. U.S.A. **77:** 1311-1315.

Jackson, T. K., A. I. Salhamck, J. Elovson, M. L. Diechman, and J. M. Amatruda. 1990. Insulin regulates apolipoprotein B turnover and phosphorylation in rat hepatocytes. J. Clin. Invest. **86:** 1746-1751.

Kaibuchi, K., Y. Takai, M. Sawamura, M. Hoshijima, T. Fujikura, and Y. Nishizuka. 1983. Synergistic functions of protein phosphorylation and calcium mobilization in platelet activation. J. Biol. Chem. **258:** 6701-6704.

Knott, T. J., S. C. Rall, Jr., T. L. Innerarity, S. F. Jacobson, M. S. Urdea, Levy-Wilson, L. M. Bog Powell, R. J. Pease, R. Eddy, H. Nakai, M. Byers, L. M. Priestly, E. Robertson, L. B. Rall, C. Betsholtz, T. B. Shows, R. W. Mahley, and J. Scott. 1985. Human apolipoprotein B: structure of carboxyl-terminal domains, sites of gene expression, and chromosomal localization. Science **230:** 37-43.

Knowles, B. B., C. C. Howe, and D. P. Aden. 1980. Human hepatocellular carcinoma cell lines secrete the major plasma proteins and hepatitis B surface antigen. Science. **209:** 497- 499.

Laemmli, U. K. 1970. Cleavage of structural proteins during the assembly of the head of bacteriophage T_4. Nature (London). **227:** 680- 685.

Lee, P., and W. C. Breckenridge. 1967. The carbohydrate composition of human apo low density lipopoprotein from normal and type II hyperlipoproteinemic subjects. Can. J. Biochem. **52:** 42-49.

Mendez, A. J., J. F. Oram, and E. L. Bierman. 1991. Protein kinase C as a modulator of high density lipoprotein receptor- dependent efflux of intracellular cholesterol J. Biol. Chem. **266:** 10104-10111.

Muller, D., and H. Holzer. 1981. Regulation of fructose-1,6- Disphosphatase in yeast by phosphrylation/dephosphorylation. Biochem.Biophys. Res. Commun. **103:** 926-933.

Osborne, J. C., and H. B. Brewer, Jr. 1977. The plasma lipoproteins. Adv. Protein Chem, **31:** 253-337.

Parker, R. A., S. J. Miller, and D. M. Gibson. 1984. Phosphorylation of microsomal HMG-CoA reductase increases susceptibility to proteolytic degradaion **in vitro**. Biochem. Biophys. Res. Commun. **125:** 629-635.

Parker, R. A., S. J. Miller, and D. M. Gibson. 1989. Phosphorylation Of native 97-kDa 3-hydroxy-3-methylglutaryl coenzyme A reductase from Rat liver: impact on activity and degradation of the enzyme. J. Biol. Chem. **264:** 4877-4887.

Pontremoli, S., E. Melloni, M. Michetti, B. Sparatore, F. Salamino, O.Sacco, and B. L. Horecker, 1987. Phosphorylation and proteolytic modification of specific cytoskeletal proteins in human neutrophils stimulated by phorbol 12-myristate 13- acetate. Proc. Natl. Acad. Sci. USA. **84:** 3604-3608.

Powell, M. A., and J. R. Glenrey. 1987. Regulation of calpactin I phospholipid binding by calpactin I light chain binding and phospho- rylation by $P60^{v\text{-}src}$. Biochem. J. **247:** 321-328.

Sato, R. T. Imanaka, A. Takatsuki, and T. Takano. 1990. Degradation of newly synthesized apolipoprotein B-100 in a pre- Golgi compartment. J. Biol. Chem. **265:** 11880-11884.

Sparks, J D., C E. Sparks, A. M. Roncone, And J. M. Amatruda. 1988. Secretion of high and low molecular weight phosphorylated apolipoprotein B by hepatocytes from control and diabetic rats. J. Biol. Chem. **263:** 5001-5004.

Stoffel, W., E. Kruger, and R. DEutzmann. 1983. Cell-free translation of human liver apolipoprotein A-I and A-II mRNA: processing of primary translation products. HoppeSeyler's Z. Physiol. Chem. **364:**227-237.

Swaminathan, N., and F. Aladjem. 1976. The monosaccharide composition and sequence of the carbohydrate moiety of human serum low density lipoproteins. Biochemistry. **15:** 1516-1522.

Wang, S.-Y., and D. L. Williams. 1982. Biosynthesis of vitellgenin: identification and characterization of non-phosphorylated precursors to avain vitellogenin I and vitellogenin II. J. Biol. Chem. **257:** 3837-3846.

Woodgett, J. R., and T. Hunter. 1987. Isolation and characterization of two distinct forms of protein kinase C. J. Biol. Chem. **262:** 4863-4843.

Yasuda, I., A. Kishimoto, S. Tanaka, M. Tominga, A. Sakurai, and Y. Nishizuka. 1990. A synthetic peptide substrate for selective assay of protein kinase C. Biochem. Biophys. Res. Commun. **166:** 1220-1227.

Zannis, V.I., and J. L. Breslow. 1981. Human very low density lipopoprotein E isoprotein polymorphism is explained by genetic variation and post-translational modification. Biochemistry **20:** 1033-1041

Zannis, V. I., S. K. Karathanasis, H. Keutmann, G. Goldberger, and J. L. Breslow. 1983. Intracellular and extracellular processing of human apolipoprotein A-I isoprotein 2 is a propeptide. Proc. Natl. Acad.Sci. USA **80:** 2574-2578.

32

MIXED LINEAGE KINASES

A New Family of Protein Kinases Containing a Double Leucine Zipper Domain, a Basic Motif and a SH3 Domain

Donna S. Dorow,[1] Lisa Devereux,[1] and Richard J. Simpson[2]

[1] Research Division
The Peter MacCallum Cancer Institute
Melbourne, Victoria 3000 Australia
[2] The Joint Protein Structure Laboratory
Ludwig Institute for Cancer Research (Melbourne Branch) and
The Walter and Eliza Hall Institute for Medical Research
Parkville, Victoria 3050 Australia

INTRODUCTION

Protein kinases play critical roles in the regulation of cellular processes. They control many of the pathways leading to the biochemical and morphological changes associated with cellular growth and division (Dunphy and Newport, 1988; Morgan *et al.*, 1989). They also serve as growth factor receptors and signal transducers and have been implicated in cellular transformation and malignancy (reviewed by Hunter and Karin, 1992; Posada and Cooper, 1992; Birchmeier *et al.*, 1993).

While protein kinases vary widely in their primary structures, each contains a catalytic domain of 250-300 amino acids, consisting of 11 highly conserved motifs or subdomains separated by areas of reduced conservation (Hanks *et al.*, 1988). The presence of these motifs within a new sequence is strongly predictive of protein kinase activity. Specificity of a protein kinase can also be predicted by the sequence of two of the motifs (VIB and VIII) in which different residues are conserved in either the tyrosine or serine/threonine specific kinases (Hunter, 1991). Within the two subgroups, protein kinases with similar substrates or modes of activation cluster into families (Hanks *et al.*, 1988) whose members share a higher degree of catalytic domain sequence identity with each other than with the entire protein kinase specificity class. Recently, several protein kinases with specificity for both tyrosine and serine/threonine have been reported (eg. Featherstone and Russell, 1991; Stern *et al.*, 1991; Ben-David *et al.*, 1991), however, they do not form a distinct family grouping on the basis of catalytic domain sequence identity.

Most protein kinase family members also share other structural features which reflect their particular cellular roles. These include regulatory domains that control protein kinase activity or interaction with other proteins (Hanks, 1991). Two regulatory elements, originally

Methods in Protein Structure Analysis, Edited by M. Z. Atassi and E. Appella
Plenum Press, New York, 1995

identified as conserved sequences in members of the *src* related kinase family, are the *src* homology 2 (SH2) and 3 (SH3) domains (Mayer *et al.*, 1988; reviewed by Koch *et al.*, 1991). These domains have been found in a variety of proteins in intracellular signalling pathways where they link activated cell surface receptors to the G-protein signalling pathway (reviewed by Pawson and Gish, 1992). Furthermore, it has recently been shown that SH3 domains also participate in regulating the activity of GTPase effector proteins (Gout *et al.*, 1993) which are vital for the passage of G-protein signals.

A second type of regulatory domain, which is usually found in transcription factors such as the oncogenes *fos, jun* and *myc*, is the "leucine zipper" (Landschultz *et al.*, 1988). In this motif, leucine or isoleucine residues are repeated with a heptad periodicity over a stretch of at least 22 amino acids. This sequence, which has a higher than average content of charged amino acids, is postulated to form a helix with a hydrophobic "stripe" or ridge of leucines down one face. In the transcription factors, the zipper motif is preceded by a stretch of basic residues that constitute the DNA binding region. There is strong evidence that zipper motifs promote dimerization (O'Shea *et al.*, 1991) through hydrophobic interactions between heptad leucines. Such dimerization appears to activate DNA binding by orientating the basic side chains of the DNA binding residues to enable correct contact with the DNA (Vinson *et al.*, 1989).

We have recently identified, in human epithelial tumour cells, two members of a new family of protein kinases that are unique in having a SH3 domain as well as a novel double leucine zipper and basic domain within their sequences. In addition, their catalytic domain structures display similarity to both the tyrosine and serine/threonine specific kinases and are related to many of the oncogenic protein kinases. Because of this unusual mixture of domain structures, the kinases have been named 'mixed lineage kinases' (MLK's). The sequence of a third member of the MLK family has now been published by others (Ezoe *et al.*, 1994; Gallo *et al.*, 1994; Ing *et al.*, 1994). In the present report we will discuss characterisation and structure of this new group of cellular control molecules.

CHARACTERISATION OF THE MLK's

Isolation, Expression and Chromosomal Localisation

A cDNA fragment of MLK1 was first isolated from human squamous epithelial carcinoma cell line mRNA by reverse transcriptase PCR. Clones for human MLK1 and MLK2 were then isolated from colonic epithelial cDNA libraries. The cDNA sequence determined for MLK1 codes for 394 amino acids containing a kinase catalytic domain, two leucine zippers and a basic domain and a short C-terminal peptide (Dorow *et al.*, 1993). A full-length MLK2 clone has recently been isolated from a human brain cDNA library. As well as the kinase catalytic, double leucine zipper and basic domains homologous to those of MLK1, the MLK2 clone encodes an N-terminal SH3 domain and an extended C-terminal tail rich in proline and serine (see Figure 1). A human MLK1 genomic fragment, isolated by hybridisation to a cDNA probe from the MLK2-SH3 domain coding region, is at present being sequenced. In Northern analysis of more than thirty carcinoma cell lines, mRNAs for human MLK's 1 and 2 were shown to be expressed at low levels in epithelial cell lines of breast and colonic origin and in some melanoma cell lines (data not shown). In human tissues, MLK 1 and 2 mRNA expression was highest in brain and skeletal muscle tissue. Other tissues tested, including heart, placenta, lung, liver, kidney and pancreas, showed extremely low levels of MLK2-mRNA expression and MLK1-mRNA was undetectable (data not shown).

We have also isolated and sequenced MLK1 clones from a murine brain cDNA library. These clones code for the kinase catalytic, double leucine zipper and basic domains

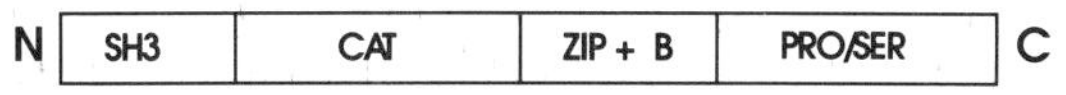

N | SH3 | CAT | ZIP + B | PRO/SER | C

Figure 1. Schematic representation of the arrangement of structural domains of the MLK proteins. Designations are SH3 (*src* homology 3 domain), CAT (kinase catalytic diomain), ZIP+B (double leucine zipper and basic domain) and PRO/SER (proline and serine/threonine rich C-terminal domain).

as well as part of an extended C-terminal domain (data not shown). There is 98% amino acid identity between human and mouse MLK's 1 within the catalytic, zipper and basic domains (only 2 out of the 6 substitutions in a 368 amino acid overlap are non-conservative). The homology between mouse and human MLK's 1, however, ends just after the basic domain where mouse MLK1 has a serine and proline rich C-terminal domain similar to that of human MLK's 2 and 3 (discussed in the C-terminal domain section below).

Using a panel of human, mouse and Chinese hamster hybrid cell lines, the MLK1 gene was mapped to human chromosome 14 (Dorow *et al.*, 1993). *In situ* hybridisation of human chromosomes (Choo *et al.*, 1990) was then used to further localise the MLK1 gene to 14q24.3-31. This area of chromosome 14 has been shown to be involved in translocation in a large number of human malignancies (Testa, 1990). Other genes assigned to this region of human chromosome 14 include the 70kD heat shock cognate protein-2 (14q22-24 [Harrison, *et al.*, 1987]), transforming growth factor beta-3 (14q24 [Barton *et al.*, 1988]) and the c-fos oncogene (14q24.3 [Ekstrand and Zech, 1987]).

In the last few months, three different laboratories have reported the sequence of a third member of the MLK family, called variously PTK1 (Ezoe *et al.*, 1994), SPRK (Gallo *et al.*, 1994) and MLK3 (Ing *et al.*, 1994). These three identical protein sequences were predicted from nucleotide sequences of melanocyte (Ezoe, *et al*, 1994), haematopoietic cell (Gallo *et al.*, 1994) and thymus (Ing *et al*, 1994) cDNA clones. Results presented in each of these reports indicate that this protein, referred to here as MLK3, is much more widely expressed than either MLK's 1 or 2, being found in a very large range of human tissues and cell lines. Gallo *et al.* (1994) used recombinant MLK3 expressed in mammalian cells in *in vitro* kinase assays to show that the expressed protein auto-phosphorylated on serine and threonine residues. Ing *et al.* (1994) have mapped the MLK3 gene to human chromosome 11q13.1-13.3. As discussed in that report, amplifications of this area of chromosome 11 have been observed with varying frequencies in malignancies of the breast, lung, oesophagus, bladder, head and neck and in melanomas. Furthermore, Ezoe *et al.* (1994) showed that anti-sense oligonucleotides to the region of the MLK3 initiator methionine were able to inhibit the growth of human melanocytes in culture.

MLK STRUCTURAL DOMAINS

The SH3 Domain

SH2 and SH3 domains are conserved sequences present in a number of proteins involved in intracellular signalling, including members of the src related family of protein kinases (Mayer *et al.*, 1988; Koch *et al.*, 1991). SH3 domains are also found in cytoskeletal proteins (Rodaway *et al.*, 1989). SH2 domains are comprised of about 100 amino acids and SH3 about 60, each with several highly conserved motifs that form consensus sequences (Koch *et al.*, 1991). Both domains are involved in protein-protein interactions in the passage of intracellular signals from activated growth factor receptors (Anderson *et al.*, 1990; Cantley *et al.*, 1991). SH2 domains bind to specific phosphorylated tyrosines in the cytoplasmic tails of the activated receptors (Mayer *et al.*, 1991; Cantley *et al.*, 1991) while SH3 domains bind to proline rich sequences in their target molecules (Ren *et al.*, 1993; Yu *et al.*, 1994). Many of the proteins to which SH3 domains bind are involved in the control of GTPase activity and thus participate in passage of signals through G-protein pathways (Booker *et al.*, 1993; Egan *et al.*, 1993). Several adaptor proteins have been described which consist of only SH2 and SH3 domains. These proteins link receptor kinase signals to the G-protein pathway. One such adaptor protein, GRB2, is comprised of an SH2 domain flanked by N- and C-terminal SH3 domains (Lowenstein *et al.*, 1992). GRB2 binds activated EGF and PDGF receptors and connects them to the *ras* signalling pathway. Of the protein kinases that contain SH3 domains, the MLK's are the only ones so far described which do not also contain a SH2 domain.

Structural studies of SH3 domains of a number of proteins have shed light on the mechanism by which these domains bind their target sequences. Crystal structures for SH3 domains from α-spectrin (Musacchio *et al.*, 1992), c-Fyn (Noble *et al.*, 1993), and Lck (Eck *et al.*, 1994) and solution structures for those of c-*src* (Yu *et al.*, 1992), phosphatidylinositol-3' (PI-3') kinase (Booker *et al.*, 1993; Koyama *et al.*, 1993) and phospholipase C-γ (Khoda *et al.*, 1993) have been published. Sequence alignments based on structural data revealed a series of highly conserved aromatic residues (see Figure 2) located in β-strands joined by variable loops (Koyama *et al.*, 1993). In the folded structure, conserved residues line a binding pocket into which the aromatic side chains protrude (Booker *et al.*, 1993; Yu *et al.*, 1994). Part of one variable loop forms an end of the binding pocket, leading to the speculation that residues within this loop may contribute to the fine specificity of the domain. The SH3 domains of the PI-3' kinase and the neuronal form of the *src* oncogene (N-*src*), each have an insert in this loop containing 15 residues in the case of PI-3' kinase and 6 for N-*src*. The placement of these inserts near the binding pocket suggests that they may play a role in target recognition by these SH3 domains.

A comparison of the SH3 domains of MLK's 2 and 3, with a number of related SH3 domains, is shown in Figure 2. The degree of amino acid conservation between the MLK-SH3 domains is very high with only 4 of 20 substitutions in 61 residues being non-conservative. There is however, one 7 residue peptide, located in a variable region between the last two consensus motifs (boxed in Figure 2), in which there are 4 substitutions and one deletion in MLK3 compared to MLK2. While all but one of the substitutions is conservative, the deletion may cause a more striking difference in conformation between the two molecules in this peptide than in the rest of the domain.

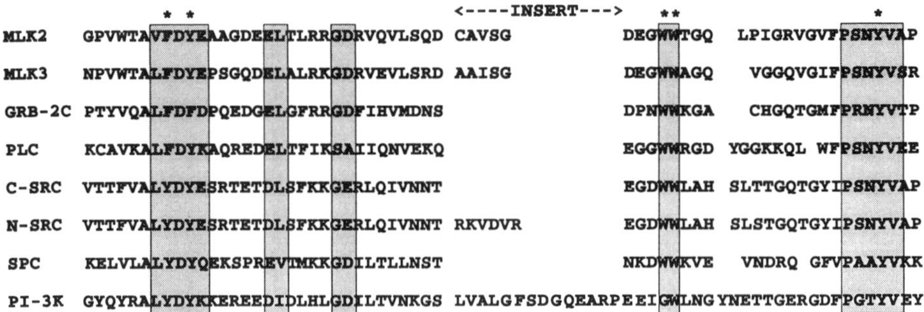

Figure 2. Alignment of the SH3 domain amino acid sequences of MLK's 2 and 3 with SH3 domains of selected signalling and cytoskeletal proteins. Alignments were done relative to the MLK2 sequence with gaps introduced to maximise matches. SH3 domains are from GRB-2C (C-terminal SH3 domain of the growth-factor receptor binding protein 2), PLC (phospholipase C γ), C-src (the cellular *src* oncogene), N-src (the neuronal form of *src*), SPC (the cytoskeletal protein spectrin) and PI-3K (the p85 subunit of the phosphatidylinosotol 3′ kinase).

Among other SH3 containing proteins, the SH3 domain of the MLK's is most closely related to the C-terminal SH3 domain of the adaptor GRB2. The MLK-SH3 domains, however, differ from those of GRB2 in that they contain a 5 residue insert, analogous to that of N-*src*. While 4 of the 6 N-*src* insert residues are charged, however, those of the MLK inserts are mainly hydrophobic, suggesting substantial differences of specificity between the SH3 domains of these molecules.

The Kinase Catalytic Domain

The predicted amino acid sequence of MLK's 1-3 catalytic domains is shown in Figure 3 Each sequence contains all of the amino acid residues conserved in the 11 kinase catalytic subdomains described by Hanks *et al.* (1988) and postulated to be necessary for protein kinase activity. While in subdomain VIb the MLK sequences contain a lysine residue thought to be diagnostic of serine/threonine specificity, their overall catalytic domain structure is more closely related to the tyrosine, rather than the serine/threonine kinases. In several conserved motifs, the sequence of the MLK's are related only to the tyrosine kinases. In particular, two tryptophan residues in subdomain IX and a motif within subdomain XI (Cys-Trp-X-X-Asp/Glu-Pro-X-X-Arg-Pro-X-Phe) are highly conserved in the tyrosine protein kinases, the raf/mos oncogene kinases and the MLK's (see Figure 3), but are not found in other members of the serine/threonine specific class (Hanks *et al.*, 1988).

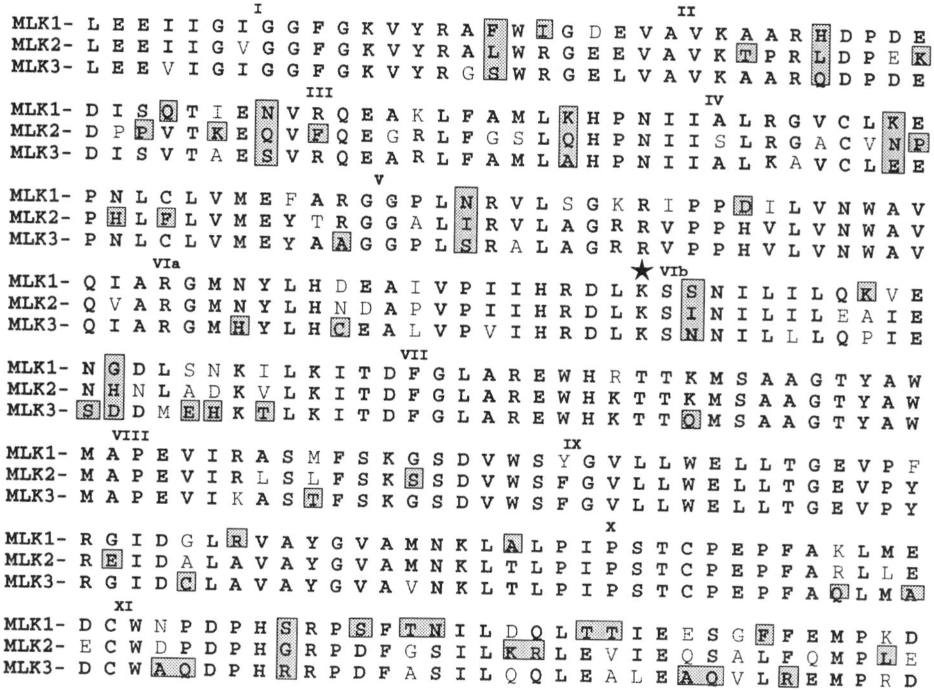

Figure 3. Alignment of the predicted catalytic domain amino acid sequences of MLK's 1, 2 and 3. Residues conserved in all three proteins are in bold type, conservative replacements in normal type and non-conservative replacements are in shaded boxes. Roman numerals refer to protein kinase catalytic subdomains as delineated by Hanks *et al.* (1988). The lysine residue conserved in members of the serine/threonine specific class is marked with a star.

Table 1. Relationship of the MLK1 catalytic domain to other human protein kinases - pairwise similarity scores

Tyr PKs		Ser/Thr PKs	
c-Ros	73	c-Raf	63
c-Abl	71	PKC	55
EGFR	69	DSRNA	54
c-Src	68	c-Mos	52
TRK	68	cGMPDK	50
PDGFR	63		
InsR	61		

Pairwise similarity scores are derived from the sequence alignment program CLUSTAL (Higgins and Sharp, 1988) and represent the numbers of absolute identities between two sequences minus a penalty for gaps introduced to maximise the alignment of the sequences Scores are from comparisons between MLK1 and human c-Ros, c-Abl, c-Src, c-Raf c-Mos (proto-oncogenes), TRK (colon carcinoma oncogene product), EGFR (epidermal growth factor receptor), PDGFR (platelet derived growth factor receptor), InsR (insulin receptor), PKC (protein kinase C), DSRNA (double stranded RNA activated protein kinase) and cGMPDK (cyclic GMP dependent kinase). Table reprinted from Dorow et al., 1993

Comparison of the catalytic domain amino acid sequences of the MLK's (Figure 4) reveals that the three family members share about 75% identity (85% if conservative substitutions are considered) with no gaps needed to completely align the sequences. The spatial arrangement of catalytic domain motifs, therefore, provides support for the notion that MLK1 represents a new and distinct family of protein kinases. As the MLK3 protein has been shown to auto-phosphorylate on serine and threonine (Gallo *et al.*, 1994), the high degree of amino acid conservation within the family suggests that MLK's 1 and 2 are likely to also have serine/threonine specificity. As discussed by Gallo *et al.* (1994), MLK3 is the first SH3 domain containing kinase to have Ser/Thr specificity. Along with the close structural similarity of the MLK catalytic domain to that of the tyrosine kinases, it is unlike that of any previously described family of serine/threonine specific kinases.

The MLK1 protein kinase catalytic domain sequence was aligned with a series of other human protein kinase domain sequences using the alignment program, CLUSTAL (Higgins and Sharp, 1988). The similarities between MLK1 and each of the other protein kinase domain sequences are presented as pairwise similarity scores in Table 1. These data show that the strongest similarities of the MLK1 protein kinase catalytic domain are to the tyrosine kinases c-Ros, c-Abl, EGFR and c-Src. Among the serine/threonine protein kinases, MLK1 is most closely related to c-Raf. Most members of the protein kinase families to which the MLK's show strongest catalytic domain homology are oncogenes with transforming ability, or growth factor receptors.

Alignment of the MLK1 catalytic domain with that of human Ros, EGFR and Raf is shown in Figure 4. At approximately 35% of positions in the MLK1 catalytic domain, amino acid residues are identical to Ros, 32% to EGFR and 27% to Raf. When conservative amino acid substitutions are considered, this becomes 47%, 45% and 41% similarity to Ros, EGFR and Raf, respectively. It can be seen by the number of gaps needed to align the sequences, however, that the spacing between the conserved areas varies extensively from one sequence

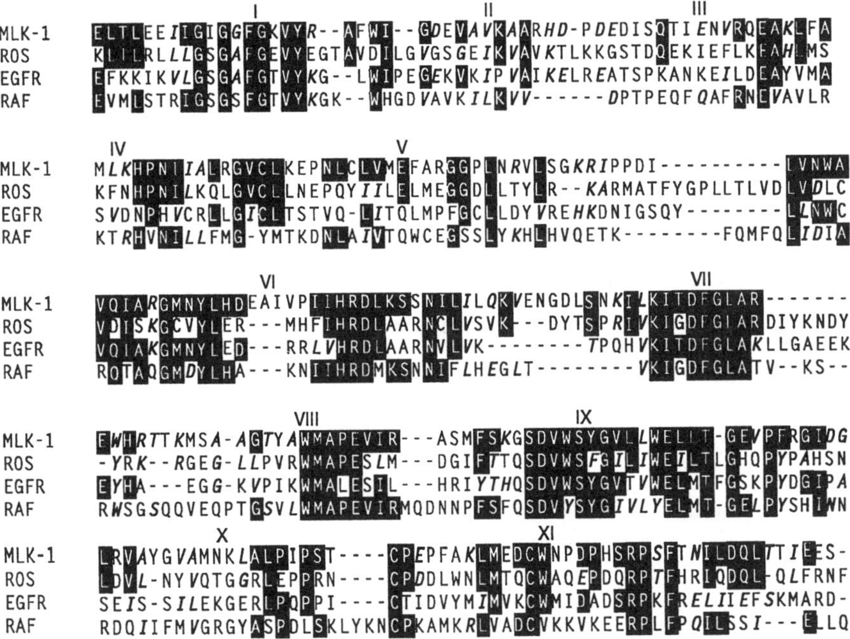

Figure 4. Alignment of the catalytic domain amino acid sequence of MLK1 with those of human Ros, EGFR and Raf. Alignments were done by the sequence alignment program CLUSTAL (Higgins and Sharp, 1988). Conserved amino acids are in shaded boxes and conservative replacements are in italics. Gaps, introduced to maximise the alignment, are denoted by dashes. Figure reprinted from Dorow et al., 1993.

to another. The alignment of amino acids within protein kinase catalytic domains is, with some exceptions, conserved among members of each kinase family (Hanks *et al.*, 1988).The catalytic domain sequence arrangement, therefore, supports the notion that the MLK's form a new and distinct family of protein kinases.

The Double Leucine Zipper and Basic Domain

A further notable feature of the MLK protein sequences is the presence of two closely spaced leucine/isoleucine zipper motifs followed by a basic domain C-terminal to the catalytic domain. This double leucine zipper region has a novel structure not previously reported for any protein. In the classic leucine zipper, first described by Landschultz *et al.* (1988) leucine residues are repeated with a heptad periodicity over a stretch of at least 22 amino acids. This sequence, which has a higher than average representation of charged residues, is postulated to form a helix with a hydrophobic ridge of leucines down one face. This motif, which is most commonly found in transcription factors, promotes dimerization (O'Shea *et al.*, 1991) through hydrophobic interactions between heptad leucines. Such dimerization appears to activate DNA binding by orienting the basic side chains of DNA binding residues for correct contact with the DNA. The strongly helix breaking amino acids, proline and glycine, are under-represented in zipper motifs (McKnight, 1991). The leucine/isoleucine zipper is a variation in that isoleucine replaces leucine at some positions in the heptad repeat (Atkinson *et al.*, 1991).

An alignment of the leucine zipper domains of the MLK's is shown in Figure 5. Each zipper motif contains 22 amino acids and the two motifs are separated by a 13 residue

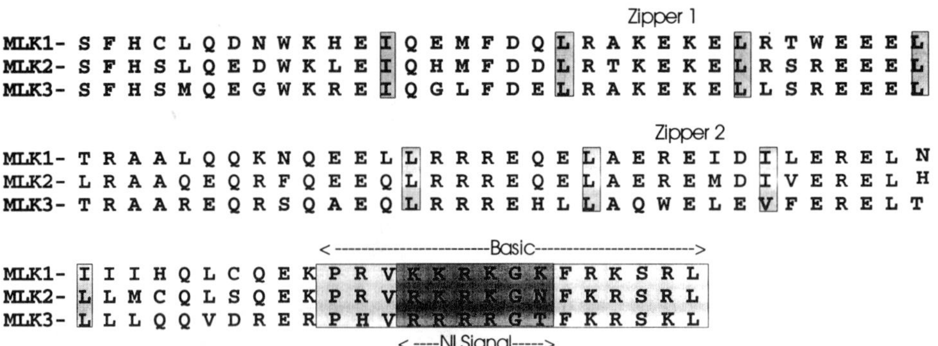

Figure 5. Alignment of the leucine zipper motifs, spacer and basic domains of MLKs 1, 2 and 3. Heptad leucine/isoleucine residues and basic domain residues are in shaded boxes. A possible nuclear localisation signal is double shaded.

"spacer" segment. The entire structural domain covers a region of 57 amino acids from which proline and glycine are totally absent in MLK's 1 and 2 while in MLK3 there is but one glycine located in zipper motif #1. MLK3 also has a valine substitution at the third heptad position of zipper #2, while all other heptad positions in both zipper motifs of the three proteins are occupied by either leucine or isoleucine. In each of the three proteins, charged amino acids occupy about 50% of positions within the double zipper domain. Among the charged amino acids, acidic residues are the most prevalent, yielding a net negative charge within the double zipper domain.

The degree of amino acid conservation between the MLK proteins is more varied within the double zipper domain than the catalytic domain (see Figure 5) While the zipper domain sequences of MLK's 1 and 2 share 74% identity, similar to that of their catalytic domains, residues in this domain are less highly conserved in MLK3 with 58% identity to MLK1 and 68% to MLK2. Within zipper motif #2, MLK's 1 and 2 are particularly similar with 19 of 22 positions having identical amino acids, and two of the three substitutions being conservative. There is also a very high degree of conservation of charged residues in the double zipper domain. Of 27 charged amino acids within the double zipper domain of MLK1 there are but two replacements in MLK2 and one of these is conservative. This is slightly reduced for MLK3 with 5 non-conservative substitutions of charged residues compared to MLK's 1 and 2. This high degree of amino acid conservation attests to the probable importance of the charged residues to the activity of the MLK proteins.

When arrayed on helical wheel templates (Schiffer and Edmundson, 1967), the two zipper motifs of MLK1 (Figure 6) differ slightly from one another in the positioning of charged amino acids. In motif #1, one side of the helix, at wheel positions 2, 5 and 6, has a preponderance of charged residues, creating a highly charged surface on that entire side of the putative helix. Position 5 is acidic, 2 is basic and 6 contains both basic and acidic amino acids. In motif #2, the leucine/isoleucine stripe is flanked on one side by three arginine residues and on the other by three glutamic acid residues. This creates basic, hydrophobic and acidic stripes along the length of the putative helix.

The positioning of charged residues within the MLK zipper motifs is similar to the zipper region of the cGMPDK (Landgraf *et al.*, 1990; Atkinson *et al.*, 1991). This is more striking for MLK zipper motif #2 in which the acidic (Gln-Glu-Glu-Glu) and basic (His-Arg-Arg-Arg) stripes flank the ridge of heptad leucine/isoleucine residues in the helical configuration (see Figure 6). In the cGMPDK zipper sequences, there are similarly placed acidic (Ala-Glu-Glu-Glu) and basic/hydrophobic (Lys-Leu-Lys-Leu) stripes (Atkinson *et*

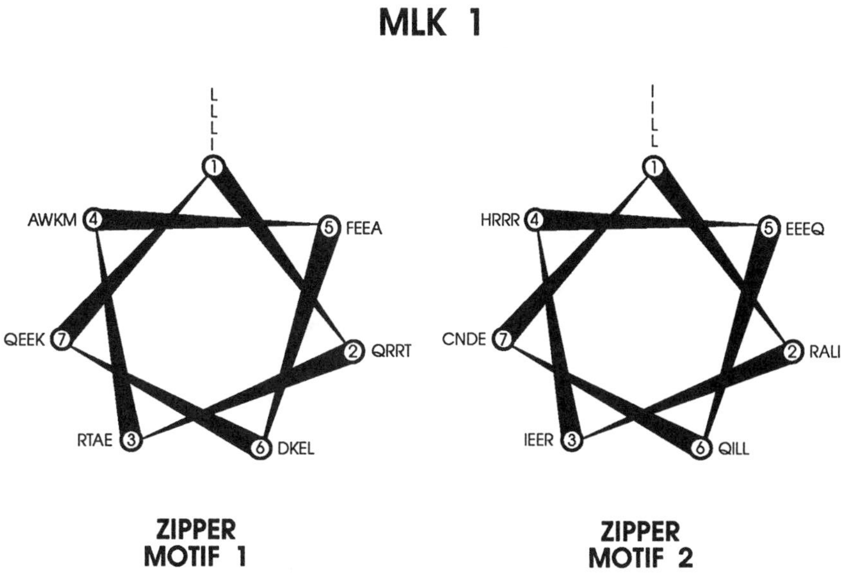

Figure 6. Helical wheel representation of the leucine zipper motifs of MLK1. The spokes of the wheel show schematically the relative positions of the amino acids of the zipper motifs in an idealised alpha-helix. The seven positions correspond to a typical alpha-helix with seven residues to every 2 helical turns. The most N-terminal heptad position is the residue closest to the center at the top spoke. Positions proceed around the wheel to the right skipping every second spoke. Residues in each succeeding 2 turns of the helix are in the second, third and fourth positions out from the center at each spoke. Figure reprinted from Dorow et al., 1993.

al., 1991). The cGMPDK leucine/isoleucine zipper sequence forms a helix in solution which is more stable to heat or acid conditions than that of the transcription factor-type leucine zipper, as measured by circular dichroism (Atkinson *et al.*, 1991). Furthermore, the stripes of charged residues within the cGMPDK zipper are thought to contribute to electrostatic stabilisation of the helical conformation (Landgraf *et al.*, 1990).

Another motif which has recently been described in DNA binding proteins is the helix-loop-helix (HLH) (Murre *et al.*, 1989). This motif is characterised by two putative amphipathic alpha-helices, separated by a short loop region, in a span of approximately 60 amino acids. It has been identified in a number of proteins which are involved in gene regulation. This HLH domain has some similarity to the MLK's with their two closely spaced but distinct zipper motifs. Based on the similarity of the MLK zipper domains to the cGMPDK leucine/isoleucine zipper, with its well characterised helical conformation (Landgraf *et al.*, 1990; Atkinson *et al* 1991), the MLK zippers would be expected to form stable helices. The 13 residue spacer sequence between the two helical segments suggests a possible break in the helical nature of this domain. We therefore used a structural prediction algorithm (Chou and Fasman, 1978) to define possible secondary structures for the MLK1 zipper domain sequence. In this analysis the two zipper motifs were indeed predicted to be helical in nature, with a turn conformation predicted in the spacer region (Dorow *et al.*, 1993). If this arrangement does occur in the folded protein, the two zipper helices within the one MLK protein could interact with each other in an anti-parallel "zipper-turn-zipper" configuration analogous to the HLH helices.

A further region of interest in the MLK structures is an extremely basic sequence that begins 9 residues C-terminal to the final heptad isoleucine of the double zipper domain (see Figure 5). In this 15 amino acid stretch there are 10 basic (Lys or Arg), 4 hydrophobic and no acidic amino acids. Like the double zipper domain, the basic region shows a high degree of charge conservation between the sequences of the MLK's. While there are several replacements within the 15 amino acids, there is but one substitution (Lys to Asn or Thr) which changes the basic nature of the amino acid side chain. While this basic sequence has some similarity to the DNA binding domains of the transcription factors (Vinson *et al.*, 1989), it also has several significant differences including it's location on the C-terminal, rather than the N-terminal side of the leucine zipper. Furthermore, the MLK basic region contains the sequence Val-Lys/Arg-Lys-Arg-Lys-Gly which is very similar to the nuclear localisation signal of the SV40 large T antigen (Pro-Lys-Lys-Lys-Arg-Lys-Val [Kalderon *et al.*, 1984]).

The C-terminal Region

Both MLK's 2 and 3 have C-terminal domains which are rich in proline and serine/threonine (data not shown). This C-terminal domain is not coded in the MLK1 cDNA clone from human colon, however, it is present in mouse brain MLK1 and in human MLK2. Because of it's presence in the mouse MLK1 sequence, this domain would be expected to exist in the human MLK1 protein as well. It is possible that truncated or rearranged human colonic MLK1 cDNAs represent RNAs that have been spliced or edited in human colonic cells. It cannot be ruled out, however, that they may be an artefact that occurred in preparation of the colonic cDNA library.

While there is considerable similarity within the C-terminal domain among three of the MLK proteins (mouse brain MLK1, human colon and brain MLK2 and human MLK3 from several sources), it is not as striking as in the SH3, catalytic, zipper, and basic domains. Sequences within this domain, therefore, may be involved in conferring specificity on these three similar molecules. The C-terminal domain of mouse MLK1 contains consensus sequences for phosphorylation by a number of known protein kinases, including casein kinase II, protein kinase C, and cyclic AMP dependent protein kinase. The C-terminal domain of MLK3 also contains sequences similar to the proline rich consensus for SH3 domain binding (Gallo *et al.*, 1994; Ing *et al.*, 1994). These features could possibly allow the C-terminal domain to regulate protein interactions involving the MLK-SH3 domain.

SUMMARY

In this report, we have described the structure of three members of the MLK's, a new family of human protein kinases. These are the first kinases described with an SH3 domain in the absence of an SH2 domain and MLK3 is the first SH3 domain containing kinase to be shown to have serine/threonine specificity. In addition, the MLK's have a unique double leucine zipper/basic domain that has not been found in any other protein to date.

In the current model for many signal transduction pathways, activated receptors recruit protein kinases and adaptor molecules that contain SH2 and SH3 domains. This is accompanied by a general rise in protein kinase activity and interaction among proteins residing near the plasma membrane and/or bound to cytoskeletal elements. Many of these proteins contain SH3 domains. Among other effects, this leads to G protein activation and guanine-nucleotide exchange, generating a signal which translocates into the nucleus to trigger complex formation of transcription factors through their leucine zippers. This activates DNA binding and gene transcription. Phosphorylation plays a role at each step of this multi-faceted process. As well as a novel type of kinase catalytic domain structure, the MLK's contain both a SH3 domain, leucine zippers and a possible nuclear localisation signal. This makes them unusual among the protein kinases in that they may act in several compartments of the pathway through their distinct domains. Definition of the role of this new family of biological control molecules in signal transduction pathways may yield new insights into the regulation of cellular events.

ACKNOWLEDGMENTS

The work on MLK's 1 and 2 was supported by a grant from the National Health and Medical Research Council of Australia.

REFERENCES

Anderson, D., Koch, C.A., Grey, L., Ellis, C., Moran, M. & Pawson, T.(1990). Binding of SH2 domains of phospholipase C 1, GAP, and Src to activated growth factor receptors. Science 250, 979-982.

Atkinson, R.A., Saudek, V., Huggins, J.P. & Pelton, J.T. (1991). 1H NMR and circular dichroism studies of the N-terminal domain cyclic GMP dependent protein kinase: a leucine/isoleucine zipper. Biochemistry 30, 9387-9395.

Barton, D.E., Foellmer, B.E., Du, J., Tamm, J., & Derynck, R. (1988). Chromosomal mapping of genes for transforming growth factors beta 2 and beta 3 in man and mouse: dispersion of the TGF-beta gene family. Oncogene Res. 3, 323-331.

Ben-David, Y., Letwin, K., Tannok, L., Bernstein, A. & Pawson, T.(1991). A mammalian protein kinase with potential for serine/threonine and tyrosine phosphorylation is related to the cell cycle regulators. EMBO J. 10, 317-325.

Birchmeier, C., Sonnenberg, E., Weidner, K.M. & Walter, B. (1993) Tyrosine kinase receptors in the control of epithelial growth and morphogenesis during development. BioEssays 15, 185-189.

Booker, G.W., Gout, I., Downing, A.K., Driscoll, P.C., Boyd, J., Waterfield, M.D. & Campbell, I.D. (1993). Solution structure and ligand-binding site of the SH3 domain of the p85α subunit of phosphatidylinositol 3-kinase. Cell 73, 813-822.

Cantley, L.C., Auger, K.R., Carpenter, C., Duckworth, B., Graziani, A., Kapeller, R. & Soltoff, S. (1991). Oncogenes and signal transduction. Cell 64, 281-302.

Choo, K.H., Brown, R.M. & Earle, E. (1990) *In situ* hybridization of chromosomes. In: *Protocols in Human Molecular Genetics*, Vol 7, Ed. C.Matthews. Humana Press, USA.

Chou, P.Y. & Fasman, G.D.(1978). Empirical predictions of protein conformation. Ann. Rev. Biochem. 47, 251-276.

Dorow, D.S., Devereux, L., Dietzsch, E. and deKretser, T. (1993). Identification of a new family of human epithelial protein kinases containing two leucine/isoleucine zipper domains. Eur J Biochem 231, 701-710.

Dunphy, W.G. & Newport, J.W.(1988). Unravelling of mitotic control mechanisms. Cell 55, 925-928.

Eck, M.J., Atwell, S.K., Shoelson, S.E. & Harrison, S.C. (1994) Structure of the regulatory domains of the Src-family tyrosine kinase Lck. Nature 368, 764-769.

Egan, S.E., Giddings, B.W., Brooks, M.W., Buday, L., Sizeland, A.M., & Weinberg, R.A. (1993). Association of Sos Ras exchange protein with Grb2 is implicated in tyrosine kinase signal transduction and transformation. Nature 363, 45-51.

Ekstrand, A.J. & Zech, L.(1987). Human c-fos proto-oncogene mapped to chromosome 14 band q24.3-31. Possibilities for oncogene activation by chromosomal rearrangements in human neoplasms. Exp Cell Res 169, 262-266.

Ezoe, K., Lee, S-t., Strunk, K. and Spritz, R.A. (1994). PTK1, a novel protein kinase required for proliferation of human melanocytes. Oncogene 9, 935-938.

Featherstone, C. & Russell, P.(1991). Fission yeast P107*wee1* mitotic inhibitor is a tyrosine/serine kinase. Nature 349, 808-811.

Gallo, K.A., Mark, M.R., Scadden, D.T., Wang, Z., Gu, Q. & Godwoski, P.J. (1994) Identification and characterization of SPRK, a novel src-homology 3 domain-containing proline-rich kinase with serine/threonine kinase activity. J Biol Chem 269, 15092-15100.

Gout, I., Dhand, R., Hiles, I.D., Fry, M.J., Panayotou, G., Das, P., Truong, O., Totty, N.F., Hsuan, J., Booker, G.W., Campbell, I.D. & Waterfield, M.D. (1993). The GTPase dynamin binds to and is activated by a subset of SH3 domains. Cell 75, 25-36.

Hanks, S.K., Quinn, A.M. & Hunter, T.(1988). The protein kinase family: conserved features and deduced phylogeny of the catalytic domains. Science 241, 42-52.

Hanks, S.K. (1991). Eukaryotic protein kinases. Current Opinion in Structural Biology 1, 369-383

Harrison, G.S., Drabkin, H.A., Kao, F.T., Hartz, J., Chu, E.H., Wu, B.J. & Morimoto, R.I.(1987). Chromosomal location of human genes encoding major heat-shock protein HSP70. Somat. Cell Mol. Genet. 13, 119-130.

Higgins, D.G. & Sharp, P.M.(1988). Clustal: a package for performing multiple sequence alignments on a microcomputer. Gene 73, 237-244.

Hunter, T.(1991). Protein kinase classification. Methods Enzymol 200, 3-37.

Hunter, T. & Karin, M.(1992). The regulation of transcription by phosphorylation. Cell 70, 375-387.

Ing, Y.L., Leung, I.W.L., Heng, H.H.Q., Tsui, L-C., Lassam, N.J. (1994). MLK-3: identification of a widely-expressed protein kinase bearing an SH3 and leucine zipper-basic region domain. Oncogene 9, 1745-1750.

Kalderon, D., Richardson, W.D., Markham, A.T. & Smith, A.E. (1984). Sequence requirements for nuclear localization of simian virus large-T antigen. Nature 311, 33-38.

Khoda, D., Hatanaka, H., Odaka, M., Mandiyan, V., Ullrich, A., Schlessinger, J & Inagaki, F. (1993). Solution structure of the SH3 domain of phospholipase C-γ. Cell 72, 953-960.

Koch, C.A., Anderson, D., Moran, M.F., Ellis, C. & Pawson, T.(1991). SH2 and SH3 domains: Elements that control interactions of cytoplasmic signalling proteins. Science 252, 668-674.

Koyama, S., Yu, H., Dalgarno, D.C., Shin, T.B., Zydowsky, L.D. & Schreiber, S.L. (1993). Structure of the PI3K SH3 domain and analysis of the SH3 family. Cell 72, 945-952.

Landgraf, W., Hofmann, F., Pelton, J.T. & Huggins, J.P. (1990). Effects of cyclic GMP on the secondary structure of cyclic GMP dependent protein kinase and analysis of the enzyme's amino-terminal domain by far-ultraviolet circular dichroism. Biochemistry 29, 9921-9928.

Landschultz, W.H., Johnson, P.F. & McKnight, S.L. (1988). The leucine zipper: a hypothetical structure common to a new class of DNA binding proteins. Science 240, 1759-1764.

Lowenstein, E.J., Daly, R.J., Batzer, A.G., Li, W., Margolis, B., Lammers, R., Ullrich, A., & Schlessinger, J. (1992). The SH2 and SH3 domain-containing protein GRB2 links receptor tyrosine kinases to ras signaling. Cell 70, 431-442.

Mayer, B.J., Hamaguchi, M. & Hanafusa, H. (1988). A novel viral oncogene with structural similarity to phospholipase C. Nature 332, 272-275.

Mayer, B.J., Jackson, P.K., & Baltimore, D. (1991). The noncatalytic *src* homology region 2 segment of *abl* tyrosine kinase binds to tyrosine-phosphorylated cellularproteins with high affinity. Proc. Natl. Acad. Sci. (USA) 88, 627-631.

McKnight, S.L.(1991). Molecular zippers in gene regulation. Scientific American 4, 32-39.

Morgan, D., Kaplan, J.M, Bishop, J.M. & Varmus, H.A.(1989). Mitosis-specific phosphorylation of $p60^{c\text{-}src}$ by $p34^{cdc2}$-associated protein kinase. Cell 57, 775-786.

Murre, C., McCaw, P.S. & Baltimore, D.(1989). A new DNA binding and dimerization motif in immunoglobulin enhancer binding, *daughterless*, *MyoD*, and *myc* proteins. Cell 56, 777-783.

Musacchio, A., Noble, M., Pauptit, R., Wierenga, R and Saraste, M. (1992). Crystal structure of a Src-homology 3 (SH3) domain. Nature 359, 851-855.

Noble, M.E.M., Musacchio, A., Saraste, M., Courtneige, S.A. and Wierenga, R.K. (1993). Crystal structure of the SH3 domain in human Fyn; comparison of the three-dimensional structures of SH3 domains in tyrosine kinases and spectrin. EMBO J. 12, 2617-2624.

O'Shea, E.K., Klemm, J.D., Kim, P.S. & Alber, T.(1991). X-ray structure of the GCN4 leucine zipper, a two-stranded, parallel coiled coil. Science 254, 539-544.

Pawson, T. & Gish, G.D. (1992). SH2 and SH3 domains: from structure to function. Cell 71, 359-362.

Posada, J., & Cooper, J.A. (1992). Molecular signal integration. Interplay between serine, threonine, and tyrosine phosphorylation. Mol. Biol. Cell. 3, 583-592.

Ren, R., Mayer, B.J., Cicchetti, P., Baltimore, D. (1993). Identification of a ten-amino acid proline-rich SH3 binding site. Science 259, 1157-1161.

Rodaway, A.R.F., Sternberg, M.J.E., & Bentley, D.L. (1989). Similarity in membrane proteins. Nature 342, 624.

Schiffer, M. & Edmundson, A.B.(1967). The use of helical wheels to represent the structures of proteins and to identify segments with helical potential. Biophysical J. 7, 121-135.

Stern, D.F., Zheng, P., Beider, D.R. & Zerillo, C. (1991). Spk1, a new kinase from *Saccharomyces cerevisiae*, phosphorylates proteins on serine, threonine and tyrosine. Mol. Cell. Biol. 11, 987-1001.

Testa, J.R.(1990). Chromosome translocations in human cancer. Cell Growth Differ. 1, 97-101.

Vinson, C.R., Sigler, P.B. & McKnight, S.L.(1989). Scissors-grip model for DNA recognition by a family of leucine zipper proteins. Science 246, 911-916.

Yu, H., Rosen, M., Shin, T.B., Seidell-Dugan, C., Brugge, J.S. and Schreiber, S.L. (1992). Solution structure of the SH3 domain of src and identification of it's ligand binding site. Science 258, 1665-1668.

Yu, H. Chen, J.K., Feng, S., Dalgarno, D.C., Brauer, A.W. and Schreiber, S.L. (1994) Structural basis of binding of proline-rich peptidees to SH3 domains. Cell 76, 933-945.

33

AGONIST-INDUCED INTERNALIZATION AND DEGRADATION OF γ-AMINOBUTYRIC ACID$_A$ (GABA$_A$) RECEPTOR POLYPEPTIDES FROM THE NEURONAL SURFACE

Eugene M. Barnes, Jr.* and Patricia A. Calkin

Verna and Marrs McLean Department of Biochemistry
Baylor College of Medicine
Houston, Texas 77030

INTRODUCTION

In the vertebrate brain, GABA$_A$ receptors on postsynaptic membranes are the major transducers of fast inhibitory neurotransmission. These receptors are hetero-pentameric proteins that provide specific binding sites for GABA, benzodiazepines, barbiturates, and anesthetic steroids as well as an integral chloride channel (Macdonald and Olsen, 1994). Chloride channel openings are gated by GABA and allosterically potentiated by benzodiazepines and other anxiolytic and hypnotic drugs. The exceptionally rich pharmacology associated with GABA$_A$ receptors has evoked considerable interest in their structure and function. Chronic administration of many GABA$_A$ergic compounds in humans and animals produces syndromes of dependence and tolerance which limit their clinical value. Since the development of tolerance to GABA$_A$ergic drugs is attributed to functional rather than pharmacokinetic accomodation (Greenblatt and Shader, 1986), attention has focused on use-dependent modifications of GABA$_A$.receptors. Indeed, there is general agreement that chronic exposure of rodents to benzodiazepines produces a decline in GABA$_A$ receptor function which coincides with the onset of tolerance (Miller et al., 1988; Marley and Gallager, 1989; Lewin et al., 1989). However, the molecular events which underlie this loss of receptor function are not well defined.

Our understanding of the use-dependent regulation of GABA$_A$ receptors has been greatly aided by studies of cortical neurons in tissue culture. GABA$_A$ergic ligands bind readily to receptor domains on the surface of these cells, permitting biophysical and biochemical characterization of receptor function. After the binding of GABA$_A$ receptor

* Corresponding Author: Dr. E.M. Barnes, Biochemistry Department, Baylor College of Medicine, One Baylor Plaza, Houston, TX 77030. Tel: (713)798-4523; Fax: (713)798-7854.

Methods in Protein Structure Analysis, Edited by M. Z. Atassi and E. Appella
Plenum Press, New York, 1995

agonists, the most rapid regulatory process is desensitization. Desensitization of these receptors occurs within seconds and is rapidly reversible following agonist removal. Because desensitization can be demonstrated in isolated membrane patches at room temperature (Hamill et al., 1983; Weiss, 1988), the mechanism probably involves a reduction in intrinsic $GABA_A$ receptor channel activity rather than receptor removal from the surface. After exposure of cortical neurons to GABA or the benzodiazepine clonazepam for one hr at 37°C, an increase in the intracellular fraction of ^{3}H-flunitrazepam binding sites has been reported (Tehrani and Barnes, 1991). That the sequestration of $GABA_A$ receptors occurs in vivo is suggested by ligand binding to clathrin-coated vesicles (Tehrani and Barnes, 1993). Following administration of lorazepam to mice, the level of $GABA_A$ receptors on clathrin-coated vesicles increases while that on synaptic membranes decreases (Tehrani and Barnes, 1994). Although it appears likely the sequestered $GABA_A$ receptors are derived from the neuronal surface, this has not yet been demonstrated.

It is well known that chronic (several days) exposure of cortical neurons to GABA reduces the density of $GABA_A$ receptor ligand binding sites (Maloteaux et al., 1987; Tehrani and Barnes, 1988; Roca et al., 1990). This process, referred to as down-regulation, is accompanied by persistent losses of spontaneous inhibitory postsynaptic currents as well as chloride currents evoked by applied GABA (Hablitz et al., 1989). An agonist-dependent reduction in receptor number could be explained by a decrease in receptor biosynthesis or by an increase in degradation. Since $GABA_A$ receptor subunit mRNAs are also subject to down-regulation by GABA (Montpied et al., 1991; Baumgarter et al., 1994; Mhatre and Ticku, 1994), changes in the synthesis or stability of receptor transcripts appear to be a part of the control mechanism. However, after administration of GABA, the reduction in ligand binding sites precedes that of the subunit mRNAs (Baumgartner et al., 1994), leading to the hypothesis that translational or post-translational regulation may be important in the initial phase of down-regulation.

In order to investigate the down-regulation of $GABA_A$ receptor polypeptides from the neuronal surface, we have utilized the impermeant cleavable labeling reagent ^{125}I-DPSgt (3,3'-dithiopropionyl 1-sulfosuccinimidyl 1'-glycyltyrosine) (Bretscher and Lutter, 1988) in combination with quantitative immunoprecipitation (Calkin and Barnes, 1994). We report here the application of this technique to examine the agonist-induced sequestration and subsequent degradation of $GABA_A$ receptor polypeptides.

METHODS AND RESULTS

Effects of Chronic Exposure to $GABA_A$ Receptor Agonists

Neuronal cell cultures from the embryonic chick cerebral cortex were prepared as described by Tehrani and Barnes (1991). Living neurons were chronically treated with GABA or other agonists by addition of a single dose to the culture medium and returning the cells to the incubator for 5 days. Neurons from the same preparation but without agonist addition were used as controls. The cell monolayers on Petri dishes were washed and incubated with [^{125}I]DPSgt at 4°C (Calkin and Barnes, 1994). Analysis of the major ^{125}I-peptides by SDS-PAGE showed that the GABA treatment produced no detectible difference in the labeling pattern. Furthermore, the total incorporation of ^{125}I into cellular protein was not changed by exposure of the neurons to GABA or to the $GABA_A$ receptor agonists, isoguvacine and THIP (4,5,6,7,-tetrahydroisoazolo[5,4-c]pyridin-3-ol). Consistent with the structure and known properties of [^{125}I]DPSgt (Bretcher and Lutter, 1988), washing the intact cells with GSH (glutathione) buffer removed essentially all of the detectible ^{125}I. This shows that the ^{125}I label is initially confined to polypeptides on the neuronal cell surface.

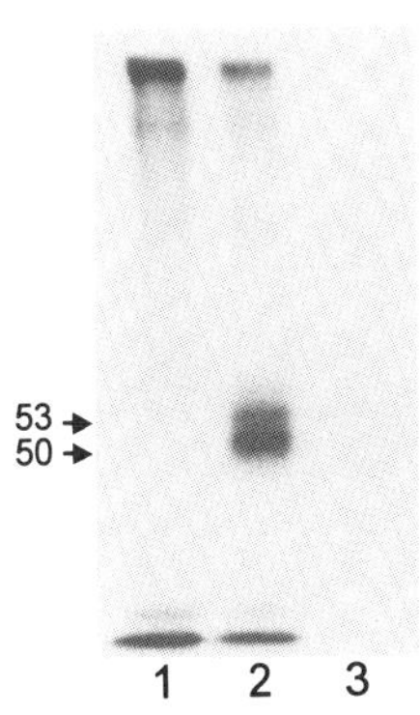

Figure 1. Labeling of surface GABA$_A$ receptor polypeptides with ^{125}I-DPSgt. Cultured cortical neurons were washed and labeled with ^{125}I-DPSgt for 30 min at 4°C. The ^{125}I medium was removed and the cells were extracted with Tris-buffered saline containing 1% Triton X-100, 0.1% SDS, and protease inhibitors (Calkin and Barnes, 1994). In *lane 3*, the cells were washed at 4°C with buffer containing 100 mM GSH before extraction. The clarified extracts were incubated with 3 μl of preimmune serum (*lane 1*) or antiserum RB4 (*lanes 2 and 3*) and then mixed and incubated further with 40 μl *Staphlococcus* A cells (10% w/v). The immunoprecipitates were run on 10% polyacrylamide-SDS gels which were then dried and autoradiographed.

GABA$_A$ receptor polypeptides with ^{125}I-labeled surface domains were isolated by Triton X-100 extraction and immunoprecipitation with polyclonal antiserum RB4, an antibody directed against the native receptor. Antiserum RB4 quantitatively precipitates GABA$_A$ receptor binding sites for ^{3}H-muscimol and ^{3}H-flunitrazepam from Triton extracts of cultured neurons and cross-reacts with 50-54-kDa subunits of the affinity-purified receptor (Calkin and Barnes, 1994). The RB4 immunoprecipitates from extracts of ^{125}I-DPSgt-labeled neurons contained labeled 50- and 53-kDa polypeptides which were not found in preimmune controls (Fig. 1). The mass of these polypeptides is similar to the major RB4 cross-reactive subunits from the GABA$_A$ receptor antigen. When the labeled cells were washed with GSH buffer prior to extraction, essentially all of the radioactivity was removed from these proteins (Fig. 1). Thus, the 50- and 53-kDa ^{125}I-polypeptides arise from GABA$_A$ receptor subunits which contain domains exposed at the outer surface of the cells.

After chronic (5 day) treatment of a set of cultures with agonists (100 mM final concentration in the growth medium), washed intact neurons were labeled with ^{125}I-DPSgt as before. This exposure to GABA caused a decline in the surface 50- and 53-kDa GABA$_A$ receptor ^{125}I-subunits compared to the untreated controls (Fig. 2). Since the specific GABA$_A$ receptor antagonist, R5135 (3α-hydroxy-16-imino-5β-17-aza-androstan-11-one) prevented this decline, the GABA$_A$ receptor appears to have a role in signaling its own down-regulation. Similar chronic treatments were carried out with the specific GABA$_A$ receptor agonists, isoguvacine and THIP. The combined autoradiographic density of the 50- and 53-kDa ^{125}I-polypeptides was determined and used to quantify the extent of GABA$_A$ receptor down-regulation produced by these agents (Table 1). Both GABA and isoguvacine caused a substantial decrease in the amount of labeled proteins, while THIP was much less effective. This is in accord with the effects of these agents on GABA$_A$ receptor channels in our

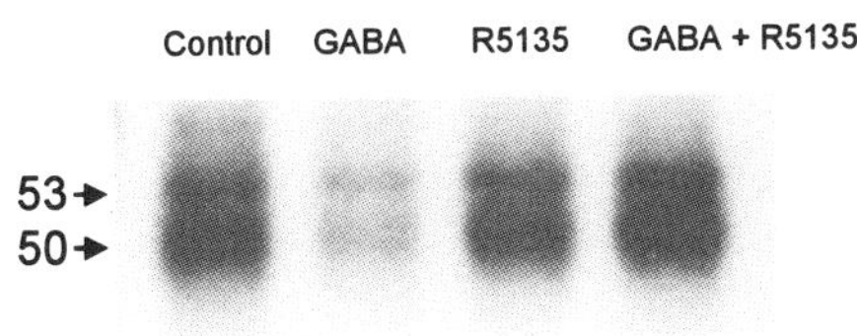

Figure 2. Levels of surface GABA$_A$ receptor polypeptides on neurons chronically exposed to GABA and R5135. Where indicated, GABA (100 μM) and R5135 (1 μM) were added to the culture medium and the cells returned to the incubator for 5 days. These additions were omitted for the samples in the control lane. The cells were then washed and labeled with ^{125}I-DPSgt and the GABA$_A$ receptor polypeptides were analyzed using RB4 immunoprecipitation as in Fig. 1.

Table 1. Effect of chronic agonist exposure on $GABA_A$ receptor surface polypeptides

	^{125}I-Receptor peptides	
Treatment	%	n
None	100 ± 10.5	7
GABA	*37.5 ± 3.2	7
Isoguvacine	*52.4 ± 4.9	4
THIP	84.2	2

Experiments were carried out as described for Fig. 2. Cells were treated for 5 days with the compound indicated at a 100 μM final concentration and then labeled with ^{125}I-DPSgt. Extracts were immunoprecipitated and analyzed on gels. Regions of the autoradiographs corresponding to 50-53 kDa (Fig. 2) were quantified as a single band by digital optical analysis. The data are expressed as a percentage of controls without agonist and represent the mean ± S.E. of the number of experiments indicated. $*p < 0.01$ compared to control. The difference between GABA and isoguvacine treatments was not statistically significant.

preparations. Application of GABA and isoguvacine induces robust chloride currents but the responses to THIP are much weaker (Mistry and Hablitz, 1990). We have also examined the down-regulation of the total cellular pool of $GABA_A$ receptor polypeptides by DPSgt iodination of membranes from 100,000 g pellets of neuronal homogenates. The membrane level of 50-53-kDa ^{125}I-subunits in the GABA- and isoguvacine-treated cells represented 36% and 53%, respectively, that in the untreated controls. Comparable results were obtained when isolated membranes were iodinated using chloramine T. This suggests that during chronic agonist exposure the down-regulated surface receptor subunits are not retained in an intracellular pool.

In order to compare these results with the down-regulation of $GABA_A$ receptor binding sites, we first measured the binding of ^{3}H-flunitrazepam and ^{35}S-TBPS (t-butylbicyclophosphorothionate), a ligand for the $GABA_A$ receptor channel. After chronic treatment of the neurons with GABA or isoguvacine, the binding of both ligands to isolated membranes was substantially reduced (Table 2). As before, THIP had little effect. To examine the intracellular binding sites after chronic treatments, we labeled intact neurons with ^{3}H-fluni-

Table 2. Effect of chronic agonist treatment on $GABA_A$ receptor ligand binding to isolated membranes

	^{3}H-Flu binding		^{35}S-TBPS binding	
Treatment	%	n	%	n
None	100 ± 4.0	16	100 ± 6.3	7
GABA	*68.8 ± 1.4	8	*55.5 ± 3.3	7
Isoguvacine	*57.0 ± 3.5	6	*47.4 ± 4.2	7
THIP	95.8 ± 5.0	4	95.2 ± 5.5	4

Cells were treated with agonist as in Table 1. The monolayers were washed and crude membranes were isolated and assayed for radioligand binding as described by Calkin and Barnes (1994). Results are expressed as a percentage of controls without agonist and represent the mean ± S.E. from the indicated number of experiments. $*p < 0.01$ relative to untreated control.

Table 3. Effect of chronic agonist treatment on ^{3}H-flunitrazepam binding to intact neurons

Treatment	Total receptor		Intracellular receptor		
	fmol/mg	n	fmol/mg	n	Intracellular/total
None	67.9 ± 1.2	15	5.07 ± 0.65	15	0.075
GABA	*42.0 ± 2.4	15	6.57 ± 1.09	14	0.156
Isoguvacine	*35.1 ± 1.7	14	5.27 ± 1.30	15	0.150

Cells were treated with agonist as in Table 1. The monolayers were washed and incubated with 1 nM ^{3}H-flunitrazepam. Nonspecific binding was determined using 1 μM benzodiazepine 1012-S and intracellular binding with 1 μM SPTC-1012S as described by Tehrani and Barnes (1991). The results are expressed per mg cell protein and represent the mean ± S.E. of the indicated number of experiments. *$p < 0.01$ relative to untreated control.

trazepam, a membrane-permeant ligand, and displaced the surface radioactivity with an impermeant benzodiazepine, SPTC-1012S (Tehrani and Barnes, 1991). The internal receptor binding sites determined in this manner represent 7.5% of the cellular total (Table 3). Consistent with the ^{125}I-labeling experiments, the levels of intracellular receptors (measured as fmol ligand bound/mg protein) did not change significantly after chronic agonist exposure. Nevertheless, the treatments doubled the fraction of internal/total receptor binding sites. This is a consequence of the decline in surface receptors.

Effects of Acute Application of Agonists

The DPSgt labeling procedure was also employed to study the fate of surface $GABA_A$ receptor subunits during acute exposure of the neuronal cultures to agonists. Cells grown in normal medium were ^{125}I-labeled with DPSgt at 0°C, incubated in culture with 200 μM GABA for 2 or 4 hr at 37°C, and then washed with GSH buffer. Extracts of the neurons were analyzed as before. Labeled 50- and 53-kDa receptor polypeptides that were protected from GSH cleavage were recovered in significant amounts from cells acutely exposed to GABA but not from the untreated controls (Fig. 3). Densitometric analysis of the autoradiographs revealed that 16.3 ± 2.4% (n = 3) of the surface polypeptides were internalized (protected) during the 2 hr GABA treatment (Fig. 4). Much lower amounts of the labeled subunits (<3% of those remaining at the surface) were sequested by cells which were incubated with GABA for 2 hr at 4°C (Fig. 4). From cells exposed to GABA plus R5135 for 2 hr at 37°C or controls without GABA, internalized polypeptides were also barely detectable (not shown). Because

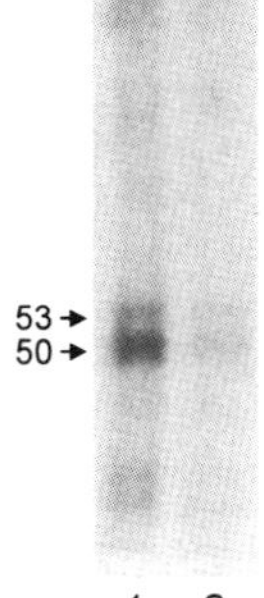

Figure 3. Sequestration of $GABA_A$ receptor polypeptides by acute GABA application. Neurons were cultured for 6 days without exogenous GABA, labeled with ^{125}I-DPSgt at 4°C as in Fig. 1, and then incubated in the presence (*lane 1*) or absence (*lane 2*) of 200 μM GABA for 2 hr at 37°C. The cells were washed with GSH buffer and the $GABA_A$ receptor polypeptides were analyzed as in Fig. 2.

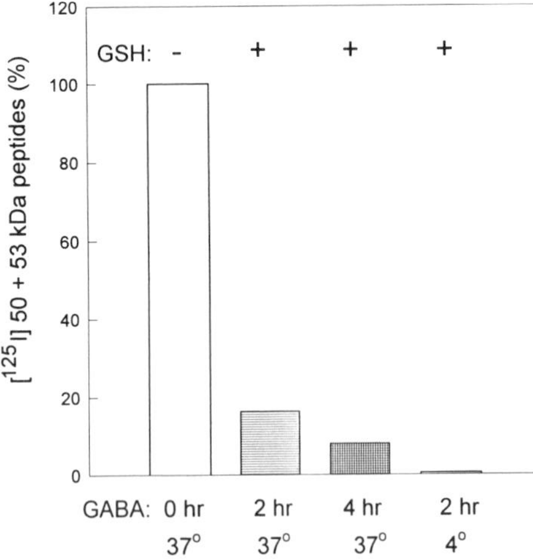

Figure 4. Acute sequestration and degradation of surface-derived $GABA_A$ receptor polypeptides. Experiments were carried out as described for Fig. 3. Cells were labeled with ^{125}I-DPSgt, incubated with 200 μM GABA under the conditions shown, and then washed with GSH buffer where indicated. $GABA_A$ receptor polypeptides were analyzed as in Table 1. The results are expressed as a percentage of controls in which the GABA treatment and GSH wash were omitted.

receptor internalization is unlikely to occur at 4°C, it is probable that the small amount of polypeptide recovered under these conditions is due to residual surface label which was not removed by the GSH wash.

We consistently found that the amount of internalized $GABA_A$ receptor polypeptides was greater after a 2 hr than after a 4 hr exposure of the neurons to GABA. Quantitation of a typical film revealed that 7.9% of the surface polypeptides were recovered in the intracellular fraction after the 4 hr treatment compared to 16% after 2 hr (Fig. 4). Since the surface subunits which are subject to chronic down-regulation are not retained by the neurons, it appears likely that the loss of sequestered polypeptides found in the 4 hr GABA treatment is due to intracellular degradation. A possible role for lysosomal proteases in this process was evaluated by the addition of 50 μM chloroquine during the acute GABA treatment. Since chloroquine had no detectable effect on the amount of internalized receptor polypeptides, lysosomes appear not to be involved in the degradation.

CONCLUSIONS

We have shown previously that $GABA_A$ receptor ligand binding sites and gated chloride channels are down-regulated during chronic exposure of cortical neurons to GABA (Tehrani and Barnes, 1988; Hablitz et al., 1989). This has been independently confirmed in a number of other laboratories. Using the DPSgt labeling procedure (Calkin and Barnes, 1994), we have demonstrated that these reductions in $GABA_A$ receptor function can be accounted for by a corresponding loss of the receptor subunit polypeptides from the neuronal surface. The down-regulation of these subunits was also induced by isoguvacine, a $GABA_A$ receptor-specific agonist, and could be completely prevented by the specific antagonist R5135. This rules out involvement of other known GABA binding proteins, such as $GABA_B$ receptors or GABA transporters, and indicates that the $GABA_A$ receptor provides the agonist site for its own down-regulation.

The DPSgt labeling procedure also permitted us to examine the fate of $GABA_A$ receptor polypeptides during their down-regulation from the cell surface. By stripping the intact cells with GSH, the label associated with exterior domains of these proteins was removed, revealing a fraction of the subunits (approximately 16%) which had become sequestered as a consequence of acute GABA application at 37°C. No detectable sequestration occurred in the absence of GABA or in the presence of GABA plus R5135. It was shown previously that acute exposure of these neuronal preparations to GABA or clonazepam increases the fraction of internal/total receptor binding sites (Tehrani and Barnes, 1991). The current studies indicate that this increase is probable due to $GABA_A$ receptors derived acutely from the surface. Although the vehicle for this sequestration in vitro is not known, clathrin-coated vesicles are strongly implicated by $GABA_A$ receptor binding studies in rodent brain (Tehrani and Barnes, 1993). Since administration of lorazepam to mice increases receptor binding in clathrin-coated vesicles, while reducing that in synaptic membranes (Tehrani and Barnes, 1994), it appears that agonist-dependent sequestration of $GABA_A$ receptors also occurs in vivo.

As the acute GABA treatment in culture progressed from 2 to 4 hr, the amount of internalized receptor polypeptides declined, suggesting that they were degraded intracellularly. This is consistent with the results obtained from the chronic agonist treatment which show that $GABA_A$ receptors do not accumulate within the cells after the initial down-regulation. Since the entire cellular pool of receptor polypeptides and ligand binding sites decreases by 50-60% during the chronic treatment, degradation appears to be a likely mechanism for down-regulation. However, reduction of receptor biosynthesis, an important mechanism in the down-regulation of β-adrenergic receptors (Collins et al., 1992), is an alternative. Indeed, chronic exposure of cortical preparations to GABA reduces the levels of $GABA_A$ receptor α1, α2, and α3 subunit mRNAs (Montpied et al., 1991; Mhatre and Ticku, 1994). Studies in our laboratory are in accord with these findings (Baumgartner et al., 1994). Quantitative RT-PCR analysis reveals that the α1, β2, β4, γ1, and γ2 subunit mRNAs are all reduced by a similar degree (47-65%) by a 7-day exposure of the cells to GABA. A more detailed examination of the decline of the α1-subunit transcript revealed that no significant change was produced during the first 4 days of GABA treatment. However, after 4 days of exposure there is a 50% reduction in the density of $GABA_A$ ligand binding. Since the attenuation of $GABA_A$ receptor subunit mRNAs appears to be a relatively slow process when compared to that for subunit polypeptides and ligand binding sites, we propose that translational or post-translational mechanisms are responsible for the initial stages of receptor down-regulation. The studies reported here suggest that agonist-induced receptor sequestration and degradation of $GABA_A$ receptor subunits from the neuronal surface may play roles in this process.

REFERENCES

Baumgartner, B.J., Harvey, R.J., Darlison, M.G., and Barnes, E.M., Jr. (1994) Developmental up-regulation and agonist-dependent down-regulation of $GABA_A$ receptor subunit mRNAs in chick cortical neurons. *Mol. Brain Res.*, **26**, 9-17.

Bretcher, M.S., and Lutter, R. (1988) A new method for detecting endocytosed proteins. *EMBO J.* **7**, 4087-4092.

Calkin, P.A., and Barnes, E.M., Jr. (1994) γ-Aminobutyric acid-A ($GABA_A$) agonists down-regulate $GABA_A$ /benzodiazepine receptor polypeptides from the surface of chick cortical neurons. *J. Biol. Chem.* **269**, 1548-1553.

Collins, S., Caron, M.G., and Lefkowitz, R.J. (1992) From ligand binding to gene expression: New insights into the regulation of G-protein-coupled receptors. *Trends Biochem. Sci.* **17**, 37-39.

Greenblatt, D.J., and Shader, R.I. (1986) Long-term administration of benzodiazepines: pharmacokinetic versus pharmacodynamic tolerance. *Psychopharmacol. Bull.* **22**, 416-423.

Hablitz, J.J., Tehrani, M.H.J., and Barnes, E.M., Jr. (1989) Chronic exposure of developing cortical neurons to GABA down-regulates GABA/benzodiazepine receptors and GABA-gated chloride currents. *Brain Res.* **501**, 332-338.

Hamill, O.P., Bormann, J., and Sakmann, B. (1983) Activation of multiple-conductance state chloride channels in spinal neurones by glycine and GABA. *Nature* **305**, 805-808.

Lewin, E., Peris, J., Bleck, V., Zahniser, N.R., and Harris, R.A. (1989) Diazepam sensitizes mice to FG 7142 and reduces muscimol-stimulated $^{36}Cl^-$ flux. *Pharmacol. Biochem. Behav.* **33**, 465-468.

Macdonald, R.L., and Olsen, R.W. (1994) $GABA_A$ receptor channels. *Ann. Rev. Neurosci.* **17**, 569-602.

Maloteaux, J.-M., Octave, J.-N., Gossuin, A., Laterre, C. and Trouet, A. (1987) GABA induces down-regulation of the benzodiazepine-GABA receptor complex in the rat cultured neurons. *Eur. J. Pharmacol.* **144**, 173-183.

Marley, R.J. and Gallager, D.W. (1989) Chronic diazepam treatment produces regionally specific changes in GABA-stimulated chloride influx. *Eur. J. Pharmacol.* **159**, 217-223.

Mhatre, M.C. and Ticku, M.K. (1994) Chronic GABA treatment downregulates the $GABA_A$ receptor α2 and α3 subunit mRNAs as well as polypeptide expression in primary cultured cerebral cortical neurons. *Mol. Brain Res.* **24**, 159-165.

Miller, L.G., Greenblatt, D.J., Barnhill, J.G., and Shader, R.I. (1988) Chronic benzodiazepine administration. I. Tolerance is associated with benzodiazepine receptor downregulation and decreased γ-aminobutyric $acid_A$ receptor function. *J. Pharmacol. Exp. Therap.* **246**, 170-176.

Mistry, D.K., and Hablitz, J.J. (1990) Activation of subconductance states by γ-aminobutyric acid and its analogs in chick cerebral neurons. *Eur. J. Physiol.* **416**, 454-461.

Montpied, P., Ginns, E.I., Martin, B.M., Roca, D., Farb, D.H. and Paul, S.M. (1991), γ-Aminobutyric acid (GABA) induces a receptor-mediated reduction in $GABA_A$ receptor subunit messenger RNAs in embryonic chick neurons in culture. *J. Biol. Chem.* **266,** 6011-6014.

Roca, D.J., Rozenberg, I., Farrant, M. and Farb, D.H. (1990) Chronic agonist exposure induces down-regulation and allosteric uncoupling of the γ-aminobutyric acid/benzodiazepine receptor complex. *Mol. Pharmacol.* **37**, 37-43.

Tehrani, M.H.J. and Barnes, E.M., Jr. (1988) GABA down-regulates the GABA/benzodiazepine receptor complex in developing cerebral neurons. *Neurosci. Lett.* **87**, 288-292.

Tehrani, M.H.J. and Barnes, E.M., Jr. (1991) Agonist-dependent internalization of γ-aminobutyric $acid_A$/benzodiazepine receptors in chick cortical neurons. *J. Neurochem.* **57**, 1307-1312.

Tehrani, M.H.J. and Barnes, E.M., Jr. (1993) Identification of $GABA_A$ /benzodiazepine receptors on clathrin-coated vesicles from rat brain. *J. Neurochem.* **60**, 1755-1761.

Tehrani, M.H.J. and Barnes, E.M., Jr. (1994) Chronic administration of lorazepam to mice promotes transfer of $GABA_A$/benzodiazepine receptors to clathrin-coated vesicles. *Soc. Neurosci. Abs.* **20**, 496.

Weiss, D.S. (1988) Membrane potential modulates the activation of GABA-gated channels. *J. Neurophysiol.* **59**, 514-527.

34

FUNCTION AND STRUCTURE OF HUMAN LEUCOCYTE COLLAGENASE

H. Tschesche,[1] V. Knäuper,[1] T. Kleine,[1] P. Reinemer,[2] S. Schnierer,[1] F. Grams,[2] and W. Bode[2]

[1] Lehrstuhl für Biochemie
Fakultät Chemie, Universität Bielefeld
33615 Bielefeld, Germany
[2] Max-Planck-Institut für Biochemie
82152 Martinsried, Germany

Human leucocyte collagenase is one member of the growing protein family of matrix metalloproteinases (MMPs) [Knäuper et al., 1990]. It is a calcium-containing Zn-endoproteinase (MMP-8) that cleaves preferentially interstitial native triple-helical type I but also type II and type III collagen into one-quarter and three quarter fragments of the native chain length. If thus differs from the fibroblast interstitial collagenase that preferentially cleaves type III. About one-third of its mass of 65 kDa (for active enzyme) is carbohydrates in contrast to the homologous interstitial collagenase from fibroblasts which carries only a small carbohydrate portion [Tschesche et al., 1992]. The enzyme is stored in the specific granules of granulocytes and is released as a proenzyme, also designated latent enzyme, upon stimulation of the cells by various chemotactic agents, such as formylpeptides, LTB_4, C5a, $F_{2\alpha}$ and Zymosan amongst others, [Tschesche et al., 1989 and 1991]. Extracellular activation is then achieved by various different proteinases, such as trypsin, kallikrein, chymotrypsin, cathepsin G [Tschesche et al., 1992] or stromelysin [Knäuper et al., 1993]. However, the physiological process of activation is not yet fully understood, since activation was also observed by isolated leucocyte membranes [Tschesche unpublished].

The enzyme is composed of a multidomain structure as are the other members of the MMP family. The hydrophobic signal peptide sequence, as deduced from the cDNA sequence, is not present in the secreted proenzyme. The secretory precursor form starts with the N-terminal propeptide domain of about 80 residues, which provides latency of the enzyme. The following domain of 163 residues bears the catalytic machinery with the reactive site residues and the zinc binding site. A hemopexin-like C-terminal domain of 188 residues is linked by a 16 residue hinge region to the catalytic domain, which was shown to be crucial for the substrate specificity of the leucocyte collagenase [Schnierer et al., 1993]. While the truncated catalytic domain itself is an enzyme exhibiting substrate specificity for cleaving peptides [Diekmann and Tschesche, 1994] and globular proteins, such as the serpins, α_1-proteinase inhibitor [Knäuper et al, 1990], C1-esterase inhibitor, and α_2-antiplasmin [Knäuper et al., 1991], it has no helicase activity in cleaving triple-helical type I, II or

Methods in Protein Structure Analysis, Edited by M. Z. Atassi and E. Appella
Plenum Press, New York, 1995

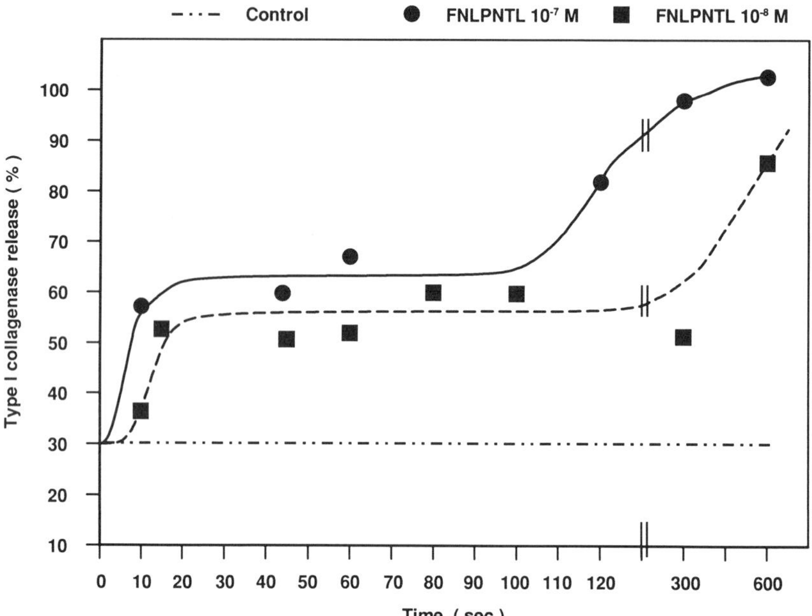

Figure 1. Secretion of collagenase from 1 x 10^6 human leucocytes unstimulated (– · · – · · –) and stimulated by 10^{-8} M (-■-■-) and 10^{-7} M (-●-●-) FNLPNTL.

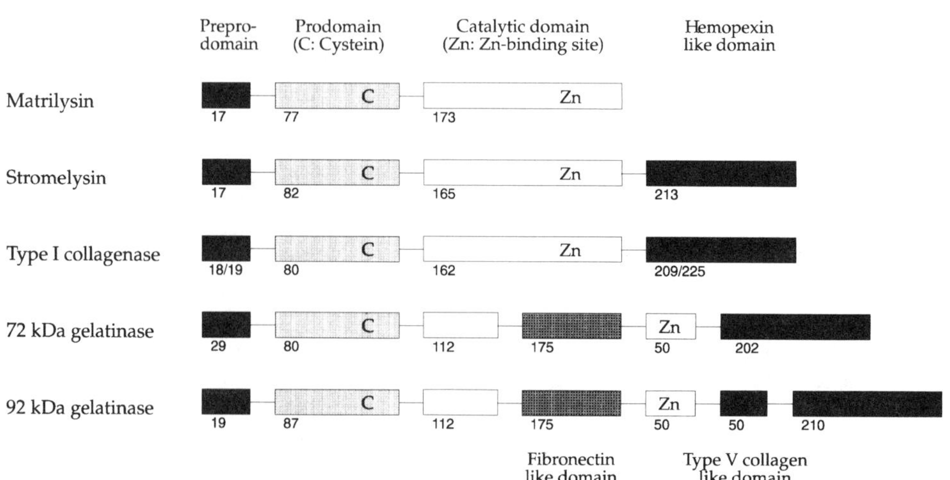

Figure 2. Schematic representation of the domain structure of the family of matrix metalloproteinases (MMPs). C denotes a free cysteine residue in the conserved PRCGVPD sequence motif responsible for chelating the catalytic zinc and maintaining latency of the proenzyme form. The catalytic zink is denoted by Zn. The number of amino acid residues per domain is indicated below each block.

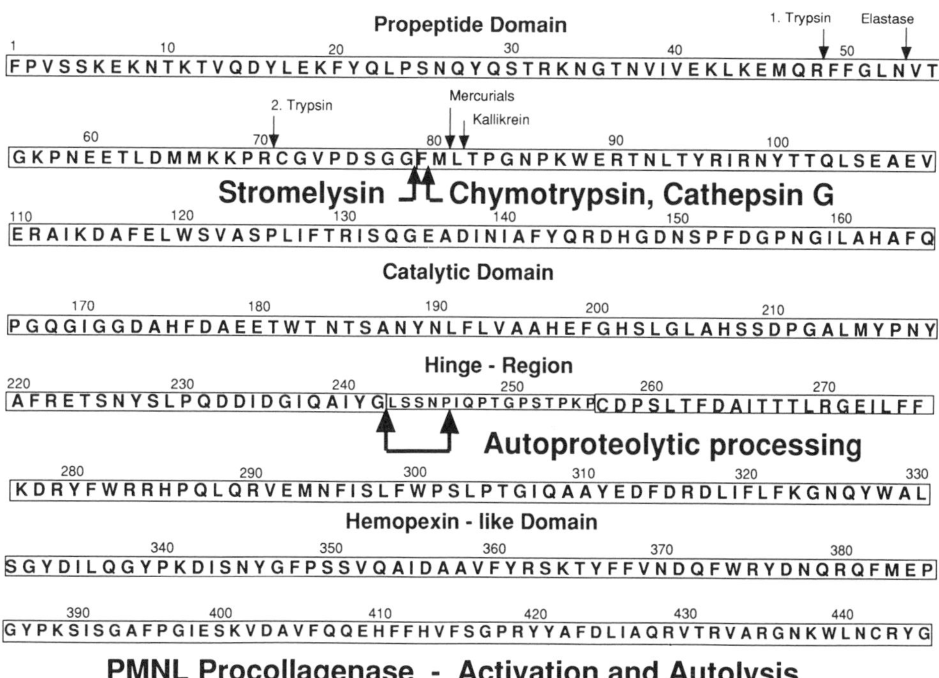

Figure 3. Sequence of amino acid residues (one letter code) in the full length leucocyte collagenase. Activation sites by stromelysin (Phe^{79}-Gly^{242} form) and by serine proteinases (Met^{80}-Gly^{242} form) are indicated as are the autocatalytic cleavage sites separating the catalytic and the hemopexin-like domain.

III collagen. Only the full length enzyme cleaves triple-helical collagen into the characteristic one-quarter and three-quarter fragments [Schnierer et al., 1993].

Activation of the latent precursor form requires removal of the propeptide domain, either by proteolytic enzymes or by autoactivation after molecular rearrangement [Knäuper et.a., 1990]. Depending on the enzyme used for activation a 78 (stromelysin activation) or a 79 residue (cathepsin G activation) propeptide is cleaved from the N-terminus. The single unpaired Cys of the strongly conserved PRCGVPD sequence motif within the propeptide domain is assumed to provide the fourth coordination ligand of the active site zinc. Activation requires replacement of the coordinating Cys moiety by a water molecule. This opening of the reactive site induced by molecular rearrangement or proteolysis has been generally accepted as the cysteine switch activation hypothesis [Van Wart and Birkedal-Hansen, 1990]. It was interesting to find that the stromelysin activated enzyme with N-terminal Phe^{79} was about three to four times more active than the trypsin, chymotrypsin or cathepsin G activated forms with N-terminal Met^{80} [Knäuper et al., 1993].

The three-dimensional structure of the catalytic domain of human leucocyte interstitial collagenase was solved at 2.0 Å after crystallisation of the recombinant protein expressed in *E.coli* [Bode et al., 1994]. The spherical molecule contains a flat active site cleft separating the smaller C-terminal part from the larger N-terminal part, which is built of a central, highly twisted five-stranded β-sheet, flanked by an S-shaped double loop and two additional bridging loops on its convex side and two long α-helices on its concave side. The catalytic zinc ion is located at the bottom of the active site cleft and is coordinated by the N-atoms of the three His within the His^{197}-Glu^{198}-X-X-His^{201}-X-X-Gly^{204}-X-X-His^{207} zinc binding consensus sequence. The active site helix contains His^{197}, Glu^{198} and His^{201} and extends to

Gly^{204}, where the polypeptide chain turns away from the helix axis towards the third zinc ligand, His^{207}. Besides the "catalytic" zinc ion a second "structural" zinc ion is sandwiched between the surface S-shaped double loop Arg^{145}—Leu^{160} and the surface of the β-sheet. It is tetrahedrally coordinated by His^{147}, Asp^{149}, His^{162} and His^{175} while a structural calcium ion is octahedrally coordinated by Asp^{154}, Gly^{155}, Asn^{157}, Ile^{159}, Asp^{177} and Glu^{180}. A second structural calcium ion is located on the convex side of the β-sheet. It is also octahedrally coordinated by Asp^{137}, Gly^{169}, Gly^{171}, Asp^{173} and two water molecules.

The small C-terminal domain exhibits a largely irregular folding with a right-handed loop followed by an α-helix. The loop is stabilised by a tight 1.4 turn Ala^{213}-Leu-Met-Tyr^{216} known as the "Met turn" [Gomis-Rüth et al., 1993; Bode et al., 1993], a conserved topological element in the "metzinkins" [Bode et al., 1993] providing a hydrophobic base for the catalytic zinc ion and the three His residues which ligate the catalytic zinc.

For stabilisation of the proteinase catalytic domain a zinc chelating inhibitor, Pro-Leu-Gly-hydroxamate, was co-crystallised. The inhibitor lies antiparallel to the edge strand (β4) with Pro^{I_1} residing in a hydrophobic groove formed by the side chains of His^{162}, Phe^{164} and Ser^{151}; Leu^{I_2} forms two inter-main chain hydrogen bonds to Ala^{163} and its side chain is situated in a small opening lined by His^{201}, Ala^{206} and His^{207}, while Gly-NHOH is oriented towards Glu^{198} with the carbonyl oxygen and the hydroxyl oxygen complexing the "catalytic" zinc. A characteristic feature of the X-ray structure of the Met^{80}-Gly^{242} catalytic domain of human leucocyte collagenase is that electron density is observed only from the seventh residue Pro 86 onwards. While the structure of the Phe^{79} variant reveals that the N-terminal heptapeptide segment, Phe^{79}-Met-Leu-Thre-Pro-Gly-Asn^{85} binds to the concave surface at the bottom of the molecule between Pro^{86} and Ser^{209} [Reinemer et al., 1994]. The side chain of Trp^{88} slots into a hydrophobic groove formed by the side chains of Leu^{93}, Ile^{138}, Pro^{166} and Gly^{172} and Thr^{82} fits into a small hydrophobic groove lined by Pro^{86}, Gln^{165}, Gly^{204} and Ala^{206}. N-terminal to Thr^{82} the chain turns downward to the C-terminal helix αC crossing over with the strand following helix αB at residue Leu^{205} to form the only regular inter-main chain hydrogen bond.

The N-terminal ammonium group of Phe^{79} forms a salt bridge with the carboxylate moiety of the strictly conserved Asp^{232} (Fig. 2a and 2b). The side chains of Leu^{81} and Phe^{79}

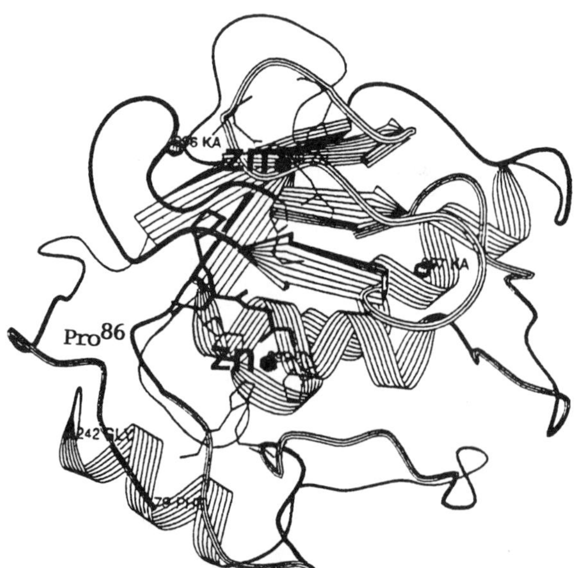

Figure 4. Representation of the leucocyte collagenase catalytic domain in a ribbon plot structure with the two zinc and the two calcium atoms. In the Met^{80}-Gly^{242}-form the N-terminal heptapeptide FMLTPGN prior to Pro^{86} is disordered but packs against a concave hydrophobic surface of the enzyme made by the C-terminal helix in the Phe^{79}-Gly^{242} form (see text).

are oriented towards the C-terminal helix with the former packing against the side chains of Ile^{240} (αC) and Asn^{85} and the latter packing against Gly^{236} (which is strictly conserved) and Ala^{239} of the C-terminal helix αC, while the side chain of Met^{80} points towards the bulk solvent and seems to be disordered, as no significant electron density could be observed for this residue. Interestingly, the N-terminus locks two water molecules in a cavity created by the C-terminal helix and the strand following the active-site helix and lined by the side chains of Val^{205}, Trp^{120}, Met^{215}, Asp^{232}, Asp^{233} and Gly^{236}. Both internal solvent molecules are hydrogen bonded to Asp^{233} $O_{\delta 1}$ and to Met^{80} NH.

The disorder-order transition of the N-terminal segment in the two structural forms (i.e. the Phe^{79}-Gly^{242} and the Met^{80}-Gly^{242} catalytic domain) must in some way be significant to activity enhancement. The formation of the salt-bridge between the N-terminal ammonium group of Phe^{79} and the side chain carboxylate of Asp^{232} seems to lead to stabilisation of the active site via the neighbouring Asp^{233}. The latter residue, which is strictly conserved as is Asp^{232}, has its side chain buried and forms a hydrogen bond to the Met turn residues Leu^{214} N and Met^{215} N, thus stabilising the active site basement. This is in accordance with the finding that this residue is essential for catalytic activity [Hirose et al., 1993]. Stabilisation of the active site might be a prerequisite for that of the transition states.

The specific 'triple-helicase' activity of the full length enzyme containing the C-terminal hemopexin-like domain obviously requires at least partial unfolding of the collagen triple helix around the active site [Bode et al., 1994]. The repetitive Pro-X-Gly segment of one strand of a regular collagen triple-helix could probably be arranged in such a way, that the glycyl carbonyl group approached the catalytic zinc. However, the P_1'-proline side chain would then not adequately fill the S_1'-subsite pocket of the enzyme and the 15 Å diameter collagen triple helix [Yonath and Traub, 1969; Fraser et al., 1979] would not fit properly through the opening at the S_2' and S_3' subsites.

Elucidation of the X-ray structure of the leucocyte interstitial collagenase catalytic domain certainly allows a better understanding of the catalytic properties of the enzyme and facilitates a design for an alignment of small molecular weight enzyme inhibitors [Grams et al., 1994].

ACKNOWLEDGMENTS

We thank P. Widawka and I. Mayr-Kröner for excellent help in cloning human PMNL-CL catalytic domain cDNA and crystallisation, respectively. The financial support of the SFB of the University München and of the Fonds der Chemischen Industrie to W.B., of the BAYER AG (PF-F/Biotechnology, Monheim, Germany) to P.R. and of the SFB 223 of the Universität Bielefeld and of the Fonds der Chemischen Industrie to H.T. is gratefully acknowledged.

REFERENCES

1. Bode, W., Gomis-Rüth, F.-X. & Stöcker, W. FEBS Lett., **331,** 134-140 (1993)
2. Bode, W., Reinemer, P., Huber, R., Kleine, T., Schnierer, S. & Tschesche, H. EMBO J. 13, no. **6** pp. 1263-1269 (1994)
3. Diekmann, O. & Tschesche, H. Brazilian. J. of Med. & Biol. Research, **27**, 1865-1876 (1994)
4. Gomis-Rüth, F.-X., Kress, L.F. & Bode, W. EMBO J, 12, 4151-4157 (1993)
5. Grams, F., Reinemer, P., Powers, J.C., Kleine, Th., Pieper, M., Tschesche, H., Haber, R. & Bode, W. Eur. J. Biochem **228**, 830-834 (1995)
6. Hirose, T., Patterson, C., Pourrmotabbed, T., Mainardi, C.L. & Hasty, K.A. Proc. Natl. Acad. Sci. USA, **90**, 2569-2573 (1993)

7. Knäuper, V., Krämer, S., Reinke, H. & Tschesche, H. Eur. J. Biochem. **189**, 295-300 (1990)
8. Knäuper, V., Reinke, H. & Tschesche, H. FEBS Lett. **263**, 355-357 (1990)
9. Knäuper, V., Triebel, S., Reinke, H. & Tschesche, H. FEBS Lett. **290**, 99-102 (1991)
10. Knäuper V., Wilhelm, S.M., Seperack, P.K., DeClerck, Y.A., Langley, K.E., Osthues, A. & Tschesche, H. Biochem. J. **295**, 581-586 (1993)
11. Reinemer, R., Grams, F., Huber, R., Kleine, T., Schnierer, S., Pieper, M., Tschesche H., & Bode, W. FEBS Lett. **338**, 227-233 (1994)
12. Schnierer, S., Kleine, T., Gote, T., Hillemann, A. & Tschesche,H. Biochem. Biophys. Res. Comm. **191**, 319-326 (1993).
13. Tschesche, H., Bakowski, B., Schettler, A., Knäuper, V. & Reinke, H. Biomed. Biochim. Acta 50, 755-761 (1991)
14. Tschesche, H., Knäuper, V., Krämer, S., Michaelis, J., Oberhoff, R. & Reinke, H. Matrix Supplement 1, 245-255 (1992)
15. Tschesche, H., Schettler, A., Thorn, H., Bakowski, B., Knäuper, V., Reinke H. & Jockusch, B. M. Proceedings of the 8th Winter School, KFA Jülich GmbH, 31-36 (1989)
16. Van Wart, H. & Birkendal-Hansen, H. Proc. Natl. Acad. Sci. USA, **87**, 5578-5582 (1990).
17. Yonath, A., & Traub, W. J. Mol. Biol. **43**, 461-477 (1969)

35

THE HUMAN DNA-ACTIVATED PROTEIN KINASE, DNA-PK: SUBSTRATE SPECIFICITY

Carl W. Anderson,[1] Margery A. Connelly,[1] Susan P. Lees-Miller,[2] Lauri G. Lintott,[2] Hong Zhang,[1] John D. Sipley,[1] Kazuyasu Sakaguchi,[3] and E. Appella[3]

[1] Biology Department, Brookhaven National Laboratory
Upton, New York 11973

[2] Department of Biological Sciences, University of Calgary
2500 University Drive, N.W., Calgary, Alberta T2N 1N4, Canada

[3] Laboratory of Cell Biology, Building 37, National Institutes of Health
Bethesda, Maryland 20892

INTRODUCTION

Agents that cause damage to DNA (DNA damage- inducing (DDI) agents) arrest cell cycle progression in all eukaryotes from yeast to humans at positions in late G_1 and G_2 that have become known as "checkpoints" (Hartwell and Weinert, 1989; Murray, 1992; Sheldrick and Carr, 1993; Weinert and Lydall, 1993), presumably to allow time for DNA repair. Otherwise the DNA damage would become irreversibly fixed as a consequence of DNA replication in S phase, or through cell division at mitosis (M phase). The mammalian G_2 checkpoint mechanism is not yet well characterized (O'Connor and Kohn, 1992), but the key observation that tumor cells with mutant p53 were unable to arrest in G_1 (Kastan et al., 1991) quickly led to an outline of the mammalian G_1 checkpoint mechanism (Hunter, 1993; Appella and Anderson, 1994; Appella et al., this volume). The p53 tumor suppressor gene is a transcription factor that normally is relatively inactive because it is rapidly degraded (Levine, 1993). However, in response to exposure to ultraviolet light, ionizing radiation, and other DDI agents, the p53 protein is transiently stabilized, accumulates in the cell nucleus, and induces the expression of several genes including *WAF1* and *GADD45* (El-Deiry et al., 1993; El-Deiry et al., 1994). The 21 kDa product of *WAF1* is a potent inhibitor of the cyclin-dependent protein kinases that are needed for the transition from G_1 to S phase and for continued DNA replication in S (Dulić et al., 1994; Harper et al., 1993). Although the G_1 checkpoint mechanism probably is much more complex, the induction of *WAF1* provides a simple explanation of how cell cycle progression can be arrested. In addition to *WAF1*, about 50 other genes are known to be induced in mammalian cells after exposure to DDI agents (Fornace, 1992; Herrlich and Rahmsdorf, 1994). Recent studies indicate that some genes are induced as a consequence of the effects of DDI agents on other cellular molecules and not

Methods in Protein Structure Analysis, Edited by M. Z. Atassi and E. Appella
Plenum Press, New York, 1995

necessarily as a consequence of damage to DNA (Anderson, 1994; Herrlich, and Rahmsdorf, 1994; Sachsenmaier et al., 1994). Such exposures activate cytoplasmic signaling mechanisms that operate through protein kinase cascades initiated at or near the plasma membrane; in turn, these kinase cascades activate several transcription factors including AP1 and NF-κB. Nevertheless, there is strong evidence that the p53-dependent induction of *WAF1* is a direct consequence of the production of DNA strand breaks (Nelson and Kastan, 1994) and that DNA strand breaks are the signals for activation of the G_1 checkpoint(s) in yeast (Siede et al., 1994).

Although much has been learned about the structure and function of p53 and the probable sequence of subsequent events that lead to cell cycle arrest, little is known about how DNA damage is detected and the nature of the signal that is generated by DNA damage. Circumstantial evidence suggests that protein kinases may be involved. Indeed, several yeast kinase genes were identified by screening for mutants defective in their ability to arrest cell cycle progression after exposure to DDI agents (Walworth et al., 1993; Weinert et al., 1994; Allen et al., 1994; Anderson, 1994). In mammalian cells, the situation is less clear; however, in hamster cells, 2-aminopurine overrides the G_1, S, and G_2 checkpoints, and 2-aminopurine and H7, another protein kinase inhibitor, block DDI-agent induction of the *GADD45* gene (Andreassen and Margolis, 1992; Luethy and Holbrook, 1994).

Two moderately abundant nuclear enzymes have been described in mammalian cells that recognize DNA strand-breaks and transmit signals to other proteins. One is poly(ADP-ribose) polymerase (de Murcia and de Murcia, 1994; Satoh and Lindahl, 1992); this enzyme is activated by binding to nicks, and it ribosylates histones, other chromosomal proteins, and itself. It may be responsible for altering chromatin structure near sites of DNA damage and also may signal the presence of damage through transient changes in NAD levels. A second DNA structure-signaling enzyme is DNA-PK (Anderson, 1993). DNA-PK is activated by binding to DNAs with nicks, gaps, or double-strand breaks, and it may function, at least in part, as a detector of DNA strand- breaks.

THE STRUCTURE OF DNA-PK

DNA-PK is believed to consist of a very large polypeptide, DNA-PK_c (or Prkdc for *pr*otein *k*inase, *D*NA-activated, *c*atalytic component), that probably contains the catalytic site, and a DNA binding/targeting and regulatory subunit, which can be the Ku autoantigen (Dvir et al., 1992; Gottlieb and Jackson, 1993; Anderson, 1993). The large DNA-PK_c polypeptide and DNA-activated kinase activity co-purify (Lees- Miller et al., 1990; Carter et al., 1990). The size of the DNA-PK_c polypeptide was initially estimated from SDS-polyacrylamide gels electrophoresis to be 300-350 kDa; however, the size of the nascent polypeptide, estimated from preliminary sequence analysis of the ~13 kbp cDNA, is close to 450 kDa. The difference is attributable to the lack of good molecular weight markers in this size range. Although nearly ten times the size of many protein kinase catalytic subunits, several findings are consistent with an assignment of the catalytic site to the 450 kDa DNA-PK_c polypeptide. DNA-PK_c binds ATP and can be labeled by the ATP analogues fluorosulfonylbenzoyladenosine (FSBA) and azido-ATP; FSBA inhibits DNA-PK kinase activity (Lees-Miller et al., 1990). A monoclonal antibody specific for the DNA-PK_c polypeptide depleted DNA-dependent kinase activity from HeLa extracts (Carter et al., 1990). Finally, sequence analysis of the cDNA has revealed a segment with homology to other kinase catalytic domains. In the human cell lines that have been examined, DNA- PK_c is moderately abundant and predominantly nuclear (Anderson and Lees-Miller, 1992). We estimated its abundance to be about 50,000 molecules per cell, but our estimate involved several assumptions and an accurate measurement has not been made. DNA- PK activity is approximately 100-fold more abundant in extracts of human and monkey cell lines compared to extracts of rodent and insect cell lines; thus, DNA-PK activity and the DNA-PK_c polypeptide are difficult to detect with current assays in unfractionated extracts of non-primate cells (Anderson and Lees- Miller, 1992).

Ku was first recognized as a heterodimeric (p70/p80), nuclear, phosphoprotein that reacted with sera from patients suffering from the autoimmune diseases lupus erythematosus and scleroderma polymyositis (Mimori et al., 1981; Reeves, 1992). HeLa cells contain ~400,000 molecules of Ku per cell (Mimori et al., 1986), but Ku also appears to be less abundant in non- primate cells and was not detected in mouse L-929 cells using mouse-specific monoclonal antibodies (Wang et al., 1993). *In vitro* Ku binds initially to the ends of linear DNA fragments but then can translate along the DNA in an ATP-independent manner (de Vries et al., 1989). Ku also recognizes DNAs with nicks and gaps, as well as DNAs with single- to double-strand transitions (Blier et al., 1993; Falzon et al., 1993), and these structures activate DNA-PK (Morozov et al., 1994). cDNAs for the two Ku polypeptides have been cloned and sequenced from both human and mouse cells (Chan et al., 1988; Reeves and Sthoeger, 1989; Yaneva et al., 1989; Mimori et al., 1990; Griffith et al., 1992; Porges et al., 1990; Falzon and Kuff, 1992), and the human genes recently were mapped to chromosomes 2 (p80) and 22 (p70) (Cai et al., 1994). The location of the p80 Ku subunit gene corresponds to the location of the human gene (*XRCC5*) that complements ionizing-radiation sensitivity in group 5 hamster cells. Transfection of the cDNA for the human Ku p80 subunit into group 5 hamster cells that are defective in repairing double-strand breaks restored their X-ray sensitivity to normal levels and corrected the defect in site-specific recombination (Rathmell and Chu, 1994a,b; Getts and Stamato, 1994; Taccioli et al., 1994). Thus, Ku is likely to play a role in these processes. It may protect DNA ends from exonucleolytic degradation; however, by activating DNA-PK, it also might have other signaling functions. Recently, Ku was shown to have DNA helicase activity and to be identical to a previously described activity, human DNA helicase II (Tuteja et al., 1994); thus, Ku probably also performs functions that are independent of DNA-PK_c. It is not known if other targeting proteins can substitute for Ku to activate DNA- PK_c, perhaps in response to other signals.

DNA-PK SUBSTRATE SPECIFICITY

In vitro, human DNA-PK phosphorylates a variety of nuclear DNA-binding, regulatory proteins including the tumor suppressor protein p53, the single- stranded DNA binding protein RPA, the heat shock protein hsp90, the large tumor antigen (TAg) of simian virus 40, a variety of transcription factors including Fos, Jun, serum response factor (SRF), Myc, Sp1, Oct-1, TFIID, E2F, the estrogen receptor, and the large subunit of RNA polymerase II (reviewed in Anderson, 1993; Jackson et al., 1993). However, for most of these proteins, the sites that are phosphorylated by DNA-PK are not known.

To determine if the sites that were phosphorylated *in vitro* also were phosphorylated *in vivo* and if DNA-PK recognized a preferred protein sequence, we identified the sites phosphorylated by DNA-PK in several substrates by direct protein sequence analysis. Table 1 shows an alignment of known DNA-PK phosphorylation sites. Each phosphorylated serine or threonine is followed immediately by glutamine in the polypeptide chain; at no other positions are the amino acid residues obviously constrained.

Two forms of hsp90, designated α and β, are found in human cells, and these proteins are 97% identical; however, only hsp90 α is phosphorylated by DNA-PK (Lees-Miller and Anderson, 1989b). The phosphorylated sites are two threonines at the amino terminus of hsp90 α in the sequence PEETQTQDQPM[11]; these residues are not present in hsp90 β. SV40 TAg was shown to be phosphorylated at four sites, serines 120, 665, 667, and 677 (Chen et

Table 1. Protein phosphorylation sites recognized by human DNA-PK

Substrate protein[a]	DNA-PK Site	Local Amino Acid Sequence
		- * -
Hsp90[α] (human)	Thr4	P-E-E-**T-Q**-T-Q-D-Q-P-M-E-E^{13}
	Thr6	P-E-E-T-Q-**T-Q**-D-Q-P-M-E-E-E-E^{15}
SV40 Large tumor antigen	Ser120	E-A-T-A-D-**S-Q**-H-S-T-P-P-K-K-K^{129}
	Ser665	E-T-G-I-D-**S-Q**-S-Q-G-S-F-Q-A-P^{674}
	Ser667	G-I-D-S-Q-**S-Q**-G-S-F-Q-A-P-Q-S^{676}
	Ser677	Q-A-P-Q-S-**S-Q**-S-V-H-D-H-N-Q-P^{686}
c-Jun transcription factor (human)	Ser249	P-I-D-M-E-**S-Q**-E-R-I-K-A-E-R-K^{258}
Serum response factor (human)	Ser435	V-L-N-A-F-**S-Q**-A-P-S-T-M-Q-V-S^{444}
	Ser446	M-Q-V-S-H-**S-Q**-V-Q-E-P-G-G-V-P^{455}
p53 Tumor suppressor		
(mouse)	Ser4	M-E-E-**S-Q**-S-D-I-S-L-E-L-P^{13}
(mouse)	Ser15	L-E-L-P-L-**S-Q**-E-T-F-S-G-L-W-K^{24}
(human)	Ser15	V-E-P-P-L-**S-Q**-E-T-F-S-D-L-W-K^{24}
		- * -

[a]Phosphorylation sites (-*-) were identified by: hsp90, Lees-Miller and Anderson (1989a); SV40 TAg, Chen et al. (1991); c-Jun, Bannister et al. (1993); serum response factor (SRF), Liu et al. (1993), and p53 (Lees-Miller et al. (1992).

al., 1991). Isolation of the phosphopeptides containing these residues was accomplished using iron-affinity chromatography in conjunction with conventional reverse phase HPLC. The phosphorylated serines then were identified by direct sequence analysis after converting the phosphoserine to S- ethylcysteine. Each phosphorylated TAg serine is followed immediately by glutamine (Table 1). Serines 120 and 677 are phosphorylated *in vivo*, but serines 665 and 667 have not been shown to be *in vivo* sites of phosphorylation. TAg serine 639 is phosphorylated *in vivo*, and although this serine is followed by glutamine in the TAg polypeptide, it was not phosphorylated by DNA-PK *in vitro*. This finding suggests that either Ser639 is phosphorylated by a different kinase or that its phosphorylation by DNA-PK requires a particular TAg conformation that was not present in the *in vitro* reaction.

Subsequent to our initial work, Bannister et al. (1993) used a genetic approach to identify serine 249 in the DNA-binding region of Jun as the likely site of phosphorylation by DNA-PK. This site is phosphorylated *in vivo*, but it also can be phosphorylated *in vitro* by casein kinase II. Jun residue 250 is glutamine; changing it to alanine largely prevented DNA-dependent Jun phosphorylation, again suggesting that glutamine is important for substrate recognition by DNA-PK. Changing glutamic acid 251 to alanine decreased the rate of Jun phosphorylation about twofold, indicating that the residue in this position also may contribute to substrate recognition. Glutamic acid also is present at the +2 position with respect to the serine 15 site of human and mouse p53, and aspartic acid (D) is present at +2 after the second TQ site in hsp90 (Table 1, see below), but neither glutamic or aspartic acid are at this position in sites from SV40 TAg, or the human serum response factor (SRF). Two serines followed by glutamine in a peptide derived from the carboxy- terminal transactivation domain of SRF are phosphorylated by DNA- PK, and they also appear to be phosphorylated in serum-stimulated cells (Liu et al., 1993).

To determine whether an adjacent glutamine is important for phosphorylation site recognition by DNA-PK, we examined its ability to phosphorylate synthetic peptides corresponding to segments of the human p53 protein sequence. These peptides covered all of the known phosphorylation sites in human p53, including serines 9, 15, 33, 315, and 392, and all -SQ- or -QS- sites (i.e. S15, -LSQE-; S37, -PSQA-, S99, -PSQK-, S166, QSQH, and S376, GQSTS) (Lees-Miller et al., 1992). Peptides containing three of the five SQ or QS motifs were phosphorylated by DNA-PK; one of these, serine 15, is in a highly conserved region and is phosphorylated *in vivo* (Ullrich et al., 1993). The sequence requirements for phosphorylation at this site were examined in more detail using a series of synthetic peptides (Table 2). Shortening the sequence to less than six residues on the carboxy-terminal side dramatically reduced the rate of peptide phosphorylation, but shortening the sequence on the amino-terminal side actually increased the rate. Changing the glutamine (Q) at the position corresponding to p53 residue 16 to tyrosine (Y), asparagine (N), or lysine (K) decreased the rate of peptide phosphorylation, and inverting the glutamine and the following glutamic acid (SQE -> SEQ) essentially abolished peptide phosphorylation. Km values for the peptide substrates varied from about 0.2 to 0.7 mM; these values are not remarkably good compared to peptide substrates for several other kinases (Kemp and Pearson, 1991). A glutamine or glutamic acid immediately before the SQ motif and a glutamic acid immediately following it (i.e. QSQE or ESQE, as in the c-Jun site, see above) gave a slight improvement in the apparent association constant (compare peptides 4, 11, and 15 in Table 2); about a twofold further improvement was obtained by removing all but two residues from the amino-terminal side of the phosphorylation site (Table 1, Figure 1). We also noticed that the version of peptide 4 (EPPLSQEAFADLWKK) ending with an amide is a slightly better substrate than the same peptide ending with a carboxyl group, perhaps suggesting that the carboxy-terminal extension of six residues is not quite optimal. To date, the best substrate peptide for human DNA-PK is P*ESQE*AFADLWKK, while the similar peptide P*ESEQ*A-FADLWKK is not appreciably phosphorylated (Table 1, Figure 1).

A second substrate determinant for recognition by DNA-PK is the ability to bind DNA. Jackson et al. first observed that quantitative phosphorylation of Sp1 required a DNA template with a GC-box DNA binding element (Jackson et al., 1990). Subsequently it was found that efficient phosphorylation required that both DNA-PK and Sp1 be bound to the same DNA molecule (Gottlieb and Jackson, 1993). Similar findings for murine p53 and a protein containing the POU DNA binding domain of human Oct- 1 were made by Lees-Miller and Anderson (Lees-Miller et al., 1992; Anderson and Lees-Miller, 1992). DNA both activates DNA-PK and increases the local concentration of substrate in the vicinity of activated kinase. Short duplex oligonucleotides that bound kinase or the substrate but could not bind both simultaneously gave much lower rates of phosphorylation for p53 or the POU domain protein than long DNA fragments; however, the rates of phosphorylation of peptide substrates or of hsp90, a protein that does not bind DNA, were independent of DNA length

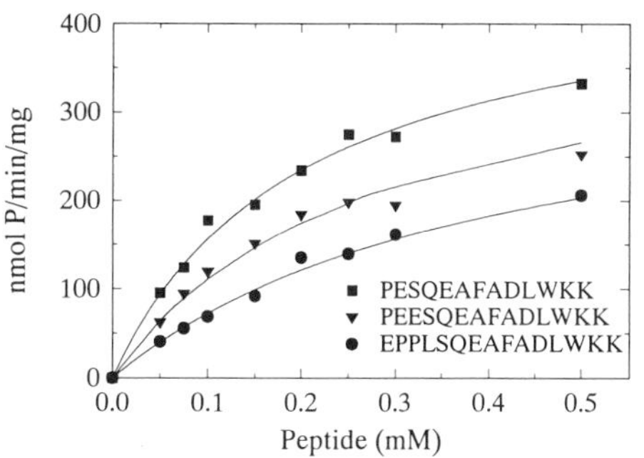

Figure 1. Phosphorylation of synthetic peptides by human DNA-PK. Rates of peptide phosphorylation are shown as a function of peptide concentration. Reactions were at 30°C for 10 min as described (Lees-Miller et al., 1992); the data represent an average of three independent experiments for each peptide. Apparent Km and Vmax values were determined from Lineweaver-Burk plots. These values were respectively: (■) $PESQEAFADLWKK_{COOH}$, 0.20 mM, 470 nmol/min/mg; (▼) $PEESQEAFADLWKK_{COOH}$, 0.27 mM, 410 nmol/min/mg; (●) $PEESQEAFADLWKK_{amide}$, 0.4 mM, 370 nmol/min/mg. Note that the Km value for (●) is about half that of its carboxyl group terminated equivalent peptide (Table 2, peptide 4). Peptides EPPL*SEQ*AFDLWKK and PESEQAFADLWKK were not significantly phosphorylated (data not shown).

Table 2. Phosphorylation of synthetic p53 peptide substrates by human DNA-PK

Pept. no.	Peptide sequence	Apparent Km (mM)	Apparent Vmax (nmol/min/mg)	Activity PO_4/min/mg at 0.2 mM
p53	11 15 20 24			
1	E P P L **S Q** E T F S D L W K-K	0.35	360	130
2	- - - - - **S Q** - - - KK	ND	ND	15
3	- - - - - **S Q** - - - - - KK	ND	ND	12
4	- - - - - **S Q** - A - A - - - - K	0.76	380	83
5	- - - - - **S Q** - A - A - - L - K	0.56	160	36
6	- - - - - S Y - A - A - - L - K	ND	ND	2.3
7	- - - - - S E - A - A - - - - K	ND	ND	0.4
8	- - - - - S N - A - A - - - - K	ND	ND	0.1
9	- - - - - S E Q A - A - - - - K	ND	ND	0
10	- - - - - **S Q** K A - A - - - - K	1.0	70	~14
11	- - - Q S L - A - A - - - - K	0.4	310	99
12	- - - Q **S Q** - A - A - - - - K	0.3	390	161
13	- - - - - **T Q** - A - A - - - - K	0.67	460	116
14	- - - D **S Q** - A - A - - - - K	ND	ND	48
15	P E E **S Q** - A - A - - - - K	0.27	410	180
16	P E **S Q** - A - A - - - - K	0.20	470	220
17	P E S E Q A - A - - - - K	ND	ND	0

Dash (-) indicates amino acid is same as in control peptide #1, p53(11- 24)K; ND = not determined. Phosphorylation reactions were performed as described in Lees-Miller et al. (1992).

or concentration. Thus, DNA length (above 18 bp) did not affect kinase activation, nor did the interaction of the substrates with non-specific DNA sequences change their conformation in a manner that significantly affected the rate of phosphorylation.

It seems likely that DNA binding also may be important for substrate recognition *in vivo*. Many site-specific DNA binding proteins, including p53 and SV40 TAg, bind DNA in a non-sequence- specific manner, and this ability may help proteins scan chromatin for their sequence-specific recognition sites. Presumably activated DNA-PK is fixed in space by binding to chromatin, although it may be able to slide in either direction from the initial binding site. Thus, we imagine that substrates become phosphorylated as they scan along chromatin strands via their non-sequence specific DNA binding mode and collide with DNA-PK. Since both substrate and kinase will be physically constrained if both are associated with the same chromatin segment, the specific location of potential sites within the three dimensional structure of the substrate also may be important. Most proteins that function as DNA-PK substrates *in vitro* are DNA binding proteins (Anderson, 1993). If DNA binding ability accounts for a substantial fraction of substrate specificity and recognition, then DNA-PK may not have a highly specific sequence or structure recognition ability. This situation would account for the relatively poor Km values that have been measured for substrate peptides.

If chromatin binding plays an important role in DNA-PK substrate recognition, then reaction conditions in the test tube are likely to be far removed from those within the nucleus of a cell. Thus, some putative substrates or sites that are phosphorylated *in vitro* may not be phosphorylated *in vivo*. One apparent example is the Oct-1 POU expression construct mentioned earlier. This protein (T7HPOU1) consists of the POU domain from human Oct-1 with a 17 amino acid amino-terminal extension derived from vector sequences (Figure 2).

```
                5         10        15        20        25        30
  1 M A S M T G H H H H H H G M S G G M E E P S D L E E L E Q F
 31 A K T F K Q R R I K L G F T Q G D V G L A M G K L Y G N D F
 61 S Q T T I S R F E A L N L S F K N M C K L K P L L E K W L N
 91 D A E N L S S D S S L S S P S A L N S P G I E G L S R R R K
121 K R T S I E T N I R V A L E K S F L E N Q K P T S E E I T M
151 I A D Q L N M E K E V I R V W F C N R R Q K E K R I N P *
```

Figure 2. Sequence of the Oct-1/Pou expression construct T7HPOU1. The predicted amino acid sequence of the human Oct-1 POU-domain construct from plasmid pT7HPOU1 (Anderson and Lees-Miller, 1992) is given in the single letter code. The POU domain extends from residue 19 to 178; residues 1 to 18 are derived from vector sequences and include a six-histidine tag for affinity purification. The POU domain has a TQ motif at residues 44/45 and a SQ motif at residues 61/62 (bold). Serine 3 (bold) is phosphorylated by DNA-PK (see text); the initiating methionine (residue 1) is removed in *E. coli*.

The POU domain has two sites that resemble other identified DNA-PK phosphorylated sites, a TQ at residues 44/45 and an SQ at residues 61/62. Phosphoamino acid analysis revealed only phosphoserine (data not shown); thus, the TQ sequence is not a DNA-PK phosphorylation site. After digestion with trypsin and CNBr, none of the phosphopeptides were retained on a C18 reverse phase column, but after digestion with trypsin alone, a major, modestly hydrophobic phosphopeptide was retained by the reverse phase column (Figure 3). Amino-terminal sequence analysis showed that this peptide was derived from the amino terminus of the expressed protein and that the serine at position 3 was partially phosphorylated. These findings are consistent with the HPLC data shown in Figure 3; after digestion with CNBr and trypsin, serine 3 would be in the tripeptide Ala- Ser-Met* (Met* = homoserine). We cannot formally exclude serine 22 as a possible second site of phosphorylation; after digestion with both CNBr and trypsin, it would be present in the tetrapeptide Ser-Gly-Gly-Met*. However, another Oct-1/POU derivative containing serine 22 but lacking the serine 3 site was not phosphorylated by DNA-PK. Serine 3 lies outside the POU domain and is not followed by a glutamine, but, in this case, the non-specific DNA binding ability of the POU domain may be sufficient to drive phosphorylation of this non-physiological, vector-encoded site. Distinguishing between physiological and non-physiological substrates and sites may well be even more difficult for DNA-PK than for other protein kinases.

Our results with the POU domain protein raise the issue of whether *in vivo* DNA-PK also phosphorylates serines or threonines that are not followed by glutamine. One such putative target is the carboxy-terminal domain (CTD) of the large subunit of RNA polymerase II. The CTD consists of about 50 conserved repetitions of the consensus heptad motif YSPTSPS, and, although the CTD contains no SQ or TQ sites, it is phosphorylated efficiently by DNA-PK when coupled to a GAL4 DNA binding domain (Peterson et al., 1992). The

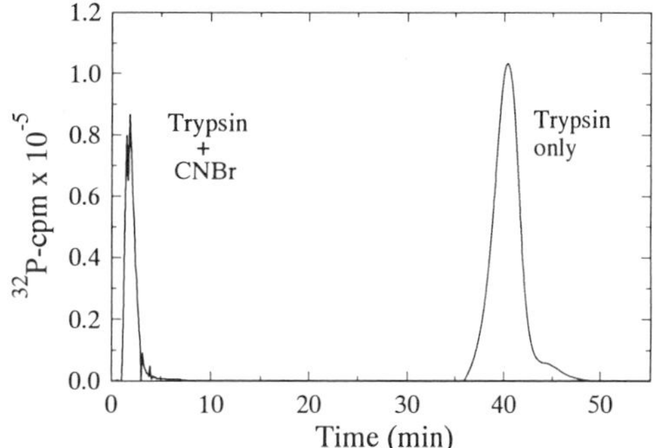

Figure 3. Reverse phase fractionation of phosphopeptides from the Oct-1 POU domain polypeptide after phosphorylation by human DNA-PK. Purified Oct-1/POU-domain construct (T7HPOU1, see Figure 3) was incubated with DNA-PK, calf thymus DNA, and [^{32}P]ATP as described (Anderson and Lees-Miller, 1989b), digested with trypsin and CNBr (solid line) or with trypsin alone (dashed line), and the resulting peptides were fractionated by reverse phase (C18) HPLC essentially as described (Chen et al., 1991).

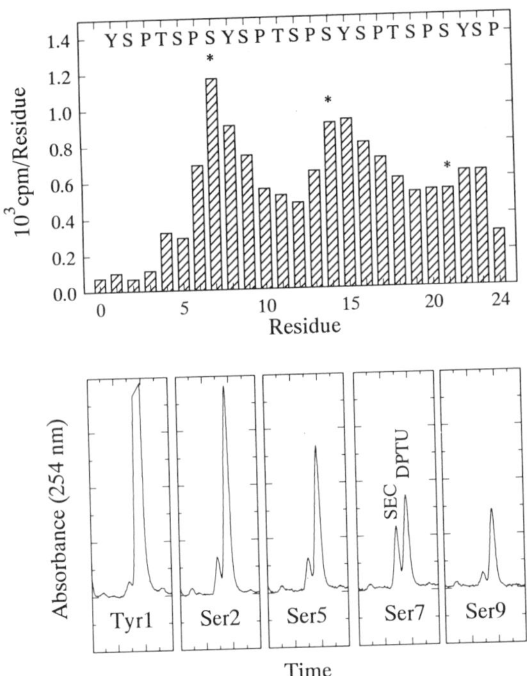

Figure 4. Sequence analysis of an RNA polymerase II CTD peptide phosphorylated by human DNA-PK. Top: The RNA polymerase II CTD peptide (YSPTSPS)$_4$RRR, containing 4 heptad repeats, was labeled by incubation at 1 mM with purified human DNA-PK, calf thymus DNA, and [^{32}P]ATP as described (lees-Miller et al., 1992). After desalting on BioGel P4, the labeled peptide (100,000 cpm) was applied to a Beckman 890M sequencer together with polybrene (3 mg) and apomyoglobin (0.1 mg). The radioactivity released after each Edman cycle is shown; the peptide sequence is given at the top of the figure. The major sites of phosphorylation were serines 7, and 14 (*). A substantial lag in the release of phosphate occurs with this method of sequence analysis. **Bottom:** The CTD peptide was phosphorylated as described above, and the phosphorylated peptide was enriched by iron-affinity chromatography (Lees-Miller and Anderson, 1989a). Before sequence analysis, phosphoserine was converted to S-ethylcysteine (SEC) by incubation with ethanethiol in 0.1 M barium hydroxide (Lees-Miller et al, 1989a). Shown are limited regions of the PTH-amino acid chromatogram near the diphenylthiourea (DPTU) peak from sequencer cycles 1, 2, 5, 7, and 9; SEC elutes immediately before DPTU. An Applied Biosystems 470A protein sequencer equipped for on-line PTH detection was used for this analysis.

CTD also becomes hyperphosphorylated *in vivo* during the initiation of transcription, and several putative CTD kinases have been described (Dhamus and Dynan, 1992; O'Brien et al., 1994).

A CTD peptide consisting of four heptad repeats, (YSPTSPS)$_4$RRR, is phosphorylated by purified human DNA-PK, although the Km for this peptide substrate is above 1 mM. The CTD peptide was phosphorylated exclusively on serine (data not shown), and sequence analysis (Figure 4) revealed that the major site of phosphorylation was the serine at position 7, or the equivalent serine in the second (and probably at the third and fourth) repeat. These serines are followed in the repeat by tyrosine; however, tyrosine did not efficiently substitute for glutamine in the p53-derived sequence (Table 2). The GAL4-CTD protein was reported to be phosphorylated at both serine and threonine (Dvir et al., 1993); thus, the number of repeats, the DNA template, or other factors may influence the recognition of sites by DNA-PK.

Another consideration with regard to substrate recognition is that some sites phosphorylated *in vivo* may not be phosphorylated *in vitro*. We alluded to a possible case in SV40 TAg. Serine 639 is phosphorylated *in vivo* and resembles other DNA-PK sites, but no kinase has been identified that phosphorylates this site *in vitro* (Chen et al., 1991). *In vivo* TAg assembles into a dodecameric structure at the SV40 origin of replication, whereas, in our *in vitro* experiments, calf thymus DNA was used as the DNA-PK activator, and TAg most likely binds in a non-sequence specific manner to fragments of calf thymus DNA primarily as monomers. It remains to be determined if the pattern of TAg residues phosphorylated by DNA-PK would be different if the Tag had been assembled on replication origins.

CONCLUSIONS

Most of the DNA-PK in normal, unsynchronized, cultured human fibroblasts and lymphoid cells is not tightly associated with DNA; thus, presumably it is inactive (Anderson and Lees-Miller, 1992). *In vitro* assays indicated that total cellular DNA-PK activity changes only slightly with stage of the cell cycle, and no significant increase in activity was observed after HeLa cells were treated with DNA damage-inducing agents. These observations suggest that DNA-PK normally may be inactive and that synthesis and degradation are not important mechanisms of regulating its activity. We presume that DNA-PK activation occurs as a consequence of changes to chromatin that create entry sites for the Ku DNA-binding, regulatory subunit. Such changes may be caused by normal cellular activities including transcription, replication or recombination, or they may be a consequence of damage to DNA that is caused by endogenous or exogenous agents. Further regulation may occur through autophosphorylation (Lees-Miller et al., 1990), other posttranslational events, or through the synthesis or activation of inhibitors (An et al., 1994). Because the methods used to disrupt cells in preparing extracts inevitably fragments chromatin, it has not been possible to determine the state of activation of DNA-PK with existing *in vitro* assays. Furthermore, because decay of the ^{32}P-label required to detect substrate phosphorylation *in vivo* also produces DNA strand breaks that may activate DNA-PK, it has not been possible to correlate changes in the phosphorylation of putative substrates with kinase activation. Thus, a major challenge will be to develop assays suitable for measuring DNA-PK activity and activation *in vivo*. A detailed knowledge of the factors that affect DNA-PK substrate recognition *in vitro* undoubtedly will be helpful for developing improved assays for DNA-PK activity.

ACKNOWLEDGMENTS

We thank S. Lamb and J. Wysocki for technical assistance. This work was supported by the Office of Health and Environmental Research of the U. S. Department of Energy (CWA) and by grants from the National Science and Engineering Research Council (of Canada) and the Alberta Cancer Board (SPLM). Margery A. Connelly is an Alexander Hollaender Distinguished Postdoctoral Fellow.

REFERENCES

Allen, J. B., Zhou, Z., Siede, W., Friedberg, E. C. and Elledge, S. J., 1994 The SAD1/RAD53 protein kinase controls multiple checkpoints and DNA damage induced transcription in yeast. *Genes Dev.* 8:2416.

An, S., Chen, Y., Wu, W. and Wu, J. M., 1994 Demonstration of a double-stranded DNA-stimulated protein kinase (DNA-PK) inhibitory activity in human HL-60 leukemia cells. *Biochem. Biophys. Res. Comm.* 202:1530.

Anderson, C. W., 1993 DNA damage and the DNA-activated protein kinase. *Trends Biochem. Sci.* 18:433.

Anderson, C. W., 1994 Protein kinases and the response to DNA damage. *Semin. Cell Biol.* (in press).

Anderson, C. W. and Lees-Miller, S. P., 1992 The human DNA- activated protein kinase, DNA-PK. *Crit. Rev. Eukaryotic Gene Express.* 2:283.

Andreassen, P. R. and Margolis, R. L., 1992 2-Aminopurine overrides multiple cell cycle checkpoints in BHK cells. *Proc. Natl. Acad. Sci. USA* 89:2272.

Appella, E. and Anderson, C. W., 1994 Tumor suppressor protein p53: Response to DNA damage, cell cycle control, and mechanisms of inactivation in cancer. *Biochemica in Italia* 1:19.

Appella, E., Sakaguchi, K., Sakamoto, H., Lewis, M. S., Omichinski, G. J, Gronenborn, A. M., Clore, G. M. and Anderson, C. W. 1994 The p53 tumor suppressor protein: biophysical characterization of the carboxy-terminal oligomerization domain. *in*: Methods in Protein Structure Analysis - 1994, Atassi, M. Z., ed., Plenum Press, New York (NB: p 407).

Bannister, A. J., Gottlieb, T. M., Kouzarides, T. and Jackson, S. P., 1993 c-Jun is phosphorylated by the DNA-dependent protein kinase *in vitro*; definition of the minimal kinase recognition motif. *Nucleic Acids Res.* 21:1289.

Blier, P. R., Griffith, A. J., Craft, J. and Hardin, J. A. 1993 Binding of Ku protein to DNA. Measurement of affinity for ends and demonstration of binding to nicks. *J. Biol. Chem.* 268:7594.

Cai, Q.-Q., Plet, A., Imbert, J., Lafage-Pochitaloff, M., Cerdan, C. and Blanchard, J.-M., 1994 Chromosomal location and expression of the genes coding for Ku p70 and p80 in human cell lines and normal tissues. *Cytogenet. Cell Genet.* 65:221.

Carter, T., Vancurova, I., Sun, I., Lou, W. and DeLeon, S., 1990 A DNA-activated protein kinase from HeLa cell nuclei. *Mol. Cell. Biol.* 10:6460.

Chan, J. Y. C., Lerman, M. I., Prabhakar, B. S., Isozaki, O., Santisteban, P., Notkins, A. L. and Kohn, L. D., 1988 Cloning and characterization of a cDNA that encodes a 70-kDa novel human thyroid autoantigen. *J. Biol. Chem.* 264:3651.

Chen, Y.-R., Lees-Miller, S. P., Tegtmeyer, P. and Anderson, C. W., 1991 The human DNA-activated protein kinase phosphorylates SV40 T-antigen at amino- and carboxy-terminal sites. *J. Virol.* 10:5131.

de Murcia, G. and de Murcia, J. M., 1994 Poly(ADP-ribose) polymerase: a molecular nick-sensor. *Trends Biochem. Sci.* 19:172.

de Vries, E., van Driel, W., Bergsma, W. G., Arnberg, A. C. and van der Vliet, P. C., 1989 HeLa nuclear protein recognizing DNA termini and translocating on DNA forming a regular DNA-multimeric protein complex. *J. Mol. Biol.* 208:65.

Dahmus, M. E. and Dynan, W. S., 1992 Phosphorylation of RNA polymerase II as a transcriptional regulatory mechanism, *in*: Transcriptional Regulation, McKnight, S. L. and Yamamoto, K. R., eds., Cold Spring Harbor Laboratory Press, NY, pp. 109-128.

Dvir, A., Peterson, S. R., Knuth, M. K., Lu, H. and Dynan, W. S., 1992 Ku autoantigen is the regulatory component of a template- associated protein kinase that phosphorylates RNA polymerase II. *Proc. Natl. Acad. Sci. USA* 89:11920.

Dvir, A., Stein, L. Y., Calore, B. L. and Dynan, W. S., 1993 Purification and characterization of a template-associated protein kinase that phosphorylates RNA polymerase II. *J. Biol. Chem.* 268:10440.

Dulić, V., Kaufmann, W. K., Wilson, S. J., Tisty, T. D., Lees, E., Harper, J. W., Elledge, S. J. and Reed, S. I., 1994 p53- dependent inhibition of cyclin-dependent kinase activities in human fibroblasts during radiation-induced G1 arrest. *Cell* 76:1013.

El-Deiry, W. S., Tokino, T., Velculescu, V. E., Levy, D. B., Parsons, R., Trent, J. M., Lin, D., Mercer, W. E., Kinzler, K. W. and Vogelstein, B., 1993 *WAF1*, a potential mediator of p53 tumor suppression. *Cell* 75:817.

El-Deiry, W. S., Harper, J. W., O'Connor, P. M., Velculescu, V. E., Canman, C. E., Jackman, J., Pietenpol, J. A., Burrell, M., Hill, D. E., Wang, Y., Wiman, K. G., Mercer, W. E., Kastan, M. B., Kohn, K. W., Elledge, S. J., Kinzler, K. W. and Vogelstain B., 1994 *WAF1/CIP1* is induced in p53-mediated G_1 arrest and apoptosis. *Cancer Res.* 54:1169.

Falzon, M. and Kuff, E. L., 1992 The nucleotide sequence of a mouse cDNA encoding the 80 kDa subunit of the Ku (p70/p80) autoantigen. *Nucleic Acids. Res.* 20:3784.

Falzon, M., Fewell, J. W. and Kuff, E. L., 1993 EBP-80, a transcription factor closely resembling the human autoantigen Ku, recognizes single- to double-stranded transitions in DNA. *J. Biol. Chem.* 268:10546.

Fornace Jr., A. J., 1992 Mammalian genes induced by radiation; activation of genes associated with growth control. *Annu. Rev. Genet.* 26:507.

Getts, R. C. and Stamato, T. D., 1994 Absence of a Ku-like DNA end binding activity in the *xrs* double-stranded DNA repair- deficient mutant. *J. Biol. Chem.* 269:15981.

Gottlieb, T. M. and Jackson, S. P., 1993 The DNA-dependent protein kinase: requirement for DNA ends and association with Ku antigen. *Cell* 72:131.

Griffith, A. J., Craft, J., Evans, J., Minori, T. and Hardin, J. A., 1992 Nucleotide sequence and genomic structure analysis of the p70 subunit of the human Ku autoantigen: evidence for a family of genes encodeing Ku (p70)-related polypeptides. *Mol. Biol. Rep.* 16:91.

Harper, J. W., Adami, G. R., Wei, N., Keyomarsi, K. and Elledge, S. J., 1993 The p21 Cdk-interacting protein Cip1 is a potent inhibitor of G1 cyclin-dependent kinases. *Cell* 75:805.

Hartwell, L. and Weinert, T. A., 1989 Checkpoints: controls that ensure the order of cell cycle events. *Science* 246:629-634.

Herrlich, P. and Rahmsdorf, H. J., 1994 Transcriptional and post- transcriptional responses to DNA-damaging agents. *Curr. Opin. Cell Biol.* 6:425.

Hunter, T., 1993 Braking the cycle. *Cell* 75:839-841.

Jackson, S. P., MacDonald, J. J., Lees-Miller, S. and Tjian, R., 1990 GC box binding induces phosphorylation of Sp1 by a DNA- dependent protein kinase. *Cell* 63:155.

Jackson, S., Gottlieb, T., and Hartley. K., 1993 Phosphorylation of transcription factor Sp1 by the DNA-dependent protein kinase. *Adv. Second Messenger Phosphoprotein Res.* 28:279.

Kastan, M. B., Onyekwere, O., Sidransky, D., Vogelstein, B. and Craig, R. W., 1991 Participation of p53 protein in cellular response to DNA damage. *Cancer Res.* 51:6304.

Kemp, B. E. and Pearson, R. B., 1991 Design and use of peptide substrates for protein kinases. *Meth. Enzymol.* 200:121.

Lees-Miller, S. P. and Anderson, C. W., 1989a Two human 90-kDa heat shock proteins are phosphorylated *in vivo* at conserved serines that are phosphorylated *in vitro* by casein kinase II. *J. Biol. Chem.* 264:2431.

Lees-Miller, S. P. and Anderson. C. W., 1989b The human dsDNA- activated protein kinase phosphorylates the 90-kDa heat shock protein, hsp90α, at two N-terminal threonine residues. *J. Biol. Chem.* 264:17275.

Lees-Miller, S. P., Chen, Y.-R. and Anderson, C. W., 1990 Human cells contain a DNA-activated protein kinase that phosphorylates simian virus 40 T antigen, mouse p53, and the human Ku autoantigen. *Mol. Cell. Biol.* 10:6472.

Lees-Miller, S. P., Sakaguchi, K., Ullrich, S., Appella, E., and Anderson, C. W., 1992 Human DNA-activated protein kinase phosphorylates serines 15 and 37 in the amino-terminal transactivation domain of human p53. *Mol. Cell. Biol.* 12:5041.

Levine, A. J., 1993 The tumor suppressor genes. *Annu. Rev. Biochem.* 62:623.

Liu, S.-H., Ma, J.-T., Yueh, A. Y., Lees-Miller, S. P., Anderson, C. W. and Ng, S.-Y., 1993 The carboxy-terminal transcription activation domain of human serum response factor is phosphorylated by DNA-PK, the DNA-activated protein kinase. *J. Biol. Chem.* 268:21147.

Luethy, J. D. and Holbrook, N. J., 1994 The pathway regulating *GADD153* induction in response to DNA damage is independent of protein kinase C and tyrosine kinases. *Cancer. Res.* 54:1902s.

Mimori, T., Akizuki, M., Yamagata, H., Inada, S., Yoshida, S. and Homma, M., 1981 Characterization of a high molecular weight acidic nuclear protein recognized by antibodies in sera from patients with polymyositis-scleroderma overlap. *J. Clin. Invest.* 68:611.

Mimori, T., Hardin, J. A. and Steitz, J. A., 1986 Characterization of the DNA-binding protein antigen Ku recognized by autoantibodies from patients with rheumatic disorders. *J. Biol. Chem.* 261:2274.

Mimori, T., Ohosone, Y., Hama, N., Suwa, A., Akizuka, M., Homma, M., Griffith, A. J. and Hardin, J. A. , 1990 Isolation and characterization of cDNA encoding the 80-kDa subunit protein of the human autoantigen Ku (p70/p80) recognized by autoantibodies from patients with scleroderma-polymyositis overlap syndrome. *Proc. Natl. Acad. Sci. USA* 87:1777.

Morozov, V. E., Falzon, M., Anderson, C. W. and Kuff, E. L., 1994 DNA-dependent protein kinase is activated by nicks and larger single-stranded gaps. *J. Biol. Chem.* 269:16684.

Murray, A. W., 1992 Creative blocks: cell-cycle checkpoints and feedback controls. *Nature* 359:599.

Nelson, W. G. and Kastan, M. B., 1994 DNA strand breaks: the DNA template alterations that trigger p53-dependent DNA damage response pathways. *Mol. Cell. Biol.* 14:1815.

O'Brien, T., Hardin, S., Greenleaf, A. and Lis, J. T., 1994 Phosphorylation of RNA polymerase II C-terminal domain and transcriptional elongation. *Nature* 370:75.

O'Connor, P. M. and Kohn, K. W., 1992 A fundamental role for cell cycle regulation in the chemosensitivity of cancer cells? *Semin. Cell Biol.* 3:409.

Porges, A. J., Ng, T. and Reeves, W. H., 1990 Antigenic determinants of the Ku (p70/p80) autoantibody are poorly conserved between species. *J. Immunol.* 145:4222.

Rathmell, W. K. and Chu, G., 1994a A DNA end-binding factor involved in double-strand break repair and V(D)J recombination. *Mol. Cell. Biol.* 14:4741.

Rathmell, W. K. and Chu, G., 1994b Involvement of the Ku autoantigen in the cellular response to DNA double-strand breaks. *Proc. Natl. Acad. Sci. USA* 91:7623.

Reeves, W. H., 1992 Antibodies to the p70/p80 (Ku) antigens in systemic lupus erythematosus. *Rheumatic Dis. Clinics of N. America* 18:391.

Reeves, W. H. and Sthoeger, Z. M., 1989 Molecular cloning of cDNA encoding the p70 (Ku) lupus autoantigen. *J. Biol. Chem.* 264:5047.

Sachsenmaier, C., Radler-Pohl, A., Müller, A., Herrlich, P., and Rahmsdorf, H. J. 1994 Damage to DNA by UV light and activation of transcription factors. *Biochem. Pharmacol.* 47:129.

Satoh, M. S. and Lindahl, T., 1992 Role of poly(ADP-ribose) formation in DNA repair. *Nature* 356:356.

Sheldrick, K. S. and Carr, A. M., 1993 Feedback controls and G2 checkpoints: fission yeast as a model system. *Bioessays* 15:775.

Siede, W., Friedberg, A. S., Dianova, I. and Frieberg, E. C., 1994 Characterization of G_1 checkpoint control in the yeast *Saccharomyces cerevisiae* following exposure to DNA-damaging agents. *Genetics* 138:271.

Taccioli, G. E., Gottlieb, T. M., Blunt, T., Priestley, A., Demengeot, J., Mizuta, R., Lehmann, A. R., Alt, F. W., Jackson, S. P. and Jeggo, P. A., 1994 Ku80: Product of the *XRCC5* gene and its role in DNA repair and V(D)J recombination. *Science* 265: 1442.

Tuteja, N., Tuteja, R., Ochem, A., Taneja, P., Huang, N. W., Simoncsits, A., Susic, S., Rahman, K., Marusic, L., Chen, J., Zhang, J., Wang, S., Pongor, S. and Falaschi, A., 1994 Human DNA helicase II: a novel DNA unwinding enzyme identified as the Ku autoantigen. *EMBO J.* 13:4991.

Ullrich, S. J., Sakaguchi, K., Lees-Miller, S. P., Fiscella, M., Mercer, W. E., Anderson, C. W. and Appella, E., 1993 Phosphorylation at serine 15 and 392 in mutant p53s from human tumors is altered compared to wild-type p53. *Proc. Natl. Acad. Sci. USA* 90:5954.

Walworth, N., Davey, S. and Beach, D., 1993 Fission yeast *chk1* protein kinase links the *rad* checkpoint pathway to cdc2. *Nature* 363:368.

Wang, J., Chou, C.-H., Blankson, J., Satoh, M., Knuth, M. W., Eisenberg, R. A., Pisetsky, D. S. and Reeves, W. H., 1993 Murine monoclonal antibodies specific for conserved and non-conserved antigenic determinants of the human and murine Ku autoantigens. *Mol. Biol. Rep.* 18:15.

Weinert, T. and Lydall, D., 1993 Cell cycle checkpoints, genetic instability and cancer. *Semin. Cancer Biol.* 4:129.

Weinert, T. A., Kiser, G. L. and Hartwell, L. H., 1994 Mitotic checkpoint genes in budding yeast and the dependence of mitosis on DNA replication and repair. *Genes Dev.* 8:652.

Yaneva, M., Wen, J., Ayala, A. and Cook, R., 1989 cDNA-derived amino acid sequence of the 86-kDa subunit of the Ku antigen. *J. Biol. Chem.* 264:13407.

36

THE P53 TUMOR SUPPRESSOR PROTEIN

Biophysical Characterization of the Carboxy-Terminal Oligomerization Domain

Ettore Appella,[1] Kazuyasu Sakaguchi,[1] Hiroshi Sakamoto,[1]
Marc S. Lewis,[2] James G. Omichinski,[3] Angela M. Gronenborn,[3]
G. Marius Clore,[3] and Carl W. Anderson[4]

[1] Laboratory of Cell Biology, National Cancer Institute, National Institutes of Health
Bethesda, Maryland 20892

[2] Biomedical Engineering and Instrumentation Program, National Center for Research Resources
Bethesda, Maryland 20892

[3] Laboratory of Chemical Physics, National Institute of Diabetes and Digestive and Kidney Diseases, National Institutes of Health
Bethesda, Maryland 20892

[4] Biology Department, Brookhaven National Laboratory
Upton, New York 11973

INTRODUCTION

In response to damaged DNA, mammalian cell growth is arrested at cell cycle checkpoints in G1, near the border of S phase, or in G2, before mitosis (Murray, 1992; Hunter, 1993; Weinert and Lydall, 1993). In some circumstances, DNA damage initiates apoptosis, a program that results in cell death. Recent studies have shown that the p53 tumor suppressor protein is an essential component of the G1 checkpoint pathway (Kastan et al., 1991); it also modulates the initiation of apoptosis (Oren, 1994). The arrest of cell cycle progression provides time for DNA damage to be repaired, whereas apoptosis may insure the death of more severely damaged cells that are at risk of loss of growth control through genome rearrangements. Thus, these functions account, at least in part, for the importance of p53 in suppressing or eliminating preneoplastic or neoplastic cells in the human and other vertebrate species. In turn, p53 function is mediated through its physical characteristics, and these may be modulated by post-translational mechanisms (Ullrich et al., 1992; Meek, 1994). Thus, biophysical studies of p53 and its functional domains are fundamental to an understanding of those properties that are important for normal p53 function.

p53 is a nuclear phosphoprotein with a short half-life, and its normal concentration in the nucleus is low (Levine, 1993). However, p53 protein accumulates in the nuclei of cells

Methods in Protein Structure Analysis, Edited by M. Z. Atassi and E. Appella
Plenum Press, New York, 1995

exposed to UV irradiation, ionizing radiation, and other DNA damage-inducing agents (Maltzman and Czyzyk, 1984; Kuerbitz et al., 1992; Lu and Lane, 1993; Nelson and Kastan, 1994), and this accumulation is thought to result from a transient stabilization of the p53 protein through undefined post-transcriptional mechanisms. Overexpression of wild-type p53 protein through transient transfection or stable integration of a wild-type p53 cDNA driven by an inducible promoter was shown to arrest cell growth at or near the G1/S border (Mercer et al., 1990). However, cell cycle arrest did not occur when mutant p53s were overexpressed, nor did it occur in tumor cells without a functional p53 gene after they were exposed to DNA damage-inducing agents (Kastan et al., 1991).

p53 is a sequence specific activator of transcription. When the wild-type p53 protein was fused to the yeast GAL4 DNA binding domain, it activated transcription from GAL4 reporter plasmids in both yeast and mammalian cells (Fields and Jang, 1990; Raycroft et al., 1990). The fragment containing the first 73 residues of p53, which are rich in acidic amino acids, was sufficient to confer transcriptional activation with the GAL4 domain. Thus, it was surprising that several missense mutants, with lesions in the central domain, failed to transactivate when coupled to GAL4 (Unger et al., 1992). Subsequently, wild-type p53 was shown to bind to a specific consensus DNA sequence consisting of two copies of the 10 base-pair motif: 5'-PuPuPuC(A/T)(T/A)GPyPyPy-3' (Funk et al., 1992; El-Deiry et al., 1992; Halazonites et al., 1993). Each 10 base-pair element is palindromic, with two five base-pair motifs in opposite orientations. p53 DNA binding sites have been identified in the 5' regions of several natural genes, including *WAF1/CIP1, MDM2, GADD45,* and *MCK* (muscle creatine kinase), and in these natural sites, the 10 base-pair elements are separated by 0 to 13 base-pairs (Bargonetti et al., 1991; Zambetti et al., 1992; Kastan et al., 1992; El-Deiry et al., 1993). These natural p53 binding elements confer p53 inducibility to their associated transcription units and, when placed upstream, to reporter genes *in vivo* and *in vitro* (Farmer et al., 1992; Funk et al., 1992; Kern et al., 1992; Zambetti et al., 1992). *WAF1/CIP1* encodes a 21 kDa inhibitor of several cyclin-dependent kinases that are required for entry into S phase (El-Deiry et al., 1993, 1994; Xiong et al., 1993). Thus, induction of Waf1 synthesis in response to elevated levels of nuclear p53 provides a mechanism that accounts for the arrest of cell cycle progression after DNA damage or after the engineered overexpression of p53 protein in cultured cells.

Two approaches, partial proteolysis and the expression of recombinant deletion constructs, were used to demonstrate that the central, highly conserved domain of p53, encompassing about 200 amino acids from residue 100, is responsible for sequence-specific DNA binding (Pavletich et al., 1993; Bargonetti et al., 1993; Wang et al., 1993; Halazonetis and Kandil, 1993). This domain includes four of the five most highly conserved segments of the p53 protein (Soussi et al, 1990); most p53 mutations from human tumors affect residues in this region (Hollstein et al., 1991). X-ray diffraction analysis of co-crystals of the central core DNA binding domain and a double-stranded recognition site oligonucleotide revealed the three dimensional structure of the complex at a resolution of 2.2 Å (Cho et al., 1994). The core DNA-binding domain consists of two anti-parallel β sheets, of four and five strands, respectively, that anchor the DNA binding elements, two large loops and a loop-sheet-helix motif, on one face of the structure. The positions of the two loops are further stabilized by a tetrahedrally coordinated zinc atom. The most frequently detected mutations in human tumors cluster in the gene segments that encode the large loops and the loop-sheet-helix motif, suggesting that these elements are essential for p53's role as a tumor suppressor.

Wild-type p53 protein forms tetramers and higher order oligomeric structures in solution (Stenger et al., 1992), and the domain responsible for oligomerization was localized to the carboxyl terminus (Wang et al., 1993). p53 with its 47 carboxy-terminal amino acids deleted is monomeric (Milner and Medcalf, 1991). Subsequent studies showed that amino acids 315 to 360 were sufficient for the formation of recombinant protein tetramers (Wang et al., 1993, 1994); proteolytic digestion of human p53 yielded a 53 residue fragment, which contained residues 311 to 363 and formed tetramers (Pavletich et al., 1993). Although the role of tetramerization in p53 function is not completely understood, it is required for efficient site-specific DNA binding and contributes to p53's ability to activate transcription from natural promoters. In addition, the tetramerization domain of mouse p53, residues 315 to 360, was shown to be sufficient for cooperation with an activated *ras* oncogene in cell transformation (Reed et al., 1993). Formation of heterotetramers between wild-type and mutant p53 *in vivo* may inactivate wild-type p53 function, thus potentiating tumor development.

Carboxy-terminal p53 sequences appear to have an inhibitory effect on site-specific DNA binding (Hupp et al., 1992). Wild-type p53 made in *E. coli* binds relatively poorly to the p53 consensus recognition sequence, but its binding is enhanced by several treatments that affect the carboxyl terminus. These include deletion of the 30 carboxy-terminal residues, complex formation with antibodies or other proteins that recognize carboxy-terminal sequences, and phosphorylation by casein kinase II (CKII). Casein kinase II phosphorylates the penultimate residues of mouse and human p53, Ser^{389} and Ser^{392}, respectively (Meek et al., 1990; Meek, 1994). The last 27 residues of human p53 are encoded by a separate exon, suggesting that this segment may comprise a distinct functional domain (Lamb and Crawford, 1986). The carboxyl terminus is rich in basic residues and provides non-sequence specific DNA binding and strand annealing activities that are independent of the sequence-specific DNA binding central domain (Wang et al., 1993; Pavletich et al., 1993; Oberosler et al., 1993; Bakalkin et al., 1994).

Wild-type p53 is phosphorylated *in vivo* at several amino-terminal and carboxy-terminal sites (Ullrich et al., 1993; Meek, 1994), and several of these sites can be phosphorylated *in vitro* by known serine/threonine protein kinases. Serine 389 and 392, which are homologous residues in mouse and human p53, respectively, are phosphorylated *in vivo*, and, as noted above, *in vitro* by CKII. Serine 312 (murine) and serine 315 (human) also are phosphorylated *in vivo*, and these residues can be phosphorylated *in vitro* by the $p34^{cdc2}$ cyclin dependent kinase (Addison et al., 1990; Bischoff et al., 1990). Several sites in the amino-terminal region are phosphorylated; these include serines 4, 6, 9, 15, and 32, and threonines 78 and 88 in murine p53, and serines 9, 15, and 33 in human p53 (Wang and Eckhart, 1992; Lees-Miller et al., 1992; Ullrich et al., 1993; Milne et al., 1994; Meek, 1994). In murine p53, serines 4, 6, and 9 are phosphorylated *in vitro* by a casein kinase I-like enzyme (Milne et al., 1992), serines 4 and 15 can be phosphorylated by DNA-activated protein kinase (Wang and Eckhart, 1992; Lees-Miller et al., 1992), and threonines 78 and 88 are phosphorylated by MAP kinase (Milne et al., 1994). Serine 15 of human p53 is phosphorylated by the DNA-activated protein kinase (Lees-Miller et al., 1992), and serine 33 may be phosphorylated by MAPK or JNK1, the Jun N-terminal kinase (De'rijard et al., 1994). Phosphorylation is an important mechanism for regulating the activity of several transcription factors, including Jun, NF-κB, and SRF (Jackson, 1992; Hunter and Karin, 1992), and the fact that the amino acid sequences at the five major sites phosphorylated in human p53 are conserved in most vertebrates (Soussi et al., 1990) suggests that phosphorylation plays an important role in regulating p53 function. However, mutant p53s with nonphosphorylatable residues at individual phosphorylation sites are similar to wild-type p53 in the properties that have been examined. Even p53 with a mutant CKII site was found to be indistinguishable from wild-type p53 with respect to the *in vivo* functions of transactivation, growth arrest, and suppression of cell transformation by *ras* and *E1A* (Fiscella et al., 1994). Thus, to date, no clear role for p53 phosphorylation has emerged. The basic structure of human p53 protein is summarized in figure 1.

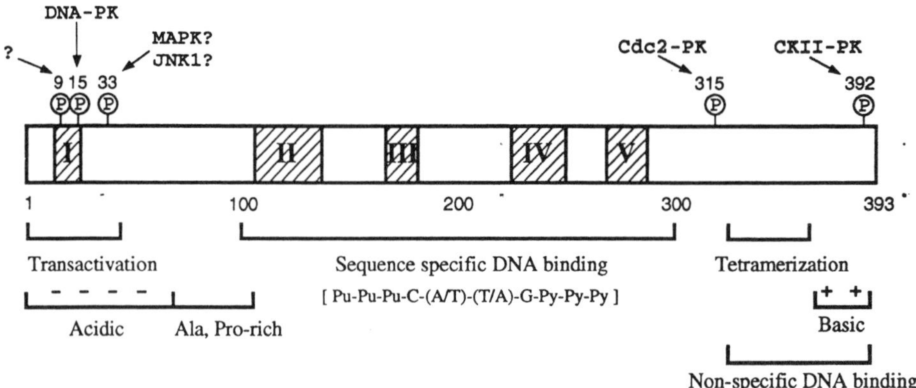

Figure 1. Schematic representation of the 393 amino acid human p53 polypeptide illustrating properties of its domains. The amino terminus of p53 contains an acidic transactivation domain (residues 1–73) followed by an alanine-rich segment. Residues 102 to 286 contain the site-specific DNA binding domain. Carboxy-terminal residues 319–360 provide for tetramerization, and this region is followed by a segment rich in basic residues; the carboxy-terminal segment provides non-specific DNA binding and strand annealing functions. At least five human p53 sites, serines 9, 15, 33, 315, and 392 are phosphorylated; kinases that phosphorylate three of these sites *in vitro* have been identified. The five segments that are evolutionarily highly conserved are indicated. References are given in the text.

Structure of the Tetramerization Domain of Human p53

To investigate the structure of the region responsible for tetramer formation, nine peptides, corresponding to carboxy-terminal segments of human p53 were chemically synthesized by solid phase methods using Fmoc chemistry. The larger peptides, S303-D393, K319-D393, R335-D393, and K319-D393($A^{323,327,330}$) were synthesized by a segment condensation method using peptide thioesters (Hojo and Aimoto, 1992). Each peptide was studied by equilibrium analytical ultracentrifugation using a Beckman XL-A analytical ultracentrifuge and analyzing the data by fitting to appropriate mathematical models using MLAB (Sakamoto et al. 1994). Figure 2 shows the concentration distribution of p53(319-360) at equilibrium fit as a monomer, as a monomer-dimer equilibrium, and as a monomer-tetramer equilibrium. The equilibrium concentrations of peptides S303-D393, K319-D393 and K319-G360 were entirely consistent with the formation of peptide tetramers in equilibrium with monomers. Residues between K319 and G334 are essential for tetramer formation since peptide R335-D393 was observed only as a monomer, as was the variant peptide, K319-D393($A^{323,327,330}$), in which three hydrophobic amino acids, Leu^{323}, Tyr^{327}, and Leu^{330}, were changed to alanine. Deletion of the 33 carboxy-terminal residues of p53, G361-D393, had no effect on tetramer formation, but deletion of another nine residues, to K351, reduced the association constant for tetramer formation, and deletion of another four residues, to A347, allowed only a minimal association to form tetramers. Thus, the segment from residues K319 to G360 forms a core domain that exhibits a strong propensity to form tetramers; no evidence was seen for dimer formation.

The thermodynamic parameters describing p53 tetramerization were obtained by the method of Clarke and Glew (1966). This procedure involves plotting R lnK_A as a function of absolute temperature and fitting the data with a mathematical model derived from basic thermodynamic principles. An examination of the thermodynamic characteristics of p53 peptides containing the tetramerization domain was revealing (Figure 3). A comparison of peptides S303-D393 and K319-D393 indicated that the increased propensity of K319-D393

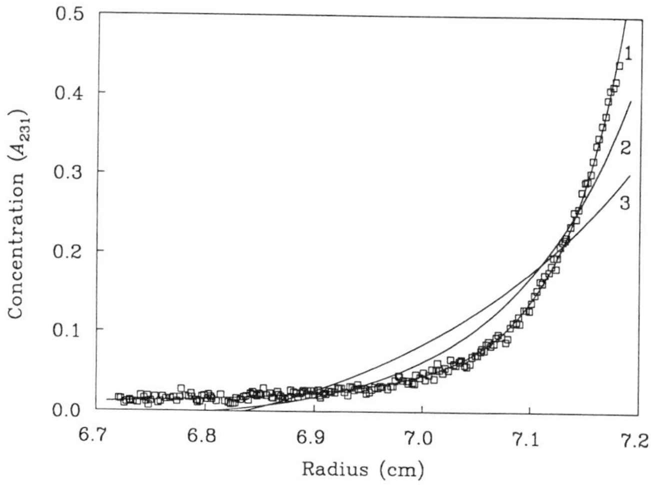

Figure 2. Concentration distribution of p53 (319-360), expressed as absorbance at 231 nm, as a function of radial position in the ultracentrifuge cell at equilibrium at 32000 rpm and 25°C. The fitting lines are best fits for the following models of association: (1) a monomer–tetramer equilibrium, (2) a monomer–dimer equilibrium, and (3) monomer alone. The failure of the latter two models is readily apparent. When a monomer–dimer–tetramer equilibrium model was used for fitting, a zero value for K_{12} was obtained, indicating that dimer was not present in detectable quantities.

to form tetramers (smaller dissociation constant, K_D, Figure 3) is a consequence of an increase in the magnitude of the change of the standard free energy (ΔG°) of 0.6 kcal mol^{-1}. A similar comparison of S303-G360 with K319-G360 revealed an even greater increase in ΔG° with removal of the amino-terminal segment, this time by 1.3 kcal mol^{-1}. Thus, the amino acid segments on both the carboxyl- and amino-terminal sides of the tetramerization domain adversely affect the ability of the core domain to form tetramers. Comparable data for the intact p53 protein currently are not available, but a comparison of its thermodynamic parameters with those of the model peptides would be of interest, as would be the comparison of these parameters from unmodified and post-translationally modified p53 or p53 peptides. As noted above, serines 315 and 392 are modified by phosphorylation, and one or more additional residues in the carboxy-terminal region may be modified by phosphorylation or glycosylation.

Circular dichroism (CD) was used to investigate the secondary structure of the p53 peptides. From the CD spectrum, it is possible to estimate the fraction of the residues that form α helices or β sheets; in particular, α helices are associated with a large negative ellipticity at 222 nm (Johnson, Jr., 1988). All of the peptides that failed to form tetramers, e.g. K319-D393($A^{323,327,330}$), R335-D393 (Figure 3), also had small ellipticity values ($\theta_{222} > -2000$). In contrast, K319-G360 and K319-D393, both of which are nearly 90% tetramer at the concentrations used for the CD measurements, had high negative ellipticity values ($\theta_{222} < -5000$). The ellipticity of K319-G360, $\theta_{222} = -9580$, was greater that that of K319-D393, which implies that the carboxy-terminal segment distal to the tetramerization domain has a low helical content. However, the core tetramerization domain cannot be a typical coiled-coil four helical bundle since the ellipticity of this segment is not indicative of the required α helical content. Substitution of three hydrophobic amino acids with alanine within the core domain prevented tetramer formation and the acquisition of an α helical conformation. Thus, these residues are critical for forming the correct secondary structure that, in turn, is necessary for oligomerization. Interestingly, these residues, Leu^{323}, Tyr^{327},

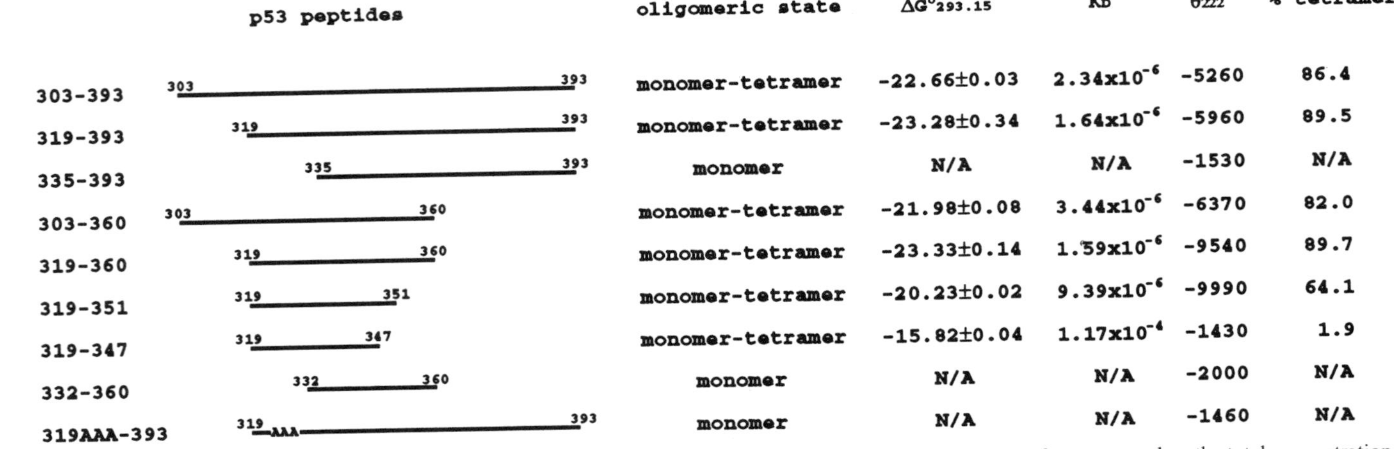

p53 peptides	oligomeric state	$\Delta G^{\circ}_{293.15}$	K_D	θ_{222}	% tetramer
303-393 (303–393)	monomer-tetramer	-22.66±0.03	2.34×10^{-6}	-5260	86.4
319-393 (319–393)	monomer-tetramer	-23.28±0.34	1.64×10^{-6}	-5960	89.5
335-393 (335–393)	monomer	N/A	N/A	-1530	N/A
303-360 (303–360)	monomer-tetramer	-21.98±0.08	3.44×10^{-6}	-6370	82.0
319-360 (319–360)	monomer-tetramer	-23.33±0.14	1.59×10^{-6}	-9540	89.7
319-351 (319–351)	monomer-tetramer	-20.23±0.02	9.39×10^{-6}	-9990	64.1
319-347 (319–347)	monomer-tetramer	-15.82±0.04	1.17×10^{-4}	-1430	1.9
332-360 (332–360)	monomer	N/A	N/A	-2000	N/A
319AAA-393 (319–AAA–393)	monomer	N/A	N/A	-1460	N/A

Figure 3. Oligomerization status and ellipticity of p53 peptides. K_D: K_D is commonly defined as the molar concentration of monomer when the total concentration is such that the molar concentration of monomer equals the molar concentration of oligomer. With this definition, for the monomer-tetramer equilibrium, $K_D = K_{14}^{-1/3}$. $\theta 222$: The ellipticity (deg cm^2 $dmol^{-1}$) of peptides at a wavelength of 222 nm at 20 mM concentration and 20°C. All concentrations are expressed in terms of molarity of monomer. % tetramer: The percent tetramer was calculated by solving the equation $C_T = C_1 + 4C_4 = C_1 + 4K_{14}C_1^4$ for C_1. When the total concentration, C_T = 20 mM, and all concentrations are expressed in terms of molarity of monomer, then the % tetramer = $100(C_T - C_1)/C_T$. N/A: data not available.

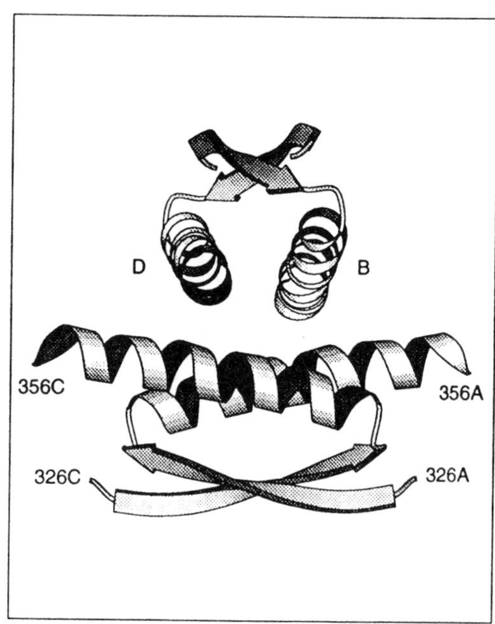

Figure 4. Schematic ribbon drawing of the structure of the tetramerization domain of human p53, residues 319-360. The tetramer is represented as a dimer of dimers with the dimers at approximately right angles to each other. This view is an expanded representation of the tetramerization model shown in figure 5.

and Leu[330], lie outside of the segment, residues 331-360, most likely to adopt an α helical structure.

Recently, the three dimensional structure of the core tetramerization domain was determined using multidimensional NMR spectroscopy (Clore et al., 1994). NMR spectroscopy was carried out on unlabeled samples, on peptides uniformly labeled with $^{15}N/^{13}C$, and on mixed tetramers comprised of equal amounts of the unlabeled and isotopically labeled peptides. Inter-proton distance constraints were derived from isotope-edited and filtered 3D and 4D NOE spectra (Bax et al., 1993; Clore and Gronenborn, 1994). With this methodology, it is possible to determine the positions of amino acid residues within each subunit as well as the relative positions of the different subunits within the tetramer. Figure 4 shows a schematic ribbon drawing of the structure of the tetramerization domain as determined by NMR. Each 42 amino acid peptide monomer consists of a short turn (D324-G325) followed by a strand (E326-G334), another short turn (R335-E336), and an α helix (R337-A355). The first residues at the amino terminus, K319-L323, and the four carboxy-terminal residues, K357-G360, were disordered in solution. Two monomer peptides form a dimer in which the α helices and β strands are antiparallel and interact. Two dimers interact to form a tetramer through their α helices; in this case the interacting helices are aligned in a antiparallel manner. The β strands lie on the outside of the tetramer on opposite faces. Thus, the tetramer is best described as a dimer of dimers, with the two dimers interacting through their α helices, which form a four helix bundle. In view of the fact that we have been unable to observe the presence of dimers in the ultracentrifugal analysis, it is implicit that the association constant for tetramer formation from dimers is much greater than the association constant for dimer formation from monomers.

The crucial structural element for tetramer formation appears to reside in the β-strand structures of corresponding dimers as is shown by the number of NOEs between the β-strands as compared to the helices of the adjacent subunits. A recombinant p53 protein has been produced in which the carboxy terminus of p53 distal to residue 333, including the turn and α helical region of the tetramerization domain, was replaced by a leucine zipper dimerization domain of yeast transcription factor GCN4 (Pietenpol et al, 1994). This hybrid p53 protein cannot form tetramers, suggesting that the small loop connecting the β strand to the helical segment also is important for tetramer formation. In addition to hydrophobic interactions between helices in the four helix bundles, several salt bridges contribute to stabilization of dimer-dimer interactions in the tetramers. Most p53 mutations from human tumors are located in regions encoding the sequence-specific DNA binding domain; however, four, which cause the substitution of His for Leu^{330}, Val for Gly^{334}, Cys for Arg^{337}, and Asp for Glu^{349}, lie within sequences encoding the oligomerization domain (Greenblatt et al., 1994). These residues are located at the interface of the dimers; thus, they may partially disrupt interactions which are required for tetramer formation.

The specific dimer-of-dimers topology of the p53 tetramerization domain has not been observed previously in other tetrameric protein structures. This topology places some constraints on the relative locations of the four amino-terminal segments of the p53 tetramer, which includes the central site-specific DNA binding domain containing residues T102 to K292 (Cho et al., 1994). The topology suggests that DNA-binding domains of adjacent subunits may be in closer proximity than the DNA binding domains for non-adjacent subunits, thus permitting a close apposition in the dimer of the sequence specific DNA binding domain with the carboxy-terminal, basic, non-sequence specific DNA binding tail segment (residues 361-393) (Figure 5). This apposition may explain how alterations to the carboxy terminus of p53, including phosphorylation of serine 292 and deletion of the carboxy-terminal 30 residues, affect sequence-specific DNA binding (Hupp et al., 1992). Based on this predicted structural arrangement (Figure 5), we envision each monomer of adjacent dimer subunits (AC) contacting one-quarter of the consensus sites in one 10 base-pair palindrome while the other dimer (BD) contacts equivalent sites in the second 10 base-pair palindrome. This disposition of binding site interactions might be expected to

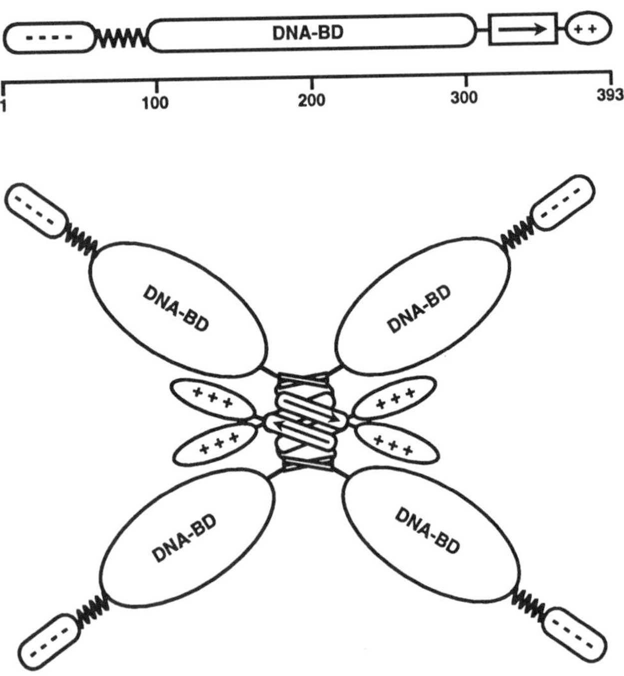

Figure 5. A. Primary structure of the human p53 protein showing the location of the various domains along the polypeptide chain. B. Model of the tetrameric arrangement of the human p53 protein illustrating the proposed positions of the DNA-binding domains with respect to the tetramerization domains.

induce a sizable bend in the DNA. No such bending is observed in the co-crystal structure reported by Cho et al. (1994); however, in this complex, only two binding domain monomers were associated with consensus site oligonucleotide, and the tetramerization domain was not present.

CONCLUSIONS

Sedimentation equilibrium, a quantitative physical technique, was used to show that peptides derived from the carboxy-terminal oligomerization region of human p53 form stable tetramers that are in equilibrium with peptide monomers in aqueous solution. Although less commonly used today than gel electrophoretic techniques (Stenger et al., 1994), the behavior of interacting components during sedimentation equilibrium is rigorously based on reversible thermodynamics. Thus, the quantitative measurements that are readily made with modern instrumentation can be meaningfully interpreted in terms of association constants and changes in the free energy of association. Using this technique in conjunction with modern methods for the efficient synthesis of longer peptides, we have defined the optimal segment of p53 for the formation of tetramers. CD, an optical technique for estimating secondary structure content, was then used to predict the secondary structure of the tetramerization domain. Subsequently, the three dimensional structure of the tetramerization domain was determined by high resolution multidimensional NMR techniques. This structural determination not only confirmed that the segment of p53 between residues K319 and G360 forms tetramers, but also that both the amino-terminal β sheet element, Y327-G334, and the carboxy-terminal α helical region, R337-A355, contribute to the structure. The structure is best described as a dimer-of-dimers, with the opposing dimers held together principally through interactions between their α helices.

The topology of the tetramerization domain has important implications relating to the biological properties of p53. Some constraints are placed on relative positions of the sequence specific DNA binding domain of each monomer within the tetramer. The DNA binding domain is amino-terminal to the tetramerization domain and is connected to it by a linker of about 30 residues that may be flexible. The basic tail of p53, residues G361 through the carboxy terminus, modulates sequence specific DNA binding. The topology of the tetramerization domain suggests that this tail, which in conjunction with the tetramerization domain has non-specific DNA binding properties, may modulate the conformation of the site-specific DNA binding domain within adjacent dimers, thus preventing the latter domain from adopting an optimal structure for sequence-specific DNA recognition. Although the tetramerization domain facilitates the interaction of p53 with DNA, it may also place the carboxy-terminal tail in position to regulate p53 binding to its consensus sites. The juxtaposition of DNA binding domains in the tetramer also suggests that p53 may bend DNA. Strong bending of DNA has been observed for other transcription factor DNA-complexes and may be a property that is important for the activation of transcription (Harrington and Winicov, 1994). Thus, the arrangement of domains, provided by the topology of the dimer-of-dimers tetramerization element, may provide an adaptable DNA reading mechanism that expedites specific interactions with p53 response elements in response to DNA damage. The speculations raised here undoubtedly will be tested in the near future when new high resolution p53 structures become available. A precise understanding of the relationship of the domains in the p53 tetramer may suggest ways to manipulate the interaction of wild-type and mutant p53s with each other and with DNA. The discovery of drugs that circumvent the loss of p53 function is an exciting prospect that would have wide implications for cancer treatment.

ACKNOWLEDGMENTS

C.W.A. is supported by the Office of Health and Environmental Research of the U.S. Department of Energy.

REFERENCES

Addison, C., Jenkins, J. R. and Stürzbecher, H.-W., 1990. The p53 nuclear-localization signal is structurally linked to a p34cdc2 kinase motif. *Oncogene* 5: 423.

Bakalkin, G., Yakovleva, T., Selivanova, G., Magnusson, K. P., Szekely, L.. Kiseleva, E., Klein, G., Terenius, L. and Wiman, K. G., 1994. p53 binds single-stranded DNA ends and catalyzes DNA renaturation and strand transfer. *Proc. Natl. Acad. Sci. USA* 91: 413.

Bargonetti, J., Friedman, P. N., Kern, S. E., Vogelstein, B. and Prives, C., 1991. Wild-type but not mutant p53 immunopurified proteins bind to sequences adjacent to the SV40 origin of replication. *Cell* 65: 1083.

Bargonetti, J., Manfredi, J. J., Chen, X., Marshak, D. R. and Prives, C., 1993. A proteolytic fragment from the central region of p53 has marked sequence-specific DNA-binding activity when generated from wild-type but not from oncogenic mutant p53 protein. *Genes Dev.* 7: 2565.

Bax, A. and Grzesiek, S., 1993. Methodological advances in protein NMR. *Acc. Chem. Res.* 26: 131.

Bischoff, J. R., Friedman, P. N., Marshak, D. R., Prives, C., and Beach, D., 1990. Human p53 is phosphorylated by p60-cdc2 and cyclin B-cdc2. *Proc. Natl. Acad. Sci. USA* 87: 4766.

Cho, Y., Gorina, S., Jeffrey, P. D. and Pavletich, N. P., 1994. Crystal structure of a p53 tumor suppressor-DNA complex: understanding tumorigenic mutations. *Science* 265: 346.

Clarke, E. C. W. and Glew, D. N., 1966. Evaluation of thermodynamic function from equilibrium constants. *Trans. Farady Soc.* 62: 539.

Clore, G. M. and Gronenborn, A. M., 1994. Multidimensional heteronuclear magnetic resonance of proteins. *Methods Enzymol.* 239: 349.

Clore, G. M., Omichinski, J. G., Sakaguchi, K., Zambrano, N., Sakamoto, H., Appella, E. and Gronenborn, A. M., 1994. High-resolution structure of the oligomerization domain of p53 by multidimensional NMR. *Science* 265: 386.

De'rijard, B., Hibi, M., Wu, I.-H., Barrett, T., Su, B., Deng, T., Karin, M. and Davis, R. J., 1994. JNK1: a protein kinase stimulated by UV light and Ha-Ras that binds and phosphorylates the c-Jun activation domain. *Cell* 76: 1025.

El-Deiry, W. S., Tokino, T., Velculescu, V. E., Levy, D. B., Parsons, R., Trent, J. M., Lin, D., Mercer, W. E., Kinzler, K. W. and Vogelstein, B., 1993. *WAF1*, a potential mediator of p53 tumor suppression. *Cell* 75: 817.

El-Deiry, W. S., Harper, J. W., O'Connor, P. M., Velculescu, V. E., Canman, C. E., Jackman, J., Pietenpol, J. A., Burrell, M., Hill, D. E., Wang, Y., Wiman, K. G., Mercer, W. E., Kastan, M. B., Kohn, K. W., Elledge, S. J., Kinzler, K. W. and Vogelstain B., 1994. *WAF1/CIP1* is induced in p53-mediated G_1 arrest and apoptosis. *Cancer Res.* 54: 1169.

Farmer, G., Bargonetti, J., Zhu, H., Friedman, P., Prywes, R. and Prives, C., 1992 Wild-type p53 activates transcription *in vitro*. *Nature* 358: 83.

Fields, S. and Jang, S. K., 1990. Presence of a potent transcription activating sequence in the p53 protein. *Science* 249: 1046.

Fiscella, M., Zambrano, N., Ullrich, S. J., Ungar, T., Lin, D., Cho, B., Mercer, W. E., Anderson, C. W. and Appella, E., 1994. The carboxy-terminal serine 392 phosphorylation site of human p53 is not required for wild-type activities. *Oncogene* 9: 3249.

Funk, W. D., Pak, D. T., Karas, R. H., Wright, W. E. and Shay, J. W., 1992. A transcriptionally active DNA-binding site for human p53 protein complexes. *Mol. Cell. Biol.* 12: 2866.

Greenblatt, M. S., Bennett, W. P., Hollstein, M. and Harris, C. C., 1994. Mutations in the p53 tumor suppressor gene: clues to cancer etiology and molecular pathogenesis. *Cancer Res.* 54: 4855.

Halazonetis, T. D., Davis, L. J. and Kandil, A. N., 1993. Wild-type p53 adopts a 'mutant'-like conformation when bound to DNA. *EMBO J.* 12: 1021.

Halazonetis, T. D. and Kandil, A. N., 1993. Conformational shifts propagate from the oligomerization domain of p53 to its tetrameric DNA binding domain and restore DNA binding to select p53 mutants. *EMBO J.* 12: 5057.

Harrington, R. E. and Winicov, I., 1994. New concepts in protein-DNA recognition: sequence-directed DNA bending and flexibility. *Prog .Nucleic Acid Res. Mol. Biol.* 47: 195.

Hojo, H. and Aimoto, S., 1992. Protein synthesis using *S*-alkyl thioester of partially protected peptide segments. synthesis of DNA-binding protein of *Bacillus sterothermophilus*. *Bull. Chem. Soc. Jpn.* 65: 3055.

Hollstein, M., Sidransky, D., Vogelstein, B. and Harris, C. C. 1991. p53 mutations in human cancers. *Science* 253: 49.

Hunter, T., 1993. Braking the cycle. *Cell* 75: 839.

Hunter, T., and Karin, M., 1992. The regulation of transcription by phosphorylation. *Cell* 70: 375.

Hupp, T. R., Meek, D. W., Midgley, C. A. and Lane, D. P., 1992. Regulation of the specific DNA binding function of p53. *Cell* 71: 875.

Jackson, S. P., 1992. Regulating transcription factor activity by phosphorylation. *Trends Cell Biol.* 2: 104.

Johnson, Jr., W. C., 1988. Secondary structure of proteins through circular dichroism spectroscopy. *Annu. Rev. Biophys. Biochem.* 17: 145.

Kastan, M. B., Onyekwere, O., Sidransky, D., Vogelstein, B. and Craig, R. W., 1991. Participation of p53 protein in cellular response to DNA damage. *Cancer Res.* 51: 6304.

Kastan, M. B., Zhan, Q., El-Deiry, W. S., Carrier, F., Jacks, T., Walsh, W. V., Plunkett, B. S., Vogelstein, B. and Fornace, Jr., A. J., 1992. A mammalian cell cycle checkpoint pathway utilizing p53 and *GADD45* is defective in ataxia-telangiectasia. *Cell* 71: 587.

Kern, S. E., Pietenpol, J. A., Thiagalingam, S., Seymour, A., Kinzler, K. W. and Vogelstein, B., 1992. Oncogenic forms of p53 inhibit p53-regulated gene expression. *Science* 256: 827.

Kuerbitz, S. J., Plunkett, B. S., Walsh, W. V. and Kastan, M. B., 1992. Wild-type p53 is a cell cycle checkpoint determinant following irradiation. *Proc. Natl. Acad. Sci. USA* 89: 7491.

Lamb, P. and Crawford, L., 1986. Characterization of the human p53 gene. *Mol Cell Biol* 6: 1379.

Lees-Miller, S. P., Sakaguchi, K., Ullrich, S. J., Appella, E. and Anderson, C. W., 1992. Human DNA-activated protein kinase phosphorylates serines 15 and 37 in the amino-terminal transactivation domain of human p53. *Mol Cell Biol* 12: 5041.

Levine, A. J., 1993. The tumor suppressor genes. *Annu. Rev. Biochem.* 62: 623.

Lu, X. and Lane, D. P., 1993. Differential induction of transcriptionally active p53 following UV or ionizing radiation: defects in chromosome instability syndromes? *Cell* 75: 765.

Maltzman, W. and Czyzyk, L., 1984. UV irradiation stimulates levels of p53 cellular tumor antigen in nontransformed mouse cells. *Mol. Cell. Biol.* 4: 1689.

Meek, D., 1994. Post-translational modification of p53. *Semin. Cancer Biol.* 5: 203.

Meek, D. W., Simon, S., Kikkawa, U. and Eckhart, W., 1990. The p53 tumor suppressor protein is phosphorylated at serine 389 by casein kinase II. *EMBO J.* 9: 3253.

Mercer, W. E., Shields, M. T., Amin, M., Sauve, G. J., Appella, E., Romano, J. W. and Ullrich, S. J., 1990. Negative growth regulation in a glioblastoma tumor cell line that conditionally expresses human wild-type p53. *Proc. Natl. Acad. Sci. USA* 87: 6166.

Milne, D. M., Palmer, R. H., Campbell, D. G. and Meek, D. W., 1992. Phosphorylation of the p53 tumor-suppressor protein at 3 N-terminal sites by a novel casein kinase I-like enzyme. *Oncogene* 7: 1361.

Milne, D. M., Campbell, D. G., Caudwell, F. B. and Meek, D. W., 1994. Phosphorylation of the tumor suppressor protein p53 by mitogen-activated protein kinases. *J. Biol. Chem.* 269: 9253.

Milner, J., and Medcalf, E. A., 1991. Cotranslation of activated mutant p53 with wild-type drives the wild-type p53 protein into the mutant p53 conformation. *Cell* 65: 765.

Murray, A. W., 1992. Creative blocks: cell-cycle checkpoints and feedback controls. *Nature* 359: 599.

Nelson, W. G. and Kastan, M. B., 1994. DNA strand breaks: the DNA template alterations that trigger p53-dependent DNA damage response pathways. *Mol. Cell. Biol.* 14: 1815.

Oberosler, P., Hloch, P., Ramsperger, U. and Stahl, H., 1993. p53-catalyzed annealing of complementary single-stranded nucleic acids. *EMBO J.* 12: 2389.

Oren, M., 1994. Relationship of p53 to the control of apoptotic cell death. *Semin. Cancer Biol.* 5: 221.

Pavletich, N. P., Chambers, K. A. and Pabo, C. O., 1993. The DNA-binding domain of p53 contains the four conserved regions and the major mutation hot spots. *Genes Dev.* 7: 2556.

Pietenpol, J. A., Tokino, T., Thiagalingam, S., El-Deiry, W. S., Kinzler, K. W. and Vogelstein, B., 1994. Sequence-specific transcriptional activation is essential for growth suppression by p53. *Proc. Natl. Acad. Sci. USA* 91: 1998.

Raycroft, L., Wu, H. and Lozano, G., 1990. Transcriptional activation by wild-type but not transforming mutants of the p53 anti-oncogene. *Science* 249: 1049.

Reed, M., Wang, Y., Mayr, G., Anderson, M. E., Schwedes, J. F., and Tegtmeyer, P., 1993. p53 domains: suppression, transformation, and transactivation. *Gene Expression* 3: 95.

Sakamoto, H., Lewis, M. S., Kodama, H., Appella, E., and Sakaguchi, K., 1994. Specific sequences from the carboxy-terminus of human p53 form anti-parallel tetramers in solution. *Proc. Natl. Acad. Sci. USA* 91: 8974.

Soussi, T., Caron de Fromentel, C. and May, P., 1990 Structural aspects of the p53 protein in relation to gene evolution. *Oncogene* 5:945.

Stenger, J. E., Mayr, G. A., Mann, K. and Tegtmeyer, P., 1992. p53 forms stable homotetramers and multiples of tetramers. *Mol. Carcinog.* 5: 102.

Ullrich, S. J., Anderson, C. W., Mercer, W. E. and Appella, E., 1992. The p53 tumor suppressor protein, a modulator of cell proliferation. *J. Biol. Chem.* 267: 15259.

Ullrich, S. J., Sakaguchi, K., Lees-Miller, S. P., Fiscella, M., Mercer, W. E., Anderson, C. W. and Appella, E., 1993. Phosphorylation at serine 15 and 392 in mutant p53s from human tumors is altered compared to wild-type p53. *Proc. Natl. Acad. Sci. USA* 90: 5954.

Unger, T., Nau, M. M., Segal, S. & Minna, J.D., 1992. p53: a transdominant regulator of transcription whose function is ablated by mutations occurring in human cancer. *EMBO J.* 11: 1383.

Wang, Y., and Eckhart, W., 1992. Phosphorylation sites in the amino-terminal region of mouse p53. *Proc. Natl. Acad. Sci. USA* 89: 4231.

Wang, Y., Reed, M., Wang, P., Stenger, J. E., Mayr, G., Anderson, M. E., Schwedes, J. F. and Tegtmeyer,P., 1993. p53 domains: identification and characterization of two autonomous DNA-binding regions. *Genes Dev.* 7: 2575.

Wang, Y., Reed, M., Wang, Y., Mayr, G., Stenger, J. E., Anderson, M. E., Schwedes, J. F. and Tegtmeyer, P., 1994. p53 domains: structure, oligomerization, and transformation. *Mol. Cell. Biol.* 14: 5182.

Weinert, T. and Lydall, D., 1993. Cell cycle checkpoints, genetic instability and cancer. *Semin. Cancer Biol.* 4: 129.

Xiong, Y., Hannon, G. J., Zhang, H., Casso, D., Kobayashi, R. and Beach, D., 1993. p21 is a universal inhibitor of cyclin kinases. *Nature* 366: 701.

Zambetti, G. P., Bargonetti, J., Walker, K., Prives, C. and Levine, A. J., 1992. Wild-type p53 mediates positive regulation of gene expression through a specific DNA sequence element. *Genes Dev.* 6: 1143.

DISTINCTIVE CLASS RELATIONSHIPS WITHIN VERTEBRATE ALCOHOL DEHYDROGENASES

Lars Hjelmqvist, Mats Estonius, and Hans Jörnvall

Department of Medical Biochemistry and Biophysics
Karolinska Institutet
S-171 77 Stockholm, Sweden

INTRODUCTION

Mammalian alcohol dehydrogenases (ADH) constitute a well-studied enzyme system composed of sub-forms at different levels of multiplicity. The family has diverged into a number of different enzymes. At the next level (Fig. 1), fairly different forms ("classes") of alcohol dehydrogenase, with distinct structural and enzymatic properties, occur. The subsequent level constitutes still more similar forms ("isozymes") with gradual differences in properties and fewer residue exchanges.

At the class level, five different classes have thus far been characterized in humans. Considering also other mammals, the number of classes in mammalian alcohol dehydrogenases appears to be minimally six (Jörnvall and Höög, 1995). Knowledge is by far most extensive for the class I, III, and IV enzymes, class I being the classical liver alcohol dehydrogenase with considerable ethanol dehydrogenase activity, class III the ubiquitous glutathione-dependent formaldehyde dehydrogenase, and class IV the stomach enzyme with high ethanol and retinol dehydrogenase activity (Vallee and Bazzone, 1983; Koivusalo et al., 1989; Yang et al., 1993; Parés et al., 1994).

CLASSES: DISTINCT PROPERTIES

Distinct properties of the classes, forming more or less separate enzymes, are summarized in Table 1 regarding the three classes thus far best characterized, I, III, and IV. The distinctions concern both functional and structural properties, as well as molecular building units, origins and expressions. Class I is the classical liver alcohol dehydrogenase, a major enzyme in metabolism of ingested ethanol, for which methylpyrazole is a strong competitive inhibitor. Class III is identical to glutathione-dependent formaldehyde dehydrogenase, has a limited ethanol dehydrogenase activity only detectable at high substrate concentrations and with virtually no sensitivity to methylpyrazole. Class IV has a limited

Methods in Protein Structure Analysis, Edited by M. Z. Atassi and E. Appella
Plenum Press, New York, 1995

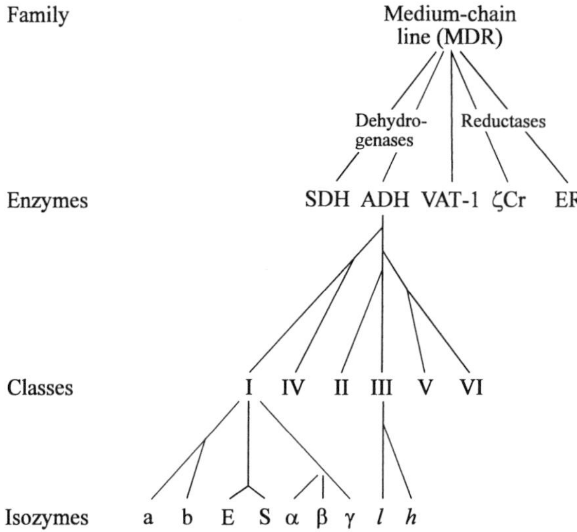

Figure 1. Different levels of multiplicity representing separate stages of gene duplications. Branch points and relative branch lengths at the class and isozyme levels are as presently estimated from available structures, where especially order of interconnections of classes II, V, and VI are still considered tentative in detail because of the limited number of such structures known. MDR, medium-chain dehydrogenases/reductases. SDH, sorbitol dehydrogenase. ADH, alcohol dehydrogenase. VAT-1, protein in synaptic vesicles of *Torpedo* electric organs. ζCr, ζ-crystallin. ER, enoyl reductase of fatty acid synthase. I-VI, classes of ADH. Bottom letters indicate isozyme subunits of *Uromastix* class I ADH (a,b). horse ADH I (E,S), human ADH I (α,β,γ), and cod ADH III (*h*,*l*).

organ distribution (epithelia with emphasis on the stomach) and constitutes the class with highest ethanol activity, but also with considerable retinol dehydrogenase activity.

Structurally, the classes of alcohol dehydrogenase are clearly related and of similar overall conformation as deduced both by modelling (Eklund et al., 1990) and recent crystallographic investigations (cf. El-Ahmad et al., 1995). Nevertheless, also structurally, they exhibit distinct properties. Rates of structural change are different (cf. below), with an approximately 3-fold difference between the "constant" class III enzyme and the "variable" class I enzyme, and with the class IV form in between (Kaiser et al., 1989; Farrés et al., 1994). Also internally, the proteins differ in variable segments and their positions (cf.

Table 1. Properties of classes I, III, and IV alcohol dehydrogenases, illustrating distinct differences

	I	III	IV
Active site	EtOH/MePyr	HCHO-GSH	EtOH/$k_{cat}\uparrow$
Overall structure	Variable	Constant	Intermediate
Molecular segments	$V1\text{-}3_I$	$V1\text{-}2_{III}$	
Origin	From III (subpiscine)	Ancestral (subvertebrate)	Common with I (subamphibian)
Expression	Liver	Ubiquitous	Epithelial

$V1\text{-}3_I$ indicate variable segments 1-3 in the class I enzyme, $V1\text{-}2_{III}$ those in the class III enzyme (Danielsson et al., 1994a). MePyr denotes inhibition by methylpyrazole, EtOH denotes ethanol, and GSH glutathione. $k_{cat}\uparrow$ indicates a high k_{cat} with ethanol.

Patterns). The classes all derive from gene duplications (below, cf. Jörnvall et al., 1995). Finally, expression sites (Table 1) and amounts differ. In short, the classes resemble different enzymes. Class I and III can even have different EC numbers. Considering the long-chain alcohol dehydrogenase activity, both are EC 1.1.1.1, but considering the more separate activities, the ethanol-active class I enzyme is still EC 1.1.1.1, while the formaldehyde-active class III is EC 1.2.1.1. They would hardly have been initially considered as merely classes (rather than enzymes) had they not had a few substrates in common, in particular long-chain alcohols, most easily detectable through a common octanol dehydrogenase activity (cf. Danielsson et al., 1994a).

ORIGIN: GENE DUPLICATIONS

As with other protein families, the common structures derive from a set of gene duplications correlated with subsequent mutational events. Class III is the parent form, present in more or less unaltered form through prokaryotes, yeasts, plants, insects and other invertebrates, and vertebrates. Details of formation regarding the other classes are thus far limited, but available evidence suggests that the class I line was the first to branch off, at early vertebrate times, as suggested both by evaluation of evolutionary speed (Cederlund et al., 1991), absence in lines originating before bony fish (Danielsson et al., 1994b) and the presence of a class I/III-mixed form in the line where class I first occurs (bony fish, Danielsson and Jörnvall, 1992). Similarly, calculation of phylogenetic trees suggests that class IV has a common origin with class I, at a somewhat later stage although still fairly early in vertebrate evolution as evidenced by its presence also in amphibians (unpublished together with Parés et al.). In summary, the class system appears to accompany the evolution of vertebrates, through a series of gene duplications starting early and with class III as the ancestral form.

PATTERNS: TYPICAL AND ATYPICAL PROPERTIES

In many respects, class I and class III illustrate different principles, class III being a "typical" protein of basic metabolism, and class I an "atypical" protein with unexpected properties. This is clearly visible in both the rate and positions of evolutionary changes.

Regarding rate, class III is "constant", which means a rate closely identical to that observed by glycolytic enzymes like glyceraldehyde-3-phosphate dehydrogenase and enolase (Danielsson et al., 1994a). Similarly, its segments of maximal variability affects two superficial regions (called 1 and 2 in the right panel of Fig. 2), both situated at the surface away from the entrance to the active site. This pattern, with little variability overall, and with those regions that do differ situated away from the active site, is the pattern expected for a basic enzyme of central importance in cellular metabolism, and hence a property common to many proteins in general.

In contrast, the class I enzyme has a faster evolutionary rate overall, and exhibits three variable segments, all affecting functionally important parts. Thus, one of the variable segments (1 in the left panel of Fig. 2) affects part of the entrance to the active site. Another (2 in Fig. 2, left) affects the segment around one of the zinc atoms of this metalloenzyme, and the third (3 in Fig. 2, left) affects the major area of subunit interactions. Therefore, the class I enzyme exhibits an atypical pattern, with variability at functionally important segments, and with few strictly conserved properties.

Remaining classes are at present difficult to judge in this detail, since they are thus far only known from single species, whereas class I and III both are established in close to

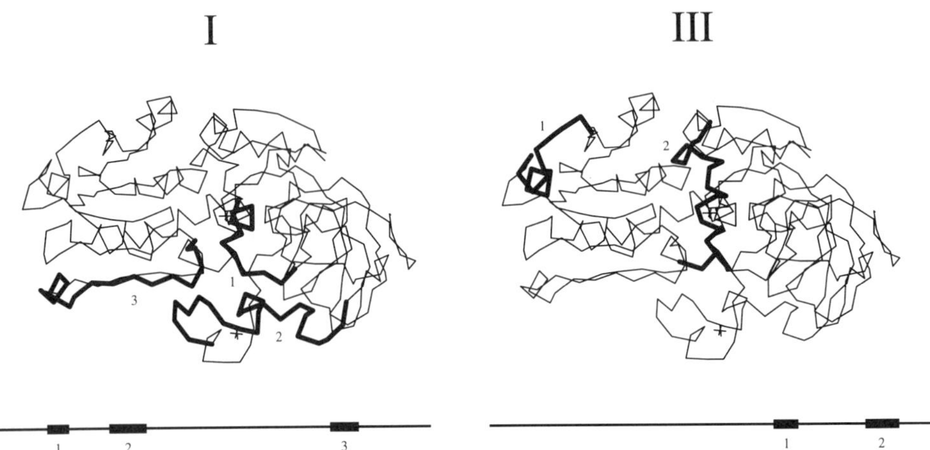

Figure 2. Fundamental differences in molecular patterns between the class I and III alcohol dehydrogenases. Top models show the subunit conformations and bottom lines the primary structures, in both cases with the most variable segments (species variants) represented by thick lines numbered to correspond to $V1\text{-}3_I$ and $V1\text{-}2_{III}$ in Table 1, respectively, and remaining parts by thin lines. As noted, the variability patterns differ, affecting superficial and "non-functional" sites in class III, as typical of highly conserved proteins in basic functions, but functional sites in class I of an atypical, non-conserved pattern, interpreted to suggest emergence of novel functions and interactions.

twenty species, showing the properties as stated. However, recent evaluation of species divergence in class IV (from the rat and human proteins) shows that class IV is likely to occupy a position in between the constant class III and variable class I forms (Farrés et al., 1994). Similarly, preliminary data on the variability of remaining classes appear to suggest that at least some of them are still more variable. Consequently, the alcohol dehydrogenase system encompasses an impressive set of enzymes ranging from constant, typical, basic enzymes to variable forms, with in addition intermediately variable as well as possibly super-variable forms. Nevertheless, in spite of this inter-class variability, these properties within each class are kept largely in the same manner over long time periods, identifying also the evolutionary variability as a dinstinct property in each case.

CLASS-MIXED PROPERTIES: EMERGENCE OF NOVEL FORMS

The alcohol dehydrogenase system with its multiplicity of forms derived through gene duplications at several levels also offers examples of what appears to be emergence of novel forms. Thus far, two such occasions have been discerned at the class level. In these cases, the novel form originating through a duplication and evolving via subsequent mutations, appears to have acquired mutations to get novel class-distinct properties but still not a sufficient number of mutations to loose the derivation from its original class. The first such example in this family was described in the form of the ethanol-active enzyme from bony fish, represented by the analysis of the cod enzyme (Danielsson and Jörnvall, 1992). Thus, this protein, as summarized in Table 2, is overall structurally more related to the class III enzyme than to the class I enzyme, as shown by its relationships to these two human proteins (64% identity versus 55%), yet its enzymatic properties toward ethanol are like a class I enzyme (cf. K_m-values, Table 2). Thus, it appears as if changes at active site residues have given rise to class I properties, while the origin from class III is still visible by the

Table 2. Class-mixed properties of a piscine enzyme (class I/III mixed properties) and an avian enzyme (class I/II mixed properties)

"Hybrid enzyme"	Compared with	
	Human I	Human III
Cod I		
Structure	55%	**64%**
Function (K_m, ethanol)	**1.2/1.1**	1.2/NS
	Human I	Human II
Ostrich II		
Structure	61%	**69%**
Function (K_m, ethanol)	**0.7/1.1**	0.7/120

The cod enzyme is called Cod I, denoting the class nomenclature after its functional assignment, while the ostrich enzyme is called Ostrich II after its structural assignment, since there is also another, true class I enzyme in the ostrich (and other avian species). As shown by the bold values, the two enzymes are structurally most closely related to one class and functionally to another. Values in the structure lines refer to residue identities with the classes of the human enzymes, while those in the function lines refer to K_m values (in mM) with ethanol as substrate. NS, non-saturable.

overall, remaining residue identities with that class. It therefore appears as if we even now, in much later, divergent lines, can still observe the emergence or enzymogenesis (Danielsson and Jörnvall, 1992) of the novel enzyme type (class I) from the parent form (class III).

Recently, yet unpublished, we have detected the same overall pattern in relation to the duplication giving rise to class II. In this case, the class-mixed form was detected in the avian line, more closely in a ratite liver, as a class II (structurally) protein from ostrich (Hjelmqvist et al., unpublished). Also here, the protein has acquired class I ethanol dehydrogenase activity, but has an overall structural relationship to class II (Table 2). In this case, since the origin of class II is not well established (cf. legend Fig. 1), the exact branching points are thus far unknown, but independent of order of events, the ostrich enzyme has class-mixed properties (Table 2).

In conclusion, the classes are distinct with separate properties over long periods of time, but with emergence of novel forms post-duplicationary, giving rise to enzymogenesis of new activity while keeping traces of the parent form. Furthermore, the class-mixed properties, once locked in an animal line, appear to survive until present times, although of distant origin, as exemplified by the present day cod enzyme reflecting the early vertebrate class I/III duplication. It is to be expected that further elucidation of all the mammalian and human alcohol dehydrogenase classes and their origins may trace still further acquirements of novel functions.

ISOZYME DIVERGENCE: MULTIPLE EVENTS

To some extent, also the development of isozymes establishes the emergence of mixed properties: the isozymes, being of more recent duplicatory origin than the classes (Fig. 1), have kept many substrates in common (hence are true isozymes), but still have acquired novel substrate distinctions. This was early noticed in for example the horse E (for ethanol-active) and S (for steroid-active) liver alcohol dehydrogenases. Furthermore, the three class I lines where isozymes have thus far been deteceted, humans and other primates, horse, and

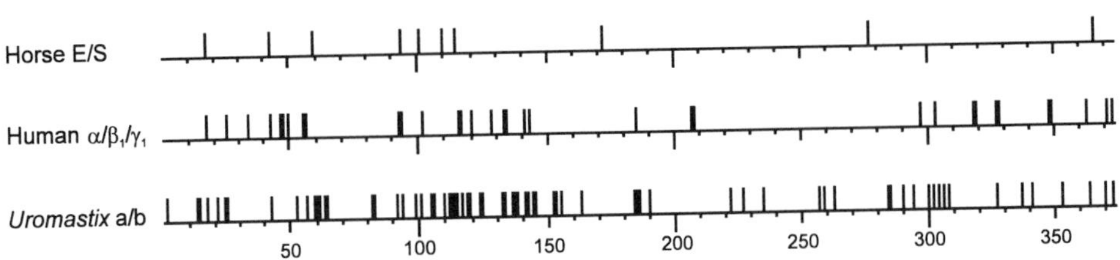

Figure 3. Positions of residue differences between isozyme subunits of the three species with characterized class I alcohol dehydrogenase isozyme differences. As shown, 30% of the differences in the horse enzyme (at positions 17, 43 and 94) coincide exactly with differences at those positions also in the other two isozyme systems.

Uromastix lizard, all have the corresponding gene duplications at separate positions in the phylogenetic tree (Jörnvall et al., 1995), suggesting that isozyme formation has occurred repeatedly during vertebrate evolution through multiple duplicatory events, some fairly recent (like in the primate line), some more distant (like in the lizard line). In spite of the multiple events, it is noteworthy that positions affected by the isozyme differences are in part identical in the different lines. Thus, no less than three of the totally only ten positions with mutational differences between the isozyme subunits in the horse enzyme also are affected by isozyme differences in the human and *Uromastix* enzymes (Fig. 3). This coincidence again suggests that some regions are especially variable and that this is noticable also at the isozyme level.

Also in class III, isozymes have recently been detected (Danielsson and Jörnvall, 1992). They differ in specific activity (composed of *h* and *l* chains, from high and low activity forms, respectively) again illustrating apparent emergence of novel properties.

In conclusion, isozyme development has occurred repeatedly, establishing that the gene duplications have multiple origins with known, emerging isozyme patterns in different lineages.

COMMON PROPERTIES

In view of the many distinct properties established within both the isozyme and class development patterns, it should be stressed, however, that the overall conformational properties of the enzymes have been kept over wide time periods. This is visible already in the residue conservation pattern of alcohol dehydrogenases: glycine is by far the most conserved residue in the MDR family, as illustrated by the three columns in Table 3, taken from summaries at three different times: 1977, when just two alcohol dehydrogenases (horse and yeast) were known (first column); in 1993, when five different enzyme lines were compared (second column); and in 1994, when ten-odd dehydrogenases/reductases were combined within the MDR family (third column). In all cases, and clear already at the first comparison, the conserved nature of glycine stands out. Apart from illustrating the value of comparisons at any stage, these facts establish the unique nature of the small (without side-chain) glycine residues.

As shown in the bottom figure insert of Table 3, the conserved glycine positions largely correspond to reverse turns in the conformation of mammalian alcohol dehydrogenase class I, establishing that these turns, and hence the overall conformation, is largely conserved. Of course, as long as conformation and functional propeties are conserved, the glycine conservation is expected to be observed in a protein family.

Table 3. Conserved residues, illustrating the excess of glycine among such residues (top columns) and their conformational positions

Amino acid residue	Horse ADH + yeast ADH	ADH + SDH + TDH + XDH + Zcr	All MDR proteins
Gly	20	13	3
Val	9	2	—
Ala	9	—	—
Lys	6	2	—
Asp	6	1	—
Cys	6	—	—
Glu	5	1	—
Leu	5	—	—
Pro	3	1	—
Ile	3	—	—
Arg	2	—	—
His	2	—	—
Phe	2	—	—
Ser	2	—	—
Thr	2	—	—
Tyr	2	—	—
Asn	1	—	—
Gln	1	—	—
Sum	86	20	3
Gly/sum	20/86 (23%)	13/20 (65%)	3/3 (100%)

The columns are taken from three previous comparisons of the family as discerned early (left), 1993 (middle) and now (right, cf. Persson et al., 1994). Black indicates the positions of the three strictly conserved residues in the rightmost column, dense and light stippling remaining glycine residues in the middle and left columns, respectively.

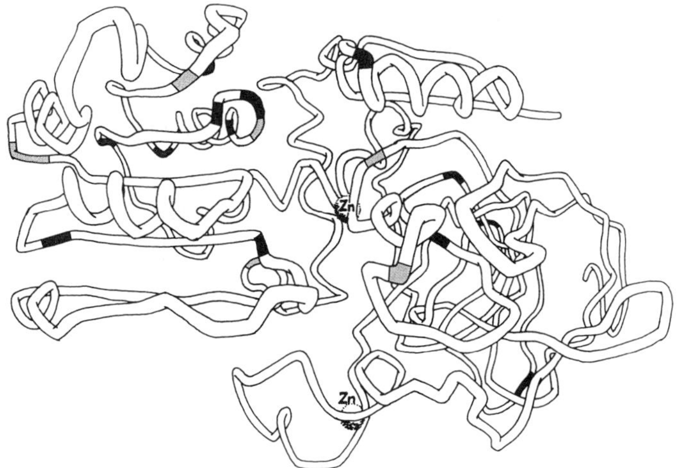

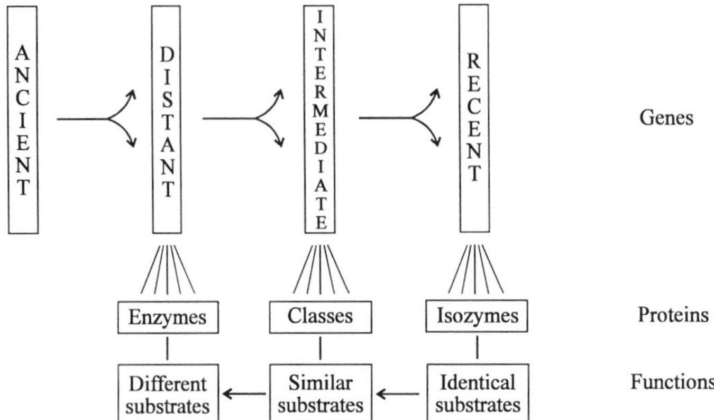

Figure 4. Evolutionary pattern illustrating the continuity and time dependence on accumulation of mutational differences in the development of novel functions.

CONCLUSION

Combined, the different structures, related through three series of gene duplications as illustrated in Fig. 1, may be interpreted to relate isozymes, classes, and enzymes in a continuous flow of structural divergence and functional emergence. Thus, as given in Fig. 4, repeated gene duplications, starting at distant times and progressing all through recent times have given rise to multiple forms which, through further mutational events, are now traceable as present-day enzymes, classes, and isozymes, respectively. The patterns of origin of enzymes, classes and isozymes are highly similar as revealed by the middle portion of Fig. 4, the major difference being essentially the time of emergence, and hence time of accumulation of subsequent mutations. The most recently originating forms are still fairly similar in function, the most distantly ones most different, with the intermediate ones just similar. In this manner, the pattern as discerned from present-day structures and phylogenetic constructions reflect successive changes and functional divergence through time. This is well observed in the MDR enzymes of the alcohol dehydrogenase type which nicely illustrate the pattern also discernible in several other well-studied protein families. Through these observations on native forms, exact positions governing class distinctions and isozyme distinction, may now be further scrutinized by site-directed mutagenesis in tests to confirm the critical roles of particular positions in class and isozyme functions. This has, in the case of alcohol dehydrogenases, established roles of positions 57 and 115 in the class I/III distinctions (Engeland et al., 1993; Estonius et al., 1994), as well as of position 48 in the human isozyme distinctions (Höög et al., 1992), illustrating the value of confirmation by mutagenesis of patterns traced by observations of native forms.

ACKNOWLEDGMENT

This work was supported by grants from the Swedish Medical Research Council (13X-3532), Karolinska Institutet, and Peptech (Australia).

REFERENCES

Cederlund, E., Peralba, J. M., Parés, X., and Jörnvall, H., 1991, Amphibian alcohol dehydrogenase, the major frog liver enzyme. Relationships to other forms and assessment of an early gene duplication separating vertebrate class I and class III alcohol dehydrogenases, *Biochemistry* 30:2811-2816.

Danielsson, O. and Jörnvall, H., 1992, "Enzymogenesis": Classical liver alcohol dehydrogenase origin from the glutathione-dependent formaldehyde dehydrogenase line. *Proc. Natl. Acad. Sci. USA* 89:9247-9251.

Danielsson, O., Atrian, S., Luque, T., Hjelmqvist, L., Gonzàlez-Duarte, R. and Jörnvall, H., 1994a, Fundamental molecular differences between alcohol dehydrogenase classes. *Proc. Natl. Acad. Sci. USA* 91:4980-4984.

Danielsson, O., Shafqat, J., Estonius, M., and Jörnvall, H., 1994b, Alcohol dehydrogenase class III contrasted to class I. Characterization of the cyclostome enzyme, existence of multiple forms as for the human enzyme, and distant cross-species hybridization, *Eur. J. Biochem.* 225:1081-1088.

Eklund, H., Müller-Wille, P., Horjales, E., Futer, O., Holmquist, B., Vallee, B. L., Höög, J.-O., Kaiser, R., and Jörnvall, H., 1990, Comparison of three classes of human liver alcohol dehydrogenase. Emphasis on different substrate-binding pockets, *Eur. J. Biochem.* 193:303-310.

El-Ahmad, M., Ramaswamy, S., Danielsson, O., Karlsson, C., Höög, J.-O., Eklund, H., and Jörnvall, H., 1995, Crystallizations of novel forms of alcohol dehydrogenase, *in*: Enzymology and Molecular Biology of Carbonyl Metabolism 5, Weiner, H., Holmes, R, and Flynn, T. G., eds, Plenum, New York, pp. 365-371.

Engeland, H., Höög, J.-O., Holmquist, B., Estonius, M., Jörnvall, H., and Vallee, B. L., 1993, Mutation of Arg-115 of human class III alcohol dehydrogenase, a binding site required for formaldehyde dehydrogenase activity and fatty acid activation, *Proc. Natl. Acad. Sci. USA* 90:2491-2494.

Estonius, M., Höög, J.-O., Danielsson, O., and Jörnvall, H., 1994, Residues specific for class III alcohol dehydrogenase. Site-directed mutagenesis of the human enzyme. *Biochemistry* 33:15080-15085.

Farrés, J., Moreno, A., Crosas, B., Peralba, J.M., Allali-Hassani, A., Hjelmqvist, L., Jörnvall, H., and Parés, X., 1994, Alcohol dehydrogenase of class IV ($\sigma\sigma$-ADH) from human stomach. cDNA sequence and structure/function relationships. *Eur. J. Biochem.* 224:549-557.

Höög, J.-O., Eklund, H., and Jörnvall, H., 1992, A single-residue exchange gives human recombinant $\beta\beta$ alcohol dehydrogenase $\gamma\gamma$ isozyme properties, *Eur. J. Biochem.* 205:519-526.

Jörnvall, H., Danielsson, O., Hjelmqvist, L., Persson, B., and Shafqat, J., 1995, The alcohol dehydrogenase system, *in*: Enzymology and Molecular Biology of Carbonyl Metabolism 5, Weiner, H., Holmes, R, and Flynn, T. G., eds, Plenum, New York, pp. 281-294.

Jörnvall, H. and Höög, J.-O., 1995, Nomenclature of alcohol dehydrogenases, *Alcohol and Alcoholism*, 30, 153-161.

Kaiser, R., Holmquist, B., Vallee, B. L., and Jörnvall, H., 1989, Characterization of mammalian class III alcohol dehydrogenases, an enzyme less variable than the traditional liver enzyme of class I, *Biochemistry* 28:8432-8438.

Koivusalo, M., Baumann, M., and Uotila, L., 1989, Evidence for the identity of glutathione-dependent formaldehyde dehydrogenase and class III alcohol dehydrogenase, *FEBS Lett.* 257:105-109.

Parés, X., Cederlund, E., Moreno, A., Hjelmqvist, L., Farrés, J., and Jörnvall, H., 1994, Mammalian class IV alcohol dehydrogenase (stomach ADH): structure, origin and correlation with enzymology. *Proc. Natl. Acad. Sci. USA* 91:1893-1897.

Persson, B., Zigler, J. S. Jr, and Jörnvall, H., 1994, A super-family of medium-chain dehydrogenases/reductases (MDR). *Eur. J. Biochem.* 226:15-22.

Vallee, B. L. and Bazzone, T. J., 1983, Isozymes of human liver alcohol dehydrogenase, *Isozymes* 8:219-244.

Yang, Z. N., Davis, G. J., Hurley, T. D., Stone, C. L., Li, T.-K., and Bosron, W. F., 1993, Catalytic efficiency of human alcohol dehydrogenase for retinol oxidation and retinol reduction, *Alcohol. Clin. Exp. Res.* 17:496.

38

PRIMARY STRUCTURE OF A NOVEL STYLAR RNASE UNASSOCIATED WITH SELF-INCOMPATIBILITY IN A TOBACCO PLANT, *NICOTIANA ALATA*

S. Kuroda,[1] S. Norioka,[1] M. Mitta,[2] I. Kato,[2] and F. Sakiyama[1]

[1] Institute for Protein Research
Osaka University
Suita, Osaka 565, Japan
[2] Biotechnology Research Laboratory
Takara Shuzo Co.
Otsu, Shiga 520-21, Japan

INTRODUCTION

Self-incompatiblity is a system for segregating self pollen from non-self pollen in the female organ of flowering plants (de Nettancourt, 1977). This system is not caused by malfunction of reproductive organs since self-incompatibility does not appear when the pistil accepts pollen with different S-alleles. *Nicotiana alata* has gametophytic self-incompatiblity in which pollen bearing the same S-allele as one of those in the pistil is rejected by arresting pollen tube growth. This rejection takes place in the style where S-allele specific glycoproteins (S-glycoproteins) responsible possibly for both segregation of self and non-self pollens and arrest of pollen tube growth are synthesized prior to anthesis (Anderson et al.,1986). Recently, we have found that S-glycoproteins associated with self-incompatibility are RNases in the RNase T_2 family (McClure et al.,1989), based on predictions made by amino acid sequence comparisons and chemical modification experiments (Kawata et al.,1990). Subsequently, a number of studies have been reported for the primary structures of S-glycoproteins mainly from *Solanaceae*. These studies have revealed that all the proteins contain two conserved amino acid sequences, including two histidine residues corresponding to those existing in the active site of RNase T_2. According to very recent investigations (Huang et al.,1994; Lee et al.,1994; Murfett et al.,1994), RNase activity is indispensable for the function of S-glycoprotein in petunia and tobacco. However, the presence of this enzyme activity in the style is not sufficient for the appearance of self-incompatibility. In fact, we detected the same RNase activity in the style (with the stigma) of either

Methods in Protein Structure Analysis, Edited by M. Z. Atassi and E. Appella
Plenum Press, New York, 1995

a self-incompatible variant (Nijisseiki) of Japanese pear (*Pyrus pyrifolia*) or its self-compatible mutant (Osa-Nijisseiki). *Arabidopsis thaliana*, a substantially self-compatible plant also synthesizes enzymes of the RNase T_2 type in the style (Taylor and Green, 1991). Moreover, two or more RNases have been detected in the style of heterozygous petunia (Lee et al.,1992) and tobacco (McClure et al.,1989), implying that the style of self-incompatible plants synthesize RNase that is not associated with self-incompatibility. Our investigation was undertaken to characterize the RNase in the style of *N.alata* in order to seek a structure motif(s) specific for S-RNase or non-S-RNase. We then separated three stylar RNases, assigned individual RNases to S-RNase and non-S-RNase by analyzing their appearance in the style during flower development and elucidated the amino acid sequences of these RNases. This paper summarizes the result of some of these experiments and analyses of the structural information for RNase MS1, a novel non-S-RNase.

SEPARATION OF THREE MAJOR RNASES FROM THE STYLE OF *N. Alata*

The extract of the style of *N.alata* was chromatographed on a Mono S column. This yielded three major peaks containing RNase activity which were named MS1, MS2 and MS3 according to elution order (Fig.1).

A single RNase [MSl (29kD), MS2 (31kD) or MS3 (30kD)] was purified by reverse phase HPLC from each of fractions MS1, MS2 and MS3, respectively. The N-terminal and internal amino acid sequences of these RNase were analyzed, revealing that all of them are of the RNase T_2 type and that they are structurally similar to each other.

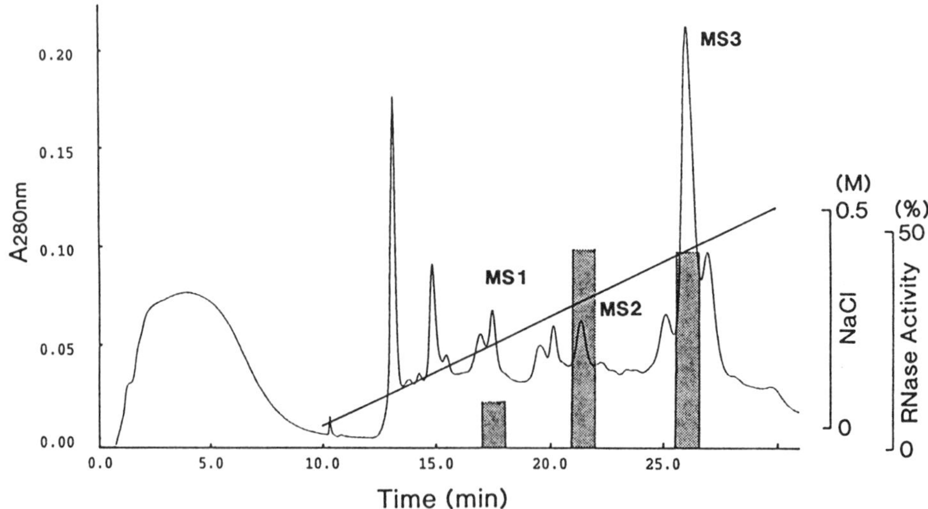

Figure 1. Separation of fractions MS1, MS2 and MS3 by Mono S column chromatography. The proteins precipitated by 40-90% saturation with ammonium sulfate were chromatographed on a Mono S column (1.6 x 50 mm) using a gradient of 0 - 0.5M NaCl in 50 mM sodium acetate buffer, pH 5.0, at a flow rate of 100 µl/min and at 8°C. Shaded columns indicate RNase activity as the percentage of the total RNase activities for fractions MS1, MS2 and MS3. The protein bearing RNase activity in each fraction was eluted as a single peak by reverse phase HPLC, collected and used for sequence analysis.

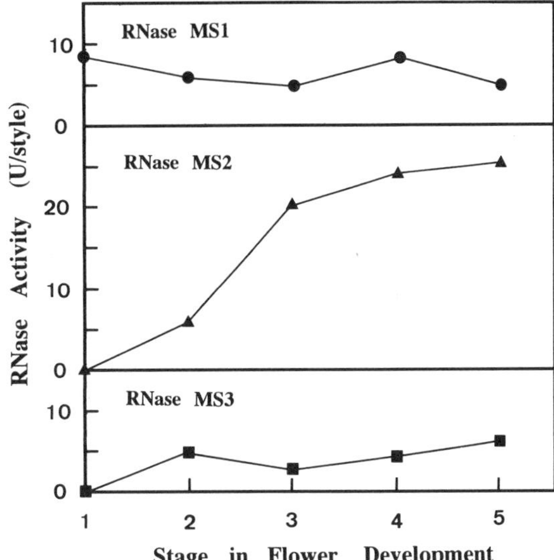

Figure 2. Changes in RNase activity of fractions MS1, MS2 and MS3 during flower development. The extract of five or six styles at each stage was chromatographed on a Mono S column as described in the legend of Fig.1 and each fraction separated was assayed for RNase activity. Lengths of styles at stages 1 to 5 were 0.5-1.5, 1.5-2.0, 2.5-3.5, 5.5-6.5 and 7 cm, respectively. Anthesis took place at stage 5.

ASSIGNMENT OF MS1, MS2 AND MS3 TO S-RNase AND NON-S-RNase

To examine whether these three RNases are associated with self-incompatibility, the appearance of these proteins in the style during flower development was followed by assaying the RNase activity for individual RNase peaks separated by Mono S chromatography (Fig.2).

At stage 1 (green bud around a week before anthesis), RNase activity was clearly detected in the style of *N.alata* and was exclusively associated with fraction MS1. No detectable activity was found in fractions MS2 and MS3. Fraction MS1 retained its enzyme activity at the same level throughout all stages until anthesis. In contrast, the activity of fraction MS2 was detected first at stage 2 (green bud around 5-6 days before anthesis) and then the activity was rapidly increased to a plateau upon anthesis, amounting to about 60% of total RNase activity. RNase activity of fraction MS3 was also detected at stage 2 and its increase in subsequent stages was marginal. Since *N.alata* gains self-incompatibility when S-RNase is present in the style (Anderson et al.,1986), the observed time-dependence of appearance of individual stylar RNases suggests that MS1 is a non-S-RNase (RNase MS1) unassociated with self-incompatibility and both MS2 and MS3 are S-RNases associated with it.

THE AMINO ACID SEQUENCE OF RNase MS1

To elucidate the amino acid sequence of the stylar RNase MS1 purified above, two oligonucleotides corresponding to the conserved sequences involving either one of the two essential histidine residues of RNase T_2 were synthesized and used for RT-PCR with mRNA from a mixture of styles collected from bud and mature flower. Three cDNAs [*ms1*(657 bp), *ms2*(654 bp) and *ms3*(654 bp)] were eventually cloned. The nucleotide sequence of *ms1* was determined and its deduced amino acid sequence was verified by sequencing peptides obtained by digestion with *Achromobacter* protease I, V8 protease and Asp-N of RNase MS1.

This protein is composed of 197 amino acid residues and its pI was calculated as 6.1. There is a potential N-glycosylation site (Asn 27) where PTH-Asn was not identified by Edman degradation. Similarly, the amino acid sequences of MS2- and MS3-RNases were elucidated by sequencing their cloned genes and verifying as described above (data not shown). All three stylar RNases hold several conserved amino acid sequences compared with *Solanaceae* S-RNases and were classified as enzymes in the RNase T_2 family.

AMINO ACID SEQUENCE COMPARISON BETWEEN RNase MS1 AND S-RNases FROM Solanaceae

The amino acid sequence of RNase MS1 was compared with those of known S-glycoproteins from N.*alata* and revealed several interesting observations (Fig.3).

Firstly, the sequence identity between a pair of any two proteins of MS1, MS2 and MS3 is 45-52%, showing that these three RNases are homologous proteins even though RNase MS1 is a non-S-RNase. Secondly, RNase MS1 is highly homologous (95%) to S_a-RNase when compared with the published amino acid sequence in which approximately 20 of the N-terminal residues are not known (Keyrs-Pour et al., 1990). Thirdly, MS2-RNase has a 74% sequence identity with S_3-RNase. Fourthly, MS3-RNase has a sequence identical to S_{F11}-RNase.

As described earlier, sequence homology between RNase MS1 and S_a-RNase is unusually high. It should be possible to detect a structural motif characteristic of S-RNase or non-S-RNase when the above two proteins are compared. The small 5% sequence difference is localized to a short stretch of six consecutive residues at positions 135- 140. If the unknown amino acid sequence of the N-terminal portion of S_a-RNase is the same as that established for RNase MS1, it may be possible to identify an amino acid residue(s) responsible for distinguishing S-RNase from non-S-RNase. Unfortunately, no particular features were detected when we carefully compared the short sequence of RNase MS1,

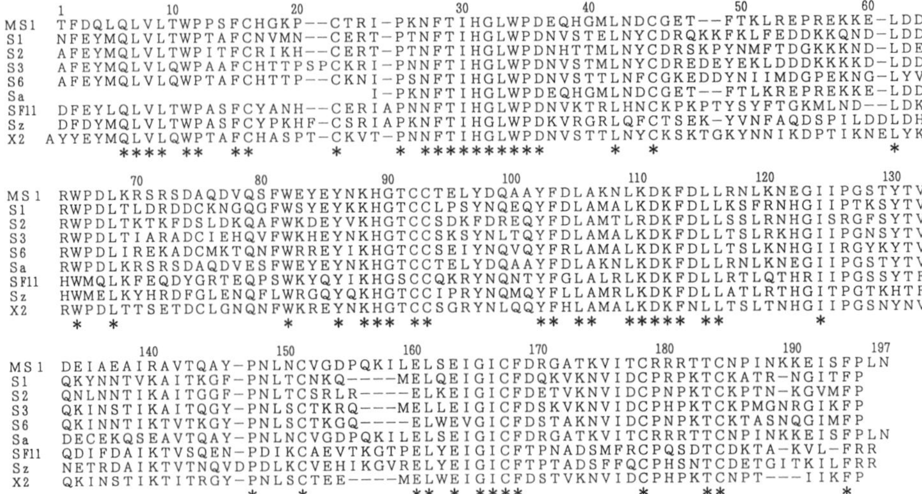

Figure 3. Alignment of the amino acid sequences of RNase MS1, RNase X2 and S-RNases. S_1 - S_Z represent enzymes reported as S-RNases in *N.alata*. X2 is RNase X2 that is not linked to S-locus in *P.inflata*. Conserved amino acid residues among all of listed RNases are marked with asterisks. Amino acid sequences were reported by McClure et al., (1989) (S_2 - S_6), Kheyr-Pour et al., (1990) (S_1, S_a, S_{F11} and S_Z) and Lee et al.,(1992)(X2).

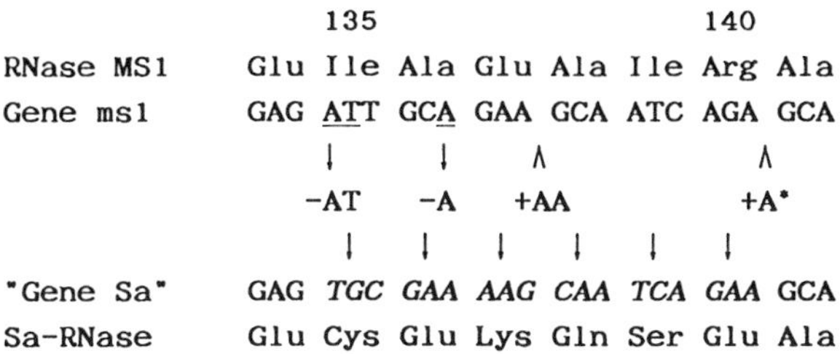

Figure 4. Possible changes in the nucleotide sequences encoding residues 135 - 140 observed between RNase MS1 and S_a-RNase. The putative nucleotide sequence of the "gene S_a", the putative gene encoding S_a-RNase, is shown in italics. Either A or G may be inserted at position marked by asterisk.

Ile-Ala-Glu-Ala-Ile-Arg, with that of its counterpart of S_a-RNase, Cys-Glu-Lys-Gln-Ser-Glu. However, an interesting correlation was found at the nucleotide sequence level (Fig.4).

If the first two base (AT) in the ATT codon for Ile and the third base (A) in the GCA codon for Ala in *ms1* were deleted and instead the AA sequence and A or G were inserted between the GAA (for Glu) and GCA (for Ala) and after the AGA codon for Arg, respectively, the nucleotide sequence coding for the amino acid sequence of RNase MS1 would be revised to code for the counter-sequence of six residues in S_a-RNase. This finding is very interesting because structural differences of these nucleotide sequences may be responsible for functional difference between RNase MS1 and S_a-RNase. However, it is unclear whether such mutations at the extremely limited region of the nucleotide sequence is possible. In any event, the following will be needed to clarify whether this structural difference is responsible for the identity of S-RNase in *N.alata*: verification of the ambiguous amino acid sequence of this limited region at the protein level; elucidation of the as yet unidentified N-terminal sequence and reconfirmation of association with self-incompatibility of S_a-RNase.

Petunia inflata, a self-incompatible plant and a member of the *Solanaceae* family, also produces a non-S-RNase in the style and the enzyme is called RNase X2 (Lee et al.,1992). Sequence identity between RNase MS1 and RNase X2 is 48.7%, which is in the range of average similarity among S-RNases. Moreover, there is no particular local sequence similarity except the common ones for S-RNases and no special structural motif characteristic of non-S-RNase was found by comparing these two non-S-RNases. RNS2 is a stylar RNase in the RNase T_2 family and has been isolated from *Arabidopsis thaliana*, a plant devoid of self-incompatibility (Taylor et al.,1993). RNase MS1 holds 31% sequence identity to this scenesence associated RNase. Again, a particular motif indicating structural relatedness as a non-S-RNase to RNS2 was not found in the primary structure.

CONCLUSION

The present investigation demonstrates that the style of heterozygous *N. alata* contains a novel RNase T_2-type enzyme. The enzyme is thought to be a constitutive protein, indicating that all pistil RNases of RNase T_2 type in self-incompatible plants are not associated with self-incompatibility. It was difficult to distinguish RNase MS1 from coexisting S-RNases based on the primary structure. Search for a structural factor(s) which is responsible for the functional difference between constitutive and S-locus related RNases is under progress in our laboratory.

REFERENCES

Anderson, M.A., Cornish, E.C., Mau, S.L., Williams, E.G., Hoggart, R., Atkison, A., Bonig, I., Grego, B., Simpson, R., Roche, P.J., Haley, J.D., Penschow, J.D., Niall, H.D., Tregear, G.W., Cochran, J.P., Crawford, R.J., and Clarke, A.E., 1986, Cloning of cDNA for a stylar glycoprotein associated with expression of self-incompatibility in *Nicotiana alata*., *Nature* 321: 38-44

de Nettancourt, D., 1977, Incompatibility in Angiosperms, Springer-Verlag, Berlin.

Huang, S.-S., Lee, S.-H., Karunanandaa, B., and Kao, T.-H., 1994, Ribonuclease activity of *Petunia inflata* S proteins is essential for rejection of self-pollen., *Plant Cell* 6: 1021-1028.

Kawata, Y. Sakiyama, F., Hayashi, F., and Kyogoku, Y., 1990, Identification of two essential histidine residues of ribonuclease T_2 from *Aspergillus oryzae*., *Eur.J.Biochem.*, 187: 255-262.

Kheyr-Pour, A., Bintrim, S.B., Ioerger, T.R., Remy, R., Hammond, S.A., and Kao, T.-H., 1990, Sequence diversity of pistil S-proteins associated with gametophytic self-incompatibility in *Nicotiana alata*., *Sex Plant Reprod.*, 3: 88-97.

Lee, H.-S., Singh, A., and Kao, T.-H., 1992, RNase X2, a pistil-specific ribonuclease from *Petunia inflata*, shares sequence similarity with solanaceous S proteins., *Plant Mol.Biol.*, 20: 1131-1141.

Lee, H.-S., Huang, S., and Kao, T.-H., 1994, S proteins control refection of incompatible pollen in *Petunia inflata*, *Nature*, 367: 560-563.

McClure, B.A., Haring, V., Ebert, P.R., Anderson, M.A., Simpson, R.J., Sakiyama, F., and Clarke, A.E., 1989, Style self-incompatibility gene products of *Nicotiana alata* are ribonucleases., *Nature*, 342: 955-957.

Murfett, J., Atherton, T.L., Mou, B., Gasser, C.S., and McClure, M.A., 1994, S-RNase expressed in transgenic *Nicotiana* causes S-allele-specific pollen rofiction., *Nature*, 367: 563-566.

Taylor, C.B. and Green, P., 1991, Genes with homology to fungal and S-gene RNases are expressed in *Arabidopsis thaliana*., *Plant Physiol.*, 96: 980-984.

Taylor, C.B., Bariola, P.A., Delcardayre, S.B., Raines, R.T., and Green P., 1993, RNS2: A senescence-associated RNases of *Arabidopsis* that diverged from the S-RNases before speciation., *Proc.Natl.Acad.Sci.USA*, 90: 5118-5122.

39

HUMAN FIBRINOGEN OCCURS AS OVER 1 MILLION NON-IDENTICAL MOLECULES

Agnes H. Henschen-Edman

Department of Molecular Biology and Biochemistry
University of California, Irvine
Irvine, California 92717-3900

INTRODUCTION

Traditionally, proteins have been regarded as well-defined, uniform molecules, where one molecule is virtually identical to the next. This notion has been supported by the fact that the highly efficient protein primary structure analysis by prediction from the DNA sequence will result in a well-defined, unique amino acid sequence, containing no direct indication of any kind of modification or processing. However, information is accumulating about protein heterogeneity, co- or post-translational modification and processing as well as about the functional implications of the structural variation (Krishna and Wold, 1993; Graves et al., 1994). Human fibrinogen may serve as an extreme example of a protein existing in a multitude of structural forms, many of which have been demonstrated to differ in functional properties (Henschen and McDonagh, 1986; Henschen, 1993). In the following, the various, so far recognized structural variations and their possible function effects, together with some relevant identification procedures will be described.

FIBRINOGEN STRUCTURE

Fibrinogen is a central protein in the blood coagulation system (Henschen and McDonagh, 1986). The fibrinogen molecule is composed of three pairs of non-identical peptide chains denoted Aα, Bβ and γ. The overall structure can thus be described as (Aα, Bβ, γ)$_2$. During blood clotting thrombin proteolytically cleaves two pairs of peptide chains releasing fibrinopeptides A and B to form fibrin monomer with the structure (α, β, γ)$_2$. The fibrin monomer can polymerize in an ordered fashion. The human fibrinogen chains Aα, Bβ and γ contain 610, 461 and 411 amino acid residues respectively in their most commonly occurring forms. The peptide chains are interconnected both within each half and between the halves of the molecule by a total of 29 disulfide bridges. The molecular weight of the human protein is 340 000. Its covalent structure was first elucidated by protein sequence analysis (see Henschen and McDonagh, 1986; Henschen, 1993). The work was completed

Methods in Protein Structure Analysis, Edited by M. Z. Atassi and E. Appella
Plenum Press, New York, 1995

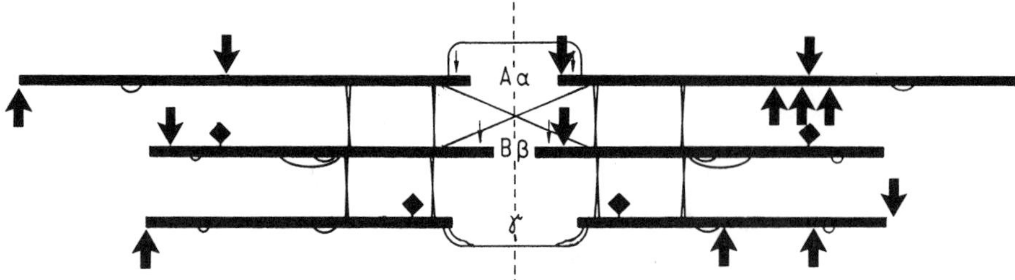

Figure 1. Human fibrinogen, a model of the covalent structure. The chains are aligned according to homology with the N-termini in the center. The thin, connecting lines represent disulfide bonds, the diamonds carbohydrate sidechains and the thin arrows thrombin cleavage sites. On the left side, the bold arrows point upwards to the sites for alternative processing during biosynthesis and downwards to the polymorphic sites. On the right side, the bold arrows point upwards to the sites for proteolytic processing and downwards to the phosphorylation site in the Aα chain, the proline-hydroxylation site in the Bβ chain and the sulfation site in the longer γ chain.

in 1979 and somewhat later confirmed and extended by DNA sequence analysis (see Chung et al., 1990). A model of human fibrinogen is shown in Fig. 1.

FIBRINOGEN FUNCTION

It is generally assumed that the most fundamental biological role of fibrinogen lies in its ability to form the skeleton of the blood clot and thereby prevent blood leakage. Furthermore, fibrinogen is believed to play a significant role in many additional pathophysiological processes, such as those related to wound healing, defense mechanisms and tumor growth and metastasis. Evidence has been presented by many research groups that fibrinogen may be one of the most important risk factors for cardiovascular disease (Humphries et al., 1987; Kannel et al., 1987). In order to fulfill all these roles fibrinogen has to interact in highly specific ways with itself and with a large number of other proteins as well as cells and lower-molecular- weight ionic components as listed in Table 1. Each type of interaction is expected to be due to the structure of one or more functional sites in the fibrinogen molecule.

FIBRINOGEN HETEROGENEITY

Human fibrinogen has since long time been described as a highly heterogeneous protein (see Henschen and McDonagh, 1986; Henschen, 1993), the heterogeneity being evident already from the variations in solubility properties, ion-exchange chromatography behavior of total fibrinogen and of its peptide chain components, as well as the gel-electrophoretic behavior of the peptide chains. Over the years, a considerable number of sites or sections of the molecule have been shown to exist as several structurally alternative forms which at least partly explain the overall heterogeneity of the total molecule. The regional variants in normal human fibrinogen can belong to either of two main categories, those which are non-inherited and may be present in all individuals and those which are inherited and therefore present only in certain individuals. Additional regional structural variants are caused by several types of non-genetic and genetic diseases. The so far identified types of variants in normal human fibrinogen are listed in Table 2. The positions of the more unique

Table 1. Functional sites in fibrinogen

Intrinsic	Thrombin cleavage
	Polymerization
	Crosslinking
	Plasmin cleavage
Protein interaction	Thrombin
	Factor XIII
	Plasmin(ogen)
	Plasminogen activators
	Fibronectin
	α_2-Antiplasmin
	Thrombospondin
	Albumin
	Collagen
	Lipoprotein(a)
Cell interaction	Platelets
	Erythrocytes
	Monocytes
	Macrophages
	Endothelial cells
	Fibroblasts
	Staphylococci, streptococci
Ion binding	Heparin
	Calcium
	Zinc
	Citrate
	EDTA

Table 2. Types of variants in human fibrinogen

Non-inherited	Alternative processing of Aα/γ chains
	Phosphorylation of Aα chain
	Sulfation of γ chain
	Proline-hydroxylation of Bβ chain
	Glutamine-cyclization in Bβ chain
	Methionine-oxidation in Aα/Bβ/γ chains
	Desamidation of Aα/Bβ/γ chains
	Glycosylation of Bβ/γ chains
	Proteolysis of Aα/γ chains
Inherited	Polymorphism in Aα/Bβ chains
	Mutation in Aα/Bβ/γ chains

Normal variant combinations: over 1 million.

Table 3. Posttranslationally modified amino acids in human fibrinogen

Modification	Peptide chain	Residue	Position	Sequence
Phosphorylation	Aα	S	3	ADSGEGD
	Aα	S	345	NPGSSER
Sulfation	Long γ	Y	418	ETEYDSL
	Long γ	Y	422	DSLYPED
Hydroxylation	Bβ	P	31	SLRPAPP
Cyclization	Bβ	Q	1	ZGVNDN
Glycosylation	Bβ	N	364	MGENRTM
	γ	N	52	QVENKTS

sites within the protein structure are indicated in Fig. 1 and corresponding sequences in Table 3.

Non-inherited Variants

There are three principal types of non-inherited regional variants present in mammalian fibrinogen. They are caused by alternative processing during biosynthesis, posttranslational modification of specific amino acid residues and proteolytic degradation.

Alternative Processing. Both the Aα and the γ chain occur in two different forms due to alternative biosynthetic processing. The two forms differ at the C-terminal end of the peptide chains depending on the utilization or non-utilization of an additional exon (Fig. 1). In the γ chain, the last four of the 411 amino acid residues of the sequence are replaced in about 10% of the chains by a stretch of 20 amino acids (Chung and Davie, 1984; Fornace et al., 1984). The two types of γ chains differ in their functional properties in the way that only the shorter ones can mediate the ability of fibrinogen to interact with platelets. However, both forms can be crosslinked by the transglutaminase, factor XIII, and thus participate in clot stabilization. In the Aα chain, only 2% of the chains seem to occur in the longer form, but here the chain carries a 236 amino acid residue extension, the function of which is not yet known (Fu and Grieninger, 1994).

Phosphorylation. All three peptide chains are posttranslationally modified at certain amino acid residues (Table 1). The Aα chain is partially phosphorylated at two serine residues (Seidewitz et al., 1984), one in fibrinopeptide A, i.e. position 3 of the chain, and the other in the middle part of the chain, in position 345. Both serine residues occur in a Ser-Xaa-Glu sequence, i.e. a coding sequence for casein kinase II. It is assumed that both positions are completely phosphorylated during biosynthesis, but that the phosphate groups subsequently are removed by a phosphatase in the blood so that only about 20% are left. However, during an acute phase reaction, with increased synthesis giving rise to a higher level of fibrinogen, and in the fetus or newborn, up to 70% phosphorylation is observed (Seidewitz et al., 1984). Phosphorylation often serves as an important signal in several biological processes, but the functional relevance in fibrinogen is unclear as the rate of thrombin-induced fibrinopeptide release is independent of the degree of phosphorylation. There is, however, some indication that phosphorylation protects against proteolytic degradation (see below).

The presence of a phosphorylated residue can often be surmised when, after fractionation by reversed-phase high-performance liquid chromatography (HPLC), two peptide fragments turn out to contain the same amino acid sequence, except for a certain position where the earlier eluting fragment has a suspiciously low yield of, e.g., serine or even a gap in the sequence and the later eluting fragment a normal yield of the residue. Evidence can conveniently be obtained by digestion with alkaline phosphatase, followed by re-chromatography and re-sequencing; the earlier eluting fragment should now appear in the position of the later eluting one and the yield of the relevant amino acid residue should be normal.

Sulfation. Many mammalian fibrinogens contain sulfated tyrosine residues, and typically, these residues are found in highly acidic sequence environments. In the human protein, only the longer γ chain splice variant is sulfated (Farrell et al., 1991). The two tyrosines, in positions 418 and 422, were both fully sulfated in the samples so far analyzed (Henschen, 1993). The functional relevance of the modification is unknown.

Sulfation of tyrosine residues may often escape attention, as the modification is highly acid-labile, the sulfate group being hydrolyzed off during sequencing and unmodified tyrosine appearing in the PTH amino acid chromatogram. In order to identify modified tyrosine residues, a procedure was developed which utilizes the protective effect of the sulfate (or phosphate) group against nitration by tetranitromethane. When peptide fragments are nitrated before sequencing, the originally unmodified tyrosines will be completely converted into nitrotyrosine, which easily can be identified in high yield during standard PTH amino acid analysis (eluting between Val and Phe/Lys, but separate from DPTU), and sulfated tyrosines will appear exclusively as tyrosine, though in slightly lower yield (Henschen, 1993). Additional evidence for sulfation can be obtained by specific cleavage of the sulfate bond by arylsulfatase or by M HCl for 4 minutes at 100 degrees followed by identification of unmodified tyrosine in the sequence (phosphorylated tyrosines are unaffected by these treatments). Furthermore, sulfated peptides are eluted before the corresponding unsulfated ones on reversed-phase HPLC.

Proline-Hydroxylation. The human Bβ chain is hydroxylated at the proline residue in position 31 to about 20% (Henschen et al., 1991). The finding was quite unexpected as hydroxyproline occurs primarily in collagen-like proteins where it is of great functional importance in regulating the optimal temperature stability of the triple helix. However, in fibrinogen the function is unknown and the sequence around the modified proline differs from the collagen consensus sequence. So far only few samples of fibrinogen have been analyzed for hydroxyproline content.

The presence of 4-hydroxyproline can easily be established in a homogeneous peptide by sequence analysis, where two characteristic PTH amino acid derivatives may be observed. Also the hydroxylated fragment is eluted before the corresponding unmodified fragment on reversed-phase HPLC.

Cyclization of N-Terminal Glutamine. The Bβ chains of many mammalian fibrinogens, including the human, start with a pyroglutamic acid residue, derived from a glutamine residue according to DNA sequence analysis. In fact, the N-terminal pyroglutamic acid in human fibrinogen seems to have to have been the first residue of this kind detected in a protein (see Blombäck et al., 1966). In the human Bβ chain the conversion to the pyroglutamic acid form is complete already in the blood. The blocked N-terminal may serve as a protection against degradation by aminopeptidases (see below).

Methionine-Oxidation. Recently, it was discovered that several methionine residues in all three peptide chains of human fibrinogen exist partially in the methionine sulfoxide form (Chen and Henschen, 1994). It had been observed that certain methionine residues are in part resistant to cyanogen bromide cleavage in native, but not in mercaptolyzed and alkylated, fibrinogen. The incomplete cleavage resulted in the appearance of additional components on fractionation by reversed-phase HPLC. The extent of oxidation at the relevant positions seemed to be 10-25%. The oxidation was not caused by the commercial fibrinogen purification procedure, which often includes virus-inactivation treatment, as fibrin, quickly isolated from blood donor plasma by simple clotting with thrombin, showed the same cyanogen bromide fragmentation pattern as the commercial fibrinogen. It may be suggested that the oxidation is caused by the hypochlorite-related agents released by activated leukocytes or phagocytes in the blood. The number of oxidized methionines and the extent of oxidation could both be specifically increased by treatment with low concentrations of chloramine-T. The oxidation caused a chloramine-T-concentration dependent loss in clotting or polymerization ability. Molecular species with damaged and undamaged polymerization sites could be separated by affinity chromatography, allowing the identification of methionine residues relevant to fibrinogen function (Chen and Henschen, 1994).

The analysis of methionine sulfoxide is hampered by the labile nature both of methionine and its sulfoxide, one being easily converted into the other. Certain methionine residues seem prone to oxidation, especially in denatured proteins. Methionine sulfoxide residues appear as methionines during sequence and amino acid analysis. A procedure was therefore developed for the identification of the oxidized residues (Chen and Henschen, 1994). The positions of these residues and the extent of oxidation could be established by quantitative N-terminal sequence analysis after cyanogen bromide cleavage of specifically modified samples. Thus, in one set of samples, all methionine residues were converted into the cyanogen bromide-refractive form by alkylation and the methionine sulfoxide residues subsequently reduced to methionine by mercaptoethanol treatment. In control samples, the reduction preceded the alkylation. The results indicated that the reduction, alkylation and cyanogen bromide cleavage were quantitative under the conditions used, so that the sequencing results could be employed for the identification and quantification of the modified residues.

Amide-Ammonia Loss. Preliminary results indicate that certain asparagine and glutamine residues in all three peptide chains have been partially converted to aspartic acid and glutamic acid residues by spontaneous amide-ammonia loss. The desamidation results in the appearance of pairs of reversed-phase HPLC components with identical sequence, except for certain positions which differ in their state of amidation. The amidated versions of the peptides are eluted before the corresponding desamidated peptides. It is not know if amide-ammonia loss has any effect on the function or survival of fibrinogen.

Glycosylation. Fibrinogen is glycosylated at two different sites (Fig. 1), i.e. in the N-terminal region position 52 of the γ chain (Blombäck et al., 1973) and in the C-terminal region of the Bβ chain (Töpfer-Petersen et al., 1976). The two carbohydrate sidechains are highly similar since both are N-glycosidically linked to asparagine residues and are biantennary. The glycosylation at the two sites is complete, but heterogeneity is caused by the presence of two or one sialic acid residue per sidechain. The amount of sialic aid influences the rate of fibrin polymerization in the way that an increase in acidic charge delays the polymerization. An increased extent of sialylation, probably in a tri- or tetra-antennary sidechain, and a correspondingly impaired polymerization is found in individuals with liver disease. A different, less specific type of glycosylation is caused by the excessive glucose level in diabetics, the glucose being added to certain amino groups in the protein.

Proteolytic Degradation. The Aα and γ chains occur even in the blood of normal, healthy individuals in degraded forms. The high susceptibility to proteolytic degradation of the Aα chain leads to the well known molecular weight variants of human fibrinogen.

Full size, undegraded fibrinogen has a molecular weight of 340 kDa and accounts for 70% of the blood plasma fibrinogen. A degraded form with a molecular weight of 305 kDa accounts for 25% and one of 270 kDa for the remaining 5%. The three forms are designated HMW, LMW and LMW′, respectively, and they differ in their Aα chains. The N-terminal region of the chain is preserved in all three forms. In the LMW-form one of the two Aα chains of the molecule is lacking a C-terminal portion; in the LMW′-form both Aα chains lack their C-terminal part. The degradation leads to a heterogeneous C-terminus, some identified C-terminal residues corresponding to positions 269, 297 and 309 (Nakashima et al., 1992). The enzyme responsible for the degradation has not yet been identified, but both plasmin and leukocyte elastase can be excluded because of their characteristic cleavage patterns (Müller and Henschen, 1988). Differences in distribution among the HMW, LMW and LMW′ forms have been observed in connection with certain diseases. Fibrin clots derived from degraded fibrinogen are less stable. An additional, less extensive, C-terminal degradation of about 25% of the Aα chains produces a variant ending at position 583, the degradation presumably being caused by plasmin as it corresponds to the earliest plasmic cleavage site in fibrinogen. The Aα chain is also N-terminally degraded in normal individuals, but only the first amino acid is missing in 10% of the chains, presumably due to the action of an aminopeptidase. The first residue is lacking only in chains which have lost the phosphate group in position 3, indicating a protective effect of phosphorylation against proteolysis.

Also the γ chain is proteolytically degraded in a heterogeneous way from the C-terminal side of the chain with preserved N-terminus (Henschen and Edman, 1972). In about 7% of the γ chains 200-300 amino acid residues have disappeared. Also here the causing agent is unknown. The shorter γ chains are unable to participate in clot-stabilization by crosslinking and in platelet interaction.

Common, Inherited Variants

Two types of inherited regional variants are present in human fibrinogen, i.e. those which are common and those which are very uncommon in the population. The common genetic variants may be detected by comparing samples from many individuals in the population. These polymorphic variants give rise to sequence microheterogeneity in pooled samples. Two polymorphic sites have recently been detected in human fibrinogen. Position 312 in the Aα chain can contain threonine or alanine and position 448 in the Bβ chain lysine or arginine (Baumann and Henschen, 1993, 1994). The allele frequencies for the pair threonine-alanine were 0.76 and 0.24, those for the pair lysine-arginine 0.85 and 0.15 in California blood donors. A polymorphic variation is, in principle, expected to be unrelated to the functional properties of the protein, and nothing is yet known about a direct effect of the polymorphic variation on the properties of fibrinogen. However, the polymorphism in the Bβ peptide chain turned out to be highly correlated to some other polymorphisms in the Bβ gene (Baumann and Henschen, 1994), which previously have been reported to correlate with the level of fibrinogen in the blood, and thus, at least indirectly, with the property of fibrinogen as a risk factor in thromboembolic, cardiovascular disease (Humphries et al., 1987; Kannel et al., 1987). It now seems meaningful to question if the variation in the protein structure or the variation in the gene structure or possibly both together contribute to the fibrinogen-related risk.

Recently, a system has been developed for the analysis of the polymorphic variants in the fibrinogen protein using restriction endonuclease digestion with the enzymes Rsa I and Mnl I, respectively, after polymerase chain reaction amplification, providing detection as restriction fragment length polymorphisms (Baumann and Henschen, 1993).

Rare, Inherited Variants

Uncommon, inherited variants of human fibrinogen have so far been described as observed only in association with fibrinogen dysfunction, i.e. incorrect or insufficient function. Genetically abnormal fibrinogens have now been detected in over 300 families and the structural aberrations identified in over 80 of these (Henschen and McDonagh, 1986; Ebert, 1991). Obviously, the abnormal, dysfunctional variants can be used as highly specific probes for structure-function relationships in fibrinogen. However, most genetic variants are discovered in the hospital routine laboratory when prolonged thrombin-clotting times are noticed, and this results in a selection of those dysfunctional variants which are related to thrombin cleavage and fibrin monomer polymerization. Out of the over 80 structurally elucidated variants 55 of the structural errors were detected in the regions of the fibrinopeptides A and B and the corresponding thrombin cleavage sites, 20 of the errors were found in the primary, complementary polymerization site in the carboxyterminal region of the γ chain and only 6 were discovered in other parts of the fibrinogen structure. It is remarkable that only 22 of the structural errors are unique to a single family. However, the mutant fibrinogen genes are rare in the population and most individuals carrying these genes are therefore heterozygous.

CONCLUSION

It may be summarized that normal human fibrinogen contains at least 13 variant sites in the Aα chain, i.e. those due to alternative processing, phosphorylation, oxidation, desamidation, proteolysis and polymorphism, 9 in the Bβ chain, i.e. those due to hydroxylation, oxidation, desamidation, glycosylation and polymorphism, and 9 in the γ chain, i.e. those due to alternative processing, sulfation, oxidation, desamidation, glycosylation and proteolysis. Each of these 31 regional variants may be symmetrically or asymmetrically distributed in the fibrinogen molecules, but one or two variants may occur in genetically homozygous form. Obviously, some variants are mutually exclusive. It can be estimated that each individual would carry over one million non-identical fibrinogen molecules in the blood.

ACKNOWLEDGMENT

This work was supported by Public Health Services Grant No. HL 424121.

REFERENCES

Baumann, R.E., and Henschen, A.H., 1993, Human fibrinogen polymorphic site analysis by restriction endonuclease digestion and allele-specific polymerase chain reaction amplification: Identification of polymorphisms at positions Aα312 and Bβ448, *Blood*, 82:2117-2124.

Baumann, R.E., and Henschen, A.H., 1994, Linkage disequilibrium relationships among four polymorphisms within the human fibrinogen gene cluster, *Human Genetics*, 94:165-170.

Blombäck, B., Blombäck, M., Edman, P., and Hessel, B., 1966, Human fibrinopeptides. Isolation, characterization, and structure, *Biochim. Biophys. Acta*, 115:371-396.

Blombäck, B., Gröndahl, N. J., Hessel, B. Iwanaga, S., and Wallén, P., 1973, Primary structure of human fibrinogen and fibrin. II. Structural studies on NH_2-terminal part of γ chain, *J. Biol. Chem.*, 248:5806-5820.

Chen, N. and Henschen, A., 1994, Identification of methionine sulfoxide in native and oxidized fibrinogen, *Protein Sci.* 3, Suppl. 1:147.

Chung, D.W., and Davie, E.W., 1984, γ and γ′ chains of human fibrinogen are produced by alternative mRNA processing, *Biochemistry* 23:4232-4236.

Chung, D.W., Harris, J.E., and Davie, E.W., 1990, Nucleotide sequences of the three genes coding for human fibrinogen. In: Fibrinogen, thrombosis, coagulation and fibrinolysis. (Liu, C.Y., and Chien, S. eds.) Plenum, New York, pp. 39-48.

Ebert, R.F., 1991, Index of variant human fibrinogens, CRC Press, Boca Raton.

Farrell, D. H., Mulvihill, E.R., Huang, S., Chung, D.W., and Davie, E.W., 1991, Recombinant human fibrinogen and sulfation of the γ′ chain, *Biochemistry*, 30:9414-9420.

Fornace, A.J., Cummings, D.E., Comeau, C.M., Kant, J.A., and Crabtree, G.R., 1984, Structure of the human γ-fibrinogen gene. Alternate mRNA splicing near the 3′ end of the gene produces γA and γB forms of γ-fibrinogen, *J. Biol. Chem.* 259:12826-12830.

Fu, Y., and Grieninger, G., 1994, Fib 420: a normal human variant of fibrinogen with two extended α chains, *Proc. Natl. Acad. Sci. USA* 91:2625-2628.

Graves, D. J., Martin, B.L., and Wang, J. H., 1994, Co- and post-translational modification of proteins. Chemical Principles and biological effects. Oxford University, New York. pp. 1-348.

Henschen, A., and Edman, P., 1972, Large scale preparation of S-carboxymethylated chains of human fibrin and fibrinogen and the occurrence of γ-chain variants, *Biochim. Biophys. Acta* 263:351-367.

Henschen, A., and McDonagh, J., 1986, Fibrinogen, fibrin and factor XIII. In: Blood coagulation. (Zwaal, R.F.A., and Hemker H.C. eds.) Elsevier, Amsterdam. pp. 171-241.

Henschen, A.H., 1993, Human fibrinogen - structural variants and functional sites, *Thromb. Haem.* 70:42-47.

Henschen, A.H., 1993, Identification of tyrosine sulfate and tyrosine phosphate residues during sequence analysis, *Protein Sci.*, 2, Suppl. 1:152.

Henschen, A.H., 1994, Human fibrinogen occurs as over 1 million nonidentical molecules, *J. Protein Chem.*, 13:504-505.

Henschen, A.H., Theodor, I., and Pirkle, H., 1991, Hydroxyproline, a posttranslational modification of proline, is a constitutent of human fibrinogen, *Thromb. Haem.* 65:821.

Humphries, S.E., Cook, M., Dubowitz, M., Stirling, Y., and Meade, T.W., 1987, Role of genetic variation at the fibrinogen locus in determination of plasma fibrinogen concentrations, *Lancet*. 1452-1455.

Kannel, W.B., D'Agostino, R.B., and Belanger, A.J., 1987, Fibrinogen, cigarette smoking, and risk of cardiovascular disease: Insights from the Framingham Study, *Am. Heart J.* 113:1006-1010.

Krishna, R.G., and Wold, F., 1993, Post-translational modification of proteins, *Adv. Enzymol. Related Areas of Mol. Biol.* 67:265-298.

Müller, E., and Henschen, A., 1988, Isolation and characterization of human plasma fibrinogen molecular-size-variants by high-performance liquid chromatography and amino acid sequence analysis. In: Fibrinogen. (Mosesson, M.W., Amrani, D.L., Siebenlist, K.R., and Diorio, J.P. eds.) Elsevier, Amsterdam, pp. 279-282.

Nakashima, A., Sasaki, S., Miyazaki, K., Miyata, T., and Iwanaga, S., 1992, Human fibrinogen heterogeneity: The COOH-terminal residues of defective Aα chains of fibrinogen II, *Blood Coag. Fibrinol.* 3:361-370.

Seydewitz, H.H., and Witt, I., 1985, Increased phosphorylation of human fibrinopeptide A under acute phase conditions, *Thromb. Res.* 40:29-39.

Töpfer-Petersen, E., Lottspeich, F., and Henschen, A., 1976, Carbohydrate linkage site in the β-chain of human fibrin, *Hoppe-Seyler's Z. Physiol Chem.* 357:1509-1513.

40

MICROSEQUENCING OF PROTEINS OF THE ROUGH ENDOPLASMIC RETICULUM (rER) MEMBRANE

Regine Kraft,[1,2] Susanne Kostka,[1,2] Enno Hartmann[1]

[1] Max-Delbrück-Center for Molecular Medicine
Berlin-Buch, Germany
[2] Humboldt University
Berlin, Germany

INTRODUCTION

The ER membrane is an important organelle involved in such diverse functions as protein translocation, protein folding and phospholipid biosynthesis. It is the entry point for most membrane and soluble proteins into the secretory pathway. Most proteins destined for translocation into the ER contain a signal sequence at their N-terminus consisting of basic amino acids followed by a stretch of hydrophobic amino acids. The SRP (signal-recognition particle) binds to the signal sequence of a nascent secretory protein that is bound to a ribosome. The ribosome-SRP complex is then bound to its receptor (DP, docking protein) in the ER membrane. After release of SRP, the nascent proteins are inserted into the membrane or translocated into the lumen of the ER (Walter and Lingappa, 1986; Rapoport, 1992) by means of a number of membrane proteins like TRAMp and the Sec61p complex. In the lumen of the ER, the synthesized polypeptide may undergo ER-specific cotranslational and post-translational modifications such as cleavage of the signal peptide, disulfide bond formation, N-linked glycosylation, fatty acylation, or prolyl hydroxylation. Thus, a multitude of functions are carried out by ER proteins which either integrate into the membrane or are located in the lumen. Up to now none of these functions is completely understood because not all proteins involved in these processes have been identified.

For these reasons we have started a systematic analysis of the proteins located in canine pancreatic microsomes by using a mini-two-dimensional(2-D) PAGE technique, followed by blotting of the separated peptides onto PVDF membranes and automated sequencing.

Here we present first results from the N-terminal amino acid sequencing of twenty-one protein spots from one up to four Coomassie Blue stained PVDF membranes. The amino acid sequence comparisons were carried out using the FASTA computer program of the Genetics Computer Group. Eight protein spots could be clearly identified. From the others, only insufficient sequence information could be obtained or the N-terminus was blocked.

Methods in Protein Structure Analysis, Edited by M. Z. Atassi and E. Appella
Plenum Press, New York, 1995

The preceding fractionation of the microsomes by detergent partitioning with Triton X-114 (Bordier, 1981) into lumenal and membrane fractions provided additional information about the cellular localization of the polypeptides identified and results in an increased resolution of the individual polypeptide spots.

METHODS

Two-Phase Partitioning with Triton X-114

Ribosome-stripped microsomes were prepared by standard methods (Walter and Blobel, 1983). This microsomal preparation (about 1 mg total protein) in 200 µl buffer containing 50 mM HEPES-KOH (pH 7.8), 500 mM potassium acetate, 5 mM magnesium acetate, 5 mM DTT, and protease inhibitor (1 : 500) was centrifuged for 10 min at 70,000 rpm at 2°C in micro test tubes in a TLA 100.3 rotor.

The pellet was dissolved by incubating for 10 min at 0°C in 900 µl of the same buffer, except that 100 µl 20% Triton X-114 (TX-114, from *Sigma*) was added (final concentration 2%). After centrifugation at 14,000 rpm, insoluble material was removed and the supernatant was divided into two aliquotes. One of them was treated by ethanol to obtain a pellet containing all proteins. The other aliquot was subjected to phase separation for 10 min at 37°C followed by centrifugation for 3 min at 7000 rpm in a microfuge. This step was repeated twice. Then the membrane proteins in the detergent phase and the lumenal ones in the aqueous phase were precipitated by addition of a fourfold excess of isopropanol and ethanol, respectively, at -20°C for 48 h. After centrifugation for 10 min at 14,000 rpm and 4°C, the pellets were washed with 400 µl each of the respective alcohols. The detergent pellet and the ethanol precipitate from the aqueous phase were finally dissolved in IEF sample buffer to analyze aliqots of them together with the input sample by mini-2-D PAGE.

Two-Dimensional Electrophoresis

2-D PAGE was essentially performed according to Klose (1989), and Jungblut and Seifert (1990). Gel solutions of the first dimension were obtained ready for use from the Wittmann Institute of Technology and Analysis of Biomolecules (WITA), Technology Center Teltow.

Glass tubes (inner diameter 1.5 mm, length 9.3 cm) were filled with degassed separation gel solution consisting of 9.2 M urea, 2% Triton X-100, 4% acrylamide, 0.3% bisacrylamide, 2% Servalyt ampholytes (four parts pH 5-7, one part pH 3-10), 0.14% TEMED, and 0.02% ammonium persulfate.

Electrophoresis in the first dimension was carried out under non-equilibrium conditions (NEPHGE, non-equilibrium pH gradient electrophoresis) applying the samples at the acidic end of the gels (anodic isoelectric focusing). The cathode buffer (lower chamber) consisted of 9 M urea, 5% glycerol and 5% ethylendiamine, the anode buffer (upper chamber) of 3 M urea, 4.25% phosphoric acid. Alcohol precipitates of the total ER proteins and samples from the detergent extraction procedure were dissolved in 10 to 20 µl buffer (theoretically about 2.5-5 µg protein/µl) consisting of 9.5 M urea, 2% Triton X-100, 5% β-mercaptoethanol, and applied onto the Sephadex gel on the upper acidic end of the first dimension gel. The samples were overlayered with overlay solution. Gels were run for 3 h at 500 V. The gels were extruded from the tubes, incubated for 10 min in a solution of Tris-phosphate buffer, glycerol and SDS and stored at -70°C until use.

SDS gel electrophoresis was performed using a Mini Protean II cell from BioRad. Rod gels were placed onto the top of 10% Laemmli-gels (0.75 mm, 6 x 9 cm), embedded in 1% agarose/buffer solution. The gels run for 1 h at 30 V followed for 2.5 h at 60 mA.

Blotting and Sequencing

Immediately after 2-D separation, the proteins were electrophoretically transferred onto ProBlott™ membranes under semi-dry (SD) conditions using a Trans-Blot SD electrophoretic transfer cell from BioRad. The transfer buffer consisted of 10mM CAPS, pH 11, containing 10% methanol and 0.07% SDS (only in the cathode buffer). Blotting was performed for 2.5 h at constant current per area of 1 mA/cm^2. The membranes were stained with Coomassie Blue for 1 - 2 min, destained several times in 50% methanol and dried.

N-terminal sequencing was carried out after direct loading of the PVDF-blotted protein spots onto the blot-cartridge of a 477A pulsed-liquid phase sequencer linked on-line with an 120A phenylthiohydantoin amino acid analyzer (Applied Biosystems, Foster City, CA).

RESULTS

To analyze proteins of the endoplasmic reticulum, we employed rough microsomes from dog pancreas, which can be obtained in high purity. The ER proteins were further separated into integral membrane proteins and soluble ones by using phase separation with TX 114 (Bordier, 1981). Hydrophobic and hydrophilic proteins are known to partition into the detergent phase and aqueous phase, respectively.

The proteins present in both fractions were separated by mini-2D PAGE (Fig. 1A, TX-114 supernatant, and 2A, TX-114 detergent phase). In the first dimension, the gels were run for only a short time period (1500 volt-hours) toward the cathode site, as recommended by O'Farrell *et al.* (1975). This procedure permits the separation of proteins at higher pH values and is referred to be a nonequilibrium pH gradient electrophoresis (NEPHGE). Under the conditions chosen, even very basic proteins remained in the gel.

The results show that more proteins are found in the fraction of soluble proteins especially on the acidic left side of the gel (Figure 1A) than in that of membrane sample. The basic membrane proteins are separated as rather broad spots (Figure 2A). The surprisingly few spots of membrane proteins represent, of course, only the most abundant proteins, particularly because of the limited amount of sample that can be loaded without compromising the separation (theoretically 50 µg of the alcohol precipitated sample pellets were redissolved in the sample buffer of the first dimension gels). In addition some proteins may not be completely solubilized or failed to enter the gel (Hjelmeland, 1990). Nevertheless very different polypeptide maps were obtained from the different fractions (see Figures) and the transfer of the proteins from the gel onto the PVDF support is complete (data not shown).

As demonstrated in Table 1 twenty-one proteins were subjected to N-terminal sequence analysis by cutting out one up to four Coomassie Blue stained spots selected from identical positions. Seven proteins (spots number ***1, 3, 4, 5, 6, 7*** and ***16***) could be unambiguously identified by comparing determined sequences with these of the data base, using the FASTA computer program of the Genetics Computer Group. One protein with unambiguous sequence (spot number ***0***) could not be identified. The proteins of spot numbers ***2*** and ***15*** seem to be blocked at their N-terminus because no N-terminal sequence could be obtained despite the fact that the intensity of staining was comparable to that of other identified proteins from this region. Because of the limited amount of material, a few spots (***11, 12, 14, 18, 19*** and ***20***) yielded only preliminary sequencing results (marked by double parenthe-

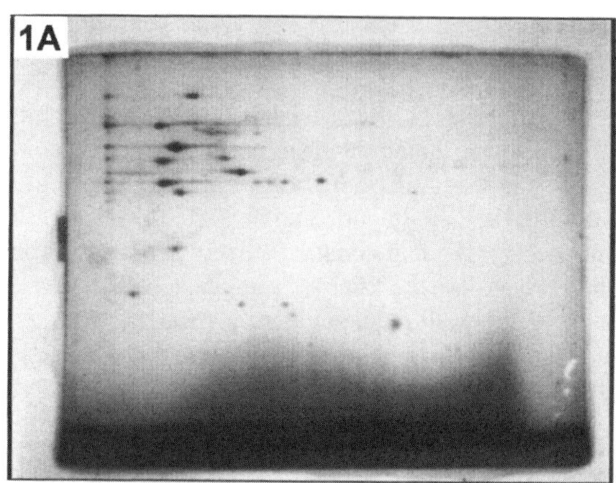

1 heat shock protein HSP 70 (73kDa)

```
   EEEDXKEDVG
   ||||:|||||
..AREEEDKKEDVGTVV..
   *    25   30
```

3 protein disulfide isomerase (PDI, 55kDa)

```
  APEEEDDVLVLNKXN
  ||||||:||||:|:|
..GAPDEEDHVLVLHKGNF..
  *  25    30
```

4 PDI family (49kDa)

```
  LYXSXDDVIELTPSX
  ||:|:|||||||||:
..SGLYSSSDDVIELTPSNFN..
  *    25   30
```

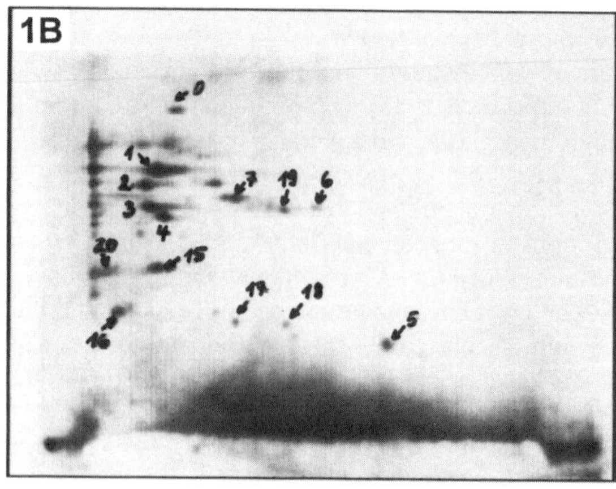

7 HIP 70 (70kDa)

```
  SDVLEXTDDNFE
  |||||:||:|||
..AASDVLELTDENFE.....
  *    30   35
```

16 EF1ß (25kDa)

```
 GFGDLKSPAGLQVLNDYLAD
 ||||||||||||||||||||
MGFGDLKSPAGLQVLNDYLADKSY...
 *       10         20
```

6 triacylglycerol-acylhydrolase (52kDa) **5 TRAP ß (22kDa)**

```
    KEVXYEQIGX
    |||:|||||:
..KAKEVXYEQIGCFSD...
  *20          30
```

Figure 1. TX-114 Supernatant (Lumenal ER Proteins).

Figure 2. TX-114 Detergent Phase (Integral ER Membrane Proteins). The following refers to both figures 1 and 2: In 1A and 2A are shown the respective gels, only to illustrate the protein pattern, gels are stained with Coomassie blue. After transfer (1B and 2B), polypeptide spots (one up to four selected from several blots) were cut out from the ProBlott paper for N-terminal micro-sequencing. The upper amino acid sequences are those determined from polypeptide spots, the lower ones represent that of known proteins obtained with the FASTA computer program from the data base. Amino acid residues are given in one-letter code. (X) means that the assignment of a phenylthio-hydantoin (PTH)-amino acid was not quite sure. * indicated the putative signal cleavage sequence site.

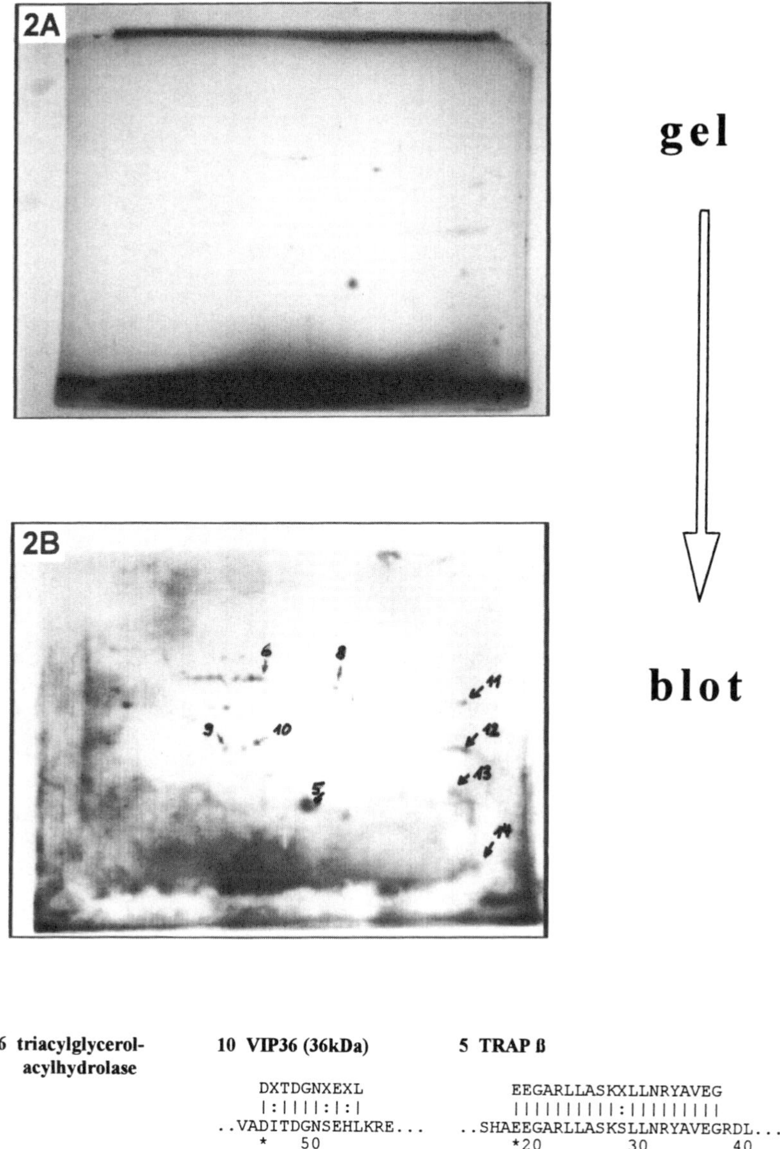

Figure 2. TX-114 Detergent Phase (Integral ER Membrane Proteins). See caption for figure 1.

sis), which were insufficient to detect homologies to other proteins. Proteins in spots numbers ***8, 9, 13*** and ***17*** could also not be sequenced. Interestingly, spot number ***10*** turned out to be VIP36, hitherto not reported to be resident in the ER-membrane.

DISCUSSION

We have initiated a systematic analysis of the proteins present in the ER membrane. This organelle contains a high number of both membrane bound and soluble proteins, which

Table 1. N-terminal sequencing results from 2D-blots

Spot number	Number of the spots sequenced (1,2,3)[a]	PTH initial yield (pmol)	Sequences	Protein (NBRF-database)
0	4 (1,3)	0.5	DAVVSED(P)G	?
1	1 (1)	3.2	EEED(K)KEDVG	heat shock protein family(HSP 70)
2	4 (1,3)		blocked	
3	3 (1,3)	4.1	APEEEDDVLVLN(K)XN	protein disulfide-isomerase (PDI)
4	4 (1,3)	2.3	LY(V)SXDDVIELTPS	canine homologous protein of the PDI family
5	2 (1,2,3)	8.3	EEGARLLASKXLLNRYAVEG	translocon-associated protein β (TRAP β)[b]
6	4 (1,2,3)	0.8	K(E)VXYEQIGX(F)	triacylglycerol-acyl-hydrolase
7	3 (1,3)	2.4	SDVLELTDDNF(E)	hormone-induced protein 70 (HIP 70) or phospholipase Cα1
8	2 (1,2)		blocked/to minor amount	
9	2 (1,2)		blocked/to minor amount	
10	2 (1,2)	2.7	(D)X(T)DGNXEXL	VIP36
11	1 (2)	0.7	((NELTQ))	?
12	1 (2)	1.9	((XXAG(S)XGGNLX))	?
13	1 (2)		blocked/to minor amount	
14	1 (2)	2.3	((G)XP(G)AX(T)(G)(L)(E))	?
15	4 (1,3)		blocked	
16	4 (1,3)	3.8	GFGDLKSPAGLQVLNDYLAD	elongation factor-1-β (EF1 β)
17	2 (3)		blocked/to minor amount	
18	3 (1,3)	1.2	((P)XGQ(E)AEEG))	?
19	4 (1,3)	0.9	((K)EVXFPXXGXX(Y)DD))	?
20	2 (3)	1.2	((NX(R)T(G)NX(D)IT))	?

[a]Blots from (1), total protein; (2), TritonX-114 pellet; (3), TritonX-114 supernatant.
[b]Görlich *et al.*, 1990; Hartmann *et al.*, 1993.

are involved in various functions, such as protein translocation (Rapoport, 1992a; Rapoport *et al.*, 1992b, Jungnickel *et al.*, 1994), protein modification (S. Hurtley and A. Helenius, 1989), phospholipid biosynthesis and detoxification reactions. The final goal of this effort is to get a full catalog of the ER resident proteins, at least of these, which are most abundant. So far, a similar endeavor has only been tried for synaptic vesicles (Sudhof *et al.*, 1993). Other systematic approaches have used total cells. The prefractionation of cells is expexted to yield increased resolution and a higher number of identificable proteins. In recent years many efforts have been made to obtain microsequence information for polypeptide spots isolated directly from 2D-PAGE gels (Vandekerckhove *et al.*, 1985; Aebersold *et al.*, 1987; Matsudaira, 1987; Rasmussen *et al.*, 1991; Celis *et al.*, 1991; Baker *et al.*, 1992; Hughes *et al.*, 1992). With a standard mini-2-D gel, a protein must constitute 0.1-1% of the loaded mixture to be detectable.

Of course, if total cellular protein is applied, only a few most abundant proteins can be detected. In our approach, we not only show the analysis with a purified cell organelle, but also additionally separated proteins into hydrophobic membrane bound ones and hydrophilic soluble ones.

By visual inspection one up to four spots from comparable positions of several blots were selected for structural characterization by automated microsequence analysis. Results are given in Table 1. In spite of the limited quantities of a few proteins, identity of seven proteins from a total of twenty-one analyzed spots could be established by comparison of the amino acid sequences obtained with sequences available from the data base. Surprisingly, elongation factor-1-β (spot number ***16***) known to be a highly abundant cytosolic enzyme (Sanders *et al.*, 1991) is found in the TX-114 supernatant implying that some cytosolic proteins have not been separated completely from the microsomal sample preparation. Two of the proteins identified, the lipases triacylglycerol-acylhydrolase (spot number ***6***) and the hormone induced protein HIP-70 (spot number ***7***), respectively, are reported to be secretory proteins (Mickel *et al.*, 1989; Mobbs *et al.*, 1990). Whether their finding in our ER supernatant preparations reflects the presence of substantial amounts of these proteins in the ER during their way from the cytosol to the cell exterior or whether they present ER resident isoforms of these proteins remains an open question.

The translocon-associated protein TRAP β (spot number ***5***) and one of the lipases (spot number ***6***), respectively, are present in both TX-114 supernatant and detergent-phase. In the latter case this behavior is probably due to the function of this enzyme binding at lipid interfaces for digestion of fats. However, the occurrence of detectable amounts of TRAP β in the TX-114 supernatant, although TRAP β is proved to be part of a larger protein complex in the ER memrane (Görlich *et al.*, 1990), is not quite clear. But it seems to be possible, that a small proportion of TX-114 remained in the aqueous phase preventing the complete separation of this major component from the other ones, particularly since TRAP β with only a single membrane-spanning region and two attached carbohydrates has also some hydrophilic character.

Three of the polypeptides found, protein disulfide-isomerases (spots number ***3*** and ***4***) and a heat shock protein (spot ***1***), respectively, were uniquely localized to the lumenal fraction. In fact, they belong to the "welcoming committee" (Hurtley and Helenius, 1989) of enzymes and factors thought to play an important role in the process of protein folding *in vivo*, like reshuffling of disulfides to accelerate a proper folding (Freedman, 1989) and associating with polypeptide intermediates to prevent them from aggregation and misfolding, respectively (Munro and Pelham, 1986).

Finally, protein of spot number ***10*** from the detergent fraction (Figure 2B) turned out to be authentically with the very recently described VIP36, a vesicular integral membrane protein (Fiedler *et al.*, 1994). Interestingly, Fiedler *et al.* reported that attempts to microsequence the N-terminus of the mature protein were unsuccessful. Our found N-terminal sequence DXTDGNXEXL (see Table 1) is in accordance to the putative signal-sequence cleavage site deduced from the cDNA as suggested by Fiedler *et al.*

VIP36 is assumed to be involved in protein sorting between the Golgi and the cell surface. Therefore, Fiedler *et al.* also analyzed the subcellular localization of VIP36 establishing its occurrence in the Golgi apparatus, endosomal and vesicular structures and the plasma membrane by immunoelectron microscopy and immuno fluorescence. The unexpected finding of this protein in our TX-114 detergent fraction, suggesting its residence also in the ER membrane remains to be verified.

ACKNOWLEDGEMENT

We thank Tom A. Rapoport for his helpful comments and critical reading of the manuscript

REFERENCES

Aebersold, R.H., Leavitt, J., Saavedra, R.A., Hood, L.E. and Kent, S.B.H., 1987, Internal amino acid sequence analysis of proteins separated by one or two dimensional gel electrophoresis after *in situ* protease digestion on nitrocellulose, *Proc. Natl. Acad. Sci. USA* 84:6970.

Baker, C.S., Corbett, J.M., May, A.J., Yacoub, M.H. and Dunn, M.J., 1992, A human myocardial two-dimensional electrophoresis database: Protein characterization by microsequencing and immunoblotting, *Electrophoresis* 13:723.

Bordier, C., 1981, Phase separation of integral membrane proteins in Triton X-114 solution, *J. Biol. Chem.* 256:1604.

Celis, J.E., Rasmussen, H.-H., Leffers, H., Madsen, P., Honore', B., Gesser, B., Dejgaard, K. and Vandekerckhove, J., 1991, Human cellular protein pattern and their link to genome DNA sequence data: usefulness of two-dimensional gel electrophoresis and microsequencing, *FASEB J.* 5:2200.

O'Farrel, P., 1975, High resolution two-dimensional electrophoresis of proteins, *J. Biol. Chem.* 250:4007.

Fiedler, K., Parton, R.G., Kellner, R., Etzold, T. and Simons, K., 1994, VIP36, a novel component of glycolipid rafts and exocytic carrier vesicles in epithelial cells, *EMBO J.* 13:1729.

Freedman, R.B., 1989, Protein disulfide isomerase: Multiple roles in the modification of nascent secretory proteins, *Cell* 57:1069.

Görlich, D., Prehn, S., Hartmann, E., Herz, J., Otto, A., Kraft, R., Wiedmann, M., Knespel, S., Dobberstein, B. and Rapoport, T.A., 1990, The signal sequence receptor has a second subunit and is part of a translocation complex in the endoplasmic reticulum as probed by bifunctional reagents, *J. Cell Biol.* 111:2283.

Hartmann, E., Görlich, D., Kostka, S., Otto, A., Kraft, R., Knespel, S., Bürger, E., Rapoport, T.A. and Prehn, S., 1993, A tetrameric complex of membrane proteins in the endoplasmic reticulum, *Eur. J. Biochem.* 214:375.

Hjelmeland, L., 1990, Solubilization of native membrane proteins, *in*: Methods in Enzymology 182:253.

Hughes, G.J., Frutiger, S., Paquet, N., Ravier, F., Pasquali, Ch., Sanchez, J.-Ch., James, R., Tissot, J.-D., Bjellqvist, B. and Hochstrasser, D.F., 1992, Plasma protein map: An update by microsequencing, *Electrophoresis* 13:707.

Hurtley, S.M. and Helenius, A., 1989, Protein oligomerization in the endoplasmic reticulum, *Annu. Rev. Cell Biol.* 5:277.

Jungblut, P., and Seifert, R., 1990, Analysis by high-resolution two-dimensional electrophoresis of differentiation-dependent alterations in cytosolic protein pattern of HL-60 leukemic cells, *J. Biochem. Biophys. Meth.* 21:47.

Jungnickel, B., Rapoport, T.A. and Hartmann, E., 1994, Minireview, Protein translocation: common themes from bacteria to man, *FEBS Lett.* 346:73.

Klose, J., 1983, High resolution of complex protein solutions by two-dimensional electrophoresis, *in*: Tschesche, H., ed., Modern Methods in Protein Chemistry - Review Articles, Walter de Gruyter, Berlin, pp.49-78.

Laemmli, U., 1970, *Nature* (London) 277:580.

Matsudaira, P.J., 1987, Sequence from picomole quantities of proteins electroblotted onto polyvinylidene difluoride membranes, *J. Biol. Chem.* 262:10035.

Mickel, F.S., Weidenbach, F., Swarovsky, B.,LaForge, K.S. and Scheele, G.A., 1989, Structure of the canine pancreatic lipase gene, *J. Biol. Chem.* 264:12895.

Mobbs, C.V., Fink, G. and Pfaff, D.W., 1990, HIP-70: a protein induced by estrogen in the brain and LH-RH in the pituitary, *Science* 247:1477.

Munro, S. and Pelham, H.R.B., 1986, An Hsp70-like protein in the ER: identity with the 78kd glucose-regulated protein and immunoglobulin heavy chain binding protein, *Cell* 46:291.

Rapoport, T.A., 1992a, Transport of proteins across the endoplasmic reticulum membrane, *Science* 258:931.

Rapoport, T.A., Görlich, D., Müsch, A., Hartmann, E., Prehn, S., Wiedmann, M., Otto, A., Kostka, S. and Kraft, R., 1992b, Components and mechanism of protein translocation across the ER membrane, *Antonie van Leewenhoek* 61b:119.

Rasmussen, H.-H., Van Damme, J., Bauw, G., Puype, M., Gesser, B., Celis, J. and Vandekerckhove, J., 1991, Protein-electroblotting and microsequencing in establishing integrated human protein databases, *in*: Jörnvall, H., Höög, J.-O. and Gustavson, A.-M., eds., Methods in Protein Sequence Analysis, Birkhauser Verlag, Basel, pp.103-114.

Sanders, J., Maasen, J.A., Amons, R. and Moeller, W., 1991, Nucleotide sequence of human elongation factor-1-beta cDNA, *Nucleic Acids Res.* 19:4551.

Sudhof, T.C., D'Camilli, P., Niemann, H. and Jahn, R., 1993, Membrane fusion machinery: insights from synaptic proteins, *Cell* 75:1.

Vandekerckhove, J., Bauw, G., Puype, M., Van Damme, J. and Van Montagu, M., 1985, Proteinblotting on Polybrene-coated glass-fiber sheets. A basis for acid hydrolysis and gas-phase sequencing of picomole quantities of protein previously separated on sodium dodecyl sulfate/polyacrylamide gel, *Eur. J. Biochem*. 152:9.

Walter, P. and Blobel, G.,1983, Preparation of microsomal membranes for cotranslational protein translocation, *Methods Enzymol.* 96:557.

Walter, P., and Lingappa, V.R., 1986, Mechanism of protein translocation across the endoplasmatic reticulum membrane, *Annu. Rev. Cell Biol*. 2:499.

DATABASE ANALYSIS, PROTEIN FOLDING AND THREE-DIMENSIONAL STRUCTURES OF PROTEINS

41

THEORETICAL STUDIES OF PROTEIN FOLDING

Harold A. Scheraga, Ming-Hong Hao, and Jaroslaw Kostrowicki

Baker Laboratory of Chemistry
Cornell University
Ithaca, New York 14853-1301

ABSTRACT

This paper summarizes some fundamental statistical mechanical aspects of protein folding and the prediction of protein structure. In order to predict the native structure of a protein, it is necessary to understand the physical conditions that determine its unique and thermodynamically-stable native structure, and to surmount the numerous local energy minima to arrive at the native structure. A statistical mechanical approach has been used to address the problem of foldability of polypeptides, and global minimization techniques have been developed to solve the multiple-minima problem. Some recent progress in these areas, made in our laboratory, is described.

INTRODUCTION

A problem of much interest in protein chemistry is to determine how interatomic interactions dictate the folding of a polypeptide chain into the three-dimensional structure of the biologically-active native protein. The underlying thermodynamic theory for an understanding of this process has been developed previously (Scheraga, 1968; Gō and Scheraga, 1969, 1976), and subsequent use of lattice models, analytical theories, and computer simulation (Taketomi et al, 1975; Dill, 1985; Dill et al, 1989; Shakhnovich and Finkelstein, 1989; Shakhnovich and Gutin, 1989, 1990, 1993; Bryngelsen and Wolynes, 1987, 1989; Skolnick and Kolinski, 1990, 1991; Kolinski and Skolnick, 1992; Covell and Jernigan, 1990; Honeycutt and Thirumalai, 1990; Comacho and Thirumalai, 1993; Leopold et al, 1992; Fukugita et al, 1993; Sali et al, 1994; Hao and Scheraga, 1994a,b, 1995) has continued to contribute to a deeper understanding of the nature of the problems involved. Two problems are currently of central importance in theoretical studies of protein folding: (1) Determination of the conditions that lead to the unique and thermodynamically-stable native structure of a protein, and (2) an accurate calculation of the native structure of a protein among its many local minimum-energy conformations. The solution of the first problem can

Methods in Protein Structure Analysis, Edited by M. Z. Atassi and E. Appella
Plenum Press, New York, 1995

greatly facilitate, but not replace, the solution to the second problem. In this article, we review the progress that has been made in our laboratory in solving these problems.

SOME STATISTICAL MECHANICAL ASPECTS

We have recently presented three papers (Hao and Scheraga, 1994a,b, 1995), designated here as papers I, II and III, dealing with some statistical thermodynamic aspects of protein folding. Paper I deals with the order of the folding transition of a protein, paper II is concerned with the effects of polypeptide sequences on their folding transitions, and paper III addresses some aspects of the influence of the potential function on the folding of a given protein. Even though each of these papers focussed on a different problem, they were all based on an identical underlying theme, i.e. defining the statistical-mechanical characteristics of the folding transitions of polypeptides in terms of their densities of states.

Clearly, knowing that a given polypeptide has a unique and thermodynamically stable folded form is the first requirement or condition for folding the protein to its native structure by theoretical methods. More generally, it is of interest to know under what conditions, in terms of either the sequence or the force fields, polypeptides will have a unique and stable folded form. A statistical mechanical formalism has been used to answer these questions. For practical applications, it is desirable to identify the quantities or features that characterize the foldability of polypeptides. A sufficient number of case studies should be carried out to test these criteria. Our papers attacked these problems directly; therefore, the work is relevant to the problem of interest here.

Much work on protein folding (e.g. Skolnick and Kolinski, 1991; Kolinski and Skolnick, 1992; Yue and Dill, 1994; Covell and Jernigan, 1990) did not address the problem of the foldability of a protein explicitly, but was concerned with a reasonably high probability to achieve the folding of a given protein to its native (or targeted) structure in simulations. Sali et al (1994) proposed that the difference of the energies between the lowest and second lowest-energy structures can be taken as the criterion of foldability of a polypeptide. This proposal, however, was based on sampling only compact conformations of simple lattice chains, and its general validity has been questioned (Bryngelson et al, 1994). Wolynes' group (Goldstein et al, 1992a,b) proposed that the ratio of the folding transition temperature to the glass transition temperature of a polypeptide can be taken as the criterion of foldability of a polypeptide. This criterion has a sound physical basis and is elegant. But, the problem is that the estimate of the glass transition temperature is often highly approximate, which reduced the accuracy of the prediction by this criterion.

We have used the complete density of states to characterize the foldability of a given polypeptide. This is a physically sound criterion; it covers the applicable ranges of the other criteria and, more important, without suffering from the inaccuracy or approximations of the previous theories. The density of states determines the complete thermodynamic properties of the protein. However, due to the complexity of the protein molecule, an analytical determination or an exact enumeration of all conformations of a protein is impossible. Fortunately, it is the relative density of states that determines the thermodynamic properties of a protein, and the relative density of states of realistic model proteins can be determined sufficiently accurately by simulation methods (see below). Figure 1 shows some typical densities of states of model polypeptides determined from our studies.

From the density of states, many thermodynamic properties of the protein, such as its average energy, heat capacity, etc. can be calculated. Of particular interest is the free energy of the protein as a function of its energy (and temperature), $F(E,T)$, which is determined by the relationship:

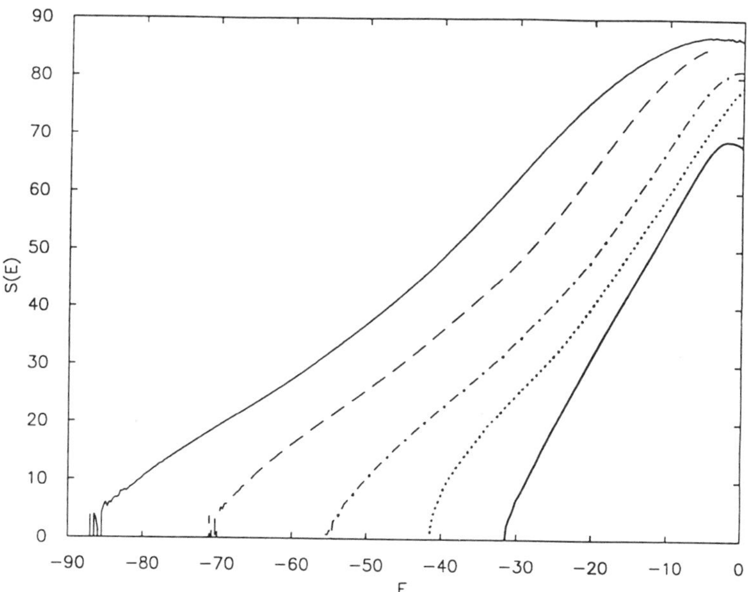

Figure 1. Illustration of different types of densities of states of polypeptides obtained from Monte Carlo simulations. $S(E)$ is the entropy of states as a function of energy E. $S(E)$ is determined by the logarithm of the density of states (Hao and Scheraga, 1995).

$$F(E,T) = E - TS \tag{1}$$

where E is the energy and $S \equiv k\ln[n(E)]$, with $n(E)$ being the density of states. Figure 2 compares the probability of occurrence for the conformations of two different polypeptides near the transition temperatures T_c. It can be seen that, while the dominant conformational distributions of the two polypeptides are similar below or above the transition temperatures, the two molecules exhibit quite different behavior at the transition temperature. The one on the left shows a bimodel distribution and, therefore, follows a first-order transition; the one on the right shows a flat distribution and is expected to follow a continuous transition.

On the basis of this approach, we have studied a number of specific problems of protein folding. Some of our findings are (1) even relatively small polypeptides can undergo a first-order folding transition; (2) depending on the sequence and potential function, there are three possible types of folding transitions: first-order, continuous, and glass-like; (3) short-range potentials and long-range interactions have different effects on the features of the folding transition of polypeptides; and (4) we demonstrate that the density of states is a sensitive indicator of the behavior of protein folding. The characteristics of a good folding sequence under a proper force field are reflected in a curve of the density of states, which has the character of a first-order transition (i.e. with a concave segment), the folded states are well separated from the unfolded states in the density of states, and there are discrete states in the lowest-energy regions.

To describe the density of states (from simulations) quantitatively, we have developed an analytical formalism (paper III). Our theory, following the spirit of Bryngelson and Wolynes (1987), is based on two components: (1) a mean-field representation of the energies of the protein and (2) a random distribution of energies in a subset of conformations defined by the fraction of native residues. The probability of an average residue with energy e is expressed as

$$P(e) = \frac{1}{(2\pi)^{1/2}\delta(\rho)} \exp\left\{-\frac{[e - \bar{\varepsilon}(\rho)]^2}{2\delta^2(\rho)}\right\} \tag{2}$$

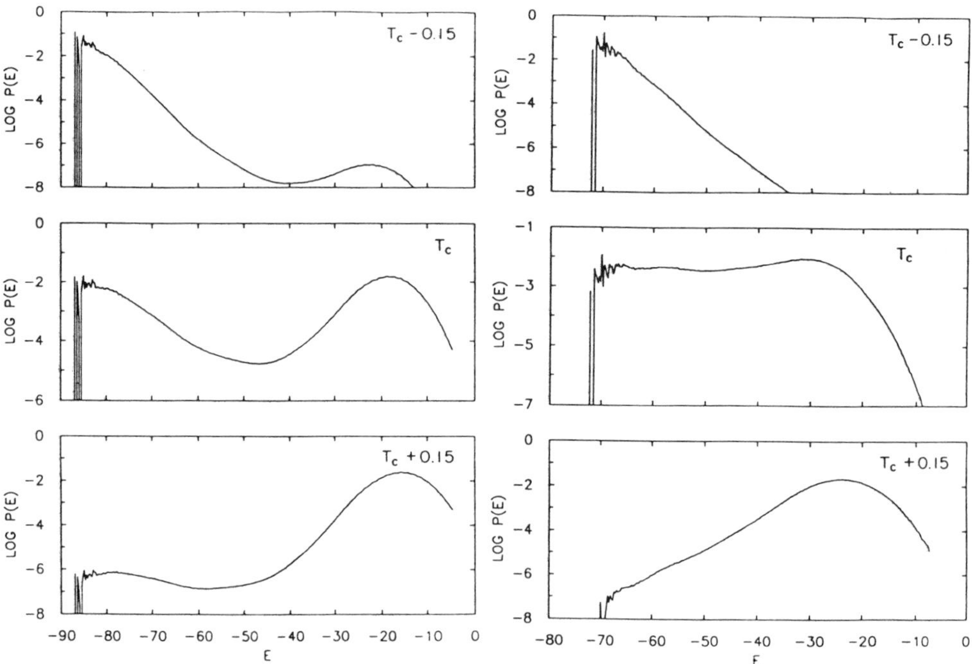

Figure 2. Comparison of the probability of occurrence of conformations $P(E)$ of two different polypeptides above the transition temperature (bottom), at the transition temperature (middle), and below the transition temperature (top) (Hao and Scheraga, 1994b).

where ρ is the fraction of native residues, $\bar{\varepsilon}(\rho)$ and $\delta(\rho)$ are defined by

$$\bar{\varepsilon}(\rho) = \varepsilon_0 + \varepsilon_1\rho + \varepsilon_2\rho^2 \tag{3}$$

$$\delta(\rho) = [\delta_0 - \delta_1\rho - \delta_2\rho^2]^{1/2} \tag{4}$$

Other properties of interest can be derived on the basis of the above expression. In this theory, each polypeptide is defined by three parameters ε_0, ε_1 and ε_2 and three distribution parameters δ_0, δ_1 and δ_2. Once these parameters are determined, the folding behavior of the polypeptide is completely defined by the theory. We have developed a procedure to extract the theoretical parameters from the simulated density of states for a given polypeptide (paper III). In this form, the folding characteristics of a protein can be defined quantitatively, with added insight from the theory about the nature of the interactions that determine such characteristics. It has been shown that this theory fits the simulation data very well. The work can be used as a basis for practical applications such as sequence design, the refinement of force fields, and folding a given protein.

We now briefly describe the simulation method for determining the density of states for proteins; this is obviously not a trivial problem. The so-called entropy sampling Monte Carlo (ESMC) method (Lee, 1993) played a key role in our simulations (it might be noted that the name 'ESMC' is likely to be changed in the future on the basis of arguments in paper III). The essence of the method is a Monte Carlo simulation with a probability distribution based on a scaling function; this scaling function is iterated in a series of simulations until it achieves uniformity when it converges. The biased sampling technique that we introduced into the ESMC method (Hao and Scheraga, 1994a) is a critical element that makes the ESMC method work for simulating proteins. Through our experience in simulating proteins with the ESMC method, we have summarized three advantages of the ESMC method with respect to the conventional Monte Carlo method in determining the density of states of proteins, i.e. the ESMC method has a clearly defined criterion of convergence, it has a self feed-back mechanism, and it is much less prone to being trapped in a particular local-energy minimum. We also introduced a jump-walking procedure into ESMC simulations and employed parallel programming. The experience gained in using the ESMC method to simulate the density of states of model polypeptides will be useful in our further efforts to fold realistic proteins.

Since the native structure emerges as the system is cooled, efforts have been devoted to computing it accurately with empirical potential energy functions. For this purpose, it is necessary to (i) generate an arbitrary starting conformation, (ii) compute its conformational energy (with entropy and hydration contributions included), and (iii) locate the global minimum of the conformational energy. Adequate procedures are available for steps (i) and (ii) and for minimizing the conformational energy (Scheraga, 1992). The minimization procedure, however, leads only to the local minimum closest to the starting conformation, rather than to the global minimum; this is the multiple-minima problem (Gibson and Scheraga, 1988; Scheraga, 1992). It is, therefore, necessary to have efficient methods for searching conformational space to locate the lowest minimum among all those in the whole space. The next section describes one of our methods that has been developed to facilitate such a search of conformational space. Details of other methods are discussed elsewhere (Scheraga, 1992; Vásquez et al, (1994).

THE DIFFUSION EQUATION METHOD (DEM)

The DEM is based on the use of the diffusion equation to deform the complex energy hypersurface in successive stages so that higher-energy minima disappear, and only a descendant of the global minimum remains. A reversal of the deformation procedure then recovers the global minimum of the original potential function (Piela et al, 1989). This procedure has been applied to a variety of simple mathematical functions (Piela et al, 1989), to a series of clusters of Lennard-Jones particles (Kostrowicki et al, 1991), to water clusters (Wawak et al, 1992), and to terminally-blocked alanine and the pentapeptide Met-enkephalin (Kostrowicki and Scheraga, 1992). In the application to Lennard-Jones particles (Kostrowicki et al, 1991), the Lennard-Jones potential function was expressed as a sum of Gaussians, for which an analytical solution of the diffusion equation is also a sum of Gaussians, but with modified coefficients. Calculations were carried out for various cluster sizes n = 5,6,7,...,55. For n = 55, there are ~10^{45} local minima, the global minimum being the MacKay icosahedron. This global minimum was found by the diffusion equation method (Kostrowicki et al, 1991) in ~400 seconds on an IBM 3090 supercomputer.

In calculations on oligopeptides (Kostrowicki and Scheraga, 1992), the DEM found the global minimum for the alanine compound in <1 min and for the pentapeptide in ~10 min, using one processor of an IBM 3090 supercomputer. Since the DEM scales as n^3, where n is the number of residues, then it should take (10 min) (10^3), or 10^4 min or ~7 days, to scale up by a factor of 10, i.e. to go from a pentapeptide to a 50-residue protein, using one processor of the IBM 3090 computer. Hence, using all 6 processors of the computer, the computation should take ~1-7/6 days.

The DEM is illustrated in Figure 3 by a simple one-dimensional function with two minima. The original function, $f(x)$, can be deformed, in the first iteration, to $f^{[1]}(x)$ by adding its second derivative, $f''(x)$, which is zero at the inflection points, viz.

$$f^{[1]}(x) = f(x) + \beta f''(x) = \left[1 + \beta \frac{d^2}{dx^2} \right] f(x) \tag{5}$$

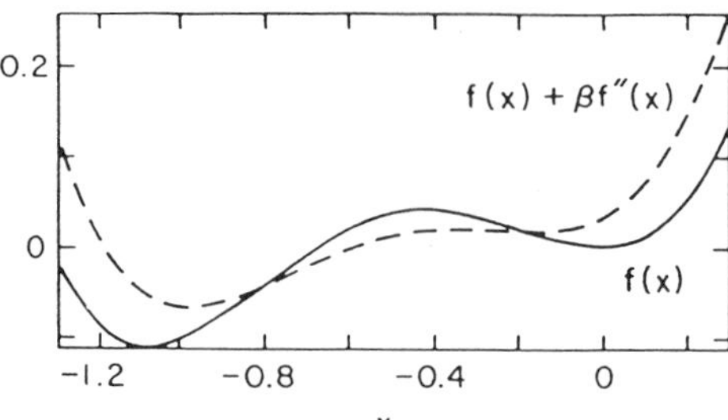

Figure 3. Original double-minimum potential energy curve $f(x)$ (solid line) transformed, according to eq. 5, into a curve with only a single minimum (dashed line). The values of the transformed function at the inflection points do not change. The particular function used in this Figure is $f(x) = x^4 + ax^3 + bx^2$, with a and b equal to 2 and 0.9, respectively. When $\beta = 0.02$, one obtains $f(x) + \beta f''(x)$, which exhibits only one minimum (Piela et al, 1989).

where β is a small positive constant. Repeated applications of this procedure lead to the following result in the Nth iteration:

$$f^{[N]}(x) = \left(1 + \frac{t}{N}\frac{d^2}{dx^2}\right)^N f(x) \tag{6}$$

where t/N has been written for β, with the parameter t being positive. Destabilization of the surface is most effective when $N \to \infty$. Taking this limit, we may write

$$\mathrm{F}(x,t) = \lim_{N\to\infty}\left(1 + \frac{t}{N}\frac{d^2}{dx^2}\right)^N f(x) = \exp\left(t\frac{d^2}{dx^2}\right)f(x) \tag{7}$$

It can be shown that, equivalently, F(x,t) is a solution of the diffusion equation

$$\frac{\partial^2 \mathrm{F}}{\partial x^2} = \frac{\partial \mathrm{F}}{\partial t} \tag{8}$$

where the parameter t takes on the meaning of "time", with the initial condition being $\mathrm{F}(x,0) = f(x)$. Equations 5-8 and Figure 3 serve only to show how the DEM was originally derived (Piela et al, 1989); in current applications, the diffusion equation is itself the starting point for the computations.

In higher dimensions, d^2/dx^2 is replaced by the Laplacian, $\Delta = \Sigma_i\, \partial^2/\partial x_i^2$, so that the diffusion equation becomes

$$\Delta\mathrm{F} = \frac{\partial \mathrm{F}}{\partial t} \tag{9}$$

The successive deformations of the one-dimensional function of Figure 3 from $t = 0$ to $t = t_o = 0.25$, and the reversal from $t = 0.25$ to $t = 0$, is illustrated in Figure 4. It can be seen how the global minimum of the original function is achieved.

In the diffusion equation method, the original potential surface is the analogue of a varying concentration which becomes uniform as $t \to \infty$. Thus, as $t \to \infty$, all minima would disappear, and the surface would become uniformly flat. However, for a sufficiently large, finite time, t_o, only one minimum (a descendant of the global minimum) remains.

Up to now, the DEM has been implemented in the space of Cartesian coordinates of the atoms, and an analytical solution of the diffusion equation was obtained by symbolic evaluation of the Fourier-Poisson integral, with the potential function expressed as a sum of Gaussians or cut Gaussians and a cut $1/r$ potential. When the DEM is applied to chain molecules, it is necessary to introduce constraints to limit the bond lengths and bond angles to acceptable values; i.e., although the diffusion equation is solved in Cartesian coordinates,

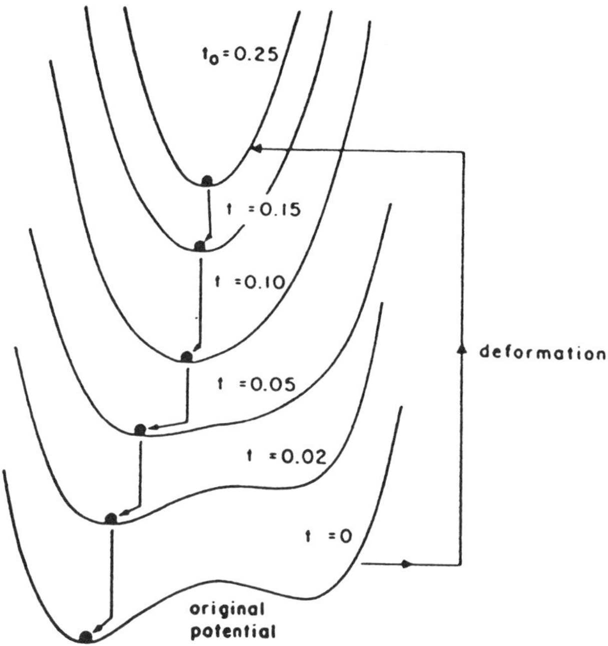

Figure 4. Illustration of the deformation of the original potential $f(x)$ (the same as in Fig. 3), and of the reversing procedure. The deformation at $t_0 = 0.25$ leads to the curve with the unique minimum that is achievable from any point of the space by a simple minimization. Then, the reversing procedure (shown by the arrows directed downward) is applied by considering a sequence of the deformed curves at $t = 0.15, 0.10, 0.05, 0.02$, and finally 0, where the original function is recovered. Each step of the procedure is followed by a minimization symbolized in the Figure by a ball moving downhill from the minimum position of the upper curve and always reaching the position of the minimum in the lower curve. In the final step, the global minimum is found (Piela et al, 1989).

its solution is examined on the manifold of fixed bond lengths and bond angles. However, even for a smoothed function having only one minimum in the whole space, it is still possible to have more than one constrained minimum for fixed bond lengths and bond angles.

This problem can be circumvented by solving the diffusion equation in internal rather than Cartesian coordinates. However, when using internal coordinates, it has not yet been possible to obtain an analytical solution. Instead, the energy function is expressed as a Fourier series, and the solution of the diffusion equation for such an energy function is also a Fourier series with coefficients related to those of the original function (Kostrowicki et al, 1995). It appears that a good approximate solution of the diffusion equation can be obtained for large t by considering only the lowest-order Fourier coefficients. These coefficients are evaluated by separation of the geometrical and physical (energetical) part of the integral $\int \cos\theta_k \, V_{ij}(\theta_1,...,\theta_n) \, d\,\theta_1 ... d\,\theta_n$. The geometrical part of the integral is a generalization of the end-to-end distance distribution function, and is derived iteratively from the chain geometry (Kostrowicki and Scheraga, 1994). This function is then integrated numerically with the distance-dependent pair interaction function $V_{ij}(r_{ij})$. In the time reversal, the solution of the diffusion equation at small t (ultimately to $t = 0$) is expressed as an analytical solution of the diffusion equation in the linear subspace tangent to the manifold of geometries with fixed bond lengths and bond angles. This approach is currently undergoing testing and evaluation.

ACKNOWLEDGMENTS

This work was supported by NIH Grant No. GM-14312 and NSF Grant No. DMB 90-15815. Support was also received from the Association for International Cancer Research and the Cornell Biotechnology Center. Many of the computations were carried out on the IBM 3090 and 9000 supercomputers of the Cornell National Supercomputer Facility (CNSF), a resource of the Cornell Theory Center, which receives major funding from the

National Science Foundation and the IBM Corporation, with additional support from New York State and members of its Corporate Research Institute. Some of the simulations were carried out on the KSR parallel computer of the CNSF, which is funded, in part, by the National Institutes of Health.

REFERENCES

Bryngelson, J. D. and Wolynes, P. G. (1987) *Proc. Natl. Acad. Sci. U.S.A.* **84**, 7524-7528.

Bryngelson, J. D. and Wolynes, P. G. (1989) *J. Phys. Chem.* **93**, 6902-6915.

Bryngelson, J. D., Onuchic, J. N., Socci, N. D. and Wolynes, P. G. (1995) *Proteins: Structure, Function and Genetics* **21**, 167-195.

Camacho, C. J. and Thirumalai, D. (1993) *Proc. Natl. Acad. Sci. U.S.A.* **90**, 6369-6372.

Covell, D. G. and Jernigan, R. L. (1990) *Biochemistry* **29**, 3287-3294.

Dill, K. A. (1985) *Biochemistry* **24**, 1501-1509.

Dill, K. A., Alonso, D. O. V., and Hutchinson, K. (1989) *Biochemistry* **28**, 5439-5449.

Fukugita, M., Lancaster, D. and Mitchard, M. G. (1993) *Proc. Natl. Acad. Sci. U.S.A.* **90**, 6365-6368.

Gibson, K. D. and Scheraga, H. A. (1988) in "Structure & Expression: Vol. 1: From Proteins to Ribosomes", Eds. R.H. Sarma & M.H. Sarma, Adenine Press, Guilderland, N.Y., p. 67-94.

Gō, N and Scheraga, H. A. (1969) *J. Chem. Phys.* **51**, 4751-4767.

Gō, N. and Scheraga, H. A. (1976) *Macromolecules* **9**, 535-542.

Goldstein, R. A., Luthey-Schulten, Z. A., and Wolynes, P. G. (1992a) *Proc. Natl. Acad. Sci. U.S.A.* **89**, 4918-4922.

Goldstein, R. A., Luthey-Schulten, Z. A., and Wolynes, P. G. (1992b) *Proc. Natl. Acad. Sci. U.S.A.* **89**, 9029-9033.

Hao, M.-H. and Scheraga, H. A. (1994a) *J. Phys. Chem.* **98**, 4940-4948.

Hao, M.-H. and Scheraga, H. A. (1994b) *J. Phys. Chem.*, **98**, 9882-9893.

Hao, M.-H. and Scheraga, H. A. (1995) *J. Chem. Phys.*, **102**, 1334-1348.

Honeycutt, J. D. and Thirumalai, D. (1990) *Proc. Natl. Acad. Sci. U.S.A.* **87**, 3526-3529.

Kolinski, A. and Skolnick, J. (1992) *J. Chem. Phys.* **97**, 9412-9426.

Kostrowicki, J., Piela , L., Cherayil, B. J. and Scheraga, H. A. (1991) *J. Phys. Chem.* **95**, 4113-4119.

Kostrowicki, J. and Scheraga, H. A. (1992) *J. Phys. Chem.* **96**, 7442-7449.

Kostrowicki, J., Oberlin, D. M. and Scheraga, H. A. (1995) to be submitted.

Kostrowicki, J. and Scheraga, H. A. (1995) Computational Polymer Science, **5**, 47-55.

Lee, J. (1993) *Phys. Rev. Let.* **71**, 211-214.

Leopold, P. E., Montal, M. and Onuchic, J. N. (1992) *Proc. Natl. Acad. Sci. U.S.A.* **89**, 8721-8725.

Piela, L., Kostrowicki, J. and Scheraga, H. A. (1989) *J. Phys. Chem.* **93**, 3339-3346.

Sali, A., Shakhnovich, E. I., and Karplus, M. (1994) *Nature* **369**, 248-251.

Scheraga, H. A. (1968) *Adv. Phys. Org. Chem.* **6**, 103-184.

Scheraga, H. A. (1992) in *Reviews in Computational Chemistry*, Vol. 3, Eds. Lipkowitz, K. B. and Boyd, D. B., VCH Publ., New York, pp. 73-142.

Shakhnovich, E., Farztdinov, G., Gutin, A. M. and Karplus, M. (1991) *Phys. Rev. Lett* **67**, 1665-1669.

Shakhnovich, E. I. and Finkelstein, A. V. (1989) *Biopolymers* **28**, 1667-1680.

Shakhnovich, E. I. and Gutin, A. M. (1989) *Biophys. Chem.* **34**, 187-199.

Shakhnovich, E. I. and Gutin, A. M. (1990) *Nature* **346**, 773-775.

Shakhnovich, E. I. and Gutin, A. M. (1993) *Proc. Natl. Acad. Sci. U.S.A.* **90**, 7195-7199.

Skolnick, J. and Kolinski, A. (1990) *Science* **250**, 1121-1125.

Skolnick, J. and Kolinski, A. (1991) *J. Mol. Biol.* **221**, 499-531.

Taketomi, H. Ueda, Y. and Gō, N. (1975) *Int. J. Peptide Protein Res.* **7**, 445-459.

Vásquez, M., Némethy, G. and Scheraga, H. A. (1994) *Chem. Revs.*, **94**, 2183-2239.

Wawak, R. J., Wimmer, M. M. and Scheraga, H. A. (1992) *J. Phys. Chem.* **96** 5138-5145.

Yue, K. and Dill, K. A. (1994) *Phys. Rev. E.* **48**, 2267-2278.

42

PROTEIN SEQUENCE ANALYSIS, STORAGE AND RETRIEVAL

J. B. C. Findlay,[1] D. Akrigg,[1] T. K. Attwood,[2] M. J. Beck,[1] A. J. Bleasby,[3] A. C. T. North,[1] D. J. Parry-Smith,[1*] and D. N. Perkins[1]

[1] Department of Biochemistry and Molecular Biology,
The University of Leeds
Leeds LS2 9JT, United Kingdom
[2] Department of Biochemistry, University College London
Gower Street, London, United Kingdom
[3] SEQNET, Daresbury
Warrington, United Kingdom

INTRODUCTION

Studies of the structures and amino-acid sequences of proteins, and the relationships between structure and function, are increasing rapidly in terms of the quantity of information available, the number of groups engaged in the field and the areas in which the resultant knowledge is being applied. An obvious impetus results from the projects to determine the entire genetic constitution of humans and other species, which demands interpretation in terms of the proteins which the chromosomal nucleic acid encodes. Many such proteins have been characterised solely as the translations of open reading frames of the nucleic acid code and are likely to be of unknown function and 3-dimensional structure. It seems clear, however, that protein molecules and their constituent domains belong to families that have evolved from a common ancestor and that there may well only be between one and two thousand such families. This number should be compared to the number of proteins for which sequences are so far known (over 80000 in the current release of the OWL composite, non-redundant database (Akrigg *et al.*.1992; Bleasby & Wootton, 1990) and the 500 or so different proteins whose 3-dimensional structures are known.

The relative difficulty of determining experimentally both the functions and the 3-D structures of proteins means that there will be an ever-increasing number of proteins of potential interest, for which sequence, but not structural, information is available. While in some cases (for example the same protein from different species, or proteins that are very closely related in function) a family relationship is readily revealed by comparison of sequences, it is clear that proteins belonging to the same family, as evidenced by closely

* now at Pfizer Ltd., Sandwich, Kent, UK.

Methods in Protein Structure Analysis, Edited by M. Z. Atassi and E. Appella
Plenum Press, New York, 1995

similar chain folds, may have such dissimilar sequences that the relationship is obscure.or at least statistically tentative. Nevertheless, there may be evidence of the relationship through the occurrence of one or more sequence motifs that together constitute a "fingerprint" or "signature" that is characteristic and thus diagnostic of family membership. An example is provided by the lipocalin family of proteins. The half dozen members of this family for which 3-dimensional structures have been determined experimentally are found to share a similar chain topology the main feature of which is in 8-stranded β-barrel. Any two members of the family, however, may well have only about 20% sequence identity. Within their sequences, however, are embedded three regions that include a pattern of invariant or similar residues which, once they have been located, are seen to be characteristic of the family. Such motifs may have a functional or a structural role.

Further properties of the amino-acid sequences of proteins that may be indicative of family relationships include the periodicities of amino-acid type that are characteristic of secondary structure and the patterns of hydrophobic and hydrophilic side-chains that are indicative of internal or external environment or of an intramembranous location.

While a number of automatic alignment procedures have been devised, these are in general successful only when sequence similarity is fairly high. When the similarity is lower than about 20% identity, with the implication that there may be insertions or deletions in one or other sequence, automatic alignment becomes problematic. Manual alignment procedures may nevertheless be successful and we have found that colour-coding according to amino acid class may reveal patterns that the eye can pick up even when overall similarity is low.

When a 3-dimensional structure is available for one or more members of a protein family, alignment of sequences is greatly facilitated by the knowledge that insertions and deletions usually occur in surface loops and that structure-dependent properties of the aligned sequences should be consistent.

The above considerations demand computer software that allows for the input and interactive alignment of protein sequences, the pictorial display of 3-dimensional structure, the graphical display of amino-acid properties as a function of position within the sequence and access to databases of sequences, structures and motifs for the retrieval and deposition of data as required. This contribution describes the development of two intercalated databases and associated software which make up a storage, interrogation and retrieval system for the analysis of protein sequence and structure.

The increasing number of scientists in academic and industrial laboratories, whose research would be facilitated by the software include many whose use of computers is only occasional and who therefore require software that is easy and straightforward. The use of windows-based and menu-driven software has now become widespread in place of command-line driven programs and this has governed our approach in the development of the VISTAS suite described in this paper. The software has been designed on a modular basis and in such a way that it can be implemented on a variety of hardware platforms. The illustrations in this paper are based on the Silicon Graphics Iris implementation, but it is also operational on Digital Alpha AXP range and it can readily be ported to other Unix-based systems.

DATABASES

At the core of our system are two data-storage elements which together with associated software can be accessed at the UK EMBnet Node at Daresbury (SEQNET), the anonymous FTP address for which is s-ind2.dl.sc.UK. The first is a non-redundant composite protein sequence database (OWL) which at its last update contained nearly 80,000 entries, drawn from SWISS-PROT (Bairoch & Boeckman, 1991), NBRF-PIR 1 (George *et al.*, 1986),

NBRF-PIR 2, -PIR 3, NRL-3D (Namboodiri *et al* 1989) and GenBank (translation) (Burks *et al* 1989; Fickett, 1986). The source databases are assigned a priority with respect to sequence validation and their contents are amalgamated. Redundant/trivially different entries are eliminated according to defined criteria using the COMPO suite of programs (Bleasby & Wootton, 1990). OWL is in the NBRF format for compatibility with established search software and is interrogated using the query language DELPHOS to allow retrieval of sequence and textual material in the database.

Other modules in the system include SWEEP which incorporates best-local and complete sequence alignment algorithms based on the approach of Lipman & Pearson (1985) and which allow database searches with complete sequences. ADSP (Parry-Smith & Attwood, 1991) is an associated package which permits multiple sequence alignment and manipulation, local similarity detection and the development of discriminating sequence-based features. ADSP was designed to permit rigorous, iterative development of pattern-recognition discriminators which are diagnostic of the structural and functional characteristics of proteins or protein domains. These can then be compiled into a new second-generation biological database (PRINTS) containing the discriminators along with all the information (references, scan histories, etc.) and commentaries relevant to each PRINT entry. This new database currently contains 250 entries almost all of which encompass multiple discriminators (2000), i.e. the set of sequence motifs characteristic of the protein. The information is rapidly addressed and retrieved by the query language SMITE (Bleasby, unpublished) which shares a common syntax with DELPHOS. The database differs from PROSITE in that protein families are usuall characaterized by more than one motif.

The discovery of multiple discriminators and the development of an associated search system arose from the study of membrane-bound G-protein linked receptors. Even though sequence identity is well below the statistically significant level in these proteins, it is clear that the substitution patterns for particular positions and regions in the various sequences fall within fairly strictly defined limits. Thus it is possible, based on the primary structure data alone, to compile a series of 7 discriminators each describing a transmembrane segment (Attwood & Findlay, 1994). This analysis reveals that each of the transmembrane segments have their own special identity which clearly indicates features of structural importance within the super-family of receptors. There are within this overall class of receptor, sub-families which have their own distinctive elements. Furthermore, there are 7-transmembrane proteins which form their own special sub-groups related or completely unrelated to the G-protein linked receptors, e.g. the secretin sub-family.

The database now contains a large number of protein families and domains which possess multiple discriminators. In many cases individual motifs had been identified previously. These discriminators describe areas which may be of either structural or functional significance. In the case of the lipocalins mentioned earlier, we have discovered other members of the family such as a protein known hitherto as the "mouse 24p3 oncogene product", to which we are now able to ascribe both membership of a structural family and a putative function (Flower *et al.*, 1991, 1993).

All this information has been amalgamated into a single interactive piece of software VISTAS which is described below.

OPERATION AND DESIGN OF THE VISTAS PROGRAM

The principal facilities of the program are menu-driven, with selection of and from the menus being through use of the mouse buttons. The only occasions when keyboard entries are required are for the initial selection of data and parameter files and for communication with external programs for the retrieval and storage of data.

The two main types of data that VISTAS must manipulate are those related to structural and sequence information. Data for the pictorial display of 3-D structure are maintained in a pre-defined 3-D array corresponding to the scaled x, y and z co-ordinates of the atoms. As the data required to draw van der Waals and full-bond representations are potentially very memory-consuming, a linked list is used, with each structure being assigned when needed by using the dynamic memory allocation facilities of the C programming language.

Sequence data are stored in pre-defined structures rather than as linked lists. While this may be wasteful of memory and potentially limit the length and number of sequences that may be manipulated, it allows the program very quickly to locate selected sequences and residues within them without having to move along a linked list.

SOURCE DATA FORMATS AND DEFAULT PARAMETERS

Sequence data are handled in the standard NBRF/PIR format and may be read in as single sequences or as aligned sets of sequences. Coordinates of 3-D structures are read in standard PDB format. The program can handle protein molecules with one or more chain and can also represent ligands. Default files are available for colouring residue types on the screen and on postscript output. As default, the program uses the colours we have previously adopted for classifying residues, *viz.*:

- grey: alkyl chains (Ala, Val, Ile, Leu, Met)
- purple: aromatic residues (Phe, Tyr, Trp)
- red: acidic residues (Asp, Glu)
- blue: basic residues (Arg, His, Lys)
- green: polar uncharged residues (Ser, Thr, Asn, Gln)
- yellow: sulphydryl and disulphide residues (CysH, Cys)
- brown: conformational exceptions (Gly, Pro)

Users may specify any number of alternative colour schemes if they wish. Other default parameters for which alternatives can be specified include: colours to be used to indicate variations in properties such as hydrophobicity; substitution matrix values; secondary structure propensity data etc.

THE SEQUENCE WINDOW

This window allows the display and manipulation of up to 20 sequences at a time out of up to 500 stored in the computer. The sequences are colour-coded, initially according to side-chain property. The sequences may be scrolled vertically or horizontally by use of the l and r keys.

The mouse cursor may be used to mark positions and control insertions or deletions, thus allowing sequences to be edited and aligned. The colour coding is found to be valuable in achieving alignments, which frequently depend upon the recognition of similarities of property rather than of identities.

Sequences may be grouped so that insertions or deletions are incorporated similarly in all members of the group, and anchor points may be defined to allow insertions or deletions to be made while leaving established parts of the alignment intact. The order in which sequences appear in the window may be changed to facilitate comparison. New sequences may be added to the alignment, unwanted ones deleted and alignments stored for future use.

Sequence motifs may be defined by use of the mouse and then stored, with their locations shown also on the structure and graph window. The selected motifs may also be used to scan the OWL database to check for their occurrence in other proteins.

Integration of the scanning procedures with VISTAS yields two major benefits: first, any additional protein sequences that are identified as being of interest may immediately be displayed in the sequence window and analysed; second, newly identified motifs may be fed into the PRINTS definition and refinement modules, thereby updating the PRINTS database. The "Find motif" menu command allows the user to type in a short motif sequence; VISTAS then carries out a fuzzy search for the occurrence of the motif within the displayed sequences.

Automatic alignment procedures may be invoked as an alternative to manual alignment through interfaces to appropriate programs, including CLUSTAL, and the global sequence searching programs FASTA and SWEEP may be used to extract and display sequences related to those previously selected.

Navigation through the sequences is facilitated by a ruler which may be set to indicate the residue numbers corresponding to any one of the individual sequences displayed or to the alignment as a whole.

THE STRUCTURE WINDOW

This window allows the display of the 3-D structure of the protein corresponding to one of the sequences, pre-selected on entry to VISTAS. In designing VISTAS, no attempt was made to emulate the many comprehensive modelling packages that are now widely available, but a number of basic manipulative features are provided: they include scaling, clipping , translation and rotation. Alternative styles of display include Cα trace, skeletal, Cα ball-and-spoke, all atoms ball-and-spoke, space-filling and van der Waals atoms. Protein ligands may be displayed and manipulated independently from the protein itself. Ligand atoms may be depicted with double van der Waals radii, a feature that is useful in searching for close contacts with the protein, which is represented in skeletal form. This is useful for pinpointing the residues involved with ligand interactions.

The structures can be coloured by residue type (in conformity with the sequence window) or, as described below, with selected motifs or specified properties. The matrix corresponding to a chosen view of the molecule can be stored for later use.

THE GRAPHICS WINDOW

This window, which is invoked optionally, allows the display of a variety of properties as a function of position in the sequence. The properties include:

1. Positional variability, i.e. the number of different types of amino acid that occur at a particular position in an aligned set of sequences. This may be a simple count of different residue types or it may be a weighted sum, with weights derived from a specified substitution matrix (*e.g.* that of Risler *et al.* (1988)).
2. Secondary structure propensity, as determined by the Garnier, Osguthorpe & Robson (1978) algorithm. The graphical display shows the computed propensities for the four canonical conformations (turn, coil, α helix, β strand) and the sequence alignment and structure may be coloured accordingly. Although secondary structure prediction remains notoriously unreliable, the use of an alignment allows the joint propensities to be evaluated, which improves prediction, and,

taken together with the known structure of at least one member of a family, is an aid to homology modelling.

3. Hydropathy, as determined by one of three methods, those of Kyte & Doolittle (1982), Sweet & Eisenberg (1983) or Engelman *et al.* (1986). Again, the calculation can be derived from a single sequence or from the aligned set, and the sequences and structure may be coloured to show the calculated value.
4. Solvent accessible area parameter, using the scale derived by Rose *et al.* (1985) by calculating the mean solvent accessible area of each residue type in 23 proteins of known structure, and evaluated with a window length of five residues.
5. Flexibility, using a scale derived by Ragone *et al.* (1989) of the mobility of residues (as evidenced by thermal parameter, B) in proteins of known structure, evaluated with a window length of ten residues.

Clearly, the properties hydropathy, solvent-accessible area and flexibility are closely correlated, but each different aspect of amino-acid properties is an aid to sequence alignment and the detection of homology.

INTEGRATION OF WINDOWS

As indicated in preceding sections, the sequence and structure displays are both initially coloured by residue type; when a graph of properties in invoked, sequence and structure displays will normally become coloured to represent the variation in those properties. It is, however, straightforward to use different colour schemes in all three windows - for example, the sequence could be coloured according to residue type, and the structure according to residue variability while the graph displayed hydrophobicity.

Further features of the program include the ability to point to a residue, or a sequence of residues, in any of the windows and it will then be high-lighted in the other windows. Thus, an invariant residue (colour-coded as having zero variability in the graph window) may be located in the structure window, or a region shown in the graph window as expected to be very flexible can be located in the sequence and structure windows. Motifs selected in any of the windows will be high-lighted on the others and their sequences may be written to a file.

INTERFACES TO OTHER PROGRAMS

Reference has already been made to the ways in which the display features of VISTAS interact with other procedures. These fall into two categories. First, the PRINTS database scanning module is an integral part of VISTAS. Motifs selected by the user of VISTAS can be submitted directly to a motif database-scanning routine with similar functionality to that of ADSP (Parry-Smith & Attwood, 1991), which adopts an iterative procedure to refine a matrix describing the motifs in terms of the frequency of occurrence of the 20 types of amino acid at each position of each motif. At each stage of the refinement, a hit-list is generated of the protein sequences that include the motifs. The amino acids occurring in any newly-identified family members are then incorporated to modify the matrix appropriately. The SCAN and COMPARE algorithms of VISTAS allow the user to search the existing entries in the PRINTS database for any potential motif identified by inspection of the sequence, structure and graphics windows and then to update existing entries, or create new entries, in PRINTS by invoking the scanning and analysis procedures.

Second, the DELPHOS software used to scan the OWL database may be called directly from VISTAS in a separate window; DELPHOS allows OWL to be scanned, either by obligatory or fuzzy searches, for the occurrence of whole or partial sequences, bibliographic or textual information in the databases. Boolean procedures allow the combination of queries - for example, for the occurrence of the sequence "A*CDEF" in a protein from "Homo sapiens" in a paper by "Smith & Jones".

Third, it is possible for VISTAS to initiate as background jobs external programs such as the FASTA and SWEEP global sequence searching programs.

THE ALIGN PROGRAM

ALIGN is essentially a version of VISTAS that is designed to offer the sequence and graphical windows of VISTAS but without the structure display facilities. It is aimed at studies of sequences and sequence alignments in cases where no 3-dimensional structure information is available. The omission of the structure display window allows the interactive display of a much larger number of sequences.

COMPARISON WITH OTHER SOFTWARE WITH RELATED FUNCTIONALITY

There exist several other systems that offer some of these facilities, but none, to our knowledge, provide the combination provided by VISTAS for an integrated and user-friendly analysis of protein sequences, structure and properties.

ACKNOWLEDGMENTS

We gratefully acknowledge the support of the SERC Protein Engineering Club in the early phases of this project.

REFERENCES

Akrigg, D., Attwood, T.K., Bleasby, A.S., Findlay, J.B.C., North, A.C.T., Maughan, N., Parry-Smith, D.J. & Perkins, D.N. (1992) *CABIOS* **8**, 295-296.

Attwood, T.K. & Findlay, J.B.C. (1994) *Protein Engng.* **7**, 195-203.

Bairoch, A. & Boeckmann, B. (1991) *Nucleic Acids Res.* **19**, Suppl., 2247-2249.

Barton, G.J. & Sternberg, M.J.E. (1990) *J. Mol. Biol.* **212**, 389-402.

Bleasby A.J. & Wootton, J.C. (1990) *Protein Engng* **3**, 153-159.

Burks, C., Fickett, J.W., Goad, W.B., Kanehisa, M., Lewitter, F.I., Rindone, W.P., Swindell, C.D., Tung, C-S. & Bilofsky, H.S. (1986) *Comput. Applic. Biosci.* **1**, 225-233.

Engelman, D.M., Steitz, T.A. & Goldman, A. (1986) *Ann. Rev. Biophys. & Biophys. Chem.* **5**, 321-325.

Fickett, J.W. (1986) *Tends Biochem. Sci.* **11**, 190.

Flower, D.R., North, A.C.T. & Attwood, T.K. (1992) *Biochem. Biophys. Res. Comm.* **180**, 65-74.

Flower, D.R. North, A.C.T. & Attwood, T.K. (1993) *Protein Science* **2**, 753-761.

Garnier, J., Osguthorpe, D.J. & Robson, B. (1978) *J. Mol. Biol.* **120,** 97-120.

George, D.G. Hunt, L.T. (1986) *Nucleic Acids Res.* **14**, 11-15.

Kyte, J. & Doolittle, R.F. (1982) *J. Mol. Biol.* **157,** 105-132.

Lipman, D.J. & Pearson, W.R. (1985) *Science* **227,** 1435-1441.

Namboodiri, K., Pattabiraman, N., Lowrey, A., Gaber, B., George, D.G. & Barker, W.C. (1989), *NRL-3D PIR Newslett* **8**, 5.

North, A.C.T. (1989) *Int. J. Biol. Macromol.* **11**, 56-58.
Parry-Smith, D.J. & Attwood, T.K. (1991) *Comput. Applic. Biosci.* 7, 233-235.
Person, W.R. & Lipman, D.J. (1988) *Proc. Natl. Acad. Sci. USA* **85**, 2444-2448.
Ragone, R., Facchiano, F., Facchiano, A., Pacchiano, A.M. & Colonna, G. (1989) *Protein Engng* **2**, 497-504
Risler, J.L., Delorme, M.O., Delacroix, H, Henaut, A. (1988) *J. Mol. Biol.* **204**, 1019-1029.
Rose, G.D., Geselowitz, A.R., Lesser, G.J., Lee, R.M.& Zehfus, M.H.. (1985) *Science,* **229**, 834-838.
Sweet, R.M. & Eisenberg, D. (1983) *J. Mol. Biol.* **171**, 479-488.

43

SUPERFAMILY AND DOMAIN

Organization of Data for Molecular Evolution Studies

Winona C. Barker,[1] Friedhelm Pfeiffer,[2] and David G. George[1]

[1] Protein Information Resource
National Biomedical Research Foundation
Washington DC, 20007
[2] Martinsried Institute for Protein Sequences
Max Planck Institute for Biochemistry
Martinsried, Germany

THE ORIGINAL SUPERFAMILY CONCEPT

In the mid 1970's, Dayhoff proposed that all naturally occurring proteins would cluster into families and superfamilies whose members have diverged from common ancestral forms (Dayhoff et al., 1975; Dayhoff, 1976). A similar proposal was made by Emil Zuckerkandl (1975). Estimates of the number of protein superfamilies were in the low thousands. Recently this estimate has been reassessed. Using a variety of criteria for superfamily membership, estimates of the same order of magnitude as the original Dayhoff estimate have been reported (Green et al., 1993; Gonnet et al., 1992; Chothia, 1992).

Although superfamily relationships in some cases would be so ancient as to preclude recognition solely on the basis of sequence similarity, our group used sequence similarity as the main criterion for partitioning the Protein Sequence Database into independent, nonoverlapping groups. The nearly 500 completely sequenced proteins then known were each assigned to one of 116 superfamilies (Dayhoff et al., 1976). At that time, there were no examples in the database of complete precursor sequences, of polyproteins, or of products of alternative splicing of mRNA. Most of the known sequences were of mature forms of peptides or proteins. The longest completely sequenced proteins were 500-600 residues (serum album, prothrombin, immunoglobulin mu heavy chains, glutamate dehydrogenase). There were a few examples of the existence of sequence regions (domains) in some members of a superfamily but not in others. Prothrombin clearly had a large amino-terminal extension with respect to trypsinogen and other known members of this superfamily and that unique region showed evidence of internal duplication. There were also several examples of members of a superfamily containing variable numbers of such related sequence segments, which we called "homology regions." Pseudomonas rubredoxin contained two domains homologous to the other known rubredoxins, with an unrelated segment in between. Immunoglobulin C regions contained from one to four homologous domains; parvalbumin contained two domains that were clearly related to the four domains of troponin C and myosin light chains; and the known apolipoproteins contained from four to 18 repeats of an 11-residue pattern (Barker & Dayhoff, 1977).

The superfamily classification scheme provided an effective architecture for the intercomparison, correlation, and analysis of the nonsequence data associated with sequences within any particular homology class. Members of a superfamily were partitioned into closely related groups (families) of proteins, usually homologs in various species or products of more recent gene duplications; these could reasonably be expected to share many structural and functional characteristics. Indeed, much of the biological information concerning protein sequences reported in the published literature has been inferred by homology with closely related sequences. These observations have recently been confirmed by a systematic study (Sander & Schneider, 1991).

Homology provides a sound basis for the verification of information in the Protein Sequence Database and for induction of new knowledge concerning these data. Given that there is justification for such inference among members of homology classes, new sequences can directly inherit annotation information associated with existing homologous sequences. This provides a mechanism for comprehensive and consistent annotating throughout the database. Moreover, as new experimental information becomes available, it can be applied uniformly to entire classes of homologous sequences.

PROTEIN DOMAINS: LIMITATIONS OF THE ORIGINAL SUPERFAMILY CONCEPT

Within a few years of the introduction of the superfamily concept, it became evident that many protein sequences contained regions of local similarity with otherwise unrelated proteins. In many cases such domains were clearly responsible for the similar properties (such as calcium binding, DNA binding, or catalytic activity) shared by diverse proteins. In other cases the properties associated with a particular domain remained to be discovered. Evidence from X-ray crystallography and chemical studies revealed that these often corresponded with compact regions of the structure or with easily cleaved fragments. More surprising was the discovery that the genes for many proteins contain noncoding regions (introns) that divide the protein coding region into exons that sometimes, but not always, approximately correspond with the domains as defined by structure or protein sequence. "Exon shuffling" among genes is now recognized as an important mechanism in the evolution of "new" proteins.

In the literature, terms such as "the immunoglobulin superfamily" came to mean the collection of all proteins that contain one or more immunoglobulin-related domains (Hunkapiller & Hood, 1986; Williams & Barclay, 1988). Initially the members that were not classical immunoglobulins (beta-2 microglobulin, T-cell receptor chains, poly-Ig receptor) contained only various numbers of immunoglobulin-like domains, but later such domains were found associated with various other domains including protein kinase and protein-tyrosine-kinase domains. As the term "superfamily" is commonly used, these and other multidomain proteins would be placed into (at least) two superfamilies. In such a usage, the superfamily concept fails to partition the sequence data. For the general scientific public, this may be of little concern, but for those trying to organize sequence data it creates dilemmas.

PROTEIN DOMAINS AND EVOLUTIONARY STUDIES

Because different domains in proteins may have different evolutionary origins and histories, a complete and accurate understanding of the evolution of a given protein may require treating each domain of the protein as a separate entity. Only when it can be shown that the evolutionary history of each domain is "congruent" (Nakayama et al., 1992) with the history of the entire molecule is one justified in constructing a phylogeny encompassing the entire molecule. Unfortunately, many domains are rather small so that derived topologies are often based on very few useful and unambiguously informative sites. This principle is well illustrated by the extensive studies of the evolution of proteins containing EF-hand domains (Moncrief et al., 1990; Nakayama et al., 1992). These proteins, which contain from two to eight repeats of the domain, have been clustered by these workers into 29 types. Of ten types that each contain four repeats of the domain, eight were judged to have evolved from a common four-domain ancestor, which itself was formed by duplication of a two-domain precursor. These include the well-known calmodulin, troponin C, myosin essential light chain, and myosin regulatory light chain, as well as caltractin (Chlamydomonas) and CDC31 (Saccharomyces), squidulin from Loligo, call from Caenorhabditis, and calcium-dependent protein kinase from soybean. Two other types, calpain and sarcoplasmic calcium-binding protein, have each evolved independently into the four-domain form from a single-domain ancestor. It would be misleading to derive an evolutionary model based on the entire sequences (four domains) that includes these noncongruent forms.

REVISED SUPERFAMILY CONCEPT

Because of the fundamental role played by the superfamily concept in organizing the Protein Sequence Database, we have recently developed a formal model that encompasses most common usages of the term superfamily and provides an architecture for partitioning the database into domain superfamilies based on sequence homology. This architecture provides a mechanism for systematic analysis and refinement of information induced by homology.

In this model, the concepts of superfamily and family are generalized to encompass any scheme for classifying proteins (or regions within proteins) that partitions the proteins (or protein regions) into hierarchically nested sets that are closed under transitivity, i.e., if members A and B are in the same set and B and C are in the same set, then A and C are also in the same set. Formally, a superfamily is a union over families. Families are sets within the superfamily hierarchy for which the members meet a threshold level of relatedness. The threshold concept of families is based on empirical evidence that a threshold can be established whereby closely related proteins can be inferred to share common biological properties. Beyond this threshold more distantly related forms may diverge in these properties (although clear evidence of their relatedness remains).

We have applied this model to establish a classification of protein sequence homology domains. We define a homology domain as a subsequence that is related by common evolutionary ancestry to other sequence domains of the same homology class. Domains may encompass the entire protein sequence, in which case they are denoted as "homeomorphic" domains. Note that domains need not be contiguous and that the classification is based on the conceptually complete protein sequence, i.e., when only fragmentary data are available, these data may be classified provided that there is sufficient data available to allow the assumption that the missing data conform with the established relationships. This provides a natural mechanism for handling domains broken by intervening "loops" and for classifying fragmentary data. In particular, we consider conceptually complete homeomorphic domains

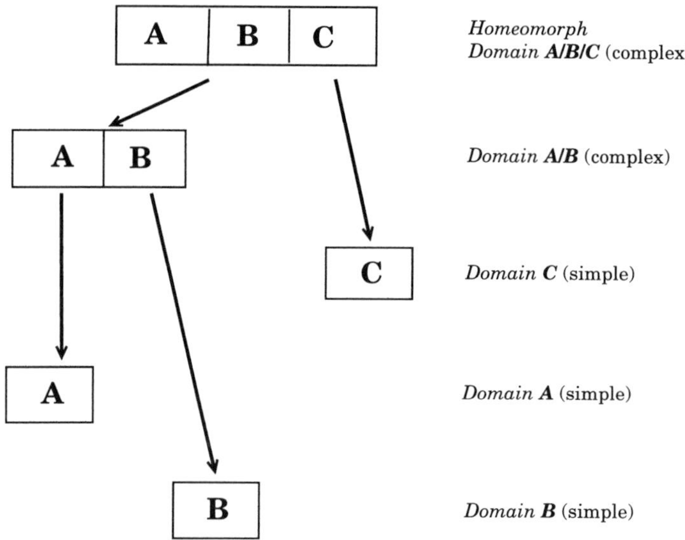

Figure 1. The homeomorphic domain corresponds with the complete sequence and may contain overlapping complex and simple domains.

to be those that can be expected to comprise the entire protein sequence even though the sequence data may be fragmentary and therefore such homology cannot be established directly. Moreover, relatively small intervening "loops" can be ignored provided that there is no evidence that these loops have established an independent evolutionary identity, i.e., they cannot be recognized in other related protein sequences.

Partitioning is achieved by independently treating homology domains containing overlapping regions. It has been observed that protein domains often coalesce, forming a composite domain that evolves as a unit after the concatenation of the original evolutionarily independent domains. Moreover, examples of the original domain may remain in isolated form and may continue to evolve independently. We denote domains that do not contain other domains as "simple" and denote composite domains as "complex." Simple and complex domains are classified along their entire extent. Composite domains are classified independently of the simple (and/or complex) domains from which they are composed. For example, the complex domain A/B is recognized as a homology class that is independent of the two simple domain classes A and B (see Figure 1). Superfamilies are constructed for all three of these classes. The overlap is not ignored; however, it is not considered within the superfamily classification scheme. A separate classification will be developed to characterize the relationships among overlapping domains.

This approach allows organization of protein sequences by homeomorphic family and superfamily while simultaneously characterizing the data explicitly by domain. The Superfamily record in a database entry contains a list of the names of the homology domain superfamilies into which the domains identified in the sequence have been classified. These names are descriptive and contain the word "homology." The name of the homeomorphic superfamily into which the conceptual complete sequence has been classified appears first on the list and does not contain the word "homology." Previously described methods to classify proteins without specifically addressing their domain architecture and effectively handling fragmentary data do not partition the data rigorously (Harris et al., 1992; States et al., 1993; Henikoff & Henikoff, 1991, 1993).

ORGANIZATION OF SEQUENCE DATA

Using this model, we are converting the placement groups in the database into homeomorphic protein superfamilies and families and are developing protocols to assign sequences into families. In addition, we are attempting to identify and annotate all homology domains. Previously, we classified only the best characterized and most fully annotated entries in the database. The project to classify all of the sequences, regardless of their state of annotation, has proceeded in stages. For practical reasons, we have changed our approach from one of classifying sequences into superfamilies, which were then further subdivided, to one of classifying sequences into families, which can then be grouped into superfamilies. Dr. Friedhelm Pfeiffer and other collaborators at the PIR-International database center at the Martinsried Institute of Protein Sequences (MIPS) embarked upon the family classification project. The procedure followed is iterative and began from the family structure in Release 36 of the database. Having first identified the unclassified sequences that belong in already defined families, the remaining unclassified sequences were grouped beginning with the longest sequences.

So that the family classification procedure can be highly automated, we adopt a working definition of a protein family as a set of conceptually complete sequences that can be aligned end-to-end without major discrepancy by standard multiple sequence alignment methods. In practice, such sequences will have the same domain architecture and an overall sequence identity of at least approximately 50%. This is virtually the same range for which the three-dimensional structure for a protein can be confidently predicted from that of a homolog whose structure has been determined (Sander & Schneider, 1991).

The classification project is facilitated by the FASTA Database, which contains all scores above a certain threshold from FASTA searches (Pearson & Lipman, 1988) of each sequence against all others. The database is updated each time the sequence database is updated (currently weekly or biweekly) and an interactive retrieval system allows annotators to query it (see Figure 2). The FASTA Database is used for the selection of candidates for classification into a family. The sequences are then aligned automatically and the aligned pairs are screened for congruence of length and threshold level of similarity. Those that meet

```
$ fastadb

     Please, enter ENTRY CODE or e to exit
   > a29002

ENTRY A29002

   ENTRY                  TITLE                              initn initl   opt
PIR2:A29002     phospholamban - dog                           275    275   275
PIR3:S05540     phospholamban - pig                           275    275   275
PIR2:A26805     phospholamban - dog                           275    275   275
PIR2:S00249     phospholamban - rabbit                        265    265   274
PIR2:B40424     phospholamban - rabbit                        265    265   274
PIR3:A49057     phospholamban - mouse                         265    265   274
PIR3:S37638     phospholamban - rat                           265    265   274
PIR2:A40424     phospholamban - human                         264    264   273
PIR2:A39535     phospholamban - chicken                       236    236   245
PIR2:A24818     phospholamban - dog (fragment)                207    207   207
PIR2:A25307     phospholamban - dog (fragment)                183    183   183

         found 11 entries with opt. score > 80
```

Figure 2. Retrieval of scores from the FASTA database. On the basis of this interactive display, the sequences from dog and rabbit (respectively) were merged to produce single entries. The resulting seven sequences were aligned and assigned placement numbers. The original papers were then reviewed and annotation made consistent throughout this family.

rather stringent requirements are routinely classified; others are examined and classified by scientific staff. As of Release 41, over 70% of the sequences in the database had been analyzed. Over 90% of those were classified as belonging to a family or as being the sole representative of a new family. About 5% of all entries are considered not classifiable by this method, generally because the sequences are too short or fragmentary.

The detection of homology domains within sequences and their classification can be approached in a similar way. The group at MIPS periodically partitions all of the sequences into recognized homology domains (annotated as features in the database) and unclassified subsequences. A database is created that contains FASTA scores (above a threshold) of all of these subsequences against each other. Pairwise local alignments are constructed and used to select the boundaries of provisional domains. These assignments are later checked and refined on the basis of multiple sequence alignments. Once the known domains are substantially annotated in the database, FASTA scores of unclassified segments searched against all other unclassified segments will be used to reveal additional domain homologies between sequences classified into different families.

INFERENCE BASED ON HOMOLOGY

Assigning sites of biological interest by homology requires the construction of a multiple sequence alignment. Mathematically rigorous algorithms for multiple sequence alignment cannot guarantee biologically realistic alignments, particularly for more distantly related sequences. Nevertheless, for sequences and subsequences that are longer than about 50 residues and that are at least 35-50% identical, the major features of an alignment are reproduced by many algorithms. Within this realm, alignments derived by comparison of three-dimensional structures also agree well with those derived by sequence comparison methods (Sander & Schneider 1991).

Sequence homology among domains does not guarantee close structural homology or preservation of function. For example, some calmodulin repeat homology domains may not adopt the E-F hand conformation or bind calcium. Nevertheless, a combination of homology and other biological or chemical knowledge frequently allows properties of domains to be predicted. This in turn may allow prediction of functional characteristics of a multidomain protein even when it is the first sequenced example of its type. The homology domain superfamilies that have been identified and annotated in the PIR-International Protein Sequence Database, Release 42 (September 1994) are listed in Table 1.

Table 1. Homology domain superfamilies (September 1994)

(S)-2-hydroxy-acid oxidase homology	alanine dehydrogenase homology
3-dehydroquinate dehydratase homology	alpha-actinin actin-binding domain homology
3-dehydroquinate synthase homology	alpha-amylase core homology
3-hydroxyacyl-CoA dehydrogenase homology	animal Kunitz-type proteinase inhibitor homology
3-hydroxyisobutyrate dehydrogenase homology	ankyrin repeat homology
3-oxoadipate CoA-transferase alpha chain homology	annexin repeat homology
3-oxoadipate CoA-transferase beta chain homology	antileukoproteinase repeat homology
3-phosphoshikimate 1-carboxyvinyltransferase homology	apple homology
6-phosphofructokinase 1 homology	aspartate kinase homology
acetate—CoA ligase homology	aspartate/ornithine carbamoyltransferase homology
adenylate cyclase homology	astacin homology
ADP,ATP carrier protein repeat homology	Bacillus dihydroorotase homology
agrin inhibitor-like repeat homology	Bacillus phosphoribosylamine—glycine ligase homology

Table 1. *Continued*

barley yellow dwarf virus RNA-directed RNA polymerase homology	hemopexin repeat homology
Berne virus hemagglutinin homolog homology	herpesvirus tegument protein homology
beta-lactamase OXA2 homology	herpesvirus thymidine kinase homology
Bowman-Birk inhibitor repeat homology	hevein chitin-binding domain homology
BUD5 protein homology	hexokinase homology
C-type lectin homology	hisI bifunctional enzyme homology
C1r/C1s repeat homology	hisI protein homology
cadherin repeat homology	histidine—tRNA ligase homology
calmodulin repeat homology	histidinol dehydrogenase homology
calpain catalytic domain homology	HMG box homology
carbamoyl-phosphate synthase (ammonia) homology	homeobox homology
carbamoyl-phosphate synthase (glutamine-hydrolyzing) large chain homology	homoserine dehydrogenase homology
cellular retinaldehyde-binding protein homology	human herpesvirus 1 UL35 protein homology
Chalara lysozyme homology	imidazoleglycerol-phosphate dehydratase homology
cholinesterase homology	immunoglobulin homology
cold shock domain homology	influenza C virus nonstructural protein NS1/NS2 homology
complement factor H repeat homology	Kazal proteinase inhibitor homology
cpl repeat homology	kringle homology
crk transforming protein homology	lactaldehyde reductase homology
cystatin homology	large structural phosphoprotein homology
cytochrome b5 core homology	LDL receptor ligand-binding repeat homology
cytochrome b6 homology	LDL receptor YWTD-containing repeat homology
cytochrome c homology	LDL receptor/EGF precursor homology
cytochrome c3 homology	leucine-rich alpha-2-glycoprotein repeat homology
cytochrome-b5 reductase homology	leukocyte common antigen cytosolic domain homology
cytokine receptor homology	LIM metal-binding repeat homology
desulforedoxin homology	lipocalin homology
discoidin I N-terminal homology	LTE1 protein homology
dnaJ N-terminal domain homology	malK protein homology
EGF homology	MAP2/tau repeat homology
elongation factor Tu homology	methylated-DNA—protein-cysteine S-methyltransferase homology
endozepine homology	methylphosphotriester-DNA methyltransferase homology
equine herpesvirus 1 glycoprotein homology	motor domain homology
erbA transforming protein homology	myb DNA-binding repeat homology
ets DNA-binding domain homology	myc transforming protein homology
eubacterial ribosomal protein L27 homology	myosin head homology
eubacterial ribosomal protein S15 homology	NAD(P)+ transhydrogenase (B-specific) alpha chain homology
ferredoxin homology	NAD(P)+ transhydrogenase (B-specific) beta chain homology
ferroxidase repeat homology	NGF receptor repeat homology
fibrinogen beta/gamma homology	nifA central domain homology
fibronectin type I repeat homology	orotate phosphoribosyltransferase homology
fibronectin type II repeat homology	orotidine-5'-phosphate decarboxylase homology
fibronectin type III repeat homology	osteonectin homology
flavodoxin homology	parathyroid hormone homology
gelsolin repeat homology	peptidylglycine monooxygenase I homology
glutamate receptor homology	phage T4 DNA topoisomerase (ATP-hydrolyzing) medium chain homology
gramicidin S synthetase I repeat homology	phage T4 lysozyme homology
guanylate cyclase catalytic domain homology	phosphoprotein phosphatase homology
guanylate kinase homology	phosphoribosylaminoimidazole carboxylase carbon dioxide-fixation chain homology
H+-transporting ATP synthase alpha chain homology	
Helicobacter urease alpha chain homology	
Helicobacter urease beta chain homology	
hemolysin A homology	

Table 1. *Continued*

phosphoribosylaminoimidazole carboxylase catalytic chain homology	shikimate kinase homology
phosphoribosylformylglycinamidine cyclo-ligase homology	sigma factor katF homology
phosphoribosylglycinamide formyltransferase homology	sigma factor region 1 homology
phosphotransferase system glucose-specific enzyme II, factor II homology	spectrin/dystrophin repeat homology
phosphotransferase system glucose-specific enzyme II, factor III homology	statherin/histatin signal sequence homology
phosphotransferase system mannitol-specific enzyme II/III homology	subtilisin homology
plastoquinol—plastocyanin reductase 17K protein homology	sucrase/isomaltase homology
pleckstrin repeat homology	sucrose/sucrose-phosphate synthase homology
potato leaf roll virus coat protein homology	sulfite oxidase homology
POU domain homology	thioredoxin homology
protein 4.1 membrane-binding domain homology	thymidylate synthase homology
protein kinase C C2 region homology	trefoil homology
protein kinase C zinc-binding repeat homology	trpC homology
protein kinase homology	trpD homology
protein kinase regulatory chain nucleotide-binding repeat homology	trpD-trpG homology
protein-glutamate O-methyltransferase homology	trpF homology
protein-tyrosine-phosphatase homology	trpG homology
rel homology	trypsin homology
response regulator homology	tryptophan synthase alpha chain homology
ribonucleoprotein repeat homology	tryptophan synthase beta chain homology
rRNA N-glycosidase homology	type I dihydrofolate reductase homology
rubredoxin homology	ubiquinol—cytochrome-c reductase 11K protein homology
S-locus-specific glycoprotein homology	ubiquitin homology
serum albumin repeat homology	V/P protein homology
SH2 homology	vaccinia virus 13.6K HindIII-C protein homology
SH3 homology	vaccinia virus 27.4K HindIII-C protein homology
shikimate dehydrogenase homology	vaccinia virus 8.6K HindIII-C protein homology
	vaccinia virus 8.8K HindIII-C protein homology
	VH1-type protein-tyrosine-phosphatase homology
	villin headpiece homology
	virB4.1 protein homology
	virB4.2 protein homology
	187 superfamilies found

ACKNOWLEDGMENTS

We are grateful to the staff of PIR-International, whose dedication makes possible the implementation of the ideas presented here, especially to Susanne Leibl of MIPS for implementation of FASTA databases and to Drs. Lai-Su Yeh and Geetha Srinavasarao of PIR for their work in defining and annotating homology domains. We also thank Dr. Lois Hunt for helpful comments on the manuscript and Margaret C. Blomquist and Kathryn Sidman for assistance in manuscript editing and preparation. This work was partially supported by grant LM05206 from the National Library of Medicine.

REFERENCES

Barker, W.C., and Dayhoff, M.O. (1977) Comp. Biochem. Physiol. 57B, 309-315

Chothia, C. (1992) Nature 357, 543-544

Dayhoff, M.O. (1976) Fed. Proc. 35, 2132-2138

Dayhoff, M.O., Barker, W.C., and Hunt, L.T. (1976) in Atlas of Protein Sequence and Structure 1976, Vol. 5, Suppl. 2, Dayhoff, M.O., ed., pp. 9-19, National Biomedical Research Foundation, Silver Spring, MD

Dayhoff, M.O., McLaughlin, P.J., Barker, W.C., and Hunt, L.T. (1975) Naturwissenschaften 62, 154-161

Gonnet, G.H., Cohen, M.A., and Benner, S.A. (1992) Science 256, 1443-1445

Green, P., Lipman, D., Hillier, L., Waterson, R., States, D., and Claverie, J.-M. (1993) Science 259, 1711-1716.

Harris, N., Hunter, L., and States, D.J. (1992) in Proceedings of the Tenth National Conference on Artificial Intelligence (AAAI-92), MIT Press, Cambridge, MA, pp. 837-842.

Henikoff, S., and Henikoff, J.G. (1991) Nucl. Acids Res. 19, 6565-6572

Henikoff, S., and Henikoff, J.G. (1992) Proc. Natl. Acad. Sci. USA 89, 10915-10919

Hunkapiller, T., and Hood, L. (1986) Nature 323, 15-16

Moncrief, N.D., Kretsinger, R.H., and Goodman, M. (1990) J. Mol. Evol. 30, 522-562

Nakayama, S., Moncrief, N.D., and Kretsinger, R.H. (1992) J. Mol. Evol. 34, 416-448

Pearson, W.R., and Lipman, D.J. (1988) Proc. Natl. Acad. Sci USA 85, 2444-2448

Sander, C., and Schneider, R. (1991) Proteins: Struct. Funct. Genet. 9, 56-68

States, D.J., Harris, N.L., and Hunter, L. (1993) in Proceedings: First International Conference on Intelligent Systems for Molecular Biology, AAAI Press, Menlo Park, CA, pp. 387-394

Williams, A.F., and Barclay, A.N. (1988) Ann. Rev. Immunol. 6, 381-405

Zuckerkandl, E. (1975) J. Mol. Evol. 7, 1-57

44

MUTATIONAL ANALYSIS OF THE BPTI FOLDING PATHWAY

David P. Goldenberg, Jose A. Mendoza,* and Jian-Xin Zhang

Department of Biology
University of Utah
Salt Lake City, Utah 84112

INTRODUCTION

Over the past three decades, considerable effort has been focused on elucidating the mechanisms by which polypeptide chains fold into well-defined three-dimensional structures (Kim & Baldwin, 1990; Creighton, 1992a; Mattthews, 1993). Major goals of these studies include the identification and characterization of partially folded intermediates and the analysis of transition states that represent the energetic barriers in the folding mechanism. Recently, there has been great progress in the structural analysis of folding intermediates by high resolution NMR spectroscopy of intermediate analogs and native proteins that have been isotopically-labeled during refolding. Structural analysis alone, however, is not sufficient to determine why particular intermediates form or what types of interactions stabilize their conformations. By their very nature, transition states are even more difficult to characterize directly. Questions about folding energetics and the roles of individual interactions in determining the folding mechanism can often be addressed by studying the folding of protein variants that differ by relatively small perturbations of the covalent structure. Recently-developed genetic techniques have made it possible to alter virtually any amino acid residue in a protein, and mutational methods have now been used to study the folding mechanisms of several proteins (Fersht et al., 1992; Goldenberg 1992a; Jennings et al., 1992). We describe here some of our recent work using amino acid replacements to study the folding of a particularly well-characterized protein, bovine pancreatic trypsin inhibitor (BPTI).

BPTI is a small protein, composed of 58 amino acid residues, that folds into a single compact domain. The native conformation is stabilized by three disulfide bonds, and the protein can be unfolded by reducing the disulfides. Disulfide-bonded intermediates accumulate during the oxidative refolding of the reduced protein, and these intermediates can be chemically trapped, physically separated and characterized individually. The disulfide-coupled refolding pathway (Figure 1) for BPTI was first characterized by T.E. Creighton in the

* Present address: Department of Chemistry, California State University, San Marcos, CA 92096-0001.

Methods in Protein Structure Analysis, Edited by M. Z. Atassi and E. Appella
Plenum Press, New York, 1995

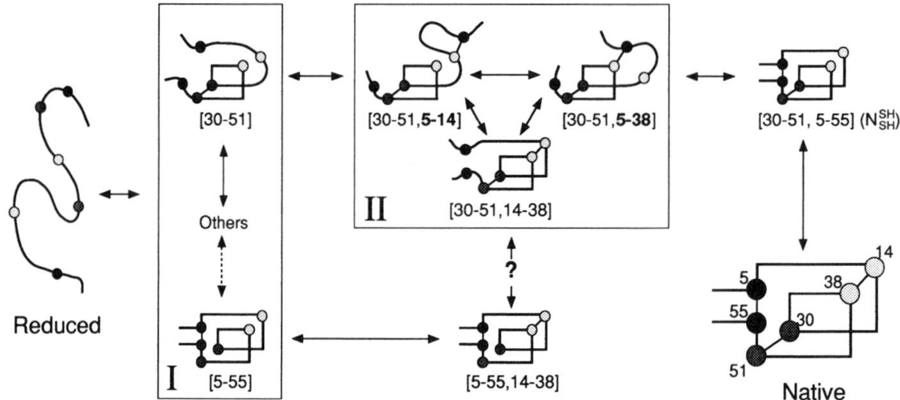

Figure 1. The BPTI folding pathway. Intermediates are identified by the disulfide bonds they contain, and the approximate path of the polypeptide chain in the intermediates is shown schematically. Intermediates grouped together in the boxes labeled I and II interconvert rapidly on the time scale of the folding experiments, except [5-55], which rearranges with the other one-disulfide intermediates relatively slowly. Reproduced, with permission, from Goldenberg, 1992b.

1970s and has continued to be the subject of extensive analysis (Creighton, 1978; Creighton & Goldenberg, 1984; Weissman & Kim 1991; Creighton, 1992b; Goldenberg, 1992b). During the past few years, new methods have been developed to trap and isolate the intermediates (Weissman & Kim, 1991), analogs of the intermediates have been analyzed by high-resolution NMR (Staley & Kim, 1992; van Mierlo et al., 1991ab, 1993, 1994), and mutational methods have been used to assess the energetic contributions of individual amino acid residues (Goldenberg et al, 1989, 1992; Coplen et al., 1990; Zhang & Goldenberg, 1993; Mendoza et al., 1994).

While the recent advances in characterizing the BPTI folding intermediates have led to a much better understanding of this pathway, the new results have also been accompanied by considerable controversy. Two of the more controversial aspects concern the origin of the intramolecular rearrangements in the pathway and the nature of the major transition states. As described in the following sections, we have attempted to address these questions by analyzing the folding mechanisms and kinetics of genetically modified BPTI variants (Zhang & Goldenberg, 1993; Mendoza et al., 1994).

KINETIC TRAPS AND THE ORIGIN OF INTRAMOLECULAR REARRANGEMENTS

One of the most striking aspects of the BPTI folding pathway is the role of intramolecular rearrangements in forming the three-disulfides of the native protein (Creighton, 1977a). Of the various two-disulfide intermediates that accumulate during folding, only one, containing the 30-51 and 5-55 disulfides and designated N_{SH}^{SH}, readily forms a third disulfide. But, this species does not form readily from the population of one-disulfide intermediates. Instead, the kinetically preferred pathway involves the formation of other two-disulfide intermediates, which then undergo intramolecular rearrangements to yield N_{SH}^{SH} (Creighton, 1977a; Goldenberg, 1988). Although N_{SH}^{SH} can, under certain conditions, form directly, the rate constant for the intramolecular step in forming the second disulfide is approximately 1,000-fold lower than for the competing reactions.

Because the thiol-disulfide exchange reactions involve a nucleophilic displacement of one sulfur atom in an existing disulfide by a thiol sulfur atom, the rearrangement(s) that produce N_{SH}^{SH} must involve at least one non-native disulfide. Two species with one non-native disulfide each, [30-51,5-14] and [30-51,5-38] have been detected in refolding reactions, and it seems likely that one or both of these species plays this role. Weissman and Kim (1991), using improved methods for trapping and isolating the intermediates, recently found that the non-native intermediates are somewhat less stable than earlier experiments had indicated. While this result might, at first glance, suggest that the non-native intermediates are less important than originally thought, the evidence for the rearrangement mechanism remains solid, and it is generally agreed that at least one non-native species must act as an intermediate, though it may accumulate to only low levels.

Two other intermediates that accumulate during refolding, [30-51,14-38] and [5-55,14-38], also contain two of the three native disulfides and have folded conformations very similar to that of the native protein, but these species do not readily form a third disulfide. In one case, [5-55,14-38], the failure to form the third disulfide is due to the inaccessibility of the remaining two cysteine thiols to the disulfide reagents (such as the disulfide forms of glutathione or dithiothreitol) used as oxidants (States et al., 1984; Creighton & Goldenberg, 1984). The other case, [30-51,14-38] is somewhat more complex, however, since the thiols are at least partially accessible to disulfide reagents (Creighton, 1977b). Here, it appears that at least part of the reason that the third disulfide does not form is that there are steric constraints that inhibit formation of the transition state for the intramolecular transition required for direct disulfide formation. At neutral or slightly alkaline pH, [5-55,14-38] accumulates for very long times and acts as a kinetic trap during folding, limiting the fraction of molecules that form a third disulfide even after several hours (States et al., 1984; Creighton & Goldenberg, 1984). The other species, [30-51,14-38], can rearrange with other two-disulfide intermediates at pH 8.7, but at pH 7.3 these rearrangements are very unfavorable and [30-51,14-38] also acts as a kinetic trap (Weissman & Kim, 1991).

Since intramolecular rearrangements are necessary to convert [5-55,14-38] or [30-51,14-38] to N_{SH}^{SH}, it might appear that the rearrangements in the BPTI pathway arise only because of the stabilities of the former two species and their inability to directly form a third disulfide (Weissman & Kim, 1992). We have recently described a mutant form of BPTI, in which Tyr 35 is replaced by Leu, that we believe helps clarify this issue (Zhang & Goldenberg, 1993). This variant is one of eight aromatic → Leu mutants we have constructed to examine the roles of the four Phe and four Tyr residues of the wild-type protein. Like most of the other aromatic residues in BPTI, Tyr 35 is largely buried in the native protein (Figure 2) and its replacement with Leu severely destabilizes the folded conformation. The folded mutant protein has circular dichroism spectra similar to those of the wild-type protein and is an active trypsin inhibitor, suggesting that the mutation has not altered the overall conformation of the protein.

The distribution of folding intermediates for the Y35 variant is shown in Figure 3, along with that for the wild-type protein. In this experiment, the proteins were unfolded by reducing the three disulfides, refolding was initiated at pH 7.3 by adding 0.1 mM oxidized glutathione (GSSG), and the reactions were quenched at the indicated times by acidification. The trapped intermediates were separated by reversed phase HPLC as described by Weissman and Kim (1991), and the disulfides in the isolated intermediates were determined by peptide mapping. The most pronounced effect of the mutation is the elimination of two intermediates, [30-51,14-38] and [5-55,14-38]. These are the two species that act as kinetic traps during refolding of the wild-type protein, and their absence greatly increases the rate at which the native protein appears.

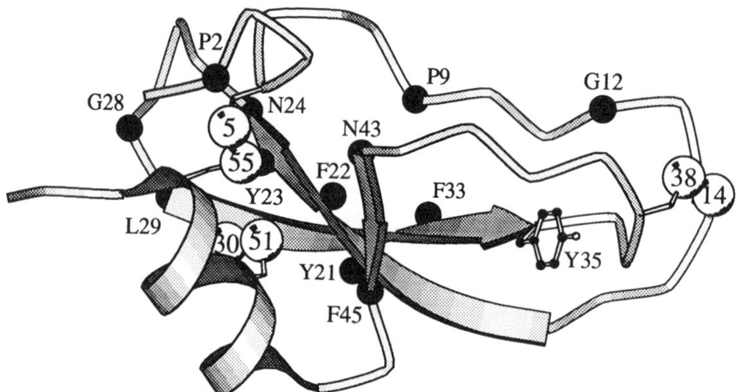

Figure 2. Schematic representation of the native structure of BPTI, showing the Tyr 35 side-chain and the sites of the amino acid replacements used to analyze the major transition states for unfolding and folding. Adapted from Figure 1 of Mendoza et. al, 1994, and drawn with the program Molscript by Per Kraulis.

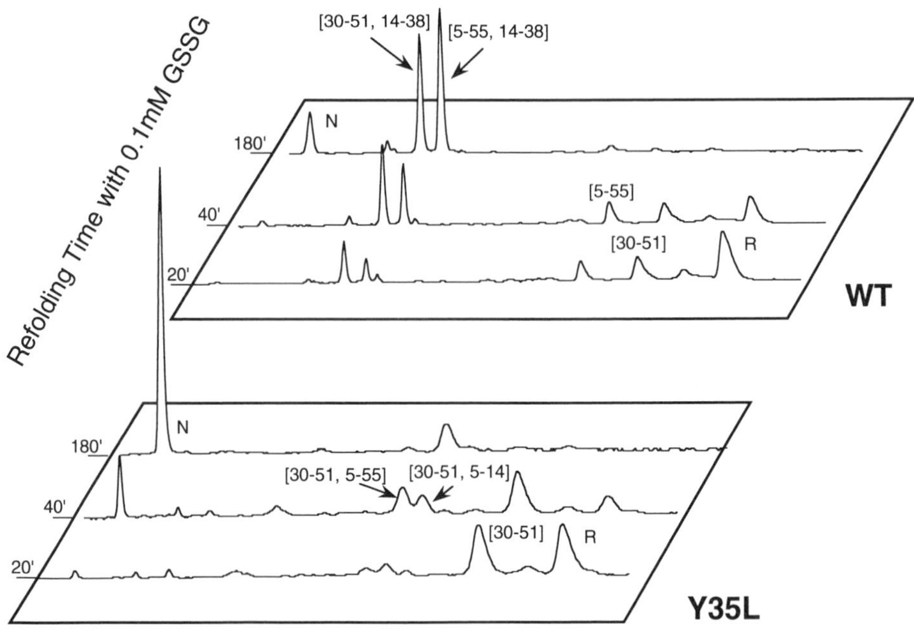

Figure 3. HPLC analysis of intermediates trapped at various times during the refolding of wild-type and Y35L BPTI. Refolding reactions were carried out at pH 7.3 in the presence of 0.1 mM GSSG. At the indicated times, the reactions were quenched by the addition of formic acid to a final concentration of 5%. The trapped intermediates were applied to a Vydac C_{18} column and eluted with a gradient of acetonitrile in 0.1% trifluoroacetic acid as absorbance of the eluent at 229 nm was monitored. The identities of the intermediates were determined by peptide mapping. Reproduced, with permission, from Zhang & Goldenberg, 1993.

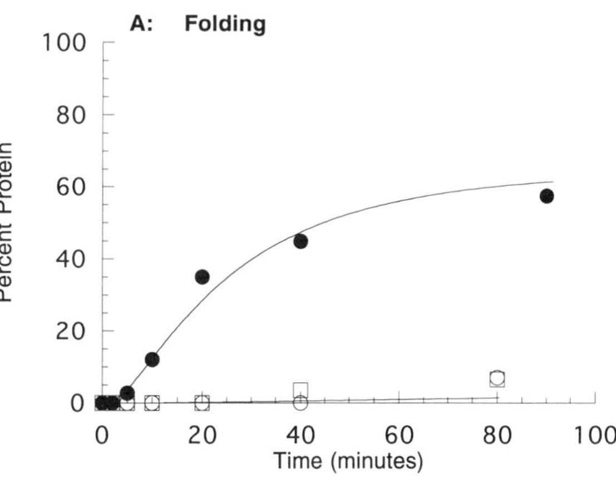

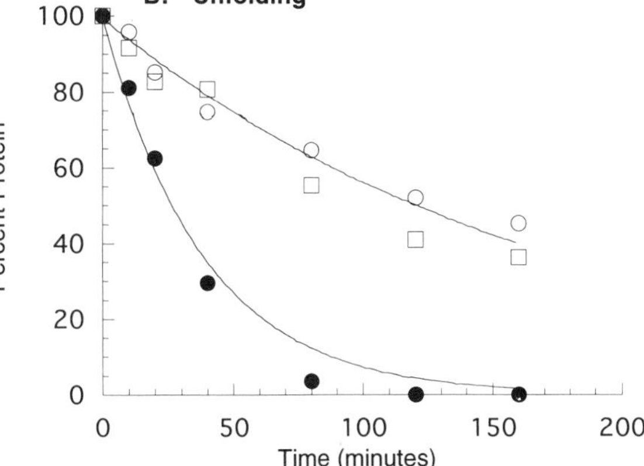

Figure 4. Kinetics of folding and unfolding of unmodified Y35L BPTI (filled circles) and modified forms in which the thiols of Cys 14 and 38 were blocked with iodoacetamide (open squares) or iodoacetate (open circles). Refolding was carried out in the presence of 80 mM oxidized dithiothreitol at pH 8.7, 25° C, while unfolding reactions contained 2 mM dithiothreitol but were carried out under conditions that were otherwise the same as for the refolding experiments. The refolding reactions were monitored by gel electrophoresis of samples trapped by reaction with iodoacetate at the indicated times. For the modified protein, the indicated percent folded represents the concentration of [30-51,5-55], while for the unmodified protein the percent folded represents the sum of [30-51,5-55] (N_{SH}^{SH}) and N. Reproduced, with permission, from Zhang & Goldenberg, 1993.

The major intermediates that accumulate during refolding of the Y35G protein are [30-51], [30-51,14-38] (N_{SH}^{SH}) and a smaller amount of [30-51,5-14]. Since the two kinetically-trapped intermediates are eliminated, it might be expected that the mutant protein would fold by a more direct pathway, such as

$$R \rightarrow [30\text{-}51] \rightarrow N_{SH}^{SH} \rightarrow N,$$

where R and N represent the fully reduced and native proteins. In order to test this possibility, we prepared modified forms of the mutant protein in which the 14-38 disulfide was selectively reduced to produce N_{SH}^{SH} and the resulting thiols alkylated with either iodoacetamide or iodoacetate. If the mutant protein folds primarily via the direct mechanism shown above, then the proteins lacking the Cys14 and 38 thiols should be able to form N_{SH}^{SH} as readily as the unmodified mutant protein. If, on the other hand, intramolecular rearrangements play a role in the formation of N_{SH}^{SH}, blocking these thiols should greatly reduce the rate of forming this species, as has been demonstrated previously for the wild-type protein (Creighton, 1977a; Goldenberg, 1988).

The folding and unfolding kinetics of the modified (open symbols) and unmodified (filled symbols) forms of the mutant protein are shown in Figure 4. The unmodified protein both folds and unfolds more rapidly than the forms in which the Cys 14 and 38 thiols have been alkylated. The rate constant for the intramolecular step in the formation of a second disulfide bond is approximately 200-fold greater for the unmodified protein than it is in the absence of the two thiols. These results indicate that during the refolding of the unmodified Y3 BPTI the preferred mechanism involves formation of two-disulfide intermediates other than N_{SH}^{SH}, followed by intramolecular rearrangements to produce this species. As in the folding of the wild-type protein, these rearrangements must involve at least one species containing a non-native disulfide, such as [30-51,5-14].

These results argue that the predominance of the rearrangement mechanism cannot be accounted for by the stabilities of the kinetically-trapped intermediates, since these species do not accumulate during the refolding of the mutant protein. Further, the low rate of forming N_{SH}^{SH} is not due to inaccessibility of the Cys 5 and 55 thiols in the [30-51] intermediate, since the thiols have been shown to react readily with oxidized glutathione (Creighton, 1977a; Goldenberg, 1988). Rather, the rearrangement mechanism is kinetically preferred because of the very low rate of the intramolecular step in forming N_{SH}^{SH} versus the other two-disulfide intermediates. What, then, is structural basis for this kinetic preference? The answer to this question must lie in the energetic and structural differences between the one-disulfide intermediates and the transition state for directly forming N_{SH}^{SH}. The structures of the intermediates have been characterized through NMR spectroscopy of analogs (van Mierlo et al., 1991a; 1993; Staley & Kim 1992), but that of the transition state can only be inferred from kinetic experiments, as described in the following section.

MUTATIONAL ANALYSIS OF THE MAJOR TRANSITION STATES

During the refolding of reduced BPTI, the slowest intramolecular processes are associated with the formation of N_{SH}^{SH}, by either direct disulfide formation from the one-disulfide intermediates, as discussed above, or by rearrangement of other two disulfide intermediates (II in Figure 1). The reverse of these reactions are also very slow; during reductive unfolding of the native protein, the 14-38 disulfide is rapidly reduced to produce N_{SH}^{SH}, but further reduction and unfolding takes several hours. Because the refolding and unfolding experiments are carried out under identical conditions, except for the concentrations of thiol and disulfide reagents present, the transition states for forming N_{SH}^{SH} during folding are expected to be equivalent to those for breaking down this species during unfolding. Thus, the transition states for forming N_{SH}^{SH} can be characterized by analyzing the kinetics of unfolding, which are experimentally more accessible than the folding kinetics.

In order to characterize these transition states, we have examined the effects of a series of destabilizing amino acid replacements on the kinetics of direct reduction of N_{SH}^{SH} and rearrangement of this species (Mendoza et al., 1994). The rationale of these experiments is illustrated in Figure 5. As shown in the figure, N_{SH}^{SH} can either undergo intramolecular rearrangements to generate other two-disulfide intermediates, such as [30-51,5-14], or can be directly reduced by reaction with a thiol reagent, shown as R-SH in the figure. In either reaction, formation of the transition state requires attack of one of the buried disulfides in N_{SH}^{SH} by a thiol, and at least part of the molecule must undergo an unfolding transition. It is possible, however, that different regions of N_{SH}^{SH} would have to unfold during the two processes. If this is the case, then substitutions that destabilize some regions of the protein would be expected to enhance selectively the rate of direct reduction, while other replacements might preferentially enhance the rate of rearrangement. Thus, it should be possible to characterize the two transition states by looking for amino acid replacements that selectively enhance one rate or the other.

Toward this end, we measured the rates of direct reduction and intramolecular rearrangement for 18 BPTI variants that were known to increase the overall rate of reductive

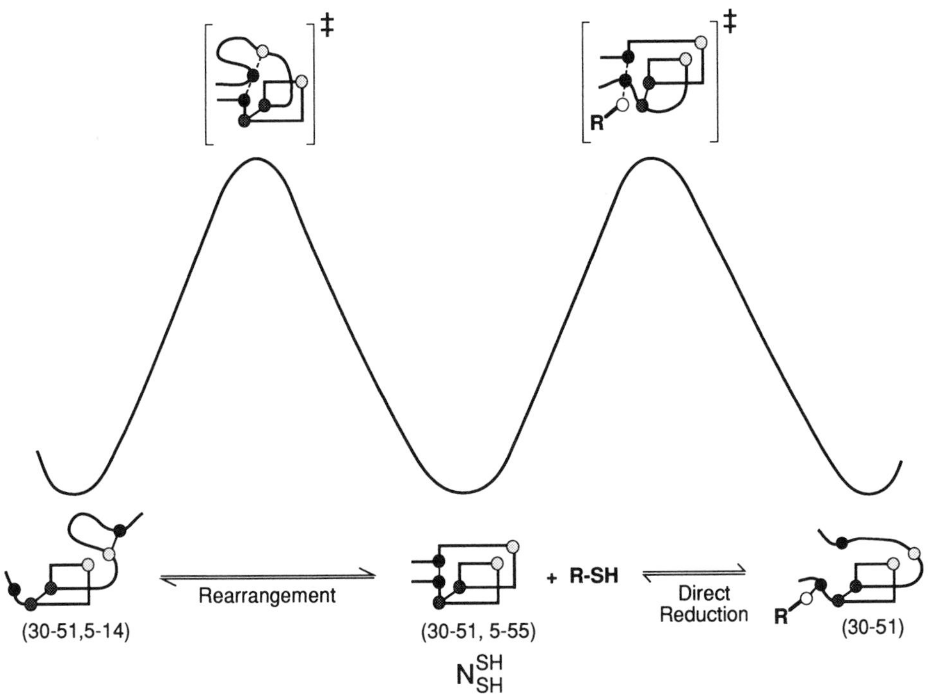

Figure 5. Rationale of a mutational experiment to characterize the transition states for direct reduction of N^{SH}_{SH} and intramolecular rearrangement of this species. As shown by the schematic representations, both processes are expected to require at least partial unfolding of N^{SH}_{SH}, but the two processes may involve disruption of different parts of the folded conformation. If the two corresponding transition states contain residual stabilizing interactions, amino acid replacements at different sites might be expected to selectively enhance the rate of one reaction or the other.

unfolding. As illustrated in Figure 2, these replacements are located at 13 sites throughout the protein.

To determine the rate constants for direct reduction and intramolecular rearrangement of N^{SH}_{SH}, the kinetics of reductive unfolding were measured in the presence of varying concentrations of DTT^{SH}_{SH}. As for the wild-type protein, the native mutant proteins were quickly converted to a native-like two-disulfide intermediate, which was then reduced further at a rate that depended upon the concentration of DTT^{SH}_{SH}. The rate constant for the intramolecular rearrangement, k_{rearr}, was estimated from the value of the apparent rate constant for disappearance of N^{SH}_{SH} extrapolated to zero DTT^{SH}_{SH}, and the rate constant for direct reduction, k_{dir}, was determined from the slope of the apparent rate versus DTT^{SH}_{SH} concentration. All of the amino acid replacements examined were found to increase both rate constants, in some cases by as much as 100,000-fold.

In Figure 6, the rate constants for the two process are plotted against one another on logarithmic scales. As shown, there is a remarkably good correlation between the logarithms of the two rate constants; the line shown in the figure has a slope of 1.3 and a correlation coefficient of 0.97 for data covering five orders of magnitude. Since the logarithms of the rate constants are proportional to the free energy differences between the common ground state (N^{SH}_{SH}) and the two transition states, the correlation illustrated in Figure 6 implies an energetic similarity between the transition states for rearrangement and direct reduction.

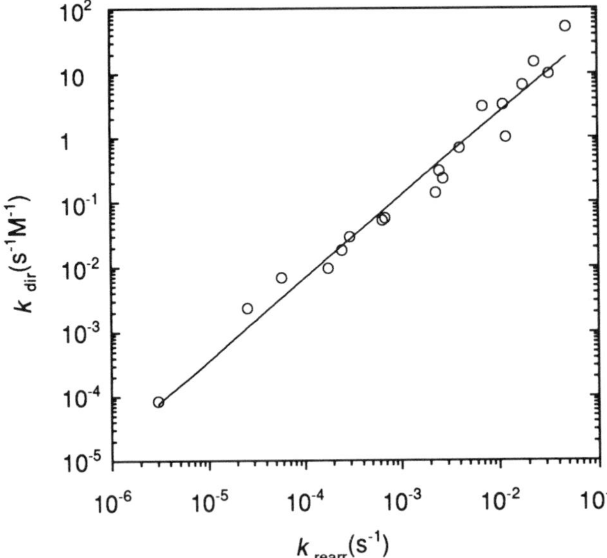

Figure 6. Correlation between the rate constants for direct reduction of N_{SH}^{SH} and intramolecular rearrangement of this species for different BPTI variants. Reproduced, with permission, from Mendoza et al., 1994.

We believe that the simplest interpretation of these results is that all of the amino acid replacements examined destabilize interactions in N_{SH}^{SH} that must be disrupted or weakened in both of the transition states. Because these substitutions alter a variety of different types of residues located throughout much of the folded protein, it seems most likely that the structure of N_{SH}^{SH} is extensively disrupted in both transition states, with few if any interactions other than the disulfide bonds stabilizing specific conformations. This conclusion is consistent with a variety of other evidence about the transition states (Weissman & Kim, 1991,1992; Mendoza et al., 1994).

These results can help explain why direct formation of N_{SH}^{SH} is disfavored during folding and why intramolecular rearrangements play a prominent role in the BPTI folding pathway. Since the transition states for direct reduction or rearrangement of N_{SH}^{SH} are expected to be equivalent to those for the reverse reactions required to form N_{SH}^{SH}, any stable interactions present in intermediates preceding this species during folding are probably disrupted in the transition states. NMR studies of analogs of the major one-disulfide intermediates, [30-51] and [5-55], indicate that both of these species contain significant native-like structure. Analogs of [5-55] have folded conformations that are nearly indistinguishable from, though less stable than, that of the native protein (van Mierlo et al., 1991a; Staley & Kim, 1992). An analog of [30-51] has been shown to contain about two thirds of the structure of the native protein, including the central β-sheet and α-helix, but some segments of the polypeptide chain, including the N-terminal 15 residues, are disordered (van Mierlo et al., 1993). Our mutational analysis of the transition states suggests that the structure present in the one-disulfide intermediates probably has to be disrupted during direct formation of N_{SH}^{SH}, thus contributing to the high energy barrier observed for this reaction. Unfortunately, it is not currently known how stable the structure in the one-disulfide intermediates is, and it is difficult to conclude whether this effect is sufficient to account entirely for the observed rate reduction, but it is likely to be a significant factor.

The other native-like two-disulfide intermediates, [5-55,14-38] and [30-51,14-38], are believed to be formed by direct formation of the 14-38 disulfide in the corresponding one-disulfide intermediates. Because the 14-38 disulfide is located on the surface of the native protein and can be both reduced and formed readily in the native conformation, native-like structure in the one-disulfide intermediates is expected to enhance, rather than impede, the rate of forming these species.

Analogs of the non-native two-disulfide intermediates, [30-51,5-14] and [30-51,5-38], have also been studied by high resolution NMR (van Mierlo et al., 1994). These species have conformations very similar to those of the [30-51] intermediate, in spite of the non-native disulfide between Cys 5 and either Cys 14 or 38. Because the N-terminal 15 residues of [30-51] are disordered, it appears that the non-native disulfides can form without significant disruption of structure already present in [30-51], consistent with kinetic results indicating that these species are formed directly from the population of one-disulfide intermediates (Creighton, 1977a).

Thus, it appears that all of the significant two-disulfide intermediates, except N_{SH}^{SH}, can be formed from the one-disulfide intermediates without disrupting structure that is already present. These two-disulfide species, therefore, form in preference to N_{SH}^{SH}. Once formed, however, these intermediates cannot directly form a third disulfide to yield the native protein; two of the intermediates contain non-native disulfides, and the other two, [5-55,14-38] and [30-51,14-38] contain so much native-like structure that direct formation of the remaining buried disulfide is very slow. Thus, each of these species must undergo one or more intramolecular rearrangements to yield N_{SH}^{SH} before a third disulfide can be incorporated. Our mutational studies, and other results, indicate that these rearrangements require extensive unfolding of the structure present in the intermediates. As a consequence, both the direct disulfide rearrangement pathway and the rearrangement pathway require disruption of structure already present in the protein. The rearrangement pathway is preferred kinetically because at the point where the pathways diverge, i.e. formation of a second disulfide, direct disulfide formation is particularly slow.

SUMMARY

The experiments described here illustrate how mutational analysis can be used to probe the energetics of a protein folding pathway. Studies with a mutant for which native-like intermediates are destabilized have demonstrated that the rearrangements observed in the BPTI pathway are not due simply to the stabilities of these kinetically-trapped species. Measurements of the effects of mutations on unfolding kinetics indicate that the major transition states for folding and unfolding are extensively unfolded. These results, together with structural studies of the intermediates, suggest that the rearrangement mechanism seen in the BPTI folding pathway is a consequence of steric constraints in the major one-disulfide intermediates.

Results such as those described here also illustrate the high degree of cooperativity among the individual interactions that stabilize protein conformations and determine folding mechanisms; a single amino acid replacement can dramatically change the distribution of folding intermediates (Figure 3) or increase the rate of reducing a protein disulfide by as much as 100,000-fold (Figure 6). Understanding the physical basis of this cooperativity is now one of the major challenges in the study of protein structure, folding and function. Further analysis of genetically modified proteins, using thermodynamic, kinetic, structural and computational methods, is likely to be an important means of addressing these problems.

ACKNOWLEDGMENTS

This work has been supported by grant no. GM42494 from the National Institutes of Health.

REFERENCES

Coplen, L. J., Frieden, R. W., and Goldenberg, D. P., 1990, A genetic screen to identify variants of bovine pancreatic trypsin inhibitor with altered folding energetics, *Proteins: Struct. Funct. Genet.* 7:16-31.

Creighton, T. E., 1977a, Conformational restrictions on the pathway of folding and unfolding of the pancreatic trypsin inhibitor., *J. Mol. Biol.* 113:275-293.

Creighton, T. E., 1977b, Energetics of folding and unfolding of pancreatic trypsin inhibitor., *J. Mol. Biol.* 113:295-312.

Creighton, T. E., 1978, Experimental studies of protein folding and unfolding., *Prog. Biophys. Mol. Biol.* 22:221-298.

Creighton, T. E.,1992a, Protein Folding, New York: W.H. Freeman.

Creighton, T. E., 1992b, Protein folding pathways determined using disulfide bonds, *BioEssays* 14:195-199.

Creighton, T. E., and Goldenberg, D. P., 1984, Kinetic role of a meta-stable native-like two-disulphide species in the folding transition of bovine pancreatic trypsin inhibitor., *J. Mol. Biol.* 179:497-526.

Fersht, A. R., Matoushchek, A., and Serrano, L., 1992, The folding of an enzyme. I. Theory of protein engineering analysis of stability and pathway of protein folding, *J. Mol. Biol.* 224:771-782.

Goldenberg, D. P., 1988, Kinetic Analysis of the Folding and Unfolding of a Mutant Form of Bovine Pancreatic Trypsin Inhibitor Lacking the Cysteine-14 and -38 Thiols, *Biochemistry* 27:2481-2489.

Goldenberg, D. P., 1992a, Mutational analysis of protein folding and stability, in Protein Folding, T. E. Creighton (ed.), pp. 353-403, New York: W.H. Freeman.

Goldenberg, D. P., 1992b, Native and non-native intermediates in the BPTI folding pathway, *Trends Biochem. Sci.* 17:257-261.

Goldenberg, D. P., Berger, J. M., Laheru, D. A., Wooden, S., and Zhang, J. X., 1992, Genetic dissection of pancreatic trypsin inhibitor, *Proc. Natl. Acad. Sci., USA* 89:5083-5087.

Goldenberg, D. P., Frieden, R. W., Haack, J. A., and Morrison, T. B., 1989, Mutational analysis of a protein folding pathway, *Nature* 338:127-132.

Jennings, P. A., Saalau-Bethell, S. M., Finn, B. E., Chen, X. W., and Matthews, C. R., 1991, Mutational analysis of protein folding mechanisms, *Methods Enzymol.* 202:113-126.

Kim, P. S., and Baldwin, R. L., 1990, Intermediates in the folding reactions of small proteins, *Annu. Rev. Biochem.* 59:631-660.

Matthews, C. R., 1993, Pathways of protein folding, *Annu. Rev. Biochem.* 62:653-683.

Mendoza, J. A., Jarstfer, M. B., and Goldenberg, D. P., 1994, Effects of amino acid replacements on the reductive unfolding kinetics of pancreatic trypsin inhibitor, *Biochemistry* 33:1143-1148.

Staley, J. P., and Kim, P. S., 1992, Complete folding of bovine pancreatic trypsin inhibitor with only a single disulfide bond, *Proc. Natl. Acad. Sci., U.S.A.* 89:1519-1523.

States, D. J., Dobson, C. M., Karplus, M., and Creighton, T. E., 1984, A new two-disulphide intermediate in the refolding of reduced bovine pancreatic trypsin inhibitor, *J. Mol. Biol.* 174:411-418.

van Mierlo, C. P. M., Darby, N. J., Neuhaus, D., and Creighton, T. E., 1991a, (14-38,30-51) Double-disulfide intermediate in folding of bovine pancreatic trypsin inhibitor: A two-dimensional ^{1}H nuclear magnetic resonance study, *J. Mol. Biol.* 222:353-371.

van Mierlo, C. P. M., Darby, N. J., Neuhaus, D., and Creighton, T. E., 1991b, Two-dimensional ^{1}H nuclear magnetic resonance study of the (5-55) single-disulfide folding intermediate of bovine pancreatic trypsin inhibitor, *J. Mol. Biol.* 222:373-390.

van Mierlo, C. P. M., Darby, N. J., Keeler, J., Neuhaus, D., and Creighton, T. E., 1993, Partially folded conformation of the (30-51) intermediate in the disulfide folding pathway of bovine pancreatic trypsin inhibitor: ^{1}H and ^{15}N resonance assignments and determination of backbone dynamics form ^{15}N relaxation measurements, *J.Mol.Biol.* 229:1125-1146.

van Mierlo, C. P. M., Kemmink, J., Neuhaus, D., Darby, N. J., and Creighton, T. E., 1994, ^{1}H NMR analysis of the partly-folded non-native two-disulfide intermediates (30-51,5-14) and (30-51,5-38) in the folding pathway of bovine pancreatic trypsin inhibitor, *J. Mol. Biol.* 235:1044-1061.

Weissman, J. S., and Kim, P. S., 1991, Reexamination of the folding of BPTI: Predominance of native intermediates, *Science* 253:1386-1393.

Weissman, J. S., and Kim, P. S., 1992, Kinetic role of nonnative species in the folding of bovine pancreatic trypsin inhibitor, *Proc. Natl. Acad. Sci., USA* 89:9900-9904.

Zhang, J. X., and Goldenberg, D. P., 1993, Amino acid replacement that eliminates kinetic traps in the BPTI folding pathway, *Biochemistry* 32:14075-14081.

45

THREE- AND FOUR-DIMENSIONAL HETERONUCLEAR NMR

G. Marius Clore and Angela M. Gronenborn

Laboratory of Chemical Physics, Building 5
National Institute of Diabetes and Digestive and Kidney Diseases
National Institutes of Health
Bethesda, Maryland 20892

INTRODUCTION

The principal source of geometric information used to solve three dimensional structures of macromolecules by NMR resides in short (< 5Å) approximate interproton distance restraints derived from nuclear Overhauser enhancement (NOE) measurements (1-5). In order to extract this information it is essential to first completely assign the ^{1}H spectrum of the macromolecule in question and then to assign as many structurally useful NOE interactions as possible. The larger the number of NOE restraints, the higher the precision and accuracy of the resulting structures (5-7). Indeed, with current state-of-the-art methodology it is now possible to obtain NMR structures of proteins at a precision and accuracy comparable to 2 Å resolution crystal structures (7-9)

For proteins of 100 residues or less, conventional homonuclear 2D NMR methods can be applied with a considerable degree of success (1-4, 10, 11). As the number of residues and molecular weight increases beyond 100 and 12 kDa, respectively, two main obstacles present themselves which made it necessary to extend the 2D NMR techniques to higher demensions and develope new approaches. First, the increased spectral complexity arising from the presence of a larger number of protons results in extensive chemical shift overlap and degeneracy, rendering the 2D spectra uninterpretable. Second, the rotational correlation time increases with molecular weight resulting in large ^{1}H linewidths and a concomitant severe decrease in the sensitivity of correlation experiments based on intrinsically small three-bond ^{1}H-^{1}H couplings. These obstacles can be overcome by increasing the dimensionality of the spectra to resolve problems associated with spectral overlap and by simultaneously making use of heteronuclear couplings that are larger than the linewidths to circumvent limitations in sensitivity (5, 6). This approach necessitates uniform ^{15}N and/or ^{13}C labeling of the macromolecule under consideration.

The concept of increasing spectral dimensionality to extract information can perhaps most easily be understood by analogy (5). Consider for example the encyclopedia Britannica. In a one-dimensional representation, all the information (i.e. words and sentences arranged

Methods in Protein Structure Analysis, Edited by M. Z. Atassi and E. Appella
Plenum Press, New York, 1995

in a particular set order) present in the encyclopedia would be condensed into a single line. If this line were expanded to two-dimensions in the form of a page, the odd word may be resolved but the vast majority would still be superimposed on each other. When this page is expanded into a book (i.e. three-dimensions) comprising a set number of lines and words per page, as well as a fixed number of pages, some pages may become intelligible, but many words will still lie on top of each other. The final expansion to the multi-volume book (i.e. four dimensions) then makes it possible to extract in full all the information present in the individual entries of the encyclopedia.

APPLICATION OF MULTIDIMENSIONAL HETERONUCLEAR NUCLEAR MAGNETIC RESONANCE SPECTROSCOPY TO PROTEIN STRUCTURE DETERMINATION

The design and implementation of higher dimensionality NMR experiments can be carried out by the appropriate combination of 2D NMR experiments, as illustrated schematically in Fig. 1.

A 3D experiment is constructed from two 2D pulse schemes by leaving out the detection period of the first experiment and the preparation pulse of the second (12). This results in a pulse train comprising two independently incremented evolution periods t_1 and t_2, two corresponding mixing periods M_1 and M_2, and a detection period t_3. Similarly, a 4D experiment is obtained by combining three 2D experiments in an analogous fashion. Thus, conceptually *n*-dimensional NMR can be conceived as a straightforward extension of 2D NMR. The real challenge, however, of 3D and 4D NMR is two-fold: first, to ascertain which 2D experiments should be combined to best advantage; and second, to design the pulse sequences in such a way that undesired artifacts, which may severely interefere with the interpretation of the spectra, are removed. This task is far from trivial.

Heteronuclear 3D and 4D NMR experiments exploit a series of large one-bond heteronuclear couplings for magnetization transfer through-bonds which are summarized in Fig. 2.

This, together with the fact that the ^1H nucleus is always detected, renders these experiments very sensitive. Indeed, high quality 3D and 4D heteronuclear-edited spectra can easily be obtained on samples of 1-2 mM uniformly labeled protein in a time frame that is

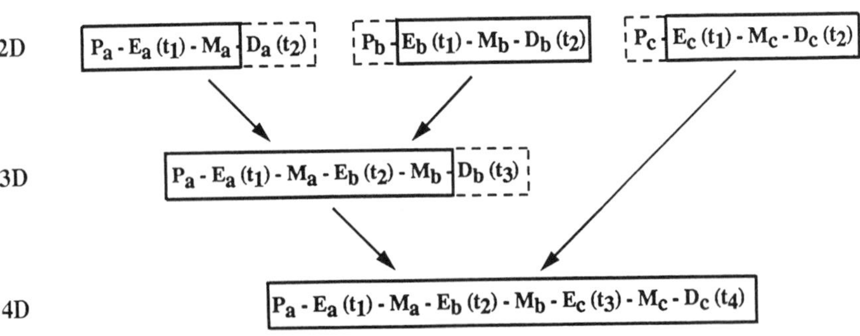

Figure 1. General representation of pulse sequences used in multi-dimensional NMR illustrating the relationship between the basic schemes used to record 2D, 3D and 4D NMR spectra. Note how 3D and 4D experiments are constructed by the appropriate linear combination of 2D ones. Abbreviations: P, preparation; E, evolution; M, mixing; and D, detection. In 3D and 4D NMR, the evolution periods are incremented independently.

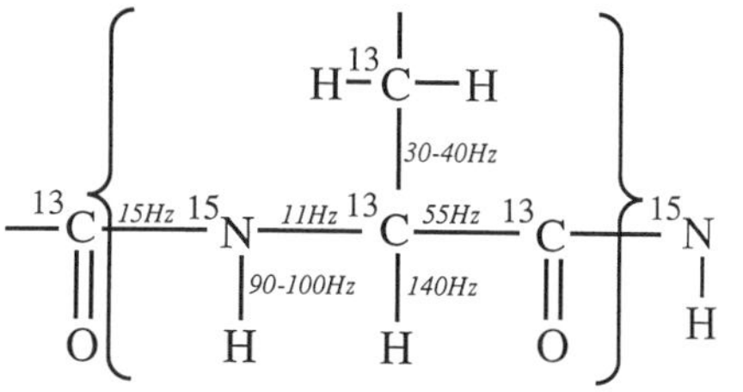

Figure 2. Summary of the one-bond heteronuclear couplings along the polypeptide chain utilized in 3D and 4D NMR experiments.

limited solely by the number of increments that have to be collected for appropriate digitization and the number of phase cycling steps that have to be used to reduce artifacts to an acceptably low level. Typical measurement times are 1 to 3 days for 3D experiments and 2.5 to 5 days for 4D ones. A detailed technical review of heteronuclear multi-dimensional NMR has been provided by Clore & Gronenborn (13) and Bax & Grzesiek (14).

Many of the 3D and 4D experiments are based on heteronuclear-editing of ^{1}H-^{1}H experiments so that the general appearance of conventional 2D experiments is preserved and the total number of cross-peaks present is the same as that in the 2D equivalents (5, 6, 13, 14). The progression from a 2D spectrum to 3D and 4D heteronuclear-edited spectra is depicted schematically in Fig. 3.

Consider, for example the cross-peaks involving a particular ^{1}H frequency in a 2D NOESY spectrum, a 3D ^{15}N or ^{13}C-edited NOESY spectrum, and finally a 4D ^{15}N/^{13}C or ^{13}C/^{13}C-edited NOESY spectrum. In the 2D spectrum a series of cross peaks will be seen from the originating proton frequencies in the F_1 dimension to the single destination ^{1}H frequency along the F_2 dimension. From the 2D experiment it is impossible to ascertain whether these NOEs involve only a single destination proton or several destination protons with identical chemical shifts. By spreading the spectrum into a third dimension according to the chemical shift of the heteronucleus attached to the destination proton(s), NOEs involving different destination protons will appear in distinct ^{1}H-^{1}H planes of the 3D spectrum. Thus each interaction is simultaneously labeled by three chemical shift coordinates along three orthogonal axes of the spectrum. The projection of all these planes onto a single plane yields the corresponding 2D spectrum. For the purposes of sequential assignment, heteronuclear-edited 3D spectra are often sufficient for analysis. However, when the goal of the analysis is to assign NOEs between protons far apart in the sequence, a 3D ^{15}N- or ^{13}C-edited NOESY spectrum will often prove inadequate. This is because the originating protons are only specified by their ^{1}H chemical shifts, and more often than not, there are several protons which resonate at the same frequencies. For example, in the case of the 153 residue protein interleukin-1β, there are about 60 protons which resonate in a 0.4 ppm interval between 0.8 and 1.2 ppm. Such ambiguities can then be resolved by spreading out the 3D spectrum still further into a fourth dimension according to the chemical shift of the heteronucleus attached to the originating protons, so that each NOE interaction is simultaneously labeled by four chemical shift coordinates along four orthogonal axes, namely those of the originating and destinations protons and those of the corresponding heteronuclei directly bonded to these protons (15-17). The result is a 4D spectrum in which each plane of the 3D spectrum constitutes a cube in the 4D spectrum.

For illustration purposes it is also useful to compare the type of information that can be extracted from a very simple system using 2D, 3D and 4D NMR. Consider a molecule with only two NH and two aliphatic protons in which only one NH proton is close to an

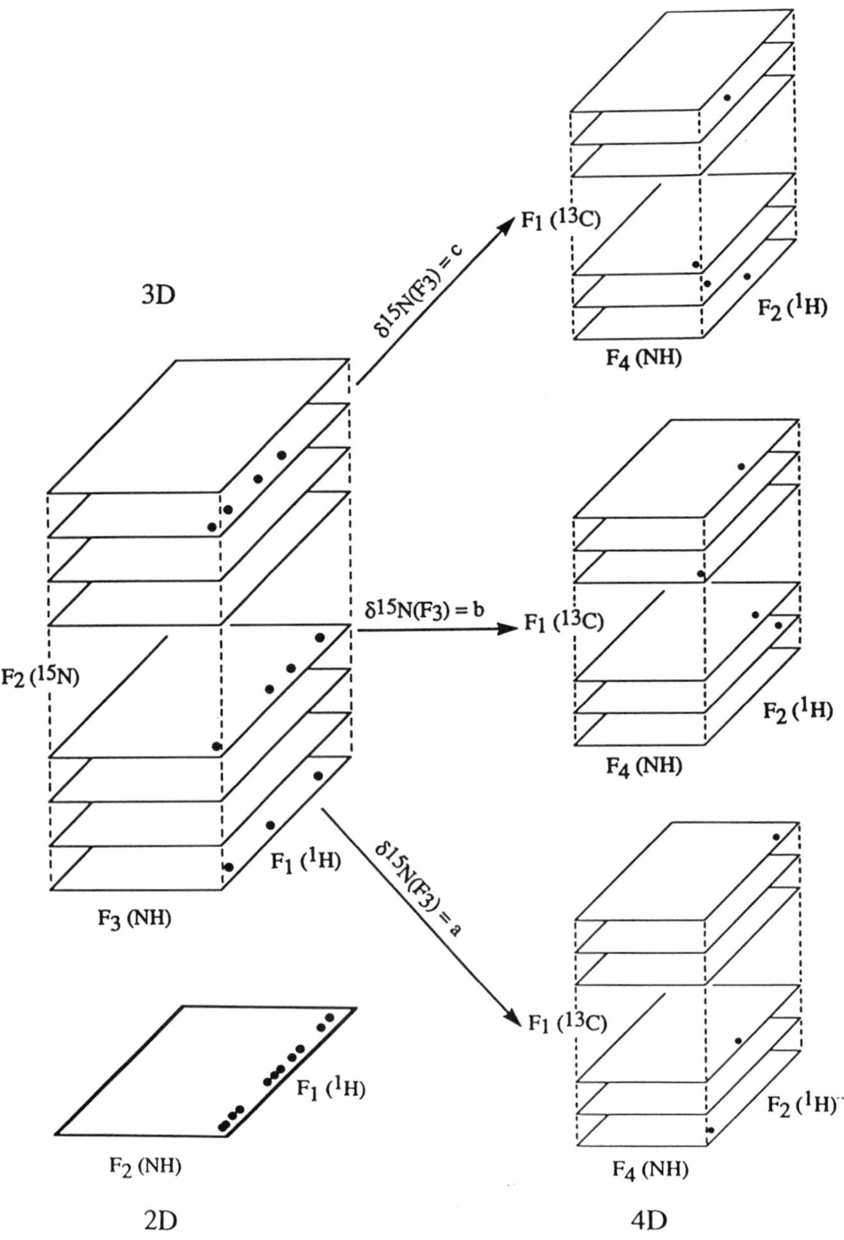

Figure 3. Schematic illustration of the progression and relationship between 2D, 3D and 4D heteronuclear NMR experiments. The closed circles represent NOE cross peaks. In the example shown there are 11 NOEs originating from 11 different protons in the F_1 dimension to a single frequency position in the F_2 dimension. In the 2D spectrum, it is impossible to ascertain whether there is only one destination proton or several in the F_2 dimension. By spreading the spectrum into a third dimension (labeled F_2), according to the chemical shift of the heteronucleus attached to the destination proton, it can be seen that the NOEs now lie in three distinct $^1H(F_1)$-$^1H(F_3)$ planes, indicating that three different destination protons are involved. However, the ^1H chemical shifts still provide the only means of identifying the originating protons. Hence the problem of spectral overlap still prevents the unambiguous assignment of these NOEs. By extending the dimensionality of the spectrum to four, each NOE interaction is labeled by four chemical shifts along four orthogonal axes. Thus, the NOEs in each plane of the 3D spectrum are now spread over a cube in the 4D spectrum according to the chemical shift of the heteronucleus directly attached to the originating protons. Adapted from ref. (15).

aliphatic proton. In addition, the chemical shifts of the NH protons are degenerate, as are those of the aliphatic protons, so that only two resonances are seen in the one-dimensional spectrum. In the 2D NOESY spectrum, an NOE will be observed between the resonance position of the NH protons and the resonance position of the aliphatic protons, but it will be impossible to ascertain which one of the four possible NH-aliphatic proton combinations gives rise to the NOE. By spreading the spectrum into a third dimension, for example by the chemical shift of the ^{15}N atoms attached to the NH protons, the number of possibilities will be reduced to two, provided, of course, that the chemical shifts of the two nitrogen atoms are different. Finally, when the fourth dimension corresponding to the chemical shift of the ^{13}C atoms attached to the aliphatic protons is introduced, a unique assignment of the NH-aliphatic proton pair giving rise to the NOE can be made.

Fig. 4A presents a portion of the 2D ^{15}N-edited NOESY spectrum of interleukin-1β (153 residues) illustrating NOE interactions between the NH protons along the F_2 axis and the C$^{\alpha}$H protons along the F_1 dimension. Despite the fact that a large number of cross-peaks can be resolved, it can be seen that many of the cross peaks have identical chemical shifts in one or other dimensions. For example, there are 15 cross peaks involving NH protons at a $F_2(^1H)$ chemical shift of ~9.2 ppm. A single $^1H(F_1)$-$^1H(F_3)$ plane of the 3D ^{15}N-edited NOESY spectrum of interleukin-1β at $\delta^{15}N(F_2)$ = 123.7 ppm is shown in Fig. 4B. Not only is the number of cross-peaks in this slice small, but at $\delta^1H(F_3)$ ~ 9.2 ppm there is only a single cross peak involving one NH proton. The correlations observed in the ^{15}N-edited NOESY spectrum are through-space ones. Intraresidue correlations from the NH protons to the C$^{\alpha}$H and C$^{\beta}$H protons can similarly be resolved using a 3D ^{15}N-edited HOHAHA spectrum in which efficient isotropic mixing sequences are used to transfer magnetization between protons via three-bond ^{1}H-^{1}H couplings.

The 3D ^{15}N-edited NOESY and HOHAHA spectra constitute only one of several versions of a 3D heteronuclear-edited spectrum. Many alternative through-bond pathways can be utilized to great effect. Consider for example, the delineation of amino acid spin systems which involves grouping those resonances which belong to the same residue. In 2D NMR, correlation experiments are used to delineate either direct or relayed connectivities via small three-bond ^{1}H-^{1}H couplings. Even for proteins of 50-60 residues, it can be difficult to delineate long chain amino acids such as Lys and Arg in this manner. In heteronuclear 3D NMR an alternative pathway can be employed which involves transferring magnetization first from a proton to its directly attached carbon atom via the large $^1J_{CH}$ coupling (~130 Hz), followed by either direct or relayed transfer of magnetization along the carbon chain via the $^1J_{CC}$ couplings (~30-40 Hz), before transferring the magnetization back to protons (18-20). An example of such a spectrum is the so called HCCH-TOCSY shown in Fig. 4C. The $^1H(F_1)$-$^1H(F_3)$ plane at $\delta^{13}C(F_2)$ = 59 ppm illustrates both direct and relayed connectivities along various side chains originating from C$^{\alpha}$H protons. As expected, the resolution of the spectrum is excellent and there is no spectral overlap. Just as importantly, however, the sensitivity of the experiment is extremely high and complete spin systems are readily identified in interleukin-1β even for long side chains, such as those of two lysine residues shown in the figure. Indeed, analysing spectra of this kind, it was possible to obtain complete ^{1}H and ^{13}C assignments for the side chains of interleukin-1β (21).

3D NMR also permits one to devise experiments for sequential assignment which are based solely on through-bond connectivities via heteronuclear couplings (13, 14, 22, 23) and thus do not rely on the NOESY experiment. This becomes increasingly important for larger proteins, as the types of connectivites observed in these correlation experiments are entirely predictable, whereas in the NOESY spectrum which relies solely on close proximity of protons, it may be possible to confuse sequential connectivities with long range ones. These 3D heteronuclear correlation experiments are of the triple resonance variety and make use of one-bond ^{13}CO(i-1)-^{15}N(i), ^{15}N(i)-^{13}C$^{\alpha}$(i), ^{13}C(i)-^{13}C(i) and ^{13}C$^{\alpha}$(i)-^{13}CO(i) couplings,

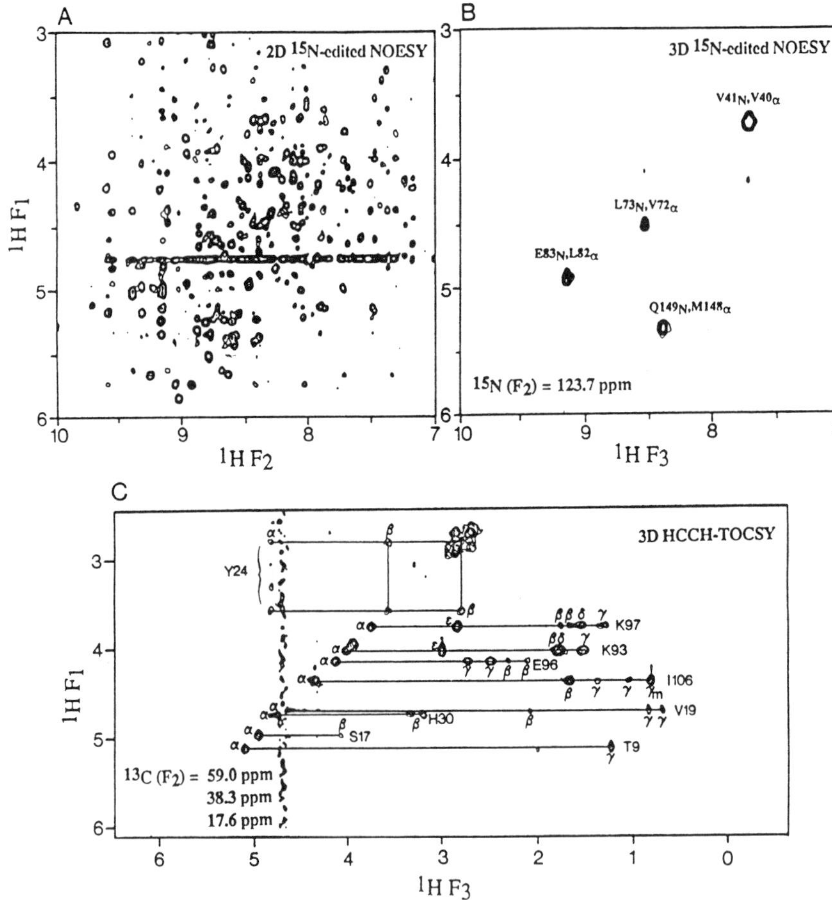

Figure 4. Example of 2D and 3D spectra of interleukin-1β recorded at 600 MHz. The 2D spectrum in panel A shows the NH(F_2 axis)-$C^{\alpha}H$(F_1 axis) region of a 2D ^{15}N-edited NOESY spectrum. The same region of a single NH(F_3)-^{1}H(F_1) plane of the 3D ^{15}N-edited NOESY at $\delta^{15}N(F_2) = 123.7$ ppm is shown in panel B. The actual 3D spectrum comprises 64 such planes and projection of these on a single plane would yield the same spectrum as in (A). Panel C shows a single ^{1}H(F_3)-^{1}H(F_1) plane of the 3D HCCH-TOCSY spectrum at $\delta^{13}C(F_2) = 38.3 \pm nSW$ (where SW is the spectral width of 20.71 ppm in the ^{13}C dimension) illustrating both direct and relayed connectivities originating from the $C^{\alpha}H$ protons. Note how easy it is to delineate complete spin systems of long side chains such as Lys (i.e. cross peaks to the $C^{\beta}H$, $C^{\gamma}H$, $C^{\delta}H$ and $C^{\epsilon}H$ protons are observed) owing to the fact that magnetization along the side chain is transferred via large $^1J_{CC}$ couplings. Several features of the HCCH-TOCSY spectrum should be pointed out. First, extensive folding is employed which does not obscure analysis as ^{13}C chemical shifts for different carbon types are located in characteristic regions of the ^{13}C spectrum with little overlap. Second, the spectrum is edited according to the chemical shift of the heteronucleus attached to the originating proton rather than the distination one. Third, multiple cross checks on the assignments are readily made by looking for the symmetry related peaks in the planes corresponding to the ^{13}C chemical shifts of the destination protons in the original slice. Adapted from ref. (5).

as well as two-bond $^{13}C^{\alpha}(i\text{-}1)\text{-}^{15}N(i)$ couplings. In this manner multiple independent pathways for linking the resonances of one residue with those of its adjacent neighbour are available, thereby avoiding ambiguities in the sequential assignment.

In practice, only a limited number of 3D triple and double resonance experiments need to be performed to obtain complete assignments. In our experience, the following eight

3D experiments not only provide all the information required, but are also characterized by high sensitivity and can be recorded in as little as two weeks of measuring time. Specifically, the 3D CBCA(CO)NH and HBHA(CBCACO)NH experiments (24, 25) are used to correlate the chemical shifts of $C^{\alpha}(i)/C^{\beta}(i\text{-}1)$ and $H^{\alpha}(i)/H^{\beta}(i\text{-}1)$, respectively, of residue i-1 with the $^{15}N(i)/NH(i)$ chemical shifts of residue i; and the complementary the 3D C(CO)NH and H(CCO)NH experiments (26) are used to correlate the chemical shifts of the aliphatic side chain ^{13}C and ^{1}H resonances, respectively, of residue i-1 with the $^{15}N(i)/NH(i)$ chemical shifts of residue i. In the first two experiments, magnetization originating on C^{β} is transferred to C^{α} by a COSY mixing pulse, while in the second pair of experiments, magnetization is transferred from a side chain C along the carbon chain to C^{α} via isotropic mixing. Intraresidue correlations to the NH group can be obtained from the 3D CBCANH (27) and ^{15}N-edited HOHAHA (28) experiments. The 3D CBCANH experiment correlates the chemical shifts of $C^{\alpha}(i)/C^{\beta}(i)$ (as well as those of $C^{\alpha}(i\text{-}1)/C^{\beta}(i\text{-}1)$ which invariably give rise to weaker cross peaks), with the $^{15}N(i)/NH(i)$ chemical shifts of residue i; the 3D ^{15}N-separated HOHAHA experiment correlates the chemical shifts of the side chain protons of residue i with the $^{15}N(i)/NH(i)$ chemical shifts of residue i. Finally, the 3D HCCH-COSY and HCCH-TOCSY experiments (21, 22)can be used to confirm and obtain complete ^{1}H and ^{13}C assignments of the side chains.

The power of 4D heteronuclear NMR spectroscopy for unraveling interactions that would not have been possible in lower dimensional spectra is illustrated in Fig. 5 by the $^{13}C/^{13}C$-edited NOESY spectrum of interleukin-1β (16).

Fig. 5A shows a small portion of the aliphatic region between 1 and 2 ppm of a conventional 2D NOESY spectrum of interleukin-1β. The overlap is so great that no single individual cross peak can be resolved. One might therefore wonder just how many NOE interactions are actually superimposed, for example, at the ^{1}H chemical shift coordinates of the letter X at 1.39 (F_1) and 1.67 (F_2) ppm . A $^{1}H(F_2)$-$^{1}H(F_4)$ plane of the 4D spectrum at $\delta^{13}C(F_1)$, $\delta^{13}C(F_3)$ = 44.3, 34.6 ppm is shown in panel B and the square box at the top right hand side of this panel encloses the region between 1 and 2 ppm. Only two cross peaks are present in this region, and the arrow points to a single NOE between the $C^{\gamma}H$ and $C^{\beta}H$ protons of Lys-77 with the same ^{1}H chemical shift coordinates as the letter X in panel A. All the other NOE interactions at the same ^{1}H chemical shift coordinates can be determined by inspection of a single $^{13}C(F_1)$-$^{13}C(F_3)$ plane taken at $\delta^{1}H(F_2)$, $\delta^{1}H(F_4)$ = 1.39, 1.67 ppm. This reveals a total of 7 NOE interactions superimposed at the ^{1}H chemical shift coordinates of the letter X. Another feature of the 4D spectrum is illustrated by the two $^{1}H(F_2)$-$^{1}H(F_4)$ planes at different F_1 and F_3 ^{13}C frequencies shown in panels C and B. In both cases, there are cross-peaks involving protons with identical or near chemical shifts, namely that between Pro-91($C^{\alpha}H$) and Tyr-90($C^{\alpha}H$), diagnostic of a cis-proline, in panel C, and between Phe-99($C^{\beta b}H$) and Met-95($C^{\gamma}H$) in panel D. These interactions could not be resolved in either a 2D spectrum or a 3D ^{13}C-edited spectrum as they would lie on the spectral diagonal (i.e.the region of the spectrum corresponding to magnetization that has not been transferred from one proton to another). In the 4D spectrum, however, they are easy to observe, provided, of course, that the ^{13}C chemical shifts of the directly bonded ^{13}C nuclei are different.

Because the number of NOE interactions present in each $^{1}H(F_4)$-$^{1}H(F_2)$ plane of 4D $^{13}C/^{15}N$ or $^{13}C/^{13}C$-edited NOESY spectra is so small, the inherent resolution in a 4D spectrum is extremely high, despite the low level of digitization. Indeed, spectra with equivalent resolution can be recorded at magnetic field strengths considerably lower than 600 MHz, although this would obviously lead to a reduction in sensitivity. Further, it can be calculated that 4D spectra with virtual lack of resonance overlap and good sensitivity can be obtained on proteins with as many as 400 residues. Thus, once complete ^{1}H, ^{15}N and ^{13}C assignments are obtained, analysis of 4D spectra should permit the automated assignment of almost all NOE interactions.

CONCLUSION

In this chapter we have summarized the recent developments in heteronuclear 3D and 4D NMR which have been designed to extend the NMR methodology to medium sized proteins in the 15-30 kDa range. The underlying principle of this approach consists of extending the dimensionality of the spectra to obtain dramatic improvements in spectral resolution while simultaneously exploiting large heteronuclear couplings to circumvent problems associated with larger linewidths. A key feature of all these experiments is that they do not result in any increase in the number of observed cross peaks relative to their 2D counterparts. Hence, the improvement in resolution is achieved without raising the spectral complexity, rendering data interpretation straightforward. Thus, for example, in 4D heteronuclear-edited NOESY spectra, the NOE interactions between proton pairs are not only labeled by the ^{1}H chemical shifts but also by the corresponding chemical shifts of their directly bonded heteronuclei in four orthogonal axes of the spectrum. Also important in terms of practical applications is the high sensitivity of these experiments which makes it feasible to obtain high quality spectra in a relatively short time frame on 1-2 mM protein samples uniformly labeled with ^{15}N and/or ^{13}C.

Figure 5. Comparison of 2D and 4D NMR spectra of interleukin-1β recorded at 600 MHz (16). The region between 1 and 2 ppm of the 2D NOESY spectrum is shown in (A). ^{1}H(F$_2$)-^{1}H(F$_4$) planes at several ^{13}C(F$_1$) and ^{13}C(F$_3$) frequencies of the 4D ^{13}C/^{13}C NOESY spectrum are shown in panels B to D. No individual cross peaks can be observed in the 2D spectrum and the letter X has ^{1}H coordinates of 1.39 and 1.67 ppm. In contrast, only two cross peaks are observed in the boxed region in panel B betwen 1 and 2 ppm, one of which (indicated by an arrow) has the same ^{1}H coordinates as the letter X. Further analysis of the complete 4D spectrum reveals the presence of 7 NOE cross peaks superimposed at the ^{1}H coordinates of the letter X. This can be ascertained by looking at ^{13}C(F$_1$)-^{13}C(F$_3$) plane taken at the ^{1}H coordinates of X. True diagonal peaks corresponding to magnetization that has not been transferred from one proton to another, as well as intense NOE peaks involving protons attached to the same carbon atom (i.e. methylene protons), appear in only a single ^{1}H(F$_2$)-^{1}H(F$_4$) plane of each ^{13}C(F$_1$), ^{1}H(F$_2$), ^{1}H(F$_4$) cube at the carbon frequency where the originating and destination carbon atoms coincide (i.e. at $F_1 = F_3$). Thus, these intense resonances no longer obscure NOEs between proton with similar or degenerate chemical shifts. Two examples of such NOEs can be seen in panels C (between the C$^{\alpha}$H protons of Pro-91 and Tyr-90) and D (between one of the C$^{\beta}$H protons of Phe-77 and the methyl protons of Met-95). These various planes of the 4D spectrum also illustrate another key aspect of 3D and 4D NMR, namely the importance of designing the pulse scheme to optimally remove undesired artifacts which may severely interfere with the interpretation of the spectra. Thus, while the 4D ^{13}C/^{13}C-edited NOESY experiment is conceptually analogous to that of a 4D ^{13}C/^{15}N-edited one, the design of a suitable pulse scheme is actually much more complex in the ^{13}C/^{13}C case. This is due to the fact that there are a large number of spurious magnetization transfer pathways that can lead to observable signals in the homonuclear ^{13}C/^{13}C case. For example, in the 4D ^{15}N/^{13}C-edited case there are no "diagonal peaks" which would correspond to magnetization that has not been transferred from one hydrogen to another, as the double heteronuclear filtering (i.e. ^{13}C and ^{15}N) is extremely efficient at completely removing these normally very intense and uninformative resonances. Such a double filter is not available in the ^{13}C/^{13}C case so that both additional pulses and phase cycling are required to suppress magnetization transfer through these pathways. This task is far from trivial as the number of phase cycling steps in 4D experiments is severely limited by the need to keep the measurement time down to practical levels (i.e less than 1 week). The most efficient way of obtaining artefact free spectra is through the incorporation of pulse field gradients to suppress undesired coherence transfer pathways (29). Indeed, inclusion of 6 pulse field gradients into the original pulse scheme of Clore et al. (16) reduces the phase cycle from eight to two steps (30). The results of such care in pulse design can be clearly appreciated from the artifact free planes shown in panels B-D. However, when a 4D ^{13}C/^{13}C-edited NOESY spectrum is recorded with the same pulse scheme as that used in the 4D ^{15}N/^{13}C experiment (with the obvious replacement of ^{15}N pulses by ^{13}C pulses), a large number of spurious peaks are observed along a pseudo-diagonal at $\delta^{1}H(F_2) = \delta^{1}H(F_4)$ in planes where the carbon frequencies of the originating and destination protons do *not* coincide. As a result, it becomes virtually impossible under these circumstances to distinguish artifacts from NOEs between protons with the same ^{1}H chemical shifts, as was possible with complete confidence in panels C and D.

Just as 2D NMR opened the application of NMR to the structure determination of small proteins of less than about 100 residues, 3D and 4D heteronuclear NMR provide the means of extending the methodology to medium sized proteins in the 150 to 300 residue range. Indeed, the determination in 1991 of the first high resolution structure of a protein in the 15-20 kDa range, namely the cytokine interleukin-1β (153 residues and 18 kDa), using 3D and 4D heteronuclear NMR (31) demonstrated beyond doubt that the technology is now available for obtaining the structures of such medium sized proteins at a level of accuracy and precision that is comparable to the best results attainable for small proteins. Subsequently, a number of other medium sized protein structures have been determined using these method, including interleukin-4 (32-34), glucose permease IIA (35), a complex of calmodulin with a target peptide (36), a complex of cyclophilin and cyclosporin A (37), a specific complex of the transcription factor GATA-1 (38) with its DNA target site, human macrophage inflammatory protein 1β (39), and the oligomerization domain of p53 (40).

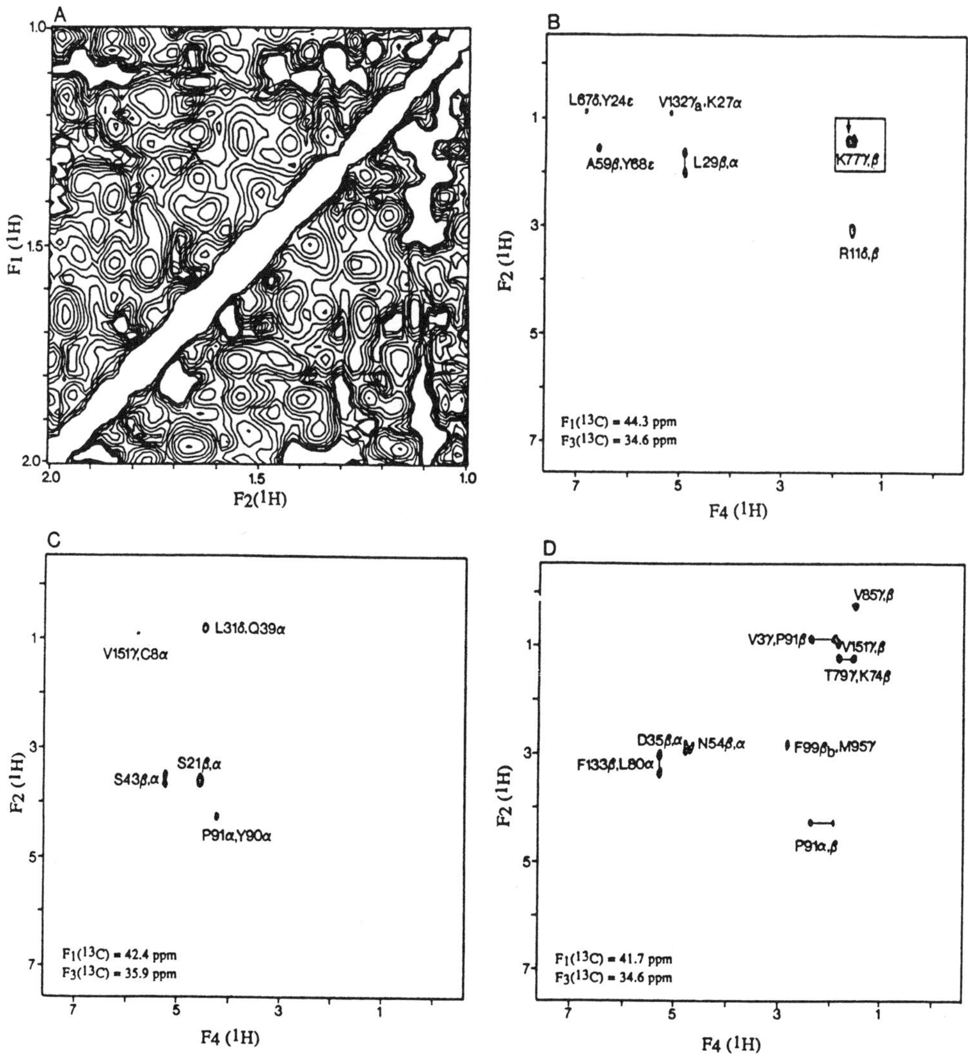

ACKNOWLEDGEMENTS

We thank Ad Bax for many stimulating discussions. This work was supported in part by the AIDS Targeted Anti-Viral Program of the Office of the Director of the National Institutes of Health. This chapter has been previously published in G.M. Clore and A.M. Gronenborn, *Methods Enzymol.* 239: 349-363 (1994) which itself was adapted and updated from G.M. Clore and A.M. Gronenborn, *Science* 252: 1390-1399 (1991).

REFERENCES

1. Wüthrich, K., 1986, *NMR of Proteins*, Wiley, New York
2. Clore, G.M., and Gronenborn, A.M., 1989, Determination of three-dimensional structures of proteins in solution by nuclear magnetic resonance spectroscopy. *Prot. Eng.* 1: 275-288.
3. Clore, G.M., and Gronenborn, A.M., 1989, Determination of three-dimensional structures of proteins and nucleic acids in solution by nuclear magnetic resonance spectroscopy. *CRC Crit Rev. Biochem. Mol. Biol.* 24: 479-564.
4. Bax, A., 1989, Two-dimensional NMR and protein structure, *Ann Rev. Biochem.* 58: 223-256.
5. Clore, G.M., and Gronenborn, A.M., 1991, Structures of larger proteins in solution: three- and four-dimensional hetronuclear NMR spectroscopy. *Science* 252: 1390-1399.
6. Clore, G.M., and Gronenborn, A.M., 1991, Two, three and four dimensional NMR methods for obtaining larger and more precise three-dimensional structures of proteins in solution. *Ann. Rev. Biophys. Biophys. Chem.* 20: 29-63.
7. G.M. Clore, and A.M. Gronenborn, 1991, Comparison of the solution nuclear magnetic resonance and X-ray crystal structures of human recombinant interleukin-1β. *J. Mol. Biol.* 221: 47-53.
8. Clore, G.M., Robien, M.A., and Gronenborn, A.M., 1993, Exploring the limits of precision and accuracy of protein structures determined by nuclear magnetic resonance spectroscopy. *J. Mol. Biol.* 231: 82-102.
9. Shaanan, B., Gronenborn, A.M., Cohen, G.H., Gilliland, G.L., Veerapandian, B., Davies, D.R., and Clore, G.M., 1992, Combining experimental information from crystal and solution studies: joint X-ray and NMR refinement, *Science* 257: 961-964.
10. Dyson, H.J., Gippert, D.A., Case, D.A., Holmgren, A., and Wright, P.E., 1990, Three-dimensional solution structure of the reduced form of *Escherichia coli* thioredoxin determined by nuclear magnetic resonance spectroscopy, *Biochemistry* 29: 4129-4136.
11. Forman-Kay, J.D., Clore, G.M., Wingfield, P.T., and Gronenborn, A.M., 1991, The high resolution three-dimensional structure of reduced recombinant human thioredoxin in solution, *Biochemistry* 30: 2685-2698.
12. Oschkinat, H., Griesinger, C., Kraulis, P.J., Sørensen, O.W., Ernst, R.R., Gronenbor, A.M., and Clore, G.M., 1988, Three-dimensional NMR spectroscopy of a protein in solution, *Nature (Lond.)* 332: 374-376.
13. Clore, G.M., and Gronenborn, A.M., 1991, Applications of three- and four-dimensional heteronuclear NMR spectroscopy to protein structure determination, *Progr. Nucl. Magn. Reson. Spectrosc.* 23: 43-92.
14. Bax, A., and Grzesiek, S., 1993, Methodological Advances in Protein NMR, *Acct. Chem. Res.* 26: 131-138.
15. Kay, L.E., Clore, G.M., Bax, A., and Gronenborn, A.M., 1990, Four-dimensional heteronuclear triple resonance NMR spectroscopy of interleukin-1β in solution, *Science* 249: 411-414.
16. Clore, G.M., Kay, L.E., Bax, A., and Gronenborn, A.M., 1991, Four-dimensional $^{13}C/^{13}C$-edited nuclear Overhauser enhancement spectroscopy of a protein in solution: application to interleukin-1β. Biochemistry 30, 12-18.
17. Zuiderweg, E.R.P., Petros, A.M., Fesik, S.W., and Olejniczak, E.T., 1991, Four-dimensional [^{13}C, ^{1}H, ^{13}C, ^{1}H} HMQC-NOE-HMQC NMR spectroscopy: resolving tertiary NOE distance restraints in spectra of larger proteins, *J. Am. Chem. Soc.* 113: 370-372.
18. Bax, A., Clore, G.M., Driscoll, P.C., Gronenborn, A.M., Ikura, M., and Kay, L.E., 1990, Practical aspects of proton-carbon-carbon-proton three-dimensional correlation spectroscopy of ^{13}C-labeled proteins. *J. Magn. Reson.* 87: 620-628.
19. Bax, A., Clore, G.M., and Gronenborn, A.M., 1990, ^{1}H-^{1}H correlation via isotropic mixing of ^{13}C magnetization: a new three-dimensional approach for assigning ^{1}H and ^{13}C spectra of ^{13}C-enriched proteins. *J. Magn. Reson.* 88, 425-431.

20. Fesik, S.W., Eaton, H.L., Olejniczak, E.T., Zuiderweg, E.R.P., McIntosh, L.P., and Dahlquist, F.W., 1990, 2D and 3D NMR spectroscopy employing ^{13}C-^{13}C magentization transfer by isotropic mixing: spin system identification in large proteins, *J. Am. Chem. Soc.* 112, 886-888.
21. Clore, G.M., Bax, A., Driscoll, P.C., Wingfield, P.T., and Gronenborn, A.M., 1990, Assignment of the side chain ^{1}H and ^{13}C resonances of interleukin-1β using double and triple resonance hetronuclear three-dimensional NMR spectroscopy. *Biochemistry* 29: 8172-8184.
22. Ikura, M., Kay, L.E., and Bax, A., 1990, A novel approach for sequential assignment of ^{1}H, ^{13}C, and ^{15}N spectra of larger protens: heteronuclear triple-resonance NMR spectroscopy: application to calmodulin, *Biochemistry* 29, 4659-4667.
23. Powers, R., Gronenborn, A.M., Clore, G.M., and Bax, A., 1991, Three-dimensional triple resonance NMR of $^{13}C/^{15}N$ enriched proteins using constant-time evolution. *J. Magn. Reson.* 94: 209-213.
24. Grzesiek, S., and Bax, A., 1992, Correlating backbone amide and sidechain resonances in larger proteins by multiple relayed triple resonance NMR, *J. Am. Chem. Soc.* 114: 6291-6293.
25. Grzesiek, S., and Bax, A., 1993, Amino acid type determination in the sequential assignment procedure of uniformly $^{13}C/^{15}N$ enriched proteins, *J. Biomol. NMR* 3: 185-204.
26. Grzesiek, S., Anglister, J., and Bax, A., 1993, Correlation of backbone amide and aliphatic sidechain resonances in $^{13}C/^{15}N$-enriched proteins by isotropic mixing of ^{13}C magnetization, *J. Magn. Reson. Series B* 101: 114-119.
27. Grzesiek, S., and Bax, A., 1992, An efficient experiment for sequential backbone assignment of medium sized isotopically enriched proteins, *J. Magn. Reson.* 99: 201-207.
28. Clore, G.M., Bax, A., and Gronenborn, A.M., 1991, Stereospecific assignment of b-methylene protons in larger proteins using three-dimensional ^{15}N-separated Hartmann-Hahn and ^{13}C-separated rotating frame Overhauser spectroscopy. *J. Biomol. NMR* 11: 13-22.
29. Bax, A., and Pochapsky, S.J., 1992, Optimized recording of heteronuclear multi-dimensional NMR spectra using pulsed field gradients, *J. Magn. Reson.* 99: 638-643.
30. Vuister, G.W., Clore, G.M., Gronenborn, A.M., Powers, R., Garrett, D.S., Tschudin, R, and Bax, A., 1993, Increased resolution and improved spectral quality in four-dimensional $^{13}C/^{13}C$ separated HMQC-NOE-HMQC spectra using pulsed field gradients, *J. Magn. Reson. Series B* 101: 210-213.
31. Clore, G.M., Wingfield, P.T., and Gronenborn, A.M., 1991, High resolution three-dimensional structure of interleukin-1β in solution by three and four dimensional nuclear magnetic resonance spectroscopy. *Biochemistry* 30: 2315-2323.
32. Powers, R., Garrett, D.S., March, C.J., Frieden, E.A., Gronenborn, A.M., and Clore, G.M., 1992, Three-dimensional solution structure of interleukin-4 by multi-dimensional heteronuclear magnetic resonance spectroscopy, *Science* 256: 1673-1677.
33. Smith, L.J., Redfield, C., Boyd, J., Lawrence, G.M.P., Edwards, R.G., Smith, R.A.G., and Dobson, C.M., 1992, Human interleukin-4: the solution structure of a four helix bundle protein. *J. Mol. Biol.* 224: 900-904.
34. Powers, R., Garrett, D.S., March, C.J., Frieden, E.A., Gronenborn, A.M., and Clore, G.M., 1993, The high resolution three-dimensional solution structure of interleukin-4 determined by multi-dimensional heteronuclear magnetic resonance spectroscopy, *Biochemistry* 32: 6744-6762.
35. Fairbrother, W.J., Gippert, G.P., Reizer, J., Saier, M.J., and Wright, P.E., 1992, Low resolution structure of the *Bacillus subtilis* glucose permease IIA domain derived from heteronuclear three-dimensional NMR spectroscopy, *FEBS Lett.* 296: 148-152.
36. Ikura, M., Clore, G.M., Gronenborn, A.M., Zhu, G., Klee, C.B., and Bax, A., 1992, Solution structure of a calmodulin-target peptide complex by multi-dimensional NMR, *Science* 256: 632-638.
37. Thierault, Y., Logan, T.M., Meadows, R., Yu, L., Olejniczak, E.T., Holzman, T.F., Sikmmer, R.L., and fesik, S.W., 1993, Solution structure of the cyclosporin A/cyclophilin complex by NMR, *Nature (Lond.)* 361, 88-91.
38. Omichinski, J.G., Clore, G.M., Schaad, O., Felsenfeld, G., Trainor, C., Appella, E., Stahl, S.J., and Gronenborn, A.M., 1993, NMR structure of a specific DNA complex of Zn-containing DNA binding domain of GATA-1, *Science* 261: 438-446.
39. Lodi, P.J., Garrett, D.S., Kuszewski, J., Tsang, M.L.S., Weatherbee, J.A., Leonard, W.J., Gronenborn, A.M., and Clore, G.M., 1994, High resolution solution structure of the β chemokine hMIP-1β by multi-dimensional NMR, *Science* 263: 1762-1767.
40. Clore, G.M., Omichinski, J.G., Sakaguchi, K., Zambrano, N., Sakamoto, H., Appella, E., and Gronenborn, A.M., 1994, High-resolution solution structure of the oligomerization domain of p53 by multi-dimensional NMR, *Science* 265: 386-391.

46

PROTEIN DOCKING IN THE ABSENCE OF DETAILED MOLECULAR STRUCTURES

I. A. Vakser and G. V. Nikiforovich

Center for Molecular Design
Washington University
Box 1099, St. Louis, Missouri 63130

INTRODUCTION

The design of new computational procedures to predict molecular complexes is a fast developing area stimulated by the growing demands of researchers working in various fields of molecular biology and looking for more powerful tools for their investigations. The problem for molecular recognition (docking) approaches may be shortly formulated as following: how to match two molecules with known 3D structures in order to predict the configuration of their complex? In the general case, no additional prior knowledge on binding sites is assumed to be available.

The algorithms for molecular recognition in a "ligand-receptor" system (for a review, see Refs. 1-4) include, and sometimes combine, approaches which concentrate mostly on energetic considerations (5-10), and procedures based on a search for steric fit (11-19) including those which make use of physico-chemical surface complementarity (20,21). The "rigid body" approach is justified in most cases of known macromolecular 3D structures (22,23). However, procedures which explicitly take into account the ligand (e.g. oligopeptide) flexibility (24) have also started to appear (9,25).

The problem of an inherent inaccuracy in 3D structures of the molecules is one of the most serious obstacles which docking procedures have to overcome. This inaccuracy has both a "natural" origin (internal flexibility) and a "technical" reason (poor quality of the X-ray data), which often is the consequence of the same flexibility. This problem has been treated by introducing a certain tolerance to the surface of molecules (14,17), reducing atom-atom interactions to residue-residue ones (22), or truncating certain amino-acid sidechains (16). However, even such "radical" methods as elimination of non-hydrophobic atoms (21), apparently will not help, when conformational changes upon complex formation are really substantial, or the X-ray data on one or both (macro)molecules is not available and the structure, based on alternative sources (NMR, modeling), is not well defined.

In order to address the problem of poorly determined structures in molecular recognition, we designed a direct computer experiment with molecules totally deprived of any structural features smaller than 7Å. For this purpose, we modified our previously

Methods in Protein Structure Analysis, Edited by M. Z. Atassi and E. Appella
Plenum Press, New York, 1995

developed docking procedure (17) which predicts complex configurations on the basis of surface complementarity. The modified procedure was applied on various known protein complexes taken from the Brookhaven Protein Data Bank. In most cases, except antigen - antibody complexes, a pronounced trend towards the correct complex configuration was clearly indicated and the real binding sites were predicted. The distinction between the prediction of the antigen - antibody complexes and the other molecular pairs may reflect important differences in the principles of complex formation (26,27).

METHODS

Basic Docking Algorithm

The primary molecular recognition algorithm is described in detail elsewhere (17). Briefly, the 3D atomic structures of "ligand" and "receptor" molecules are projected on a 3D grid. The surface of the receptor molecule is represented by a layer of small positive numbers and the inner part is represented by large negative numbers. The ligand molecule is represented by positive numbers only. Everywhere on the grid outside the two molecules there are zero values. Thus, when the projected images of the molecules are translated one relative to the other, and point by point multiplication all over the grid is taken, the absence of contact between projection points of the molecules contributes zero, the contact between the ligand and the surface of the receptor contributes a small positive number, and the contact between the ligand and the inner part of the receptor contributes a large negative number (penalty for penetration). All possible orientations of the ligand are sampled with a given angle interval. The resulting highest score list of the ligand positions relative to the receptor, gives the configurations of the molecular complex with largest areas of contact between the ligand and the receptor. The algorithm was used both as described above and in its "hydrophobic" modification (21) to predict the complex configurations for a number of molecules with known high resolution X-ray structures and proved to be a reliable tool for docking studies.

Modified Algorithm

One of the most important parameters for the described molecular images is the interval of the grid, which sets up the resolution, or the accuracy, of molecules representation. In a regular docking procedure, the grid step is in the range of 0.7-1.7Å (17,21). Such a grid-step represents an ultimate threshold for details of the molecular structure. No detail smaller than the grid-step is reflected in the molecule 3D grid projection, and consequently, no such detail is taken into account in the search for intermolecular fits. Thus, a natural quality of grid representation, that is, the possibility to vary the grid step, can be used to study resolution dependencies in molecular recognition.

To reveal a possibility to dock low-resolution structures, we set up an ultra large grid-step of 7Å (much larger than an atom radius). The Fig. 1 shows a typical cross section through various molecule representations. As can be seen, the high-resolution discrete molecular image preserves many atom-size shape features of the "realistic" van der Waals representation. This is not the case of the image obtained on a grid with an ultra large step (low-resolution representation). In this case, it is hard to find any specific shape characteristics at all. As a consequence, use of the surface recognition algorithm, which yields such molecular images, doesn't produce any reasonable, different from noise, results (Vakser, unpublished results).

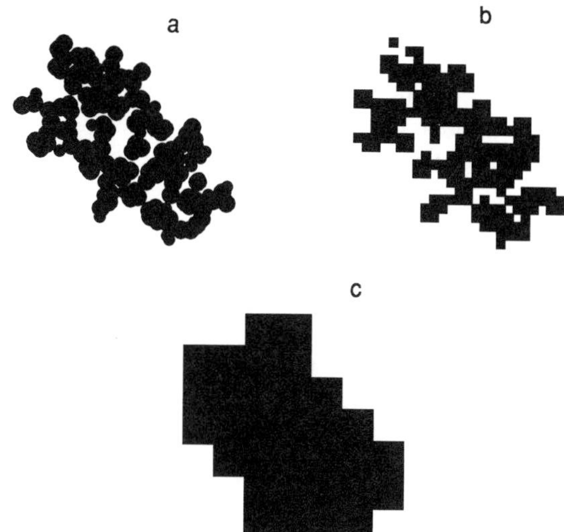

Figure 1. A cross section through different representations of the β-subunit of human hemoglobin. In (a) the molecule is represented by van der Waals spheres. In (b) and (c) the same cross section is shown for 3D grid projections with high-resolution 1.7Å and low-resolution 7Å grid-steps respectively.

In order to make the low-resolution images more informative, not breaching the main assumption of no structure details below the ultra large grid-step size, we modified the procedure of projecting molecules on the grid. The details of the modified algorithm as well as its implementation are described elsewhere (28). Briefly, the numbers which represent the density of atoms within a volume cube (element of the grid) were introduced. The density numbers are naturally smaller at the molecular surface than inside the molecule, because at the surface only part of the cube is occupied by atoms. The surface values tend to be larger within deep cavities and smaller at pronounced convexities. The rest of the docking algorithm remains exactly as in the basic procedure.

As follows from this modification, the dependence on atomic density is digitized on a low-resolution grid. Thus, our basic assumption of no structural features below the low-resolution grid-step still holds. To use an analogy with regular vision, the atom density modification means that a subject with "low-resolution" vision may distinguish not only between black and white, but also between different colors as well as their densities.

RESULTS

The high-resolution molecular recognition algorithms, at least in theory, predict the "exact" position of the ligand in complex with the receptor. It means that the six parameters, three translations and three rotations of the rigid-body ligand, have to be determined. Traditionally, these parameters are considered together, in one set, which is quite justifiable for high-resolution docking. Indeed, when the molecules are represented with high precision, there is little room, for example, to rotate the ligand after it had been translated to the correct position. There are, however, certain indications of multiplicity, at least for predicted ligand rotations at the binding site of the receptor (16). As our experience showed, this multiplicity increases dramatically when the low resolution is applied. Such increase of multiplicity is accompanied by a decrease in reliability of the multiple-value parameter(s) determination. In other words, the difference between the high-resolution and the low-resolution occurs as follows. In typical results of the high-resolution docking, there might be present multiple false-positive matches, the one match which is "fully correct", and few (or none) "partially correct" matches (e.g. those which correctly predict same five out of the total of six parameters). In low-resolution recognition, also along with false positives, there are usually no "fully correct" matches and multiple "partially correct" matches. This characteristic feature of the low-resolution docking means that it is reasonable to consider the six parameters of the rigid-body ligand translation and rotation separately in the following sequence: three translations (the correct values are equivalent to the prediction of the receptor binding site), two angles of rotation which determine the orientation of the ligand binding site toward the receptor (the correct values are equivalent to the prediction of the ligand binding site), and one spin angle around the axis which connects the two binding sites already in contact (determines the final "lock" of the two molecules).

We applied the procedure to different molecular complexes from the Brookhaven Protein Data Bank (29). Some of the molecules are shown on Fig. 2. The complexes were selected so that, while the "receptor" was always represented by a macromolecule (more than 1000 atoms), the size of the "ligand" varies, from medium (β subunits of human deoxy- and horse met-hemoglobins, lysozyme), to small proteins (trypsin and chymotrypsin inhibitors, ovomucoid third domain), and further to peptides and tyrosinyl adenylate (all in the receptor-bound conformation). From the point of view of their nature, the chosen examples represent multisubunit proteins (hemoglobins), enzyme-inhibitor complexes (trypsin, chymotrypsin, and subtilisin with the inhibitors), antigen-antibody complexes (Fab with

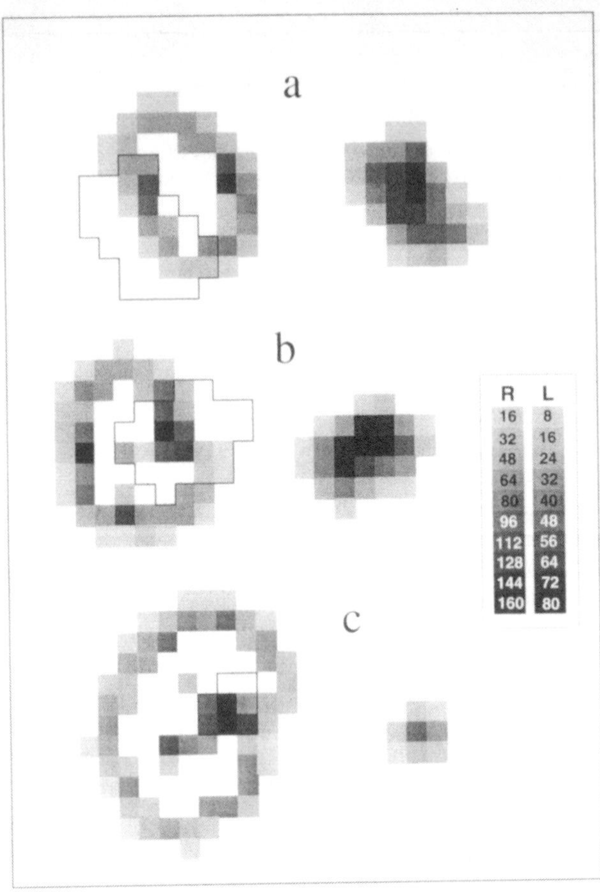

Figure 2. Cross sections through 3D grid images of (a) human hemoglobin α (left) and β (right) subunits; (b) trypsin (left) and trypsin inhibitor (right); (c) acid proteinase (left) and its peptide inhibitor (right). The step of the grid is 7Å. Different shades of gray represent various degrees of atom density within the corresponding cube of the grid for the "receptor" (R) and the "ligand" (L). White areas inside the "receptor" molecules indicate negative values (to avoid deep intermolecular penetrations, Refs. 17,28). Solid line contours show the actual position of the ligands in the cocrystallized complexes.

lysozyme and with a peptide), and such complexes as tRNA synthetase - tyrosinyl adenylate and the MHC I molecule with a peptide.

All results were analyzed in two stages, first for the three translational (prediction of the receptor's binding site) and then, in configurations with the correct translation, for the two rotational coordinates (prediction of the ligand's binding site). The axis P for the third angle of rotation was chosen between centers of gravity of the entire ligand molecule and its binding site. In cases when the whole ligand molecule is in contact with the receptor (e.g. peptides in complex with aspartic proteinase or MHC) we chose the main axis of the ligand. No molecular complex showed any significant preference toward the correct values of the sixth coordinate (spin angle around the binding site axis).

For the human deoxyhemoglobin, α - β subunits (2HHB, Ref. 30), the low-resolution representation of the molecules is shown in Fig. 2a. The results of an exhaustive search through all six docking coordinates (Fig. 3a) show that the correct values of the translational coordinates (translation of β subunit to α subunit) are determined quite unequivocally. The rotational coordinates are resolved much less distinctively, however the trend towards the correct binding site orientation of the "ligand" (β subunit) is clearly indicated.

In case of horse methemoglobin, α - β subunits (2MHB, Ref. 31), the translation of the β subunit is predicted with even better accuracy than for 2HHB (Fig. 3b). The orientation coordinates show a moderate trend to correct values.

The molecular images of trypsin and trypsin inhibitor (2PTC, Ref. 32) are presented in Fig. 2b. The trypsin binding site was predicted with remarkable accuracy (Fig. 3c). All first 100 highest score positions had the correct translational values. Actually, the first configuration with a wrong translation of the ligand appeared as number 274 in the sorted list of predicted positions. However, the prediction of the ligand orientation is poor, though the left part of the distribution (correct directions) slightly prevails. Such non-distinctive ligand's orientation may be correlated with the strong results in the translation prediction. Indeed, the less sensitive are the low-resolution molecular images to the ligand orientation, the more configurations with the correct translations and different orientations will be found.

For chymotrypsin and ovomucoid third domain (1CHO, Ref. 33), the "absolute" translational results (Fig. 3d) are similar to these of the trypsin - trypsin inhibitor complex. However, contrary to the case of 2PTC, the orientation sampling reveals a distinctive peak of the correct values in the angles distribution.

The results on subtilisin - chymotrypsin inhibitor (2SNI, Ref. 34) give a high-quality prediction of the subtilisin binding site (Fig. 3e). However, not all first 100 matches are correct, as in the case of other enzyme - inhibitor (small protein) complexes (2PTC, 1CHO). The trend toward correct ligand orientation is weak, as in the case of 2PTC.

The complex between acid proteinase and peptide inhibitor (3APR, Ref. 35), shown on Figure 2c, yielded perfect results on the ligand translation (Fig. 3f). Contrary to all previous complexes, the orientation prediction is very good. The angle distribution has certain symmetry, where the values near 0º and 180º strongly dominate over those around 90º, which clearly reflects the symmetry of the ligand (an elongated peptide). Of course, the difference between the C- and N-terminals (0º and 180º) at the ultra-low resolution is negligible.

In the case of tRNA synthetase - tyrosinyl adenylate (3TS1, Ref. 36), the translational and rotational distributions (Fig. 3g) are similar to those of the 3APR complex. Their character is even more pronounced. Just as in the previous case, the rotational distribution reflects the natural symmetry of the ligand.

The histograms in Fig. 3h show the results of a model 9-residue peptide docking on the MHC I molecule (1HSA, Ref. 37). In our computer experiment, we retained the $\alpha 1$ and $\alpha 2$ subunits which contain the binding site on MHC I. The pattern of distributions is similar to those of 3APR and 3TS1, however, the results are weaker than in both these cases. The

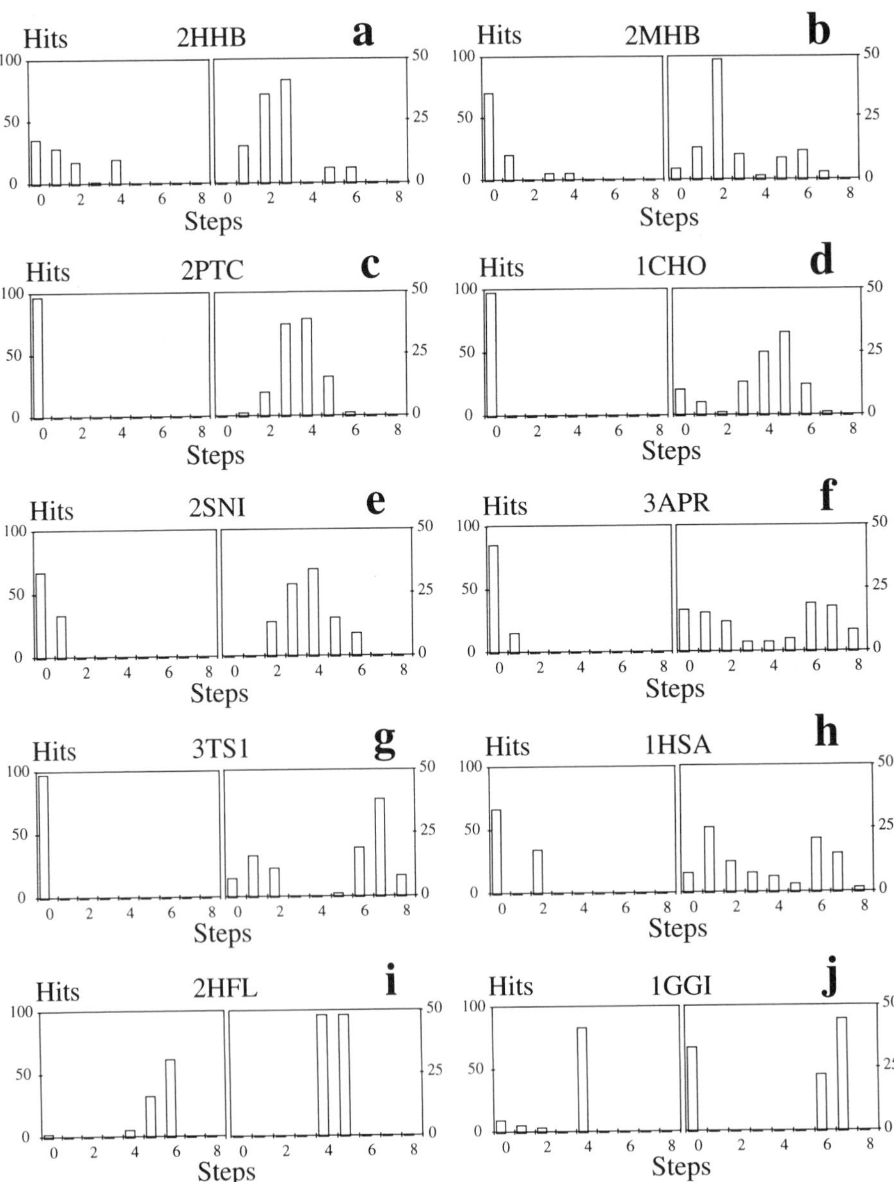

Figure 3. Docking results for different molecular complexes (a-j). For each complex the left panel represents the number of hits at various distances from the correct ligand position. Only the first 100 highest score hits were analyzed. Distances are shown in translation (grid) steps of 7Å, thus the 0 step corresponds to the correct position (within the ultimate highest accuracy of 7Å). Right panels show relative distribution of ligand orientations for the hits with the correct translations only. The orientation is calculated as an angle between the axis P (see the text) of the given and the correct orientations, in steps of 20°, thus the 0 step corresponds to the correct orientation (within 20° accuracy). The left part of the histogram (steps 0-3) corresponds to the ligand's binding site orientations which prefer the receptor direction.

reason might be that in 3APR and 3TS1 complexes the ligands are deeply immersed into the receptor molecule, while in 1HSA the receptor site is shallow. Along with matching the peptide from 1HSA complex with α1-α2 subunits of MHC, we also tested docking modes on the entire MHC molecule. The resulting distributions (not shown) appeared quite disordered. This may be attributed to a considerably large size and complicated surface structure of the entire MHC I molecule, which, combined with the shallow character of the binding site, creates highly competitive false positive matches for the ligand at the ultra-low resolution.

The results on Fab fragment - lysozyme (2HFL, Ref. 38) complex (Fig. 3i) are quite different from all the previous ones. The ligand's translation was basically not predicted (a small value at the 0-step is negligible). The same applies to the rotational distribution which was not very informative at all, since it was based on very poor statistics.

For Fab fragment and a peptide (1GGI, Ref. 39), the docking of a peptidic ligand (instead of a small protein in 2HFL) to an antibody yielded somewhat better results (Fig. 3j). The translation part is still much weaker than in the rest of the cases. The correct translations, however, are represented better than in 2HFL. The correct rotational values (including the opposite to 0º, "symmetric" part of the distribution) are very distinctive and look much more dominant than in any other complex. The character of both distributions suggests high "specificity" of the complex (small number of configurations with correct translations when most of them have also correct orientations of the ligand), as opposed to low "specificity" of complexes like 2PTC (large number of configurations with correct translations and different orientations of the ligand).

DISCUSSION AND CONCLUSIONS

The distributions of the ligand positions in the 10 tested complexes show different patterns for different groups of molecules. The complexes between medium size proteins, such as subunits of hemoglobins, which are characterized by large intermolecular interfaces with no distinct "global" concavity at the "receptor" site, demonstrate good translation predictions and moderate orientation preferences. The enzyme-inhibitor complexes, when the inhibitor is represented by a small protein, are characteristic of very strong translation predictions, which, at the same time, are not very specific to the ligand orientation. The complexes with small oligopeptide-size ligands are also good on the translation prediction (except the antigen-antibody complex). In addition to that, they demonstrate an exceptional quality of the orientation prediction, which is symmetric due to the similarity between the two ends of elongated ligands at the ultra-low resolution. The results on the antigen-antibody complexes tested are, in general, different from the rest of the cases. They are characterized by very low predictions with correct translations which are very specific on the ligand orientation (in case of the lysozyme-antibody complex, this feature might be lost due to a non-representative statistics in the rotational distribution).

The translational and rotational histograms represent a convenient and simple tools for examination of the docking results at ultra-low resolution. However, they are, by far, not sufficient for more detailed analysis. The reduction of 3 dimensional (translations) and 2 dimensional (rotations) results to single-dimension representations helps to reveal very clearly the existing trend to the correct configurations. At the same time, it makes impossible to employ such methods as cluster analysis of 3D and 2D results. If such clusters exist, their examination could contribute to a more objective view of the predictions. For example, in case of the antigen-antibody complex 2HFL, the predominant high-score false-positive matches (Fig. 3i), if found belonging to the same "wrong" binding site, could be reduced to a single, most representative match, thus allowing the analysis of matches with lower scores.

Such histograms of "clusters" rather than the histograms of "individual matches" which we used, might be quite helpful in dealing with cases as 2HFL, when the correct binding mode does not dominate. We leave this analysis for the future, as this present paper concentrates on the principal trends in molecular recognition at ultra-low resolution. For this purpose, the simple 1D analysis helps to reveal the generalities which could have been less explicit otherwise.

Our approach gives an important instrument for the practical docking studies of molecules whose structures are too uncertain to fit into high-resolution docking procedures. The low-resolution docking may be successfully used as the first, preliminary stage, which will be followed by a "regular" high-resolution procedure. This may be helpful to preselect potential areas of ligand binding in case of local conformational changes upon complex formation (which are of little importance at ultra-low resolution). We will also use this approach as a stand-alone procedure for molecules with low-resolution structures (e.g. NMR, modeled structures). The resolution of ~7Å is more than enough to accommodate inaccuracies of many low-resolution experimental and modeled structures, which will open new opportunities for investigation of molecular mechanisms.

ACKNOWLEDGMENTS

The authors wish to thank Prof. Garland R. Marshall for his support and valuable comments. We are grateful to Prof. Ephraim Katchalski-Katzir and all members of the molecular recognition group at the Weizmann Institute for encouraging discussions at the initial stages of the work. The authors acknowledge the NIH for financial support (NIH GM24483).

REFERENCES

1. Kuntz, I.D., Meng, E.C., and Shoichet, B.K., 1994, Structure-based molecular design, *Acc. Chem. Res.* 27:117-123.
2. Kollman, P.A., 1994, Theory of macromolecule-ligand interactions, *Curr. Opin. Struct. Biol.* 4:240-245.
3. Blaney, J.M., and Dixon, J.S., 1993, A good ligand is hard to find: automated docking methods, *Perspec. Drug Disc. Des.* 1:301-319.
4. Cherfils, J., and Janin, J., 1993, Protein docking algorithms: simulating molecular recognition, *Curr. Opin. Struct. Biol.* 3:265-269.
5. Goodford, P.J., 1985, A computational procedure for determining energetically favorable binding sites on biologically important macromolecules, *J. Med. Chem.* 28:849-857.
6. Warwicker, J., 1989, Investigating protein-protein interaction surfaces using a reduced stereochemical and electrostatic model, *J. Mol. Biol.* 206:381-395.
7. Goodsell, D.S., and Olson, A.J., 1990, Automated docking of substrates to proteins by simulated annealing, *Proteins* 8:195-202.
8. Yue, S.-Y., 1990, Distance-constrained molecular docking by simulated annealing, *Protein Engng.* 4:177-184.
9. Caflisch, A., Niederer, P., and Anliker, M., 1992, Monte Carlo docking of oligopeptides to proteins, *Proteins* 13:223-230.
10. Hart, T.N., and Read, R.J., 1992, A multiple-start Monte Carlo docking method, *Proteins* 13:206-222.
11. Kuntz, I.D., Blaney, J.M., Oatley, S.J., Langridge, R., and Ferrin, T.E., 1982, A geometric approach to macromolecule-ligand interactions, *J. Mol. Biol.* 161:269-288.
12. Connolly, M.L., 1986, Shape complementarity at the hemoglobin alpha1-beta1 subunit interface, *Biopolymers* 25:1229-1247.
13. DesJarlais, R.L., Sheridan, R.P., Seibel, G.L., Dixon, J.S., Kuntz, I.D., and Venkataraghavan, R., 1988, Using shape complementarity as an initial screen in designing ligands for a receptor binding site of known three-dimensional structure, *J. Med. Chem.* 31:722-729.

14. Jiang, F., and Kim, S.H., 1991, "Soft docking": matching of molecular surface cubes, *J. Mol. Biol.* 219:79-102.

15. Norel, R., Fischer, D., Wolfson, H.J., and Nussinov, R., 1994, Molecular surface recognition by a computer vision-based technique, *Protein Engng.* 7:39-46.

16. Shoichet, B.K., and Kuntz, I.D., 1991, Protein docking and complementarity, *J. Mol. Biol.* 221:327-346.

17. Katchalski-Katzir, E., Shariv, I., Eisenstein, M., Friesem, A.A., Aflalo, C., and Vakser, I.A., 1992, Molecular surface recognition: determination of of geometric fit between proteins and their ligands by correlation techniques, *Proc. Natl. Acad. Sci. U.S.A.* 89:2195-2199.

18. Helmer-Citterich, M., and Tramontano, A., 1994, PUZZLE: a new method for automated protein docking based on surface shape complementarity, *J. Mol. Biol.* 235:1021-1031.

19. Ho, C.M.W., and Marshall, G.R., 1993, SPLICE: a program to assemble partial query solutions from three-dimensional database searches into novel ligands, *J. Comput. Aided Mol. Des.* 7:623-647.

20. Shoichet, B.K., and Kuntz, I.D., 1993, Matching chemistry and shape in molecular docking, *Protein Engng.* 6:723-732.

21. Vakser, I.A., and Aflalo, C., Hydrophobic docking: a proposed enhancement to molecular recognition techniques, *Proteins*, in press.

22. Wodak, S.J., and Janin, J., 1978, Computer analysis of protein-protein interaction, *J. Mol. Biol.* 124:323-342.

23. Janin, J., and Chothia, C., 1990, The structure of protein-protein recognition sites, *J. Biol. Chem.* 265:16027-16030.

24. Marshall, G.R., 1992, 3D structure of peptide-protein complexes: implications for recognition, *Curr. Opin. Struct. Biol.* 2:904-919.

25. Leach, A.R., 1994, Ligand docking to proteins with discrete side-chain flexibility, *J. Mol. Biol.* 235:345-356.

26. Tello, D., Goldbaum, F.A., Mariuzza, R.A., Ysern, X., Schwarz, F.P., and Poljak, R.J., 1993, Tree-dimensional structure and thermodynamics of antigen binding by anti-lysozyme antibodies, *Biochem. Soc. Trans.* 21:943-946.

27. Lawrence, M.C., and Colman, P.M., 1993, Shape complementarity at protein/protein interfaces, *J. Mol. Biol.* 234:946-950.

28. Vakser, I.A., 1995, Protein docking for low-resolution structures, *Protein Engng.* 8:371-377.

29. Abola, E.E., Bernsein, F.C., Bryant, S.H., Koetzle, T.L., and Weng, J., 1987, Protein Databank, *in:* Crystallographic Databases - Information Content, Software Systems, Scientific Applications. Allen, F.H., Bergerhoff, G., and Sievers, R. eds., Data Commission of the International Union of Crystallography, Bonn, pp 107-132.

30. Fermi, G., Perutz, M.F., Shaanan, B., and Fourme, R., 1984, The crystal structure of human deoxyhaemoglobin at 1.74 A resolution, *J. Mol. Biol.* 175:159-174.

31. Ladner, R.C., Heidner, E.G., and Perutz, M.F., 1977, The structure of horse methaemoglobin at 2.0 angstroms resolution, *J. Mol. Biol.* 114:385-414.

32. Marquart, M., Walter, J., Deisenhofer, J., Bode, W., and Huber, R., 1983, The geometry of the reactive site and of the peptide groups in trypsin, trypsinogen and its complexes with inhibitors, *Acta Crystallog., Sect. B* 39:480-490.

33. Fujinaga, M., Sielecki, A.R., Read, R.J., Ardelt, W., Laskowski, M. Jr, and James, M.N.G., 1987, Crystal and molecular structures of the complex of aplpha-chymotrypsin with its inhibitor turkey ovomucoid third domain at 1.8 angstroms resolution, *J. Mol. Biol.* 195:397-418.

34. McPhalen, C.A., and James, M.N.G., 1988, Structural comparison of two serine proteinase-protein inhibitor complexes: eglin-c-subtilisin Carlsberg and CI-2-subtilisin Novo, *Biochemistry* 27:6582-6598.

35. Suguna, K., Bott, R.R., Padlan, E.A., Subramanian, E., Sheriff, S., Cohen, G.H., and Davies, D.R., 1987, Structure and refinement at 1.8 A resolution of the aspartic proteinase from Rhizopus chinensis, *J. Mol. Biol.* 196:877-900.

36. Brick, P., Bhat, T.N., and Blow, D.M., 1989, Structure of tyrosyl-tRNA synthetase refined at 2.3 angstroms resolution. Interaction of the enzyme with the tyrosyl adenylate intermediate, *J. Mol. Biol.* 208:83-98.

37. Madden, D.R., Gorga, J.C., Strominger, J.L., and Wiley, D.C., 1992, The three-dimensional structure of HLA-B27 at 2.1 angstroms resolution suggests a general mechanism for tight peptide binding to MHC, *Cell* 70:1035-1048.

38. Sheriff, S., Silverton, E.W., Padlan, E.A., Cohen, G.H., Smith-Gill, S.J., Finzel, B.C., and Davies, D.R., 1987, Three-dimensional structure of an antibody-antigen complex, *Proc. Natl. Acad. Sci. U.S.A.* 84:8075-8079.

39. Rini, J.M., Stanfield, R.L., Stura, E.A., Salinas, P.A., Profy, A.T., and Wilson, I.A., 1993, Crystal structure of an HIV-1 neutralizing antibody 50.1 in complex with its V3 loop peptide antigen, *Proc. Natl. Acad. Sci. U.S.A.* 90:6325-6329.

47

A METRIC MEASURE FOR COMPARING SEQUENCE ALIGNMENTS

Hugh B. Nicholas Jr., Alexander J. Ropelewski, David W. Deerfield II, and Joseph G. Behrmann

Pittsburgh Supercomputing Center
4400 Fifth Avenue
Pittsburgh, Pennsylvania 15213

INTRODUCTION

An important goal in sequence analysis is to identify features in sequence alignments that are reflections of the evolutionary history of the sequences rather than artifacts of the alignment method. We have developed a new measure for determining a distance between sequence alignments that will assist us in reaching this goal. We apply this new measure to alignments produced by three different sequence alignment programs and an alignment created by structural superposition.

Multiple sequence alignments are valuable tools for identifying the amino acids critical for the structural and functional integrity of a protein. Alignments of homogolous proteins also succinctly summarize the evolutionary history of the protein. In recent years, several different techniques for creating multiple alignments of proteins have been published (Barton and Sternberg, 1987; Lipman et al., 1989, Feng and Doolittle, 1990). Implicit in each alignment program are different assumptions about how to treat the evolution of sequences (Altschul, 1989; Altschul and Lipman, 1989). All multiple alignment techniques have potential theoretical shortcomings (Altschul, 1989; Altschul and Lipman, 1989). We will explore how these different assumptions lead to different alignments by applying our new measure to alignments from three multiple sequence alignment programs and a structural superposition alignment.

Our new measure for comparing alignments has several advantages over previously used methods. Most importantly, our measure is a true metric that meets the mathematical criteria for a distance between alignments. That this is a true metric is important because it allows us to compare several different alignment techniques simultaneously (Kruskal, 1983). Nonmetric measures allow only pairs of alignments to be directly compared. Hence, if several methods are contrasted with a nonmetric measure, one of the alignments must be designated as the standard alignment, assumed to be more reliable, to which the other alignments are compared.

Methods in Protein Structure Analysis, Edited by M. Z. Atassi and E. Appella
Plenum Press, New York, 1995

The second major advantage of our alignment measure is that it incorporates a more quantitative and discriminating assessment of the variations in the size and location of gaps. Methods, such as determining the percentage of amino acids that are identically aligned, measure the location and size of gaps in a dichotomous "the same location and size" or "not the same location and size" manner. Our alignment measure makes a much more quantitative and graduated measurement of gap placement. Since an alignment can be completely described by the locations and lengths of the gaps placed into each sequence, gap location and size is a critical aspect.

A third feature of our proposed alignment measure is that it has a straight forward interpretation that is easily visualized. The interpretation is that the distance between two alignments is the area between the path graphs of the two alignments.

METHODS and DATA

Multiple Sequence Alignments

The MSA (multiple sequence alignment) alignment was generated on the Cray-C90 at the Pittsburgh Supercomputing Center with the program from the National Center for Biotechnology Information described by Lipman et al., (1989). The multiple sequence alignment was generated using the PAM 250 (Dayhoff et al. 1978) similarity matrix converted to differences (Smith et al., 1981), open gap costs equal to eight, and extend gap cost of 12 (Altschul, 1989). These are the default values for these parameters. We selected the option to weight the sequences so that the alignment scores would approximate an alignment scored by summing the costs over the phylogenetic tree estimated by the program. We selected the optimal alignment rather than the heuristic alignment and elected to weight end-gaps.

The AMPS (alignment of multiple protein sequences) alignment was generated by the VAX version of the program initially described by Barton and Sternberg (1987). The alignment was generated with the PAM 250 similarity matrix and a gap score of eight. The AMPS program does not use a separate penalty for extending gaps. Thus in the AMPS alignment each gap is penalized the same cost regardless of its length. Barton (1990) has found this to be an effective strategy. The order in which sequences were joined in generating the alignment was computed by the program using normalized alignment scores from one hundred randomizations of the sequences. We used the option to generate the alignment guided by a phylogenetic tree rather than a single order alignment (Barton, 1990).

The PileUp alignment was generated by the PileUp program included release 7 of the Genetics Computer Group suite of programs (Genetics Computer Group, 1991). Again we used the PAM 250 similarity matrix to generate the alignments. The open gap penalty was minus eight and the extend gap penalty was also minus eight. The rational for this choice was to set the total costs for a gap of two amino acids equal in both the PileUp and MSA programs. An exact equivalence for all gap lengths is not possible since PileUp uses similarity scoring and MSA uses distance scoring (Smith et al., 1981). This allows us to gather information on the consequences of this difference in gap penalty functions between the two programs. The option to weight end-gaps was also chosen to remain consistent with both MSA and AMPS. Weighting end-gaps is the theoretically correct choice for the aspartyl protease sequence used in this study.

The structural superposition alignments were in an alignment database made available by Pascarella and Argos (1992) on the EMBL file server. We edited the alignments to replace the isolated insertions that Pascarella and Argos had removed from the listing after the alignment was initially generated. These changes were necessary to make the comparison with the other alignments on the same basis.

EXAMPLE SEQUENCE ALIGNMENTS

We will apply the alignment measure to four multiple sequence alignments of six aspartyl proteases. Table 1 identifies the sequences and summarizes the alignment of each pair of aligned sequences. The aligned pairs of aspartyl proteases range from twenty-five to forty-three percent identical. We expect sequences with this level of similarity to be able to select each other as related sequences when used as a query in a database search. However, this level of diversity is great enough that alignment of the sequences is challenging.

A fragment from the four multiple sequence alignments of the six aspartyl proteases (Figure 1) was selected to illustrate many of the uncertainties that arise with current multiple sequence alignment methods. Within this fragment there are five amino acids that all four alignments identify as completely conserved. Another five amino acids are identified as completely conserved in two or more of the alignments. We postulate that these residues should be aligned and accepted as completely conserved.

Table 1. Comparative statistics for six aspartyl proteases

	Carp_Y	Pepc_M	Chym_B	Catd_H	Carp_R	Pepa_A
		40%	35%	42%	34%	26%
Carp_Y	329	57%	61%	51%	60%	65%
		2%	3%	5%	5%	8%
	135		43%	40%	29%	28%
Pepc_M	190	328	52%	53%	64%	62%
	7		3%	6%	5%	9%
	118	145		43%	32%	25%
Chym_B	203	175	323	49%	63%	68%
	10	11		7%	3%	5%
	149	141	151		29%	25%
Catd_H	179	185	171	348	61%	63%
	19	22	25		9%	10%
	114	99	107	103		33%
Carp_R	204	217	210	217	325	60%
	17	20	13	32		5%
	89	97	84	92	112	
Pepa_A	224	214	230	225	203	325
	28	31	20	37	19	

Each cell in the table has three numbers that are identities at the top, differences in the middle, and gaps at the bottom in the alignment indicated by the row and column headings. Numbers to the lower left of the diagonal and on the diagonal the number of amino acids in these categories. Numbers above and to the right of the diagonal are percentages. Percentages may not add to one hundred because of rounding errors. Numbers on the diagonal are the lengths of the sequences. The statistics were computed by the MALIGNED multiple sequence alignment editor (Clark, 1992). The complete Swiss-Prot identifier and accession number corresponding to each of the six abreviated aspartyl protease names are: Carp_Y = (Carp_Yeast, P07267); Pepc_M = (Pepc_Macfu, P03955); Chym_B = (Chym_Bovin, P00794); Catd_H = (Catd_Human, P07339); Carp_R = (Carp_Rhich, P06026); and Pepa_A = (Pepa_Aspsw, P17946). Any signal and propeptide fragments were removed from the sequence in the database so that only the sequences for the mature enzymes were aligned.

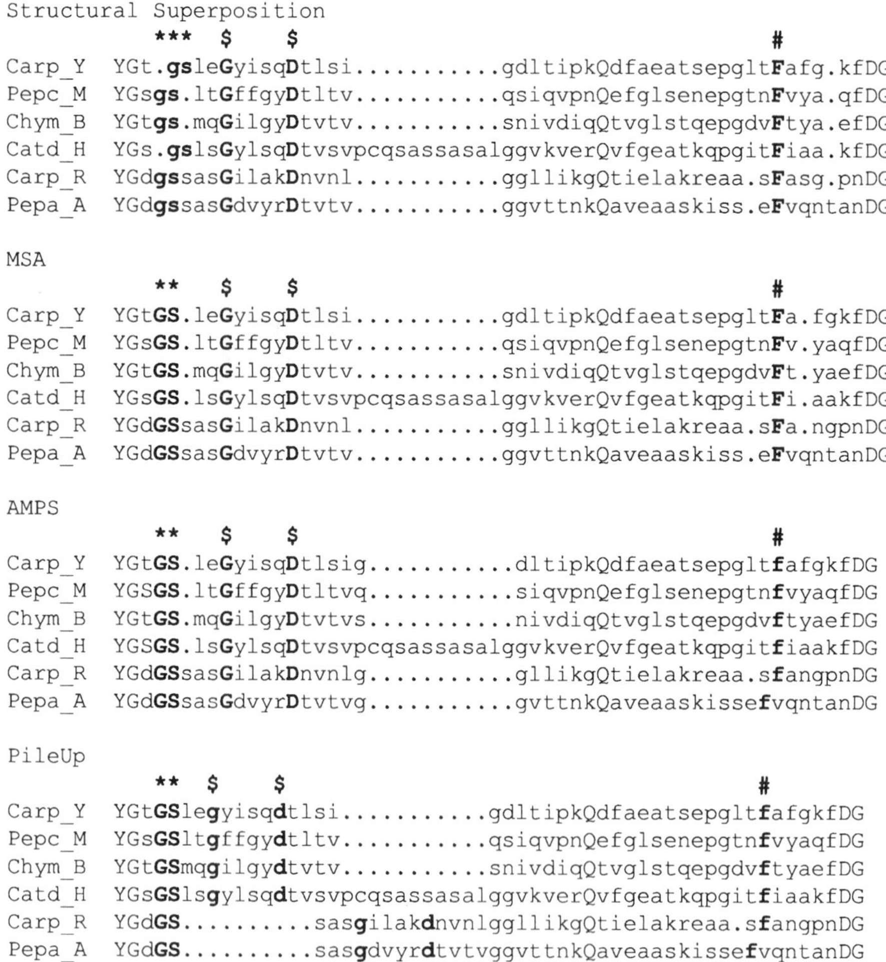

Figure 1. Multiple sequence alignments of six aspartyl proteases.

We will discuss, in turn, each of these five amino acids that. One factor that will emerge is that the differences can be succinctly and efficiently described in terms of the differences in the placements of small gaps by the different alignment methods. This suggests that, while identifying homologous amino acids in related proteins reveals the biochemically important features, the alignment process itself may be fruitfully understood in term of the placement of gaps.

Glycine Serine Dipeptide

We will first consider the Glycine Serine (GS) dipeptide marked with asterisks above the alignments. The three multiple sequence alignment programs that base their alignments entirely on phylogenetic or sequence information align this dipeptide as completely conserved, while the structural superposition alignment does not. The two unaligned GS dipeptides in the structural superposition alignment have a single amino acid gap placed in a different location from the gap placed in the other three alignments. Examination of the three dimensional structures associated with these sequences indicates that the gap placement is likely an artifact of the structural superposition alignment method rather than an accurate reflection of the evolutionary history of these sequences.

In all cases the GS dipeptide is the first two amino acids after a β Turn in a sheet-hairpin turn-sheet motif (see Brookhaven Protein Data Bank entries 1cms, 2apr, and 4pep, Abola et al., 1987). The hairpin turn is either two or three residues long with the first two residues being a β Turn. Thus, the GS is either the first two residues of the sheet (when the hairpin turn is two residues long) or the GS is the last residue of the hairpin turn and the first residue of the sheet (when the hairpin turn is three residues long). The structural superposition alignment algorithm forces the hairpin turns to be the same length; thus, for the shorter hairpin turn, a gap is inserted. The residues in the sheet-hairpin turn-sheet are on the surface of the protein. Rather than an insertion in the hairpin turn, it appears that the second strand of sheet has shifted one position because of steric interactions with core residues of the protein.

We believe that, in this instance, the sequence based alignments more accurately represent the evolutionary history of the protein than does the structural superposition alignment.

Glycine Aspartic Acid Region

The dollar signs in figure 1 mark two amino acids, a Glycine and an Aspartic acid, that are found to be conserved in all the alignments except the PileUp alignment. While it may appear that the eleven amino acid insertion in the Catd_H sequence is somehow responsible for misleading the PileUp program relative to the other programs, this is probably not the cause.

The probable cause of this difference is the PileUp gap function which contains both a penalty for opening a new gap and a penalty for extending the gap. This form of the gap penalty explicitly makes short gaps less likely. This bias against short gaps prevents PileUp from inserting a single amino acid gap either before or after the conserved GS dipeptide which is inserted by all the other programs. The MSA program inserted the single amino acid gap after the conserved GS even though it also used a gap function that includes penalties for both opening and extending gaps. The difference in these gap functions is that MSA weights the contributions from each pair of sequence so that the alignment scores approximate alignment scores summed over a phylogenetic tree rather than simple sum-of-pairs alignment score. This effectively reduces the cost of introducing this gap.

Based on these considerations, we believe that the alignments containing as conserved both the Glycine and the Aspartic acid are more likely to represent the actual evolution of these aspartyl proteases.

Phenylalanine

The final case we will examine is the pound sign marking a Phenylalanine that is completely conserved in the structural superposition and the MSA alignments but not in the alignments generated by the progressive pairwise alignment programs, AMPS and PileUp. Maintaining this Phenylalanine as completely conserved would require these two programs to introduce a pair of new gaps. One new gap is required before the Phenylalanine in the Pepa_A sequence to bring the Phenylalanines into alignment. The second gap is required after the Phenylalanine in all the sequences other than the Pepa_A sequence to keep the conserved Aspartic acid Glycine dipeptide at the C terminal end of the fragment.

This case is more difficult to access than the previous two. There are several factors that argue in favor of accepting the conservation of the Phenylalanine. First, we are inclined to give a heavy weight to the structural superposition alignment when it identifies as homologous large hydrophobic amino acids, as it does here. Second, this arrangement is a commonly observed local minimum encountered in progressive pairwise alignments.

In both the AMPS and the PileUp alignments, adding the Pepa_A sequence to the alignment of the other five sequences was the final step in the alignment process. Aligning the Pepa_A Phenylalanine with the already aligned Phenylalanines required opening two gaps, one in the Pepa_A sequence and one in the preexisting five sequence alignment. This gap barrier was too large for the pairwise programs to overcome.

We tested this hypothesis by aligning the Pepa_A sequence with the other sequences individually (alignments not shown). In these alignments the gaps necessary to align the Pepa_A Phenylalanine with the other Phenylalanines were inserted. Thus had one of these pairwise alignments been the first step in the AMPS or PileUp progressive alignment procedure all the Phenylalanines would have been aligned.

Three factors in the MSA alignment program allow it to overcome this gap penalty barrier. The first is the weighting of the sequences. The second and prehaps more important factor is that the MSA program has a much improved rationale for counting gap cost over the AMPS and PileUp programs (Altschul, 1989). The third factor is that MSA aligns all the sequences simultaneously rather than in a progressive manner. All three factors make MSA theoretically superior to both AMPS and PileUp and thus expected to give more biologically reasonable alignments. These factors argue in favor of accepting the MSA alignment over the AMPS and PileUp alignments.

Based on all these factors we believe that the Phenylalanine should be accepted as completely conserved.

Summary

The variability in the alignments of these moderate length fragments is representative of the variability in alignments that would be seen in many of the families of proteins that biochemists and molecular biologists are actively studying. Had we included alignments generated with different similarity matrices and with different gap penalties, the variability in the alignments would have been even greater. To understand how the choice of alignment algorithms, similarity scores, and gap penalties effects the alignment of the sequences we need to accurately measure the differences among alignments that result from changes in the different factors contributing to the alignment.

DISTANCES BETWEEN ALIGNMENTS

Computing the Distance

Comparing more than two different alignments of the same sequences simultaneously requires a measure that satisfies the mathematical definition of a distance (Kruskal, 1983) To be useful, the measure must also capture the important differences between alignments in a manner that is easily understood and visualized. We have developed a measure that determines a distance between sequence alignments that satisfies these criteria.

To describe the alignment of a pair of sequences X and Y, the amino acids in sequence X are denoted as x_i while those in sequence Y are denoted as y_j. Pairs of aligned amino acids in the alignment are denoted as x_i:y_j When gaps are introduced into sequence X of the alignment following amino acids x_i:y_j, the index j is incremented to denote amino acids in sequence Y that are aligned with a gap character, while the index i is not incremented until the gap has ended. Thus a two amino acid gap in sequence X and its surrounding ungaped aligned amino acids would denoted as: $x_i{:}y_j, x_i{:}y_{j+1}, x_i{:}y_{j+2}, x_{i+1}{:}y_{j+3}.$

With this notation we can write a simple summation that will yield the distance between two alignments. We denote alignment A_1 on sequences X and Y as having amino acid juxtapositions $x_i{:}y_j$ and denote alignment A_2 on sequences X and Y as having amino acid juxtapositions $x_i{:}y_k$. Then, if sequence X is N amino acids long, the distance between alignments A_1 and A_2, $D_{A1:A2}$, is given by:

$$D_{A1:A2} = \sum_{i=1}^{N} |j-k|.$$

Each term in the summation is the distance along sequence Y between amino acids y_j and y_k, the amino acids aligned with amino acid x_i in sequence X of the two alignments being compared. Since $D_{A1:A2}$ is a distance, $D_{A1:A2}$ is equal to $D_{A2:A1}$. This distance, $D_{A1:A2}$, can be easily visualized in terms of path graphs of the alignments (Figure 2). The path graph of an alignment of two sequences is a rectangular plot with sequence X listed along the horizontal axis and sequence Y listed along the vertical axis. This plot is composed of square cells, one for each pair of amino acids x_i and y_j . Each cell is filled in one of four possible ways. If the pair of amino acids x_i and y_j are aligned with each other in the alignment (either a match or mismatch) the cell is filled with a diagonal line. If either of the amino acids x_i and y_j is part of a gap (an insertion or deletion) the cell is filled with a vertical line if the gap is in sequence X or a horizontal line if the gap is in sequence Y. If the pair of amino acids x_i and y_j are not part of the alignment the cell is empty. Figure 2 shows path graphs for the alignments of the fragments Catd_H with Pepa_A generated by the MSA and PileUp programs and shown in Figure 1.

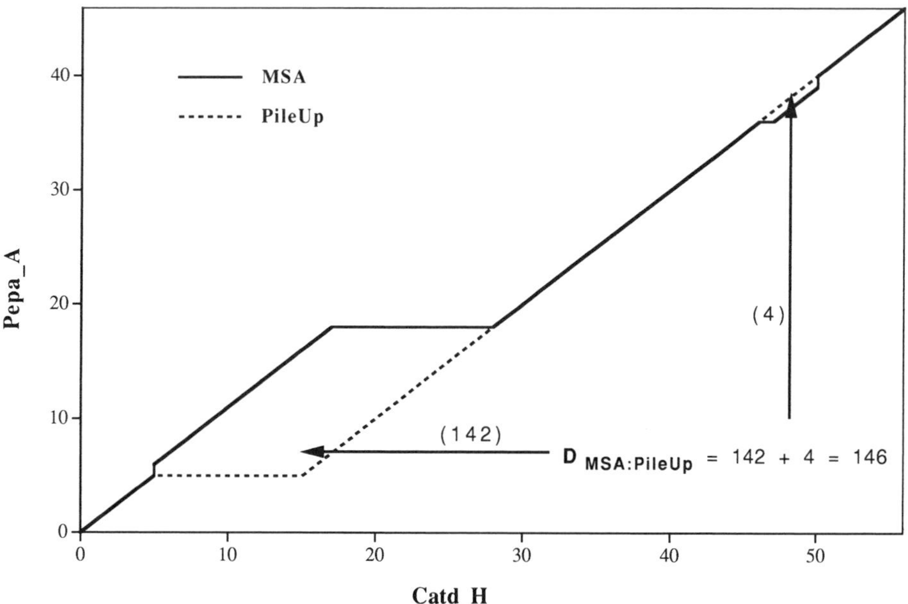

Figure 2. Path graphs for the MSA and PileUp alignments.

Table 2. Distances between alignments by different programs

(a)

	Structure	MSA	AMPS	PILEUP
		Alignments containing Carp_Y		
Structure				
Max	—	1194	1143	1385
Min	—	81	87	97
Avg	—	951.2	915	986
Median	—	1156	1108	1099
MSA				
Max	1155	—	68	197
Min	81	—	1	10
Avg	923.4	—	31	88.4
Median	1130	—	25	57
AMPS				
Max	1141	77	—	171
Min	87	1	—	15
Avg	909.8	24.4	—	85
Median	1103	18	—	55
PileUp				
Max	1227	251	237	—
Min	97	10	15	—
Avg	944.8	120.6	121.6	—
Median	1145	74	75	—
	Alignments containing Catd_H			

(b)

	Structure	MSA	AMPS	PILEUP
		Alignments containing Pepc_M		
Structure				
Max	—	1156	1126	138.5
Min	—	64	105	170
Avg	—	507	530.4	749.6
Median	—	115	177	445
MSA				
Max	1138	—	70	241
Min	75	—	12	28
Avg	509.6	—	33.8	115.6
Median	115	—	27	74
AMPS				
Max	1108	83	—	218
Min	61	68	—	5
Avg	507	74.5	—	102.2
Median	177	76	—	75
PileUp				
Max	1146	251	237	—
Min	66	102	102	—
Avg	619.6	202.8	187.8	—
Median	521	223	211	—
	Alignments containing Pepa_A			

Table 2. *Continued*

(c)

	Structure	MSA	AMPS	PILEUP
		Alignments containing Chym_B		
Structure				
Max	—	1187	1111	1285
Min	—	30	33	170
Avg	—	508.4	486.8	401.8
Median	—	115	105	288
MSA				
Max	1194	—	83	223
Min	30	—	18	28
Avg	504.8	—	37.2	98.8
Median	75	—	27	58
AMPS				
Max	1143	76	—	211
Min	33	33	—	5
Avg	500.6	52.8	—	94.4
Median	125	46	—	67
PileUp				
Max	1109	210	214	—
Min	66	102	102	—
Avg	594.4	154.2	151	—
Median	445	144	143	—
	Alignments containing Carp_R			

Table 2 is composed of six triangular subtables arranged in pairs as the upper right and lower left triangles of a square table. The three square tables are the subtables of the complete table. Each triangular subtable contains summaries of the distances between alignments generated by all pairs of the alignment programs used in this study. Each summary is based on the alignment of a single aspartyl protease sequence with each of the other five aspartyl protease sequences in the data. The single aspartyl protease sequence present in all five alignments of a subtable is indicated above the upper right subtables or below the lower left subtables. The summary includes the maximum distance, the minimum distance, the average distance, and the median distance. Giving both the average and median provides information on the asymmetry of the distribution of distances.

The distance, $D_{A1:A2}$, is the count of the number of cells between the two path graphs plotted in the same rectangular plot. Thus $D_{A1:A2}$, is easily conceptualized and visualized as the area between the path graphs. For the path graphs shown in Figure 2 the distance, $D_{MSA:PileUp}$, is 146.

Application of the Metric to Aspartyl Protease Alignments

The entire sequences of the six aspartyl proteases in the data were aligned by the MSA program, the AMPS program, the PileUp program, and structural superposition. The distances between every pair of sequence and every pair of programs was computed. These distances are summarized in Table 2 in six largely independent subtables. Each subtable summarizes five alignments and shares one pair of aligned sequences with each other subtable.

Examination of the subtables shows a consistent pattern of distances between the different methods of alignment. The MSA and AMPS programs consistently give the most similar alignments. The largest difference is between the structural superposition alignments and the alignments of any of the three sequence aligning programs. In five of the six subtables, PileUp alignments are more different from the structural superposition alignments than are the MSA and AMPS alignments.

Based on these results and the detailed analysis of the sections of alignments in figure 1 we hypothesize that in differences the gap function (i.e., the presence of gap opening and gap extending terms, or how gaps are counted) play a critical role in the variations observed among the methods of generating alignments. This hypothesis, based on a small data set from one protein family, is very tentative. However it provides a basis for further experimentation.

CONCLUSIONS

We have developed a new measure for determining the distance between alignments. This new measure has significant advantages over previously used measures (e.g., the percentage of identically aligned amino acids). The most important feature of this measure is that it meets the mathematical criteria for being a distance. This allows us to legitimately compare several different alignments simultaneously and eliminates the need to declare one of the alignments as a standard. A second important feature of our measure is that it captures the influence of both the size and locations of gaps in the sequences in a quantitative and graduated manner rather than on a binary identical or not identical scale.

These features have been gained without introducing confusing obscurity or abstractness into the discussion of alignments; rather, the measure has a simple interpretation that is readily visualized and displayed graphically. The graphical display, in fact, enhances the discussion of alignment differences by hiding the overwhelming detail of specific amino acids while illuminating the different patterns of gaps along the sequences. In terms of the sequences, each element in the summation computing the measure is simply the distance between the two different amino acids in the second sequence that are aligned with the same amino acid in the first sequence in the alignment. The total measure is the sum of all of those distances. Another advantage is that our alignment distance has an obvious and direct extension to directly comparing alignments of more than a pair of sequences.

This measure of alignment distance provides a useful tool for investigating all aspects sequence alignment. Even on this small data set it allows us to show that sequence based alignment methods give appreciably different results from methods based on structural superposition.

ACKNOWLEDGMENTS

This work was suporteded under the National Institutes of Health NCRR grant 1 P41 RR06009 to the Pittsburgh Supercomputing Center. We thank the Pittsburgh Supercomputing Center for computational resources.

REFERENCES

Abola, E.W., Bernstein, F.C., Bryant, S., Koetzle, T.F., Weng, T., 1987, in Crystallographic Databases - Information Content, Software Systems, Scientific Applications, eds., Allen, F.H., Bergerhoff, G.,

Sievers, R. Data Comission of the International Union of Crystallography, Bonn, Cambridge, Chester, pp 107-132.

Altschul, S.F., 1989. Gap Costs for Multiple Sequence Alignment. J. Theor. Biol. 138:297-309.

Altschul, S.F., Lipman, D.J., 1989. Trees, stars, and multiple biological sequence alignment. SIAM J. Appl. Math. 49:197-209.

Bairoch, A., Boeckmann, B., 1993. The Swiss-PROT protein sequence data bank, recent developments. Nuc. Acids. Res. 21:3093-3096.

Barton, G., 1990, Protein multiple sequence alignment and flexible pattern matching, Methods in Enzymology. 183:403-428.

Barton, G., Sternberg, M., 1987, Evaluation and improvements in the automatic alignment of protein sequences, Protein Engineering. 1:89-94.

Clark, S.P., 1992, Maligned: a multiple sequence alignment editor, CABIOS. 8:535-538.

Dayhoff, M., Schwartz, R., Orcutt, B., 1978, A model for evolutionary change in proteins, Atlas of Protein Sequence and Structure. 5:3:345-352.

Feng, D., Doolittle, R. 1990, Progressive alignment and phylogenetic tree construction of protein sequences, Methods in Enzymology. 183:403-428.

Genetics Computer Group, 1991, Program Manual for the GCG Package, Version 7.

Kruskal, J.B., 1983, An overview of sequence comparision I, Time Warps, String Edits, and Macromolecules: The Theory and Practice of Sequence Comparison. 1-44.

Lipman, D.J., Altschul, S.F., and Kececioglu, J.D., 1989, A tool for multiple sequence alignment Proc. Natl. Acad. Sci. USA. 86:4412-4415.

Pascarella, S., Argos, P., 1992, A data bank merging related protein structures and sequences Protein Engineering. 5:121-137.

Smith, T.F., Waterman, M.S., Fitch, W.M. 1981, Comparative biosequence metrics, J. Mol. Evol. 18:38-46

INDEX